The Chemical Biology of Nucleic Acids

The Chemical Biology of Nucleic Acids

Edited by

GÜNTER MAYER
Strathclyde Institute for Pharmacy and Biological Sciences,
University of Strathclyde, Glasgow, UK

WILEY
A John Wiley and Sons, Ltd., Publication

This edition first published 2010

Registered office
John Wiley & Sons Ltd, The Atrium, Southern Gate, Chichester, West Sussex, PO19 8SQ, United Kingdom

For details of our global editorial offices, for customer services and for information about how to apply for permission to reuse the copyright material in this book please see our website at www.wiley.com.

Library of Congress Cataloging-in-Publication Data

The chemical biology of nucleic acids / editor, Günter Mayer.
p. ; cm.
Includes bibliographical references and index.
ISBN 978-0-470-51974-5 (cloth)
1. Nucleic acids. I. Mayer, Günter, 1972–
[DNLM: 1. Nucleic Acids—chemistry. 2. Nucleic Acids—physiology. 3. Nucleic Acids—therapeutic use.
QU 58 C5165 2010]
QP620.C44 2010
572.8—dc22

2010000254

A catalogue record for this book is available from the British Library.

ISBN: 978-0-470-51974-5

Set in 10/12pt Times by Integra Software Services Pvt. Ltd., Pondicherry, India
Printed and bound in Great Britain by CPI Antony Rowe, Chippenham, Wiltshire.

Contents

Foreword

Chemists have played a key role not only in establishing the chemical structure of DNA and RNA but also in helping to understand their structure–function relationships. This fascination has continued over the years and, as the present book demonstrates, is unabated. The many chapters are great testimony to the fruitful interplay in nucleic acid chemistry and biology. As chemists have continued over the years to modify the structure and architecture of DNA and RNA for an even better insight into their biological functions, Nature has lately provided new challenges by disclosing some unanticipated novel structures and pathways. These new insights into Nature pose questions where chemistry helps to obtain a deeper understanding.

Even though the chemistry of nucleic acids has been described over the years in many textbooks and reviews, the chapters in this book describe the efforts and successes achieved in the last few years in combination with more recent developments. Hence this book comprises an appreciated update review of the chemical biology of nucleic acids as it stands today.

The book covers a wide scope, with chapters on specific topics written by experts in the particular area. Given the new horizons on the role of RNA which we have experienced over recent years, it is not surprising that several chapters describe RNA chemistry as an essential component for the deeper analysis of these phenomena. The chemical synthesis of modified RNA is, of course, fundamental and is rightly described at the beginning. RNA occurs in a surprisingly wide range of architectural motifs commensurate with the multitude of functions as a detailed account describes. The versatility of structural motifs is also apparent in the naturally occurring riboswitches. These are responsible for the control of translation or transcription by binding of small cellular metabolites. This area is still very much in flux and novel riboswitches are being discovered every year, supporting RNA as an important regulator for gene expression. MicroRNAs are a novel type of small RNAs whose mechanism of regulation of gene expression is still under investigation. So far, chemistry has not generally been applied to these, even though for medical applications this looks rather enticing. The functions of small non-coding RNAs in bacteria are attracting increasing attention and deserve the in-depth review presented. This should make chemists aware of a field which has been neglected by them so far.

RNA and DNA can actually share certain functions. This is seen in the catalytic properties which both can adopt in catalytic RNA (ribozymes) and DNAzymes, which by themselves occur in a variety of architectures. Aptamers can also consist of RNA or DNA to bind proteins and other molecules extremely specifically. The structures with these properties have been obtained by *in vitro* selection from random sequences where selective pressure plays an important role in sampling the sequence space. The examples provided demonstrate again the architectural and, with it, the functional versatility of RNA and DNA. As reviewed in several chapters, these in turn represent excellent opportunities for modulation by chemical derivatization.

Besides riboswitches, small molecules bind to RNA also in a different setting, such as to the ribosomal RNA, where aminoglycosides are the paradigm and where they exert antimicrobial effects. The fundamental features playing a role in such interactions of small molecules with nucleic acids in general is an interesting subject.

Sugar- and base-modified nucleotides have been a subject of study for many years, yet novel aspects are still being explored. One example is the use of such modified nucleotides to study the selectivity of DNA polymerases. Another is the creation of new base pairs to expand the scope of the genetic alphabet by incorporation into DNA or RNA by replication and transcription. Of the sugar-modified nucleotides, the locked nucleic acids (LNAs) have found wide application because of their strong binding to RNA. This property makes them most suitable for interaction with mRNA and microRNAs for interference with gene expression. An entirely different analogue is the peptide nucleic acids (PNAs), where the phosphate backbone and the deoxyribose are replaced by amide linkages. Even though this is a drastic change in structure, the PNAs still hybridize favourably with DNA and RNA but are, of course, completely resistant to nucleases. They represent very powerful DNA mimics with interesting properties.

The potential of nucleic acids for therapy is a well-studied area where various strategies are being explored and being examined in clinical trials. There is high potential in this applied area and we look forward to a general breakthrough for the approval of drugs.

In summary, this collection of reviews is testimony to the fruitful role that chemists can play in helping to understand structure–function relationships in nucleic acids and their mechanisms of action. Recent years have provided big surprises, particularly in the field of RNA, which indicates that the field is wide open with lots of opportunities and challenges.

Fritz Eckstein
Göttingen
2010

Preface

In early 2007, Paul Deards from Wiley approached me and asked whether I would like to edit a book on the topic of light-responsive nucleic acids. Upon this impulse we started a fruitful discussion on that topic. An initial survey of books that cover nucleic acids in general, and on light-responsive nucleic acids in particular, resulted in the finding that almost no comprehensive compendium starting with the synthesis of nucleic acids and their derivatives and also covering their biological applications was available at that time. Therefore, we decided to withdraw the initial idea of putting together a book on light-responsive nucleic acids only, and instead we set out to edit a more general book on the topic of *The Chemical Biology of Nucleic Acids*. The result of this effort finally led to what is compiled within the present book.

Fortunately, many outstanding scientists within their respective scientific fields agreed to contribute to the book. Without their efforts this compilation would not have come to life. It is amazing to learn which central role RNA, DNA and derivatives thereof play in Nature, apart from the sole storage and transmission of genetic information, and it is even more remarkable to gain knowledge on the versatile application of artificial, synthetic nucleic acids to investigate biological phenomena. Being a part within this scientific area is exciting and I am curious to learn what comes next. I am very proud that Fritz Eckstein could be won over to write the Foreword to this book. He definitely is the right and most competent person to do so – thank you very much. Of course, the book cannot cover every aspect of the complex and diverse field of nucleic acids. I apologize if we have missed any subjects that others might consider invaluable. Owing to the inevitable space constraints, we definitely had to make compromises that owing to their inherent nature will not satisfy everyone's opinions completely. Finally, I would like to thank all the authors again for their willingness to make *The Chemical Biology of Nucleic Acids* an outstanding contribution – it will certainly become an important reference in the field. I am also grateful to Paul Deards who initiated the process leading to this book, and Richard Davis, Gemma Valler and Rebecca 'Becki' Ralf at Wiley for their helpful support and encouragement during the making of this book.

Günter Mayer
Glasgow, 2009

List of Contributors

Sergio Aguirre, School of Biomedical Engineering, McMaster University, Hamilton, Ontario, Canada.

Tyler D. Ames, Department of Molecular Cellular and Developmental Biology, Howard Hughes Medical Institute, Yale University, 266 Whitney Avenue, New Haven, CT 06520, USA.

Stefan Bauer, Institut für Immunologie, Philipps-Universität Marburg, BMFZ, Hans-Meerweinstrasse 2, 35043 Marburg, Germany.

Sabine Brantl, AG Bakteriengenetik, Friedrich-Schiller-Universität Jena, Philosophenweg 12, 07743 Jena, Germany.

Ronald R. Breaker, Department of Molecular Cellular and Developmental Biology and Department of Molecular Biophysics and Biochemistry, Howard Hughes Medical Institute, Yale University, 266 Whitney Avenue, New Haven, CT 06520, USA.

Valérie Fritsch, Institut de Biologie Moléculaire et Cellulaire, Centre National de la Recherche Scientifique, Université de Strasbourg 15 rue R. Descartes, 67084 Strasbourg Cedex, France.

Alexander Heckel, Goethe University Frankfurt, Cluster of Excellence Macromolecular Complexes, Max-von-Laue-Strasse 9, 60438 Frankfurt am Main, Germany.

Ichiro Hirao, Nucleic Acid Synthetic Biology Research Team, Systems and Structural Biology Center, Yokohama Institute, RIKEN, 1-7-22 Suehiro-cho, Tsurumi-ku, Yokohama, 230-0045, Japan.

Claudia Höbartner, Max Planck Institute for Biophysical Chemistry, Research Group Nucleic Acid Chemistry, Am Fassberg 11, 37077 Göttingen, Germany.

Britta Hoehn, Department of Molecular Biology, Beckman Research Institute of the City of Hope, Duarte, CA 91010, USA.

Andres Jäschke, University of Heidelberg, Institute of Pharmacy and Molecular Biotechnology, Department of Chemistry, Im Neuenheimer Feld 364, 69120 Heidelberg, Germany.

Albert Jeltsch, School of Engineering and Science, Jacobs University Bremen, Campus Ring 1, 28759 Bremen, Germany.

Renata Z. Jurkowska, School of Engineering and Science, Jacobs University Bremen, Campus Ring 1, 28759 Bremen, Germany.

Harleen Kaur, Institute of Genomics and Integrative Biology, CSIR, Mall Road, Delhi 110 007, India.

Michiko Kimoto, Nucleic Acid Synthetic Biology Research Team, Systems and Structural Biology Center, Yokohama Institute, RIKEN, 1-7-22 Suehiro-cho, Tsurumi-ku, Yokohama, 230-0045, Japan.

Jeffrey C.F. Lam, Department of Biochemistry and Biomedical Sciences, McMaster University, Hamilton, Ontario, Canada.

Yingfu Li, Department of Biochemistry and Biomedical Sciences, McMaster University, Hamilton, Ontario, Canada.

Joe C. Liang, Department of Bioengineering, 473 Via Ortega, MC 4201, Stanford University, Stanford, CA 94305, USA.

Souvik Maiti, Institute of Genomics and Integrative Biology, CSIR, Mall Road, Delhi 110 007, India.

Andreas Marx, Department of Chemistry, Konstanz Research School Chemical Biology, Universität Konstanz, Universitätsstrasse 10, Box 726, 78457 Konstanz, Germany.

Günter Mayer, Strathclyde Institute for Pharmacy and Biological Sciences, University of Strathclyde, 27 Taylor Street, Glasgow G4 0NR, UK.

Gunter Meister, Max Planck Institute of Biochemistry, Am Klopferspitz 18, 82152 Martinsried, Germany.

Peter E. Nielsen, Department of Cellular and Molecular Medicine, Faculty of Health Sciences, The Panum Institute, University of Copenhagen, Blegdamsvej 3c, 2200 Copenhagen N, Denmark.

Michael Oberhuber, Institute of Organic Chemistry and Center for Molecular Biosciences (CMBI), University of Innsbruck, Innrain 52a, 6020 Innsbruck, Austria. Present address: Laimburg Research Centre for Agriculture and Forestry, Laimburg 6, Pfatten (Vadena), 39040 Auer (Ora), Italy.

John J. Rossi, Department of Molecular Biology, Beckman Research Institute of the City of Hope, Duarte, CA 91010, USA.

Christina D. Smolke, Department of Bioengineering, 473 Via Ortega, MC 4201, Stanford University, Stanford, CA 94305, USA.

Victor K. Tam, Department of Chemistry and Biochemistry, University of California, San Diego, La Jolla, CA 92093-0358, USA.

Yitzhak Tor, Department of Chemistry and Biochemistry, University of California, San Diego, La Jolla, CA 92093-0358, USA.

Peter J. Unrau, Department of Molecular Biology and Biochemistry, Simon Fraser University, 8888 University Drive, Burnaby, BC, V5A 1S6, Canada.

Falk Wachowius, Max Planck Institute for Biophysical Chemistry, Research Group Nucleic Acid Chemistry, Am Fassberg 11, 37077 Göttingen, Germany.

Qing Wang, Department of Molecular Biology and Biochemistry, Simon Fraser University, 8888 University Drive, Burnaby, BC, V5A 1S6, Canada.

Eric Westhof, Institut de Biologie Moléculaire et Cellulaire, Centre National de la Recherche Scientifique, Université de Strasbourg 15 rue R. Descartes, 67084 Strasbourg Cedex, France.

Maung Nyan Win, Department of Bioengineering, 473 Via Ortega, MC 4201, Stanford University, Stanford, CA 94305, USA.

Bernhard Wulffen, University of Bonn, Life and Medical Sciences Bonn, c/o Kekulé Institut für Org. Chemie und Biochemie, Gerhard-Domagk-Strasse 1, 53121 Bonn, Germany.

Yun Xie, Department of Chemistry and Biochemistry, University of California, San Diego, La Jolla, CA 92093-0358, USA.

1
Chemical Synthesis of Modified RNA

Claudia Höbartner and Falk Wachowius

1.1 Introduction

The synthesis of chemically modified RNA oligonucleotides is of paramount importance in many fields of nucleic acids research, ranging from studies of native RNA structure and function to applications in diverse areas not limited to chemistry, such as molecular biology, cell biology and medicine (Figure 1.1).

In Nature, more than 100 structurally distinguishable modified nucleotides have been identified in almost all classes of cellular RNAs [1]. The types of natural RNA modifications include simple nucleobase methylations and uridine isomerization, in addition to hypermodified nucleotides resulting from multistep biosynthetic transformations in complex RNA maturation processes [2]. A large number of structural and functional roles have been described for post-transcriptional RNA modifications, ranging from blocking or reinforcing single base pairs to increasing thermal stability and reducing conformational dynamics [3]. Changes in the physicochemical properties of RNA structures arising from nucleobase and ribose methylations have been attributed to enhanced base stacking due to increased hydrophobicity and polarizability and to changes in the hydration spheres of major and minor grooves [4]. Most insights into the roles of natural minor nucleotides come from extensive studies of highly modified tRNAs. However, the effects of individual nucleotide modifications in many other classes of RNAs are far from being fully understood. The availability of synthetic RNAs carrying natural modifications is therefore a prerequisite for studying the structural and functional contributions of individual RNA modifications. Advances in chemical synthesis of modified nucleotides will certainly lead to a more detailed understanding of natural phenomena.

In addition, artificial RNA modifications that introduce functional groups not found in Nature have proven to be useful tools for biochemical and biophysical investigations of RNA structure and function. Examples of reporter probes include fluorescent dyes to measure inter-helical distances or to report local and global changes during RNA folding [5–7], nitroxide spin probes for analysis of RNA structure and dynamics by EPR spectroscopy [8–10], disulfide crosslinks to restrict RNA helical elements for the investigation of structural models [11], selenium modifications to assist in solving the crystallographic phase problem [12–14] and

The Chemical Biology of Nucleic Acids Edited by Günter Mayer

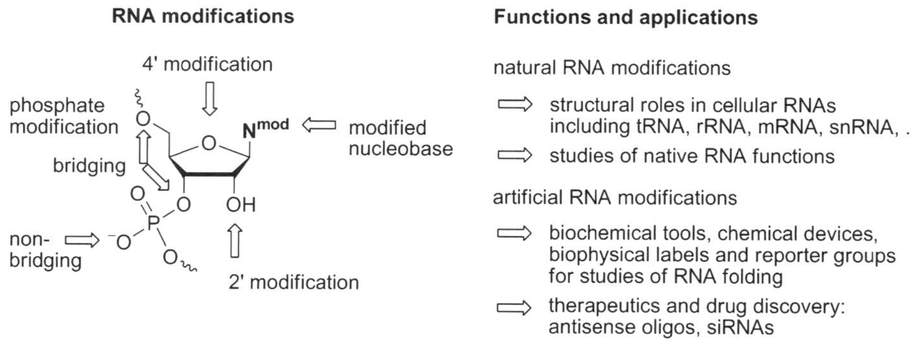

Figure 1.1 *Diagram illustrating sites for chemical modifications of ribonucleotides and versatile functions and applications of synthetic RNA*

photolabile modifications for temporal control of RNA structure and activity [15]. These are only a few examples of the large diversity of approaches reported to date.

A separate motivation for the development of novel RNA analogs and methods for their chemical synthesis is their potential use as oligonucleotide-based therapeutics. Many modified oligonucleotides have been evaluated as antisense agents against different targets, but so far only a single antisense medium has been approved for clinical use [16]. The discovery of RNA interference (RNAi) as a natural gene-silencing pathway has energized the field of oligonucleotide-based technologies. The findings that synthetic small interfering RNAs (siRNAs) can be used as *in vitro* research tools and have the potential for the modulation of various diseases have increased the demand for modified and unmodified synthetic RNA oligonucleotides. Various modifications to the ribose, phosphate and nucleobase moieties of RNA have been investigated for their efficiency to decrease the susceptibility to nuclease degradation, reduce the risk of activating the innate immune response, decrease off-target effects and improve cellular uptake and pharmacological availability. Today, no single perfect modification pattern meets the significant biomedical challenges of delivery and biodistribution. Potential drug discovery is therefore a respectable motivation for the implementation of new RNA modifications. Recent reviews document the utility of various chemical modifications for use in siRNA technology [17–19].

In this chapter we focus on RNA solid-phase synthesis using phosphoramidite chemistry and describe recent developments in protecting group strategies for RNA synthesis. We discuss general ways to introduce modifications into RNA oligonucleotides and we review selected examples of RNA modifications to illustrate various synthetic routes and give a flavor of the diversity of applications of modified RNAs.

1.2 The chemical synthesis of modified RNA

1.2.1 Solid-phase RNA synthesis using phosphoramidite building blocks

1.2.1.1 The general chemistry of oligoribonucleotide synthesis

The currently most widely used strategy for the chemical synthesis of RNA oligonucleotides applies repeated coupling of ribonucleoside phosphoramidite building blocks on a solid support (Scheme 1.1a). Similarly to solid-phase DNA synthesis, the automated preparation of RNA oligonucleotides entails a four-step reaction cycle [20]. Chain elongation is initiated in step A by 5′-deprotection of an *N*,2′,5′-*O*-protected ribonucleoside

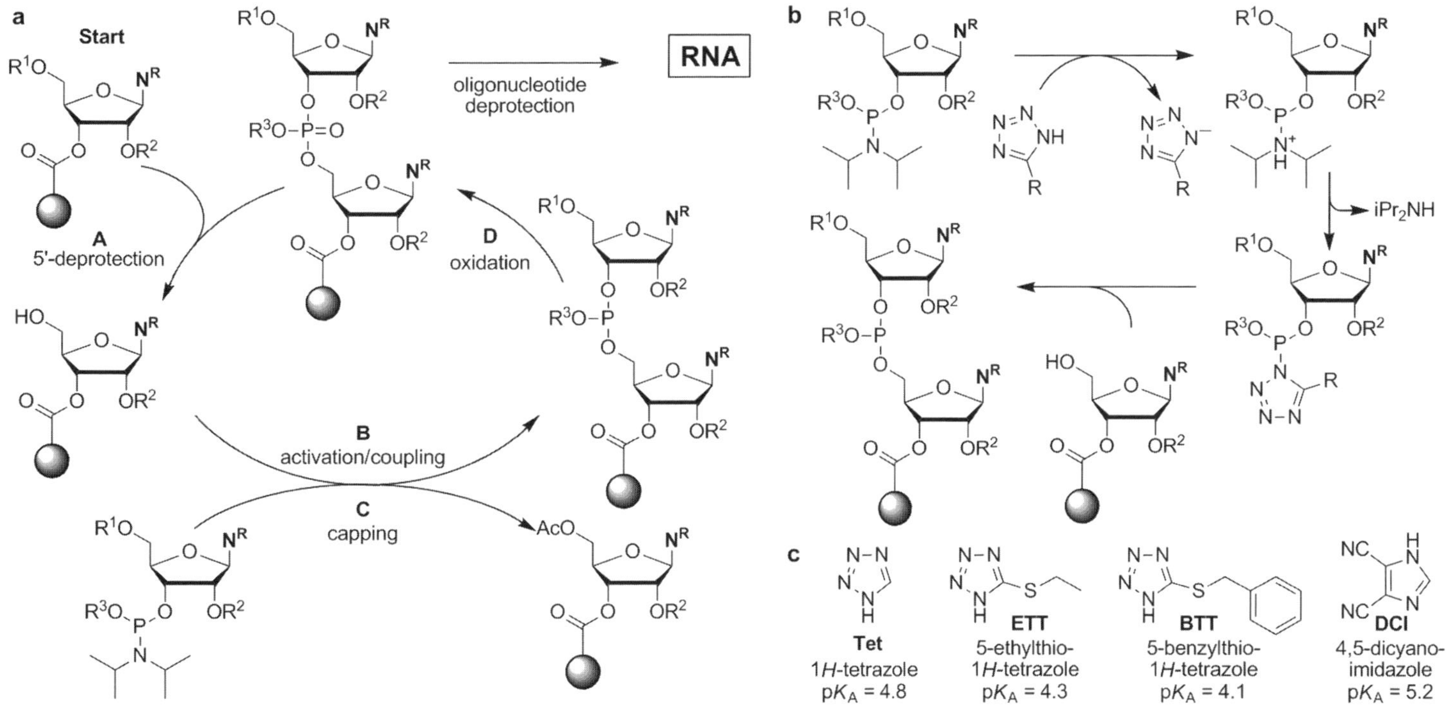

Scheme 1.1 *(a) RNA solid-phase synthesis cycle. N^R, protected nucleobase; R usually a base-labile protecting group (see Figure 1.2); R^1, transient 5′-protecting group (usually acid- or fluoride-labile); R^2, permanent 2′-protecting group, see text and Figures 1.4 and 1.5 for details; R^3, permanent phosphate protecting group, usually cyanoethyl or methyl. (b) Mechanism of phosphoramidite activation and coupling. (c). Commonly used activators*

that is attached to a solid support through a bifunctional linker. The solid support is typically composed of controlled-pore glass with a high surface density of alkylamino groups (LCAA-CPG) or of amino-functionalized polystyrene resins. The bifunctional linker is usually an aliphatic dicarboxylic acid (succinic, adipic or pimelic acid), which is connected to the surface of the support via amide bond formation and comprises an ester linkage to the 3′-hydroxyl group of the ribonucleosides. In step B of the synthesis cycle, a suitably protected phosphoramidite building block is activated with a weak acid (such as an appropriately substituted 1*H*-tetrazole or imidazole derivative) and coupled to the 5′-hydroxyl group of the support-bound ribonucleoside, which results in the formation of a dinucleoside phosphite triester in a two-step reaction. This coupling reaction, as shown in Scheme 1.1b, proceeds via protonation of the diisopropylamino group of the phosphoramidite, followed by displacement of diisopropylamine by the conjugate base of the activator to form the active species, e.g., a tetrazolide intermediate. Subsequent nucleophilic substitution of the tetrazolide by the 5′-hydroxyl of the growing oligonucleotide forms the new phosphite triester linkage. Thus, a better proton donor and/or a better nucleophile to generate the reactive intermediate will increase the rate of the coupling reaction: for example, 5-ethylthio-1*H*-tetrazole (ETT) and 5-benzylthio-1*H*-tetrazole (BTT) improve the rate of the reaction compared with 1*H*-tetrazole due to their stronger acidity; 4,5-dicyanoimidazole (DCI) improves the reaction rate, presumably because it is a better nucleophile (Scheme 1.1c). The coupling step is followed by the capping step (step C in Scheme 1.1a) which involves 5′-*O*-acetylation of unreacted 5′-termini to prevent the subsequent extension to less than full-length oligonucleotide chains. The capping reagents also accomplish the cleavage of byproducts from nucleobase phosphorylation that may have formed during the coupling reaction. The solid-phase bound dinucleoside phosphite triester is then converted into a more stable phosphate triester (= phosphotriester) by oxidation with iodine or *tert*-butyl hydroperoxide (step D). Removal of the 5′-protecting group initiates the next chain extension cycle. The four-step synthesis cycle is repeated until chain assembly of the desired oligonucleotide length is completed. The full-length oligoribonucleotide is then released from the solid support and the nucleobase and phosphate protecting groups are removed, usually under ammonolytic conditions. Finally, 2′-deprotection affords the plain oligoribonucleotide product which is analyzed and purified for further utilization.

The key to successful solid-phase RNA synthesis is the choice of a suitable combination of orthogonal transient (R^1) and permanent (R, R^2, R^3) protecting groups for the reactive functional groups in ribonucleoside phosphoramidites. The levels of orthogonality that need to be considered are as follows: deprotection of 5′-hydroxyl groups, deprotection of phosphate backbone, deprotection of exocyclic amino groups on nucleobases, release from the solid support and deprotection of 2′-hydroxyl groups. Usually, nucleobase deprotection, phosphate deprotection and cleavage from the solid support are combined in a single step, which leaves at least three levels of orthogonality that must be met by any successful approach for RNA solid-phase synthesis. Occasionally, separate deprotection of the phosphate group is required. In general, base-labile protecting groups of the acyl, amidine or carbamoyl type are used for the exocyclic amino groups of nucleobases (Figure 1.2). Ammonolytic conditions are commonly applied for nucleobase deprotection and release of the oligonucleotide from the solid support. Simultaneously, the phosphotriester is converted into a phosphodiester (a few important exceptions are discussed below). It is imperative that the 2′-protecting groups are stable under these conditions, because generation of a free 2′-OH in the presence of a basic reaction medium results in strand cleavage and is also known to promote phosphate migration, which leads to the undesired formation of unnatural 2′–5′ internucleotide linkages.

A plethora of protecting groups and solid-phase supports have been developed for the chemical synthesis of DNA and RNA oligonucleotides and a comprehensive review is beyond the scope of this chapter. Here, we describe the most commonly used and commercially available phosphoramidite building block families as a reference point and summarize recent developments in the field.

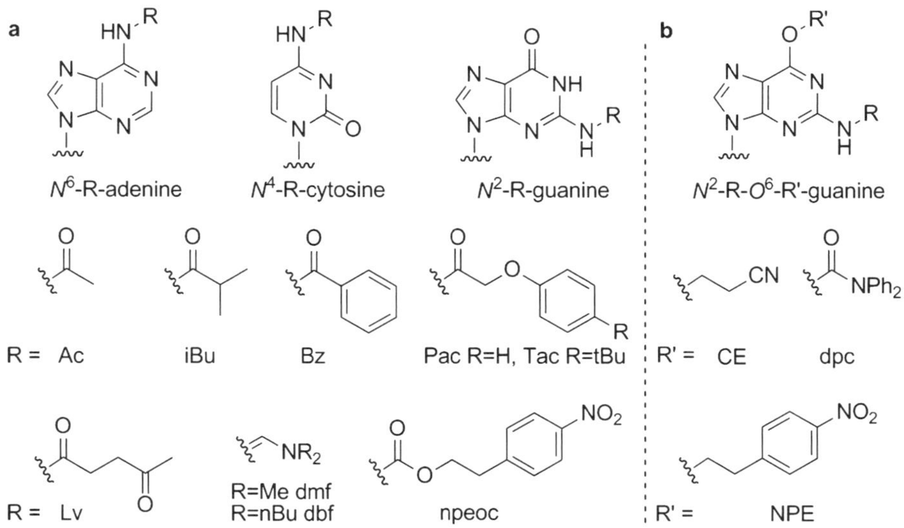

Figure 1.2 (a) Natural nucleobases adenine (A), cytosine (C) and guanine (G) that need protecting groups at their exocyclic amino groups during solid-phase synthesis. Selected examples of base-labile nucleobase protecting groups (R): Ac = acetyl, iBu = isobutyl, Bz = benzoyl, Pac = phenoxyacetyl, Tac = 4-(tert-butylphenoxy)acetyl, Lv = levulinyl, dmf = dimethylaminomethylene (same as dimethylformamidine), dbf = dibutylaminomethylene, npeoc = (4-nitrophenyl)ethyloxycarbonyl. (b) For special applications, O^6-protection of guanine is required. Examples of R′: CE = 2-cyanoethyl, dpc = diphenylcarbamoyl, NPE = (4-nitrophenyl)ethyl

1.2.1.2 Commonly used protecting group strategies for ribonucleoside phosphoramidites

The 5′-O-dimethoxytrityl-2′-O-silyl strategy

The traditional approach for orthogonal protection of 5′- and 2′-hydroxyl groups of ribonucleoside phosphoramidites is based on 5′-*O*-dimethoxytrityl (DMT)-protected 3′-β-cyanoethyl diisopropylphosphoramidites, and therefore capitalizes on the highly successful DNA synthesis strategy which routinely uses 5′-*O*-DMT protected phosphoramidites. The 5′-*O*-DMT group is rapidly removed by anhydrous acid and produces an orange-colored DMT carbocation that permits facile determination of sequential coupling yields by colorimetric detection. Various types of 2′-protecting groups have been described in combination with the 5′-*O*-DMT group, but the two most common representatives belong to the fluoride-labile class of silyl protecting groups. The 2′-*O*-*tert*-butyldimethylsilyl (TBDMS) and 2′-*O*-triisopropylsilyloxymethyl (TOM) protecting groups are described below. Conditions for phosphoramidite coupling and RNA deprotection are summarized in Figure 1.3.

The tert-butyldimethylsilyl (TBDMS) protecting group Since the 1980s, the *tert*-butyldimethylsilyl (TBDMS) group has been the most commonly used 2′-alkylsilyl protecting group for RNA solid-phase synthesis [21]. A wide variety of 5′-*O*-DMT-2′-*O*-TBDMS phosphoramidites of general formula **1** are commercially available. However, the performance of these building blocks in solid-phase RNA synthesis has not reached the level of deoxyribonucleoside phosphoramidites in DNA synthesis. The sluggish coupling kinetics (10–15 min coupling time) and relatively low coupling efficiencies (typically ~98% average coupling yield) have been attributed to steric interference of the TBDMS group with the coupling reaction [22]. For comparison, coupling times for deoxyribonucleoside phosphoramidites usually range from 0.5 to 2 min and coupling efficiencies exceed 99%. Traditionally, 1*H*-tetrazole was used as activator for 2′-*O*-TBDMS-protected RNA

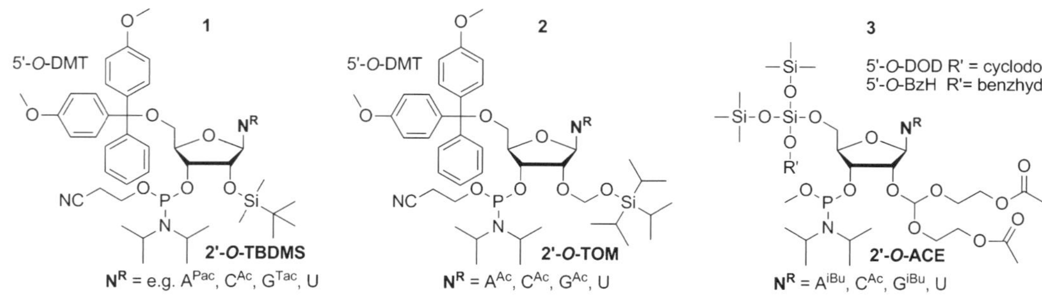

	1	2	3
used since	1980	1998	1998
activation/coupling	BTT, 6-12 min, ~98%	BTT, 2-3 min, >99%	ETT, 1 min, >99%
oxidation	I_2, pyridine/H_2O	I_2, collidine/H_2O	tBuOOH
5'-deprotection	*acid* (TCA or DCA)	*acid* (DCA)	*fluoride* (HF.NEt_3)
phosphate deprotection	*base* (NH_4OH)	*base* (NH_4OH/$MeNH_2$)	*thiolate* (S_2Na_2)
nucleobase deprotection	*base* (NH_4OH)	*base* (NH_4OH/$MeNH_2$)	*base* ($MeNH_2$)
2'-deprotection	*fluoride* (HF.NEt_3)	*fluoride* (TBAF)	*acid* (AcOH/TEMED)

Figure 1.3 *Ribonucleoside phosphoramidite building blocks for commercialized RNA synthesis methods using either the 5′-O-DMT-2′-O-silyl (**1** and **2**) or the 5′-O-silyl-2′-O-ACE chemistry (**3**). TCA = trichloroacetic acid; DCA = dichloroacetic acid; TBAF = tetrabutylammonium fluoride; S_2Na_2 = disodium-2-carbamoyl-2-cyanoethylene-1,1-dithiolate trihydrate; TEMED = tetramethylethylenediamine*

phosphoramidites, but today more powerful activators such as ETT and BTT are applied and coupling times can be reduced to 3–6 min to achieve coupling efficiencies of up to 99% [22]. The most critical issue associated with 2′-*O*-TBDMS protected ribonucleosides is the high potential of the alkylsilyl ether to migrate from the 2′- to the 3′-position during the phosphitylation step of phosphoramidite synthesis. Contamination of the 2′-*O*-TBDMS-protected 3′-phosphoramidite building blocks with the isomeric 3′-*O*-TBDMS-protected 2′-phosphoramidites will result in the undesired presence of unnatural 2′–5′ phosphodiester linkages in the final RNA product. Since the 2′-*O*-TBDMS group is not indefinitely stable under harsh ammonia deprotection conditions, migration and eventually chain cleavage might also occur during basic deprotection. The use of nucleobase protecting groups cleavable under mild conditions, such as phenoxyacetyl (Pac) and 4-(*tert*-butylphenoxy)acetyl (Tac), is therefore recommended.

The triisopropylsilyloxymethyl (TOM) protecting group Within the last decade, considerable advances in conventional 2′-silyl protection have been reported. Probably the most significant improvement in this context was the development of the 2′-*O*-triisopropylsilyloxymethyl (TOM) protecting group, reported by Wu and Pitsch in 1998 [23a]. The reduced steric demand of the TOM group compared with TBDMS during internucleotide bond formation allows for high coupling yields (>99%) of phosphoramidite building block **2** in short coupling times (2–3 min). In contrast to the TBDMS group, the acetal moiety of TOM does not undergo 2′ to 3′ isomerization during phosphitylation and therefore the risk of contamination with isomeric 2′–5′ phosphodiester linkages in the final oligoribonucleotide is eliminated. The 2′-*O*-TOM group is entirely compatible with the 5′-O-DMT and the β-cyanoethyl groups. The exocyclic amino groups of the standard nucleobases are protected with acetyl groups that can be deprotected with methylamine in aqueous ethanol at room temperature within a few hours. The removal of the 2′-*O*-TOM protecting group is achieved with tetrabutylammonium fluoride (TBAF) in THF for 5–14 h at 25–30 °C. Data for RNA synthesis in high quality and high yield have been reported for oligonucleotides of up to 84 nucleotides (nt) in length [23b]. The 2′-*O*-TOM protection strategy is also compatible with a wide variety of modified nucleosides and has been used extensively for the preparation of modified RNA [24,25].

The 5′-O-silyl-2′-O-bis(2-acetoxyethoxy)methyl orthoester (ACE) strategy

In addition to the TOM chemistry, which has been commercialized and is currently widely used in many research laboratories, a second strategy was developed in the late 1990s that has advanced to a highly powerful and commercially offered RNA synthesis method. This second strategy, the 2′-*O*-bis(2-acetoxyethoxy)methyl orthoester (ACE) RNA synthesis method, was described in 1998 by Caruthers and co-workers and is based on a complete redesign of earlier protecting group strategies [26]. The new approach was designed under the notion that mildly acidic conditions would be most desirable for the final deprotection of RNA 2′-hydroxyl groups. For this purpose, the mildly acid-labile ACE protecting group was developed. The resulting loss of orthogonality with 5′-*O*-DMT protection, which is also sensitive to acidic conditions, necessitated the development of a novel class of 5′-protecting groups. Since the 2′-protecting group in ACE chemistry is no longer silyl based, the 5′-position can now be protected with a fluoride-labile silyl group. Two examples of extensively used substituted silyl ethers are the bis(trimethylsiloxy)cyclododecyloxysilyl ether (DOD) and the benzhydryloxybis(trimethylsiloxy)silyl ether (BzH), which are rapidly removed by fluoride ions under neutral conditions. This innovative 2′-*O*-ACE setup preserves only a few aspects of the traditional DNA synthesis strategy and requires alterations to commonly applied procedures and changes to standard instrumentation. The 5′-*O*-silyl-2′-*O*-ACE building blocks **3** (Figure 1.3) are prepared as methyl diisopropylphosphoramidites (i.e. $R^3 = CH_3$) because the cyanoethyl protecting group is incompatible with fluoride treatment during repeating coupling cycles. An additional deprotection step is therefore required to cleave the methyl phosphate prior to release of the oligonucleotide from the solid support and nucleobase deprotection under alkaline conditions. During the basic deprotection step, the ACE group is converted to a 2′-*O*-bis(2-hydroxyethoxy)methyl orthoester, which is 10 times more acid labile than the acetylated form of the orthoester. The ACE RNA synthesis methodology yields a water-soluble oligonucleotide intermediate that can be rapidly 2′-deprotected (within 10 min) using mild aqueous acid immediately before use. Other advantages are the rapid coupling kinetics and high average stepwise coupling yields (>99% in less than 1 min). ACE chemistry permits RNA synthesis in excess of 70 nt in length, is easily scalable and applicable to high-throughput RNA production.

A feature that has been lacking in ACE chemistry since its development is the possibility of convenient tracking of repetitive coupling efficiencies in each cycle in analogy with the photometric detection of the dimethoxytrityl cation. This drawback has recently been overcome by the development of ACE phosphoramidite building blocks **6** that have a visible chromophore appended to the 5′-silyl protecting group via a 1,2,3-triazole linkage (Scheme 1.2) [27]. The chromophore is installed by the highly specific copper(I)-catalyzed [3 + 2] bipolar cycloaddition reaction between an alkyne on the silyl group of nucleosides **4** and an azide moiety on the chromophore **5**. The chromophore is a Disperse Red (DR) (DR = 2-{*N*-ethyl-*N*-[4-(4-nitrophenyldiazenyl)phenyl]}aminoethanol) derivative, which has similar absorption properties to the DMT cation (λ_{max} = 498 nm for DMT^+ and 470 nm for DR).

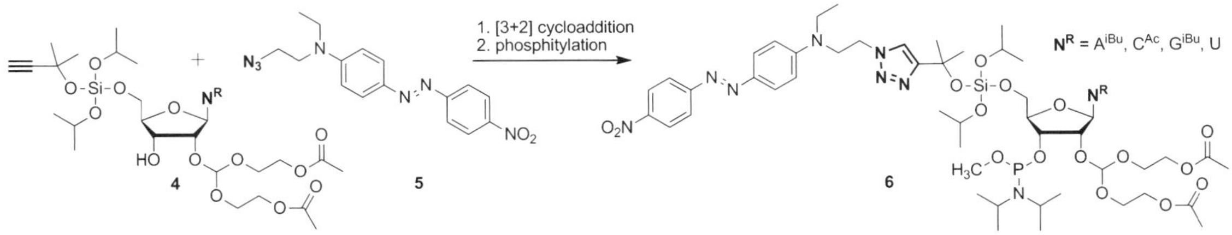

Scheme 1.2 *Synthesis of 5′-DRSil-2′-O-ACE ribonucleoside 3′-phosphoramidites **6** via 1,3-dipolar cycloaddition reaction and phosphitylation*

1.2.1.3 Recent developments in the area of 2′-protecting groups

Within the last few years, the increasing demand for synthetic oligonucleotides for RNAi applications has stimulated renewed research activities for the improvement of RNA synthesis technology beyond the capabilities of the current methods based on TBDMS, TOM and ACE chemistry. Almost all recent developments in this field involve variations of the 2′-protection strategies while maintaining orthogonality to the 5′-*O*-DMT group and compatibility with the cyanoethyl phosphate protecting group. Many of the novel 2′-protecting groups exploit a flexible formacetal moiety to take advantage of the minimized steric crowding in the vicinity of the 3′-phosphoramidite functionality, which allows for high coupling yields in short coupling times. According to the required deprotection conditions, 2′-protecting groups can traditionally be classified into the following types (Figure 1.4): fluoride-labile protecting groups [other examples apart from TBDMS and TOM are triisopropylsilyl (TIPS) (**7**) and 4-nitrobenzyloxymethyl (4-NBOM) (**8**)], photolabile nitrobenzyl protecting groups such as 2-nitrobenzyloxymethyl (2-NBOM) (**9**) and 1-(2-nitrophenyl)ethoxymethyl (NPEOM) (**10**) and acid-labile acetal/ketal protecting groups such as 1-(2-fluorophenyl)-4-methoxypiperidin-4-yl (Fpmp) (**11**) and 1-(4-chlorophenyl)-4-ethoxypiperidin-4-yl (Cpep) (**12**).

This classification is not absolutely strict and several exceptions are known. For example, some earlier reported protecting groups that are labile to UV irradiation are also cleaved by fluoride ions (e.g. 2-NBOM). More recently, base-labile and reducible protecting groups have also been found useful for RNA synthesis (Figure 1.5). A protection strategy that has been gaining increasing interest is the use of convertible or 'protected', protecting groups, in which fragmentation occurs under one set of conditions (usually base treatment) and leaves a modified protecting group that is then sensitive to a new set of conditions (usually more acid labile). This concept also applies to the design of the ACE group. Below we describe the most recent developments of fluoride-labile, base-labile, acid-labile and reducible 2′-protecting groups.

Fluoride-labile protecting groups

The 2-cyanoethoxymethyl (CEM) and 2-cyanoethyl (CE) protecting groups A recently introduced fluoride-labile, acetal-type 2′-hydroxyl protecting group is the 2′-*O*-2-cyanoethoxymethyl (CEM) group. Initially reported by Ohgi and co-workers in 2005 [28], this group showed satisfactory performance for the synthesis of up to 110 nt long RNAs [29,30] Activation of the 2′-*O*-CEM protected phosphoramidites **13** was accomplished by BTT or ETT within a coupling period of 2.5 min, resulting in >99% coupling efficiency. The CEM group is modestly sensitive to the basic conditions used for nucleobase and phosphate deprotection and it is preferably removed by treatment with TBAF in THF or DMSO. The presence of bis(2-mercaptoethyl) ether or nitromethane as an acrylonitrile scavenger is essential to prevent the formation of nucleobase adducts during fluoride-mediated CEM deprotection.

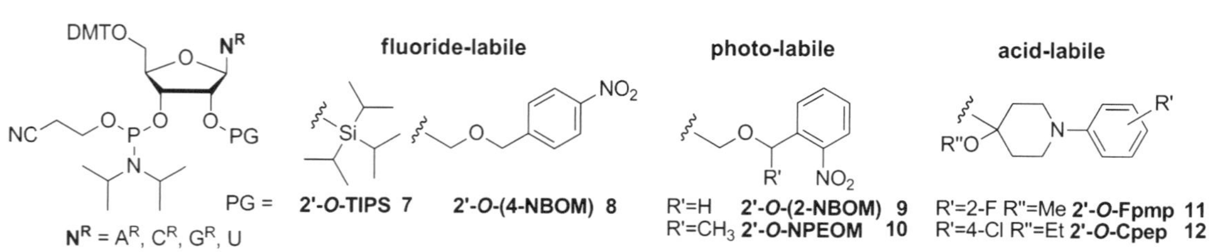

Figure 1.4 *Examples of conventional 2′-protecting groups classified according to required deprotection conditions*

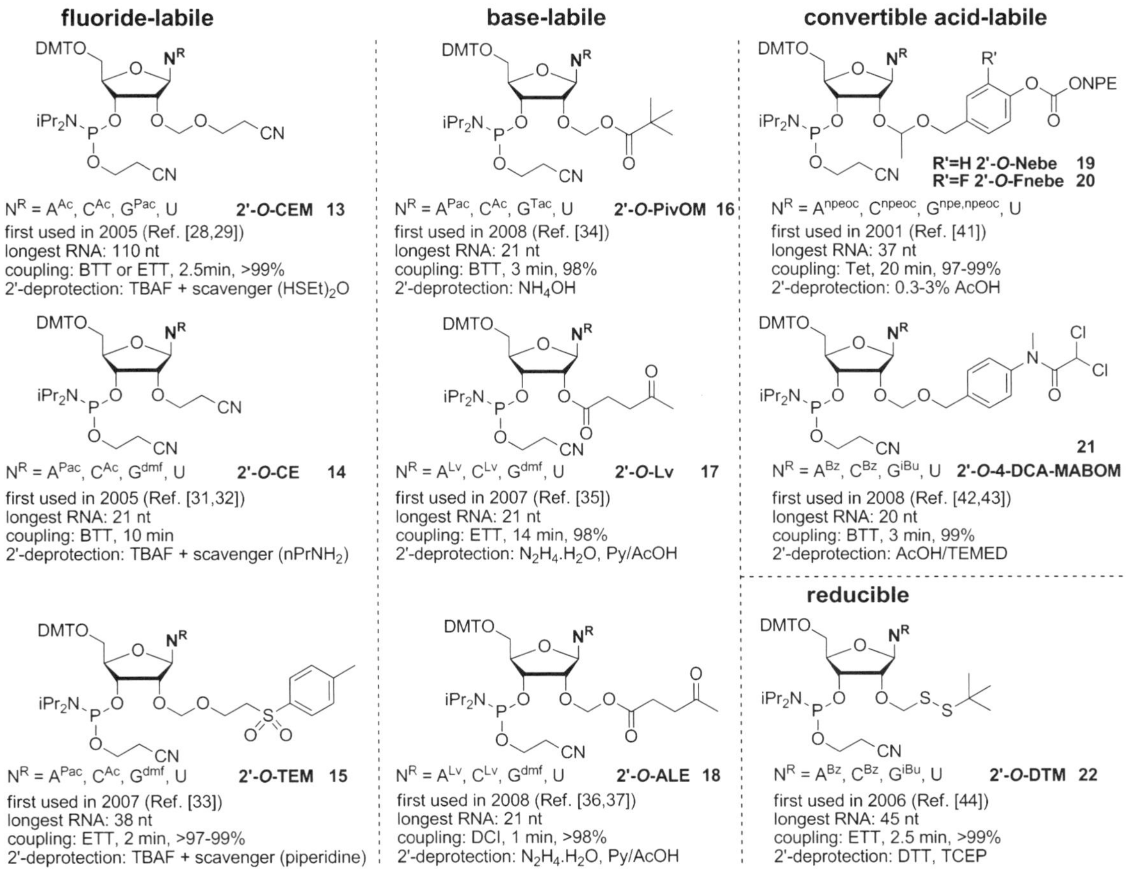

Figure 1.5 *Overview of 5′-O-DMT-protected phosphoramidites with various recently developed 2′-O-protecting groups, classified according to 2′-deprotection conditions*

The analogous 2′-*O*-cyanoethyl (CE) group without the formacetal moiety was also used as 2′-protecting group in RNA oligonucleotide synthesis [31,32] The corresponding phosphoramidites **14** were activated with BTT over a coupling time of 10 min. Complete deprotection of the 2′-cyanoethyl group was accomplished with 1 M TBAF in THF in the presence of 5% *n*-propylamine as acrylonitrile scavenger. Interestingly, the cyanoethyl group proved to be stable upon exposure to triethylamine trihydrofluoride. This property could be used to produce partially 2′-*O*-cyanoethylated RNA oligonucleotides by using a combination of 2′-*O*-CE and 2′-*O*-TBDMS phosphoramidites in solid-phase synthesis. It was proposed that partially cyanoethylated oligoribonucleotides, which were shown to form stable duplexes with RNA and DNA oligonucleotides, could be suitable candidates for RNAi experiments [32].

The 2-(4-tolylsulfonyl)ethoxymethyl (TEM) protecting group Another fluoride-sensitive 2′-acetal-type protecting group is the 2′-*O*-2-(4-tolylsulfonyl)ethoxymethyl (TEM) group that was reported by Chattopadhyaya and co-workers in 2007 [33]. The TEM group was designed to be more stable towards ammonia deprotection conditions compared with the CEM group, which reduces the risk of chain cleavage during nucleobase and phosphate deprotection. The 2′-*O*-TEM-protected phosphoramidites **15** were activated with ETT and allowed to couple for 2 min. The stepwise coupling yields ranged from 97 to 99%. RNA oligonucleotides of up to 38 nt in length have been produced. The deprotection proceeds under similar conditions as used for CEM deprotection. However, it was observed that during fluoride-assisted removal of TEM, the released 4-tolyl vinyl sulfone modified the exocyclic amino groups on nucleobases via Michael-type addition. Adduct formation was largely suppressed by using piperidine or morpholine as scavenger [33].

Base-labile protecting groups

The pivaloyloxymethyl (PivOM) protecting group The goal of reducing the number of steps for RNA deprotection after solid-phase synthesis for a rapid and efficient preparation of RNA relies on the design of base-labile protecting groups for the 2′-hydroxyl group. However, this is a major challenge due to the inherent instability of fully deprotected RNA under alkaline conditions. First success towards this goal has recently been reported by using the base-labile 2′-*O*-pivaloyloxymethyl (PivOM) protecting group for ribonucleoside phosphoramidites **16**. Debart and co-workers demonstrated the synthesis of up to 21 nt long RNAs with this base-labile acetal ester protecting group [34]. The average coupling efficiency upon activation with BTT was >99% within a 3 min coupling time. The PivOM group was designed for a fast, two-step, all-base deprotection scheme, which consists of the selective cleavage of the phosphate cyanoethyl protecting groups by a non-nucleophilic base (DBU or piperidine), followed by ammonia treatment for the release of the oligomer from the solid support, with simultaneous deprotection of nucleobases and cleavage of the 2′-*O*-PivOM groups. It has been proposed that the formacetal intermediate on the 2′-position, which is generated upon ammonolysis of the acetal ester, is stable enough to ensure that the RNA does not degrade in aqueous ammonia. Upon evaporation of the ammonia deprotection mixture, the decrease in the pH results in fragmentation of the hemiacetal to liberate the unmodified RNA. The activity of an siRNA that was prepared by the new PivOM strategy was shown to have gene silencing activity comparable to a commercial sample of the same siRNA sequence [34].

The levulinyl (Lv) and levulinyl acetal ester (ALE) protecting groups Despite the difficulties associated with using base-sensitive 2′-protecting groups, 2′-*O*-levulinyl ribonucleoside phosphoramidites **17** were recently reported for use in RNA solid-phase synthesis on a fluoride-labile hydroquinone-*O*,*O*′-diacetic acid (Q-linker) CPG support [35]. The nucleobases also carried levulinyl protecting groups on the exocyclic amino groups for adenine and cytosine and the dimethylformamidine group for guanosine. These phosphoramidites were activated with ETT and coupled for 14 min to reach an average coupling yield of 98.5%, which is comparable to the standard 2′-*O*-TBDMS phosphoramidites. Release and deprotection of 21 nt long RNA oligonucleotides proceeded in three steps: first, the cyanoethyl groups were removed by triethylamine in acetonitrile, then hydrazinolysis cleaved the nucleobase and 2′-*O*-Lv groups, and finally, the completely deprotected RNA was released from the support by 1 M TBAF in THF. The Lv group is orthogonal to the TBDMS group and can therefore be used to prepare partially 2′-protected RNA oligonucleotides that might be useful for RNAi applications. One major obstacle is the difficult synthesis of pure 2′-*O*-Lv 3′-phosphoramidites because of the tendency of the levulinyl group to undergo facile 2′ to-3′ isomerization.

To overcome the issue of protecting group migration and to improve the coupling efficiency, the analogous 2′-*O*-levulinyl acetal ester (ALE) protecting group was developed, which can also be released with buffered hydrazine solutions [36,37] The ribonucleoside phosphoramidites **18** were coupled with DCI as activator and yielded a coupling efficiency of >98% in a coupling time as short as 1 min. One additional motivation for the development of the ALE protecting group was the goal of allowing deprotection of RNA oligonucleotides that remain attached to a solid support such as a glass or chip surface [37].

Acid-labile protecting groups

A class of acid-labile acetal protecting groups that are compatible with the 5′-*O*-DMT group has also been reinvestigated recently. Early examples were the 2′-*O*-Fpmp (**11**) and 2′-*O*-Cpmp (**12**) protecting groups that showed great promise in the 1990s [38,39] The Fpmp-protected phosphoramidites **10** have been commercially available for some time, but there have been few reports of successful RNA synthesis using this method [40]. A next generation of acid-labile acetal protecting groups was proposed by Matysiak and Pfleiderer in 2001 [41]. The protected benzylacetal derivatives 2′-*O*-Nebe (**19**) and 2′-*O*-Fnebe (**20**) are also compatible with the 5′-*O*-DMT group provided that the 4-nitrophenyloxycarbonyl group is present on the acetal moiety. Upon DBU treatment for nucleobase and phosphate deprotection, the 2′-*O*-acetal is converted into a much more acid-sensitive derivative that can subsequently be released under mildly acidic conditions. Nebe-protected phosphoramidites **19** have been applied for the synthesis of up to 37 nt long RNAs, but the coupling kinetics of these building blocks were very slow (a 20 min coupling time was required for optimal performance) [41].

In 2008, Beaucage and co-workers reported the 2′-*O*-4-(*N*-dichloroacetyl-*N*-methylamino)benzyloxymethyl (4-DCA-MABOM) group for 2′-hydroxyl protection of ribonucleoside phosphoramidites **21**, which is designed according to a similar strategy for convertible acetal protecting groups [42]. It was found that a 4-aminobenzyloxymethyl (4-ABOM) group resulting from reduction of a 4-NBOM derivative was sensitive to 0.1 M acetic acid at 90 °C [43]. Deprotection presumably proceeded through formation of an iminoquinone methide intermediate and elimination of formaldehyde. It was shown that the electronic and structural parameters of the benzyl acetal critically influence the acid sensitivity and that the presence of a single electron-donating methyl group on the aminobenzyl moiety is most favorable. Accordingly, phosphoramidites **21** with 2′-*O*-4-DCA-MABOM were prepared in which the methylaminobenzyl group was protected by dichloroacetylation. Activation of **21** was achieved with BTT for 3 min and an average stepwise coupling efficiency of 99% was reported for the synthesis of a 20 nt RNA oligonucleotide [42]. Release of the oligonucleotide from the support and removal of nucleobase and phosphate protecting groups were achieved by incubation with ammonia for 10–16 h at 55 °C. Under these conditions, the dichloroacetyl group of 4-DCA-MABOM was also cleaved. The residual 2′-*O*-4-MABOM group was removed in a TEMED-buffered acetic acid solution at pH 3.8 within 30 min at 90 °C. This final deprotection step is essentially identical with the release of the orthoester moiety in ACE chemistry. So far, the 4-DCA-MABOM-protected phosphoramidites have only been used for the synthesis of a model 20-mer oligoribonucleotide. It remains to be demonstrated if the combination of the advantageous features from 5′-*O*-DMT and acid-labile 2′-*O*-acetals will become a powerful alternative to the currently most widely used RNA synthesis strategies.

The reducible 2-tert-butyldithiomethyl (DTM) protecting group

Ribonucleoside phosphoramidites **22** protected with the 2′-*O*-*tert*-butyldithiomethyl (DTM) protecting group were developed by Kwiatkowski and co-workers in 2006 [44]. Upon activation of phosphoramidites **22** with ETT, coupling efficiencies of up to 99.8% were achieved within a coupling time of 2.5 min. The longest RNA synthesized with the 2′-*O*-DTM protecting group consisted of 45 nt. Release of the oligonucleotide and deprotection of nucleobase and phosphate groups proceeded under standard ammonolytic conditions. Cleavage of the DTM group was achieved with 1,4-dithiothreitol (DTT) or tris(2-carboxyethyl)phosphine (TCEP) in a buffered pH 7.6 solution at 55 °C. The moderate stability of 2′-*O*-DTM-protected phosphoramidites in solution at room temperature (sufficiently stable within 12–24 h) currently seems to be a limitation of this otherwise attractive method.

Most of these newer approaches have not yet been evaluated extensively in the RNA research community and their suitability for widespread applications has yet to be demonstrated. The 2′-protecting groups that are orthogonal to the traditional TBDMS and TOM silyl protecting groups hold special promise to become useful for the preparation of partially protected RNAs that might have important features for siRNA delivery as they increase hydrophobicity and might facilitate cell uptake. In this context, it has been proposed that RNAs modified with novel biolabile 2′-protecting groups should enable RNAi experiments in which the finally unprotected RNAs would be liberated only after administration inside the cells [45].

1.2.2 Synthetic strategies for RNA modification

The site-specific incorporation of nucleoside modifications by RNA solid-phase synthesis generally follows one of two common strategies. The most widespread and versatile strategy is the direct incorporation of nucleotide analogs by replacing standard phosphoramidites with modified derivatives during solid-phase synthesis (see Section 1.2.2.1). Alternatively, the standard oxidation and/or capping solutions can be substituted by specialized reagents, which results in the synthesis of backbone-modified RNA (e.g. phosphorothioate or phosphoroselenoate RNA). The second common strategy for nucleoside functionalization involves post-synthetic modification, for which various strategies are feasible (see Section 1.2.2.2). Other important strategies for the synthesis of modified RNA involve combinations of chemical and enzymatic methods (see Section 1.2.2.3).

1.2.2.1 Solid-phase synthesis of modified RNA via phosphoramidite chemistry

Nucleobase and ribose modification via nucleoside analog phosphoramidites

Modified phosphoramidites for 5′-O-DMT-2′-O-silyl chemistry (i) Commercially available phosphoramidites The foremost requirement for successful incorporation of nucleoside analogs via solid-phase phosphoramidite chemistry is that the desired modification is chemically compatible with all reactions and reagents used in the synthesis cycle and also remains unchanged under the deprotection conditions. Various modified phosphoramidites compatible with TBDMS and TOM chemistries are commercially available from different sources (e.g. GlenResearch, ChemGenes, Berry Associates). Figures 1.6 and 1.7 show most of the currently obtainable nucleobase- and ribose-modified RNA building blocks according to the 5′-O-DMT-2′-O-silyl protection scheme; the nucleoside analogs are shown with appropriate protecting groups. The following classes of nucleobase modifications are represented:

- Nucleosides containing natural RNA modifications, such as alkylated nucleobases [m^3U (**23**), m^5U (**24**), m^3m^5U (**25**), m^5C (**26**), m^1A (**27**), i^6A (**28**)], pseudouridine [Ψ (**29**)] and thio-substituted nucleobases [s^6G (**30**), s^4U (**31**)], are mainly used to mimic native systems closely and to study biochemical and biophysical properties of modified RNAs. Thio-substituted nucleosides are also used for photo-crosslinking studies or for derivatization via disulfide or thioether bonds.
- Nucleosides with altered patterns of exocyclic functional groups as compared with their natural counterparts [2AP (**32**), DAP (**33**), isoG (**34**), isoC (**35**)] and other derivatives entirely missing certain exocyclic functional groups [inosine (**36**), nebularine (**37**), zebularine (**38**)] are used to study the specific roles of nucleobase amino and carbonyl groups in RNA folding and catalysis.
- Nucleoside analogs with altered patterns of ring nitrogen atoms [c^7A (**39**), c^7G (**40**), 8-aza-c^7A (**41**)] are primarily used for structural and mechanistic studies of RNAs.
- Halogenated nucleosides, such as Br^5U (**42**), Br^5C (**45**) and Br^8A (**47**), can be used for heavy atom isomorphous replacement studies in X-ray crystallography. 5-Halopyrimidine nucleosides are photoreactive and have also been used for RNA–protein crosslinking studies. The iodinated nucleosides I^5U (**43**) and I^5C (**46**) are useful for further derivatization by Pd-catalyzed cross-coupling reactions (see Section 1.2.2.2.2), whereas F^5U (**44**) is mainly used as a structural probe for studies on enzymes.
- Fluorescent nucleoside analogs such as etheno-A (**48**), pyrrolo-C (**49**) and 2AP (**32**) allow for monitoring of conformational changes during RNA folding.
- Convertible nucleosides, such as 4-triazolyluridine (**50**), O^4-(4-chlorophenyl)-U (**51**), O^6-(4-chlorophenyl)-I (**52**) and F^2I (**53**) and amino-tethered nucleosides such as 5-aminoallyl-U (**54**) are used to prefunctionalize RNA oligonucleotides for post-synthetic derivatization (see Section 1.2.2.2.1).

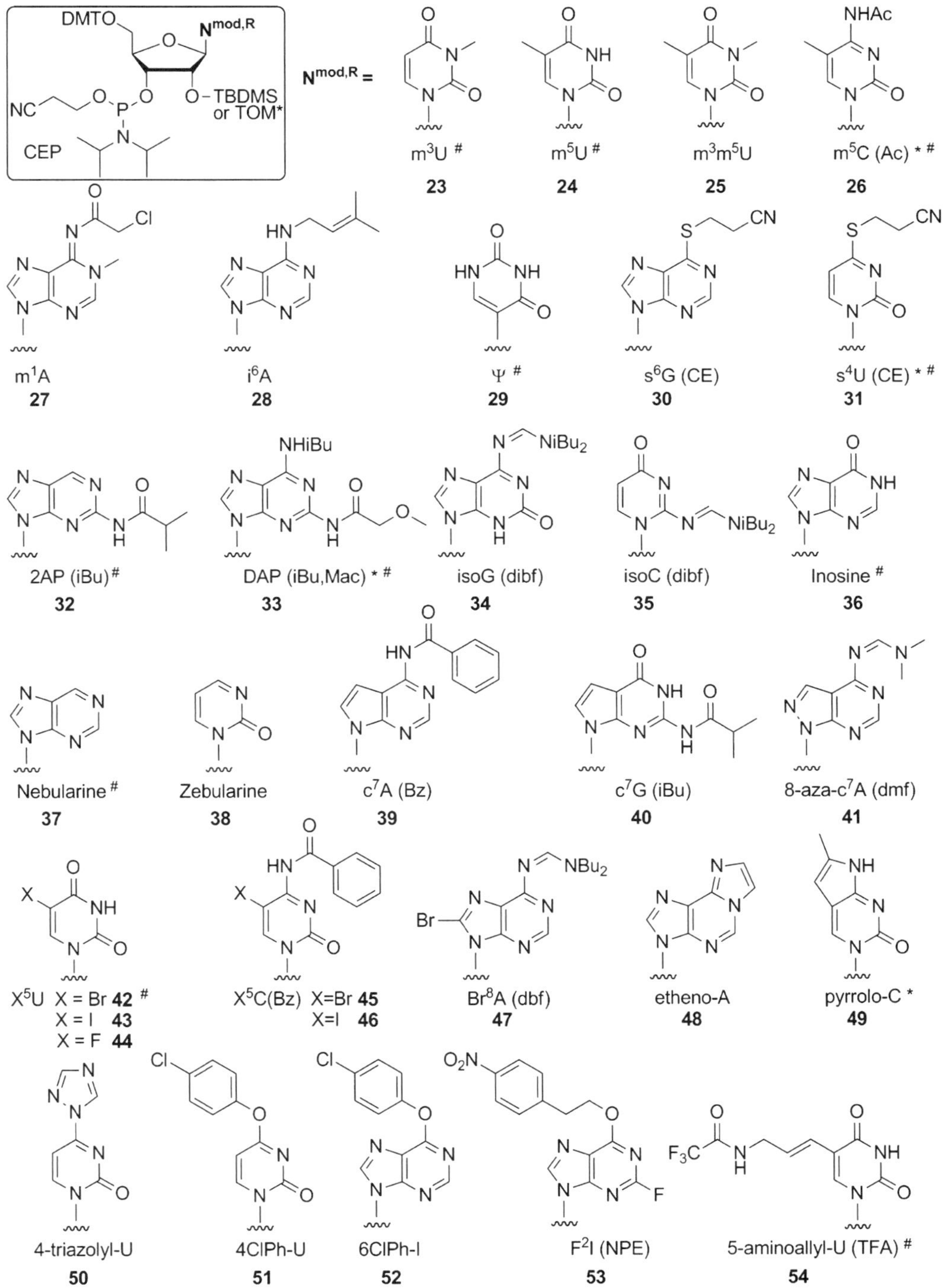

Figure 1.6 *Collection of commercially available modified phosphoramidites compatible with the 5′-O-DMT-2′-O-silyl protection scheme. The modified nucleobases are shown in their protected forms. The asterisk (*) indicates commercial availability as 2′-O-TOM-protected phosphoramidite; # denotes availability via custom synthesis service using 2′-O-ACE chemistry. Mac = methoxyacetyl, dibf = diisobutylaminomethylene, TFA = trifluoroacetyl, ClPh = chlorophenyl*

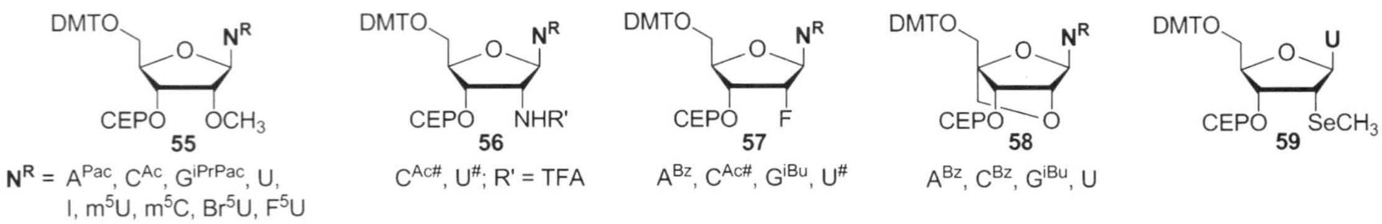

Figure 1.7 *Selection of commercially available 2′-ribose-modified phosphoramidites compatible with the 5′-O-DMT-2′-silyl protection scheme. The symbol # denotes availability via custom synthesis service using ACE chemistry. iPrPac = 4-isopropylphenoxyacetyl*

The most important classes of commercially available 2′-modified RNA phosphoramidites:

- 2′-*O*-Methyl modified nucleoside phosphoramidites **55**. 2′-OMe groups are common natural ribose modifications found in all classes of RNA. In addition to the four standard nucleosides A, C, G and U, various nucleobase-modified 2′-OMe-derivatized ribonucleoside analogs are available as phosphoramidite building blocks.
- 2′-Amino-2′-deoxy-modified nucleoside phosphoramidites **56**, of which only uridine and cytidine monomers are currently commercially available, but the analogous adenosine and guanosine phosphoramidites have been reported [46,47]. 2′-Amino-modified RNA was used in studies of the ribosomal peptidyl transferase reaction mechanism [48] and for thermodynamic analysis of ribozyme mechanisms [49]. The 2′-amino functionality can easily be derivatized with any biophysical label that can be supplied as activated carboxylic acid, isocyanate or isothiocyanate (see Section 1.2.2.2.4).
- 2′-Deoxy-2′-fluoro-modified nucleoside phosphoramidites **57**. 2′-Fluoro-modified oligonucleotides have been used as nuclease-resistant RNA analogs for antisense and RNAi applications [17], and as probes for ^{19}F NMR in the determination of RNA conformational equilibria and ligand binding [50,51].
- LNA nucleosides **58**, in which the 2′-oxygen and the 4′-carbon atoms are linked with a methylene unit to lock the ribose in the C3′-*endo* conformation. This artificial ribose modification has been designed for improved base pairing behavior to complementary RNA targets as compared with unmodified DNA or RNA strands and it confers nuclease stability to the oligonucleotide [52].
- 2′-Methylseleno RNA phosphoramidites **59**, of which currently only 2′-SeMe-U is purchasable, but synthetic strategies for the nucleosides C, A and G are known [53,54].

In addition to the nucleobase- and ribose-modified nucleoside analogs depicted in Figures 1.6 and 1.7, several non-nucleoside phosphoramidites are commercially available and can be used to incorporate internucleotide spacers (mono-, tri- or hexaethylene glycol units) or abasic site analogs into oligoribonucleotides. Moreover, phosphoramidites of fluorophores (e.g. fluorescein and its derivatives, tetramethylrhodamine, cyanine dyes), quenchers, biotin, acridine, psoralen and cholesterol are available for conjugation to the 5′- or 3′-termini of oligonucleotides. Most of these modifications are available as DMT-protected analogs, but several are also offered for combination with ACE chemistry.

(ii) Chemical synthesis of modified nucleoside phosphoramidites Although the number of commercially obtainable nucleoside analogs is constantly increasing, many modifications desired by researchers for specific experiments are not easily accessible and it remains the task of chemists to develop efficient synthetic routes for the preparation of suitable phosphoramidite building blocks.

The synthesis of modified phosphoramidites usually involves installation of the desired nucleobase or ribose modification on a partially protected nucleoside, followed by protecting group manipulations that finally result in functionalization of the modified nucleoside according to an orthogonal protecting group scheme for RNA solid-phase synthesis. Two general strategies for the synthesis of nucleobase-modified 2′-*O*-TOM-protected RNA phosphoramidites are depicted in Scheme 1.3a. These two routes differ in the order of nucleobase modification and introduction of the 2′-*O*-TOM group. Route 1 entails the transformation of the nucleobase early in the synthetic scheme, with preparation of the free (i.e. ribose-unprotected) nucleoside analog as intermediate. The ribose is then functionalized in at least three more steps. Alkylation of the vicinal 2′,3′-diol, which is achieved via activation as dibutylstannylidene diacetal (Scheme 1.3b), yields a mixture of 2′-*O*- and 3′-*O*-alkylated isomers that need to be separated chromatographically (similar considerations apply to direct silylation with TBDMS-Cl or similar reagents, which also gives a mixture of 2′-*O*- and 3′-*O*-silylated isomers). In route 2, the 2′-*O*-TOM group is introduced before the nucleobase transformation is performed. This strategy generates only the desired 2′-*O*-alkylated isomers of precious modified compounds and is required for modifications that are not stable under the conditions for installation of the TOM protecting group. The majority of reported syntheses followed one of these two general routes. In principle, alternative synthetic approaches are conceivable and the exact sequence of synthesis steps is largely depending on the chemical stability of the targeted nucleobase modification.

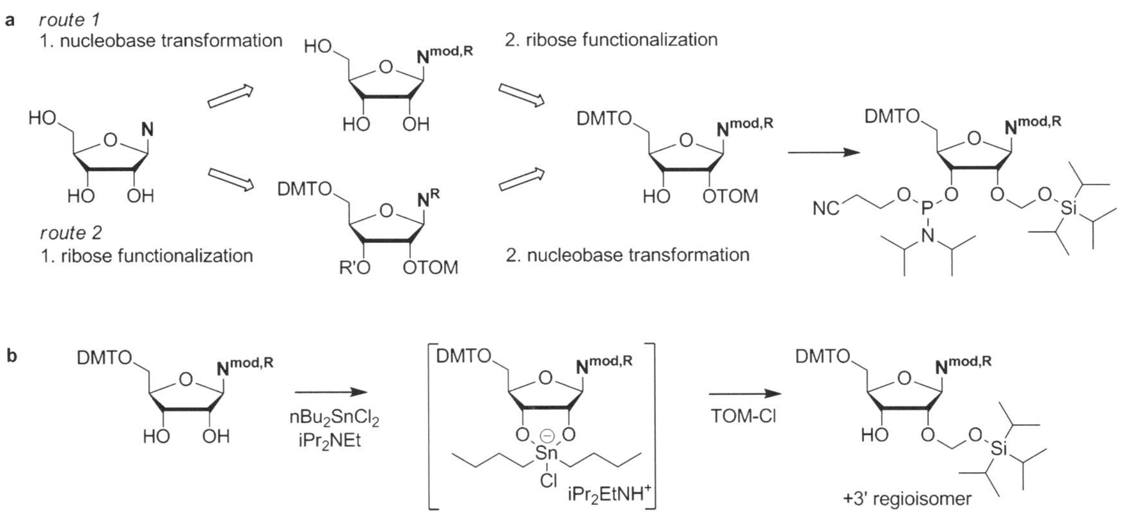

Scheme 1.3 *(a) General routes for the synthesis of nucleobase-modified 2′-O-TOM-protected phosphoramidites. (b) Introduction of 2′-O-TOM protecting group via cyclic 2′,3′-dibutylstannylidene diacetal*

To illustrate the synthetic routes outlined in Scheme 1.3a, two reported syntheses of $m^2{}_2G$ building blocks are depicted in Scheme 1.4. The phosphoramidites **65** and **70** differ only in the presence or absence of the O^6-NPE protecting group. In route 1 [24], the dimethylamino group is installed via nucleophilic aromatic substitution of the O^6-NPE-protected 2-fluoroinosine derivative **61**, which was prepared from guanosine in three steps. The resulting $m^2{}_2G$ nucleoside **62** was transformed into the phosphoramidite building block **65** by stepwise introduction of the 5′-*O*-DMT (**63**) and 2′-*O*-TOM protecting groups (**64**), followed by 3′-phosphitylation. In route 2 [25], the stepwise methylation of the 5′-*O*-DMT-2′-*O*-TOM-protected guanosine **67** was achieved by treatment with 1,3-benzodithiolylium tetrafluoroborate and subsequent reduction of the intermediate with $(Me_3Si)_3SiH$–AIBN to give the monomethylated intermediate **68**. The second N^2-methyl group on **69** was installed by repeating this reaction sequence. Again, 3′-phosphitylation as the

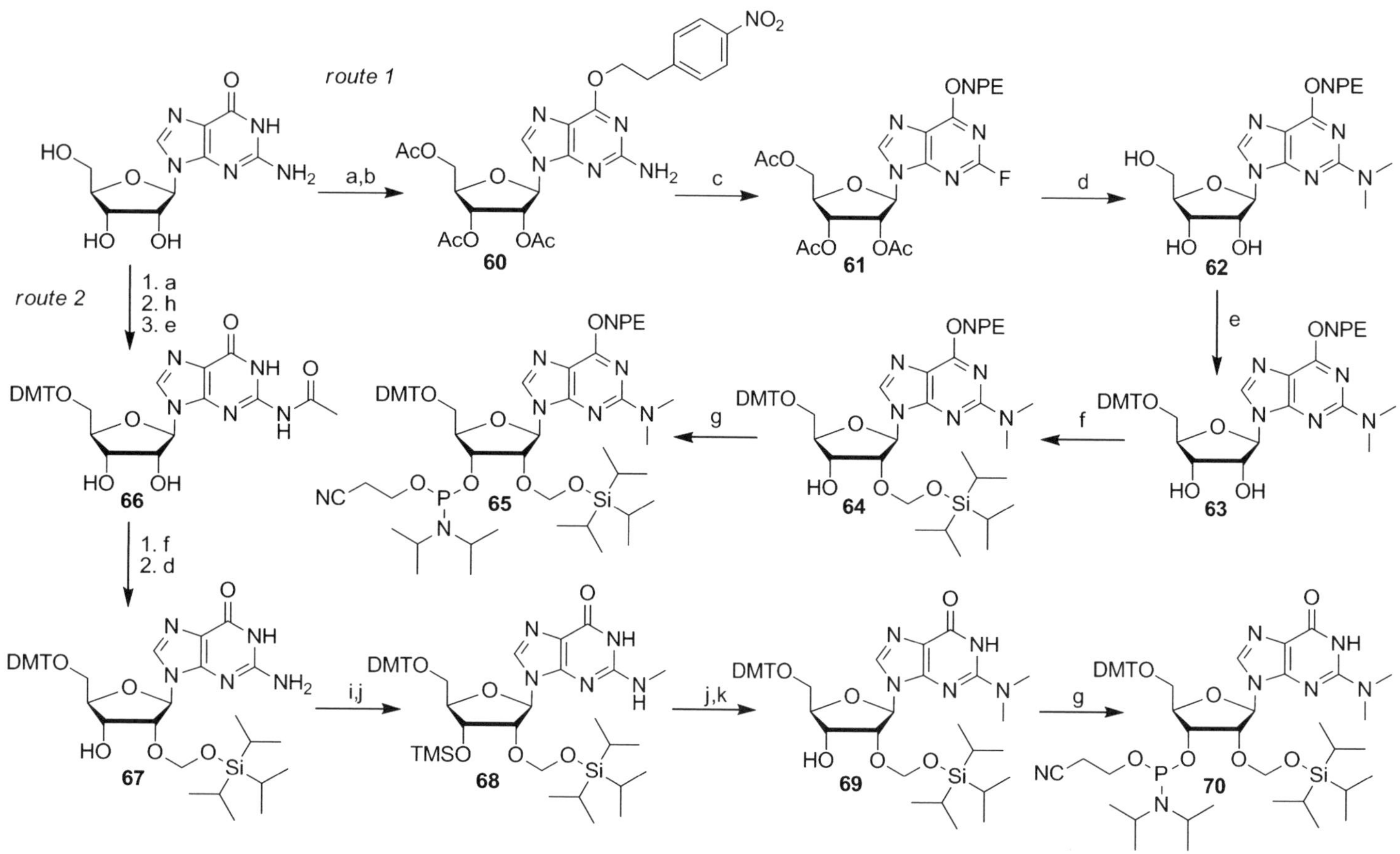

Scheme 1.4 *Two synthetic routes to 5′-O-DMT-2′-O-TOM-protected $m^2{}_2G$ phosphoramidite building blocks that differ only in their nucleobase protecting groups. (a) Ac_2O, DMF–pyridine; (b) 4-nitrophenylethanol, diisopropyl azodicarboxylate, PPh_3, dioxane; (c) HBF_4, $NaNO_2$, acetone–H_2O; (d) $MeNH_2$, EtOH–H_2O; (e) DMT-Cl, pyridine; (f) 1, Bu_2SnCl_2, iPr_2NEt; 2,. TOM-Cl, $(CH_2Cl)_2$; (g) CEP-Cl, Me_2NEt, CH_2Cl_2; (h) NaOH–MeOH–THF; (i) Me_3SiCl, in pyridine; (j) 1, 1,3-benzodithiolylium tetrafluoroborate; 2, $(Me_3Si)_3SiH$–AIBN, in benzene; (k) NH_3 in MeOH–THF*

last synthetic step yielded the phosphoramidite **70** for solid-phase RNA synthesis. Similarly, both routes have been implemented for the synthesis of a collection of nucleobase-methylated ribonucleoside phosphoramidites including m^1G, m^2G, m^1I, m^4C, m^6A and $m^6{}_2A$ [24,25]. These methylated nucleosides have been used to study the influence of nucleobase methylations on equilibria of RNA secondary structures [55–57].

The general routes 1 and 2 in Scheme 1.3a are also applicable to slightly deviating hydroxyl group protection schemes. For example, the synthesis of 3-deazaadenosine (c^3A) phosphoramidite **73** (Scheme 1.5) followed general route 1, although the fluoride-labile triisopropylsilyl (TIPS) protecting group was used instead of TOM or TBDMS (TIPS favorably gave higher 2′-regioselectivity in the silylation step). Construction of the modified nucleoside phosphoramidite **73** followed a 12-step synthetic route starting from inosine, via the crucial 5-amino-4-imidazolecarboxamide (AICA) ribonucleoside intermediate **71** [58]. The modified ribonucleoside 3-deazaadenosine **72** was obtained as free nucleoside intermediate and was further derivatized to the corresponding phosphoramidite **73** via benzoylation, tritylation, silylation and phosphitylation. The c^3A phosphoramidite **73** was incorporated into a 26 nt fragment of the 23S rRNA and was used for site-specific modification of the peptidyl transferase center in the large ribosomal subunit for mechanistic studies of ribosome-catalyzed peptide bond formation [58].

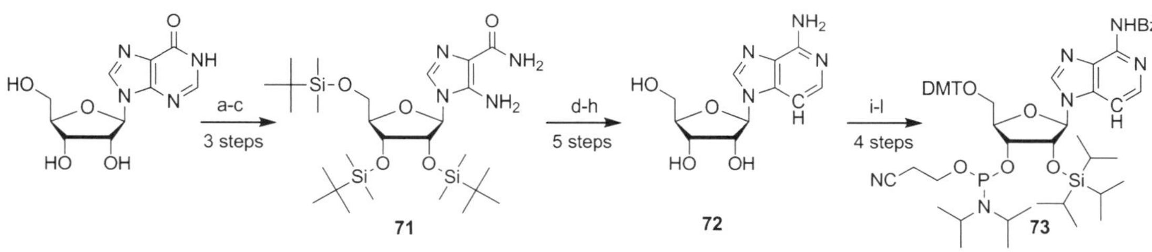

Scheme 1.5 *Synthesis of 3-deazaadenosine phosphoramidite* ***73****. (a) TBDMS-Cl, imidazole, DMF; (b) 2,4-dinitrochlorobenzene, K_2CO_3, DMF; (c) ethylenediamine; (d)* p*-toluenesulfonyl chloride, pyridine; (e) isoamyl nitrite, CH_2I_2; (f) ethinyltrimethylsilane, NEt_3, $(PhCN)_2PdCl_2$, CH_3CN; (g) NH_3, MeOH; (h) Amberlite IRA 900 fluoride form, toluene; (i) 1, Me_3SiCl, pyridine; 2, benzoyl chloride, pyridine; (j) DMT-Cl, pyridine; (k) TIPS-Cl, $AgNO_3$, pyridine–THF; (l) CEP-Cl, Me_2NEt, CH_2Cl_2*

The compatibility of the nucleoside modification with the synthesis and deprotection conditions is of vital importance for successful use in solid-phase synthesis. In this regard, the hypermodified nucleoside analogs of the wyosine family, which are often found in the anticodon loop of tRNAs, are very challenging candidates for chemical synthesis. Wyosine is known to be particularly sensitive to acidic conditions and therefore it had not been successfully incorporated into synthetic RNA oligonucleotides until recently. With the exception of 4-demethylwyosine [59], no other syntheses of wyosine analog phosphoramidite building blocks have been reported [60]. In 2008, Porcher developed a synthetic method that allows for coupling of wyosine phosphoramidite **76** at the 5′-terminus of an RNA oligonucleotide [60]. A synthetic approach related to general route 2 of Scheme 1.3a was employed for the preparation of the wyosine phosphoramidite building block **76** (Scheme 1.6), because the wyosine nucleoside was not stable under conditions for introduction of the TOM group. The 5′-*O*-DMT-2′-*O*-TOM-protected guanosine **67** was transformed into the 4-demethylwyosine intermediate **74** via alkylation of N1 with bromoacetone in the presence of potassium iodide and subsequent cyclization in the presence of molecular sieves. Detritylation and ribose acetylation afforded a diacetylated intermediate that was methylated at N5 with CH_2I_2–Et_2Zn. After deacetylation, the 5′-OH of the new wyosine building block **75** was protected with the TBDMS group and subsequent phosphitylation produced the 2′,5′-silyl-protected phosphoramidite **76**. Wyosine was incorporated at the 5′-end of a 10 nt RNA via solid-phase synthesis under optimized oxidation and deprotection conditions. The short modified RNA oligonucleotide was then enzymatically phosphorylated and used in a ligation reaction to generate 18 nt long RNA hairpins for studying the kissing-loop interaction of a fragment of the retroviral RNA of Moloney murine leukemia virus [60].

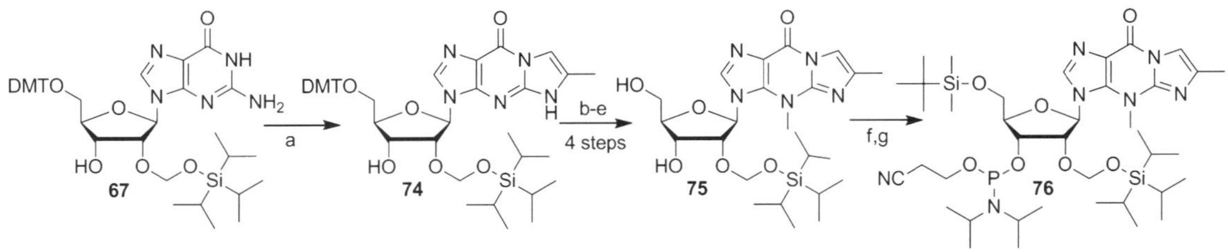

Scheme 1.6 *Synthesis of wyosine phosphoramidite **76**. (a) 1, K_2CO_3, DMF; KI, then bromoacetone; 2, molecular sieves, CH_2Cl_2; (b) dichloroacetic acid, CH_2Cl_2; (c) 1, Ac_2O, pyridine; 2, pyridine, MeOH, H_2O; (d) CH_2I_2, Et_2Zn, Et_2O; (e) NH_3, MeOH; (f) TBDMS-Cl, imidazole, DMF–CH_2Cl_2; (g) CEP-Cl, iPr_2NEt, CH_2Cl_2*

Examples of ribose-modified nucleosides not commercially available include the 2′-*O*-2-aminoethyl- and 2′-*O*-2-mercaptoethyl-modified nucleosides that are appropriately protected to be compatible with the 5′-*O*-DMT-2′-*O*-silyl chemistry (also known as 2′-tethered amine and 2′-tethered thiol, respectively). Several synthetic routes have been described for the 2′-alkylation of various nucleosides [61–63]. Silverman and co-workers reported a complete collection of 2′-tethered thiol and 2′-tethered amine building blocks for each of the four common ribonucleotides A, C, G and U [64]. The most recent approach for the synthesis of 2′-*O*-aminoethyl nucleosides involves direct alkylation of the 2′-OH with phthalimidoethyl triflate [65]. RNA analogs with 2′-tethered amino substituents are useful substrates for post-synthetic labeling [66] (see Section 1.2.2.2, *Derivatization of artificial amino groups at ribose and nucleobase residues*) and have also been studied for RNAi applications [67]. A variety of other RNAs derivatized with lipophilic and zwitterionic 2′-alkylations, such as guanidinoethyl or aminopropyl modifications, have been analyzed for their properties as antisense or siRNA agents. The improved chemical stability and higher affinity for RNA targets and also the ability partially to neutralize the negatively charged phosphate backbone furnish the modified RNAs with better physicochemical and pharmacokinetic properties for medical applications [17,18,68].

In a different area of RNA research, a diversity of 2′-modified nucleotide analogs allowed the construction of an atomic mutation cycle to determine whether 2′-hydroxyl groups donate functionally important hydrogen bonds in the catalytic mechanisms of RNA enzymes and splicing machineries [49,69,70]. In a similar context, 2′-*C*-α-hydroxyalkyl-modified cytidine analogs have recently been used to probe RNA-solvent interactions in the catalytic core of the group II intron during the spliced exon reopening (SER) reaction [71,72]. Also, phosphorothiolate linkages in RNA oligonucleotides, in which a sulfur atom replaced the 3′- or 5′-bridging oxygen atom of the phosphodiester linkage, have yielded fundamental insights into RNA functions and revealed the participation of metal ions and hydrogen bonds in catalytic mechanisms [73–75]. A recent addition to the pool of highly functionalized probes to uncover complex interaction networks within ribozyme active sites is the combination of 3′-phosphorothiolate and 2′-ribose modifications. This was illustrated by the synthesis of a 2′-*O*-methyl-3′-thioguanosine phosphoramidite and its incorporation into an oligoribonucleotide that served as a substrate for investigations of group I intron catalysis [76].

Modified phosphoramidites for 5′-O-silyl-2′-O-ACE chemistry The commercial 2′-*O*-ACE RNA synthesis service by Dharmacon (part of Thermo Fisher Scientific) now also offers certain RNA modifications to be incorporated into custom oligoribonucleotides, but the great variety of accessible analogs known from 2′-*O*-silyl chemistry has not yet been achieved with the 2′-*O*-ACE method (the modifications currently offered are marked with # in Figures 1.6 and 1.7). There have also been some recent reports on the synthesis of new nucleobase- and ribose-modified ACE-phosphoramidites and their incorporation into RNA.

Lusic *et al.* synthesized a 5-formylcytidine (f^5C) 5′-*O*-silyl-2′-*O*-ACE-protected building block (**79**, Scheme 1.7) that allowed the incorporation of f^5C into an RNA oligonucleotide for the first time [77]. The f^5C modification is found at the wobble position of human mitochondrial methionine-specific tRNA, but its function during decoding, chain initiation or chain elongation is still unknown. The recent synthesis of an f^5C-modified tRNA anticodon domain now allows the analysis of its thermodynamic properties and allows structural investigations of the modified anticodon stem–loop. The key step in the synthesis of phosphoramidite **79** was the installation of the hydroxymethylene unit into a protected cytidine nucleoside via a Baylis–Hillman-type reaction with formaldehyde. Selective oxidation of the allylic alcohol in **77** and acetonide deprotection gave the free f^5C nucleoside **78**, which was appropriately functionalized to the 2′-*O*-ACE-protected phosphoramidite **79** [77].

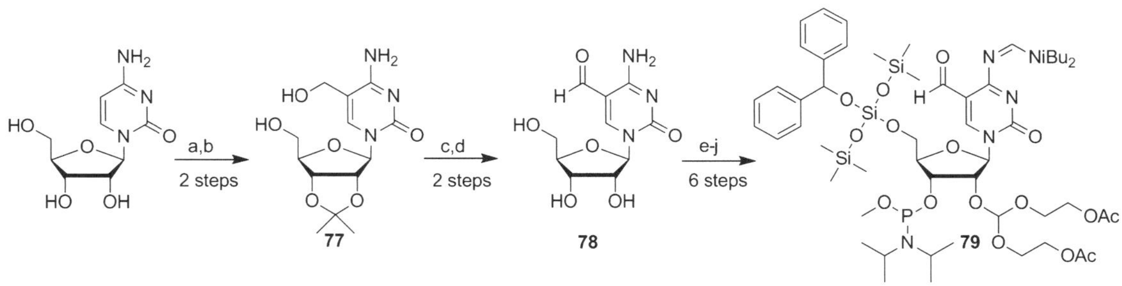

Scheme 1.7 *Synthesis of 5-formylcytidine phosphoramidite **79**. (a) Dimethoxypropane, acetone, $HClO_4$ (cat.); (b) paraformaldehyde, KOH; (c) $RuO_2{\cdot}H_2O$, dioxane; (d) 1 M HCl; (e) 1,3-dichloro-1,1,3,3-tetraisopropyldisiloxane, pyridine–DMF; (f) $iBu_2NCH(OMe)_2$, DMF; (g) tris(2-acetoxyethoxy)orthoformate, pyridinium* p-*toluenesulfonate, TBDMS-pentadione, CH_2Cl_2; (h) HF–TEMED, CH_3CN; (i) BzH-Cl, iPr_2NH, CH_2Cl_2; (j) $(MeO)P(iPr_2N)_2$, iPr_2NH, ETT, CH_2Cl_2*

Micura and co-workers described the preparation of all four (A, C, G, U) 5′-silyl protected 2′-methylseleno ribonucleoside phosphoramidites and successfully incorporated them into RNA oligonucleotides [54]. Expanding seleno-RNA synthesis to be compatible with ACE chemistry was an important contribution to making selenium-modified RNA generally accessible.

Other examples for the recent syntheses of modified ACE phosphoramidites include 2-iodoadenosine and 5-iodocytidine analogs; their application for post-synthetic RNA modification is discussed in Section 1.2.2.2, *Palladium-catalyzed cross-coupling of halogenated nucleobases* [78]. The increasing demand for modified RNAs and the growing popularity of ACE chemistry will likely result in a broader spectrum of ACE-containing modified RNA building blocks in the future.

Backbone modification via alternative oxidation reagents

Solid-phase phosphoramidite chemistry can also be used to introduce backbone modifications into RNA oligonucleotides. The most important example is represented by the phosphorothioate group, in which a non-bridging oxygen of the phosphodiester internucleotide linkage is replaced by a sulfur atom. The single-atom exchange can be classified as a conservative modification, because sulfur is situated in group VI of the periodic table right below oxygen. The van der Waals radius of sulfur is 0.3 Å larger than that of oxygen (1.5 vs 1.8 Å) and the P–S versus P–O bond length is increased by 0.5 Å (1.5 vs 2.0 Å). Phosphorothioates possess a number of chemical properties that clearly distinguish them from natural phosphodiester linkages, such as facile cleavage by iodine, their nucleophilicity and the preferred coordination of soft metal ions according to the HSAB (hard and soft acids and bases) principle. An important feature of the phosphorothioate modification is the chirality of the phosphorus center and the resulting existence of two diastereomers, denoted *R*p and *S*p (Figure 1.8a), which – in the case of short oligonucleotides – can be separated by HPLC. The availability of distinguishable isomers makes phosphorothioates ideally suited to probe enzymatic reaction mechanisms and offers the opportunity via metal ion rescue experiments (with soft metal ions such as Mn^{2+} or Cd^{2+}) to determine whether a particular phosphate is involved in metal ion interactions.

Phosphorothioate-modified oligonucleotides are the best known representatives of the first generation of antisense oligonucleotides. These analogs have been intensively investigated due to their increased nuclease resistance and attractive pharmacokinetic properties [79]. Reported shortcomings include off-target effects and observed cellular toxicity [80,81].

Interestingly, the phosphorothioate modification was the first observed naturally occurring DNA backbone modification. In 2007, phosphorothioate linkages were discovered in bacterial DNA from *Streptomyces lividans* and related bacteria [82]. This novel aspect of natural nucleic acid modification will stimulate additional investigations to understand why Nature has chosen phosphorothioates and by which biosynthetic pathways they are installed. Synthetic phosphorothioate oligonucleotides will play a significant role in such experiments.

The key step in the solid-phase synthesis of phosphorothioate oligoribonucleotides by the 2′-*O*-silyl or 2′-*O*-ACE chemistry is the replacement of the standard (iodine or *tert*-butyl hydroperoxide) oxidation solution by a sulfurizing reagent. Elemental sulfur was one of the first reagents used by Burgers and Eckstein [83]. Since then, a number of alternative sulfurizing reagents have been described. Among others, the Beaucage reagent, 3*H*-1,2-benzodithiol-3-one-1,1-dioxide (Figure 1.8b), is one of the best known backbone-modifying reagents [84]. Despite its limited stability in solution and suboptimal kinetics for the sulfurization of RNA, it has been widely used for the synthesis of phosphorothioate oligonucleotides. Other examples include 3-ethoxy-1,2,4-dithiazoline-5-one (EDITH) and [3-(dimethylaminomethylidine)amino]-3*H*-1,2,4-dithiazole-5-thione (DDTT), which feature improved performance for RNA sulfurization and extended shelf-life but are rather expensive [85]. More recently, cheaper reagents such as phenylacetyl disulfide (PADS) [86] and dimethylthiuram disulfide (DTD) [87] have been used successfully in large-scale syntheses of RNA phosphorothioates.

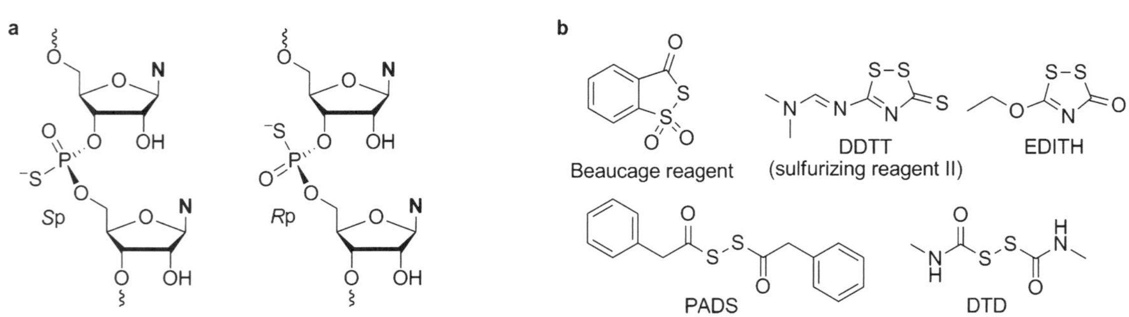

Figure 1.8 *(a) Phosphorothioate RNA diastereomers. (b) Selected sulfurizing reagents used for oligonucleotide phosphorothioate synthesis*

The stereodefined chemical synthesis of phosphorothioate oligonucleotides has been a longstanding goal, because the stereorandom synthesis of oligonucleotides containing n phosphorothioate modifications provides a mixture of 2^n diastereomers, each of which could interact differently with other biomolecules. At least two strategies have been described for the stereoselective synthesis of phosphorothioate DNA analogs. The method developed by Guga and Stec is based on 3′-O-(2-thio-1,3,2-oxathiophospholane) monomers that are condensed under strongly basic conditions [88]. This strategy is incompatible with standard coupling of phosphoramidite-based monomers, but stereodefined dinucleoside phosphoramidite building blocks with nitrobenzyl-protected sulfur atoms at the phosphorothioate junction were subsequently reported [89]. To construct any desired dinucleotide junction in either the *S*p or *R*p configuration, 32 different dinucleotide building blocks would be required, of which four have been reported so far [89]. Moreover, the necessary separation of the oxathiophospholane diastereomers is tedious and time consuming. Nevertheless, it seems possible that such dinucleoside building blocks could also be generated for the synthesis of stereodefined RNA phosphorothioates, but at present it is unclear whether such significant efforts would be worthwhile.

An alternative, phosphoramidite-based method has been investigated by the Agrawal [90,91] and Wada groups [92,93] for several years. The diastereroselective synthesis of proline-derived bicyclic 1,3,2-oxazaphospholidines as synthons for stereoselective synthesis of phosphorothioate oligonucleotides has been reported and conditions for activation and coupling have been carefully optimized [93]. The choice of activator reagent is critical because oxazaphospholidines are prone to epimerization in the presence of acidic activators. (*N*-Cyanomethyl)pyrrolidinium triflate (CMPT) was described as an optimal activator with low nucleophilicity; up to 12 nt long stereodefined phosphorothioate DNA oligonucleotides were obtained with excellent diastereoselectivities [93].

Wada's group has recently reported the successful stereodefined synthesis of phosphorothioate RNA oligonucleotides [94]. Their strategy is based on diastereomerically pure, 2′-*O*-TBDMS-protected, configurationally stable, bicyclic 1,3,2-oxazaphospholidine derivatives (*R*p)- or (*S*p)-**81** that were diastereoselectively prepared from 2-chloro-1,3,2-oxazaphospholidines L- or D-**80** (derived from L- and D-proline, respectively; Scheme 1.8). The best coupling performance of phosphoramidites **81** was observed with *N*-phenylimidazolium triflate (PhImT) as activating reagent; 99% coupling efficiency was achieved in a 15 min coupling time for formation of (*S*p)-phosphorothioate linkages and 97% for (*R*p)-phosphorothioate linkages. Sulfurization of the stereodefined phosphite triester **82** was performed with DTD. Acylation of the secondary amino group of the chiral auxiliary to afford the fully protected phosphorothioate **83** and capping

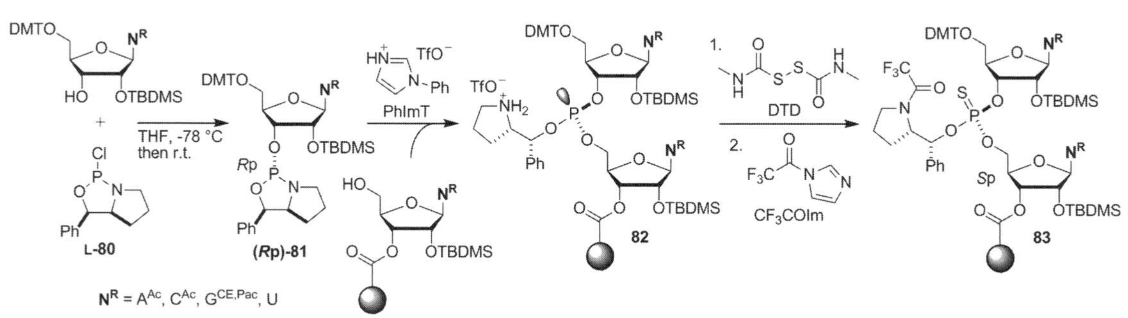

Scheme 1.8 *Synthesis of stereodefined phosphorothioate RNA **83**. Activation of 3′-O-oxazaphospholidine monomers **81** with* N*-phenylimidazolium triflate (PhImT) and coupling to the 5′-OH of a solid-phase attached (oligo)nucleotide gives phosphite intermediate **82**, which is followed by sulfurization with* N,N′*-dimethylthiuram disulfide (DTD) and capping with trifluoroacetylimidazole (CF_3COIm)*

of unreacted 5′-OH groups was achieved with trifluoroacetylimidazole. After completion of chain assembly, the stereodefined phosphorothioate RNA was cleaved from the solid support and deprotected with NH_3–EtOH and the TBDMS groups were removed with TBAF. The largest all-*R*p and all-*S*p phosphorothioate RNA oligonucleotides reported so far are only 10 nt long. Nevertheless, this recent report of successful diastereoselective synthesis of RNA phosphorothioate linkages is encouraging for future developments in the field and for syntheses of longer stereodefined RNA oligonucleotides with mixed sequences.

It should be noted that stereodefined phosphorothioates can also be prepared by *in vitro* transcription. T7 RNA polymerase accepts the *S*p diastereomers of nucleoside 5′-O-α-thiotriphosphates as substrates. The polymerization reaction leads to inversion of configuration at the α-P and produces solely *R*p phosphorothioates. However, enzymatic synthesis usually does not allow for site-specific incorporation of individual phosphorothioate modifications.

Other backbone modifications incorporated by solid-phase synthesis include phosphoroselenoates, for which similar principles apply as for phosphorothioates. In this case, the selenium modification is introduced by oxidation with KSeCN. Boranophosphates and methylphosphonates should also be mentioned as important backbone modifications. These derivatives have mainly been applied as antisense oligonucleotides [17], and will not be discussed in further detail in this chapter.

1.2.2.2 Post-synthetic RNA modification strategies

The post-synthetic modification of RNA oligonucleotides relies on the introduction of nucleoside analogs containing reactive functionalities by solid-phase synthesis and permits the site-specific attachment of a variety of reporter groups and chemical devices. Useful types of nucleoside derivatization reactions include nucleophilic aromatic substitution of appropriate leaving groups on nucleobases, palladium-catalyzed cross-coupling reactions, formation of thioether or disulfide bonds and functionalization of amino groups via formation of amide bonds or ureido groups.

Nucleophilic aromatic substitution of convertible nucleosides

A convertible nucleobase contains an appropriate leaving group that can be displaced by a nucleophile, which in turn becomes attached to that nucleobase. The most prominent examples of convertible ribonucleoside phosphoramidites, introduced by Verdine and co-workers in 1997 [95], are currently commercially available as 2′-*O*-TBDMS protected building blocks (see Figure 1.6, compounds **51–53**). Displacement of the 4-chlorophenyl leaving group with a nucleophilic primary alkylamine at uridine and inosine nucleosides leads to N^4-modified cytidine and N^6-modified adenosine analogs, respectively. Labeling of guanosines at position N2 is achieved by fluoride displacement in O^6-protected 2-fluoroinosine derivatives. These convertible nucleosides allow for attachment of novel functionalities or biophysical probes at either the major (A and C) or minor (G) groove of A-form RNA double helices while maintaining Watson–Crick base pairing capabilities of modified nucleosides (Scheme 1.9). The most prominent examples of modifications installed by this approach include disulfide crosslinking reagents, photo-crosslinking reagents and ^{15}N isotope labels. Correct incorporation of the modification needs to be verified by careful analysis of the isolated product by LC–ESI-MS or MALDI-MS. For new modifications, validation of incorporation by enzymatic digestion and base composition analysis by reversed-phase HPLC is recommended.

Palladium-catalyzed cross-coupling of halogenated nucleobases

Functionalization of the nucleobase with a biophysical reporter group can also be achieved by a carbon–carbon bond-forming reaction. The prime example is the palladium-catalyzed Sonogashira cross-coupling reaction for the derivatization of halogenated nucleosides with terminal alkynes. This strategy has been employed extensively for the derivatization of DNA oligonucleotides with ethynylpyrene [96–98].

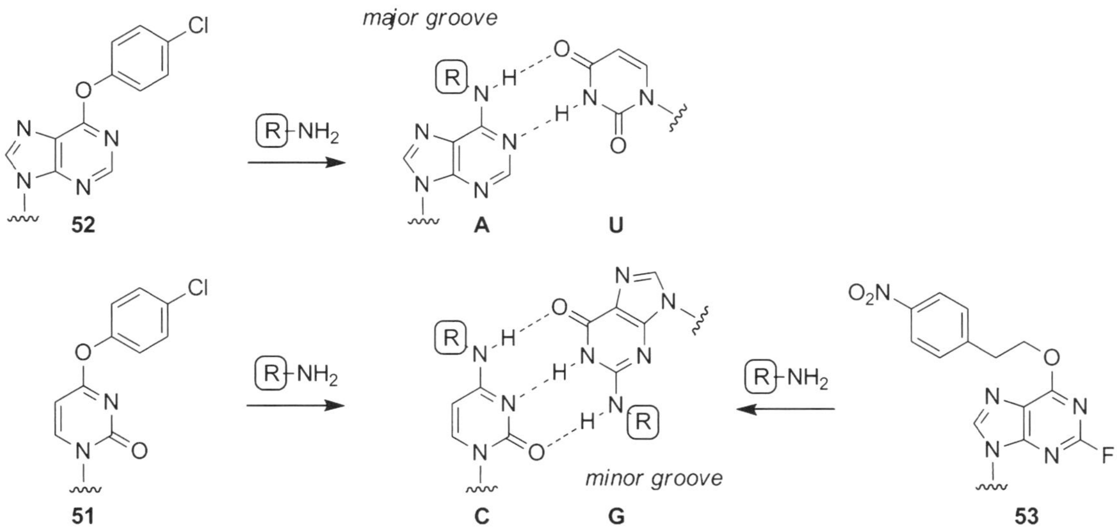

Scheme 1.9 *The convertible nucleoside approach allows installation of nucleobase modifications at the major or minor groove side in A, C and G nucleosides while maintaining Watson–Crick base pairing capability*

RNA oligonucleotides have been derivatized by the Sonogashira reaction with pyrene fluorophores and with nitroxide spin labels [78,99]. The halogenated pyrimidine or purine nucleosides are incorporated via 2′-*O*-TBDMS- or 2′-*O*-ACE-protected phosphoramidites (see Figure 1.6 and Scheme 1.10) [78,99,100]. The palladium-catalyzed derivatization reaction is usually performed while the oligonucleotide is still fully protected and attached to the solid support (Scheme 1.10c). The solid-phase synthesis is interrupted after coupling of the halogenated nucleoside phosphoramidite and the column is removed from the synthesizer. The reagent mixture for the Sonogashira coupling containing the ethynyl reagent, the palladium catalyst [e.g. $(PPh_3)_4Pd(0)$ or $(PPh_3)_2Pd(II)Cl_2$], CuI and triethylamine in dichloromethane is injected into the synthesis column. After coupling (usually 2–3 h), the solid support is washed and dried under vacuum, followed by continued chain elongation on the synthesizer to obtain the desired full-length oligonucleotide.

Other palladium-catalyzed cross-coupling reactions such as the Suzuki–Miyaura coupling of boronic acids and the Stille coupling of unsaturated stannanes have been used to derivatize solid-phase-bound

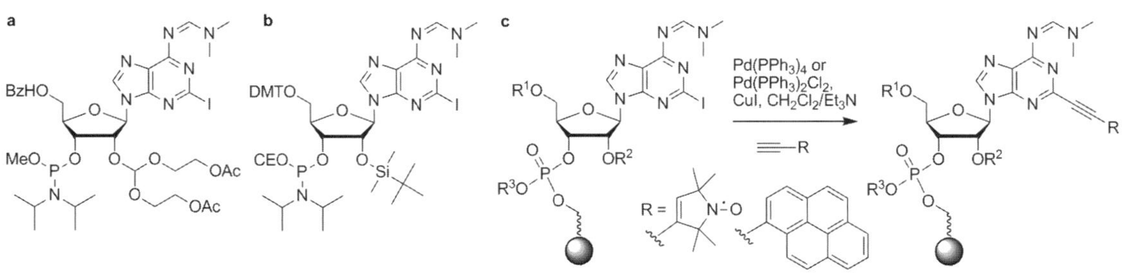

Scheme 1.10 *2-Iodoadenosine phosphoramidite building blocks (a) for 2′-O-ACE chemistry and (b) for 2′-O-TBDMS chemistry. (c) Pd-catalyzed Sonogashira cross-coupling reaction of terminal alkynes to 2-iodoadenosine on the solid support. Prominent examples of successfully coupled alkynes include alkynylpyrene derivatives and nitroxide radical-containing spin labels*

2′-deoxy-5-iodouridine nucleotides [101,102]. The major difficulties associated with these palladium-catalyzed techniques arise from the strict requirements for carefully controlled coupling conditions in dry and deoxygenated solutions under an argon atmosphere. The reaction temperatures and times (often 80–100 °C for up to 20 h) are not easily compatible with solid-phase RNA synthesis. Nevertheless, the recent success with Sonogashira coupling reactions in modifying RNA and the proof-of-concept work for Suzuki–Miyaura and Stille coupling to polystyrene-bound mononucleotides are promising examples for the development of new C–C bond-forming methods for on-column RNA derivatization.

Derivatization of sulfur-containing RNA

Post-synthetic modification of RNA oligonucleotides can be achieved by derivatization of sulfurylated RNA residues that have been incorporated during solid-phase synthesis. In contrast to the convertible nucleoside approach and to Pd-catalyzed cross-coupling reactions, derivatization reactions of thio- and amino-modified RNA are performed in solution after complete deprotection of the oligonucleotide.

One example of post-synthetic RNA backbone modification applies the reaction of phosphorothioate residues **84** with α-haloacetyl compounds (Scheme 1.11a) [103]. The nucleophilic displacement reaction leads to the formation of a phosphotriester species **85**, which is highly unstable in the presence of an adjacent 2′-hydroxyl group in standard RNA. Phosphotriesters are prone to hydrolysis in a reaction analogous to that used for phosphorothioate RNA sequencing [104] and nucleotide analog interference mapping (NAIM) [105]. It is therefore necessary to introduce a 2′-deoxyribonucleotide or a 2′-*O*-methyl derivative directly 5′ to the phosphorothioate modification. Examples of phosphorothioate labeling include derivatization of RNA with photo-crosslinking reagents and with nitroxide spin labels [103,106,107]. One of the reported complications is non-specific modification of RNA with highly reactive α-haloacetyl reagents, particularly at exocyclic amino groups of adenine and cytosine [103]. Increased selectivity for phosphorothioates is achieved by a low concentration of the derivatization reagent, low temperature and short reaction time.

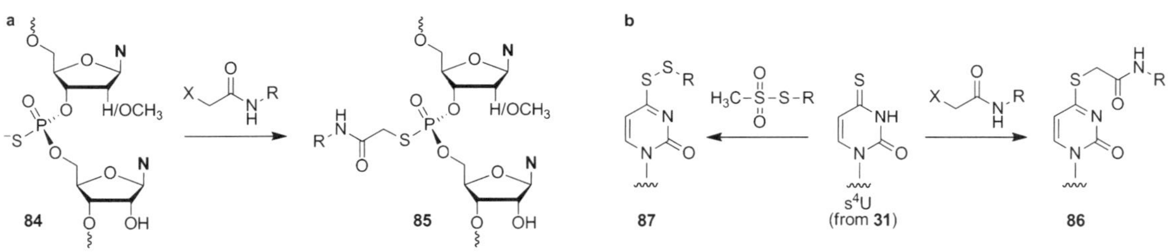

Scheme 1.11 (a) Phosphorothioate modification with an α-haloacetamide reagent leads to formation of a phosphotriester (**85**). (b) 4-Thiouridine modification with an α-haloacetyl compound gives a lactim thioester linkages (**86**) and reaction with a methanethiosulfonate reagent gives a disulfide bond (**87**). X = Cl, Br, I; R = biophysical or chemical label or reporter group, such as a fluorophore, ion complexation reagent, photo-crosslinking reagent or spin label

Post-synthetic modification can also be applied to thio-modified nucleobases, in particular to 4-thiouridine, which is introduced by solid-phase synthesis with phosphoramidite **31** (see Figure 1.6). The reaction with α-haloacetyl compounds leads to the formation of a lactim thioester linkage as in **86** (often called thioether for simplicity) [108]. A biophysical label or a chemical probe can alternatively be attached to the s^4U nucleobase via a disulfide linkage (**87**, Scheme 1.11b). In this case, the label or probe is applied as a methanethiosulfonate reagent [109].

The formation of disulfide crosslinks between two site-specifically attached thiol moieties on nucleosides also represents a useful approach for studying the structure and function of nucleic acids. Sulfhydryl groups have been directly attached to the N3 position of pyrimidine bases and were used for intrahelical disulfide crosslinking in tRNA [110]. Disulfide crosslinks involving the ribose moiety have been installed at the 2′-position via ribonucleosides containing 2′-*O*-(2-thioethyl) substitutions. Inter- and intra-helical disulfide crosslinks in tRNA were introduced by air oxidation of site-specifically incorporated 2′-thioalkyl groups on pyrimidine nucleosides [110]. All four common RNA nucleotides A, C, G and U were more recently synthesized as 2′-thioethyl-modified ribonucleoside phosphoramidites [64]. 2′-Thioethyl-modified RNA was used for the reversible connection of 5′-thiolated DNA oligonucleotides on to RNA in a disulfide exchange reaction [111]. In this manner, DNA duplexes were attached to specific 2′-positions of the P4–P6 domain of the *Tetrahymena* group I intron RNA and were used as conformational constraints to control RNA folding. The reversibility of disulfide crosslinking by reduction with DTT (1,4-dithiothreitol) was utilized as one of several methods to modulate the structural effects of DNA constraints.

Derivatization of artificial amino groups at ribose and nucleobase residues

Amino-modified nucleotides provide a very important class of functional groups for post-synthetic labeling of RNA. Similarly to labeling and crosslinking reactions of sulfur residues, the derivatization of amino groups is usually performed in solution after deprotection of the synthetic RNA. Three major types of amino-modified RNA nucleotides include 2′-amino-modified (**88**) and 2′-aminoethyl-modified (**89**) ribonucleotides and also 5-(aminoallyl)pyrimidine nucleotides (**90**) (Scheme 1.12), all of which are incorporated into RNA oligonucleotides via the corresponding 3′-phosphoramidite building blocks.

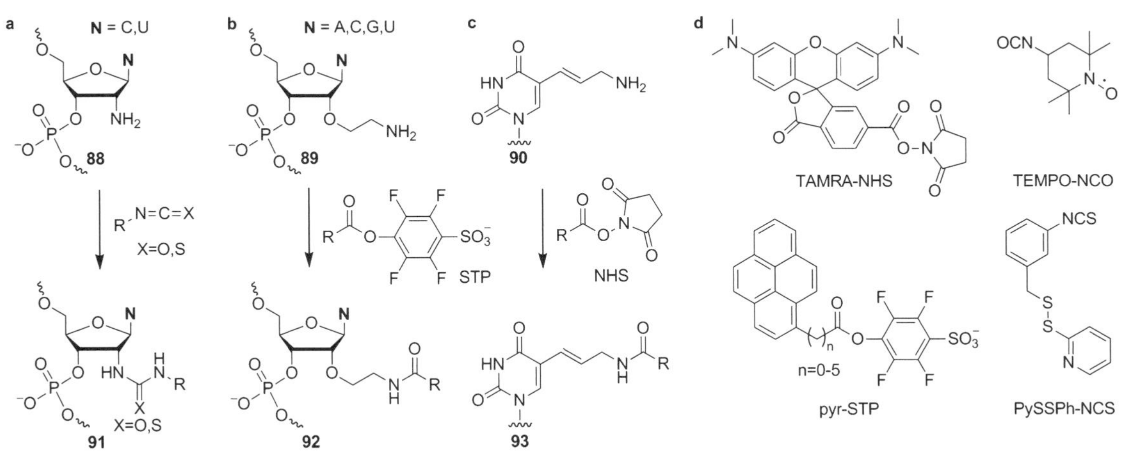

Scheme 1.12 *(a) Derivatization of 2′-amino-modified RNA* ***88*** *with isocyanate or isothiocyanate reagents to form urea or thiourea linkages as in* ***91****. (b) Reaction of 2′-aminoethyl-modified RNA* ***89*** *with STP esters. (c) Amide bond formation of 5-(aminoallyl)uridine-modified RNA* ***90*** *with NHS esters. (d) A small collection of common amino-modifying reagents, showing one example per class of NHS ester (TAMRA-NHS), STP ester (pyrene-STP esters with variable linker length), TEMPO-isocyanate for spin labeling and aromatic isothiocyanate for disulfide installation. NHS = N-hydroxysuccinimide, STP = 4-sulfonyltetrafluorophenyl, TAMRA = tetramethylrhodamine*

Site-specifically incorporated 2′-amino groups can be selectively conjugated by reactions with aromatic isothiocyanates or aliphatic isocyanates (Scheme 1.12a) to produce the corresponding urea or thiourea linkages in **91**. Primary amino groups of 2′-aminoethyl ribonucleosides and 5-(aminoallyl)pyrimidines are often conjugated via amide bond formation (**92**, **93**) with active esters (Scheme 1.12b,c) such as *N*-hydroxysuccinimide (NHS) esters and 4-sulfonyltetrafluorophenyl (STP) esters; the latter offer the advantage of higher solubility in buffered aqueous reaction media commonly used for RNA derivatization [112]. Although not explicitly shown in Scheme 1.12, it should be noted that 2′-amino-modified RNA can also react with active esters to form amide bonds and 2′-aminoethyl and 5-aminoallyl groups can be derivatized with isocyanates or isothiocyanates. Examples of different types of amino-labeling reagents are shown in Scheme 1.12d.

One notable advantage of post-synthetic RNA labeling strategies is that a relatively small number of phosphoramidites are needed to incorporate site-specifically suitable amino groups to which a wide variety of labels can easily be conjugated. Such a convergent approach facilitates the optimization of important parameters, such as the search for ideal tether lengths for the attachment of biophysical probes [66,113].

A different strategy for the derivatization of 2′-amino-modified RNA **88** involves reductive amination reactions with carbonyl groups. This approach has been used for the irreversible attachment of DNA constraints to RNA in the context of controlling RNA conformations [114]. The DNA oligonucleotide was modified on the 5′-terminus with a 1,2-diol which was converted into an aldehyde functional group upon oxidation with $NaIO_4$. The reductive amination reaction with the 2′-amino RNA **88** proceeded in the presence of $NaCNBH_3$ and $NiCl_2$ to give an engineered RNA–DNA conjugate [115].

1.2.2.3 Combined chemical and enzymatic methods

The preparation of modified RNAs up to a length of 50–60 nt can be reliably achieved by solid-phase synthesis. However, the combination of one or more protein- or nucleic acid-based enzymes is needed to incorporate chemically synthesized, modified oligonucleotide fragments into larger RNAs.

Either the required RNA segments can all be prepared by chemical synthesis, or larger fragments can be generated enzymatically by *in vitro* transcription. T7 RNA polymerase and related phage enzymes use synthetic DNA templates, double-stranded DNA fragments generated by polymerase chain reaction (PCR) or sections of linearized plasmids. Transcription is initiated from a 17 nt double-stranded T7 promoter region and continues to the 3′-end of the DNA template (Figure 1.9a). Several methods have been developed to circumvent problems of length heterogeneity that result from non-templated nucleotide addition at the 3′-end or, in rare cases typically involving multiple G nucleotides in a row, at the 5′-end. For example, the desired transcript can be engineered between a hammerhead ribozyme at the 5′-end and a hepatitis delta virus (HDV) ribozyme sequence at the 3′-end. [116] Upon transcription, the *cis*-acting ribozyme constructs will precisely excise the desired RNA by self-cleavage. It should be noted that hairpin and HDV ribozyme can also be used independently of each other, if processing of only the 5′- or 3′-end of the RNA transcript is required. Alternatively, deoxyribozymes (= DNA enzymes) can be used for the site-specific cleavage of RNA transcripts to prepare homogenous termini for further utilization in chemical or enzymatic reactions [117,118].

Although T7 RNA polymerase accepts certain modified NTPs as substrates (for example, α-thiotriphosphates and even certain nucleobase-modified derivatives) [105], it is usually not possible to introduce modified nucleotides site-specifically using standard DNA templates, unless, in rare cases, the particular nucleotide in question is present only once in the entire transcript. Another exception is the extreme 5′-terminus at which modifications can be placed by using special dinucleotide constructs (XpG) as initiators. Moreover, several research groups have developed unnatural base pair systems to attempt direct site-specific modification by RNA polymerases. Since the first report of the unnatural base pair isoG–isoC [119], a variety of other innovative base pair analogs have been described. A recent example is the shape-complementary, hydrophobic base pair formed by 7-(2-thienyl)imidazo[4,5]pyridine (Ds) and pyrrole-2-carbaldehyde (Pa)

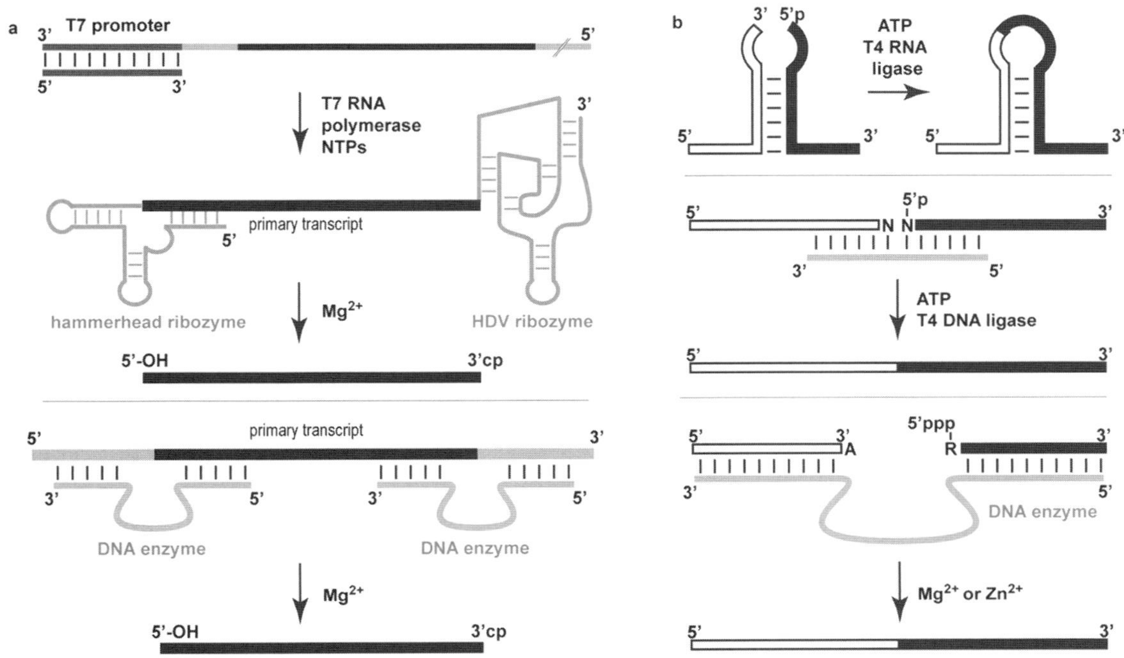

Figure 1.9 *Enzymatic methods for RNA manipulation. (a)* In vitro *transcription by T7 RNA polymerase from a synthetic DNA template with engineered 5′-hammerhead and 3′-HDV ribozyme sequences (gray) for production of desired RNA with homogenous termini (black); processing of primary RNA transcript can also be achieved with RNA-cleaving deoxyribozymes (bottom). (b) Enzymatic ligation of RNA fragments with T4 RNA ligase (top), T4 DNA ligase and splint (middle) or a deoxyribozyme (bottom)*

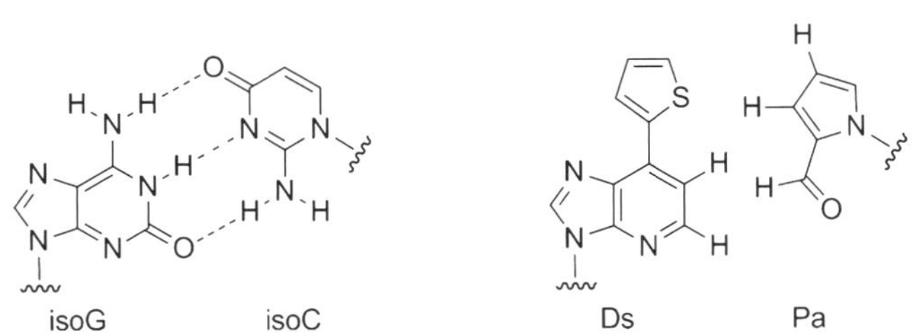

Figure 1.10 *The first unnatural base pair isoG–isoC with non-natural hydrogen bonding pattern and the recently described Ds–Pa base pair formed by specific hydrophobic shape complementation without hydrogen bonding interactions*

(Figure 1.10), which displays robust selectivity for the incorporation of these base substrates into RNA by transcription [120]. This approach is conceptually attractive, but it is still limited to highly specialized nucleoside analogs.

Another useful way for enzymatic incorporation of a modified nucleotide is the RNA ligase-catalyzed ligation of a modified nucleoside 3′,5′-bisphosphate to the 3′-terminus of an RNA transcript [121]. For example, this method has been used to incorporate s^4U and m^1A into tRNAs for the purpose of further derivatization and RNA folding studies [122].

To generate longer RNAs than are routinely achievable by direct chemical synthesis, modified and non-modified RNA fragments can be covalently joined by enzymatic ligation. Most commonly, the protein enzymes T4 DNA ligase or T4 RNA ligase are used to activate the 5′-terminal phosphate of the donor RNA fragment by adenylation and join it to the 3′-hydroxyl group of the acceptor substrate [123,124]. T4 DNA ligase catalyzes the ligation of two RNA substrates that are precisely aligned in a fully base-paired RNA–DNA heteroduplex, whereas T4 RNA ligase is used to join two single-stranded RNAs in the absence of a splint oligonucleotide (Figure 1.9b). The T4 RNA ligase-catalyzed reaction is prone to generating side products, most notably circular RNAs, unless the 3′-end of the donor RNA is blocked, for example by 3′-phosphorylation. In a different setup, T4 RNA ligase can be used with a splint that leaves the RNA termini that will be joined unpaired; this could be useful if the ligation site is not in a preformed stem–loop.

Numerous examples have been reported for the successful combination of chemical synthesis of modified RNA with enzymatic ligation methods. A recent example is the efficient preparation of 2-aminopurine-labeled riboswitch domains for fluorescence spectroscopic RNA folding studies [125–127]. Folding of the RNA tertiary structure has also been monitored by fluorescence of covalently attached pyrene. Short 2′-amino- or 2′-tethered amino-modified RNAs were chemically derivatized with pyrene and enzymatically ligated to larger RNA transcripts to provide pyrene-labeled P4–P6 RNA [66,128]. Efficient enzymatic ligation was recently also described for small and highly structured fluorophore-labeled tRNA and ribozyme fragments [129]. A combination of the convertible nucleoside approach and thiol-specific RNA labeling together with enzymatic ligation was applied for engineering of pre-mRNA and snRNA constructs [130,131]. In these studies, a site-specifically attached hydroxyl radical probe (Fe-BABE) was used to investigate the architecture of early spliceosomal complexes.

A recent addition to the repertoire of available methods for covalent ligation of RNA fragments comes from the *in vitro* selection of deoxyribozymes. Silverman and co-workers reported practically useful DNA catalysts for the ligation of a 5′-triphosphate RNA donor substrate to the 3′-hydroxyl group of a second RNA fragment [132,133].

A different technique for post-synthetic or post-transcriptional RNA labeling is the application of RNA-ligating deoxyribozymes to attach fluorophore-labeled tagging RNAs to specific RNA 2′-hydroxyl groups via 2′,5′-phosphodiester bond formation. This method has been termed DECAL (deoxyribozyme-catalyzed labeling of RNA) and has been applied for the preparation of fluorescein- and TAMRA-labeled P4–P6 RNA for studying RNA folding by fluorescence resonance energy transfer (FRET) [134].

1.3 Examples of modified RNA for the analysis and manipulation of RNA secondary and tertiary structures

Interconvertible RNA secondary and tertiary structures play key roles in natural systems for the regulation of transcription and translation events. For instance, riboswitches are mRNA-based genetic control elements that rely heavily on the interplay of alternative RNA secondary structures. Defined single-stranded regions belong to mutually exclusive stem–loop motifs in the absence or presence of dedicated riboswitch ligands [135,136]. Such structural transitions between different folding states not only may occur in large RNAs, but also alternating conformations can exist in small RNAs of only 20–40 nt [137,138]. Several research groups have focused on specific properties of chemically modified RNA oligonucleotides to analyze secondary structure equilibria and to study RNA refolding processes. Recent examples include fluorine- or aminopurine-labeled RNAs to monitor RNA secondary structure populations by NMR or fluorescence spectroscopy and caged nucleotides to manipulate deliberately RNA conformations.

1.3.1 ^{19}F NMR spectroscopy of fluorine-modified RNA

Micura and co-workers incorporated fluorine atoms site-specifically into RNA oligonucleotides and capitalized on ^{19}F NMR spectroscopy to analyze RNA secondary structure populations [50,139]. ^{19}F NMR spectroscopy offers high intrinsic sensitivity based on 100% natural abundance of the ^{19}F nucleus. The 100-fold larger chemical shift dispersion of ^{19}F compared with ^{1}H results in strongly reduced signal overlap in ^{19}F spectra compared with often severe resonance degeneracy observed in ^{1}H NMR imino proton spectra. Applications of fluorine-labeled RNA and ^{19}F NMR spectroscopy have recently been reviewed [140].

In one strategy, 2′-deoxy-2′-fluoro-modified nucleotides **94** were used as non-invasive spin labels for the analysis of RNA secondary structure equilibria. Thereby, the 2′-F-nucleotides were strategically positioned in the RNA to reside within a double helix in one RNA conformation, but to be part of a single-stranded region in an alternative conformational state (Figure 1.11a). The significantly different chemical environment experienced by the 2′-F atom in both conformations resulted in different chemical shifts for the ^{19}F resonances, with slow exchange on the NMR time-scale. Integration of the ^{19}F NMR signals allowed quantification of the equilibrium positions [50]. The combination of 2′-F-modified RNA and ^{19}F NMR spectroscopy has also been applied in a proof-of-principle study for the identification of specific RNA–ligand binding events. Upon complexation of a small-molecule target, the RNA experienced local conformational alterations in the binding pocket that were monitored in form of chemical shift changes of specific 2′-F resonances [51].

In a different approach, the nonpolar uridine analog 2,4-difluorotoluyl ribonucleoside (rF) was site-specifically incorporated into RNA (**95**) by solid-phase synthesis and was used for the analysis of RNA secondary structure populations and for the investigation of temperature-dependent conformational changes

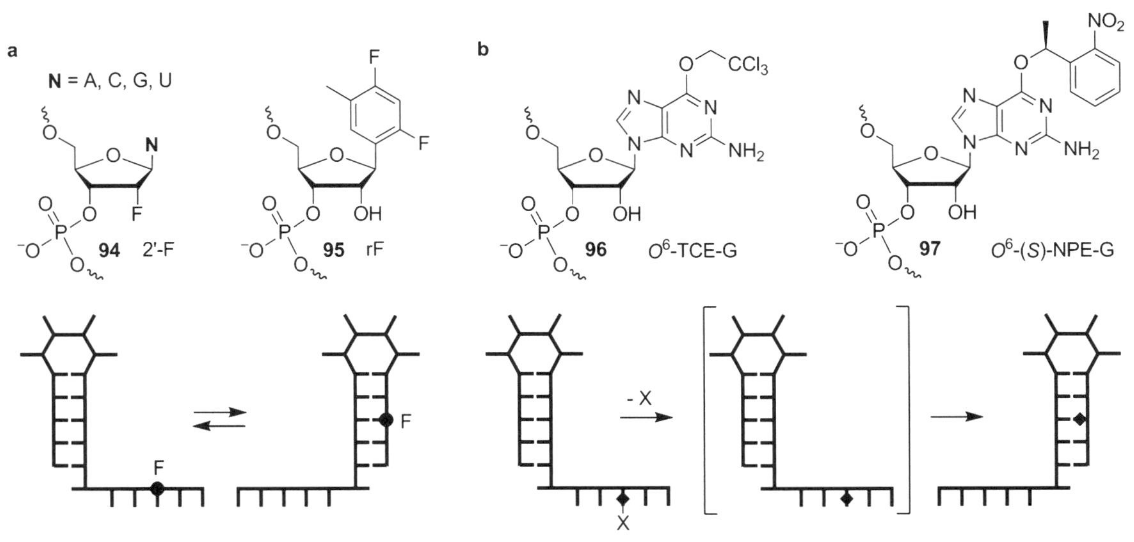

*Figure 1.11 Analysis and manipulation of RNA conformations. (a) 2′-Deoxy-2′-fluoro nucleotides **94** and 2,4-difluorotoluyl ribonucleotide **95** for characterization of interconvertible RNA secondary structures by ^{19}F NMR spectroscopy. (b) O^6-TCE-modified guanosine **96** and O^6-(S)-NPE-caged guanosine **97** for the manipulation of RNA structures. Nucleobase functionalization with X (= NPE or TCE) specifically selects one conformation. Upon release of X by reductive elimination or photolysis, the base pairing properties of the nucleobase are restored, the RNA conformation becomes metastable and is reorganized into a thermodynamically more favorable folding state*

by ^{19}F NMR spectroscopy [139]. This strategy is highly valuable for the precise determination of RNA equilibrium positions at elevated temperatures because it is not dependent on the detection of intrinsically exchangeable NH-N nuclei (in ^{1}H NMR spectroscopy, H–D exchange results in signal loss at elevated temperatures). The rF-labeling approach allows the measurement of RNA melting temperatures (T_m) with good precision at high RNA concentrations that are generally not accessible by UV melting analysis. Moreover, the rF modification has recently been incorporated into siRNAs and was shown to confer improved nuclease resistance in serum. An internal rF:A base pair did not negatively affect gene silencing activity relative to native siRNAs in HeLa cells, despite slightly reduced affinity of the rF-modified RNA for the target strand [141–143]. In combination with ^{19}F NMR spectroscopy, the rF modification could be applicable for conformational analysis of siRNAs and provide further insights for mechanistic investigations of RNAi.

1.3.2 Triggering of RNA structural transitions by functionalized nucleobases

Chemically functionalized nucleosides provide useful tools to control intentionally conformational changes of nucleic acid structures. Nucleobase modifications installed at the Watson–Crick base pairing site have been designed to prevent the formation of selected base pairs and specifically to destabilize predefined RNA secondary or tertiary structures. With respect to the site-specific incorporation into RNA, the functional labels must meet the stringent requirement of orthogonality to the protecting groups used throughout solid-phase synthesis. The first example of a functionalized RNA nucleotide described as an efficient tool for inducing defined rearrangements of RNA secondary structures is the O^6-trichloroethyl (TCE)-modified guanosine **96** [144]. The TCE group alters the hydrogen bonding donor–acceptor pattern of guanosine and therefore significantly reduces the residue's base pairing ability. Upon release of the TCE functionality under reductive conditions (Zn–AcOH), the base pairing capability is restored and the then unmodified nucleobase allows secondary structure rearrangements of the metastable conformation to occur (Figure 1.11b). In this study, alternative secondary structures were monitored by comparative imino proton NMR spectroscopy [138]. In a similar approach, Pitsch and co-workers introduced the photolabile (*S*)-1-(2-nitrophenyl)ethyl (NPE)[1] group at the O^6-position of guanosine **97** to modulate the Watson–Crick base pairing capability of defined nucleotides in bistable RNAs [145]. The kinetics of RNA conformational switching upon release of the photolabile group was studied by monitoring characteristic imino proton resonances of specific RNA conformations by time-resolved NMR spectroscopy [146–148].

For a different application, Höbartner and Silverman synthesized all four caged (*S*)-NPE-modified RNA nucleotides and placed individual caging groups into the 160-nt P4–P6 domain of the *Tetrahymena* group I intron RNA [149]. The effects on RNA folding were studied by nondenaturing (native) PAGE, which reports on global folding of P4–P6 RNA. Caging groups at selected positions were shown to disrupt key tertiary contacts by alteration of hydrogen bonding interactions, by disruption of base-stacking interactions or by introduction of steric clashes. In general, chemical interference with specific tertiary contacts, such as disruption of a key tetraloop–receptor interaction, can be used to produce stable misfolded states of large RNAs that may be of functional significance. Upon release of the caging group, the refolding into the native state can be initiated. Phototriggered RNA folding in combination with time-resolved monitoring of conformational changes by spectroscopic methods is expected to yield further insights into the mechanisms of complex RNA refolding events.

---

[1] It should be noted that the same abbreviation NPE is used in the literature for two different isomers of nitrophenylethyl groups. Here, NPE denotes the photolabile 1-(2-nitrophenyl)ethyl group, whereas in Figures 1.2 and 1.6 and Scheme 1.4 NPE at the O^6 of guanosine or inosine derivatives denotes the 2-(4-nitrophenyl)ethyl group that is cleaved by β-elimination during RNA deprotection.

1.4 Conclusion

In this chapter, we have summarized recent developments in protecting group strategies for RNA solid-phase synthesis and presented an overview of available methods for the preparation of chemically modified RNA. The diversity of chemically accessible RNA modifications is rapidly increasing, accompanied by an expanding number of applications in fundamental and applied research. Here, we could discuss only a small selection of studies demonstrating the power of oligoribonucleotide analogs for biochemical and biophysical applications.

Future research efforts will continue to yield more detailed insights into structural and mechanistic aspects of functionally important RNAs involved in various biochemical processes. The creativity and aptitude of chemists will ensure that tailor-made nucleoside modifications will be designed and synthesized to modulate RNA in various ways for multidisciplinary experiments.

Acknowledgments

The authors are grateful for support from the Max Planck Society. C.H. thanks Professor Ronald Micura, Innsbruck, for stimulating discussions. Professor Scott K. Silverman, Urbana-Champaign, and Dr. Manfred Konrad, Göttingen, are gratefully acknowledged for critical reading of the original manuscript.

References

1. Rozenski, J., Crain, P.F., McCloskey, J.A. (1999) The RNA Modification Database: 1999 update. *Nucleic Acids Res.* **27**, 196–197.
2. Dunin-Horkawicz, S., Czerwoniec, A., Gajda, M.J., Feder, M., Grosjean, H., Bujnicki, J.M. (2006) MODOMICS: a database of RNA modification pathways. *Nucleic Acids Res.* **34**, D145–D149.
3. Helm, M. (2006) Post-transcriptional nucleotide modification and alternative folding of RNA. *Nucleic Acids Res.* **34**, 721–733.
4. Agris, P.F. (1996) The importance of being modified: roles of modified nucleosides and Mg^{2+} in RNA structure and function. *Prog. Nucleic Acid Res. Mol. Biol.* **53**, 79–129.
5. Hengesbach, M., Kobitski, A., Voigts-Hoffmann, F., Frauer, C., Nienhaus, G.U., Helm, M. (2008) RNA intramolecular dynamics by single-molecule FRET. *Curr. Protoc. Nucleic Acid Chem.* **11**, Unit 11.12.
6. Walter, N.G. (2003) Probing RNA structural dynamics and function by fluorescence resonance energy transfer (FRET). *Curr. Protoc. Nucleic Acid Chem.* **11**, Unit 11.10.
7. Bevilacqua, P.C., Turner, D.H. (2002) Use of fluorescence spectroscopy to elucidate RNA folding pathways. *Curr. Protoc. Nucleic Acid Chem.* **11**, Unit 11.8.
8. Edwards, T.E., Sigurdsson, S.T. (2007) Site-specific incorporation of nitroxide spin-labels into 2′-positions of nucleic acids. *Nat. Protocols* **2**, 1954–1962.
9. Sowa, G.Z., Qin, P.Z. (2008) Site-directed spin labeling studies on nucleic acid structure and dynamics. *Prog. Nucleic Acid Res. Mol. Biol.* **82**, 147–197.
10. Schiemann, O., Piton, N., Plackmeyer, J., Bode, B.E., Prisner, T.F., Engels, J.W. (2007) Spin labeling of oligonucleotides with the nitroxide TPA and use of PELDOR, a pulse EPR method, to measure intramolecular distances. *Nat. Protocols* **2**, 904–923.
11. Sigurdsson, S.T. (2000) Site-specific sulfhydryl groups for study of RNA conformation via disulfide crosslinking. *Methods Enzymol.* **318**, 165–175.
12. Sheng, J., Huang, Z. (2008) Selenium derivatization of nucleic acids for phase and structure determination in nucleic acid X-ray crystallography. *Int. J. Mol. Sci.* **9**, 258–271.

13. Olieric, V., Rieder, U., Lang, K., Serganov, A., Schulze-Briese, C., Micura, R., Dumas, P., Ennifar, E. (2009) A fast selenium derivatization strategy for crystallization and phasing of RNA structures. *RNA* **15**, 707–715.
14. Caton-Williams, J., Huang, Z. (2008) Biochemistry of selenium-derivatized naturally occurring and unnatural nucleic acids. *Chem. Biodivers.* **5**, 396–407.
15. Tang, X., Dmochowski, I.J. (2007) Regulating gene expression with light-activated oligonucleotides. *Mol. BioSyst.* **3**, 100–110.
16. Grillone, L., Lanz, R. (2001) Fomivirsen. *Drugs Today* **37**, 245–255.
17. Behlke, M.A. (2008) Chemical modification of siRNAs for *in vivo* use. *Oligonucleotides* **18**, 305–319.
18. Rozners, E. (2006) Carbohydrate chemistry for RNA interference: synthesis and properties of RNA analogues modified in sugar–phosphate backbone. *Curr. Org. Chem.* **10**, 675–692.
19. Beaucage, S.L. (2008) Solid-phase synthesis of siRNA oligonucleotides. *Curr. Opin. Drug Discov. Devel.* **11**, 203–216.
20. Beaucage, S.L., Caruthers, M.H. (2001) Synthetic strategies and parameters involved in the synthesis of oligodeoxyribonucleotides according to the phosphoramidite method. *Curr. Protoc. Nucleic Acid Chem.* **3**, Unit 3.3.
21. Usman, N., Ogilvie, K.K., Jiang, M.Y., Cedergren, R.J. (1987) The automated chemical synthesis of long oligoribuncleotides using 2′-*O*-silylated ribonucleoside 3′-*O*-phosphoramidites on a controlled-pore glass support: synthesis of a 43-nucleotide sequence similar to the 3′-half molecule of an *Escherichia coli* formylmethionine tRNA. *J. Am. Chem. Soc.* **109**, 7845–7854.
22. Welz, R., Muller, S. (2002) 5-(Benzylmercapto)-1*H*-tetrazole as activator for 2′-*O*-TBDMS phosphoramidite building blocks in RNA synthesis. *Tetrahedron Lett.* **43**, 795–797.
23. (a) Wu, X., Pitsch. S. (1998) Synthesis and pairing properties of oligoribonucleotide analogues containing a metal-binding site attached to beta-D-allofuranosyl cytosine. *Nucleic Acids Res.* **26**, 4315–4323; (b) Pitsch, S., Weiss, P.A., Jenny, L., Stutz, A., Wu, X. (2001) Reliable chemical synthesis of oligoribonucleotides (RNA) with 2′-*O*-[(triisopropylsilyl)oxy]methyl-(2′-*O*-tom)-protected phosphoramidites. *Helv. Chim. Acta* **84**, 3773–3795.
24. Höbartner, C., Kreutz, C., Flecker, E., Ottenschläger, E., Pils, W., Grubmayr, K., Micura, R. (2003) The synthesis of 2′-*O*-[(triisopropylsilyl)oxy]methyl (TOM) phosphoramidites of methylated ribonucleosides (m^1G, m^2G, m^2_2G, m^1I, m^3U, m^4C, m^6A, m^6_2A) for use in automated RNA solid-phase synthesis. *Monatsh. Chem./Chem. Monthly* **134**, 851–873.
25. Porcher, S., Pitsch, S. (2005) Synthesis of 2′-*O*-[(triisopropylsilyl)oxy]methyl (= tom)-protected ribonucleoside phosphoramidites containing various nucleobase analogues. *Helv. Chim. Acta* **88**, 2683–2704.
26. Scaringe, S.A., Wincott, F.E., Caruthers, M.H. (1998) Novel RNA synthesis method using 5′-*O*-silyl-2′-*O*-orthoester protecting groups. *J. Am. Chem. Soc.* **120**, 11820–11821.
27. Delaney, M.O., Thomas, A., Ricketts, C., Kitchen, D.E., Kaiser, R.J. (2008) Chromophoric 5′-*O*-silyl protection of N-protected 2′-ACE ribonucleosides for solid-phase RNA synthesis. *Curr. Protoc. Nucleic Acid Chem.* **2**, Unit 2.14.
28. Ohgi, T., Masutomi, Y., Ishiyama, K., Kitagawa, H., Shiba, Y., Yano, J. (2005) A new RNA synthetic method with a 2′-*O*-(2-cyanoethoxymethyl) protecting group. *Org. Lett.* **7**, 3477–3480.
29. Shiba, Y., Masuda, H., Watanabe, N., Ego, T., Takagaki, K., Ishiyama, K., Ohgi, T., Yano, J. (2007) Chemical synthesis of a very long oligoribonucleotide with 2-cyanoethoxymethyl (CEM) as the 2′-O-protecting group: structural identification and biological activity of a synthetic 110mer precursor-microRNA candidate. *Nucleic Acids Res.* **35**, 3287–3296.
30. Ohgi, T., Kitagawa, H., Yano, J. (2008) Chemical synthesis of oligoribonucleotides with 2′-*O*-(2-cyanoethoxymethyl)-protected phosphoramidites. *Curr. Protoc. Nucleic Acid Chem.* **2**, Unit 2.15.
31. Saneyoshi, H., Seio, K., Sekine, M. (2005) A general method for the synthesis of 2′-*O*-cyanoethylated oligoribonucleotides having promising hybridization affinity for DNA and RNA and enhanced nuclease resistance. *J. Org. Chem.* **70**, 10453–10460.
32. Saneyoshi, H., Ando, K., Seio, K., Sekine, M. (2007) Chemical synthesis of RNA via 2′-*O*-cyanoethylated intermediates. *Tetrahedron* **63**, 11195–11203.
33. Zhou, C., Honcharenko, D., Chattopadhyaya, J. (2007) 2-(4-Tolylsulfonyl)ethoxymethyl (TEM) – a new 2′-OH protecting group for solid-supported RNA synthesis. *Org. Biomol. Chem.* **5**, 333–343.
34. Lavergne, T., Bertrand, J., Vasseur, J., Debart, F. (2008) A base-labile group for 2′-OH protection of ribonucleosides: a major challenge for RNA synthesis. *Chemistry* **14**, 9135–9138.

35. Lackey, J.G., Sabatino, D., Damha, M.J. (2007) Solid-phase synthesis and on-column deprotection of RNA from 2′- (and 3′-) *O*-levulinated (Lv) ribonucleoside monomers. *Org. Lett.* **9**, 789–792.
36. Lackey, J.G., Damha, M.J. (2008) The acetal levulinyl ester (ALE) group for the 2′-hydroxyl protection of ribonucleosides and the synthesis of oligoribonucleotides. *Nucleic Acids Symp. Ser.* **52**, 35–36.
37. Lackey, J.G., Mitra, D., Somoza, M.M., Cerrina, F., Damha, M.J. (2009) Acetal levulinyl ester (ALE) groups for 2′-hydroxyl protection of ribonucleosides in the synthesis of oligoribonucleotides on glass and microarrays. *J. Am. Chem. Soc* **131**, 8496–8502.
38. Reese, C.B., Thompson, E.A. (1988) A new synthesis of 1-arylpiperidin-4-ols. *J. Chem. Soc., Perkin Trans.* **1** 2881–2885.
39. Reese, C.B. (2001) Protection of 2′-hydroxy functions of ribonucleosides. *Curr. Protoc. Nucleic Acid Chem.* **2**, Unit 2.2.
40. Chaulk, S.G., MacMillan, A.M. (2007) Synthesis of oligo-RNAs with photocaged adenosine 2′-hydroxyls. *Nat. Protocols* **2**, 1052–1058.
41. Matysiak, S., Pfleiderer, W. (2001) Nucleotides, Part LXVIII, Acetals as new 2′-*O*-protecting functions for the synthesis of oligoribonucleotides: synthesis of monomeric building units and oligoribonucleotides. *Helv. Chim. Acta* **84**, 1066–1085.
42. Cieślak, J., Grajkowski, A., Kauffman, J.S., Duff, R.J., Beaucage, S.L. (2008) The 4-(*N*-dichloroacetyl-N-methylamino) benzyloxymethyl group for 2′-hydroxyl protection of ribonucleosides in the solid-phase synthesis of oligoribonucleotides. *J. Org. Chem.* **73**, 2774–2783.
43. Cieślak, J., Kauffman, J.S., Kolodziejski, M.J., Lloyd, J.R., Beaucage, S.L. (2007) Assessment of 4-nitrogenated benzyloxymethyl groups for 2′-hydroxyl protection in solid-phase RNA synthesis. *Org. Lett.* **9**, 671–674.
44. Semenyuk, A., Földesi, A., Johansson, T., Estmer-Nilsson, C., Blomgren, P., Brännvall, M., Kirsebom, L.A., Kwiatkowski, M. (2006) Synthesis of RNA using 2′-*O*-DTM protection. *J. Am. Chem. Soc.* **128**, 12356–12357.
45. Parey, N., Baraguey, C., Vasseur, J., Debart, F. (2006) First evaluation of acyloxymethyl or acylthiomethyl groups as biolabile 2′-O-protections of RNA. *Org. Lett.* **8**, 3869–3872.
46. Karpeisky, A., Sweedler, D., Haeberli, P., Read, J., Jarvis, K., Beigelman, L. (2002) Scaleable and efficient synthesis of 2′-deoxy-2′-*N*-phthaloyl nucleoside phosphoramidites for oligonucleotide synthesis. *Bioorg. Med. Chem. Lett.* **12**, 3345–3347.
47. Dai, Q., Deb, S.K., Hougland, J.L., Piccirilli, J.A. (2006) Improved synthesis of 2′-amino-2′-deoxyguanosine and its phosphoramidite. *Bioorg. Med. Chem.* **14**, 705–713.
48. Lang, K., Erlacher, M., Wilson, D.N., Micura, R., Polacek, N. (2008) The role of 23S ribosomal RNA residue A2451 in peptide bond synthesis revealed by atomic mutagenesis. *Chem. Biol.* **15**, 485–492.
49. Hougland, J.L., Sengupta, R.N., Dai, Q., Deb, S.K., Piccirilli, J.A. (2008) The 2′-hydroxyl group of the guanosine nucleophile donates a functionally important hydrogen bond in the tetrahymena ribozyme reaction. *Biochemistry* **47**, 7684–7694.
50. Kreutz, C., Kählig, H., Konrat, R., Micura, R. (2005) Ribose 2′-F labeling: a simple tool for the characterization of RNA secondary structure equilibria by ^{19}F NMR spectroscopy. *J. Am. Chem. Soc.* **127**, 11558–11559.
51. Kreutz, C., Kählig, H., Konrat, R., Micura, R. (2006) A general approach for the identification of site-specific RNA binders by ^{19}F NMR spectroscopy: proof of concept. *Angew. Chem. Int. Ed.* **45**, 3450–3453.
52. Kauppinen, S., Vester, B., Wengel, J. (2005) Technologies: locked nucleic acid (LNA): high affinity targeting of RNA for diagnostics and therapeutics. *Drug Discov. Today* **2**, 287–290.
53. Micura, R., Höbartner, C., Rieder, R., Kreutz, C., Puffer, B., Lang, K., Moroder, H. (2007) Preparation of 2′-deoxy-2′-methylseleno-modified phosphoramidites and RNA. *Curr. Protoc. Nucleic Acid Chem.* **1**, Unit 1.15.
54. Puffer, B., Moroder, H., Aigner, M., Micura, R. (2008) 2′-Methylseleno-modified oligoribonucleotides for X-ray crystallography synthesized by the ACE RNA solid-phase approach. *Nucleic Acids Res.* **36**, 970–983.
55. Micura, R., Pils, W., Höbartner, C., Grubmayr, K., Ebert, M.O., Jaun, B. (2001) Methylation of the nucleobases in RNA oligonucleotides mediates duplex–hairpin conversion. *Nucleic Acids Res.* **29**, 3997–4005.
56. Höbartner, C., Ebert, M., Jaun, B., Micura, R. (2002) RNA two-state conformation equilibria and the effect of nucleobase methylation 13. *Angew. Chem. Int. Ed.* **41**, 605–609.
57. Pallan, P.S., Kreutz, C., Bosio, S., Micura, R., Egli, M. (2008) Effects of N^2,N^2-dimethylguanosine on RNA structure and stability: crystal structure of an RNA duplex with tandem $m^2{}_2$G:A pairs. *RNA* **14**, 2125–2135.

58. Erlacher, M.D., Lang, K., Wotzel, B., Rieder, R., Micura, R., Polacek, N. (2006) Efficient ribosomal peptidyl transfer critically relies on the presence of the ribose 2′-OH at A2451 of 23S rRNA. *J. Am. Chem. Soc.* **128**, 4453–4459.
59. Ziomek, K., Kierzek, E., Biala, E., Kierzek, R. (2002) The influence of various modified nucleotides placed as 3′-dangling end on thermal stability of RNA duplexes. *Biophys. Chem.* **97**, 243–249.
60. Porcher, S. (2008) Chemical approach for the study of the ‘kissing complex’ of Moloney murine leukaemia virus. *Helv. Chim. Acta* **91**, 1219–1235.
61. Manoharan, M., Prakash, T.P., Barber-Peoc’h, I., Bhat, B., Vasquez, G., Ross, B.S., Cook, P.D. (1999) *N*-(2-Cyanoethoxycarbonyloxy)succinimide: a new reagent for protection of amino groups in oligonucleotides. *J. Org. Chem.* **64**, 6468–6472.
62. Cuenoud, B., Casset, F., Hüsken, D., Natt, F., Wolf, R.M., Altmann, K., Martin, P., Moser, H.E. (1998) Dual recognition of double-stranded DNA by 2′-aminoethoxy-modified oligonucleotides. *Angew. Chem. Int. Ed.* **37**, 1288–1291.
63. Osborne, S.D., Powers, V.E.C., Rusling, D.A., Lack, O., Fox, K.R., Brown, T. (2004) Selectivity and affinity of triplex-forming oligonucleotides containing 2′-aminoethoxy-5-(3-aminoprop-1-ynyl)uridine for recognizing AT base pairs in duplex DNA. *Nucleic Acids Res.* **32**, 4439–4447.
64. Jin, S., Miduturu, C.V., McKinney, D.C., Silverman, S.K. (2005) Synthesis of amine- and thiol-modified nucleoside phosphoramidites for site-specific introduction of biophysical probes into RNA. *J. Org. Chem.* **70**, 4284–4299.
65. Smicius, R., Engels, J.W. (2008) Preparation of zwitterionic ribonucleoside phosphoramidites for solid-phase siRNA synthesis. *J. Org. Chem.* **73**, 4994–5002.
66. Smalley, M.K., Silverman, S.K. (2006) Fluorescence of covalently attached pyrene as a general RNA folding probe. *Nucleic Acids Res.* **34**, 152–166.
67. Odadzic, D., Bramsen, J.B., Smicius, R., Bus, C., Kjems, J., Engels, J.W. (2008) Synthesis of 2′-*O*-modified adenosine building blocks and application for RNA interference. *Bioorg. Med. Chem.* **16**, 518–529.
68. Egli, M., Pallan, P.S. (2007) Insights from crystallographic studies into the structural and pairing properties of nucleic acid analogs and chemically modified DNA and RNA oligonucleotides. *Annu. Rev. Biophys. Biomol. Struct.* **36**, 281–305.
69. Das, S.R., Piccirilli, J.A. (2005) General acid catalysis by the hepatitis delta virus ribozyme. *Nat. Chem. Biol.* **1**, 45–52.
70. Hougland, J.L., Deb, S.K., Maric, D., Piccirilli, J.A. (2004) An atomic mutation cycle for exploring RNA’s 2′-hydroxyl group. *J. Am. Chem. Soc.* **126**, 13578–12579.
71. Li, N., Piccirilli, J.A. (2007) Synthesis of 2′-C-alpha-(hydroxyalkyl) and 2′-C-alpha-alkylcytidine phosphoramidites: analogues for probing solvent interactions with RNA. *J. Org. Chem.* **72**, 1198–1210.
72. Gordon, P.M., Fong, R., Deb, S.K., Li, N., Schwans, J.P., Ye, J., Piccirilli, J.A. (2004) New strategies for exploring RNA’s 2′-OH expose the importance of solvent during group II intron catalysis. *Chem. Biol.* **11**, 237–246.
73. Gordon, P.M., Fong, R., Piccirilli, J.A. (2007) A second divalent metal ion in the group II intron reaction center. *Chem. Biol.* **14**, 607–612.
74. Szewczak, A.A., Kosek, A.B., Piccirilli, J.A., Strobel, S.A. (2002) Identification of an active site ligand for a group I ribozyme catalytic metal ion. *Biochemistry* **41**, 2516–2525.
75. Thomas, J.M., Perrin, D.M. (2009) Probing general acid catalysis in the hammerhead ribozyme. *J. Am. Chem. Soc.* **131**, 1135–1143.
76. Lu, J., Li, N., Sengupta, R.N., Piccirilli, J.A. (2008) Synthesis and biochemical application of 2′-*O*-methyl-3′-thioguanosine as a probe to explore group I intron catalysis. *Bioorg. Med. Chem.* **16**, 5754–5760.
77. Lusic, H., Gustilo, E.M., Vendeix, F.A., Kaiser, R., Delaney, M.O., Graham, W.D., Moye, V.A., Cantara, W.A., Agris, P.F., Deiters, A. (2008) Synthesis and investigation of the 5-formylcytidine modified, anticodon stem and loop of the human mitochondrial tRNAMet. *Nucleic Acids Res.* **36**, 6548–6557.
78. Piton, N., Mu, Y., Stock, G., Prisner, T.F., Schiemann, O., Engels, J.W. (2007) Base-specific spin-labeling of RNA for structure determination. *Nucleic Acids Res.* **35**, 3128–3143.
79. Chan, J.H., Lim, S., Wong, W.F. (2006) Antisense oligonucleotides: from design to therapeutic application. *Clin. Exp. Pharm. Phys.* **33**, 533–540.
80. Levin, A.A. (1999) A review of the issues in the pharmacokinetics and toxicology of phosphorothioate antisense oligonucleotides. *Biochim. Biophys. Acta* **1489**, 69–84.

81. Kurreck, J. (2003) Antisense technologies. Improvement through novel chemical modifications. *Eur. J. Biochem.* **270**, 1628–1644.
82. Wang, L., Chen, S., Xu, T., Taghizadeh, K., Wishnok, J.S., Zhou, X., You, D., Deng, Z., Dedon, P.C. (2007) Phosphorothioation of DNA in bacteria by dnd genes. *Nat. Chem. Biol.* **3**, 709–710.
83. Burgers, P.M.J., Eckstein, F. (1978) Synthesis of dinucleoside monophosphorothioates via addition of sulfur to phosphite triesters. *Tetrahedron Lett.* **19**, 3835–3838.
84. Iyer, R.P., Egan, W., Regan, J.B., Beaucage, S.L. (1990) 3*H*-1,2-Benzodithiole-3-one 1,1-dioxide as an improved sulfurizing reagent in the solid-phase synthesis of oligodeoxyribonucleoside phosphorothioates. *J. Am. Chem. Soc.* **112**, 1253–1254.
85. Xu, Q., Barany, G., Hammer, R., Musier-Forsyth, K. (1996) Efficient introduction of phosphorothioates into RNA oligonucleotides by 3-ethoxy-1,2,4-dithiazoline-5-one (EDITH). *Nucleic Acids Res.* **24**, 3643–3644.
86. Ravikumar, V.T., Andrade, M., Carty, R.L., Dan, A., Barone, S. (2006) Development of siRNA for therapeutics: efficient synthesis of phosphorothioate RNA utilizing phenylacetyl disulfide (PADS). *Bioorg. Med. Chem. Lett.* **16**, 2513–2517.
87. Wang, Z., Song, Q., Sanghvi, Y.S. (2005) Dimethylthiarum disulfide: new sulfur transfer reagent in phosphorothioates oligonucleotide synthesis. *Methods Mol. Biol.* **288**, 51–64.
88. Guga, P., Stec, W.J. (2003) Synthesis of phosphorothioate oligonucleotides with stereodefined phosphorothioate linkages. *Curr. Protoc. Nucleic Acid Chem.* **4**, Unit 4.17.
89. Nawrot, B., Rebowska, B., Cieslinska, K., Stec, W.J. (2005) New approach to the synthesis of oligodeoxyribonucleotides modified with phosphorothioates of predetermined sense of P-chirality. *Tetrahedron Lett.* **46**, 6641–6644.
90. Iyer, R.P., Guo, M., Yu, D., Agrawal, S. (1998) Solid-phase stereoselective synthesis of oligonucleoside phosphorothioates: the nucleoside bicyclic oxazaphospholidines as novel synthons. *Tetrahedron Lett.* **39**, 2491–2494.
91. Yu, D., Kandimalla, E.R., Roskey, A., Zhao, Q., Chen, L., Chen, J., Agrawal, S. (2000) Stereo-enriched phosphorothioate oligodeoxynucleotides: synthesis, biophysical and biological properties. *Bioorg. Med. Chem.* **8**, 275–284.
92. Oka, N., Wada, T., Saigo, K. (2003) An oxazaphospholidine approach for the stereocontrolled synthesis of oligonucleoside phosphorothioates. *J. Am. Chem. Soc.* **125**, 8307–8317.
93. Oka, N., Yamamoto, M., Sato, T., Wada, T. (2008) Solid-phase synthesis of stereoregular oligodeoxyribonucleoside phosphorothioates using bicyclic oxazaphospholidine derivatives as monomer units. *J. Am. Chem. Soc.* **130**, 16031–16037.
94. Oka, N., Kondo, T., Fujiwara, S., Maizuru, Y., Wada, T. (2009) Stereocontrolled synthesis of oligoribonucleoside phosphorothioates by an oxazaphospholidine approach. *Org. Lett.* **11**, 967–970.
95. Allerson, C.R., Chen, S.L., Verdine, G.L. (1997) A chemical method for site-specific modification of RNA: the convertible nucleoside approach. *J. Am. Chem. Soc.* **119**, 7423–7433.
96. Mayer, E., Valis, L., Wagner, C., Rist, M., Amann, N., Wagenknecht, H. (2004) 1-Ethynylpyrene as a tunable and versatile molecular beacon for DNA. *ChemBioChem* **5**, 865–868.
97. Wagner, C., Rist, M., Mayer-Enthart, E., Wagenknecht, H. (2005) 1-Ethynylpyrene-modified guanine and cytosine as optical labels for DNA hybridization. *Org. Biomol. Chem.* **3**, 2062–2063.
98. Okamoto, A., Ochi, Y., Saito, I. (2005) Fluorometric sensing of the salt-induced B–Z DNA transition by combination of two pyrene-labeled nucleobases. *Chem. Commun*, 1128–1130.
99. Kwon, T., Piton, N., Grünewald, C., Engels, J.W. (2007) Synthesis of pyrene labeled rna for fluorescence measurements. *Nucleosides Nucleotides Nucleic Acids* **26**, 1381.
100. Grünewald, C., Kwon, T., Piton, N., Förster, U., Wachtveitl, J., Engels, J.W. (2008) RNA as scaffold for pyrene excited complexes. *Bioorg. Med. Chem.* **16**, 19–26.
101. Aucagne, V., Berteina-Raboin, S., Guenot, P., Agrofoglio, L.A. (2004) Palladium-catalyzed synthesis of uridines on polystyrene-based solid supports. *J. Comb. Chem.* **6**, 717–723.
102. El Kazzouli, S., Berteina-Raboin, S., Agrofoglio, L.A. (2007) Supported synthesis and functionnalization of 2′-deoxyuridine by Suzuki–Miyaura cross-coupling. *Nucleosides Nucleotides Nucleic Acids* **26**, 1395–1398.
103. Konarska, M.M. (1999) Site-specific derivatization of RNA with photocrosslinkable groups. *Methods* **18**, 22–28.
104. Gish, G., Eckstein, F. (1988) DNA and RNA sequence determination based on phosphorothioate chemistry. *Science* **240**, 1520–1522.
105. Ryder, S.P., Strobel, S.A. (1999) Nucleotide analog interference mapping. *Methods* **18**, 38–50.

106. Musier-Forsyth, K., Schimmel, P. (1994) Acceptor helix interactions in a class II tRNA synthetase: photoaffinity crosslinking of an RNA miniduplex substrate. *Biochemistry* **33**, 773–779.
107. Qin, P.Z., Butcher, S.E., Feigon, J., Hubbell, W.L. (2001) Quantitative analysis of the isolated GAAA tetraloop/receptor interaction in solution: a site-directed spin labeling study. *Biochemistry* **40**, 6929–6936.
108. Ramos, A., Varani, G. (1998) A new method to detect long-range protein–RNA contacts: NMR detection of electron–proton relaxation induced by nitroxide spin-labeled RNA. *J. Am. Chem. Soc.* **120**, 10992–10993.
109. Qin, P.Z., Feigon, J., Hubbell, W.L. (2005) Site-directed spin labeling studies reveal solution conformational changes in a GAAA tetraloop receptor upon Mg^{2+}-dependent docking of a GAAA tetraloop. *J. Mol. Biol.* **351**,1–8.
110. Goodwin, J.T., Osborne, S.E., Scholle, E.J., Glick, G.D. (1996) Design, synthesis and analysis of yeast tRNAPhe analogs possessing intra- and interhelical disulfide cross-links. *J. Am. Chem. Soc.* **118**, 5207–5215.
111. Miduturu, C.V., Silverman, S.K. (2006) Modulation of DNA constraints that control macromolecular folding. *Angew. Chem. Int. Ed.* **45**, 1918–1921.
112. Smalley, M.K., Silverman, S.K. (2004) Site-specific fluorescent labeling of large RNAs with pyrene. *Curr. Protoc. Nucleic Acid Chem.* **11**, Unit 11.11.
113. Sigurdsson, S.T. (1999) Thiol-containing RNA for the study of structure and function of ribozymes. *Methods* **18**, 71–77.
114. Miduturu, C.V., Silverman, S.K. (2005) DNA constraints allow rational control of macromolecular conformation. *J. Am. Chem. Soc.* **127**, 10144–10145.
115. Miduturu, C.V., Silverman, S.K. (2006) Synthesis and application of a 5′-aldehyde phosphoramidite for covalent attachment of DNA to biomolecules. *J. Org. Chem.* **71**, 5774–5777.
116. Walker, S.C., Avis, J.M., Conn, G.L. (2003) General plasmids for producing RNA in vitro transcripts with homogeneous ends. *Nucleic Acids Res.* **31**, e82.
117. Silverman, S.K. (2005) *In vitro* selection, characterization and application of deoxyribozymes that cleave RNA. *Nucleic Acids Res.* **33**, 6151–6163.
118. Cheong, H., Hwang, E., Lee, C., Choi, B., Cheong, C. (2004) Rapid preparation of RNA samples for NMR spectroscopy and X-ray crystallography. *Nucleic Acids Res.* **32**, e84.
119. Piccirilli, J.A., Benner, S.A., Krauch, T., Moroney, S.E., Benner, S.A. (1990) Enzymatic incorporation of a new base pair into DNA and RNA extends the genetic alphabet. *Nature* **343**, 33–37.
120. Hirao, I., Kimoto, M., Mitsui, T., Fujiwara, T., Kawai, R., Sato, A., Harada, Y., Yokoyama, S. (2006) An unnatural hydrophobic base pair system: site-specific incorporation of nucleotide analogs into DNA and RNA. *Nat. Methods* **3**, 729–735.
121. England, T.E., Uhlenbeck, O.C. (1978) 3′-Terminal labelling of RNA with T4 RNA ligase. *Nature* **275**, 560–561.
122. Helm, M., Giege, R., Florentz, C. (1999) A Watson–Crick base-pair-disrupting methyl group (m1A9) is sufficient for cloverleaf folding of human mitochondrial tRNALys. *Biochemistry* **38**, 13338–13346.
123. Frilander, J.M., Turunen, J.J. (2005) RNA ligation using T4 DNA ligase. In *Handbook of RNA Biochemistry*, ed. Hartmann, R.K., Bindereif, A., Schön, A., Westhof, E. Wiley-VCH Verlag GmbH, Weinheim, pp. 36–52.
124. Persson, T., Willkomm, D.K., Hartmann, R.K. (2005) T4 RNA ligase. In *Handbook of RNA Biochemistry*, ed. Hartmann, R.K., Bindereif, A., Schön, A., Westhof, E. Wiley-VCH Verlag GmbH, Weinheim, pp. 53–74.
125. Rieder, R., Lang, K., Graber, D., Micura, R. (2007) Ligand-induced folding of the adenosine deaminase A-riboswitch and implications on riboswitch translational control. *ChemBioChem* **8**, 896–902.
126. Lang, K., Micura, R. (2008) The preparation of site-specifically modified riboswitch domains as an example for enzymatic ligation of chemically synthesized RNA fragments. *Nat. Protocols* **3**, 1457–1466.
127. Lang, K., Rieder, R., Micura, R. (2007) Ligand-induced folding of the thiM TPP riboswitch investigated by a structure-based fluorescence spectroscopic approach. *Nucleic Acids Res.* **35**, 5370–5378.
128. Silverman, S.K., Cech, T.R. (1999) RNA tertiary folding monitored by fluorescence of covalently attached pyrene. *Biochemistry* **38**, 14224–14237.
129. Kurschat, W.C., Müller, J., Wombacher, R., Helm, M. (2005) Optimizing splinted ligation of highly structured small RNAs. *RNA* **11**, 1909–1914.
130. Dönmez, G., Hartmuth, K., Kastner, B., Will, C.L., Lührmann, R. (2007) The 5′ end of U2 snRNA is in close proximity to U1 and functional sites of the pre-mRNA in early spliceosomal complexes. *Mol Cell* **25**, 399–411.

131. Kent, O.A., MacMillan, A.M. (2002) Early organization of pre-mRNA during spliceosome assembly. *Nat. Struct. Biol.* **9**, 576–81.
132. Purtha, W.E., Coppins, R.L., Smalley, M.K., Silverman, S.K. (2005) General deoxyribozyme-catalyzed synthesis of native 3′–5′ RNA linkages. *J. Am. Chem. Soc.* **127**, 13124–13125.
133. Höbartner, C., Silverman, S.K. (2007) Recent advances in DNA catalysis. *Biopolymers* **87**, 279–292.
134. Baum, D.A., Silverman, S.K. (2007) Deoxyribozyme-catalyzed labeling of RNA. *Angew. Chem. Int. Ed.* **46**, 3502–3504.
135. Blouin, S., Mulhbacher, J., Penedo, J.C., Lafontaine, D.A. (2009) Riboswitches: ancient and promising genetic regulators. *ChemBioChem* **10**, 400–416.
136. Serganov, A., Patel, D.J. (2007) Ribozymes, riboswitches and beyond: regulation of gene expression without proteins. *Nat. Rev. Genet.* **8**, 776–790.
137. Micura, R., Höbartner, C. (2003) On secondary structure rearrangements and equilibria of small RNAs. *ChemBioChem* **4**, 984–990.
138. Höbartner, C., Micura, R. (2003) Bistable secondary structures of small RNAs and their structural probing by comparative imino proton NMR spectroscopy. *J. Mol. Biol.* **325**, 421–431.
139. Graber, D., Moroder, H., Micura, R. (2008) ^{19}F NMR spectroscopy for the analysis of RNA secondary structure populations. *J. Am. Chem. Soc.* **130**, 17230–17231.
140. Kreutz, C., Micura, R. (2008) Investigations on fluorine-labeled ribonculeic acids by ^{19}F NMR spectroscopy. In *Modified Nucleosides in Biochemistry, Biotechnology and Medicine*, ed. Herdewijn, P. Wiley-VCH Verlag GmbH, Weinheim, pp. 3–27.
141. Xia, J., Noronha, A., Toudjarska, I., Li, F., Akinc, A., Braich, R., Frank-Kamenetsky, M., Rajeev, K.G., Egli, M., Manoharan, M. (2006) Gene silencing activity of siRNAs with a ribo-difluorotoluyl nucleotide. *ACS Chem. Biol.* **1**, 176–183.
142. Li, F., Pallan, P.S., Maier, M.A., Rajeev, K.G., Mathieu, S.L., Kreutz, C., Fan, Y., Sanghvi, J., Micura, R., Rozners, E., Manoharan, M., Egli, M. (2007) Crystal structure, stability and *in vitro* RNAi activity of oligoribonucleotides containing the ribo-difluorotoluyl nucleotide: insights into substrate requirements by the human RISC Ago2 enzyme. *Nucleic Acids Res.* **35**, 6424–6438.
143. Somoza, A., Chelliserrykattil, J., Kool, E.T. (2006) The roles of hydrogen bonding and sterics in RNA interference 13. *Angew. Chem. Int. Ed.* **45**, 4994–4997.
144. Höbartner, C., Mittendorfer, H., Breuker, K., Micura, R. (2004) Triggering of RNA secondary structures by a functionalized nucleobase. *Angew. Chem. Int. Ed.* **43**, 3922–3925.
145. Wenter, P., Fürtig, B., Hainard, A., Schwalbe, H., Pitsch, S. (2005) Kinetics of photoinduced RNA refolding by real-time NMR spectroscopy. *Angew. Chem. Int. Ed.* **44**, 2600–2603.
146. Wenter, P., Bodenhausen, G., Dittmer, J., Pitsch, S. (2006) Kinetics of RNA refolding in dynamic equilibrium by ^{1}H-detected ^{15}N exchange NMR spectroscopy. *J. Am. Chem. Soc.* **128**, 7579–7587.
147. Wenter, P., Fürtig, B., Hainard, A., Schwalbe, H., Pitsch, S. (2006) A caged uridine for the selective preparation of an RNA fold and determination of its refolding kinetics by real-time NMR. *ChemBioChem* **7**, 417–420.
148. Furtig, B., Wenter, P., Reymond, L., Richter, C., Pitsch, S., Schwalbe, H. (2007) Conformational dynamics of bistable RNAs studied by time-resolved NMR spectroscopy. *J. Am. Chem. Soc.* **129**, 16222–16229.
149. Höbartner, C., Silverman, S.K. (2005) Modulation of RNA tertiary folding by incorporation of caged nucleotides. *Angew. Chem. Int. Ed.* **44**, 7305–7309.

2

Expansion of the Genetic Alphabet in Nucleic Acids by Creating New Base Pairs

Ichiro Hirao and Michiko Kimoto

2.1 Introduction

Contemporary DNA molecules store genetic information in a linear sequence of four different bases, A, G, C and T, and form double-stranded helices by the complementary A–T and G–C base pairing between the strands. The complementarity of the two sets of base pairs is a fundamental aspect of the central dogma of genetic information flow, comprising replication, transcription and translation. Through the complementarity, DNA molecules are amplified by DNA polymerases in replication and are copied to RNA molecules by RNA polymerases in transcription. The base sequences of the RNA molecules are then transferred into the amino acid sequences of proteins in translation. In replication and transcription, the nascent DNA and RNA molecules are synthesized by base pairing between the incoming triphosphate substrates and the bases in DNA template strands. In translation, base pairing in the codon–anticodon interaction between the mRNA and the aminoacylated tRNA is also a principal event in the ribosome machinery.

The mechanisms of genetic information flow underlie all of the recombinant techniques of modern biotechnology. Using these technologies, researchers have been generating novel nucleic acids and proteins with specific sequences of components, nucleotides for nucleic acids and amino acids for proteins, by preparing cloned DNA templates in accordance with their intended use. One serious limitation of recombinant technology is its laboriousness for the site-specific incorporation of non-standard extra components into nucleic acids and proteins. Recent progress in a wide variety of structural analyses has permitted the design of extra functional components for the creation of novel nucleic acids and proteins [1–12]. Furthermore, evolutionary engineering methods using combinatorial libraries, such as *in vitro* selection, DNA shuffling and ribosome display, have been developed and preparing libraries containing extra components will open the door to new evolution systems [13,14]. Therefore, there is an urgent need to develop a fundamental method that allows the site-specific incorporation of extra components into nucleic acids and proteins.

The Chemical Biology of Nucleic Acids Edited by Günter Mayer

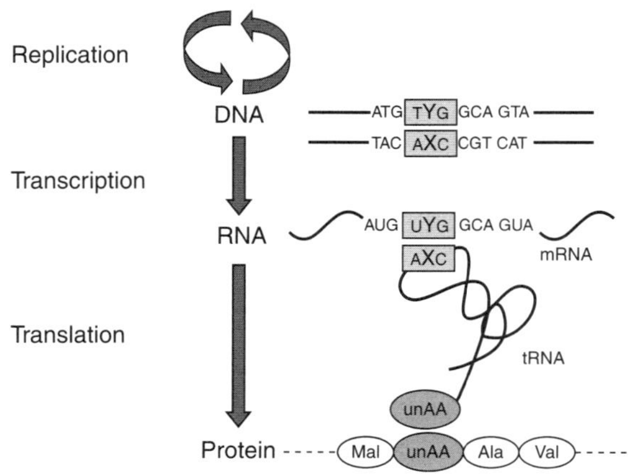

Figure 2.1 *An unnatural base pair system for the expansion of the genetic alphabet and code. The unnatural base pair (**X–Y**) that functions in the central dogma enables the site-specific incorporation of extra nucleotide analogs (**Y**) into DNA and RNA and amino acid analogs (**unAA**) into proteins*

One of the most attractive methods for this is the creation of unnatural base pair systems, in which an unnatural third base pair can expand the genetic information in the central dogma [2–7]. If DNA fragments containing the third base pair are faithfully replicated and transcribed into RNA, then the extra functional nucleotide components could be site-specifically incorporated into nucleic acids. In addition, novel codons containing unnatural bases in mRNA could be used for a translation system allowing the site-specific incorporation of amino acid analogs into proteins (Figure 2.1). The present codon table consists of 64 different combinations of triplet-base sequences, composed of the four natural bases encoding the standard 20 amino acids. By the addition of an extra base pair, the codon table would be expanded by 216 ($= 6^3$) triplet sequences and various amino acid analogs could be assigned to the extra codons. For this purpose, the extra base pair requires highly exclusive selectivity in replication, transcription and translation, to work with the natural A–T(U) and G–C pairs.

Researchers have pursued unnatural third base pairs since the late 1980s [15] and many unnatural base pairs were synthesized and tested in polymerase reactions and ribosome-mediated protein synthesis. As design prerequisites, unnatural base pairs should have similar structural, physical and chemical properties to those of the natural base pairs. For example, the distance between the positions of the *N*-glycosyl bond in the pairing bases should be around 10.7 Å, as in the natural base pairs, which is suitable to accommodate base pairing between a template base and the incoming substrate base in the closed polymerase complex. In addition, each base should have a proton acceptor residue at a specific position within the minor groove, like the N3 of purines and the 2-keto of pyrimidines, to facilitate polymerase recognition [16–18] (Figure 2.2). Needless to say, the nucleotide derivatives of the unnatural bases should be as stable in solution as the natural nucleotides. Considering these prerequisites, researchers have to create new concepts to confer faithful selectivity to the unnatural base pairs. hence so far, various concepts and their combinations, such as non-standard hydrogen bonding topology, shape complementarity and hydrophobic interfaces, have been investigated.

In this chapter, we describe a series of unnatural base pairs and their applications, studied by the research groups of Steven Benner, Harry Rappaport, Eric Kool and Floyd Romesberg, and also our group. Although many other unique nucleotide base analogs have been developed [19–26], here we will focus on the unnatural base pairs that were tested in polymerase reactions and translation systems.

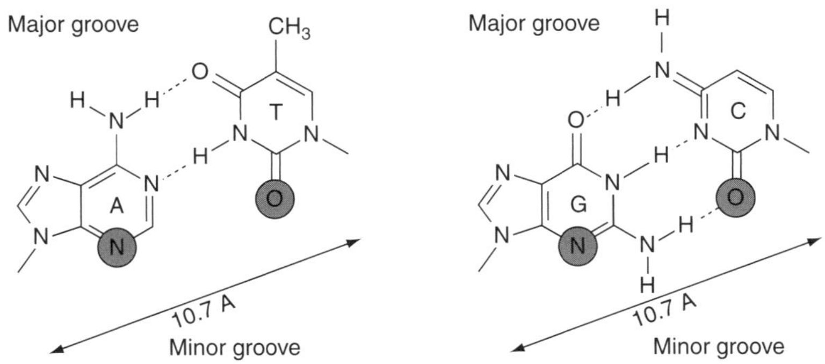

Figure 2.2 *The natural A–T and G–C pairs. Their common properties, proton acceptor residues within the minor grooves (circles) and the distance between the bases (arrows), are shown*

2.2 Procedures for unnatural base pair studies

The studies on the development of unnatural base pairs are divided two parts, the synthesis of their nucleotide derivatives and the biological and physical assessments of the pairing ability. Polymerase-mediated reactions require DNA templates and substrates, 2′-deoxyribonucleoside 5′-triphosphates (dNTPs) for replication and ribonucleoside 5′-triphosphates (NTPs) for transcription. For protein synthesis, *in vitro* cell-free translation systems are useful, in which the mRNA and the aminoacylated tRNA molecules both contain unnatural bases.

First, DNA fragments containing the unnatural bases are synthesized by phosphoramidite methods, using a DNA synthesizer and the amidite reagents of the unnatural bases. The 2′-deoxyribonucleosides and ribonucleosides are chemically synthesized by glycosidation between the sugar and base moieties and/or by base modification of nucleosides [27–29]. The nucleosides are then converted to the amidite derivatives, in which the protic residues are fully protected with appropriate protecting groups by a conventional method.

Triphosphates are also prepared from the nucleoside derivatives by chemical synthesis (Figure 2.3) [30]. Yoshikawa and co-workers invented a simple method for the selective phosphorylation of unprotected nucleosides [31,32]. The triphosphate products are purified by DEAE-cellulose column chromatography, followed by reversed-phase HPLC. Since Yoshikawa's method is performed under acidic conditions, the addition of 1,8-bis(dimethylamino)naphthalene (proton sponge) is effective for the synthesis of acid-sensitive nucleosides, such as purine-type nucleosides [33]. Another chemical synthesis is Eckstein's method, using nucleoside derivatives protected as 2′- and/or 3′-hydroxy groups [34]. This method has proved to be useful for the preparation of a wide range of triphosphate derivatives. Researchers have to monitor the purity of the triphosphates, because the triphosphates are usually used in more than a 1000-fold excess relative to the input polymerases.

The selectivity and the efficiency of unnatural base pairings are assessed by biological methods, such as single-nucleotide insertion, primer extension, polymerase chain reaction (PCR) amplification and T7 transcription, and by physical methods, such as thermal stability and structural analyses. Single-nucleotide insertion experiments are one of the useful methods for determining the kinetic parameters of cognate and non-cognate base pair selectivity in replication, by using combinations of a triphosphate and a partially double-stranded template with a labeled primer [35–37]. This method usually involves a gel-based analysis, which is tremendously laborious and time consuming for a large number of samples to collect all of the kinetic parameters of the cognate and non-cognate pairings. Thus, an improved method using an automated DNA sequencer was developed [38]. Melting temperature measurement is also a quantitative method to

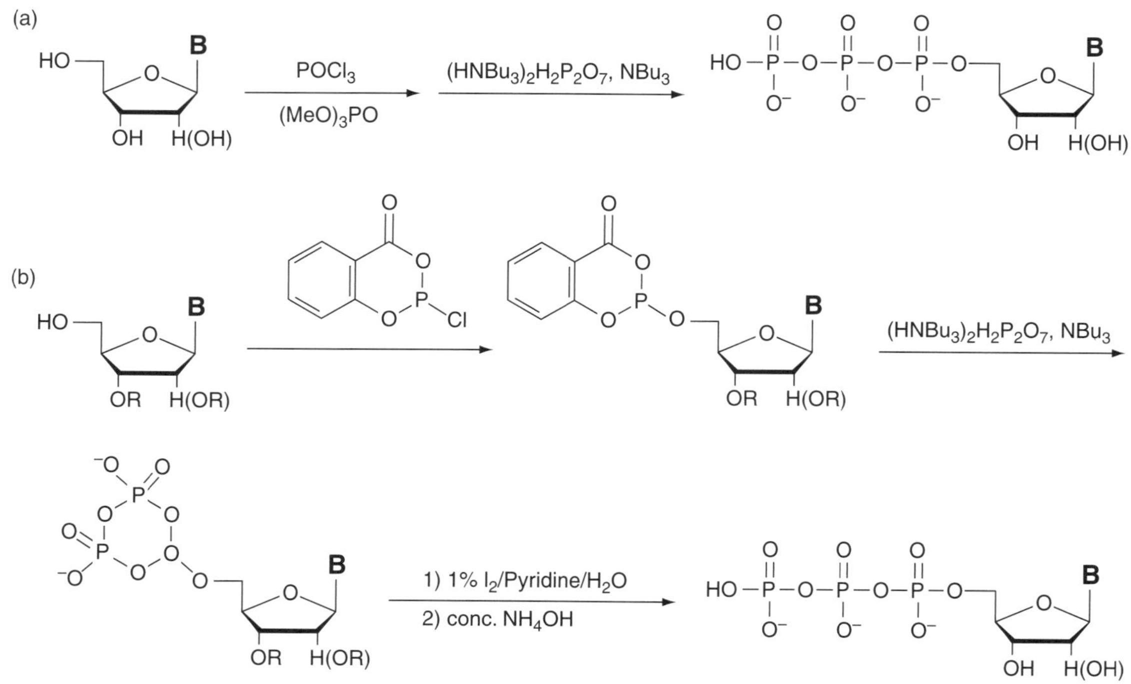

Figure 2.3 *Syntheses of nucleoside triphosphates. (a) Yoshikawa's method. (b) Eckstein's method.* ***B*** *= nucleobase,* ***R*** *= protecting groups*

assess the thermal stability of duplex DNA containing unnatural base pairs. As for other assessments, unique quantification methods for primer extension, PCR amplification and T7 transcription have been developed by each of the research groups.

2.3 Progress in unnatural base pair development

2.3.1 Benner's base pairs

In 1989, Benner and co-workers first reported a new base pair, between isoguanosine (**isoG**) and isocytidine (**isoC**) [15] (Figure 2.4a). They focused on the hydrogen bonding patterns, which are the arrangements of the proton donor–acceptor residues between pairing bases, and found that extra base pairs could be designed with non-standard hydrogen bonding patterns that differ from those of the A–T and G–C base pairs. The **isoG** substrates were incorporated into DNA and RNA, opposite **isoC** in templates, by the Klenow fragment of *Escherichia coli* DNA polymerase I and T7 RNA polymerase, respectively. They suggested that at least six different unnatural base pairs are possible. Subsequently, other unnatural base pairs, in addition to the **isoG–isoC** pair, were synthesized [39,40] and a base pair between xanthosine (**X**) and 2,6-diaminopyrimidine (**κ**) (Figure 2.4b) was also shown to be a potential candidate working in replication and transcription.

In 1992, Benner's group succeeded in the site-specific incorporation of an amino acid analog, 3-iodotyrosine, into a peptide by an *in vitro* translation system, using synthetic mRNA and tRNA molecules including the **isoG–isoC** pair [41]. The mRNA molecule contained a new (**isoC**)AG codon and encoded a peptide consisting of 16 amino acids. The aminoacylated tRNA molecule containing a CU(**isoG**) anticodon was prepared by two

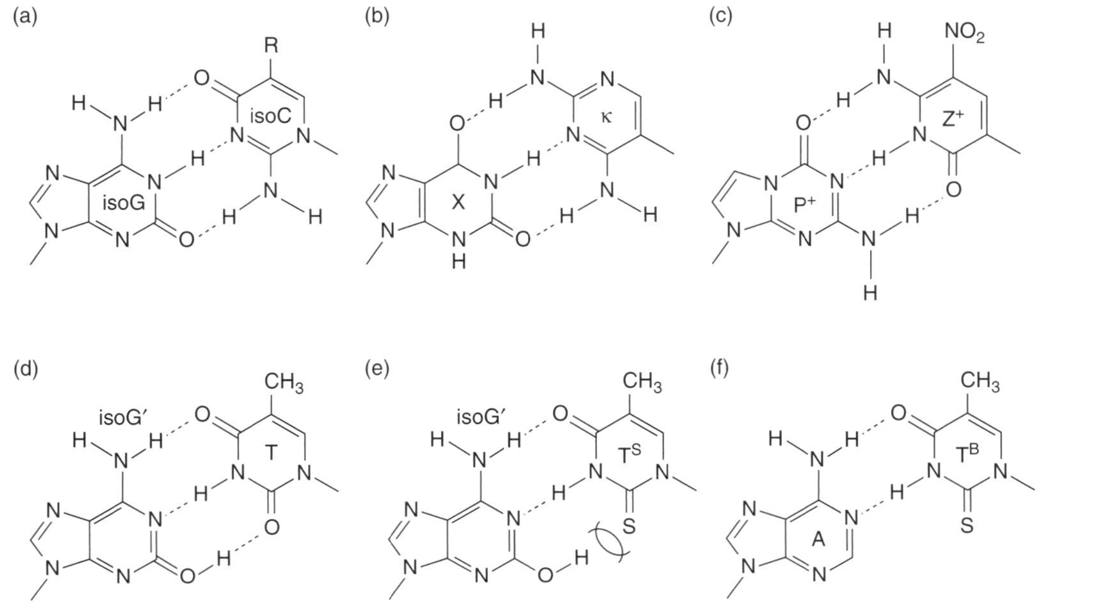

Figure 2.4 *Benner's base pairs. (a) The unnatural* ***isoG–isoC*** *pair.* ***R*** *in position 5 of isoC is H or a methyl group. (b) The unnatural* ***X–κ*** *pair. (c) The unnatural* ***P*–Z**** *pair. (d) The non-cognate pair between T and* ***isoG′*** *(the enol tautomer of* ***isoG****). (e) The non-cognate pair between the modified 2-thio-T (*$\mathbf{T^S}$*) and* ***isoG′****. (f) The unnatural A–*$\mathbf{T^S}$ *pair, instead of the natural A–T pair*

consecutive steps, developed by Hecht's [42] and Schultz's groups [43]. First, a truncated tRNA molecule missing the last two bases was chemically synthesized and then this molecule was enzymatically ligated with a dimer of 2′-deoxynucleotides linked with iodotyrosine at the 3′-end. Using these two RNA species, they observed that a peptide containing iodotyrosine was efficiently synthesized in an *E. coli* cell-free translation system.

As for transcription using the **isoG–isoC** pair, Tor and Dervan reported the site-specific incorporation of a modified **isoG**, N^6-(6-aminohexyl)isoguanosine, into an RNA fragment (10-mer) using **isoC**-containing DNA templates with T7 RNA polymerase [44]. Although a gel-based analysis of the transcripts revealed that **isoG**TP was also incorporated into RNA opposite T in a template in the absence of ATP (see below), ATP was predominantly incorporated into RNA opposite T when both ATP and **isoG**TP were utilized in the transcription reaction. The transcription efficiency was about 50% of that for transcription using the standard natural template–substrate system. The aminohexyl group in the transcripts would be used as a specific modification site after the preparation of the RNA transcripts.

These non-standard hydrogen-bonded base pairs have some shortcomings, however, including the tautomerization of **isoG**, the poor recognition of diaminopyrimidines by polymerases and the instability of **isoC**. For instance, the 2-keto form of **isoG** is in equilibrium with its enol form in solution and the enol form pairs with T, rather than with **isoC** (Figure 2.4d). As mentioned above, the **isoG** incorporation opposite T in templates could be neglected by competing with the A incorporation opposite T. However, the T incorporation opposite **isoG** would be a critical problem. In fact, in the presence of both d-**isoC**TP and dTTP, dTTP was also incorporated into DNA, opposite **isoG** in templates, by the Klenow fragment [45]. Prudent and co-workers examined the PCR amplification of DNA fragments containing the **isoG–isoC** pair using Titanium Taq DNA polymerase, in which an N-terminal region corresponding to the 5′-exonuclease activity was deleted [46]. Due to the tautomerization, the fidelity-per-round of the unnatural base pairing in PCR was ~93%, as determined by analyzing the decomposition of the incorporated **isoC** in the amplified DNA fragments.

In addition, the 2-amino groups of pyrimidine analogs, such as **isoC** and **κ**, are hardly recognized by some polymerases. For example, d**κ**TP was not incorporated into DNA by either the Klenow fragment or the bovine DNA polymerases α, β and ε [47] and **isoC**TP was not a suitable substrate for T7 RNA polymerase [45]. Furthermore, the nucleotide of **isoC** is chemically unstable in solution and about 65% of d-**isoC**TP was decomposed in an aqueous solution after being stored for 6 weeks at –20 °C [45]. This instability was improved by a methyl group modification at position 5 of **isoC** [44,48].

Recently, Benner's group overcame the tautomerization problem of **isoG** by using 2-thiothymidine triphosphate (d-$\mathbf{T^S}$TP), instead of dTTP [49]. The mispairing of the enol form of **isoG** with T was efficiently prevented by replacing the 2-keto of T with the thione, which is a less efficient hydrogen bond acceptor and spatially clashes with the 2-hydroxyl group of the **isoG** enol form (Figure 2.4e). They performed PCR amplification of DNA fragments containing the **isoG–isoC** pair with an N-terminally truncated Taq DNA polymerase (KlenTaq). As a result, the fidelity-per-round of the **isoG–isoC** pair in PCR was improved to 98%, which was higher than that in the previous attempt [46].

Other unnatural hydrogen-bonded base pairs were also examined by PCR amplification. DNA fragments containing the **X–κ** pair were amplified by variants of the reverse transcriptase from HIV-1 [50]. About 5% of the **X–κ** pair was replaced with the natural bases after five cycles of PCR. Since the reverse transcriptase is not thermostable, this system required the addition of the enzyme into the reaction mixture in every PCR cycle. Recently, PCR amplification using an unnatural pair between 2-aminoimidazo[1,2-*a*]-1,3,5-triazin-4(8*H*)one ($\mathbf{P^*}$) and 6-amino-5-nitro-2(1*H*)pyridone ($\mathbf{Z^*}$) (Figure 2.4c) was also reported and the retention per cycle of the $\mathbf{P^*}$–$\mathbf{Z^*}$ pair in the amplified products reached 97.5% with the exonuclease-deficient Vent or Deep Vent DNA polymerase [51]. In the $\mathbf{P^*}$–$\mathbf{Z^*}$ pair, the problems, such as the tautomerization, the poor recognition by polymerases and the instability of the **isoG–isoC** pair, were significantly improved.

2.3.2 Rappaport's base pairs

The use of the thione group, in place of the keto group, as part of the unnatural bases could increase the range of the unnatural base pair design, as mentioned in the discussion of Benner's 2-thiothymidine. At the early stage of the unnatural base pair studies, Rappaport created a third pair, between 6-thioguanine ($\mathbf{G^S}$) and 5-methyl-2-pyrimidinone ($\mathbf{T^H}$) (Figure 2.5a) [52,53]. These base analogs were designed by considering that the relatively large 6-thione group of $\mathbf{G^S}$ prevents the interaction with the 4-amino group of C, but fits well with the small hydrogen of $\mathbf{T^H}$. In a single-nucleotide insertion experiment using the Klenow fragment with a $3' \rightarrow 5'$ exonuclease activity, the incorporation efficiency of d$\mathbf{G^S}$TP opposite $\mathbf{T^H}$ in a template was fourfold higher than that of d$\mathbf{G^S}$TP opposite C. Although the selectivity of the $\mathbf{G^S}$–$\mathbf{T^H}$ pairing was marginal, this was the first assessment of the concept of shape complementarity in the development of unnatural base pairs.

In 2004, Rappaport reported a three-base-pair set composed of 6-thiopurine ($\mathbf{H^S}$)–$\mathbf{T^H}$ as an unnatural base pair, hypoxanthine (**H**)–C instead of G–C (Figure 2.5b and c) and A–T pairs [54]. To increase the effect of the shape complementarity of the cognate pairing, he reduced their hydrogen bonding ability by removing the 2-amino groups from G and $\mathbf{G^S}$. The selectivity of their cognate and non-cognate pairings in replication was examined with T7 DNA polymerase, which has strong $3' \rightarrow 5'$ exonuclease activity. In the system, each base pair exhibited highly exclusive selectivity in replication. The incorporation efficiency including the exonuclease process of the $\mathbf{H^S}$–$\mathbf{T^H}$ pairing was 0.01–0.1 times that of the A–T pairing and 0.001–0.2 times that of the **H**–C pairing. The misincorporations of dTTP opposite H, d$\mathbf{T^H}$TP opposite **H** and d$\mathbf{H^S}$TP opposite C were detected in the assay. However, these misincorporation efficiencies were 10^3–10^5 times lower than those of the corresponding cognate pairings.

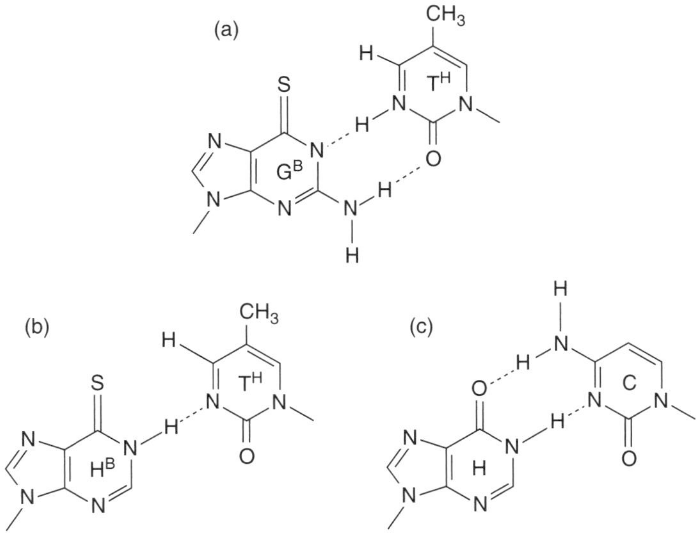

Figure 2.5 *Rappaport's base pairs. (a) The unnatural* $\boldsymbol{G^S}$*–*$\boldsymbol{T^H}$ *pair. (b) The unnatural* $\boldsymbol{H^S}$*–*$\boldsymbol{T^H}$ *pair. (c) The unnatural* ***H****–C pair, instead of the natural G–C pair*

2.3.3 Kool's base pairs

Hydrogen bonds are generally very important intra- and intermolecular forces in living cells. Therefore, since the discovery of Watson–Crick base pairs, the hydrogen bonds between pairing bases have also long been believed to be essential for both DNA duplex formation and genetic information flow. However, in the mid-1990s, Kool and co-workers studied non-hydrogen-bonded base pairs and proposed that the hydrogen bonds are less important, at least, in the fidelity of replication than commonly believed and that shape complementarity between pairing bases plays a more important role [55–58]. They created nonpolar base analogs with shapes similar to either A or T, but without strong proton acceptor or donor residues, and examined the efficiency and the selectivity of their base pairing by single-nucleotide insertion and primer extension experiments.

They first designed and synthesized 2,4-difluorotoluene (**F**), as a shape analog of T [59]. Since the thermal stability of a duplex DNA fragment containing the A–**F** pair was significantly decreased, in comparison with that containing the A–T pair, the fluoro groups in **F** might be fairly poor as hydrogen bonding acceptors [60–63]. Despite this, in single-nucleotide insertion experiments using an exonuclease-deficient Klenow fragment, the incorporation efficiency of dATP opposite **F** in a template was only four times lower than that of dATP opposite T, but it was 10^2–10^4 times higher than those of dGTP, dCTP and dTTP opposite **F** [62]. Although the incorporation efficiency of d**F**TP opposite A was 400 times lower than that of dTTP opposite A, it was 10^3–10^4 times higher than those of d**F**TP opposite G, C and T [63]. The relatively low efficiency of d**F**TP might be caused by the weak ability of the fluoro group to function as the hydrogen bond acceptor for the recognition by the Klenow fragment in the minor groove region of the forming duplex DNA [61,64]. Together, these results showed the importance of shape complementarity, rather than hydrogen bonding interactions, for the selectivity of the base pairing in replication. Furthermore, Kool's group also determined the solution structure of a duplex DNA containing the A–**F** pair by NMR spectroscopy. The 3D structure of the A–**F** pair very closely resembled that of the A–T pair [65].

In 1998, Kool's group designed and synthesized the pairing partner of **F**, 4-methylbenzimidazole (**Z**) (Figure 2.6a), which is a shape analog of A, and examined it in a replication reaction with the Klenow

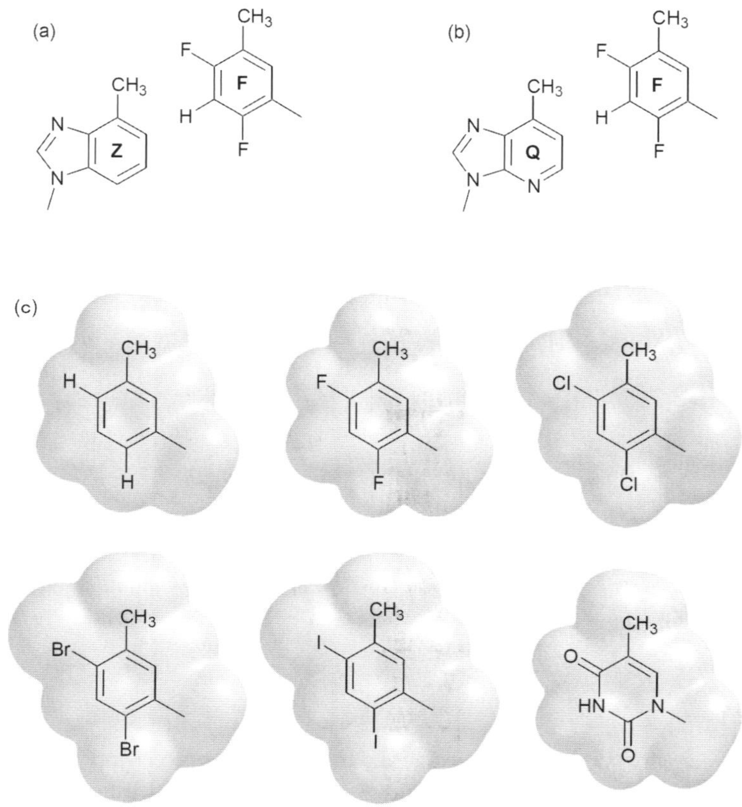

Figure 2.6 *Kool's base pairs. (a) The unnatural **Z–F** pair. (b) The unnatural **Q–F** pair. (c) The structures of nonpolar thymidine base analogs with gradually increasing size, shown in space-filling models*

fragment [66]. Again, the data indicated the importance of the shape complementarity between pairing bases. In single-nucleotide insertion experiments, the unnatural **Z–F** pair, and also the A–**F** and **Z**–T pairs, exhibited high selectivity relative to other non-cognate pairs. Extension involving dATP incorporation opposite **F** and dTTP incorporation opposite **Z** also occurred with high efficiency, although those involving the d**Z**TP incorporation opposite **F** and the d**F**TP opposite **Z** occurred with considerably lower efficiency. As in the case of d**Z**TP, **Z** also lacks a hydrogen bond acceptor at the position corresponding to N3 in purines, thus hindering efficient polymerase recognition. This was subsequently proved by synthesizing a new base, 9-methyl-1*H*-imidazo[4,5-*b*]pyridine (**Q**) (Figure 2.6b), which has the minor groove acceptor nitrogen. The incorporation efficiency of **Q** was greatly increased, in comparison with that of **Z** [67]. The solution structure of the **Z–F** pair in duplex DNA was determined by NMR spectroscopy and the geometry resembled that of the natural base pairs [68]. This was an important study to confirm their hypothesis, because there were still some arguments about the polarity of the **F** base [69].

Further support for their hypothesis was obtained from studies of another unique pair, between pyrene (**P**) and an abasic site (**Φ**) [70]. The shape of pyrene is nearly as large as that of the entire unnatural base pair, hence pyrene was expected to fit the abasic site in a shape-complementary manner in duplex DNA. Indeed, d**P**TP was efficiently incorporated into DNA opposite **Φ** in the template, by the exonuclease-deficient Klenow fragment, and the incorporation efficiency was 10^2-fold higher than those of the d**P**TP incorporation opposite the natural bases. After the d**P**TP incorporation, the extension stopped and, using this property, Kool's group demonstrated the detection of abasic sites in DNA.

The importance of shape complementarity was proved by *in vivo* experiments using the A-shape mimic, **Q**, and the T-shape mimic, **F** [71,72]. DNA fragments containing **Q** or **F** were inserted into an M13mp7(L2) single-stranded genome. These genomes were passed through *E. coli* and then their sequences were analyzed. The data revealed that **Q** and **F** were bypassed with moderate to high efficiency in the cells and with very high efficiency under damage-response (SOS induction) conditions. As expected, **Q** and **F** were replaced mainly with A and T, respectively, because these experiments were carried out without the substrates of **Q** and **F**. Kool's group also synthesized other T-shape mimics, such as 2,4-dichloro-, -dibromo- and -diiodotoluenes, in addition to **F** (Figure 2.6c), and examined their abilities under *in vitro* and *in vivo* conditions [61,72,73]. Although the dichlorotoluene base has a slightly larger shape than that of T, it was incorporated opposite A with higher efficiency than that of the **F** incorporation, suggesting that the natural base pairs appear to be smaller than optimum for polymerases and that DNA polymerases are relatively flexible in terms of the shape variation between pairing bases.

Kool's work proved that shape fitting between pairing bases was important and that non-hydrogen-bonded base pairs could work in replication. The effect of hydrogen bonds on base pairing in replication is still being debated. Other unnatural non-hydrogen-bonded bases and/or polymerases have also been used to confirm the importance of hydrogen bonding for the efficiency and the selectivity of base pairing in a total replication process [74–78]. However, Kool's work greatly contributed to the advancement of unnatural base pair creation and the concept of shape complementarity suggests that non-hydrogen-bonded base pairs could be a good candidate for a third base pair. They also developed another type of unique, unnatural base pair [79–84], shape-widened base pairs, although these are still being studied for polymerase reactions.

2.3.4 Romesberg's base pairs

The development of unnatural hydrophobic, non-hydrogen-bonded base pairs was sparked by Romesberg and co-workers. In 1999, they first reported an unnatural, hydrophobic self-pair of 7-propynylisocarbostyril (**PICS**) [85] (Figure 2.7). Surprisingly, the thermal stability of the **PICS–PICS** pair in a DNA duplex was higher than that of the G–C pair. Furthermore, in replication by the exonuclease-deficient Klenow fragment, the incorporation efficiency of d(**PICS**)TP opposite **PICS** was 20–100 times higher than those of the natural triphosphates opposite **PICS** and the misincorporation efficiency of d(**PICS**)TP opposite the natural bases was 100–5000 times lower than that of the A incorporation opposite T.

Romesberg's group have pursued hydrophobic base pairs and synthesized a variety of hydrophobic bases, such as 7-azaindole (**7AI**), isocarbostyril (**ICS**), *C3*-methylisocarbostyril (**MICS**) and pyrrolopyrizine (**PP**), shown in Figure 2.7 [86–88]. Among them, many hydrophobic base combinations of the **7AI–MICS**, **M7AI–MICS**, **7AI–7AI**, **P7AI–P7AI**, **ICS–P7AI**, **PICS–P7AI**, **MICS–P7AI** and **MICS–PP** pairings showed high incorporation efficiency in single-nucleotide insertion experiments with the Klenow fragment (exo$^-$) [87]. For example, the **MICS–PP** pair was complementarily incorporated with high efficiency and the incorporation efficiency of d(**MICS**)TP opposite **PP** was 60–240 times higher than those of d(**MICS**)TP opposite the natural bases. Interestingly, some hydrophobic self-pairs, such as trimethylphenyl (**TM**) and 2-methylnaphthyl (**2MN**), which lack a hydrogen-bond acceptor residue for polymerase recognition, also exhibited high incorporation efficiency [86,88].

Hydrophobic self-pairs, such as **PICS–PICS**, **7AI–7AI** and **P7AI–P7AI**, might be promising as the third pair. However, one problem is that these self-pairs are poorly extended after their incorporation by the Klenow fragment. Romesberg's group found that mammalian polymerase β could efficiently extend primers containing **7AI** at the 3′-end and the combination with the exonuclease-proficient Klenow fragment allowed efficient replication involving the **7AI–7AI** self-pairing with reasonable selectivity [89]. In general, Pol β is

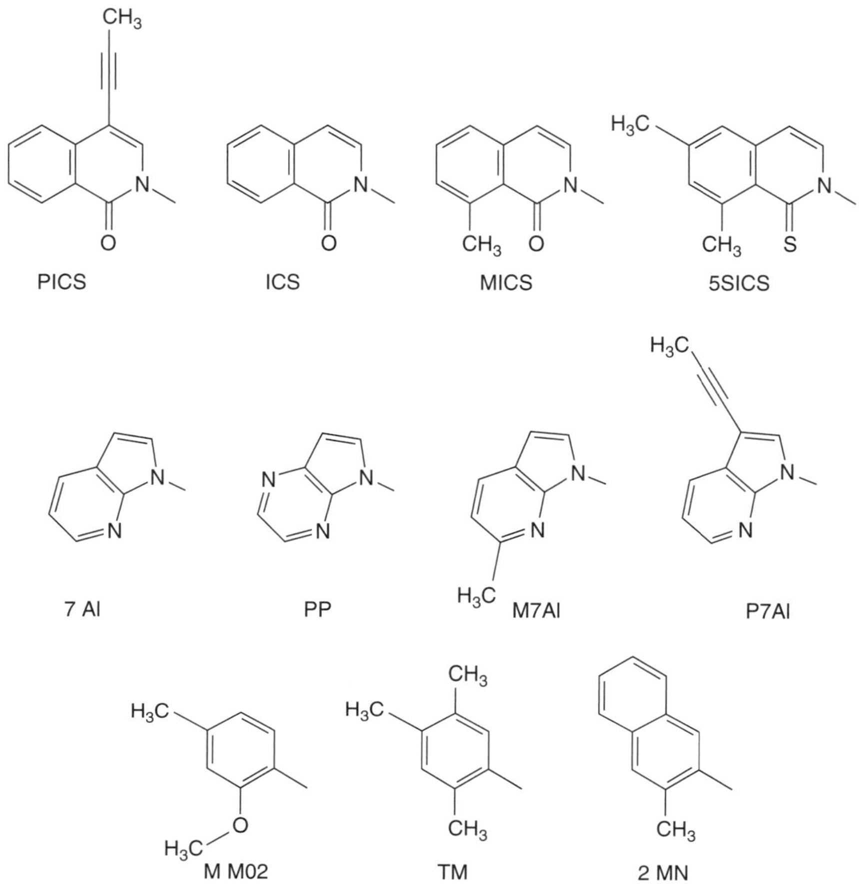

Figure 2.7 *Romesberg's unnatural bases: his group created the* ***5SICS–MMO2*** *pair, which was selected as the best candidate of the third pair, by a screening method using an unnatural hydrophobic base library*

primarily involved in DNA repair and fills single nucleotide gaps in DNA, but it lacks exonuclease activity. They also found that the exonuclease activity of the Klenow fragment was less efficient with the cognate **7AI–7AI** pair, relative to mispairs that resulted from the natural base incorporation opposite **7AI**.

Romesberg and co-workers further advanced their unnatural base pair systems by the directed evolution of DNA polymerases [90–93]. Since primer extension after the **PICS–PICS** self-pairing cannot proceed with native polymerases, such as the Klenow fragment and Taq polymerase, they evolved the Stoffel fragment of Taq polymerase by phage display [92,93]. They developed an ingenious method: first, a **PICS**-containing primer fragment (51-mer) conjugated to a 'basic peptide' was assembled, with an 'acidic peptide' expressed on the phage with the polymerase library; second, the primer fragment was hybridized with a **PICS**-containing template fragment (28-mer), in which the duplex contains the **PICS–PICS** self-pair at the 3′-terminus of the primer; and third, polymerase variants capable of extending the primer were selected by the incorporation of biotin-linked dUTP. They then isolated a polymerase variant with three mutations, which incorporates d(**PICS**)TP opposite **PICS** and then extends the primer after the **PICS–PICS** self-pairing with reasonable efficiency and fidelity. Interestingly, one of the mutation positions of the polymerase variant does not directly contact the primer–template DNA duplex, indicating that it mediates long-range interactions. The others are located at positions related to a salt bridge with the DNA phosphate backbone and to substrate recognition and catalysis.

Rational design of unnatural base pairs is still restricted, because of the limited information about the mechanisms and the lack of detailed structural analyses of replication. Thus, combinatorial methods are also useful to search for unnatural base pair candidates within a library of base analogs. Romesberg's group has synthesized a large number (more than 60) of hydrophobic base analogs, in which the pairing combinations between these base analogs could reach 3600 potential unnatural base pairs [85–88,94–104]. They then selected the best candidate, the **5SICS–MMO2** pair, from the combinations by screening methods using single-nucleotide insertion and primer extension by the Klenow fragment [105]. The incorporation efficiency of d(**5SICS**)TP opposite **MMO2** in a template was 290–10^5 times higher than those of the natural substrates opposite **MMO2** and the incorporation efficiency of d(**MMO2**)TP opposite **5SICS** was at least 7500 times higher than those of the natural substrates opposite **5SICS**. Furthermore, the **5SICS–MMO2** pairing was 10–1000 times superior to each self-pairing, **5SICS–5SICS** and **MMO2–MMO2**.

Romesberg group's extensive studies of the hydrophobic unnatural base pair systems, including polymerase mutations, have been making steady strides forward. In parallel, their studies showed that the hydrophobic interfaces between pairing bases could confer high selectivity and efficiency to the third base pair.

2.3.5 Hirao's base pairs

Our group started studying unnatural base pair development in 1997. We developed several unnatural base pairs based on the stringent shape complementarity between pairing bases, for which we designed unnatural bases with different shapes from those of the natural bases. An unnatural base pair between 2-amino-6-dimethylaminopurine (**x**) and 2-oxopyridine (**y**) (Figure 2.8a) was first designed by combining the concepts of non-standard hydrogen-bonding topology and shape complementarity [106]. Although **x** could possibly pair with T, the bulky 6-dimethylamino group of **x** was designed to clash spatially with the 4-keto group of T. The **x–y** pair functions in transcription and the triphosphate of **y** (**y**TP) is site-specifically incorporated into RNA opposite **x** in DNA templates by T7 RNA polymerase.

In 2002, another base pair, between 2-amino-6-thienylpurine (**s**) and **y** (Figure 2.8b), was developed as an improved version of the **x–y** pair [107]. The bulky dimethylamino group of **x** interferes with the stacking interactions between the unnatural and the neighboring natural base pairs in DNA duplexes and DNA–RNA hybrids. To address this problem, we replaced the 6-dimethylamino group of **x** with heterocycles, such as thiophene and furan, and designed the **s** base. The 6-thienyl group efficiently prevents the non-cognate pairing with T by clashing with its 4-keto group spatially and electrostatically (Figure 2.8e) and the incorporation efficiency and selectivity of **y**TP into RNA opposite **s** became higher than those opposite **x** [108]. In addition, several modifications at position 5 of **y** can be synthesized chemically and these modified-**y** triphosphates are site-specifically incorporated into RNA by transcription (described in a later section).

The specific transcription of the **s–y** pair was combined with an *in vitro* translation system to generate proteins containing amino acid analogs at desired positions [108]. We created an extra codon (**y**AG)–anticodon (CU**s**) interaction between mRNA and tRNA for an amino acid analog, 3-chlorotyrosine. The tRNA molecule containing the CU**s** anticodon was prepared by a combination of chemical synthesis and enzymatic ligation and its aminoacylation was performed with the help of the mischarging ability of a natural aminoacyl-tRNA synthetase [109]. The site-specific incorporation of 3-chlorotyrosine into the Ha-Ras protein (185 amino acids) was achieved by a coupled transcription–translation system with a DNA template containing **s**, **y**TP, T7 RNA polymerase, 3-chlorotyrosine-charged $tRNA_{CU\mathbf{s}}$ and an *E. coli* extract.

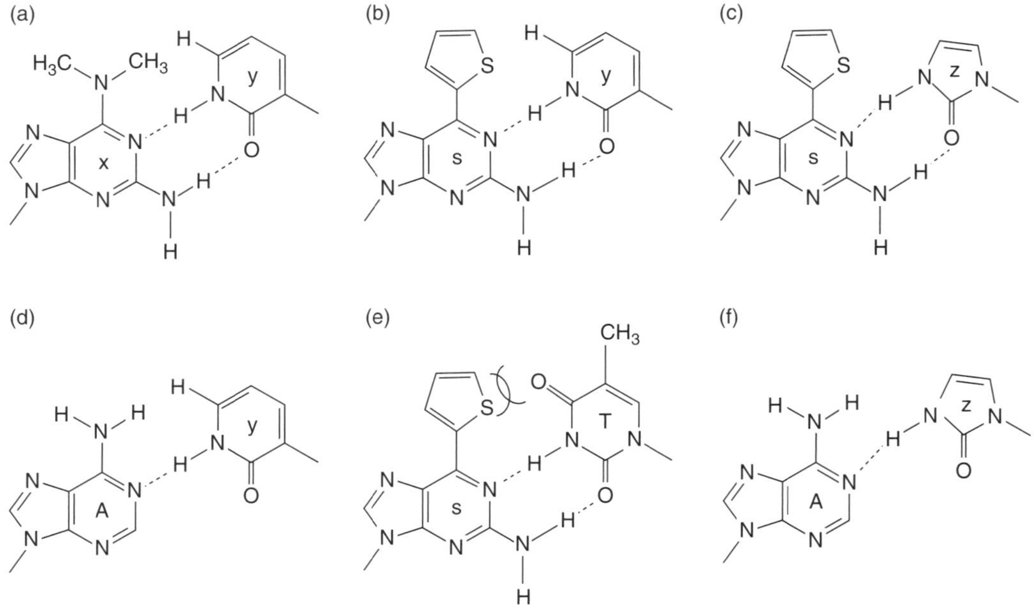

Figure 2.8 *Hirao's base pairs. (a) The unnatural **x–y** pair. (b) The unnatural **s–y** pair. (c) The unnatural **s–z** pair. (d) The non-cognate A–**y** pair. (e) The non-cognate **s**–T pair. (f) The non-cognate A–**z** pair*

We also designed imidazolin-2-one (**z**) (Figure 2.8c), as an alternative base of **y**, for an efficient pairing partner of **s** [110]. Although the site-specific **y** incorporation into RNA by transcription using **s**-containing DNA templates is useful for practical applications, the **s** incorporation into RNA opposite **y** is less selective. The bulky thienyl group of **s** efficiently prevents pairing with natural bases, whereas **y** has no such functional group to exclude non-cognate pairings with the natural bases. As a result, natural purine substrates, especially A, were partially incorporated opposite **y** in nucleic acids by polymerase reactions (Figure 2.8d). The shape complementarity of the five-membered ring of **z** with **s** is better than that of the six-membered ring of **y** with **s**. In contrast, the shape of **z** is expected to be less effective in the non-cognate pairing with A, compared with that of **s** (Figure 2.8c and f). We then chemically synthesized **z**-containing DNA templates and the triphosphate of **z**. By using the **z**-containing templates, **s**TP was more selectively incorporated into RNA opposite **z** by T7 transcription, in comparison with that opposite **y** [110]. This indicates that the five-membered ring base is useful for preventing non-cognate pairing with the natural purine bases, by virtue of its incomplete shape complementarity. However, the incorporation efficiency and selectivity of the **s–z** pairing are still insufficient in replication. This is because the hydrogen bonding residues in **s** and **z** still attract any one of the natural bases. Although these non-cognate pairings seldom occur in replication, they are sufficient to seriously reduce the unnatural base pair fidelity after repetitive replication cycles, such as PCR. Hence our design concept has been shifted to hydrophobic base pairs with specific shape complementarity.

Our first hydrophobic base was pyrrole-2-carbaldehyde (**Pa**) (Figure 2.9a) [111], which was designed to pair specifically with Kool's **Q** base (Figure 2.6b). The **Q** base was developed as a pairing partner of **F** by Kool's group [67]. The problem with the **Q–F** pair is its poor exclusivity as the third base pair, because the **Q–F** pair is an isostere of the A–T pair. Since the shape of **F** mimics that of T, **F** fits well and pairs with A in replication. Actually, the efficiency of the A–**F** pair is higher than that of the **Q–F** pair in replication, perhaps due to steric hindrance between the hydrogen atoms of **F** and **Q**. To address this problem with **F**, we designed

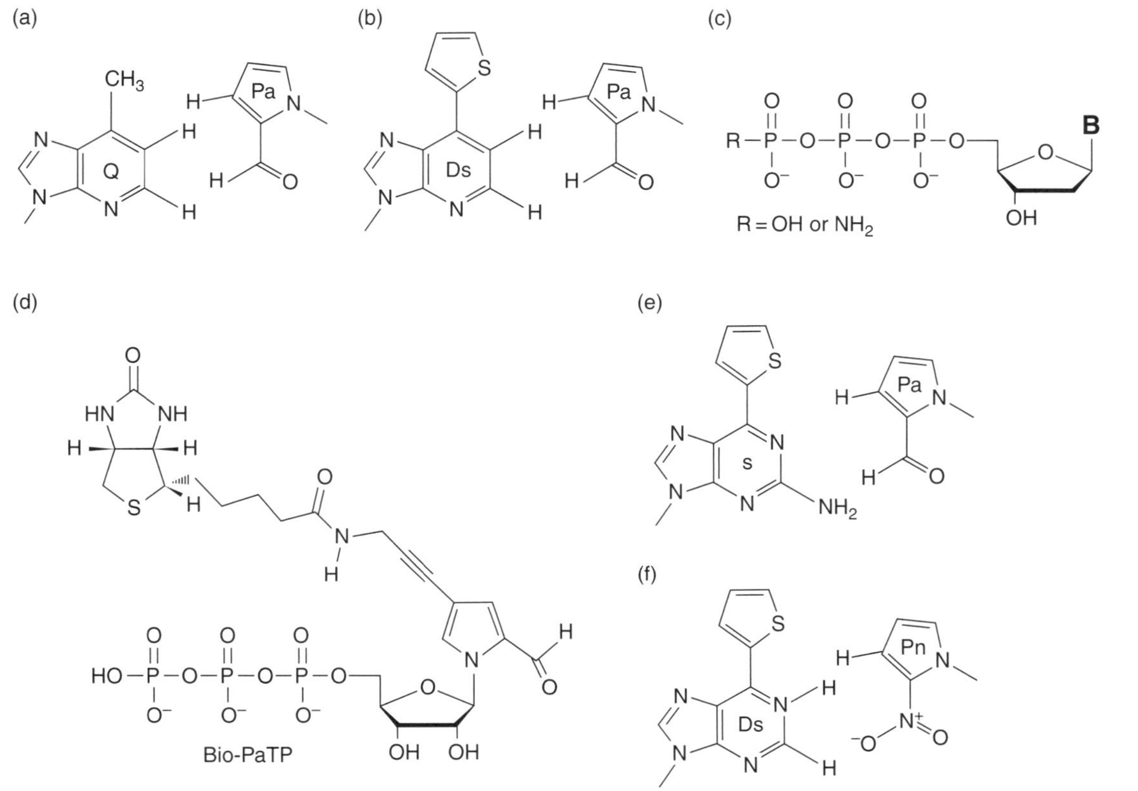

Figure 2.9 *Hirao's hydrophobic base pairs. (a) The unnatural* ***Q–Pa*** *pair. (b) The unnatural* ***Ds–Pa*** *pair. (c) Usual triphosphate (**R** = OH) and γ-amidotriphosphate (**R** = NH_2) for the unnatural* ***Ds–Pa*** *pair system. (d) The structure of Bio–**Pa**TP. (e) The unnatural* ***s–Pa*** *pair. (f) The unnatural* ***Ds–Pn*** *pair*

a hydrophobic five-membered ring base, **Pa**, by considering the stringent shape complementarity with **Q**, with the same idea of designing the **s–z** pair from the **s–y** pair. The **Pa** base has an aldehyde group to facilitate the interaction with polymerases. As a result, the selectivity of the **Q–Pa** pair was improved, relative to that of the **Q–F** pair [111]. However, **Q** is still an isostere of A and **Q** also pairs with T. Therefore, we designed a new hydrophobic base, 7-(2-thienyl)imidazo[4,5-*b*]pyridine (**Ds**) (Figure 2.9b) [112], in which the methyl group of **Q** was replaced with the thienyl group to prevent the formation of the non-cognate **Q**–T pair, as we had developed with the **s–y** pair (Figure 2.8b and e).

The hydrophobic **Ds–Pa** pair has specific shape complementarity: the shape of **Ds** is larger than those of the natural purines, A and G and the shape of **Pa** is smaller than those of the natural pyridines, C and T. We found that the use of modified triphosphates, γ-amidotriphosphates (Figure 2.9c), instead of the usual triphosphates for some bases increases the selectivity of the **Ds–Pa** pairing in replication. In the **Ds–Pa** system, the self **Ds–Ds** mispairing and the slight misincorporation of d**Pa**TP opposite A are problematic, in terms of applying the unnatural base pair to PCR amplification. The γ-amidotriphosphates of **Ds** and A are capable of reducing the non-cognate **Ds–Ds** and A–**Pa** pairings, respectively. In addition, the 3′–5′ exonuclease activity of the DNA polymerases also assists by eliminating the non-cognate base pair formations. Using the combination of the usual triphosphates of **Pa**, G, C and T and the modified triphosphates, γ-amidotriphosphates, of **Ds** and A, we succeeded in the amplification of DNA fragments containing the

Ds–Pa pair by PCR with Vent DNA polymerase, a thermophilic DNA polymerase with 3′–5′ exonuclease activity. After 10 and 20 PCR cycles, the replacements of the **Ds–Pa** pair with the natural base pairs were ~1 and 4%, respectively. The γ-amidotriphosphates can be prepared with a slight modification of the conventional chemical synthesis for triphosphates [112].

The **Ds–Pa** pair also functions complementarily in transcription. The substrates of **Ds** and **Pa** can be site-specifically incorporated into RNA, opposite **Ds** and **Pa** in the template, respectively, by T7 RNA polymerase. In addition, we chemically synthesized a modified **Pa** substrate, biotin-linked **Pa**TP (Bio-**Pa**TP) (Figure 2.9d), and incorporated it into a desired position of RNA molecules by T7 transcription. We compared the selectivity of the **Ds–Pa** pairing in transcription with that of the natural A–T(U) pairing and showed that the misincorporation percentage of Bio-**Pa**TP opposite the natural bases was lower than that of biotin-linked UTP opposite the natural bases [112].

Interestingly, **Pa** also functions as a template base for the site-specific incorporation of **s**, in addition to **Ds**, into RNA by T7 transcription (Figure 2.9e) [113]. The **s** base is strongly fluorescent (excitation 352 nm; emission 434 nm) [114] and the fluorescent intensity of **s** in RNA molecules changes according to the structural environment. The quenching of the **s** fluorescence is sensitive to its stacking with neighboring bases, hence the site-specific **s** incorporation would be a powerful tool for studying the dynamics of the local structural features of RNA molecules and their intra- and intermolecular interactions.

As an alternative to the **Ds–Pa** pair, we recently developed a new pair between **Ds** and 2-nitropyrrole (**Pn**) (Figure 2.9f) [115]. Since the nitro group of **Pn** electrostatically repels the 1-nitrogen of A, the mispairing of **Pn** with A was effectively reduced, in comparison with that of **Pa** with A. As a result, the PCR amplification involving the **Ds–Pn** pair does not require the γ-amidotriphosphate of A. Despite this, the specificity of the **Ds–Pn** pairing is higher than that of the **Ds–Pa** pairing in replication and the replacement of the **Ds–Pn** pair with the natural base pairs is ~1% after 20 cycles of PCR amplification.

Now that researchers have developed several unnatural base pairs that function in replication, transcription, and/or translation, a wide range of applications can be initiated. In the following section, we introduce some applications of unnatural base pair systems.

2.4 Applications of unnatural base pair systems

Extra unnatural base pairs could have potential in various applications for current and future biotechnologies. For DNA/RNA-based technology, there are two types of applications. First, the unnatural base pairs can expand the genetic alphabet, which augments selective molecular recognition by the base pair complementarity, thus increasing the variety of base sequences and reducing close mismatch hybridization. Second, the unnatural base pairs can mediate the site-specific incorporation of extra functional components into nucleic acids by replication and transcription, to create novel DNA and RNA molecules with increased functionality. By attaching functional groups, such as fluorophore, biotin and crosslinking reagents, to the unnatural base moiety, an ample repertoire of extra functional base substrates will be available for polymerase reactions using DNA templates containing unnatural bases.

2.4.1 Diagnosis of target DNA molecules

EraGen Biosciences and its collaborators have been fostering the commercial potential of the **isoG–isoC** pair (Figure 2.4a) in two types of molecular diagnostic testing methodologies for the detection and quantification of target gene expression [46,116–120]. One is for solution-based, quantitative real-time PCR analysis and is named MultiCode-RTx™ Systems (EraGen Biosciences, Inc.) or Plexor™ Systems (Promega Corp.) (Figure 2.10a) [117,118]. The other is for solid-phase, multiplexed end-point analysis integrated

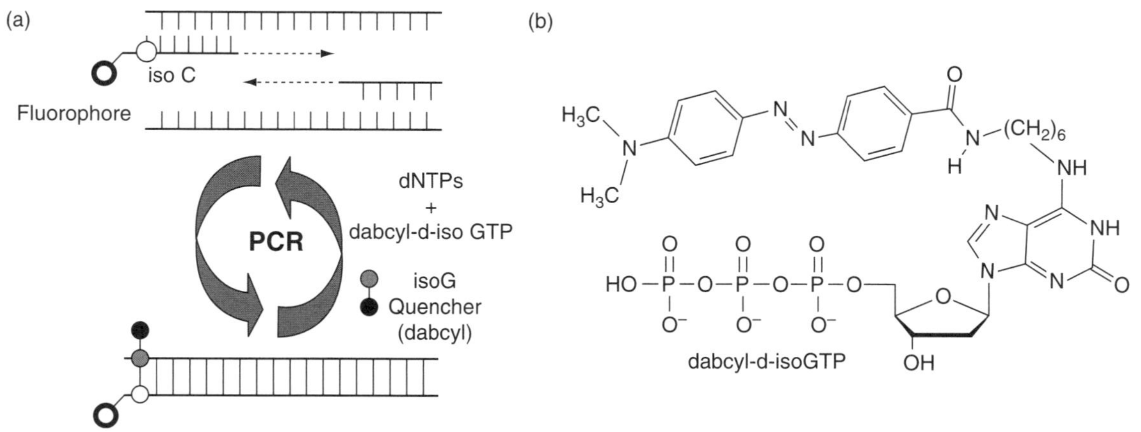

Figure 2.10 Schematic of Multicode-RTx systems. (a) Principle of the site-specific incorporation of a quencher during PCR amplification. (b) The structure of dabcyl-d-**isoG**TP

with microsphere flow cytometry and is named MultiCode-PLx™ Systems (EraGen Biosciences, Inc.) (Figure 2.11) [119,120]. Both utilize the site-specific incorporation of chemically-modified **isoG** substrates opposite **isoC** in DNA templates, for signal generation and amplification by PCR. In the MultiCode-PLx format, the selective unnatural base pair formation in the DNA duplex can be utilized to capture a target molecule among adventitious DNA molecules.

For solution-based, quantitative real-time PCR analysis, Prudent and co-workers demonstrated the simple monitoring of the amplified sequences (amplicons), by the incorporation of dabcyl-labeled d-**isoG**TP (dabcyl-d-**isoG**TP) opposite **isoC** in a fluorophore-labeled primer (Figure 2.10). The primer-derived fluorescence is quenched upon the incorporation of the dabcyl-d-**isoG**TP by PCR amplification. The quenching was proportional to the amount of the resulting amplicon, which permitted the monitoring of the PCR amplification progress in real time. Unlike the conventional real-time PCR methods, such as TaqMan, this methodology does not require an extra probe or a special primer design. Furthermore, multiplex analysis is also available by using different fluorophore-labeled primers [117].

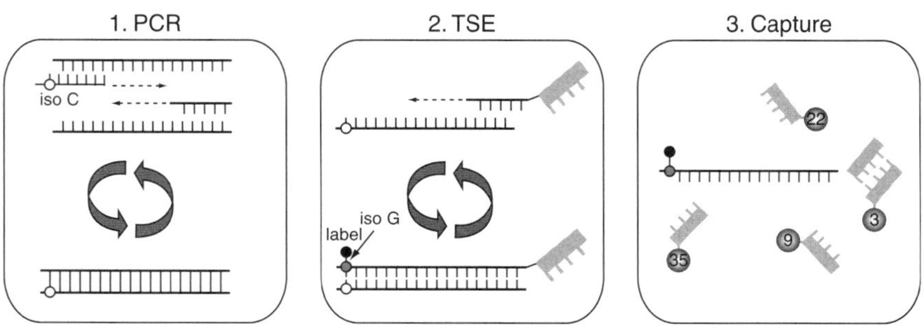

Figure 2.11 Schematic of Multicode-PLx systems. In step 1, target regions are amplified by PCR, producing the amplicon with **isoC** at the 5′ end. In step 2, target-specific labeling extension (TSE) is carried out with a tagged target-specific extender and labeled d-**isoG**TP. In step 3, Eracode-modified Luminex microspheres are added and each microsphere captures the corresponding target with labeled **isoG**

MultiCode-PLx [118], a simplified solid-phase end-point genetic analysis, comprises three steps: PCR, extension labeling and liquid decoding, which all involve the **isoG–isoC** pair (Figure 2.11). In the first step, the target DNA sequence is amplified by PCR, using a primer containing one **isoC** base at the 5′-end. In the next step, the target-specific labeling extension (TSE) of the **isoC**-containing amplicon is performed in the presence of labeled d-**isoG**TP and a tagged target-specific extender. Each tag in the extenders is a short sequence (typically 7–10 nucleotides in length) consisting of a mixture of natural bases (A, T, G and C) and unnatural bases (**isoG** and **isoC**). This tag system is called EraCode and the sequences are designed to hybridize only to their perfect complementary sequences at room temperature, without a washing procedure to remove mis-hybridization products resulting from cross-hybridization between other codes and natural DNA [116]. In the final decoding step, the tagged extenders are captured on the EraCode-addressed Luminex microspheres and are analyzed with the Luminex detection system.

2.4.2 Tools for site-specific incorporation of extra components into RNA

For the site-specific incorporation of extra, functional components into RNA, unnatural base pair systems can also be applied to *in vitro* transcription, which is widely used for preparing RNA molecules with defined sequences. Other methods, such as chemical RNA synthesis [1], are also available for the incorporation of functional groups into RNA. However, the conventional methods are restricted, in some respects. For example, chemical synthesis is accessible only for the introduction of chemically stable analogs and it is still difficult to synthesize long-chain RNA molecules with more than 100 nucleotides. Another random incorporation method, using modified natural base substrates together with unmodified substrates, is generally used for RNA labeling [121], but it is very difficult to adjust the modification site. In that context, the site-specific incorporation of functional components by transcription mediated by unnatural base pairs is a very attractive method [122].

The **s–y** pair allows the site-specific incorporation of modified **y** substrates into RNA molecules opposite **s** in DNA templates, by the standard T7 transcription method. A series of modified **y** substrates, where the functional groups were covalently attached to position 5 of the **y** base, are illustrated in Figure 2.12. These modified **y** substrates can be prepared from the iodo-**y** ribonucleoside, which is obtained by iodination of the **y** nucleoside using *N*-iodosuccinimide [123]. The iodo-**y** incorporation into RNA is useful as a photosensitive component, which is capable of crosslinking with proximal reactive residues in a target molecule that binds to the modified RNA [124]. This could be applicable for the tight binding of aptamers to target molecules by photo-crosslinking, towards diagnostic and therapeutic applications as RNA photoaptamers [125], and also for studies of RNA–RNA and RNA–protein interactions [124]. In addition, RNA molecules containing iodo-**y** at a specific position could be used as rational phasing tools for X-ray crystallography. One of the modified **y** bases, phenylethynyl-**y** (Ph-**y**), augmented the thermal stability of RNA molecules by the incorporation of Ph-**y** into RNA, resulting in increased stacking ability between Ph-**y** and its neighboring bases [126]. The incorporation of an aminohexanamide-1-propynyl-linked **y** (NH_2-hx-**y**) substrate into RNA molecules permits site-specific post-transcriptional modification [123]. The site-specific incorporation of biotin-linked **y** (Bio-**y**) into RNA molecules can facilitate their immobilization on solid supports, such as a sensor chip [127]. Site-specific fluorescent labeling of RNA molecules was also accomplished by transcription using fluorophore-linked **y** substrates, such as a FAM-**y** substrate [123].

2.4.3 Development of functional RNA molecules

The incorporation sites of functional components into RNA molecules are rationally designed by means of their structural information, to endow them with novel functionality. Thus, the unnatural base pair systems have also been applied to RNA aptamers that bind to their specific target molecules. Since most RNA aptamers change their structural conformation upon binding to their target molecules, the incorporation of the

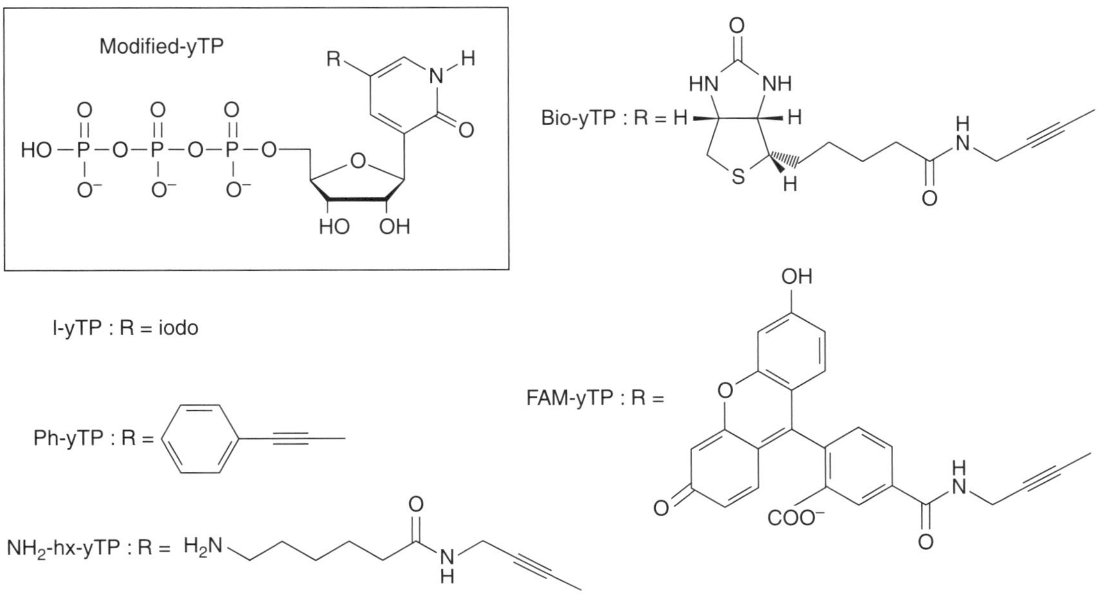

Figure 2.12 *Functional substrates of modified **y** bases, which are site-specifically incorporated into RNA by T7 transcription*

fluorophores at a specific position within aptamers provides a detection system, by the alteration of the fluorescent intensity depending on their conformational changes [128].

The fluorescent unnatural base FAM-**y** was incorporated into a specific position of an anti-theophylline aptamer [129] and the conformational change due to theophylline binding was monitored (Figure 2.13) [123]. Upon binding to theophylline, the aptamer forms a stable base triplet, U6–U23–A28, adjacent to the binding site [130,131] and U6 was replaced with FAM-**y**. Note that the 4-keto group of U6 is not involved in the U6–U23–A28 formation and, hence the (FAM-**y**)–U23–A28 triplet can also be formed. In the absence of theophylline, the FAM residue is stacked with the bases inside the aptamer structure and thus its fluorescent intensity is decreased. Upon theophylline binding, the FAM residue protrudes outside the aptamer structure with the (FAM-**y**)–U23–A28 triplet, increasing its fluorescence intensity (Figure 2.13) [123]. By using this system, 50 pmol of theophylline in a 0.1 ml of solution can be detected specifically.

2.4.4 Local structural analysis of RNA molecules

The unnatural **s** base is strongly fluorescent (excitation 352 nm; emission 434 nm; quantum yield 41%) [114]. The fluorescence intensity of **s** in RNA molecules is stronger than that of a well-known fluorescent base analog, 2-aminopurine, which is significantly quenched by stacking with neighboring bases in RNA molecules. However, the fluorescence of the **s** base is also quenched to some extent, depending on its stacking environment [113]. In contrast, a high fluorescence intensity of the **s** base in RNA molecules is observed when the **s** base is exposed outside the RNA structure. Furthermore, the fluorescent **s** base can be incorporated into desired positions on RNA molecules by transcription mediated by the **s**–**Pa** pair (Figure 2.9e) [113]. Thus, the site-specific incorporation of the **s** base into RNA is a powerful tool for the local structural analysis of long RNA molecules. We demonstrated the site-specific **s** labeling of a tRNA molecule, which provided characteristic fluorescent profiles, depending on the incorporation site, temperature and Mg^{2+} concentration [113].

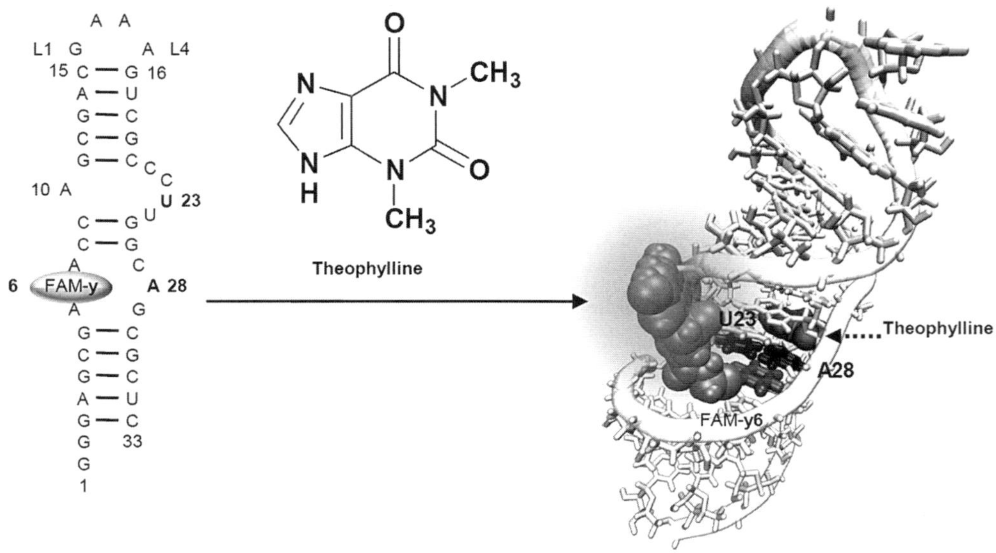

Figure 2.13 *Site-specific fluorescent labeling of an anti-theophylline aptamer. The fluorescent unnatural base, FAM-**y**, was site-specifically incorporated at position 6, in place of U6, in the aptamer. In the figure, the FAM-**y** base was superimposed at the U6 position in the aptamer–theophylline complex (PDB 1O15). The FAM-**y** at position 6 is shown in gray and the U23 and A28 are shown in black*

2.5 Future development of unnatural base pair systems

In the two decades since Benner's group reported the first unnatural base pair, researchers have achieved the development of some unnatural base pairs, such as **isoG–isoC**, **Z*–P***, **5SICS–MMO2**, **Ds–Pa** and **Ds–Pn**, that can be applied to PCR amplification. Some of them also function in transcription and translation. With the development of these unnatural base pairs, the application of these unnatural base pairs to biotechnology has just begun. At present, there has been no report about an unnatural base pair that functions in replication, transcription and translation. However, we now have versatile information to design new base pairs and therefore the advent of multipotent unnatural base pair systems might only be a matter of time. A variety of defined functional, unnatural base pairs, such as the **s–Pa** pair for fluorescent labeling of RNA, will also be developed towards specific biotechnology. In terms of the biological ramifications, our knowledge about the mechanisms of the genetic information flow has been tremendously supplemented through the efforts of unnatural base pair studies.

References

1. S. Verma and F. Eckstein, Modified oligonucleotides: synthesis and strategy for users. *Annu. Rev. Biochem.* **67**, 99–134 (1998).
2. S.A. Benner, P. Burgstaller, T.R. Battersby and S. Jurczyk, Did the RNA world exploit an expanded genetic alphabet? In *The RNA World*, 2nd edn, eds R.F. Gesteland, T. Cech and J.F. Atkins, Cold Spring Harbor Laboratory Press, Cold Spring Harbor, NY, 1999, pp. 163–181.
3. E.T. Kool, Synthetically modified DNAs as substrates for polymerases, *Curr. Opin. Chem. Biol.* **4**, 602–608 (2000).

4. A.A. Henry and F.E. Romesberg, Beyond A, C, G and T: augmenting Nature's alphabet, *Curr. Opin. Chem. Biol.***7**, 727–733 (2003).
5. J. Hunziker and G. Mathis, DNA with artificial base pairs, *Chimia* **59**, 780–784 (2005).
6. S.A. Benner and A.M. Sismour, Synthetic biology, *Nat. Rev. Genet.***6**, 533–543 (2005).
7. I. Hirao, Unnatural base pair systems for DNA/RNA-based biotechnology, *Curr. Opin. Chem. Biol.* **10**, 622–627 (2006).
8. N. Budisa, Prolegomena to future experimental efforts on genetic code engineering by expanding its amino acid repertoire, *Angew. Chem. Int. Ed.* **43**, 6426–6463 (2004).
9. T.L. Hendrickson, V. de Crécy-Lagard and P. Schimmel, Incorporation of nonnatural amino acids into proteins, *Annu. Rev. Biochem.* **73**, 147–176 (2004).
10. L. Wang, J. Xie and P.G. Schultz, Expanding the genetic code, *Annu. Rev. Biophys. Biomol. Struct.* **35**, 225–249 (2006).
11. E.A. Lemke, D. Summerer, B.H. Geierstanger, S.M. Brittain and P.G. Schultz, Control of protein phosphorylation with a genetically encoded photocaged amino acid, *Nat. Chem. Biol.* **3**, 769–772 (2007).
12. K. Wang, H. Neumann, S.Y. Peak-Chew and J.W. Chin, Evolved orthogonal ribosomes enhance the efficiency of synthetic genetic code expansion, *Nat. Biotechnol.* **25**, 770–777 (2007).
13. S.J. Wrenn and P.B. Harbury, Chemical evolution as a tool for molecular discovery, *Annu. Rev. Biochem.* **76**, 331–349 (2007).
14. M. Famulok, J.S. Hartig and G. Mayer, Functional aptamers and aptazymes in biotechnology, diagnostics and therapy, *Chem. Rev.* **107**, 3715–3743 (2007).
15. C. Switzer, S.E. Moroney and S.A. Benner, Enzymatic incorporation of a new base pair into DNA and RNA, *J. Am. Chem. Soc.*, **111**, 8322–8323 (1989).
16. D.L. Ollis, P. Brick, R. Hamlin, N.G. Xuong and T.A. Steitz, Structure of large fragment of *Escherichia coli* DNA polymerase I complexed with dTMP, *Nature* **313**, 762–766 (1985).
17. S. Doublié, S. Tabor, A.M. Long, C.C. Richardson and T. Ellenberger, Crystal structure of a bacteriophage T7 DNA replication complex at 2.2 Å resolution, *Nature* **391**, 251–258 (1998).
18. J.R. Kiefer, C. Mao, J.C. Braman and L.S. Beese, Visualizing DNA replication in a catalytically active *Bacillus* DNA polymerase crystal, *Nature* **391**, 304–307 (1998).
19. D. Loakes, Survey and summary: the applications of universal DNA base analogues, *Nucleic Acids Res.* **29**, 2437–2447 (2001).
20. X. Zhang, I. Lee and A.J. Berdis, The use of nonnatural nucleotides to probe the contributions of shape complementarity and π-electron surface area during DNA polymerization, *Biochemistry*, **44**, 13101–13110 (2005).
21. S. Hikishima, N. Minakawa, K. Kuramoto, Y. Fujisawa, M. Ogawa and A. Matsuda, Synthesis of 1,8-naphthyridine C-nucleosides and their base-pairing properties in oligodeoxynucleotides: thermally stable naphthyridine:imidazopyridopyrimidine base-pairing motifs, *Angew. Chem. Int. Ed.* **44**, 596–598 (2005).
22. M.M. Ali, M. Oishi, F. Nagatsugi, K. Mori, Y. Nagasaki, K. Kataoka and S. Sasaki, Intracellular inducible alkylation system that exhibits antisense effects with greater potency and selectivity than the natural oligonucleotide, *Angew. Chem. Int. Ed.* **45**, 3136–3140 (2006).
23. G.H. Clever, C. Kaul and T. Carell, DNA–metal base pairs, *Angew. Chem. Int. Ed.* **46**, 6226–6236 (2007).
24. J. Beckman, K. Kincaid, M. Hocek, T. Spratt, J. Engels, R. Cosstick and R.D. Kuchta, Human DNA polymerase alpha uses a combination of positive and negative selectivity to polymerize purine dNTPs with high fidelity, *Biochemistry*, **46**, 448–460 (2007).
25. A. Zahn and C.J. Leumann, Recognition properties of donor- and acceptor-modified biphenyl-DNA. *Chemistry* **14**, 1087–1094 (2008).
26. A. Ohkubo, R. Kasuya, K. Sakamoto, K. Miyata, H. Taguchi, H. Nagasawa, T. Tsukahara, T. Watanobe, Y. Maki, K. Seio and M. Sekine, 'Protected DNA probes' capable of strong hybridization without removal of base protecting groups. *Nucleic Acids Res.* **36**, 1952–1964 (2008).
27. U. Niedballa and H. Vorbrüggen, A general synthesis of *N*-glycosides. IV. Synthesis of nucleosides of hydroxy and mercapto *N*-heterocycles. *J. Org. Chem.* **39**, 3668–3671 (1974).
28. A. Marx, I. Detmer, J. Gaster and D. Summerer, Probing DNA polymerase function with synthetic nucleotides, *Synthesis* **1**, 1–14 (2004).

29. Q. Wu and C. Simons, Synthetic methodologies for C-nucleosides, *Synthesis* **10**, 1533–1553 (2004).
30. K. Burgess and D. Cook, Syntheses of nucleoside triphosphates, *Chem. Rev.* **100**, 2047–2059 (2000).
31. M. Yoshikawa, T. Kato and T. Takenishi, A novel method for phosphorylation of nucleosides to 5′-nucleotides, *Tetrahedron Lett.* **50**, 5065–5068 (1967).
32. M. Yoshikawa, T. Kato and T. Takenishi, Studies of phosphorylation. III. Selective phosphorylation of unprotected nucleosides, *Bull. Chem. Soc. Jpn.* **42**, 3505–3508 (1969).
33. T. Kovács and L. Ötvös, Simple synthesis of 5-vinyl- and 5-ethynyl–2′-deoxyuridine-5′-triphosphates, *Tetrahedron Lett.* **29**, 4525–4528 (1988).
34. J. Ludwig and F. Eckstein, Rapid and efficient synthesis of nucleoside 5′-*O*-(1-thiotriphosphates), 5′-triphosphates and 2′,3′-cyclophosphorothioates using 2-chloro-4*H*-1,3,2-benzodioxaphosphorin-4-one, *J. Org. Chem.* **54**, 631–635 (1989).
35. J. Petruska, M.F. Goodman, M.S. Boosalis, L.C. Sowers, C. Cheong and I. Tinoco, Comparison between DNA melting thermodynamics and DNA polymerase fidelity, *Proc. Natl. Acad. Sci. USA* **85**, 6252–6256 (1988).
36. M.F. Goodman, S. Creighton, L.B. Bloom and J. Petruska, Biochemical basis of DNA replication fidelity, *Crit. Rev. Biochem. Mol. Biol.* **28**, 83–126 (1993).
37. S. Moran, R.X.-F. Ren, S. Rummey and E.T. Kool, Difluorotoluene, a nonpolar isostere for thymine, codes specifically and efficiently for adenine in DNA replication, *J. Am. Chem. Soc.* **119**, 2056–2057 (1997).
38. M. Kimoto, S. Yokoyama and I. Hirao, A quantitative, non-radioactive single-nucleotide insertion assay for analysis of DNA replication fidelity by using an automated DNA sequencer, *Biotechnol. Lett.* **26**, 999–1005 (2004).
39. J.A. Piccirilli, T. Krauch, S.E. Moroney and S.A. Benner, Enzymatic incorporation of a new base pair into DNA and RNA extends the genetic alphabet, *Nature*, **343**, 33–37 (1990).
40. J.A. Piccirilli, S.E. Moroney and S.A. Benner, A C-nucleotide base pair: methylpseudouridine-directed incorporation of formycin triphosphate into RNA catalyzed by T7 RNA polymerase, *Biochemistry* **30**, 10350–10356 (1991).
41. J.D. Bain, C. Switzer, A.R. Chamberlin and S.A. Benner, Ribosome-mediated incorporation of a non-standard amino acid into a peptide through expansion of the genetic code, *Nature* **356**, 537–539 (1992).
42. T.G. Heckler, L.H. Chang, Y. Zama, T. Naka, M.S. Chorghade and S.M. Hecht, T4 RNA ligase mediated preparation of novel 'chemically misacylated' tRNAPheS, *Biochemistry* **23**, 1468–1473 (1984).
43. S.A. Robertson, C.J. Noren, S.J. Anthony-Cahill, M.C. Griffith and P.G. Schultz, The use of 5′-phospho=2 deoxyribocytidylylriboadenosine as a facile route to chemical aminoacylation of tRNA, *Nucleic Acids Res.* **17**, 9649–9660 (1989).
44. Y. Tor and P.B. Dervan, Site-specific enzymic incorporation of an unnatural base, N^6-(6-aminohexyl)isoguanosine, into RNA, *J. Am. Chem. Soc.* **115**, 4461–4467 (1993).
45. C.Y. Switzer, S.E. Moroney and S.A. Benner, Enzymatic recognition of the base pair between isocytidine and isoguanosine, *Biochemistry* **32**, 10489–10496 (1993).
46. S.C. Johnson, C.B. Sherrill, D.J. Marshall, M.J. Moser and J.R. Prudent, A third base pair for the polymerase chain reaction: inserting isoC and isoG, *Nucleic Acids Res.* **32**, 1937–1941 (2004).
47. J. Horlacher, M. Hottiger, V.N. Podust, U. Hübscher and S.A. Benner, Recognition by viral and cellular DNA polymerases of nucleosides bearing bases with nonstandard hydrogen bonding patterns, *Proc. Natl. Acad. Sci. USA* **92**, 6329–6333 (1995).
48. T. Horn, C.-A. Chang and M.L. Collins, Hybridization properties of the 5-methyl-isocytidine/isoguanosine base pair in synthetic oligodeoxynucleotides, *Tetrahedron Lett.* **36**, 2033–2036 (1995).
49. A.M. Sismour and S.A. Benner, The use of thymidine analogs to improve the replication of an extra DNA base pair: a synthetic biological system, *Nucleic Acids Res.* **33**, 5640–5646 (2005).
50. A.M. Sismour, S. Lutz, J.-H. Park, M.J. Lutz, P.L. Boyer, S.H. Hughes and S.A. Benner, PCR amplification of DNA containing non-standard base pairs by variants of reverse transcriptase from human immunodeficiency virus-1, *Nucleic Acids Res.* **32**, 728–735 (2004).
51. Z. Yang, A.M. Sismour, P. Sheng, N.L. Puskar and S.A. Benner, Enzymatic incorporation of a third nucleobase pair, *Nucleic Acids Res.* **35**, 4238–4249 (2007).
52. H.P. Rappaport, The 6-thioguanine/5-methyl-2-pyrimidinone base pair, *Nucleic Acids Res.* **16**, 7253–7267 (1988).
53. H.P. Rappaport, Replication of the base pair 6-thioguanine/5-methyl-2-pyrimidinone with the large Klenow fragment of *Escherichia coli* DNA polymerase I, *Biochemistry* **32**, 3047–3057 (1993).

54. H.P. Rappaport, The fidelity of replication of the three-base-pair set adenine/thymine, hypoxanthine/cytosine and 6-thiopurine/5-methyl-2-pyrimidinone with T7 DNA polymerase, *Biochem. J.* **381**, 709–717 (2004).
55. E.T. Kool, Replication of non-hydrogen bonded bases by DNA polymerases: a mechanism for steric matching, *Biopolymers* **48**, 3–17 (1998).
56. D. Barsky, E.T. Kool and M.E. Colvin, Interaction and solvation energies of nonpolar DNA base analogs and their role in polymerase insertion fidelity, *J. Biomol. Struct. Dyn.* **16**, 1119–1134 (1999).
57. E.T. Kool, J.C. Morales and K.M. Guckian, Mimicking the structure and function of DNA: insights into DNA stability and replication, *Angew. Chem. Int. Ed.* **39**, 990–1009 (2000).
58. E.T. Kool, Hydrogen bonding, base stacking and steric effects in DNA replication, *Annu. Rev. Biophys. Biomol. Struct.* **30**, 1–22 (2001).
59. B.A. Schweitzer and E.T. Kool, Aromatic nonpolar nucleosides as hydrophobic isosteres of pyrimidine and purine nucleosides, *J. Org. Chem.* **59**, 7238–7242 (1994).
60. B.A. Schweitzer and E.T. Kool, Hydrophobic, non-hydrogen-bonding bases and base pairs in DNA, *J. Am. Chem. Soc.* **117**, 1863–1872 (1995).
61. E.T. Kool and H.O. Sintim, The difluorotoluene debate – a decade later, *Chem. Commun.* 3665–3675 (2006).
62. S. Moran, R.X.-F. Ren, S. Rumney IV and E.T. Kool, Difluorotoluene, a nonpolar isostere for thymine, codes specifically and efficiently for adenine in DNA replication, *J. Am. Chem. Soc.* **119**, 2056–2057 (1997).
63. S. Moran, R.X.-F. Ren and E.T. Kool, A thymidine triphosphate shape analog lacking Watson–Crick pairing ability is replicated with high sequence selectivity, *Proc. Natl. Acad. Sci. USA* **94**, 10506–10511 (1997).
64. J.C. Morales and E.T. Kool, Functional hydrogen-bonding map of the minor groove binding tracks of six DNA polymerases, *Biochemistry* **39**, 12979–12988 (2000).
65. K.M. Guckian, T.R. Krugh and E.T. Kool, Solution structure of a DNA duplex containing a replicable difluorotoluene-adenine pair, *Nat. Struct. Biol.* **5**, 954–959 (1998).
66. J.C. Morales and E.T. Kool, Efficient replication between non-hydrogen-bonded nucleoside shape analogs, *Nat. Struct. Biol.* **5**, 950–954 (1998).
67. J.C. Morales and E.T. Kool, Minor groove interactions between polymerase and DNA: more essential to replication than Watson–Crick hydrogen bonds? *J. Am. Chem. Soc.* **121**, 2323–2324 (1999).
68. K.M. Guckian, T.R. Krugh and E.T. Kool, Solution structure of a nonpolar, non-hydrogen-bonded base pair surrogate in DNA, *J. Am. Chem. Soc.* **122**, 6841–6847 (2000).
69. T.A. Evans and K.R. Seddon, Hydrogen bonding in DNA-a return to the *status quo*, *Chem. Commun.* 2023–2024 (1997).
70. T.J. Matray and E.T. Kool, A specific partner for abasic damage in DNA, *Nature* **399**, 704–708 (1999).
71. J.C. Delaney, P.T. Henderson, S.A. Helquist, J.C. Morales, J.M. Essigmann and E.T. Kool, High-fidelity *in vivo* replication of DNA base shape mimics without Watson–Crick hydrogen bonds, *Proc. Natl. Acad. Sci. USA* **100**, 4469–4473 (2003).
72. T.W. Kim, J.C. Delaney, J.M. Essigmann and E.T. Kool, Probing the active site tightness of DNA polymerase in subangstrom increments, *Proc. Natl. Acad. Sci. USA* **102**, 15803–15808 (2005).
73. A.P. Silverman, Q. Jiang, M.F. Goodman and E.T. Kool, Steric and electrostatic effects in DNA synthesis by the SOS-induced DNA polymerases II and IV of *Escherichia coli*, *Biochemistry* **46**, 13874–13881 (2007).
74. I. Hirao, M. Kimoto, S. Yamakage, M. Ishikawa, J. Kikuchi and S. Yokoyama, A unique unnatural base pair between a C analogue, pseudoisocytosine and an A analogue, 6-methoxypurine, in replication, *Bioorg. Med. Chem. Lett.* **12**, 1391–1393 (2002).
75. X. Zhang, I. Lee and A.J. Berdis, A potential chemotherapeutic strategy for the selective inhibition of promutagenic DNA synthesis by nonnatural nucleotides, *Biochemistry* **44**, 13111–13121 (2005).
76. K. Kincaid, J. Beckman, A. Zivkovic, R.L. Halcomb, J.W. Engels and R.D. Kuchta, Exploration of factors driving incorporation of unnatural dNTPS into DNA by Klenow fragment (DNA polymerase I) and DNA polymerase α, *Nucleic Acids Res.* **33**, 2620–2628 (2005).
77. H.R. Lee, S.A. Helquist, E.T. Kool and K.A. Johnson, Importance of hydrogen bonding for efficiency and specificity of the human mitochondrial DNA polymerase, *J. Biol. Chem.* **283**, 14402–14410 (2008).
78. H.R. Lee, S.A. Helquist, E.T. Kool and K.A. Johnson, Base pair hydrogen bonds are essential for proofreading selectivity by the human mitochondrial DNA polymerase, *J. Biol. Chem.* **283**, 14411–14416 (2008).

79. H. Liu, J. Gao, S.R. Lynch, Y.D. Saito, L. Maynard and E.T. Kool, A four-base paired genetic helix with expanded size, *Science* **302**, 868–871 (2003).
80. H. Liu, J. Gao, L. Maynard, Y.D. Saito and E.T. Kool, Toward a new genetic system with expanded dimensions: size-expanded analogues of deoxyadenosine and thymidine, *J. Am. Chem. Soc.* **126**, 1102–1109 (2004).
81. H. Liu, S.R. Lynch and E.T. Kool, Solution structure of xDNA: a paired genetic helix with increased diameter, *J. Am. Chem. Soc.* **126**, 6900–6905 (2004).
82. J. Gao, H. Liu and E.T. Kool, Expanded-size bases in naturally sized DNA: evaluation of steric effects in Watson–Crick pairing, *J. Am. Chem. Soc.* **126**, 11826–11831 (2004).
83. J. Gao, H. Liu and E.T. Kool, Assembly of the complete eight-base artificial genetic helix, xDNA and its interaction with the natural genetic system, *Angew. Chem. Int. Ed.* **44**, 3118–3122 (2005).
84. A.H.F. Lee and E.T. Kool, A new four-base genetic helix, yDNA, composed of widened benzopyrimidine-purine pairs, *J. Am. Chem. Soc.* **127**, 3332–3338 (2005).
85. D.L. McMinn, A.K. Ogawa, Y. Wu, J. Liu, P.G. Schultz and F.E. Romesberg, Efforts toward expansion of the genetic alphabet: DNA polymerase recognition of a highly stable, self-pairing hydrophobic base, *J. Am. Chem. Soc.* **121**, 11585–11586 (1999).
86. A.K. Ogawa, Y. Wu, D.L. McMinn, J. Liu, P.G. Schultz and F.E. Romesberg, Efforts toward the expansion of the genetic alphabet: information storage and replication with unnatural hydrophobic base pairs, *J. Am. Chem. Soc.* **122**, 3274–3287 (2000).
87. Y. Wu, A.K. Ogawa, M. Berger, D.L. McMinn, P.G. Schultz and F.E. Romesberg, Efforts toward expansion of the genetic alphabet: optimization of interbase hydrophobic interactions, *J. Am. Chem. Soc.* **122**, 7621–7632 (2000).
88. A.K. Ogawa, Y. Wu, M. Berger, P.G. Schultz and F.E. Romesberg, Rational design of an unnatural base pair with increased kinetic selectivity, *J. Am. Chem. Soc.* **122**, 8803–8804 (2000).
89. E.L. Tae, Y. Wu, G. Xia, P.G. Schultz and F.E. Romesberg, Efforts toward expansion of the genetic alphabet: replication of DNA with three base pairs, *J. Am. Chem. Soc.* **123**, 7439–7440 (2001).
90. G. Xia, L. Chen, T. Sera, M. Fa, P.G. Schultz and F.E. Romesberg, Directed evolution of novel polymerase activities: mutation of a DNA polymerase into an efficient RNA polymerase, *Proc. Natl. Acad. Sci. USA* **99**, 6597–6602 (2002).
91. M. Fa, A. Radeghieri, A.A. Henry and F.E. Romesberg, Expanding the substrate repertoire of a DNA polymerase by directed evolution, *J. Am. Chem. Soc.* **126**, 1748–1754 (2004).
92. A.M. Leconte, L. Chen and F.E. Romesberg, Polymerase evolution: efforts toward expansion of the genetic code, *J. Am. Chem. Soc.* **127**, 12470–12471 (2005).
93. A.A. Henry and F.E. Romesberg, The evolution of DNA polymerases with novel activities, *Curr. Opin. Biotechnol.* **16**, 370–377 (2005).
94. M. Berger, A.K. Ogawa, D.L. McMinn, Y. Wu, P.G. Schultz and F.E. Romesberg, Stable and selective hybridization of oligonucleotides with unnatural hydrophobic bases, *Angew. Chem. Int. Ed.* **39**, 2940–2942 (2000).
95. M. Berger, S.D. Luzzi, A.A. Henry and F.E. Romesberg, Stability and selectivity of unnatural DNA with five-membered-ring nucleobase analogues, *J. Am. Chem. Soc.* **124**, 1222–1226 (2002).
96. C. Yu, A.A. Henry, F.E. Romesberg and P.G. Schultz, Polymerase recognition of unnatural base pairs, *Angew. Chem. Int. Ed.* **41**, 3841–3844 (2002).
97. S. Matsuda, A.A. Henry, P.G. Schultz and F.E. Romesberg, The effect of minor-groove hydrogen-bond acceptors and donors on the stability and replication of four unnatural base pairs, *J. Am. Chem. Soc.* **125**, 6134–6139 (2003).
98. A.A. Henry, A.G. Olsen, S. Matsuda, C. Yu, B.H. Geierstanger and F.E. Romesberg, Efforts to expand the genetic alphabet: identification of a replicable unnatural DNA self-pair, *J. Am. Chem. Soc.* **126**, 6923–6931 (2004).
99. S. Matsuda and F.E. Romesberg, Optimization of interstrand hydrophobic packing interactions within unnatural DNA base pairs, *J. Am. Chem. Soc.* **126**, 14419–14427 (2004).
100. G.T. Hwang and F.E. Romesberg, Substituent effects on the pairing and polymerase recognition of simple unnatural base pairs, *Nucleic Acids Res.* **34**, 2037–2045 (2006).
101. S. Matsuda, A.A. Henry and F.E. Romesberg, Optimization of unnatural base pair packing for polymerase recognition, *J. Am. Chem. Soc.* **128**, 6369–6375 (2006).
102. A.M. Leconte, S. Matsuda, G.T. Hwang and F.E. Romesberg, Efforts towards expansion of the genetic alphabet: pyridone and methyl pyridone nucleobases, *Angew. Chem. Int. Ed.* **45**, 4326–4329 (2006).

103. S. Matsuda, A.M. Leconte and F.E. Romesberg, Minor groove hydrogen bonds and the replication of unnatural base pairs, *J. Am. Chem. Soc.* **129**, 5551–5557 (2007).
104. S. Matsuda, J.D. Fillo, A.A. Henry, P. Rai, S.J. Wilkens, T.J. Dwyer, B.H. Geierstanger, D.E. Wemmer, P.G. Schultz, G. Spraggon and F.E. Romesberg, Efforts toward expansion of the genetic alphabet: structure and replication of unnatural base pairs, *J. Am. Chem. Soc.* **129**, 10466–10473 (2007).
105. A.M. Leconte, G.T. Hwang, S. Matsuda, P. Capek, Y. Hari and F.E. Romesberg, Discovery, characterization and optimization of an unnatural base pair for expansion of the genetic alphabet, *J. Am. Chem. Soc.* **130**, 2336–2343 (2008).
106. T. Ohtsuki, M. Kimoto, M. Ishikawa, T. Mitsui, I. Hirao and S. Yokoyama, Unnatural base pairs for specific transcription, *Proc. Natl. Acad. Sci. USA* **98**, 4922–4925 (2001).
107. T. Fujiwara, M. Kimoto, H. Sugiyama, I. Hirao and S. Yokoyama, Synthesis of 6-(2-thienyl)purine nucleoside derivatives that form unnatural base pairs with pyridin-2-one nucleosides, *Bioorg. Med. Chem. Lett.* **11**, 2221–2223 (2001).
108. I. Hirao, T. Ohtsuki, T. Fujiwara, T. Mitsui, T. Yokogawa, T. Okuni, H. Nakayama, K. Takio, T. Yabuki, T. Kigawa, K. Kodama, T. Yokogawa, K. Nishikawa and S. Yokoyama, An unnatural base pair for incorporating amino acid analogs into proteins, *Nat. Biotechnol.* **20**, 177–182 (2002).
109. S. Ohno, T. Yokogawa, I. Fujii, H. Asahara, H. Inokuchi and K. Nishikawa, Co-expression of yeast amber suppressor $tRNA^{Tyr}$ and tyrosyl-tRNA synthetase in *Escherichia coli*: possibility to expand the genetic code, *J. Biochem.* **124**, 1065–1068 (1998).
110. I. Hirao, Y. Harada, M. Kimoto, T. Mitsui, T. Fujiwara and S. Yokoyama, A two-unnatural-base-pair system toward the expansion of the genetic code, *J. Am. Chem. Soc.* **126**, 13298–13305 (2004).
111. T. Mitsui, A. Kitamura, M. Kimoto, T. To, A. Sato, I. Hirao and S. Yokoyama, An unnatural hydrophobic base pair with shape complementarity between pyrrole-2-carbaldehyde and 9-methylimidazo[(4,5)-*b*]pyridine, *J. Am. Chem. Soc.* **125**, 5298–5307 (2003).
112. I. Hirao, M. Kimoto, T. Mitsui, T. Fujiwara, R. Kawai, A. Sato, Y. Harada and S. Yokoyama, An unnatural hydrophobic base pair system: site-specific incorporation of nucleotide analogs into DNA and RNA, *Nat. Methods* **3**, 729–735 (2006).
113. M. Kimoto, T. Mitsui, Y. Harada, A. Sato, S. Yokoyama and I. Hirao, Fluorescent probing for RNA molecules by an unnatural base-pair system, *Nucleic Acids Res.* **35**, 5360–5369 (2007).
114. T. Mitsui, M. Kimoto, R. Kawai, S. Yokoyama and I. Hirao, Characterization of fluorescent, unnatural base pairs, *Tetrahedron* **63**, 3528–3537 (2007).
115. I. Hirao, T. Mitsui, M. Kimoto and S. Yokoyama, An efficient unnatural base pair for PCR amplification, *J. Am. Chem. Soc.* **129**, 15549–15555 (2007).
116. J.R. Prudent, Using expanded genetic alphabets to simplify high-throughput genetic testing, *Expert Rev. Mol. Diagn.* **6**, 245–252 (2006).
117. C.B. Sherrill, D.J. Marshall, M.J. Moser, C.A. Larsen, L. Daudé-Snow, S. Jurczyk, G. Shapiro and J.R. Prudent, Nucleic acid analysis using an expanded genetic alphabet to quench fluorescence, *J. Am. Chem. Soc.* **126**, 4550–4556 (2004).
118. M.J. Moser, D.R. Christensen, D. Norwood and J.R. Prudent, Multiplexed detection of anthrax-related toxin genes, *J. Mol. Diagn.* **8**, 89–96 (2006).
119. S.C. Johnson, D.J. Marshall, G. Harms, C.M. Miller, C.B. Sherrill, E.L. Beaty, S.A. Lederer, E.B. Roesch, G. Madsen, G.L. Hoffman, R.H. Laessig, G.J. Kopish, M.W. Baker, S.A. Benner, P.M. Farrell and J.R. Prudent, Multiplexed genetic analysis using an expanded genetic alphabet, *Clin. Chem.* **50**, 2019–2027 (2004).
120. W.M. Lee, K. Grindle, T. Pappas, D.J. Marshall, M.J. Moser, E.L. Beaty, P.A. Shult, J.R. Prudent and J.E. Gern, High-throughput, sensitive and accurate multiplex PCR-microsphere flow cytometry system for large-scale comprehensive detection of respiratory viruses. *J. Clin. Microbiol.* **45**, 2626–2634 (2007).
121. P.R. Langer, A.A. Waldrop and D.C. Word, Enzymatic synthesis of biotin-labeled polynucleotides: novel nucleic acid affinity probes, *Proc. Natl. Acad. Sci. USA*, **78**, 6633–6637 (1981).
122. I. Hirao, Placing extra components into RNA by specific transcription using unnatural base pair systems, *Biotechniques* **40**, 711, 713, 715 (2006).
123. R. Kawai, M. Kimoto, S. Ikeda, T. Mitsui, M. Endo, S. Yokoyama and I. Hirao, Site-specific fluorescent labeling of RNA molecules by specific transcription using unnatural base pairs, *J. Am. Chem. Soc.* **127**, 17286–17295 (2005).

124. M. Kimoto, M. Endo, T. Mitsui, T. Okuni, I. Hirao and S. Yokoyama, Site-specific incorporation of a photo-crosslinking component into RNA by T7 transcription mediated by unnatural base pairs, *Chem. Biol.* **11**, 47–55 (2004).
125. E.N. Brody, M.C. Willis, J.D. Smith, S. Jayasena, D. Zichi and L. Gold, The use of aptamers in large arrays for molecular diagnostics, *Mol. Diagn.* **4**, 381–388 (1999).
126. M. Endo, T. Mitsui, T. Okuni, M. Kimoto, I. Hirao and S. Yokoyama, Unnatural base pairs mediate the site-specific incorporation of an unnatural hydrophobic component into RNA transcripts, *Bioorg. Med. Chem. Lett.* **14**, 2593–2596 (2004).
127. K. Moriyama, M. Kimoto, T. Mitsui, S. Yokoyama and I. Hirao, Site-specific biotinylation of RNA molecules by transcription using unnatural base pairs, *Nucleic Acids Res.* **33**, e129 (2005).
128. S. Jhaveri, M. Rajendran and A.D. Ellington, *In vitro* selection of signaling aptamers, *Nat. Biotechnol.* **18**, 1293–1297 (2000).
129. R.D. Jenison, S.C. Gill, A. Pardi and B. Polisky, High-resolution molecular discrimination by RNA, *Science* **263**, 1425–1429 (1994).
130. G.R. Zimmermann, R.D. Jenison, C.L. Wick, J.P. Simorre and A. Pardi, Interlocking structural motifs mediate molecular discrimination by a theophylline-binding RNA, *Nat. Struct. Biol.* **4**, 644–649 (1997).
131. G.M. Clore and J. Kuszewski, Improving the accuracy of NMR structures of RNA by means of conformational database potentials of mean force as assessed by complete dipolar coupling cross-validation. *J. Am. Chem. Soc.* **125**, 1518–1525 (2003).

3

Chemical Biology of DNA Replication: Probing DNA Polymerase Selectivity Mechanisms with Modified Nucleotides

Andreas Marx

3.1 Introduction

In Nature, all DNA synthesis is catalyzed by DNA polymerases in DNA replication, repair and recombination events [1–7]. DNA polymerases catalyze DNA synthesis in a template-directed manner. Thus, all DNA synthesis in Nature depends on the ability of DNA polymerases to recognize the template and correctly insert the complementary nucleotide. DNA polymerases are presented with a pool of four structurally similar dNTPs from which the sole correct (i.e. Watson–Crick base-paired) substrate must be selected for incorporation into the growing DNA strand in order to ensure faithful transmission and maintenance of the genetic information. On the other hand, some degree of error must be allowed in order to spur evolution.

The mechanisms by which these remarkable enzymes achieve a balance between selective DNA synthesis by allowing some (the right?) degree of error have been a matter of continuing interest and intensive discussion since the discovery of the first DNA polymerase, *Escherichia coli* DNA polymerase I, by Arthur Kornberg in 1956. Enormous efforts by scientists in many disciplines have been undertaken with the aim of gaining insights into the complex mechanisms and functions of these molecular machines. Endeavors in this direction are complicated because even relatively simple organisms have more than one DNA polymerase. Currently more than a dozen human DNA polymerases and five DNA polymerases in *E. coli* are known with significantly varied selectivity [8–12]. DNA polymerases involved in DNA replication achieve selective information transfer to the offspring according to the Watson–Crick rule with intrinsic error rates as low as one mistake within one million synthesized nucleotides. On the other hand, recently discovered DNA polymerases involved in DNA repair, translesion synthesis (TLS) and somatic hypermutation exhibit strikingly low fidelity. Interestingly, it has been shown unambiguously that mutations of DNA

The Chemical Biology of Nucleic Acids Edited by Günter Mayer

polymerases which alter their selectivity or activity are involved in the development of various diseases such as cancer [13,14].

Recently, a wealth of valuable new insights into DNA polymerase mechanisms that govern selectivity were gained through application of carefully designed synthetic nucleotides and oligonucleotides in functional enzyme studies [15–17]. This chapter covers some recent approaches and highlights the mechanistic conclusions drawn from the observed experiments.

3.2 Probing the role of hydrogen bonding in DNA polymerase selectivity

A DNA polymerase is presented with a pool of four structurally similar deoxynucleotide triphosphates (dNTPs) from which it must select the sole correct (i.e. Watson–Crick base-paired) substrate for incorporation into the growing DNA strand (Figure 3.1).

Certain DNA polymerases achieve selective information transfer according to the Watson–Crick rule with error rates as low as one mistake within one million synthesized nucleotides. What are the underlying mechanisms accounting for this strikingly high selectivity? At first glance, the formation of distinct hydrogen bonding patterns between the nucleobases of the coding template strand and the incoming nucleoside triphosphate might be regarded as 'informational'. Hence it was a common perception that hydrogen bonding alone is responsible for the selectivity of DNA polymerases. Nevertheless, thermal denaturing studies of matched and mismatched DNA complexes by Echols and Goodman strongly suggest that these interactions alone are not sufficient to explain the degree of selectivity commonly observed for enzymatic DNA synthesis promoted by high-fidelity DNA polymerases [7]. Hence additional factors must contribute to the observed high fidelity. Among these factors are exclusion of water from the enzyme's active site, base stacking, solvation, minor groove scanning and steric constraints within the nucleotide binding pocket resulting in geometric selection of the nucleobase pair with the right shape and size [1–7].

In order to evaluate the participation of hydrogen bonding in DNA replication selectivity mechanisms, Kool and colleagues described a functional strategy based on chemically modified DNA polymerase substrates. They developed nucleotide analogs in which the polar natural DNA nucleobases are replaced by nonpolar aromatic molecules, which closely mimic the shape and size of the natural nucleobases but have at least significantly diminished ability to form stable hydrogen bonds (Figure 3.2) [18].

These hydrophilic nucleotide isosteres were applied as functional probes to elucidate the impact of hydrogen bonding on DNA polymerase selectivity. In first experiments, Kool and colleagues studied the insertion of dNTPs opposite **F**, the isostere of thymidine (Figure 3.2), in the template strand [19,20]. If solely hydrogen bonding drives selective nucleotide incorporation, one would expect that incorporation opposite **F** is very inefficient and unselective. Interestingly, the contrary was observed when studying a

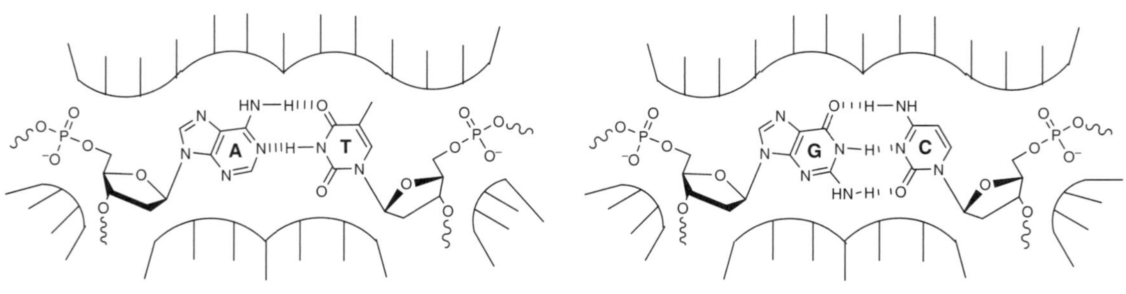

Figure 3.1 Schematic depiction of Watson–Crick base pairing in DNA polymerase active sites

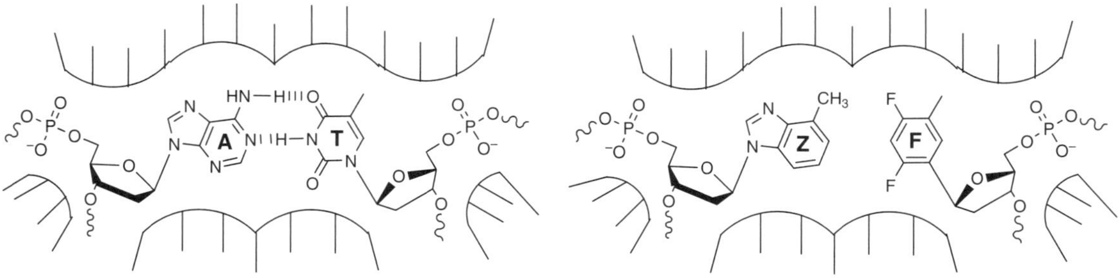

Figure 3.2 *Comparison of **A:T** nucleobase pair with the **Z:F** isosteres*

3′→5′-exonuclease-deficient mutant of the Klenow fragment of *E. coli* DNA polymerase I (henceforth just called Klenow fragment). This enzyme was able to promote efficiently **dATP** insertion opposite **F** and with high selectivity [21]. Steady-state kinetic analysis revealed that the efficiency of artificial **A–F** pair formation was only fourfold lower than that of the natural pair. Additionally, they found that **dFTP** was used as substrate by the Klenow fragment still highly efficiently and that the selectivity was as high as for the natural substrates. These results strongly suggest that, at least for Klenow fragment, the observed selectivity cannot be ascribed to hydrogen bonding alone. This hypothesis was further supported by the finding that a throughout artificial **F–Z** pair (Figure 3.2) can be efficiently processed by DNA polymerases accompanied by considerable selectivity [22]. Since these investigations clearly suggest that hydrogen bonding does not contribute substantially to DNA polymerase selectivity, other factors such as steric effects play at least a significant role in DNA polymerase selectivity processes. In a steric model for DNA polymerase fidelity, the enzyme selects the right substrate (according to the Watson–Crick rule) through editing the shape and size of the nascent base pair.

Further evidence for the steric model was derived from studying **dPTP** insertion opposite abasic sites **AP** in the template strand (Figure 3.3).

The pyrene group in **dPTP** is nearly as large as a natural base pair and obviously has no significant hydrogen bonding ability to the abasic site. The space occupied by the pyrene moiety fills in the blank of the removed base in the template strand. Remarkably, the Klenow fragment and T7 DNA polymerase are able to process **dPTP** opposite abasic sites more efficiently than opposite natural bases or another pyrene-bearing moiety in the template [23].

In further investigations, Morales and Kool compared the action of several eukaryotic and prokaryotic DNA polymerases on their action on nonpolar nucleoside isosteres [24]. When investigating interactions at

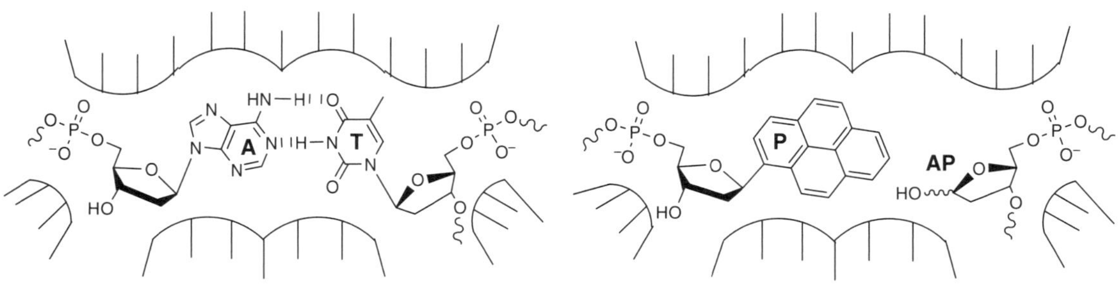

Figure 3.3 *Comparison of **A:T** nucleobase pair with the pyrene-containing nucleotide **P** opposite an abasic site **AP***

the active site (modifications include the coding nucleotide and the nucleoside triphosphate), they found that several DNA polymerases differ significantly in their action on the isosteric analogs. T7 DNA polymerase, *Thermus aquaticus* DNA polymerase and HIV-1 reverse transcriptase (RT) behave similarly to the Klenow fragment of *E. coli* DNA polymerase I, suggesting related mechanisms that govern nucleotide processing. However, different results were obtained for calf thymus DNA polymerase α and avian myeloblastosis virus RT. The results obtained suggest the existence of an energetically important hydrogen-bonded interaction between the enzyme and the incipient base pair. A third group of DNA polymerases comprising human DNA polymerase β and Moloney murine leukemia virus RT failed to replicate the artificial **F–Z** base pair, indicating that hydrogen bonds are needed at both the template and nucleoside triphosphate site.

These findings suggest the presence of hydrogen bonds between the DNA primer template complex and DNA polymerases that are important for enzyme function. Using the nonpolar isosteres, it was found that certain DNA polymerases such as *Thermus aquaticus* DNA polymerase, HIV-1 RT and the Klenow fragment of *E. coli* DNA polymerase I undergo important hydrogen bonding interactions at the first extension site in the primer strand [25,26]. Another group of enzymes comprising DNA polymerase α and β and T7 DNA polymerase fail to extend any of the non-hydrogen-bonding base pairs. Hence these studies indicate that several enzymes need hydrogen bonding interactions at both the template and primer site.

Using the nonpolar isostere **F**, eukaryotic DNA polymerase η was investigated [27]. DNA polymerase η is an extraordinary error-prone enzyme that is proficient in promoting bypass synthesis of several DNA lesions. This study reveals that the efficiency and accuracy of the incorporation of **F** by DNA polymerase η is severely impaired. Based on this finding, the authors suggest that Watson–Crick hydrogen bonding is required for DNA synthesis promoted by DNA polymerase η. Thus, this enzyme differs strikingly from high-fidelity DNA polymerases.

Taken together, the studies employing nonpolar nucleotide isosteres as probes strongly suggest that different mechanisms are employed by the various enzymes to achieve high selectivity. Studies on high-fidelity DNA polymerases (such as T7 DNA polymerase, *Thermus aquaticus* DNA polymerase, and the Klenow fragment of *E. coli* DNA polymerase I) indicate that hydrogen bonding alone does not account for the origin of DNA polymerase selectivity. Thus, steric effects within the active site of the enzyme are believed to contribute significantly, if not predominantly, to the factors that drive selectivity of DNA polymerases and canonical base-pair formation relies on the complementarities of size and shape of the nascent base pair. Needless to say, this does not imply that hydrogen bonding according to the Watson–Crick rule does not contribute to the selectivity at all. Indeed, structural and functional investigations of error-prone DNA polymerases suggest that hydrogen bonding might contribute substantially to DNA polymerase selectivity in these cases.

3.3 Varied selectivity among DNA polymerase: probing active site tightness

The results depicted in the previous section indicate that hydrogen bonding alone does not solely account for the origin of DNA polymerase selectivity. It is therefore suggested that DNA polymerase selectivity is achieved by editing nucleotide shape and size within a tight nucleotide binding pocket. However, DNA polymerase selectivity often varies significantly depending on the DNA polymerase. It is assumed that the origin of this varying error propensity stems from nucleotide binding pockets that differ in properties such as shape and tightness depending on the enzyme [1]. Thus, high-fidelity DNA polymerases are believed to form more rigid binding pockets tolerating less geometric deviation whereas low-fidelity enzymes exhibit more flexibility leading to decreased fidelity. However, this concept of varied active site tightness as the origin of different DNA polymerase selectivity needed to be tested experimentally. Recently, Marx and colleagues have introduced new functional means to investigate the effect of mainly steric constraints on the mechanism of DNA polymerase selectivity (Figure 3.4).

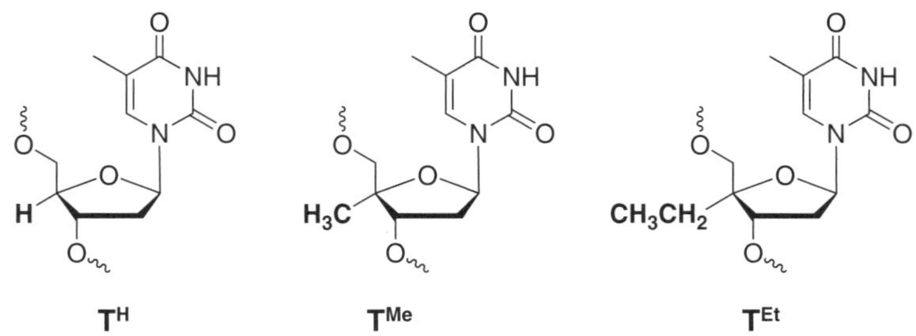

Figure 3.4 *Natural thymidine and size-augmented thymidines bearing sugar modifications with gradually increased size*

They developed steric probes by increasing the bulk of nucleoside triphosphates and substituted the standard 4′-C-hydrogen of the 2′-deoxyribose with alkyl groups that gradually increase in steric demand. Structural and functional data on DNA polymerases show that the sugar residues of the nucleotides are part of the substrate recognition process and provide the enzymes with additional paths for achieving selectivity besides inspection of nucleobase conformations. This size expansion is believed to cause less conformational flexibility within the DNA polymerase active site and, therefore, the modifications should decrease the tolerance for geometrically altered conformations of nascent nucleotide pairs. If the steric model of DNA replication selectivity holds true, this feature should result in an increase in nucleotide insertion selectivity. That was indeed found. Increasing the bulk of nucleoside triphosphate substrate through employment of the probes $\mathbf{T^{Me}TP}$ and $\mathbf{T^{Et}TP}$ led to a marked increase in nucleotide insertion selectivity catalyzed by the Klenow fragment [28–30]. Hence these results support the model that steric constraints are at least one crucial determinant of DNA polymerase selectivity.

As mentioned above, decreased tightness of enzyme active sites might be the origin of the higher error propensity observed for some DNA polymerases. One would anticipate that a more error-prone DNA polymerase (such as HIV-1 RT) would process the bulkier thymidines more efficiently than the more selective enzyme Klenow fragment. Marx and co-workers assumed that if varied active site tightness is indeed essential for differential nucleotide insertion selectivity, this should be felt by the steric probes $\mathbf{T^{R}TP}$. However, they observed that concerning the 'correct' insertion of the different $\mathbf{T^{R}TP}$ used, there is little difference between HIV-1 RT and the Klenow fragment [31,32]. Analyzing misinsertion, the two enzymes behave differently. Whereas 4′-C-methylation has little effect on the selectivity of HIV-1 RT, significant effects are observed for Klenow fragment. Thus, based on the concept of active site tightness, the results suggest that the two enzymes differ most significantly when promoting misinsertion rather than insertion opposite canonical template bases. This might be the result of differential active site conformations causing different steric constraints while promoting 'incorrect' nucleotide insertion.

In another study by Marx and co-workers, the Klenow fragment and the archaeal Y-family DinB homolog (Dbh) of *Sulfolobus solfataricus* were compared using the size-augmented sugar-modified thymidine-5′-triphosphate analogs depicted in Figure 3.4 [33]. It was found that substitution of a hydrogen atom at the 4′-position in the nucleotide analog by a methyl group reduces the maximum rate of nucleotide incorporation by about 40-fold for the Klenow fragment and about 12-fold for Dbh. Increasing the size to an ethyl group leads to a further twofold reduction in the rates of incorporation for both enzymes. Interestingly, the affinity of the Klenow fragment for the modified nucleotides is only marginally affected, which would indicate no discrimination during the binding step. Dbh even has a higher affinity for the modified analogs than it does for the natural substrate. Misincorporation of either **TTP** or $\mathbf{T^{Me}TP}$ nucleotides opposite a G template causes a drastic decline in incorporation rates for both enzymes. At the same time, the binding affinities of the Klenow fragment for these nucleotides drop by about 16- and fourfold, respectively, whereas Dbh shows only a twofold reduction. The varied proficiencies of the two enzymes in processing the size-augmented probes indicate varied flexibility of the enzymes' active sites and again support the notion of active site tightness being a criterion for DNA polymerase selectivity.

In another approach by Marx and co-workers, 4′-alkylated nucleotides and primer strands bearing 4′-alkylated nucleotides at the 3′-terminal position were employed as steric probes to investigate differential active site properties of human DNA polymerase β (Pol β) and the Klenow fragment [34]. Transient kinetic measurements indicated that both enzymes vary significantly in active site tightness at both positions. Whereas small 4′-methyl and -ethyl modifications of the nucleoside triphosphate perturb Pol β catalysis, extension of modified primer strands is only marginally affected. Just the opposite was observed for the Klenow fragment. Here, incorporation of the modified nucleotides is only slightly reduced, whereas size augmentation of the 3′-terminal nucleotide in the primer reduces the catalytic efficiency by more than 7000- and 260 000-fold, respectively. NMR studies support the notion that the observed effects derive from enzyme–substrate interactions rather than inherent properties of the modified substrates. These findings are consistent with the observed differential capability of the investigated DNA polymerases in fidelity such as the higher proficiency of Pol β to process misaligned DNA substrates. The results presented provide direct evidence for the involvement of varied steric effects among different DNA polymerases on their fidelity.

In order to probe steric effects in nucleobase recognition processes of DNA polymerases, Kool and colleagues developed an approach based on gradually expanded thymine nucleobase analogs. In these nonpolar compounds, the base size was increased incrementally over a 1.0 Å range by use of variably sized atoms (H, F, Cl, Br and I; Figure 3.5) to replace the oxygen molecules of thymine [35].

The kinetics with the Klenow fragment showed that replication efficiency opposite adenine increased through the series, reaching a peak at the chlorinated compound. The efficiency then dropped markedly as a steric tightness limit was apparently reached. Importantly, fidelity also followed this trend, with the fidelity maximum at dichlorotoluene, the largest compound that fits without apparent repulsion. The fidelity at this point approached that of wild-type thymine. Surprisingly, the maximum fidelity and efficiency were found at a base pair size significantly larger than the natural size. Interestingly, when bacteriophage T7 DNA polymerase was investigated using the same analogs, Kool and colleagues found that the enzyme strongly rejected nucleobase pairs that were smaller than the optimum by as little as 0.3 Å [36].The size preferences with T7 DNA polymerase were generally smaller and the steric rejection was greater than for the Klenow fragment. This correlates with the higher fidelity of the former enzyme. The results provide direct evidence for the importance of a tight steric fit of nucleobases to the active enzyme site on DNA polymerase fidelity.

In another set of experiments, Kool and co-workers tested steric effects in the active site of a Y-family DNA polymerase, Dpo4 [37]. It has been hypothesized that low-fidelity repair polymerases in this family more readily accept damaged or mismatched base pairs because of a sterically more open active site, which might place lower geometric constraints on the incipient pair. They tested the origin of low fidelity by use of the five nonpolar thymidine analogs depicted in Figure 3.5. The results showed that Dpo4 preferred to pair the

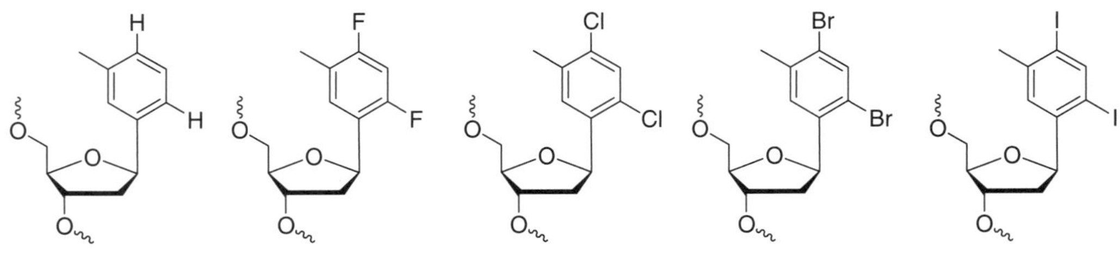

Figure 3.5 *Structures of nonpolar thymidine analogs having gradually increased size*

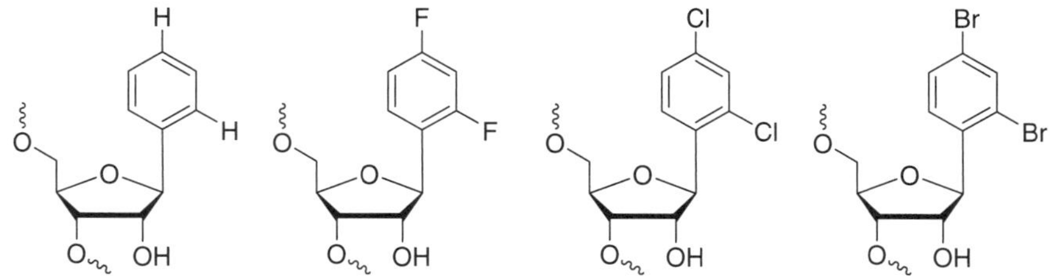

Figure 3.6 *Structures of nonpolar uridine analogs having gradually increased size*

thymidine shape mimics with adenine and, surprisingly, the preferred size was at the center of the range, the same optimum size as found recently for the high-fidelity Klenow fragment. However, the size preference with Dpo4 was fairly small, varying by a factor of only 30–35 from the most to the least efficient thymidine analog. This is in marked contrast to the Klenow fragment, which showed a rigid size preference, varying by 1100-fold from best to worst. The fidelity for the non-hydrogen-bonding analogs in pairing with A over T, C or G was much lower in Dpo4 than in the previous high-fidelity enzyme. The data establish that, unlike the Klenow fragment, Dpo4 has very low steric selectivity and that steric effects alone cannot explain the fidelity (albeit low) that Dpo4 has for a correct base pair; the findings suggest that hydrogen bonds may be important in determining the fidelity of this enzyme. The results suggest that the low steric selectivity of this enzyme is the result of a conformationally flexible or loose active site that adapts with small energetic cost to different base-pair sizes, rather than a spatially large active site. Interestingly, similar results were obtained for *E. coli* DNA polymerase IV, another translesion DNA polymerase [38].

HIV-1 RT is part of the viral replication machinery and catalyzed the DNA synthesis templated by the viral–single-stranded RNA genome. This enzyme also has the ability to catalyze DNA-templated DNA synthesis. It is a mutagenic, error-prone nucleotide polymerase. Kool and colleagues investigated whether the steric constraints in DNA- and RNA-templated DNA synthesis varies, for which purpose they synthesized nonpolar uridine analogs having gradually increased size (Figure 3.6) [39].

The analogs were incorporated into RNA templates and employed in functional studies of HIV-1 RT. They directly compared the functional flexibility of these two activities and measured the kinetics of incorporation for natural dNTPs opposite unnatural template bases in the RNA and DNA context and also for unnatural dNTPs opposite natural template bases [40]. Their results showed that the DNA-dependent polymerization by HIV-RT is highly sensitive to size, strongly biasing against both too-small and too-large base pairs, whereas, in contrast, the RNA-dependent polymerization is only biased against analogs that are too small and is much more accepting of larger base pairs. The data show greater rigidity with a DNA template as compared with an RNA template, which nicely correlates directly with the higher fidelity of the DNA-templated synthesis.

In conclusion, these studies highlight the importance of tight fitting of the nucleotide substrate within the DNA polymerase active site. The data presented provide experimental evidence that minute changes of the overall shape and size of the substrate impose significant effects on nucleotide selection. Furthermore, these studies provide strong experimental evidence that variations of steric constraints within the nucleotide binding pockets of DNA polymerases cause differences in nucleotide incorporation selectivity.

3.4 Conclusion

Numerous insights in the function of DNA polymerases have been gained through the employment of chemically modified substrates. Nevertheless, we are far from completely understanding the complex mechanisms of these enzymes, which may be regarded as complex machines. To make a detailed understanding even more difficult to achieve, often an organism expresses several DNA polymerases and crucial auxiliary factors. An understanding of how the mechanisms by which biological processes involving these proteins are orchestrated is in its infancy. Only recent studies indicate that auxiliary proteins may have a significant influence on the selectivity of processes involving DNA polymerases [41]. The author is convinced that many new and unexpected features of these fascinating enzymes can be revealed in the future.

References

1. Kool E.T. (2002) Active site tightness and substrate fit in DNA replication. *Annu. Rev. Biochem.* **71**: 191–219.
2. Patel P.H., Loeb L.A. (2001) Getting a grip on how DNA polymerases function. *Nat. Struct. Biol.* **8**: 656–659.
3. Kunkel T.A. (2004) DNA replication fidelity. *J. Biol. Chem.* **279**: 16895–16898.
4. Kunkel T.A., Bebenek K. (2000) DNA replication fidelity. *Annu. Rev. Biochem.* **69**: 497–529.
5. Kool E.T., Morales J.C., Guckian K.M. (2000) Mimicking the structure and function of DNA: insights into DNA stability and replication. *Angew. Chem. Int. Ed.* **39**: 990–1009.
6. Goodman M.F. (1997) Hydrogen bonding revisited: geometric selection as a principal determinant of DNA replication fidelity. *Proc. Natl. Acad. Sci. USA* **94**: 10493–10495.
7. Echols H., Goodman M.F. (1991) Fidelity mechanisms in DNA replication. *Annu. Rev. Biochem.* **60**: 477–511.
8. Goodman M.F. (2002) Error-prone repair DNA polymerases in prokaryotes and eukaryotes. *Annu. Rev. Biochem.* **71**: 17–50.
9. Hübscher U., Maga G., Spadari S. (2002) Eukaryotic DNA polymerases. *Annu. Rev. Biochem.* **71**: 133–163.
10. Marx A., Summerer D. (2002) Molecular insights into error-prone DNA replication and error-free lesion bypass. *ChemBioChem* **3**: 405–407.
11. Friedberg E.C., Fischhaber P.L., Kisker C. (2001) Error-prone DNA polymerases: novel structures and the benefits of infidelity. *Cell* **107**: 9–12.
12. Prakash S., Johnson R. E., Prakash L. (2005) Eukaryotic translesion synthesis DNA polymerases: specificity of structure and function. *Annu. Rev. Biochem.* **74**, 317–353.
13. Masutani C., Kusumoto R., Yamada A., Dohmae N., Yokoi M., Yuasa M. *et al.* (1999) The XPV (*Xeroderma pigmentosum* variant) gene encodes human DNA polymerase eta. *Nature* **399**: 700–704.
14. Starcevic D., Dalal S., Sweasy J.B. (2004) Is there a link between DNA polymerase β and cancer? *Cell Cycle* **3**: 998–1001.
15. Marx A., Detmer I., Gaster J., Summerer D. (2004) Probing DNA polymerase function with synthetic nucleotides. *Synthesis* 1–14.
16. Verma S., Eckstein F. (1998) Modified oligonucleotides: synthesis and strategy for users. *Annu. Rev. Biochem.* **67**: 99–134.
17. Jung K.-H., Marx A. (2005) Nucleotide analogs as probes for DNA polymerases. *Cell. Mol. Life Sci.* **62**: 2080–2091.
18. Kool E.T., Sintim H.O. (2006) The difluorotoluene debate – a decade later. *Chem. Commun.* 3665–3675.
19. Moran S., Ren R.X.-F., Rumney S. IV, Kool E.T. (1997) Difluorotoluene, a nonpolar isostere for thymine, codes specifically and efficiently for adenine in DNA replication. *J. Am. Chem. Soc.* **119**: 2056–2057.
20. Morales J.C., Kool E.T. (2000) Varied molecular interactions at the active sites of several DNA polymerases: nonpolar nucleoside isosteres as probes. *J. Am. Chem. Soc.* **122**: 1001–1007.
21. Moran S. Ren R.X.-F., Kool E.T. (1997) A thymidine triphosphate shape analog lacking Watson–Crick pairing ability is replicated with high sequence selectivity. *Proc. Natl. Acad. Sci. USA* **94**: 10506–10511.
22. Morales J.C., Kool E.T. (1998) Efficient replication between non-hydrogen-bonded nucleoside shape analogs. *Nat. Struct. Biol.* **5**: 950–954.
23. Matray T.J., Kool E.T. (1999) A specific partner for abasic damage in DNA. *Nature* **399**: 704–708.
24. Morales J.C., Kool E.T. (2000) Varied molecular interactions at the active sites of several DNA polymerases: nonpolar nucleoside isosteres as probes. *J. Am. Chem. Soc.* **122**: 1001–1007.
25. Morales J.C., Kool E.T. (1999) Minor groove interactions between polymerase and DNA: more essential to replication than Watson–Crick hydrogen bonds? *J. Am. Chem. Soc.* **121**: 2323–2324.

26. Morales J.C., Kool E.T. (2000) Functional hydrogen-bonding map of the minor groove binding tracks of six DNA polymerases. *Biochemistry* **39**: 12979–12988.
27. Washington M.T., Helquist S.A., Kool E.T., Prakash L., Prakash S. (2003) Requirement of Watson–Crick hydrogen bonding for DNA synthesis by yeast DNA polymerase η. *Mol. Cell. Biol*. **23**, 5107–5112.
28. Summerer D., Marx A. (2001) DNA polymerase selectivity: sugar interactions monitored with high-fidelity nucleotides. *Angew. Chem. Int. Ed*. **40**: 3693–3695.
29. Strerath M., Summerer D., Marx A. (2002) Varied DNA polymerase–substrate interactions in the nucleotide binding pocket. *ChemBioChem* **3**: 578–580.
30. Summerer D., Marx A. (2002) Differential minor groove interactions between DNA polymerase and sugar backbone of primer and template strands. *J. Am. Chem. Soc*. **124**: 910–911.
31. Strerath M., Cramer J., Restle T., Marx A. (2002) Implications of active site constraints on varied DNA polymerase selectivity. *J. Am. Chem. Soc*. **124**: 11230–11231.
32. Cramer J., Strerath M., Marx A., Restle T. (2002) Exploring the effects of active site constraints on HIV-1 reverse transcriptase DNA polymerase fidelity. *J. Biol. Chem*. **277**: 43593–43598.
33. Cramer J., Rangam G., Marx A., Restle T. (2008) Varied active-site constraints in the klenow fragment of *E. coli* DNA polymerase I and the lesion-bypass Dbh DNA polymerase. *ChemBioChem* **9**: 1243–1250.
34. Di Pasquale F., Fischer D., Grohmann D., Restle T., Geyer A., Marx A. (2008) Opposed steric constraints in human DNA polymerase beta and *E. coli* DNA polymerase I. *J. Am. Chem. Soc*. **130**: 10748–10757.
35. Kim T.W., Delaney J.C., Essigmann J.M., Kool E.T. (2005) Probing the active site tightness of DNA polymerase in subangstrom increments. *Proc. Natl. Acad. Sc.i USA* **102**: 15803–15808.
36. Kim T.W., Brieba L.G., Ellenberger T., Kool E.T. (2006) Functional evidence for a small and rigid active site in a high fidelity DNA polymerase: probing T7 DNA polymerase with variably sized base pairs. *J. Biol. Chem*. **281**: 2289–2295.
37. Mizukami S., Kim T.W., Helquist S.A., Kool E.T. (2006) Varying DNA base-pair size in subangstrom increments: evidence for a loose, not large, active site in low-fidelity Dpo4 polymerase. *Biochemistry* **45**: 2772–2778.
38. Silverman A.P., Jiang Q., Goodman M.F., Kool E.T. (2007) Steric and electrostatic effects in DNA synthesis by the SOS-induced DNA polymerases II and IV of *Escherichia coli*. *Biochemistry* **46**:13874–13881.
39. Silverman A.P., Kool E.T. (2007) RNA probes of steric effects in active sites: high flexibility of HIV-1 reverse transcriptase. *J. Am. Chem. Soc*. **129**:10626–10627.
40. Silverman A.P., Garforth S.J., Prasad V.R., Kool E.T. (2008) Probing the active site steric flexibility of HIV-1 reverse transcriptase: different constraints for DNA- versus RNA-templated synthesis. *Biochemistry* **47**: 4800–4807.
41. Maga G., Villani G., Crespan E., Wimmer U., Ferrari E., Bertocci B., Hübscher U. (2007) 8-Oxo-guanine bypass by human DNA polymerases in the presence of auxiliary proteins. *Nature* **447**: 606–608.

4

Nucleic Acid-templated Chemistry

Michael Oberhuber

4.1 Introduction

Conceptually, 'template' refers to objects that convey a mold or design specification, without being incorporated into the structure they define. In chemistry, templates are molecules that control the formation of particular products from reactants that, potentially, could combine in many different ways [1]. DNA is one of the most prominent examples of molecular templates, which stores and transfers genetic information via canonical Watson–Crick base pairs [2]. Nature relies on nucleic acid (NA) templating in elementary cellular processes, including replication [3], transcription [4], translation [5] and regulation of gene expression [6–8]. Molecular engineers have been inspired by these natural processes to create artificial template systems that exploit the selectivity and predictability of base-pairing interactions [1]. Watson–Crick base pairing has been used, for instance, to build complex assemblies up to nanoscale DNA objects through noncovalent interactions, giving rise to the sprawling field of DNA nanotechnology [9,10].

This chapter focuses on *chemistry* that is controlled by an NA templating effect without requiring a macromolecular catalyst. Cellular templated chemistry is typically catalyzed by enzymes and has been extensively reviewed elsewhere [3–8]. Supramolecular DNA assemblies have also been comprehensively reviewed [9,10] and will be treated here only marginally. In essence, this chapter will discuss the possibilities that NAs offer to promote and control the formation (and rarely cleavage) of chemical bonds. Particular attention will be devoted to the added value of NAs for chemistry, since several excellent reviews on NA templating have been published in recent years [11–14]. In the tradition of chemical biology [15], the chapter will discuss how the biological principles that are inherent to NAs can be used to advance chemistry, which, in turn, fuels advances in biology. Most examples will concern DNA templates, but RNA and some NA analogs will also be discussed.

The Chemical Biology of Nucleic Acids Edited by Günter Mayer

4.2 Historical overview

4.2.1 Chemical replication: nucleic acid-templated chemical synthesis of nucleic acids

Leslie Orgel's pioneering work has broadly introduced NAs as templates to mediate the ligation or polymerization of DNA, RNA or analogs to study the origin of replication and the emergence of the genetic system [16,17]. Since the goal of these studies was the replication of genetic information by synthesis of a complementary daughter strand, the reactants were typically (oligo)nucleotides with reactive groups at their end, juxtaposed on a complementary 'template strand', similar to a nicked double-stranded DNA (Figure 4.1 and Table 4.1).

The first NA-templated reaction to be reported was the chemical ligation of two hexathymidylates (dT_6) on a poly(A) RNA template in the presence of a water-soluble carbodiimide activating agent (EDC) to give a 12-mer with a native phosphodiester bond [18]. This was then rapidly followed by a series of investigations, primarily by Orgel and co-workers, which showed for the first time that the corresponding reaction of adenylates was strongly enhanced in the presence of a poly(U) RNA [19]. A poly(U), however, had no effect on the reaction of adenylates with guanylates or cytosinates [20], demonstrating that covalent bond formation was promoted by the specificity of Watson–Crick base pairing interactions. These experiments established the possibility of a nonenzymatic transfer of genetic information and, at the same time, gave birth to the field of NA-templated chemistry. These first experiments were rapidly expanded to other chemistries and RNA analogs [17,21–24], once chemical synthesis of NAs became routine in the laboratory [25]. DNA, for instance, was used as a template in similar studies regarding the nonenzymatic 'transcription' of oligo-(dC) into oligo-(G)s [26]. Soon after, von Kiedrowski reported a model of replication via carbodiimide-mediated ligations, involving a palindromic DNA hexamer that promoted the joining of two juxtaposed trimers to give another copy of itself [27]. Numerous studies

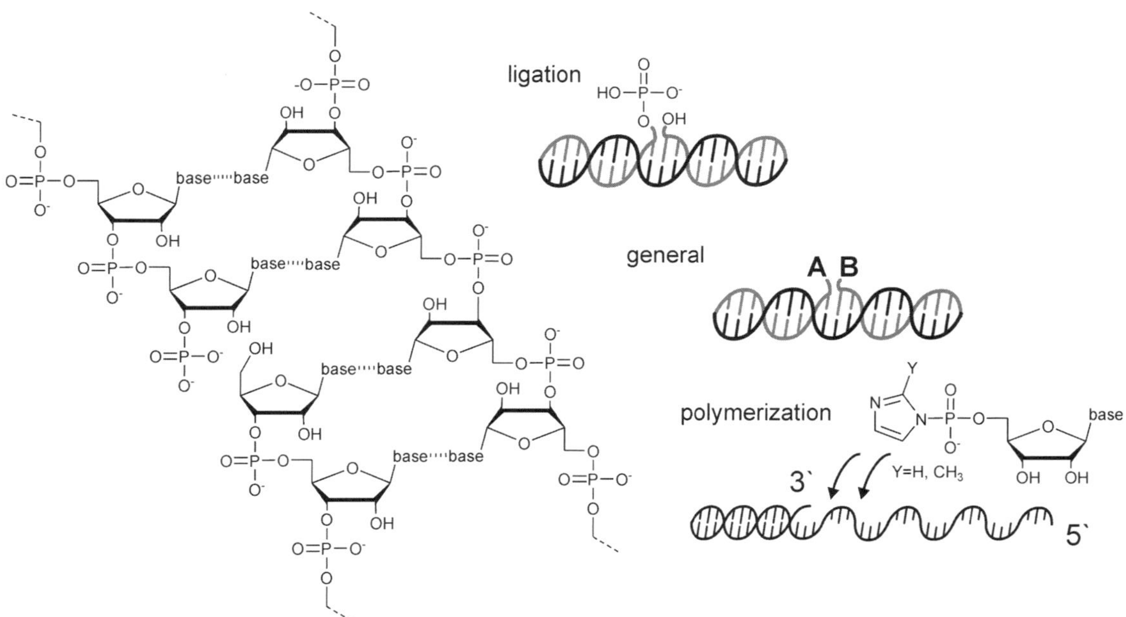

Figure 4.1 *Early examples of nucleic-acid-templated chemical reactions*

Table 4.1 *NA-templated reactions leading to structures related to the NA backbone*

Reaction type	A	B	Comments
P–X bond formation (X = O, S, Se, N)			
In situ activated ligation	-Phosphate	–OH	+ Carbodiimide (EDC), numerous examples; RNA and DNA templates [16–20,27]
In situ activated ligation	-Phosphate	–OH	+ CN–X (X = Br, imidazole), M^{2+}; numerous examples, also with Hoogsteen pairs [36–39,74,75]
Ligation	-Triphosphate	–OH	+ M^{2+}; RNA templates [79,80]
Ligation	2′,3′-Cyclic phosphate	–OH	RNA [16,17] and pyranosyl-RNA templates [29]
Autoligation	Phosphothioate or -selenoate	α-Bromoacetamido-, tosylate or I	DNA and RNA templates, numerous examples for NA detection [40–45,130]
Polymerization (ligation)	Phosphorimidazolides	–OH	RNA [21–24,70,80], DNA [26], HNA [28] and TNA [30] templates [16,17]
Phosphoramidate formation	-Phosphate	$-NH_2$	+ Carbodiimide (EDC); DNA template [34]
C–X bond formation (X = N, O, C) (see also Table 4.2)			
Amine acylation	$-NH_2$	Peptide–CO–S–DNA	DNA-templated peptide transfer [33]
Amine acylation	PNA–CO_2R	PNA–NH_2	RNA-templated PNA coupling [32]
Amine acylation	PNA–CO_2H	PNA–NH_2	For NA sensing [52,53]
Native chemical ligation	PNA–CO–SR	PNA–NH_2	For NA sensing; transthioesterification, followed by S to N acyl shift [121,122]
Imine formation – reductive amination	$-CH_2-CHO$	$-NH_2$	+ $NaCNBH_3$ (reductive amination); DNA templates; numerous examples [11,35]
Photochemical ligation via [2 + 2] cycloaddition	Thymidylate/ pyrimidine nucleotide	(Thio)thymidylate/ 5-vinylpyrimidine nucleotide	Reversible in the case of 5-vinylpyrimidines [124–126]; numerous examples with various photoreactive groups [43,46,47,127–129]

followed, including Orgel's polymerization studies on NA templates with a hexose backbone (HNA) [28], Eschenmoser's ligation studies on pyranosyl-RNA templates [29], and, more recently, polymerization studies on threose nucleic acid (TNA) [30], which has been proposed as a primitive genetic system preceding RNA and DNA [31]. Several chemistries other than phosphodiester formation have been explored as potential ligation/polymerization reactions, including Nielsen and Orgel's RNA-templated coupling of peptide nucleic acids (PNAs) [32], Joyce's DNA-templated peptide conjugation [33], von Kiedrowski's DNA-templated phosphoramidate formation [34], and Zhan and Lynn's reductive aminations [11,35]. Shabarova *et al.* used cyanogen bromide in the presence of metal ions as a condensing agent to synthesize a whole tRNA gene by chemical ligation of mutually annealed fragments [36]. This approach was then modified by Luebke and Dervan to synthesize double-stranded plasmid DNA in the presence of *N*-cyanoimidazole and Zn^{2+} [37], and by Kool and co-workers to prepare circular DNA with cyanogen bromide, imidazole and Ni^{2+} [38]. Lynn and co-workers optimized this ligation method by covalently linking imidazole and other amines to the template strand opposite the ligation site [39]. Letsinger and Kool eliminated the need for condensing agents by employing oligonucleotide-3′-phosphothioates (or -selenoates) and leaving groups on the 5′ end of a juxtaposed oligonucleotide (for more details, see below and Table 4.1) [40–45]. Oligonucleotides have also been joined via photochemical reactions. Dimerization of neighboring thymidylates (and thiothymidylates) [46,47] or other photoreactive groups [43,48,49] have been used to ligate oligonucleotides in a [2 + 2] cycloaddition reaction upon irradiation with UV light (see below and Table 4.2).

Table 4.2 *NA-templated reactions leading to structures unrelated to the NA backbone*

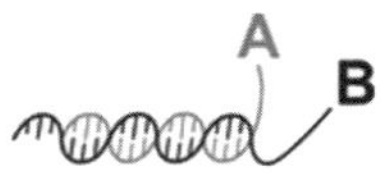

Reaction type	A[a]	B[b]	Comments
Nucleophilic displacement (see also Table 4.1)	$-CH_2SH$	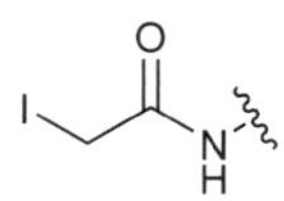	Various leaving groups [12,50,73]
Conjugate addition	$-CH_2SH$ $-NH_2$	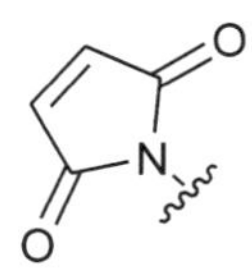	Various electrophiles tested [12,50,61]
Amine acylation (see also Table 4.1)	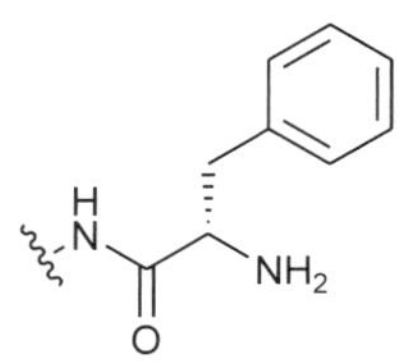	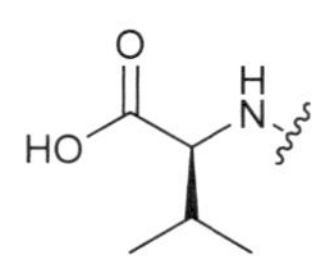	+ Coupling reagent (e.g. EDC), numerous amino acids [12,51,61,96]; also in organic solvents [136]; templated with Watson–Crick and Hoogsteen pair s[76]
Reductive amination Hydrazone formation Oxazolidine formation (see also Table 4.1)	$-NH_2$ + $NaCNBH_3$ -hydrazide –2-amino alcohol	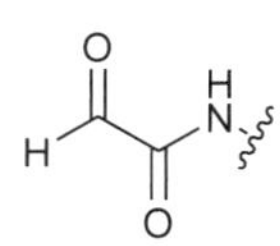	Many aldehydes [12]; also in organic solvents [136]; for ligation [11,51], polymerization [137], construction of nanoscale DNA objects [57,138] and heterocycle synthesis [51,83]
Wittig olefination	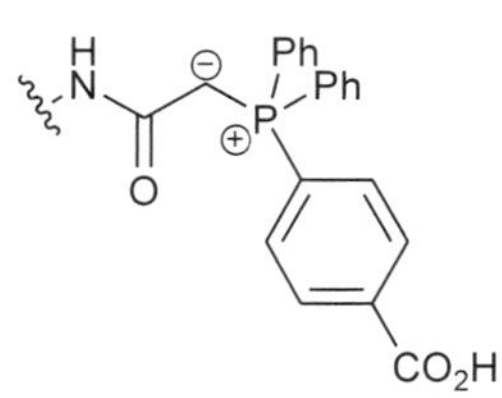	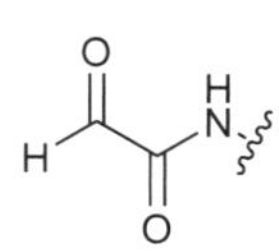	Numerous examples [12,94,96], in aqueous [51,61] and organic solutions [136]

Table 4.2 *(Continued)*

Reaction type	A[a]	B[b]	Comments
[2 + 3] Dipolar cycloadditions	$-N_3$	Alkyne	Huisgen (with DNA reagents [93] and minor groove binders [77]) and 1,3-nitrone cycloaddition [51,93]
Pd-mediated coupling reactions	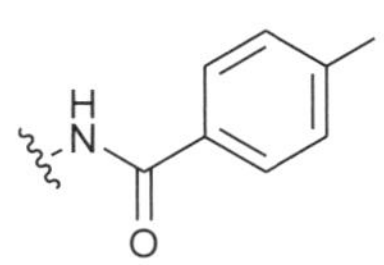	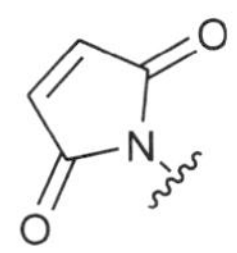	Heck, Suzuki and related chemistry [12,51,85]; also in organic solvents [136]; new coupling reaction discovered [85]
Aldol reaction	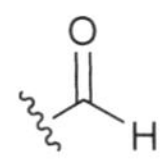	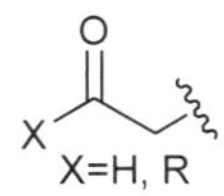	Various reactants, in aqueous [62] and organic solutions [136], enamine catalysis [62,112]
Nitro-aldol (Henry) and nitro-Michael reaction	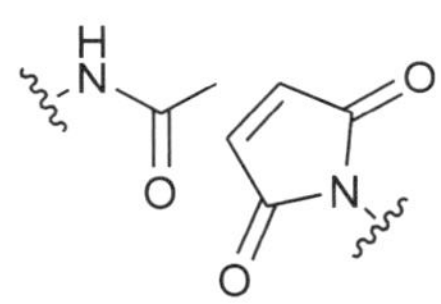	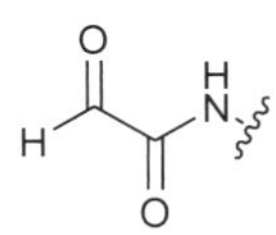	1,2 (Aldol) and 1,4 (Michael) addition [51,61]
Metallo-salen formation	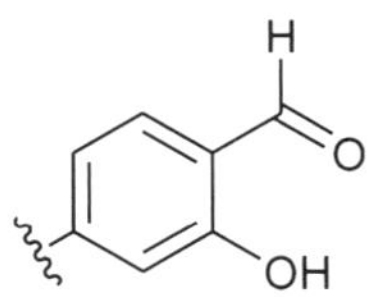	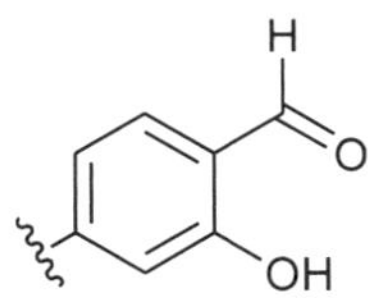	+ Ethylenediamine, Ni^{2+} or Mn^{2+} [54]; also branched [57] and inter-strand salen [139] complexes
Hydroarylation	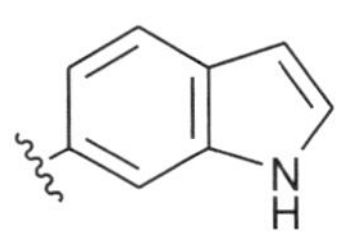	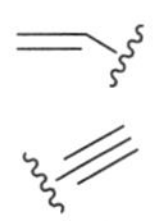	Brønsted- or Lewis acid-catalyzed [109]
Functional group transformations:			
Ester hydrolysis	$-CO_2$-R (to CO_2H)	Imidazole or metal complexes	R = *p*-NO_2-phenyl; multiple turnover [58,59]
Staudinger reaction	$-N_3$ (to NH_2)	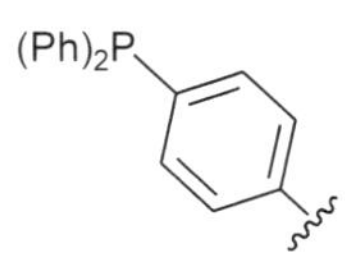	Coupled with subsequent elimination or S (O) to N acyl shift to liberate CO_2H, SH or OH, respectively [103,104]

[a,b]Representative substrate examples shown, linkers may vary.

During this first era from the 1960s to the late 1990s, the main motivation of the early reports was to study potentially ancient mechanisms of replication in the context of the origin of life. Some reports were directed at the chemical synthesis of particular NAs. A third class was aimed at the detection of NAs. The common principle in all this work is that motivation, chemistry and products structures were all closely related to the natural NA backbone itself.

4.2.2 Generalizing the concept: DNA-templated synthesis (DTS) of organic molecules

This changed profoundly with the pioneering work of Liu and co-workers, who considered NA templating to synthesize structures unrelated to the NA backbone. Liu's group added linker molecules, inserted between the reactive groups and the DNA, to facilitate conjugation and keep the reaction site more distant from the backbone (Figure 4.2a) [12,50]. Reactive groups were typically linked to ends of two hybridizing DNA strands (referred to as end-of-helix or A + A′ architecture; Figure 4.2b) [12,50], complementing earlier architectures, which typically involved ternary complexes, with two reactant–oligonucleotide conjugates juxtaposed on a complementary DNA template (A + B + A′B′ architecture) or slight variations thereof (hairpin or A + BB′A′ architecture).

In their first study, Gartner and Liu presented two reaction types with products that were topologically and structurally different from NAs: a nucleophilic displacement and conjugate addition involving typical organic chemical substrates proceeded efficiently with good to excellent yields (Table 4.2) [50]. Liu and co-workers then expanded the repertoire of NA-templated chemistry to an impressive range of chemical reactions, including [2 + 3] dipolar cycloadditions, Wittig olefinations or Pd-mediated coupling (Heck) reactions [51], but also chemistries that had already been used to create NA-related structures such as amine acylations and reductive animations (Table 4.2) [35,52,53].

Approximately at the same time, Czlapinski and Sheppard reported the first DNA-templated assembly of metallo-salen complexes from two DNA-linked salicylaldehydes, hybridized on a complementary template and free ethylenediamine in the presence of Mn^{2+} or Ni^{2+} ions [54]. Metal–NA interactions have been thoroughly studied in the context of NA catalysis and detection [40,55,56]; the salen complexes, however, stand out because their synthesis has been instructed by DNA templating. Gothelf and co-workers used this principle to build branched macromolecular DNA assemblies at the nanoscale [57]. The first application of DNA-linked metal complexes in DNA-templated catalysis was reported by Brunner and co-workers, who used a DNA-linked 3-(pyrid-2-yl)pyrazole–Cu^{2+} complex to hydrolyze a juxtaposed *p*-nitrophenyl ester-functionalized PNA [58]. This work was conceptually preceded, albeit without the use of metal complexes, by Ma and Taylor, who reported a DNA-templated hydrolysis of a nitrophenyl ester, catalyzed by a imidazolylamide moiety linked to a hybridized DNA strand. The template could dissociate after hydrolysis and enter a new cycle of catalysis [59].

Taken together, an impressive range of chemical reactions, discussed here and later in this chapter, have been successfully performed in an NA-templated format (Table 4.2). Even though the scope of the young

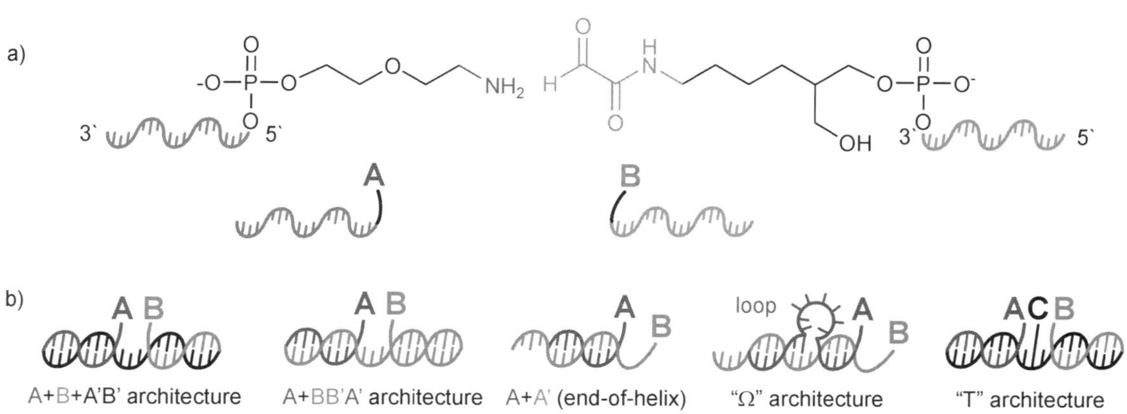

Figure 4.2 (a) Reactant examples for DNA-templated synthesis (DTS), consisting of an oligonucleotide (gray), linker (black) and reactive group (colored); (b) DTS architectures, A/A′ indicate complementary sequences, molecules are drawn in different colors and separated by '+'characters. See color plate section

field of ‘DNA-templated synthesis’ (DTS) cannot rival the generality of traditional organic synthesis [60], it enables fascinating new avenues to be explored, the principles of which are detailed below.

4.3 Enabling aspects of nucleic-acid-templated chemistry

4.3.1 Enabling aspects of Watson–Crick base pairs

4.3.1.1 Hybridization of nucleic acids

The most outstanding consequence of linking reactants to oligonucleotides is the possibility of controlling their chemical reactivity by specific NA interactions, most frequently Watson–Crick base pairs. NA-templated reactions are typically performed in very dilute solutions (low-micromolar to nanomolar reactants), where intermolecular reactions are negligibly slow. Hybridization of the tethered NAs, however, confines the reactive groups in a radius of approximately 10 bonds, dramatically increasing their local concentration in excess of 1 M (Figure 4.3a) [12]. This proximity effect is the underlying principle of NA-templated synthesis, allowing kinetic reaction control by Watson–Crick base pairing and other NA interactions.

4.3.1.2 Sequence specificity

Gartner and Liu noted that a single base pair mismatch at the center of a 10-mer DNA template was sufficient to reduce the reaction rate of a conjugate addition approximately 200-fold at room temperature [50]. Faint product fractions, derived from single-mismatch duplexes formed under these conditions, disappeared at higher temperatures. At temperatures above the calculated melting point of the fully matched duplex, no reaction products were observed, providing proof that these chemical reactions were enabled by sequence-specific Watson–Crick base pairing and not by an intermolecular mechanism [50], for which an Arrhenius-type behavior would have been expected. In other words, the outcome of a NA-templated chemical reaction is no longer dominated by the chemical reactivity of the functional groups, but rather by the sequence of the tethered oligonucleotide, allowing one to select a desired reaction partner among chemically very similar reactants. An impressive demonstration of this new dimension of selectivity, unknown in traditional organic synthesis, was reported by Liu and co-workers [61]. Three primary amines, two benzaldehyde, three maleimide, and one each carbonic acid, thiol, nitroalkane and phosphor ylid functions were linked to 12 pairwise complementary DNA sequences and allowed to react in one pot, giving rise to only six out of 28 possible products (Figure 4.3c). In a traditional synthetic reaction, the amines would be indistinguishable and cross-react with all electrophilic groups, giving a statistical product mixture. DNA templating, in contrast, directed one amine to react with its complementary benzaldehyde conjugate in a reductive amination, while the other amine reacted with its cognate carbonic acid by forming an amide bond [61]. Oberhuber and Joyce reported the first DNA-templated aldol reaction, where a DNA-linked glyceraldehyde was forced to react with a juxtaposed glycolaldehyde–DNA conjugate to give pentose sugars (Figure 4.3b) [62]. In free solution, these nearly identical aldehydes would react in a complicated network of self-aldol, isomerization and retro-aldol reactions, cluttering the observation of the desired cross-aldol reaction. This so-called ‘formose’ reaction has been extensively studied as a prebiotic pathway to ribose on the primordial Earth [63], but only the use of DNA templates allowed a level of control that permitted the first direct investigation of pentose formation [62,64].

4.3.1.3 Distance dependence

In their first report on DNA-templated synthesis, Gartner and Liu also noted that for some reactions it was not required to align the reactants in immediate proximity (~10 bonds). The rates of reaction between a

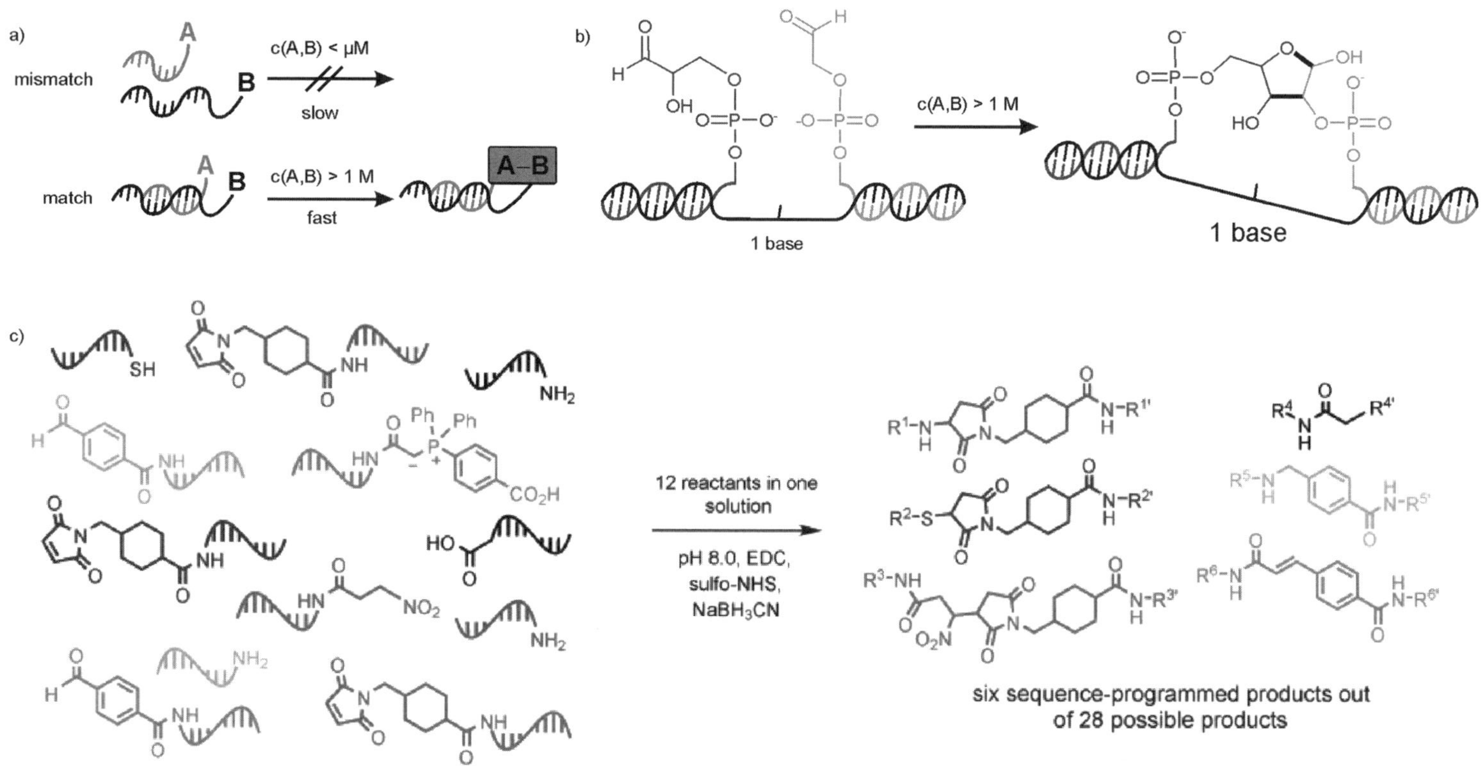

Figure 4.3 *Chemical reactivity controlled by Watson–Crick base pairing: (a) Origin of sequence specificity; (b) DNA-programmed cross-aldol reaction gives only rise to pentoses; (c) DNA-programmed reactions lead to the formation of six out of 28 possible products; matching colors indicate conjugates with complementary oligonucleotides [EDC = 1-ethyl-3-(3′-dimethylaminopropyl)carbodiimide hydrochloride; NHS = N-hydroxysuccinimide]. Part (c) reprinted from X. Li and D. R. Liu, DNA-templated organic synthesis: Nature's strategy for controlling chemical reactivity applied to synthetic molecules.* Angew. Chem. Int. Ed., ***43****, 4848–4870 (2004), with permission from Wiley-VCH Verlag GmbH & Co. KGaA. See color plate section*

DNA-linked thiol and a DNA–maleimide conjugate were comparable when the two functional groups were hybridized 1, 10 or up to 30 bases away from each other. Other reactions, in turn, showed a marked distance dependence and proceeded efficiently only when reagents were annealed in close proximity [51]. So-called 'distance independence' has been observed for conjugate additions, amine acylations, Henry reaction, Heck coupling, S_N2 reactions between thiols and iodoamides and, to a lesser extent, for Wittig olefinations. The ease of bond formation in distance-independent DNA-templated reactions, where reactants are separated by up to 200 bonds [50,51], contrasts sharply with the notorious difficulties related to the synthesis of large rings [65]. This unexpected feature of DNA-templated synthesis was rationalized with a kinetic model, where hybridization rather than bond formation is rate determining: Distance-independent reactions are fast enough to form a covalent bond, as soon as the reactive groups are brought together by DNA hybridization. Distance-dependent reactions, in contrast, are limited by bond formation rates, regardless of intervening nucleotides [12,50,51]. One notable exception, however, does not fit this model: reductive amination, an extensively studied model reaction [11,12], is highly distance dependent, yet characterized by high product formation rates that should exceed DNA hybridization [12]. In a recent investigation regarding the role of secondary structure in the intervening region, Liu and co-workers shed light on this mystery, showing that a modest degree of secondary structure in a 30-base intervening sequence significantly increased bond formation rate, for both reductive amination and amine acylation. Unstructured regions, in turn, almost completely abolished product formation for both chemistries [66]. These findings highlight the role of secondary structure in bridging the distance between reactive ends, thereby increasing the frequency of their encounter after DNA hybridization, and suggest that secondary structure demands more thorough consideration in future studies.

4.3.1.4 The aqueous milieu

Another important enabling aspect of a NA tether is that it makes organic molecules soluble in water. Although not the best solvent, water is an excellent milieu for many reactions in organic chemistry, where transitions states are typically more polar than the reactants. The scope of biological chemistry and recent observations of efficient organic transformations in aqueous suspensions ('on water') [67] suggest a potential for water that exceeds its appreciated role as a readily available and environmentally benign solvent.

4.3.1.5 Structural bias

As chiral molecules, NAs can be expected to bias the structural outcome of a templated reaction. Natural and artificial NA enzymes, for instance, achieve exquisite regio- and stereochemical control by folding into compact three-dimensional structures [55,68,69]. Regiochemical bias in simple templated systems has indeed been known for some decades. For instance, NA-templated ligations and polymerizations involving 2′,3′-cyclic phosphates as activated substrates preferably lead to unnatural 2′–5′ phosphodiester linkages. In the absence of a template, however, adenosine 2′,3′-cyclic phosphate dimerized to give predominantly 3′–5′ linkages [16,17]. DNA-templated control over the *stereochemical course* of a reaction was first reported by Joyce and co-workers, who detected chiral discrimination of L-guanosine 5′-phosphor (2-methylimidazolides) (L-2-MeImpG) over the enantiomeric D-2-MeImpG during polymerization on an RNA [poly(D-C)] template. Even though one L-2-MeImpG coupled efficiently on the D-template, the resulting product was poorly extended [70]. Also, unnatural NAs such as pyranosyl-RNA or PNA can promote reactions stereoselectively. Bolli *et al.* reported that oligomerization of D-pyranosyl-RNA tetramers on a D-ribopyranose NA template was favored over tetramers with mixed chirality [71]. Kozlov *et al.* showed that appending two distant deoxynucleotides was sufficient to induce enantioselective coupling of a D-dinucleotide on an achiral 8-mer PNA template [72]. This is a remarkable case of chiral information transfer over an achiral molecule to a distal site of another molecule, were bond formation took place. Li and Liu observed a fourfold difference in the rate of a DNA-templated nucleophilic substitution (S_N2) reaction of a thiol and a juxtaposed (*S*)-α-bromopropionamide versus the corresponding *R*-isomer, even though thioether formation took place several bonds away from the backbone. Separating the reagents by 12 intervening single-stranded nucleotides still favored the (*S*)-α-bromopropionamide by a factor of 4. Only when the intervening nucleotides were substituted with a flexible achiral linker of similar length was no difference seen between the (*R*)- and (*S*)-bromide. Most notably, the stereoselectivity was inverted when the conformation of the DNA helix was transformed from a right-handed B-form to a left-handed Z-form under high-salt conditions, demonstrating that chiral induction originated from the helical DNA structure rather than from particular chiral centers in the deoxyribose backbone [73]. These examples indicate regio- and stereochemical bias originating from NA templates, with implications for the origin of homochirality; the details, however, depend strongly on the reactants [16,17], linkers [73] and their structures, making predictions on the product stereochemistry difficult. In fact, the regio- and stereochemical details of most NA-templated reactions are unknown. The minute scale makes a thorough structural product analysis, which is routine in a traditional synthesis, a formidable challenge, essentially limiting the spectroscopic repertoire to mass spectrometry (MS) [12]. The previously cited examples are exceptions, where substrate-specific enzymes or kinetic data were available to distinguish product isomers.

4.3.2 Enabling aspects of other nucleic acid interactions

Watson–Crick base pairs are by far the most important, but not the only, interaction that has been exploited for NA templating. The earliest example of a DNA-templated reaction employing Hoogsteen base pairing was reported by Luebke and Dervan [74], who used a double-stranded DNA to template the ligation of two single-stranded oligonucleotides, bound adjacently in the major groove via Hoogsteen base pairs. The two fragments were ligated to form a 3′,5′-phosphodiester bond in a sequence-specific manner, even though the sequences were not complementary in a Watson–Crick sense [74].

Later, Li and Nicolaou reported a system for nonenzymatic replication, based on templated ligation in a triple helix [75]. A palindromic homopurine:homopyrimidine duplex templated the ligation of two adjacently annealed oligonucleotides in the presence of *N*-cyanoimidazole (Figure 4.4a). After dissociation from the double-stranded template at neutral pH, the ligation product was able to template another ligation in the classical duplex format, giving a second generation of the double-stranded palindromic sequence. Multiple rounds of replication were performed by cycling between double- and triple-strand-templated ligation reactions, triggered by a change of pH [75]. Chen and Mao reported a variant of this system, involving two amine-functionalized oligonucleotides hybridized sequentially on the 3′ end of a template bearing a carboxylic acid on its 5′ end (Figure 4.4b) [76]. The proximal amine was separated from the carboxylic acid by a stretch of 40 unstructured nucleotides and was acylated almost quantitatively in the presence of a condensing agent, as expected for a distance-independent DNA-templated reaction [51]. The 40 intervening nucleotides were not randomly chosen, but designed to hybridize intramolecularly in a triple helix at more acidic pH, bringing the carboxylic acid in proximity with the distal amine, which was acylated selectively (63% yield) when the reaction was carried out at pH 5 [76].

Sequence-specific interactions with double-stranded DNA's minor groove have also been employed for DNA templating. Polyamides containing *N*-methylpyrrole and *N*-methylimidazole moieties have been established as sequence-specific minor groove binders by Dervan's group and used by Poulin-Kerstien and Dervan [77] to template click chemistry [78], a Huisgen cycloaddition between juxtaposed azide and alkyne groups, each linked to a suited polyamide (Figure 4.4c). All reactions involving double-stranded DNA as template showed marked distance dependence, probably because the rigid duplex structures were not able to bridge the distance between separated reactants [12,51].

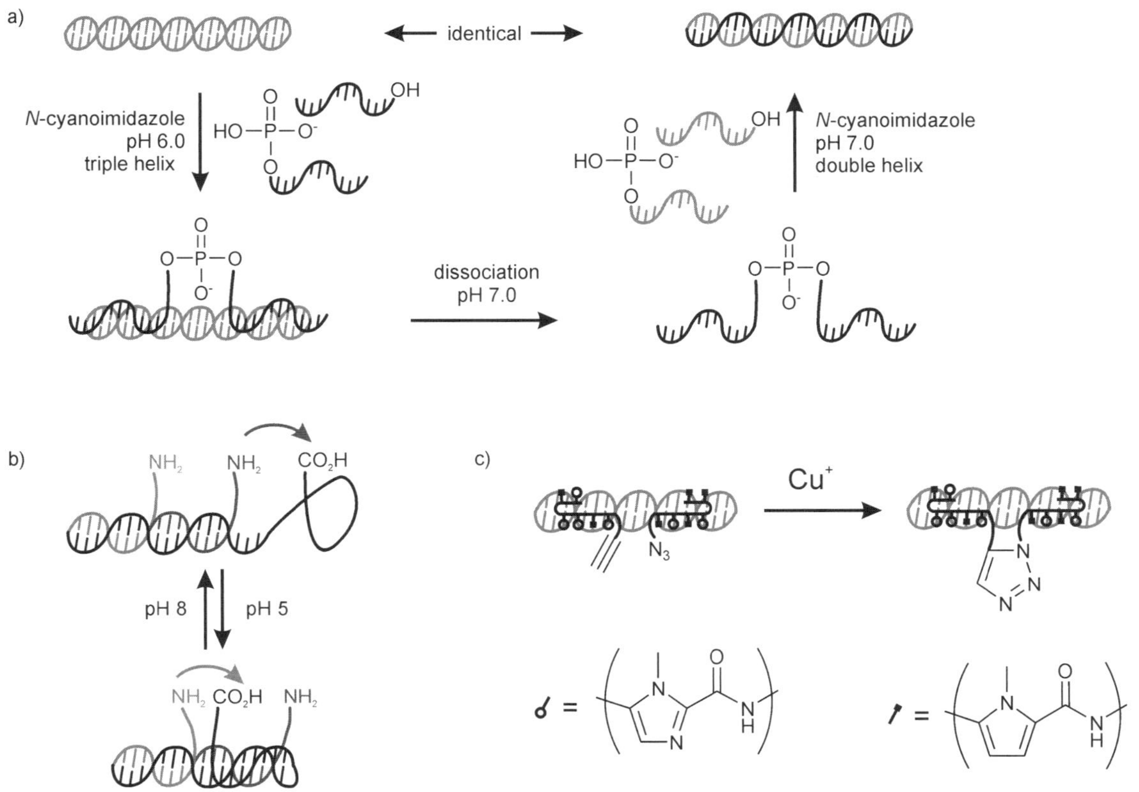

Figure 4.4 *Proximity effects mediated by other than Watson–Crick interactions: (a) double/triple helix-promoted phosphodiester bond formation, triggered by pH; (b) 'regio'selective carbodiimide-mediated amine acylation, achieved by pH-triggered double/triple helix switching; (c) Huisgen cycloaddition with minor grove-binding reactants*

4.3.3 Nucleic acids open a powerful analytical repertoire to small molecules

As a highly appreciated side-effect, linking small molecules to oligonucleotides opens a powerful analytical repertoire to traditional chemical reactions. Oligonucleotides can easily be labeled, for instance, with radioactive phosphorus or fluorescent dyes, routinely allowing the detection of femtomole amounts following chromatographic or electrophoretic separations. In addition, a vast repertoire of polymerases and NA-modifying enzymes is available to support the analytical techniques. This combination was crucial for the success of the early era. RNA-templated RNA polymerization was studied, for instance, by using a radiolabeled hairpin template (Figure 4.1), whose extension products could easily be followed by electrophoresis or paper chromatography [16,21]. The specificity of ribonuclease enzymes for native 3′,5′-phosphodiester linkages, in turn, provided the basis for analyzing the regioselectivity of templated polymerizations (e.g. [23]).

The flexibility that NA analytics adds to chemistry should not be underestimated. Standard gel electrophoresis offers sufficient resolution and sensitivity to detect even minute extents of bond formation or cleavage during DNA-templated reactions, permitting kinetic investigations of slow and difficult reactions. Rohatgi *et al*. investigated phosphodiester bond formation involving the kinetically stable 5′-triphosphate in an RNA-templated format [79,80]. Oberhuber and Joyce [62] employed NA analytics to investigate the notoriously difficult formation of pentose sugars via the formose reaction [63]. DNA templating suppressed competing reactions and allowed the detection of minute amounts of pentoses formed by this slow reaction [62]. Affinity labels such as biotin are also readily incorporated in oligonucleotides. Based on the strong and highly specific binding to streptavidin ($K_D \approx 10^{-14}$), biotinylated oligonucleotides are amenable to affinity chromatography [81], allowing an efficient and fast purification scheme. Liu and co-workers captured biotinylated product conjugates on streptavidin-coated beads, washed unreacted oligonucleotides and reagents away and released the product by cleaving the linker to enter the next reaction (see Figure 4.6) [12,82,83].

DNA microarrays – devices developed for the massive parallel analysis of genetic sequences by complementarity to immobilized oligonucleotides [84] – have been exploited by Liu and co-workers to analyze the results of 168 DNA-templated reactions, all performed simultaneously in one pot under varying conditions, achieving extraordinarily high throughput for reaction discovery (see below) [85].

Microarrays are frequently used in conjunction with the polymerase chain reaction (PCR) [86], a key technology capable of routinely amplifying less than 10^4 copies of a DNA molecule with high fidelity [87]. DNA-linked small molecules can thereby be recovered and identified even if present in small amounts in complex mixtures based on the tethered oligonucleotide which becomes a precious sequence tag, amplifiable by PCR.

4.3.4 Limitations

The fascinating enabling features of NAs come at a significant price: the ballast of a polymer that is much larger than the actual reactive group it controls. One of the main limitations of NA-templated chemistry is in fact cost. Even though DNA synthesis has become affordable at well below US$1 per base, the cost of setting up a DNA-templated synthesis exceeds substantially a traditional synthesis. The investment is therefore only justified where traditional approaches are inefficient or fail to deliver desired results.

The NA also limits the reaction scale, with a practical upper limit of a ~1 μmol, making it evident that NA-templated syntheses are not performed for preparative purposes. More important than the small molecule itself is the nucleotide tethered to it, which can be amplified or followed with sensitive techniques.

Third, reaction conditions must be – at least to some extent – compatible with the functional groups present in DNA or RNA, excluding a large part of the organic chemical repertoire. A classical DNA-templated synthesis, for instance, is limited to conditions that allow DNA hybridization: salt-containing aqueous solutions at neutral to moderately alkaline pH, temperatures from 0 to ~80 °C. Recent advances, however, are promising to expand the benefits of DNA-templated chemistry to conditions that are incompatible with hybridization (see below).

Fourth, the available analytical techniques for NAs are sensitive, but low in structural resolution. The limited amount of material makes a thorough product analysis a formidable challenge, essentially limiting the spectroscopic repertoire to MS. Molecular weight information is frequently considered sufficient for addition reactions, loss of water, etc. [12]; however, structural information is only deduced from indirect evidence and should be treated with due caution. Early replication studies, for instance, used nuclease enzymes to determine the content of natural isomers [16,17]. In our studies, we have verified that DNA-templated pentose formation – confirmed by mass spectra – depended on the presence of aldehyde functions in both aldol donor and acceptor strands. The pentose products were shown to behave like an abasic site (a ribose sugar without base) in a reverse transcriptase-based assay [62]; however, stereochemical details on the products could still not be deduced. In some cases, the reaction kinetics provided stereochemical clues to the product structure: the different reaction rates of the (*S*)- and (*R*)-bromide in Li and Liu's DNA-templated nucleophilic substitution [73] allowed conclusive predictions on the configuration of one chiral center in the product, based on the well-known stereochemical inversion during S_N2 reactions [88].

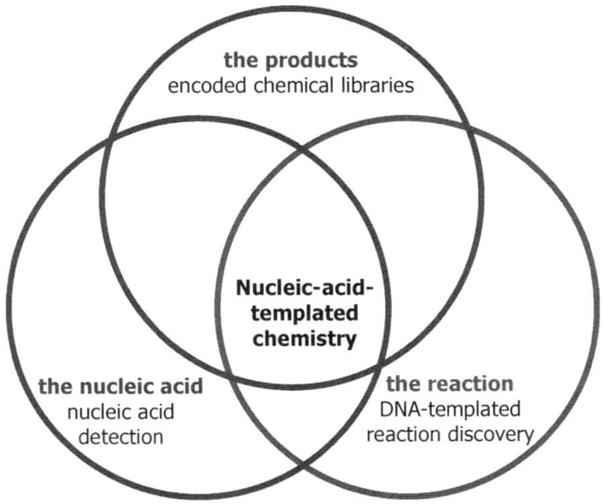

Figure 4.5 *Aspects and application of nucleic acid-templated chemistry*

NA-templated chemistry allows a level of control over the reactivity of tethered substrates that is not possible in a traditional laboratory synthesis. Most importantly, chemical reactivity can be directed by coding desired reaction partners into the sequences of oligonucleotides. The simple rules governing Watson–Crick and other base-pairing interactions allow one to design chemical reactivity control – even with nanoscale DNA objects – on the back of an envelope. NA conjugation makes a vast and powerful repertoire of instrumental and enzymatic methods available to analyze and handle reaction products, whose prime strengths are selectivity, sensitivity and high throughput. Oligonucleotides contain amplifiable sequence information that can be read out with standard techniques, even when present in a complicated mixture, but they also bring a significant molecular burden with them. Taken together, NA templating empowers chemical reactivity with properties that are characteristic of biological systems, paralleling Nature's approach for the synthesis and discovery of functional small molecules [12].

4.4 Applications

The enabling features of NA templating have provided an exciting basis for the development of new applications, exploiting all aspects of a templated reaction. Conceptually, an NA-templated reaction is an equation with three variables (Figure 4.5): (i) NA template, (ii) the chemical reaction leading to bond formation or cleavage and (iii) the product obtained from the reaction; by holding two variables constant, one can solve the equation for the third. All three faces have been exploited to empower the chemical discovery process. The next section will attempt to present applications along these lines, where NA-templated chemistry has shown its potential, but also its limitations.

4.5 Directed evolution of small-molecule libraries

4.5.1 Encoding combinatorial libraries

Combinatorial chemistry aims to improve the efficiency of the search for new functional small molecules, which are highly demanded for their ability to perturb protein function [15,89]. Libraries with hundreds of thousands of natural product-like chemical compounds are synthesized, typically in a split-and-pool approach [15]. After synthetic coupling steps on a solid support, the beads are recombined (pooled), mixed and separated (split) again for the next coupling step to give complex and structurally diverse libraries [90]. However, each compound has to be purified, identified and screened individually for a desired activity, requiring significant resources and an elaborate robotic infrastructure [12–14].

Brenner and Lerner first proposed to use oligonucleotides as sequence tags for small molecules, incorporated via alternating peptide and DNA coupling steps during library synthesis on a solid support [91]. The synthetic history of a bead would thereby be recorded in the oligonucleotide sequence tethered to it, dramatically simplifying its identification. Brenner and co-workers then demonstrated the feasibility of this concept by synthesizing the pentapeptide enkephalin and a 'barcode' oligonucleotide on the same solid support [92]. Such *encoded chemical libraries* provide a link between phenotype (the activity of the small molecule) and genotype (the sequence of the tethered oligonucleotide), making small molecules amenable to selection experiments, where all individuals are tested simultaneously and desired individuals are identified by their sequence information after physical separation of inactive molecules [12–14]. Paralleling Nature's approach to molecular discovery, selection experiments have been successfully applied to biomolecules, yielding high-affinity binders in addition to efficient catalysts from pools containing more than 10^{15} random-sequence NAs [55]. Encoded chemical libraries – in analogy with their biopolymer counterparts – hold the promise to make the power of Darwinian evolution available to small molecules [12–14].

4.5.2 DNA-templated multi-step synthesis of encoded libraries

Liu's group reported a series of studies to put this promising perspective into practice with DTS [12,14]. First, template topologies were developed that can accommodate multiple codons for the consecutive hybridization of different reagents ($A + A'$ architecture; Figures 4.2b and 4.6) [50,51]. Then, a thoughtful combination of DNA-templated reaction steps (with distance -dependence in mind), suitable linkers and a biotin-based purification scheme enabled the first DNA-templated multi-step synthesis to be achieved, involving an amine acylation, followed by a Wittig olefination and a conjugate addition in 3% overall yield [82]. Slightly later, Liu and co-workers reported new architectures for three-component reactions, involving a classical $A + B + A'B'$ architecture with a third reagent extending from a central nucleobase on the $A'B'$ strand (T architecture; Figure 4.2b). Distance-dependent reactions were made more useful by splitting the coding region for distance-dependent reagents between three and five nucleotides proximal to the reactive $5'$ end and a distal site of the template (Ω architecture; Figure 4.2b). After annealing of the reaction partner, intervening template nucleotides formed a bulge resembling an Ω, essentially eliminating distance dependence for Wittig olefinations, 1,3-dipolar cycloadditions and other chemistries [93].

These architectures brought a significant advance towards library synthesis, as demonstrated in a divergent multi-step synthesis of two different substituted *N*-acyloxazolidines in 3–7% overall yield. After an amine acylation employing a T-architecture that was shared by both pathways, one synthesis continued with oxazolidine formation and oxazolidine *N*-acylation, whereas the other involved formation of a different oxazolidine followed by *N*-acylation and Wittig olefination. In addition, different linkers and cleavage strategies were used to liberate, purify and prepare the reaction products for the next step [83]. An even more sophisticated multi-step synthesis was realized by Snyder and Liu, where the stabilities of coding duplexes were carefully balanced to allow a sequence of three Wittig olefinations (or amine acylations), to be triggered by a temperature gradient, without other intervention by the experimenter. Eliminating significant substance loss during intermediate purifications, a sequence-programmed triolefin and tripeptide were isolated in 24 and 21% yield, respectively [94].

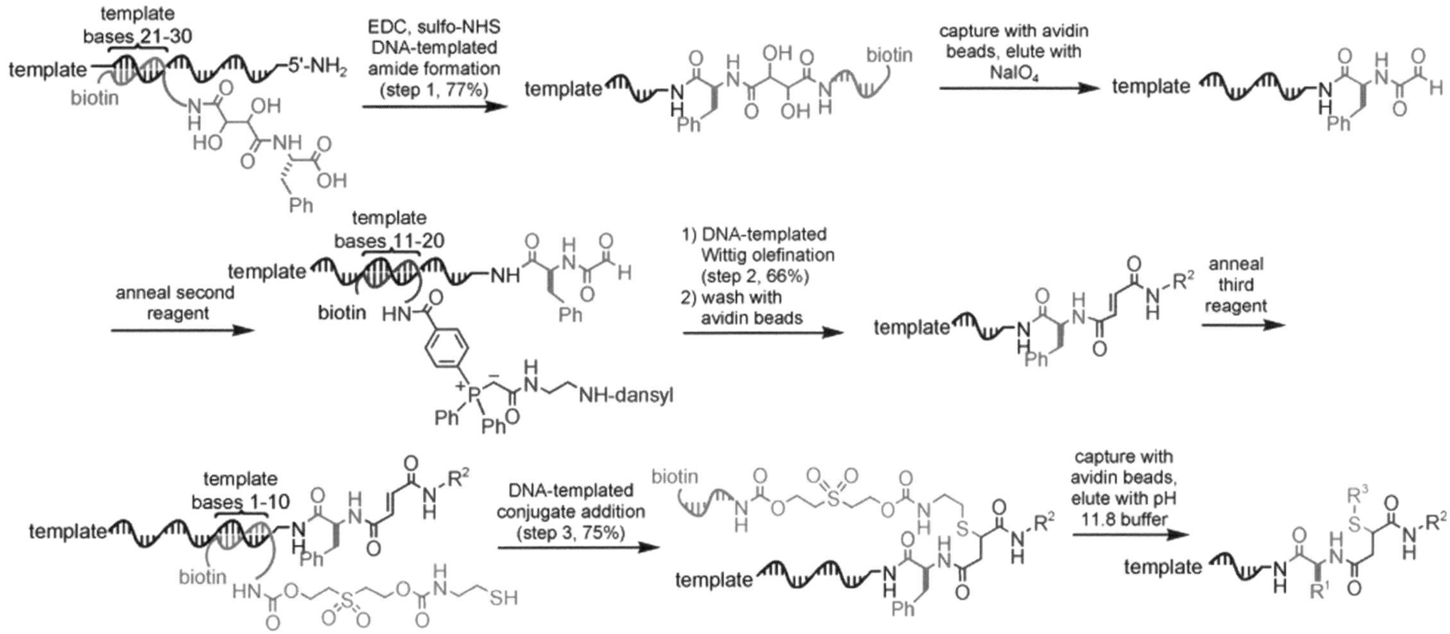

Figure 4.6 *DNA-templated multi-step synthesis; amine acylation (red), followed by Wittig olefination (blue) and conjugate addition (green) with biotin-based purification scheme. Template product conjugates are liberated during linker cleavage via periodate oxidation (red), during Wittig reaction (blue), and via β-elimination (green) (NHS = N-hydroxysuccinimide). Reprinted from X. Li and D. R. Liu, DNA-templated organic synthesis: Nature's strategy for controlling chemical reactivity applied to synthetic molecules.* Angew. Chem. Int. Ed., ***43****, 4848–4870 (2004), with permission from Wiley-VCH Verlag GmbH & Co. KGaA. See color plate section*

With these studies, DNA-templated synthesis has evolved into a promising system for molecular discovery, where DNA sequences *encode and direct* multi-step syntheses of fairly complex small molecules, in analogy with genetically encoded proteins.

4.5.3 *In vitro* evolution of small molecules

A general *in vitro* evolution scheme for small molecules was presented by Gartner and Liu [50]. Starting from an NA-tagged small-molecule library, prepared in a DNA-templated multi-step synthesis, molecules with desired activities are physically separated from inactive library members, for instance by passing the library over an immobilized protein target. Small-molecule binders are recovered by amplifying their sequence tag by PCR and, after diversification and chemical 'transcription', progeny small molecules are ready to enter the next round. In addition, all library members surviving the selection can be identified by sequencing in any round of the experiment (Figure 4.7) [50].

Liu and co-workers reported a remarkable proof-of-principle experiment, where a library of 1025 5′-maleimide-functionalized oligonucleotides with different sequences were incubated with 1025 DNA-linked thiols [50]. Only one of the thiol reagents contained a biotin group, which was captured on streptavidin-coated beads after DNA-templated conjugate addition. After extensive washing, retained oligonucleotides were amplified by PCR and the sequence encoding the reaction of the biotinylated thiol with its cognate maleimide was correctly identified based on a restriction site that had been engineered only in this particular sequence. Dideoxy sequencing [95] provided further proof that indeed the correct sequence was amplified [50].

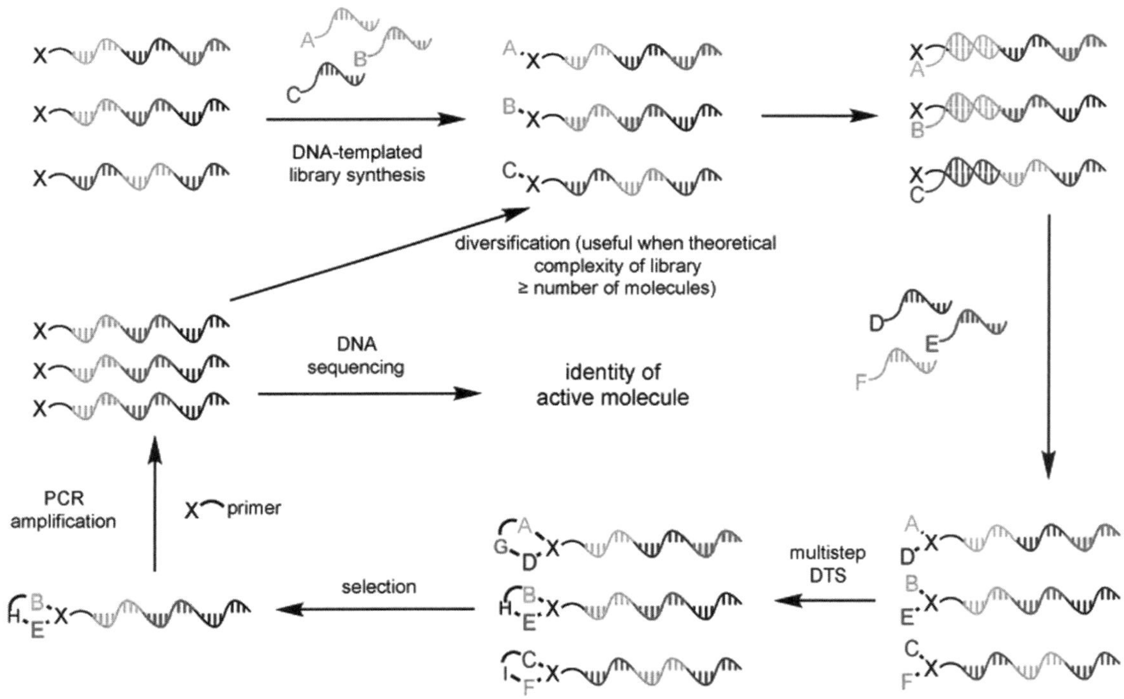

Figure 4.7 In vitro *evolution scheme for small molecules. Reprinted from X. Li and D. R. Liu, DNA-templated organic synthesis: Nature's strategy for controlling chemical reactivity applied to synthetic molecules.* Angew. Chem. Int. Ed., ***43****, 4848–4870 (2004), with permission from Wiley-VCH Verlag GmbH & Co. KGaA. See color plate section*

In a landmark experiment, Liu and co-workers then subjected an encoded small-molecule library, synthesized by DTS, for the first time to a selection experiment, aimed at identifying carboanhydrase binders [96]. Sixty-four macrocyclic fumaramides were prepared in a DTS involving four encoded chemical steps: three amine acylations followed by a Wittig olefination afforded the DNA-tagged fumaramides in 1–5% overall yield. After each step, the library was affinity purified and the major products expected from the DNA-programmed reactions were confirmed by MS. As a novel binder could not be expected in this limited library, a phenylsulfonamide-containing building block, known to bind with high affinity ($K_D \approx 1$ nM) [97], was encoded in a 65th DNA template that contained a unique *Nla*III restriction site. After two iterative cycles of selection, involving as little as 100 fmol of the library (without retranslation), the phenylsulfonamide-encoding template dominated the pool, as evident from *Nla*III restriction analysis (Figure 4.8). This work demonstrated that DNA templating permits the translation, selection and amplification of encoded small-molecule libraries [96], providing the first experimental proof that laboratory evolution is not reserved to biological macromolecules, but can also be performed on small, drug-like compounds.

4.5.4 Overcoming the limitations

Some technical issues such as distance dependence and structural bias, which are unwanted during library synthesis, have been approached with the design of new architectures [93] and the introduction of flexible linkers [73], respectively. The modest overall yields are much less severe than in a traditional synthesis, as very small amounts of material are sufficient for selection experiments. In addition, PCR can easily amplify as few as 10^4 copies of a DNA [87]. Potentially, protein binding could be accomplished with the DNA portion instead of the small molecule. DNA-dependent binding of a library member has not been detected in several

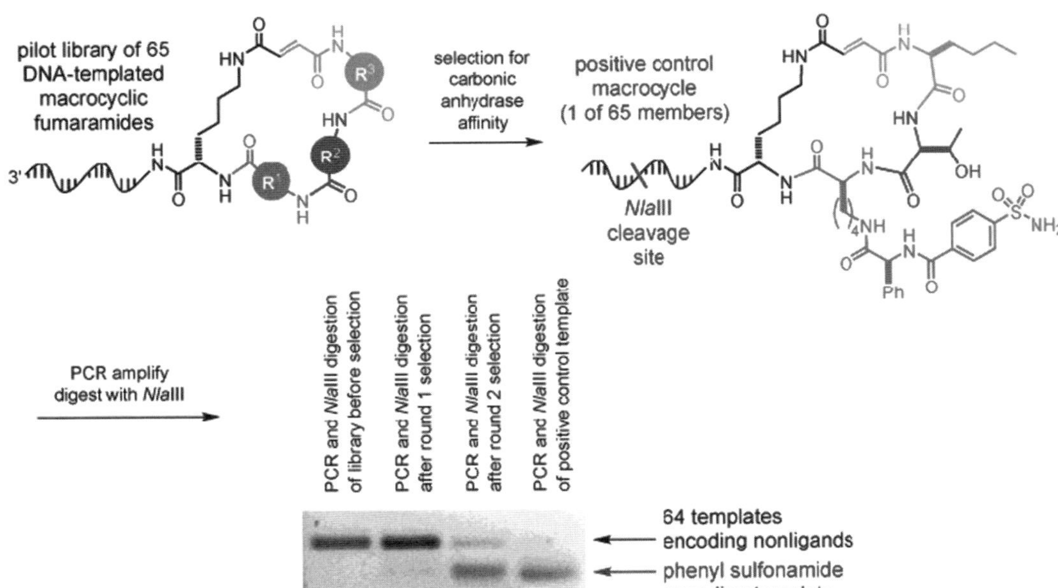

Figure 4.8 *In* vitro *selection of a carbonic anhydrase ligand from a spiked 65-membered library of DNA-templated macrocyclic fumaramides. Reprinted from X. Li and D. R. Liu, DNA-templated organic synthesis: Nature's strategy for controlling chemical reactivity applied to synthetic molecules.* Angew. Chem. Int. Ed., ***43****, 4848–4870 (2004), with permission from Wiley-VCH Verlag GmbH & Co. KGaA. See color plate section*

selections [96,98–100] but requires continued attention, as single-stranded DNA molecules, capable of specific binding to many molecular targets, including proteins, are well known to emerge from *in vitro* selections [101]. Possible countermeasures include converting the coding DNA to a rigid double strand before the selection step.

A more severe concern is the limited structural information on DTS products. Inevitably, all possible isomers of a hit structure have to be resynthesized and tested individually, but the related effort is still moderate compared with a fully traditional library screen.

Probably the most severe limitation of DTS in terms of chemistry is the requirement for reaction conditions that allow DNA hybridization, effectively excluding a significant part of the organic chemical repertoire [60]. Despite the surprising scope of DNA-templated chemistry in water and the respectably diverse structures synthesized so far, the chemical repertoire would need to be expanded significantly to build molecular structures that match the diversity of natural products [90,102]. Indeed, the largest library synthesized in a DNA hybridization-driven format contained only 65 members, orders of magnitude smaller than traditional combinatorial libraries (up to 10^6) [90] or libraries of random-sequence NAs (10^{15}) [55].

4.5.5 Improvements and latest developments

The latest developments attempted to enlarge the chemical repertoire for the synthesis of encoded libraries. Liu and co-workers separated hybridization from the bond formation step by transforming azide-containing protecting groups to a desired functional group [103] via a DNA-templated Staudinger reaction (Table 4.2) [104]. Sequence-specific functional group transformation was followed by a chemical reaction in DMF involving non-DNA-linked, water-incompatible reagents such as ethyl chloroformate. In a sequence of four sequence-specific 'deprotections', followed by acylation of the liberated amine, carbamate, urea and thiourea products were synthesized in 51% yield in a single solution [103]. Harbury and co-workers were the first to develop this concept into a system for the synthesis and selection of DNA-encoded small molecule libraries. Referring to phage and related display technologies [105], where a combinatorial library of peptides are displayed on the surface of a phage particle, Harbury and colleagues termed their approach 'DNA display', since a library of small molecules was 'displayed' on the 5′ end of a long DNA [98,106,107]. In DNA display, the role of the oligonucleotide portion is not to promote the formation of chemical bonds, but to route the DNA-tagged starting materials through a set of anticodon columns, each recognizing a unique codon stretch on the DNA. DNA hybridization is not relevant during bond formation, making a vast chemical repertoire available for the synthesis of encoded libraries [13]. DNA displayed libraries have been constructed using peptide coupling reactions, nucleophilic displacement reactions involving amines and thiols, reductive aminations, electrophilic aromatic substitutions, periodate oxidations and phosphine-based reductions [13,107,108] without detectable sign of DNA or side-chain modification, as evident from mass spectra, Edman degradation and binding characteristics to the 3-E7 antibody [106].

DNA-encoded combinatorial libraries have significant potential for molecular discovery, as they apply Nature's approach to artificial small molecules. Linking DNA molecules with unique sequences to each library member is a key aspect of encoded library synthesis, allowing one to amplify and identify minute amounts of selected members.

4.6 DNA-templated reaction discovery

In contrast to the highly organized search for new functional molecules, the discovery of previously unknown reactions is frequently left to serendipity, as exploring all possible reactive interactions under a myriad of reaction conditions is a challenging task. Liu and colleagues reported a DNA-templated system for reaction discovery [85] that exploits biological principles to advance one essential endeavor in chemistry, the discovery of new reactions.

The authors used DNA templating to set up a multi-dimensional reaction array system that allows the simultaneous evaluation of hundreds (potentially even thousands) of synthetically interesting bond-forming reactions in a single solution. Each of the DNA-linked reactants (e.g. A_1, Figure 4.9b) contained a unique coding region for its identification and second region for the annealing of a series of reaction partners B_1–B_m. A pool of reactants A_1–A_n was constructed and incubated with a pool of reaction partners B_1–B_m, to bring molecules in reactive proximity, comparable to $n \times m$ preparative reactions at high concentration. Conjugates B_1–B_m carried a biotin group for affinity purification extending from the oligonucleotide via a cleavable disulfide linker. Only bond-forming combinations lead to transfer of the biotin group to the corresponding reactant–oligonucleotide A_x, making it eligible by affinity purification after cleavage of the disulfide bond. PCR amplification of the coding DNA following capture on streptavidin-coated beads afforded DNAs encoding productive combinations A_x–B_y, which were analyzed on microarrays after being labeled with a green chromophore (Figure 4.9a) [85].

Two pools A and B were synthesized, each containing 12 substrates with simple functional groups that are commonly encountered in organic molecules. In addition, all 24 substrates were added with a sequence complementary to the original A and B conjugates to allow the detection of homocoupling reactions, bringing the total possible combinations to 168. Despite the significant effort to synthesize the pool, it provided sufficient material for 1000 different analyses, potentially evaluating 168 000 reaction conditions [85]. After some well-known reactive combinations had been correctly identified as bright green spots on the microarray, the library was tested for reactivity in the presence of Pd(II) salts, revealing several green spots that indicated bond formation between an aryl iodide and an enone, a terminal alkyne with each of an enone, a terminal alkene and an alkyne. While the first combination corresponds to the well-known Heck reaction, Pd-mediated coupling between a double and triple bond was not yet known. This oxidative coupling reaction was validated under standard synthesis conditions in organic solvents, resulting in an efficient macrocyclization reaction that gave an enone in almost quantitative yield (Figure 4.9c) [85]. This experiment demonstrated that typical organic reactions can be discovered with DNA templating (in water).

Recently, Liu and co-workers expanded this discovery system to allow reaction conditions that do not support DNA hybridization [109]. The essential difference is that the reaction partners are not brought into proximity via DNA hybridization, but a covalent enzyme-synthesized DNA bridge between oligonucleotides linked to (and coding for) A_1–A_n and corresponding B_1–B_m reaction partner conjugates, which contain a biotin and a cleavable disulfide linkage [109].

NAs enable reaction discovery by increasing the throughput by orders of magnitude. In a traditional setting, reactions must be individually prepared and analyzed in isolated vessels, requiring researchers to focus on a certain reaction product or reaction type. Recent advances involve sophisticated instrumentation to investigate bond formation of one product with many reaction partners simultaneously in an LC–MS-based approach [110]. With the addition of an NA, in contrast, hundreds or thousands of reactants can be present simultaneously in the same solution, the reaction partners being organized by proximity effects or covalent linkage to the same oligonucleotide. Despite the significant effort to link all reactants to suited DNA sequences, minute amounts of material are sufficient to assess 1000 or more different reaction conditions on DNA arrays by one researcher in parallel experiments. With emerging improvements in sequencing technologies becoming available, even more reactions involving less material will be accessible in the future.

The powerful techniques for analysis of NAs have been instrumental for discovering new reactions and deepening our understanding of known reactions. Examples include chemical replication [17], triphosphate-activated ligation [79] and aldol chemistry (see above) [62]. The last technique was used to screen amine-containing small molecules for potential catalysts. All experiments could be performed in parallel with little material, identifying lysine as an efficient catalyst of pentose formation [62], which was confirmed

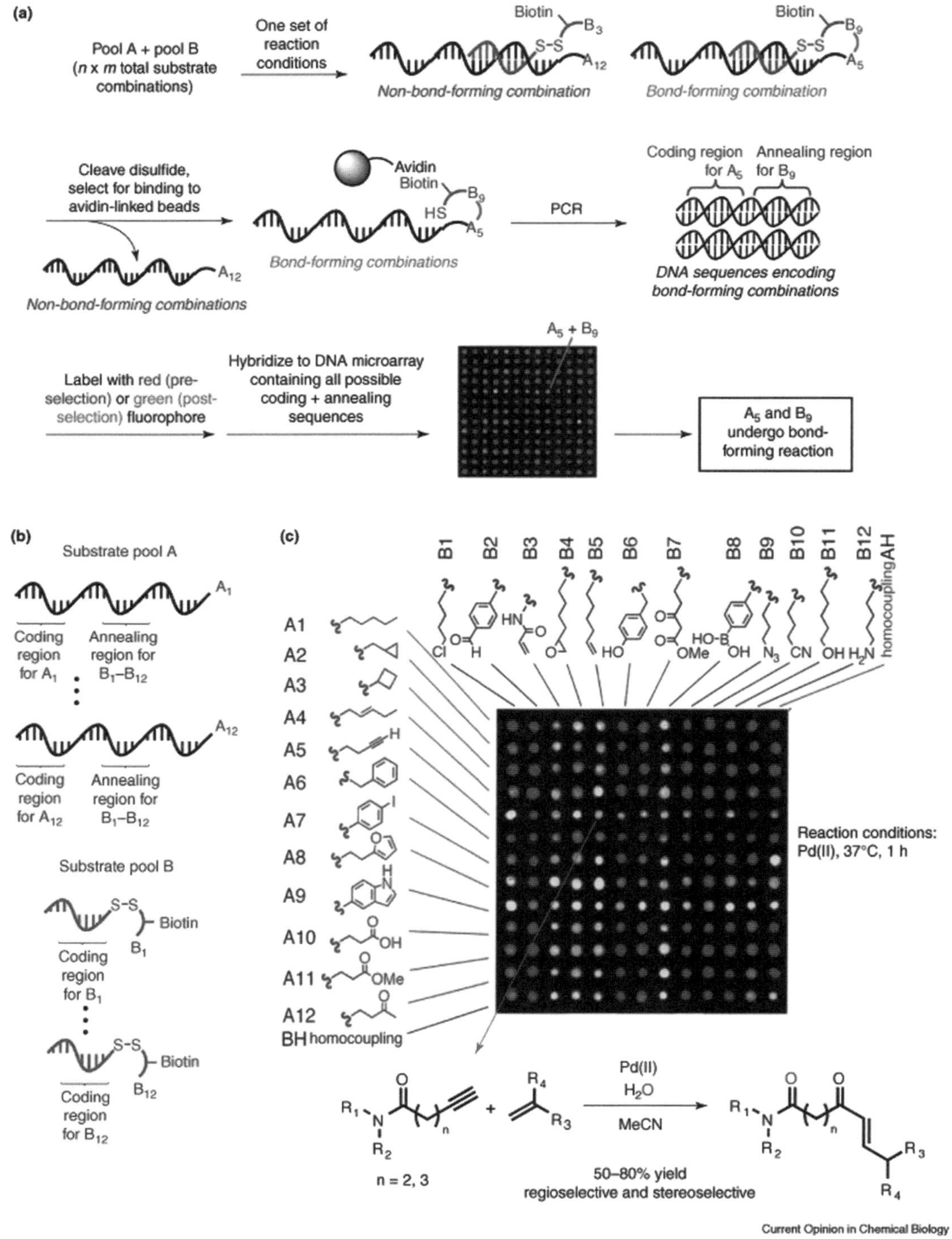

Figure 4.9 *DNA-templated reaction discovery. Reprinted from M. Rozenman, B. McNaughton and D. Liu, Solving chemical problems through the application of evolutionary principles,* Curr. Opin. Chem. Biol.*, **11**, 259–268 (2007), with permission from Elsevier © 2007. See color plate section*

independently with traditional methods [111]. Tang and Marx developed a related DNA-templated reaction system for the discovery of organocatalysts [112], small organic molecules with potential applications in industrial catalytic processes [113]. The authors used DNA-linked amines as potential organocatalysts for the aldol reaction of a juxtaposed benzaldehyde–DNA conjugate with various aldol donors in free solution. A DNA-linked diprolinamide was shown to catalyze aldol formation more than 100-fold with multiple turnover [112].

4.7 Nucleic acid sensing

Sensing and quantifying the presence a particular NA sequence has become an increasingly important endeavor in biology and medicine, with applications ranging from organism genotyping, NA sequencing and forensic medicine to the identification of microbial genes in human patients, food and environmental samples. Numerous challenges arising from these applications have fueled the development of a wide array of detection techniques, including molecular beacons [114], photophysical probes [115], quantitative PCR [116] and template-dependent enzymatic reactions [117,118], typically following isolation and PCR-amplification of NAs from body fluids or living organisms.

As these topics have been reviewed extensively (see the cited references), here the focus is on detection systems that rely on nonenzymatic NA-templated reactions which typically consist of the target NA analyte (accessible sites in cellular RNAs or PCR-amplified DNA), NA-linked reagents that carry signal molecules (frequently referred to as 'probes') and a suitable NA-templated reaction that translates the presence of a particular NA sequence into a detectable signal (Figure 4.5 and Figure 4.10; for an extensive review on this topic, see [40]). The sparsity and similarity of analytes, sometimes differing by one point mutation [for the detection of single nucleotide polymorphism (SNP)] [119], pose significant challenges to the design of the participating molecules. Methods that can detect such subtle differences in intact organisms are most desired, as they eliminate the laborious and cost-intensive isolation of the target NA analyte.

4.7.1 Nucleic acid-templated reactions for sensing nucleic acids

In principle, the whole range of NA-templated reactions can be used in a detection system, provided that the probes are stable under the assay conditions, for instance during thermal cycling in PCR applications or against cellular nucleases. Probes are typically 5 and 20 nucleotide-long DNAs or analogs such as PNA to increase serum stability [40]. Length and sequence have to be carefully designed to balance affinity with specificity at given physical conditions that include temperature, ionic strength and pH [119]. Six or seven base pairs are required for substantial hybridization at room temperature, while probes exceeding 15 nucleotides make it generally difficult to detect SNPs [119]. Finally, the significant size of the three gigabase

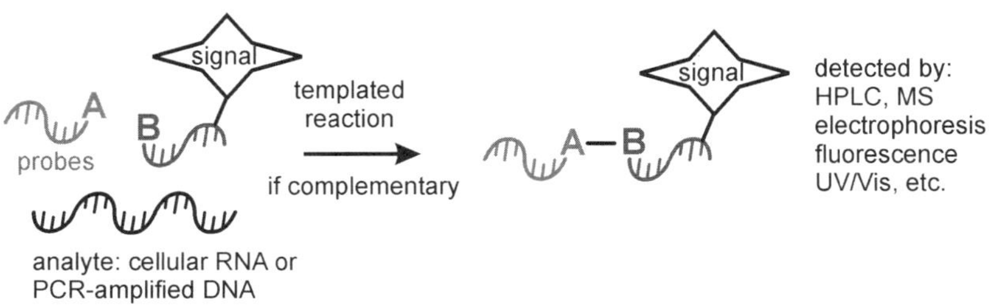

Figure 4.10 *Nucleic acid detection system*

human genome poses additional constraints, demanding at least 18 nucleotides to identify a sequence uniquely [40].

Ideally, the templated reaction is bioorthogonal (not cross-reactive with biological functional groups), quantitative or even catalytic (i.e. one template molecule can give rise to multiple products) and self-sufficient (reagents other than the oligonucleotide probes are not required) [40]. The quest for suitable NA detection chemistries has led to the emergence of numerous templated reactions, contributing significantly to today's knowledge of NA-templated chemistry (Tables 4.1 and Table 4.2). Orgel's replication systems and Shabarova *et al.*'s approach to synthesize a tRNA gene, for instance, provided the first chemical means for a nonenzymatic phosphodiester formation [17,36]. Von Kiedrowski's studies on prebiotic replication pioneered catalytic systems performing multiple turnover [27], sometimes supported by thermal cycling [34]. Gat and Lynn introduced the reversible and self-sufficient imine formation chemistry to study prebiotic replication and reported that product duplexes dissociated more easily when the ligated products were permanently linked by reduction with $NaCNBH_3$, promoting multiple turnover [11].

Letsinger and co-workers pioneered self-sufficient ligation chemistry between nucleophilic phosphothioates and oligonucleotides containing an electrophilic 5′-bromoacetyl group [41,42]. These so-called autoligation reactions proceeded rapidly without additional reagents in the presence of complementary templates; a single mismatch, in contrast, already reduced the rate 20-fold [42]. Xu and Kool replaced the bromoacetyl group with a more stable iodide leaving group, establishing a quantitative autoligation reaction with excellent selectivity [44]. After hybridization, these probes were shown to discriminate all possible mismatches in a single position on a 50-mer DNA template with selectivities between 10- and 1000-fold, rivaling established methods based on T4-DNA-ligase [45].

Mattes and Seitz established PNA probes designed to hybridize juxtaposed on a matching DNA template, promoting ligation via peptide bond formation in the presence of EDC with excellent mismatch selectivity [52,53]. Later, Seitz and co-workers introduced native chemical ligation, a reaction sequence originally developed to couple peptides [120], to eliminate the requirement for a condensing agent for their PNA-based detection system [121]. Native chemical ligation joins a peptide bond from a thioester and a juxtaposed cysteine, via a highly chemoselective transthioesterification reaction followed by an S to N acyl shift [120]. Seitz and co-workers successfully detected single nucleotide differences with >1000-fold selectivity in PCR-amplified double-stranded DNA templates with HPLC or MS [121]. Recently, Dose and Seitz improved the system to work with fluorescent signals and multiple turnover, suggesting that it could be applied in living cells [122].

The Staudinger reaction, one of the most prominent bioorthogonal chemistries [123], was first employed in a DNA-templated format by Taylor and co-workers to release a fluorescent signal in the presence of a complementary DNA. Reduction of a DNA–azide reagent by a juxtaposed DNA-linked phosphine-modified fluorescein ester was accelerated more than 30-fold in the presence of a fully matched versus mismatched DNA strand [104].

A series of photochemical ligation chemistries have been reported based on the well-known thymidine dimerization (see Table 4.1) [46]. Photochemical ligations bear many advantages of reagent-free systems provided that long-wavelength UV light can be used. Most notably, Fujimoto, Saito and co-workers presented a series of reversible DNA-templated photoligation reactions involving 5-vinylpyrimidines and related compounds capable of undergoing a [2 + 2] cycloaddition with juxtaposed nucleotides after irradiation with UV light at 366 nm [124–126]. In a recent study, Fujimoto and co-workers used 5-carboxyvinyl-2′-deoxyuridine-terminated oligonucleotides as photoreactive probes for a chip-based detection system with fluorescent readout. The probes were anchored on a solid support and reacted quantitatively with thymidine-terminated and fluorescently labeled oligonucleotide in the presence of complementary RNA templates when irradiated with UV light [126]. The chip was able to detect a series of single mismatches at different positions with 10–30-fold selectivity, demonstrating its potential as detection system for short cellular RNAs [126].

Other photochemical NA-templated reactions involving stilbene dicarboxamide [43], anthracene [127], 4-thiothymidine [47], or an azide–sensitizer pair [128,129] have also been shown to detect single mismatches in NA templates.

Finally, the chemical reaction requires to be coupled to some sort of signal for readout. There are numerous examples of DNA-templated detection systems, where bond formation gives rise to an optical signal, including generation or quenching fluorescence [40]. Most notably, Kool and co-workers quenched the fluorescence of a fluorophore-labeled probe with a dabsyl sulfonate leaving group. In the presence of a phosphothioate probe and a matching template, This so-called quenched autoligation (QUAL) successfully detected ribosomal RNA sequences inside intact cells [130] (see also [40] and references therein).

Even more sophisticated examples (reviewed in [131,132]) include electrical signals, which could be integrated into electronic devices or label-free detection systems based on MS or a quartz crystal microbalance. Mattes and Seitz, for instance, relied on MS to detect products from their DNA-templated PNA ligation label free [53].

4.8 Conclusions and outlook

Linking reactive groups to oligonucleotides adds powerful features to chemical reactions. Paralleling Nature's approach to the synthesis of small molecules, DNA templating enables a chemical synthesis to be programmed through non-covalent interactions. In recent decades, an impressive range of chemical reactions have been promoted by NA templating, giving rise to product structures both related and unrelated to the backbone of natural NAs. Many of these chemistries have been developed and exploited to study chemical replication or detect cellular NAs with exquisite sequence specificity. While the former are of paramount academic interest regarding the quest for the origin of life, the latter have broad practical implications, for instance, as analytical devices in diagnostic applications. Even though significant advances have been achieved in probe development and signal generation, sensitivity remains a significant challenge [40]. Many targets are not abundant in living cells and membranes are still a significant obstacle for the cellular delivery of NA-based probes. Probes undergoing multiple turnover combined with improved delivery and signal generation systems will be a key aspect of the development of future NA-templated diagnostic devices for real-world applications.

NA-templated synthesis has matured considerably over the past 10 years. Encompassing carbon–heteroatom and carbon–carbon bond formation reactions, ionic and pericyclic reactions and organometallic and photochemical examples, NA-templated chemistry has gained sufficient scope to allow the synthesis of complex small organic molecules. DTS is certainly not aimed at rivaling traditional synthesis in its strongholds. It can, however, complement existing chemical methods with valuable NA features, where traditional approaches are inefficient or fail completely: NAs permit highly substrate-specific chemistry with reaction partners being programmed by sequence complementarity, are amenable to sensitive analytics requiring only minute amounts of material and add amplifiable molecular sequence tags to chemical reaction products. With these features, NA templating has the potential of bringing biological solutions to chemical problems. Indeed, one of the greatest promises of NA-templated synthesis is to bring evolutionary principles to the chemical discovery process via the selection and evolution of encoded small-molecule libraries and DNA-assisted discovery of new chemical reactions [12–14,133]. Results from these efforts, in turn, are expected to fuel advances in biology, for instance, by supporting functional genomics initiatives with new inhibitors to dissect the function of the countless cellular proteins [134].

The future will show if encoded libraries can meet the high expectations. Undoubtedly, increasingly complex small-molecule libraries will be constructed using encoded techniques to obtain higher affinity binders, rivaling established display technologies [133]. Even though the latter are increasingly optimized to include unnatural structures [13], structurally diverse small-molecule libraries are still reserved for chemical synthesis. Future DTS efforts for the development of drug-like molecules [135] should aspire to the generation of diverse natural product-like structures, known for their excellent inhibitory potential [90,102,134]. However, the challenges related to DNA-encoded libraries increase with increasing library size, structural diversity and requirements for diverse chemical conditions: reaction efficiencies, yields, purification and structural characterization protocols will become increasingly important to ensure that libraries contain the desired chemical entities. DNA sequences will require computational optimization to accommodate the combined demands from sequence-specific hybridization, specialized reaction architectures and efficient parallel decoding after selection. Despite the significant achievements in recent years [13,14,133], a structurally diverse library of drug-like molecules that is sufficiently complex to obtain previously unknown binders in a selection experiment has not yet been reported. This might require the integration of complementary approaches such as DTS, DNA display and others, and also the replacement of the selection of individuals present in an initial library with true Darwinian evolution, which is characterized by the formation of new progeny species[55]; in other words, require chemical retranslation and mutation of selected 'genotypes' to prepare focused daughter libraries with enhanced fitness between rounds.

4.9 Abbreviations

DMF	dimethylformamide
DTS	DNA-templated synthesis
EDC	1-ethyl–3-(3′-dimethylaminopropyl)carbodiimide hydrochloride
ESAC	encoded self-assembling chemical library
HPLC	high performance liquid chromatography
MS	mass spectrometry
NA	nucleic acid
NHS	*N*-hydroxysuccinimide
PNA	peptide nucleic acid
THF	tetrahydrofuran

Acknowledgments

The author thanks the Austrian National Science Fund, the Max Kade Foundation, the Tyrolean Science Fund and the South Tyrolean Government for financial support, Evelyn Profanter-Oberhuber for critical reading of the original manuscript and G. F. Joyce, B. Kräutler and R. Micura for many helpful discussions.

References

1. S. Anderson and H. Anderson, Templates in organic synthesis: definitions and roles, *in* Templated Organic Synthesis, ed. P. Stang and F. Diederich, Wiley-VCH Verlag GmbH, Weinheim, 1999, pp. 1–38.
2. J. D. Watson and F. H. Crick, Molecular structure of nucleic acids; a structure for deoxyribose nucleic acid, *Nature*, **171**, 737–738 (1953).
3. A. Kornberg and T. Baker, *DNA Replication*, W. H. Freeman, New York, 1992.
4. R. Kornberg, The molecular basis of eukaryotic transcription (Nobel Lecture), *Angew. Chem. Int. Ed.*, **46**, 6956–6965 (2007).

5. M. M. Yusupov, G. Z. Yusupova, A. Baucom *et al.*, Crystal structure of the ribosome at 5.5 Å resolution, *Science*, **292**, 883–896 (2001).
6. S. McKnight and K. Yamamoto, *Transciptional Regulation*, Cold Spring Harbor Laboratory Press, Cold Spring Harbor, NY, 1992.
7. C. C. Mello and D. Conte, Revealing the world of RNA interference, *Nature*, **431**, 338–342 (2004).
8. M. Mandal, B. Boese, J. E. Barrick *et al.*, Riboswitches control fundamental biochemical pathways in *Bacillus subtilis* and other bacteria, *Cell*, **113**, 577–586 (2003).
9. U. Feldkamp and C. Niemeyer, Rational design of DNA nanoarchitectures, *Angew. Chem. Int. Ed.*, **45**, 1856–1876 (2006).
10. N. Seeman, DNA in a material world, *Nature*, **421**, 427–431 (2003).
11. Y. Gat and D. Lynn, Template-directed ligation: towards the synthesis of sequence specific polymers, in *Templated Organic Synthesis*, ed. P. Stang and F. Diederich, Wiley-VCH Verlag GmbH, Weinheim, 1999, pp. 133–158.
12. X. Li and D. R. Liu, DNA-templated organic synthesis: Nature's strategy for controlling chemical reactivity applied to synthetic molecules, *Angew. Chem. Int. Ed.*, **43**, 4848–4870 (2004).
13. S. J. Wrenn and P. B. Harbury, Chemical evolution as a tool for molecular discovery, *Annu. Rev. Biochem.*, **76**, 331–349 (2007).
14. M. Rozenman, B. McNaughton and D. Liu, Solving chemical problems through the application of evolutionary principles, *Curr. Opin. Chem. Biol.*, **11**, 259–268 (2007).
15. G. Quinkert, H. Wallmeier, N. Windhab *et al.*, Chemistry and biology – historical and philosophical aspects, in *Chemical Biology*, ed. S. Schreiber, T. Kapoor and G. Wess, Wiley-VCH Verlag GmbH, Weinheim, 2007, pp. 3–55.
16. L. E. Orgel and R. Lohrmann, Prebiotic chemistry and nucleic acid replication, *Acc. Chem. Res.*, **7**, 368–377 (1974).
17. L. E. Orgel, Unnatural selection in chemical systems, *Acc. Chem. Res.*, **28**, 109–118 (1995).
18. R. Naylor and P. T. Gilham, Studies on some interactions and reactions of oligonucleotides in aqueous solution, *Biochemistry*, **5**, 2722–2728 (1966).
19. J. Sulston, R. Lohrmann, L. E. Orgel *et al.*, Nonenzymatic synthesis of oligoadenylates on a polyuridylic acid template, *Proc. Natl. Acad. Sci. USA*, **59**, 726–733 (1968).
20. J. Sulston, R. Lohrmann, L. E. Orgel *et al.*, Specificity of oligonucleotide synthesis directed by polyuridylic acid, *Proc. Natl. Acad. Sci. USA*, **60**, 409–415 (1968).
21. B. J. Weimann, R. Lohrmann, L. E. Orgel *et al.*, Template-directed synthesis with adenosine–5′-phosphorimidazolide, *Science*, **161**, 387 (1968).
22. T. Inoue and L. E. Orgel, Oligomerization of (guanosine 5′-phosphor)-2-methylimidazolide on poly(C). An RNA polymerase model, *J. Mol. Biol.*, **162**, 201–217 (1982).
23. T. Inoue and L. E. Orgel, A nonenzymatic RNA polymerase model, *Science*, **219**, 859–862 (1983).
24. O. L. Acevedo and L. E. Orgel, Non-enzymatic transcription of an oligodeoxynucleotide 14 residues long, *J. Mol. Biol.*, **197**, 187–193 (1987).
25. M. H. Caruthers, G. Beaton, J. V. Wu *et al.*, Chemical synthesis of deoxyoligonucleotides and deoxyoligonucleotide analogs, *Methods Enzymol.*, **211**, 3–20 (1992).
26. C. B. Chen, T. Inoue and L. E. Orgel, Template-directed synthesis on oligodeoxycytidylate and polydeoxycytidylate templates, *J. Mol. Biol.*, **181**, 271–279 (1985).
27. G. von Kiedrowski, A self-replicating hexadeoxynucleotide, *Angew. Chem. Int. Ed. Engl.*, **25**, 932–935 (1986).
28. I. A. Kozlov, B. De Bouvere, A. Van Aerschot *et al.*, Efficient transfer of information from hexitol nucleic acids to RNA during nonenzymatic oligomerization, *J. Am. Chem. Soc.*, **121**, 5856–5859 (1999).
29. S. Pitsch, R. Krishnamurthy, M. Bolli *et al.*, Pyranosyl-RNA ('p-RNA'): base-pairing selectivity and potential to replicate, *Helv. Chim. Acta*, **78**, 1621–1635 (1995).
30. B. Heuberger and C. Switzer, Nonenzymatic oligomerization of RNA by TNA templates, *Org. Lett.*, **8**, 5809–5811 (2006).
31. K. U. Schoning, P. Scholz, S. Guntha *et al.*, Chemical etiology of nucleic acid structure: the alpha-threofuranosyl-(3′,2′) oligonucleotide system, *Science*, **290**, 1347–1351 (2000).
32. C. Böhler, P. E. Nielsen and L. E. Orgel, Template switching between PNA and RNA oligonucleotides, *Nature*, **376**, 578–581 (1995).

33. R. K. Bruick, P. E. Dawson, S. B. Kent *et al.*, Template-directed ligation of peptides to oligonucleotides, *Chem. Biol.*, **3**, 49–56 (1996).
34. A. Luther, R. Brandsch and G. von Kiedrowski, Surface-promoted replication and exponential amplification of DNA analogs, *Nature*, **396**, 245–248 (1998).
35. Z. Zhan and D. Lynn, Chemical amplification through template-directed synthesis, *J. Am. Chem. Soc*, **119**, 12420–12421 (1997).
36. Z. A. Shabarova, I. N. Merenkova, T. S. Oretskaya *et al.*, Chemical ligation of DNA: the first non-enzymatic assembly of a biologically active gene, *Nucleic Acids Res.*, **19**, 4247–4251 (1991).
37. K. Luebke and P. Dervan, Nonenzymatic sequence-specific ligation of double-helical DNA, *J. Am. Chem. Soc.*, **113**, 7447–7448 (1991).
38. E. Rubin, S. Rumney IV, S. Wang *et al.*, Convergent DNA synthesis: a non-enzymatic dimerization approach to circular oligodeoxynucleotides, *Nucleic Acids Res.*, **23**, 3547–3553 (1995).
39. J. Ye, Y. Gat and D. Lynn, Catalyst for DNA ligation: towards a two-stage replication cycle, *Angew. Chem. Int. Ed.*, **39**, 3641–3643 (2000).
40. A. Silverman and E. Kool, Detecting RNA and DNA with templated chemical reactions, *Chem. Rev.*, **106**, 3775–3789 (2006).
41. S. Gryaznov and R. Letsinger, Chemical ligation of oligonucleotides in the presence and absence of a template, *Angew. Chem. Int. Ed. Engl.*, **115**, 3808–3809 (1993).
42. S. Gryaznov, R. Schultz, S. Chaturvedi *et al.*, Enhancement of selectivity in recognition of nucleic acids via chemical autoligation, *Nucleic Acids Res.*, **22**, 2366–2369 (1994).
43. R. Letsinger, T. Wu and R. Elghanian, Chemical and photochemical ligation of oligonucleotide blocks, *Nucleosides Nucleotides*, **16**, 643–652 (1997).
44. Y. Z. Xu and E. T. Kool, A novel 5′-iodonucleoside allows efficient nonenzymatic ligation of single-stranded and duplex DNAs, *Tetrahedron Lett.*, **38**, 5595–5598 (1997).
45. Y. Xu, N. B. Karalkar and E. T. Kool, Nonenzymatic autoligation in direct three-color detection of RNA and DNA point mutations, *Nat. Biotechnol.*, **19**, 148–152 (2001).
46. R. Lewis and P. Hanawalt, Ligation of oligonucleotides by pyrimidine dimers: a missing 'link' in the origin of life?, *Nature*, **298**, 393–396 (1982).
47. J. Liu and J. Taylor, Template-directed photoligation of oligodeoxyribonucleotides via 4-thiothymidine, *Nucleic Acids Res.*, **26**, 3300–3304 (1998).
48. R. Letsinger and T. Wu, Control of excimer emission and photochemistry of stilbene units by oligonucleotide hybridization, *J. Am. Chem. Soc.*, **116**, 811–812 (1994).
49. J. Woo and P. Hopkins, Template directed modification of single stranded DNA by psoralen-tethered oligonucleotides: sites of photoadduct formation analyzed by sequence-specific and sequence-random cleavage, *J. Am. Chem. Soc*, **113**, 5457–5459 (1991).
50. Z. J. Gartner and D. R. Liu, The generality of DNA-templated synthesis as a basis for evolving non-natural small molecules, *J. Am. Chem. Soc.*, **123**, 6961–6963 (2001).
51. Z. J. Gartner, M. W. Kanan and D. R. Liu, Expanding the reaction scope of DNA-templated synthesis, *Angew. Chem. Int. Ed.*, **41**, 1796–1800 (2002).
52. A. Mattes and O. Seitz, Sequence fidelity of a template-directed PNA-ligation reaction, *Chem. Commun.*, 2050–2051 (2001).
53. A. Mattes and O. Seitz, Mass-spectrometric monitoring of a PNA-based ligation reaction for the multiplex detection of DNA single-nucleotide polymorphisms, *Angew. Chem. Int. Ed.*, **40**, 3178–3181 (2001).
54. J. L. Czlapinski and T. L. Sheppard, Nucleic acid template-directed assembly of metallosalen–DNA conjugates, *J. Am. Chem. Soc.*, **123**, 8618–8619 (2001).
55. G. F. Joyce, Directed evolution of nucleic acid enzymes, *Annu. Rev. Biochem.*, **73**, 791–836 (2004).
56. R. K. O. Sigel and A. M. Pyle, Alternative roles for metal ions in enzyme catalysis and the implications for ribozyme chemistry, *Chem. Rev.*, **107**, 97–113 (2007).
57. K. Gothelf, A. Thomsen, M. Nielsen *et al.*, Modular DNA-programmed assembly of linear and branched conjugated nanostructures, *J. Am. Chem. Soc.*, **126**, 1044–1046 (2004).
58. J. Brunner, A. Mokhir and R. Krämer, DNA-templated metal catalysis, *J. Am. Chem. Soc.*, **125**, 12410–12411 (2003).

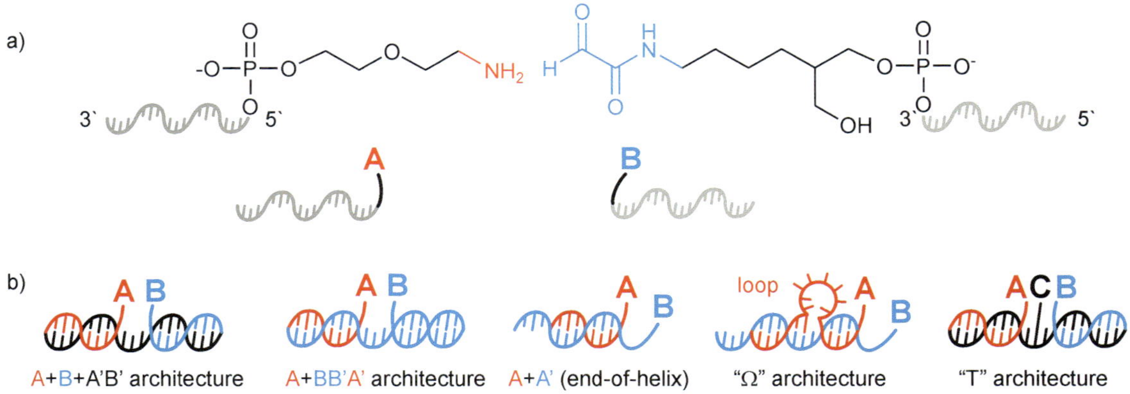

Plate 1 *(a) Reactant examples for DNA-templated synthesis (DTS), consisting of an oligonucleotide (gray), linker (black) and reactive group (colored); (b) DTS architectures, A/A′ indicate complementary sequences, molecules are drawn in different colors and separated by '+' characters. See Figure 4.2*

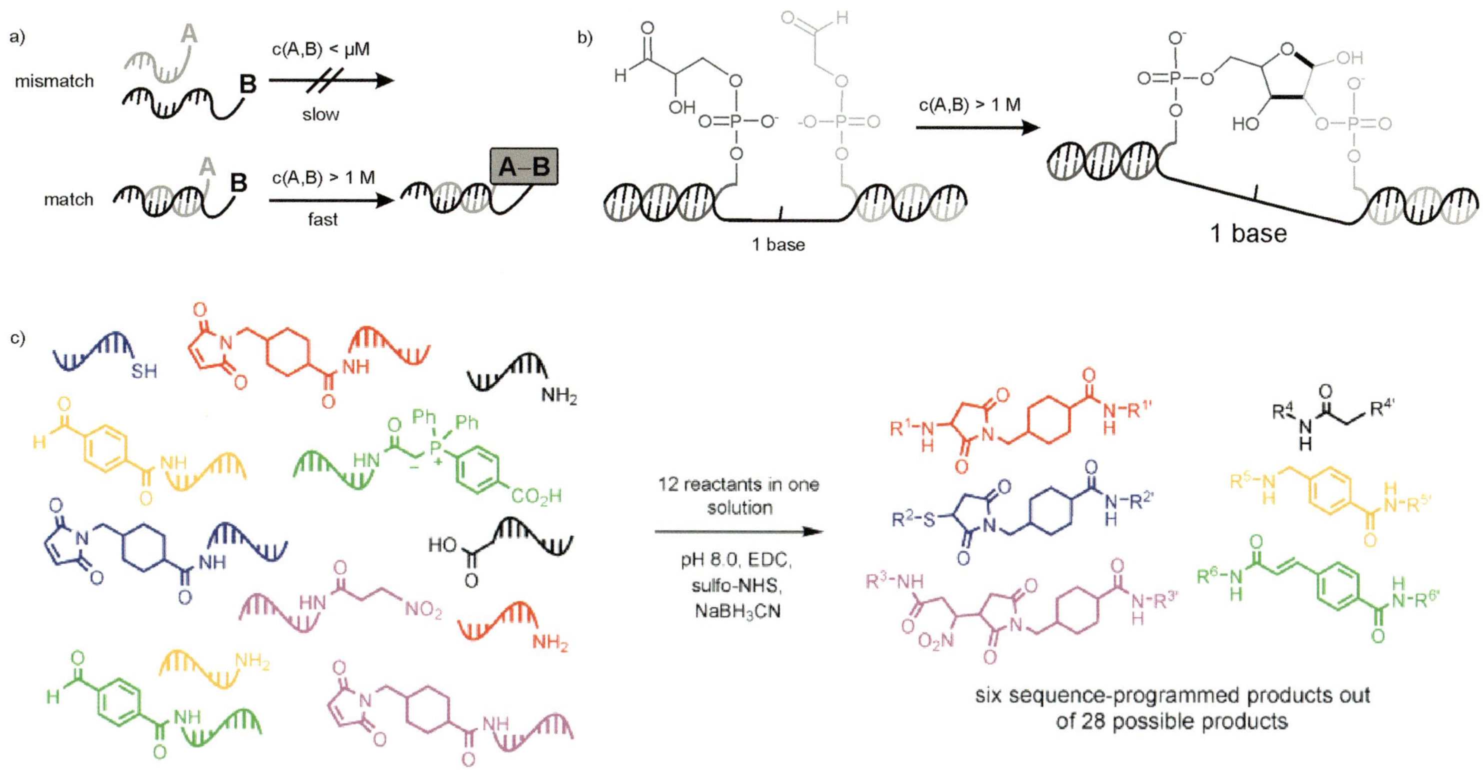

Plate 2 *Chemical reactivity controlled by Watson–Crick base pairing: (a) Origin of sequence specificity; (b) DNA-programmed cross-aldol reaction gives only rise to pentoses; (c) DNA-programmed reactions lead to the formation of six out of 28 possible products; matching colors indicate conjugates with complementary oligonucleotides [EDC = 1-ethyl-3-(3′-dimethylaminopropyl)carbodiimide hydrochloride; NHS =* N*-hydroxysuccinimide]. Part (c) reprinted from X. Li and D. R. Liu, DNA-templated organic synthesis: Nature's strategy for controlling chemical reactivity applied to synthetic molecules.* Angew. Chem. Int. Ed., ***43****, 4848–4870 (2004), with permission from Wiley-VCH Verlag GmbH & Co. KGaA. See Figure 4.3*

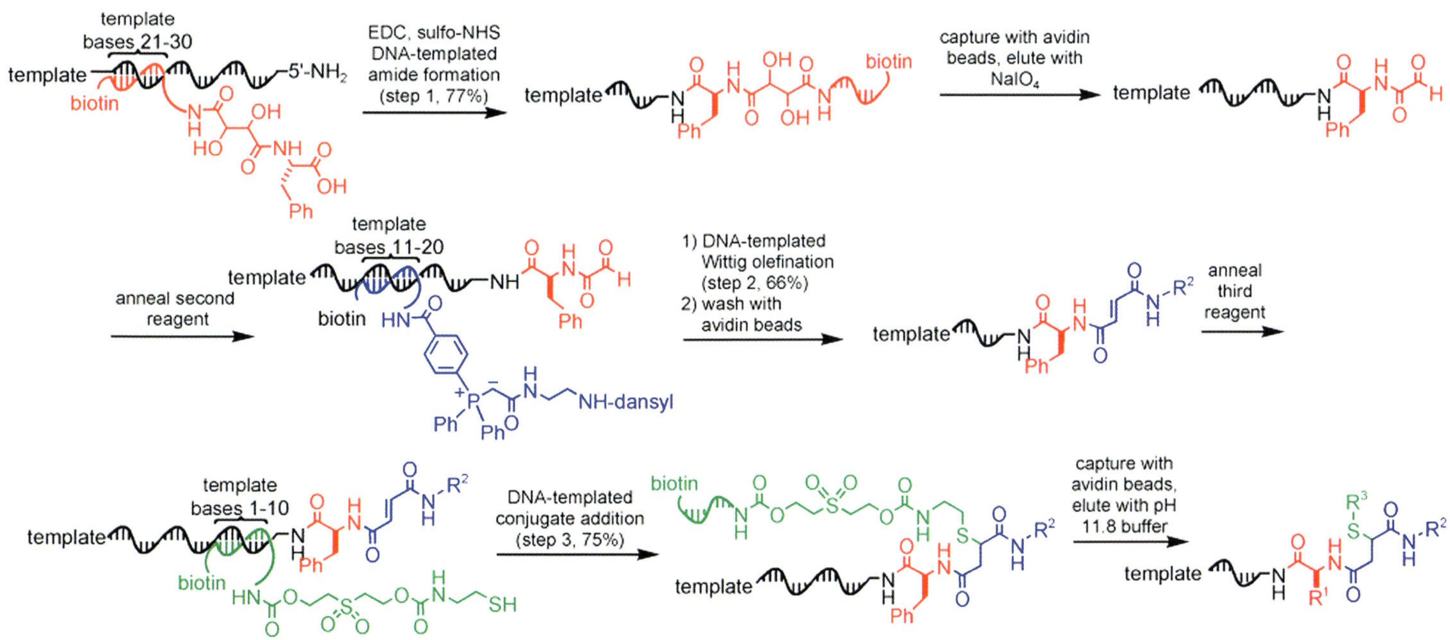

Plate 3 *DNA-templated multi-step synthesis; amine acylation (red), followed by Wittig olefination (blue) and conjugate addition (green) with biotin-based purification scheme. Template product conjugates are liberated during linker cleavage via periodate oxidation (red), during Wittig reaction (blue), and via β-elimination (green) (NHS = N-hydroxysuccinimide). Reprinted from X. Li and D. R. Liu, DNA-templated organic synthesis: Nature's strategy for controlling chemical reactivity applied to synthetic molecules.* Angew. Chem. Int. Ed., ***43***, *4848–4870 (2004), with permission from Wiley-VCH Verlag GmbH & Co. KGaA. See Figure 4.6*

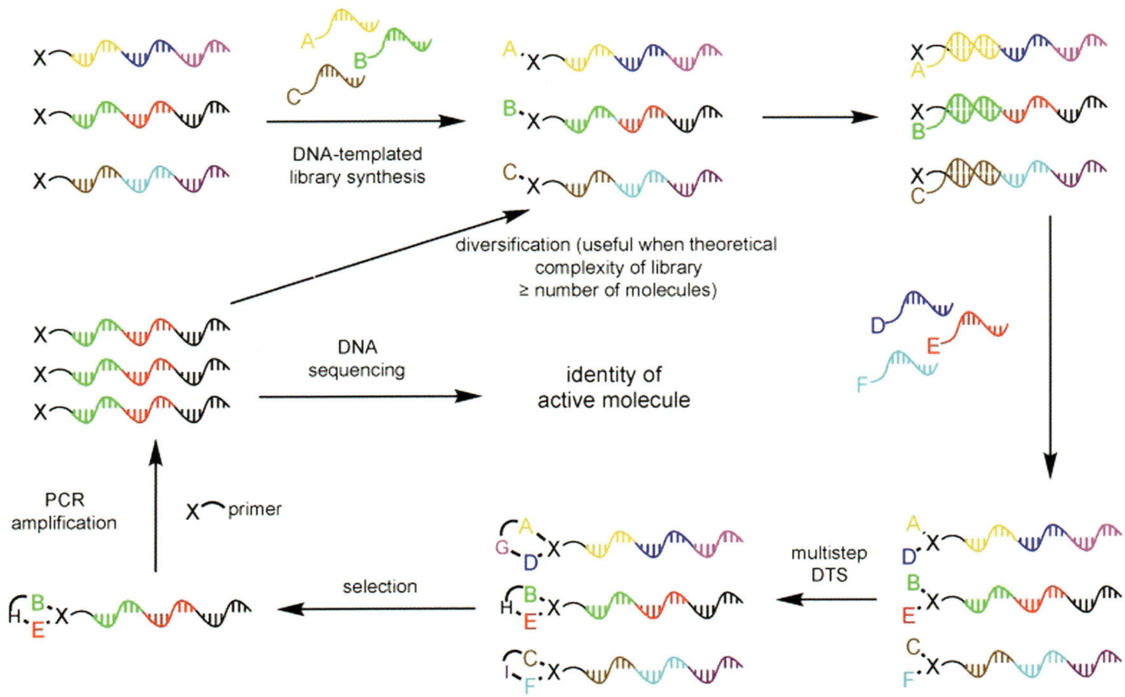

Plate 4 In vitro *evolution scheme for small molecules. Reprinted from X. Li and D. R. Liu, DNA-templated organic synthesis: Nature's strategy for controlling chemical reactivity applied to synthetic molecules.* Angew. Chem. Int. Ed.*,* ***43****, 4848–4870 (2004), with permission from Wiley-VCH Verlag GmbH & Co. KGaA. See Figure 4.7*

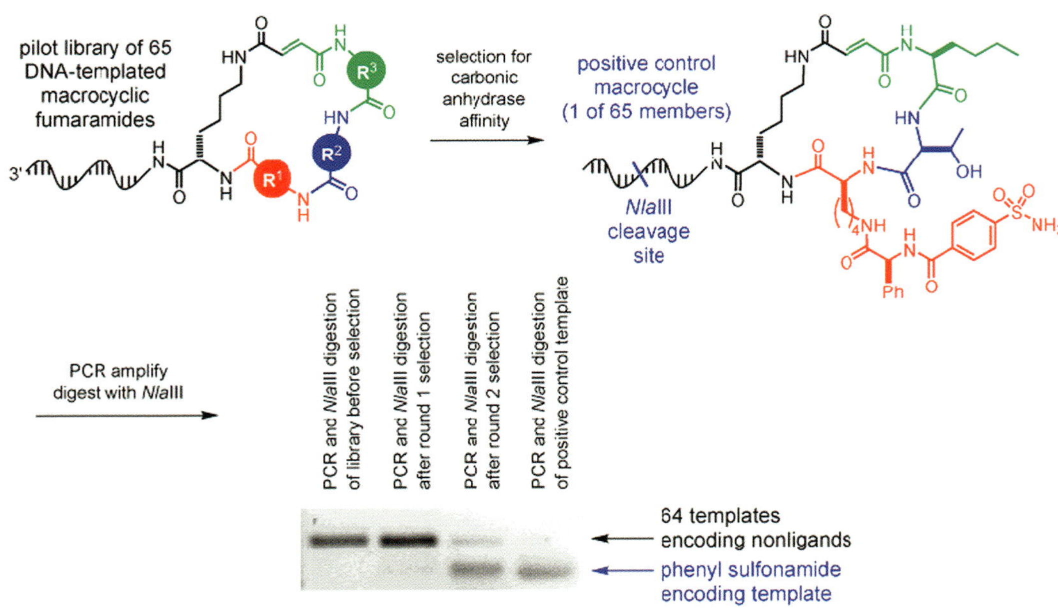

Plate 5 In vitro *selection of a carbonic anhydrase ligand from a spiked 65-membered library of DNA-templated macrocyclic fumaramides. Reprinted from X. Li and D. R. Liu, DNA-templated organic synthesis: Nature's strategy for controlling chemical reactivity applied to synthetic molecules.* Angew. Chem. Int. Ed.*,* ***43****, 4848–4870 (2004), with permission from Wiley-VCH Verlag GmbH & Co. KGaA. See Figure 4.8*

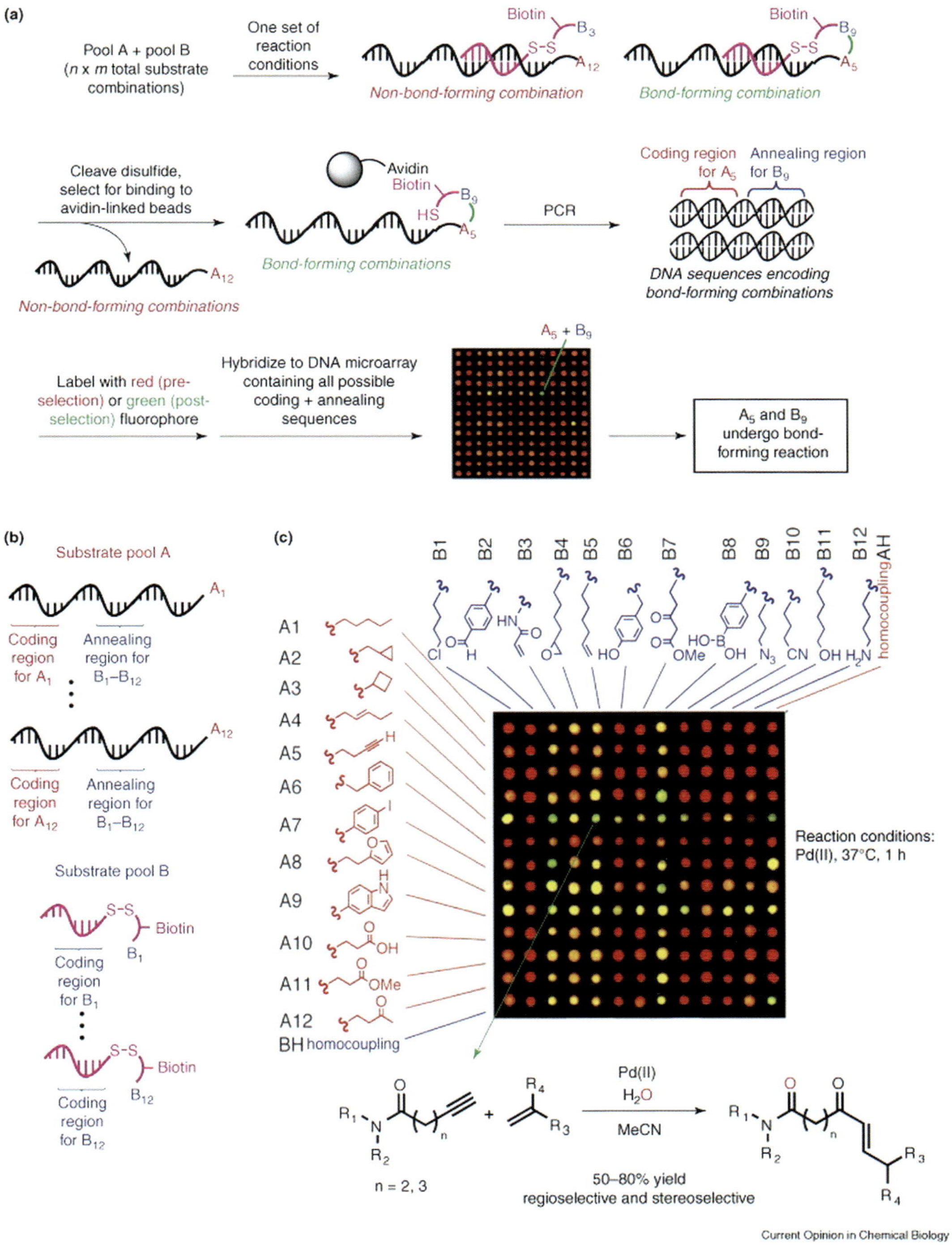

Plate 6 *DNA-templated reaction discovery. Reprinted from M. Rozenman, B. McNaughton and D. Liu, Solving chemical problems through the application of evolutionary principles,* Curr. Opin. Chem. Biol., ***11****, 259–268 (2007), with permission from Elsevier © 2007. See Figure 4.9*

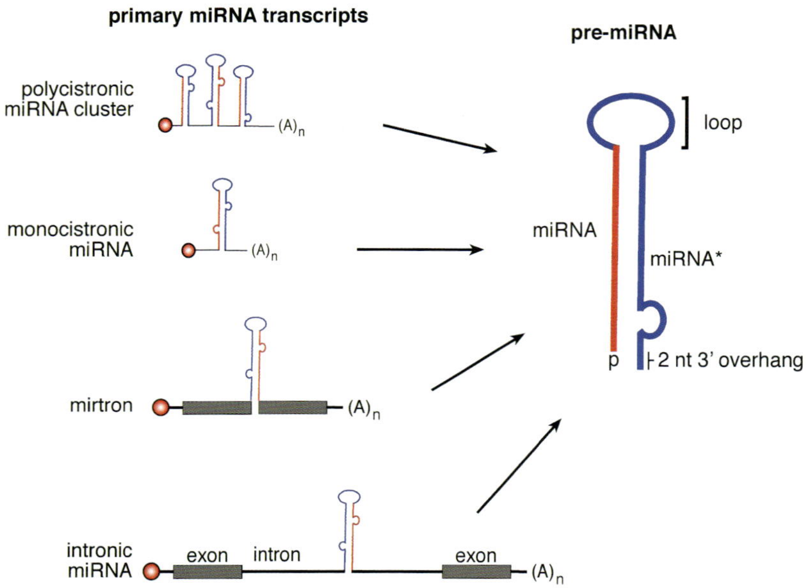

Plate 7 *Architecture of primary miRNA transcripts and miRNA precursors (pre-miRNAs). MiRNAs can be transcribed from polycistronic miRNA clusters that contain multiple miRNA genes or from a single miRNA gene (monocistronic). Many miRNA genes, however, are found in introns of protein coding genes. Mirtrons form an entire intron without any extra intronic sequences (for details, see text). Pre-miRNAs are characterized by a stem–loop structure. One strand of the stem gives rise to the mature miRNA (highlighted in red). The opposite strand forms the miRNA* sequence, which is unstable. Pre-miRNAs are further characterized by 5′ phosphates (p) and 2 nt 3′ overhangs. $(A)_n$, polyA tail. See Figure 10.1*

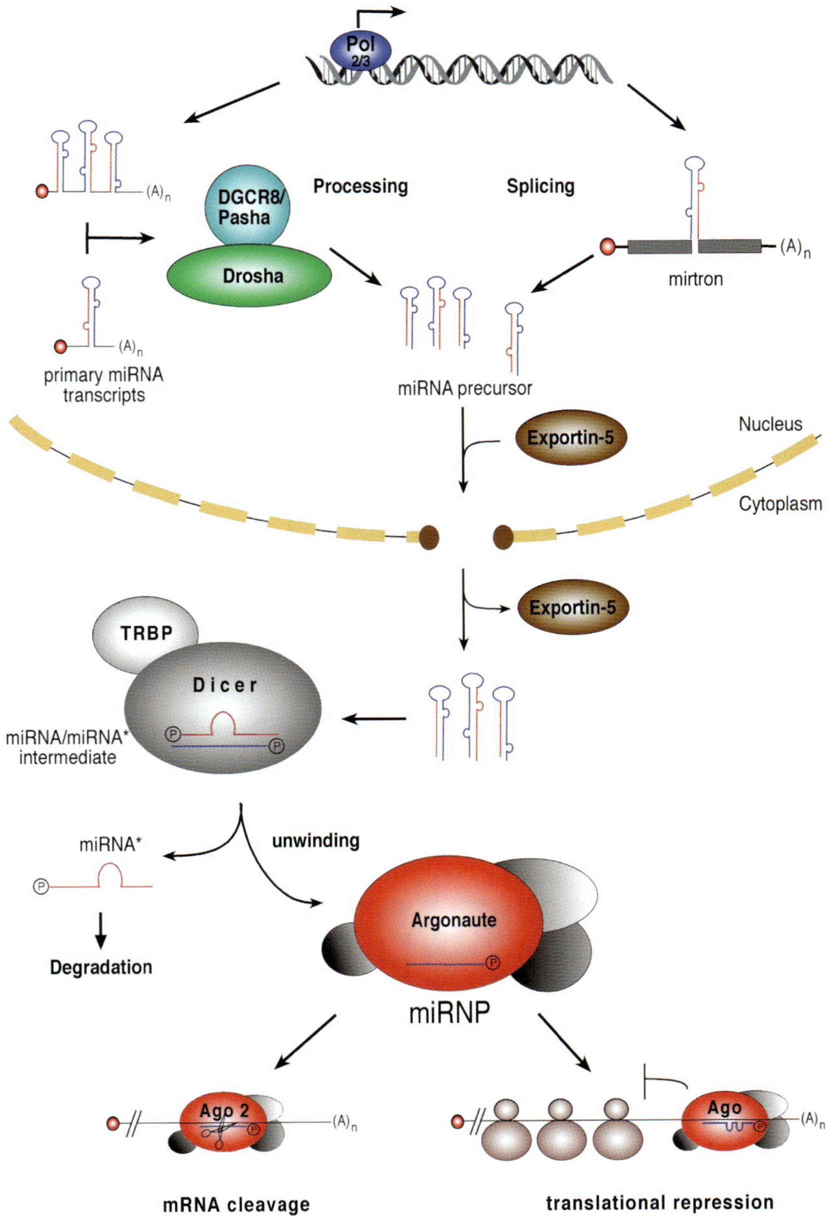

Plate 8 *Schematic representation of miRNA biogenesis and function. MiRNA genes are transcribed as primary miRNA transcripts by RNA polymerases II and III and are further processed to pre-miRNAs by the RNase III enzyme Drosha and its co-factor DGCR8/Pasha. The export receptor exportin-5 transports pre-miRNAs to the cytoplasm, where the RNase III enzyme Dicer cleaves off the loop and generates a double-stranded short RNA. Such short intermediate RNA duplexes are rapidly unwound and one strand is incorporated into protein complexes (miRNPs) as the mature miRNA. Key components of such miRNA 'effector complexes' are members of the Argonaute protein family. In plants, and very rarely in animals, miRNAs hybridize to highly complementary target sequences and guide its sequence-specific cleavage in a process identical with RNA interference (RNAi). Many miRNAs, however, bind to only partially complementary target sites typically located in the 3′ untranslated region of distinct target mRNAs. Such imperfect pairing leads either to translational repression or de-adenylation followed by degradation of the mRNA. See Figure 10.2*

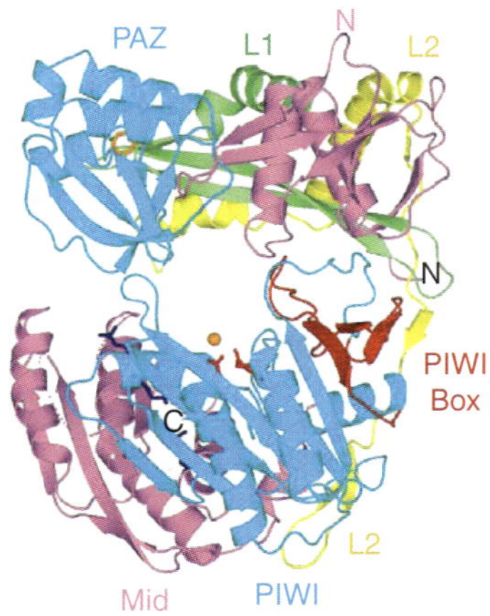

Plate 9 *Architecture of the* Aquifex aeolicus *Argonaute protein. An N-terminal domain (magenta) with so far unknown functions is linked to the PAZ domain (blue) by linker L1 (green). The PAZ domain interacts with the 3′ end of miRNAs. Linker L2 (yellow) connects the PAZ domain to the MID domain (magenta), which anchors the 5′ end of the miRNA. The MID domain is followed by the RNase H-like PIWI domain. See Figure 10.3*

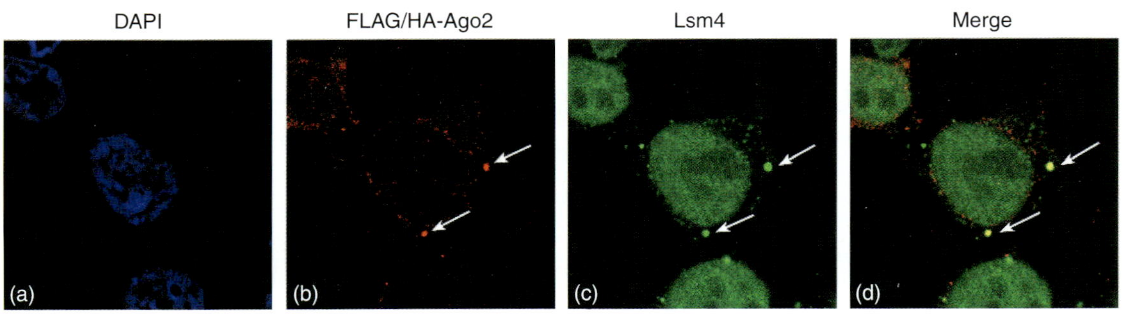

Plate 10 *Localization of Argonaute proteins in human cells revealed by indirect immunofluorescence. Cells were treated with chromophore-conjugated antibodies directed against human Ago2 (b) or LSm4 (c), a marker protein that localizes to cytoplasmic processing bodies (P-bodies). The dot-like cytoplasmic structure represents P-bodies. The nuclei of the cells are stained with DAPI (a, blue), a dye that intercalates with DNA. Panel (d) shows merged panels (b) and (c) and demonstrates that Ago2 is present in P-bodies. See Figure 10.4*

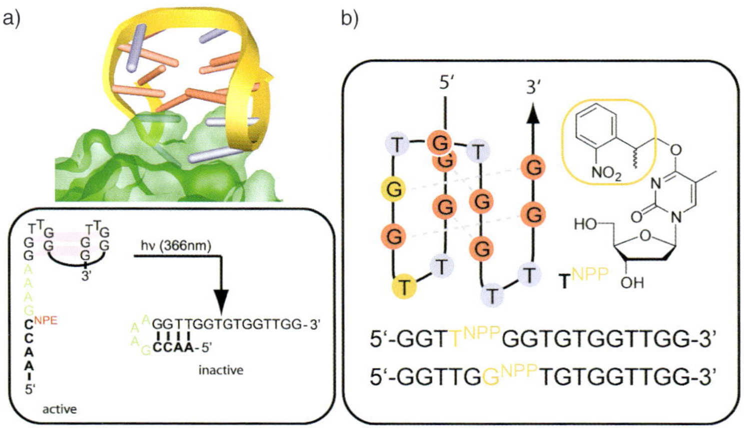

Plate 11 *Caged aptamers targeting the exosite I of human α-thrombin. The 15 nucleotide-containing aptamer was synthesized in order to obtain aptamer variants that can be deactivated (a) or activated (b) upon irradiation with light (λ = 365 nm). Light control was achieved by incorporation of* o*-nitrophenylethyl (NPE)-modified cytidine or* o*-nitrophenylpropyl (NPP)-modified thymidine and guanine nucleobases [50,51,53]. See Figure 13.11*

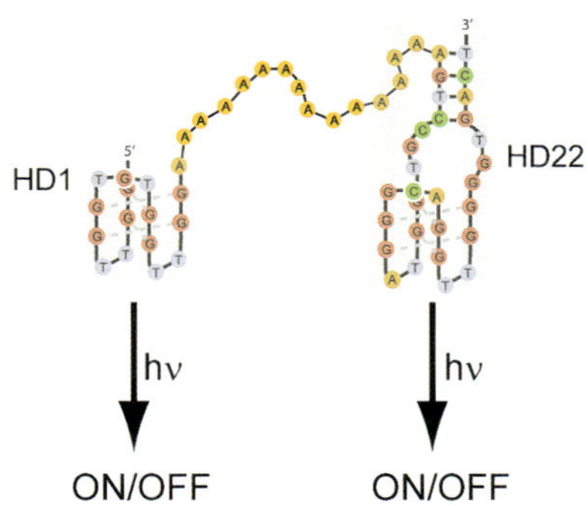

Plate 12 *Caged bivalent aptamer for differential protein subdomain regulation. A bivalent aptamer comprising the HD1 domain (targeting exosite I of thrombin) and the HD22 domain (targeting exosite II of thrombin) – both shaded gray – can be employed for being light controllable by the incorporation of caged nucleobases at strategic positions [54,55]. See Figure 13.12*

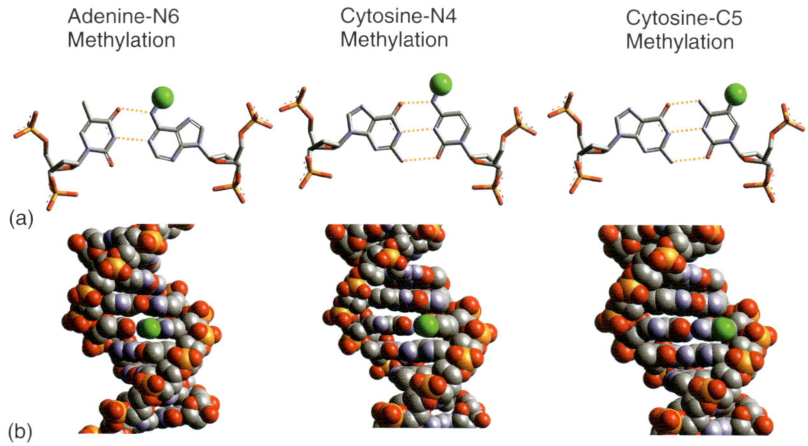

Plate 13 *Types of methylated bases observed in DNA. (a) TA and GC base pairs are shown in stick representation. The positions of methyl groups in N6-methyladenine, N4-methylcytosine and C5-methylcytosine are indicated by green balls. The methylation does not interfere with Watson–Crick base pairing in any of these cases. (b) B-DNA structures are shown in space fill representation. The methyl groups positioned in the major grove of the DNA are highlighted in green. See Figure 14.1*

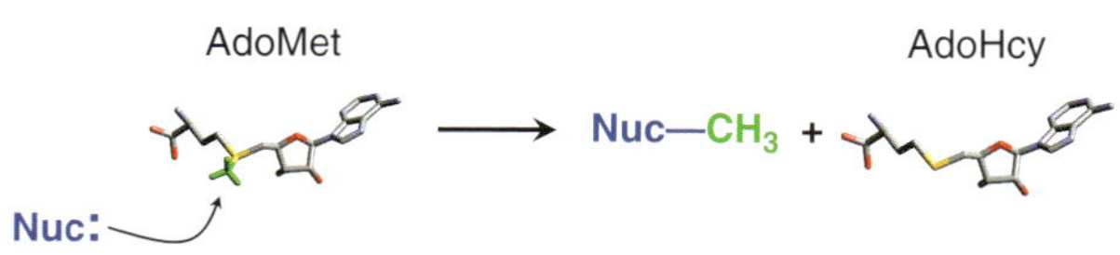

Plate 14 *General reaction catalyzed by methyltransferases. AdoMet and AdoHcy are shown in stick models colored by atom type, except that the activated methyl group is colored green. The reaction proceeds by an attack of a nucleophile on the methyl group, which leads to its transfer and the release of the cofactor product AdoHcy. See Figure 14.3*

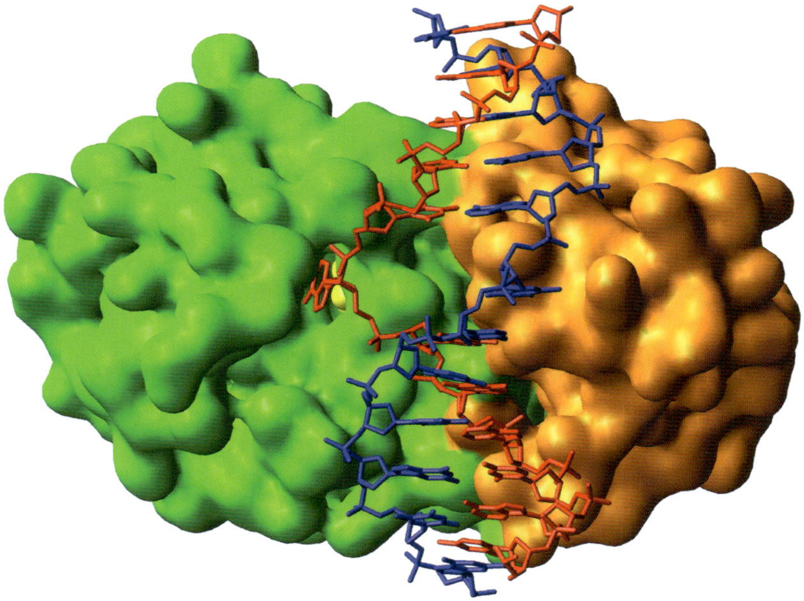

Plate 15 *Structure of the T4Dam DNA-(adenine-N6)-methyltransferase bound to a specific DNA molecule. The enzyme is shown in surface representation colored green and orange for its large and small domains. The two strands of the DNA are colored red and blue and shown as a stick model. AdoMet is rendered in space fill representation and colored yellow. Note the flipping of the central target base out of the DNA helix into a binding pocket of the enzyme, which harbors the catalytic center. See Figure 14.4*

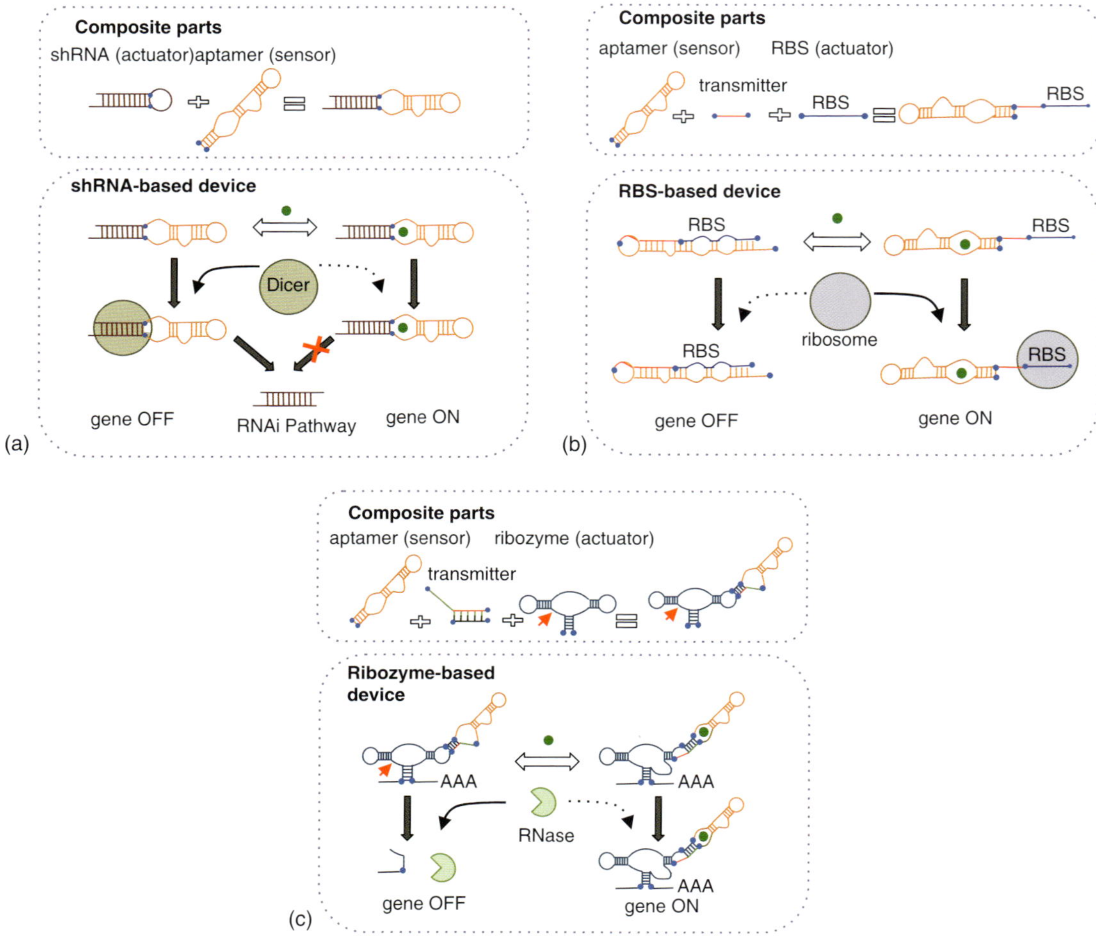

Plate 16 *RNA device design strategies. (a) A device design strategy based on the direct coupling of sensor and actuator parts. In the example shown, an shRNA actuator part is linked directly to an RNA sensor part. Binding of the molecular input to the RNA sensor modulates cleavage by Dicer and subsequent RNAi-induced gene silencing. (b) A device design strategy based on the integration of an information transmission part between the sensor and actuator parts. In the example shown, an RBS actuator part is linked to an RNA sensor part through a linker sequence that acts as a transmitter part. Binding of input to the RNA device stabilizes the input-bound conformation in which the RBS is not occluded in a secondary structure, thereby increasing ribosome access. (c) A device design strategy based on the integration of an information transmission part that permits the modular assembly of sensor, actuator and information transmitter parts. In the example shown, a hammerhead ribozyme actuator part is linked to an RNA sensor part through a transmitter sequence that provides modular coupling between the two parts. Binding of the input to the RNA device stabilizes the input-bound conformation in which the ribozyme is inactive. All of the illustrated device designs result in gene expression increasing as a function of increasing input concentrations, thereby encoding gene expression ON switches. Reproduced from Win, M.N., Liang, J.C., and Smolke, C.D. (2009). Frameworks for programming biological function through RNA parts and devices.* Chem Biol ***16****, 298–310. Reprinted with permission from Elsevier © 2009.* *See Figure 15.1*

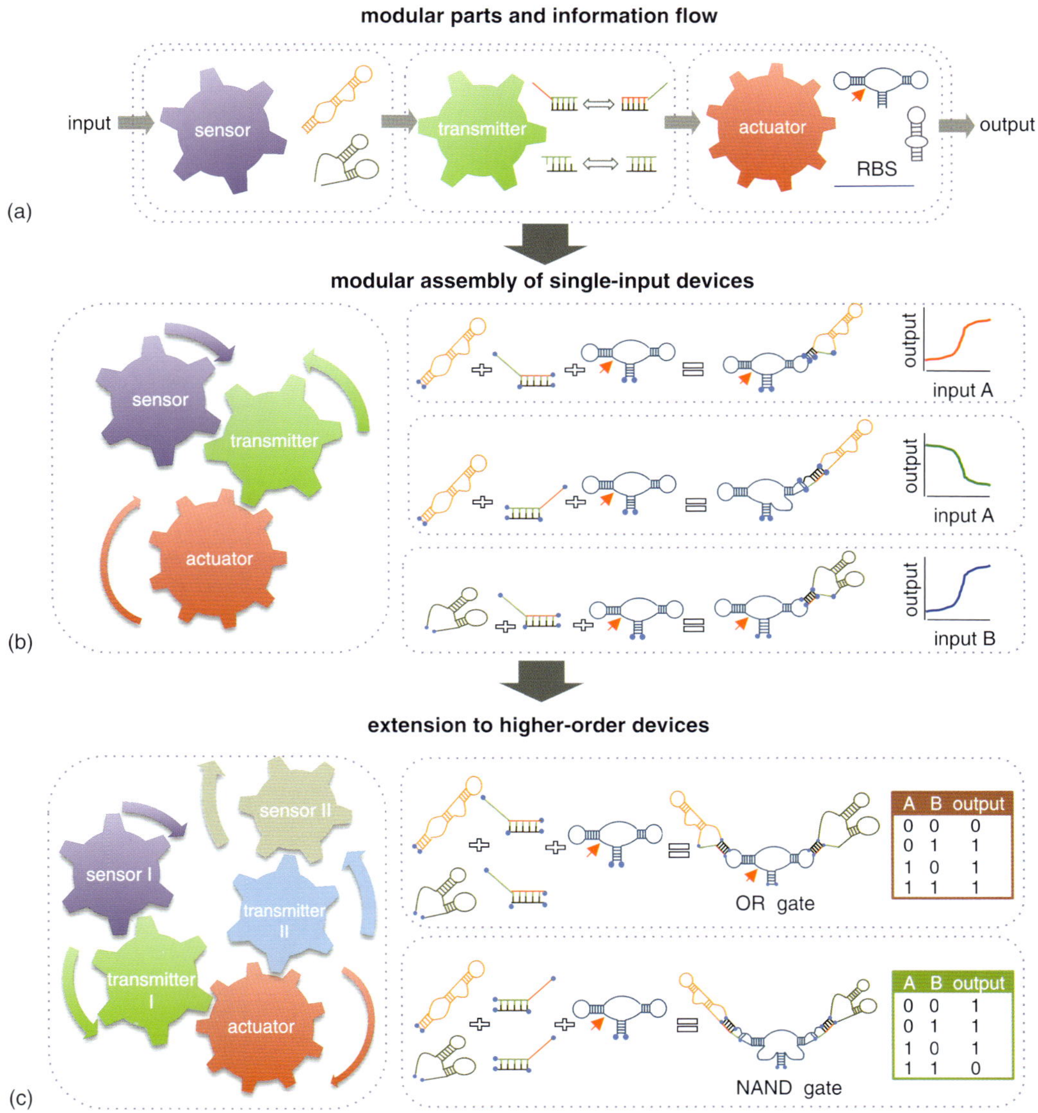

Plate 17 *Functional composition frameworks support the modular assembly of single-input, single-output RNA devices and extension to higher order devices. (a) A functional composition framework for assembling RNA devices from modular components. Information in the form of a molecular input is received by the sensor and transmitted by the transmitter to a regulated activity of the actuator, which in turn controls a gene expression event output. (b) Modular composition frameworks allow the modular assembly of single-input, single-output RNA devices from sensor, transmitter and actuator parts. In this example, the function of the RNA device is encoded in the transmitter part, such that alteration of the transmitter sequence results in devices that perform different computations. In addition, the input responsiveness of the device, or the particular input molecule to which the device responds, is encoded in the sensor part, such that direct replacement of the sensor part results in devices that process different inputs. (c) This single-input, single-output device framework can be directly extended to more complex devices encoding higher order functions. The computational operation encoded in the device is determined by the particular combination of parts (sensor, transmitter) and the device architecture. Reproduced from Win, M.N., Liang, J.C., and Smolke, C.D. (2009). Frameworks for programming biological function through RNA parts and devices.* Chem Biol ***16****, 298–310. Reprinted with permission from Elsevier © 2009.* *See Figure 15.2*

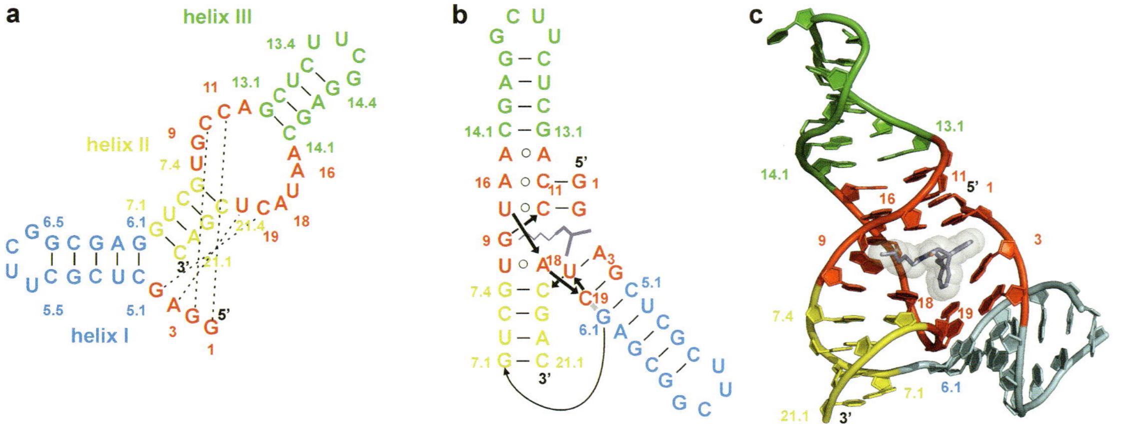

Plate 18 Crystal structure of the Diels–Alderase ribozyme. (a) Ribozyme secondary structure with helices I, II and III (cyan, yellow, green), the asymmetric bubble and the conserved 5′ end (both in red). Dotted lines represent pseudoknot Watson–Crick-type interactions. (b) Tertiary fold and (c) three-dimensional topology in the crystal structure of the ribozyme–product complex. One enantiomer of the Diels–Alder product (dark blue sticks with transparent spheres) is bound into the catalytic pocket of the ribozyme [21]. See Figure 16.5

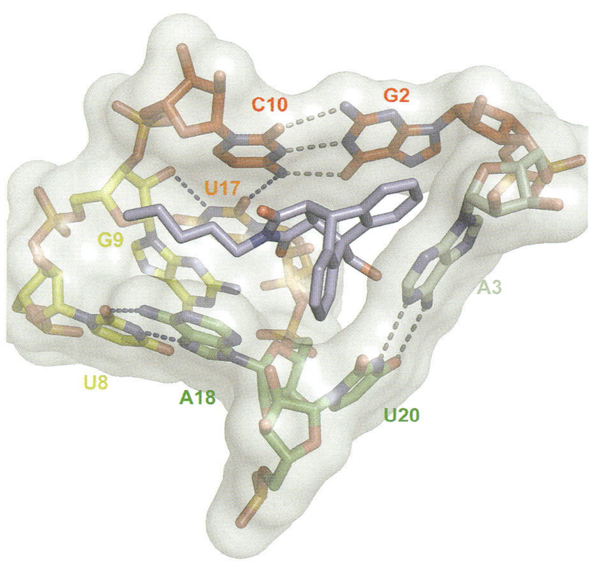

Plate 19 Surface representation of the catalytic pocket with stick representations of the nucleotides. Standard Watson–Crick type interactions are indicated by gray dotted lines; blue indicates non-standard interaction. The Diels–Alder product is shown in blue. See Figure 16.6

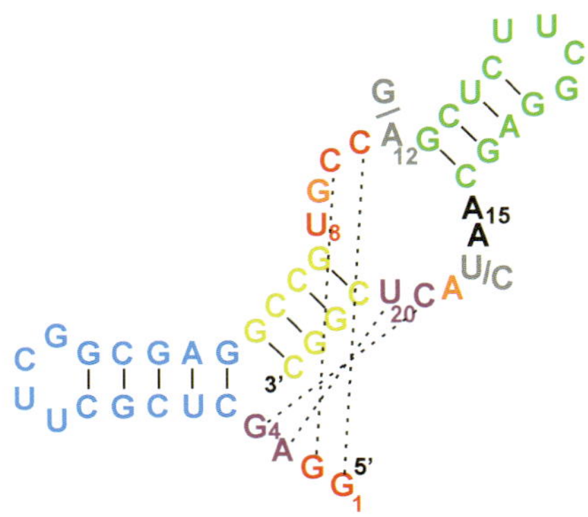

Plate 20 *Mutation analysis of the Diels–Alderase ribozyme [24]. Green and blue stem–loop structures are completely variable in size and sequence, whereas the yellow helix is variable in sequence, but conserved in size. Nucleotides in red are invariant. Orange: highly conserved (over 90% activity reduction on mutation). Purple: pairwise complementary substitutions allowed. Gray: strong preference for two nucleotides. Black: variable (less than 50% activity reduction on mutation). Black dotted lines: Watson–Crick pseudoknot base pairs. See Figure 16.9*

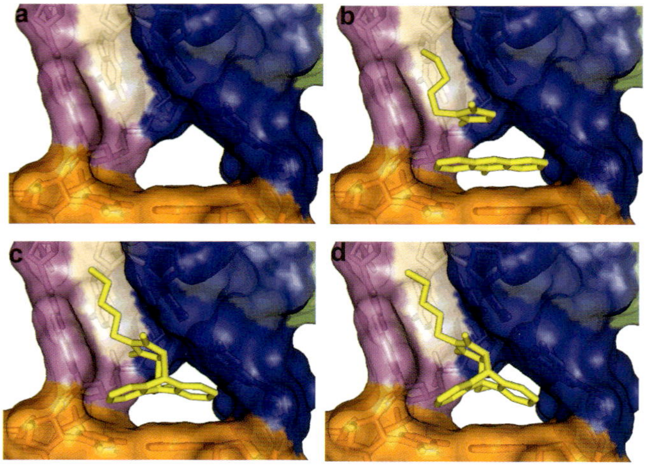

Plate 21 *Proposed model for the catalytic mechanism of the Diels–Alderase ribozyme. (a) Empty catalytic pocket. (b) Michaelis complex with both substrates bound. (c) Transition state inside the catalytic pocket. (d) Ribozyme-bound product. Panels (a) and (d) are derived directly from the crystallographic data; panels (b) and (c) were obtained by manually docking the two substrates and the transition state, respectively, into the pocket. Coloring scheme: orange, A3:U20 tertiary Watson–Crick base pair; blue, G2:C10:U17 base triple; violet, U8:A18 reversed Hoogsteen base pair; wheat, unpaired G9 [21]. See Figure 16.10*

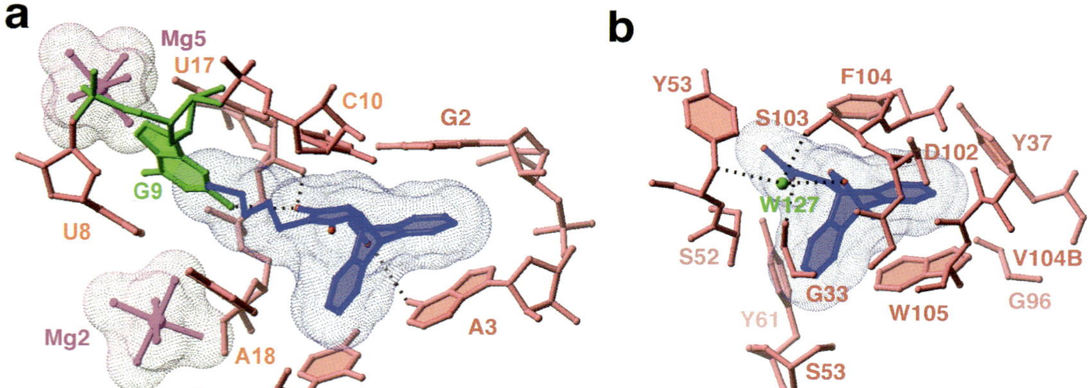

Plate 22 *Comparison of RNA and protein catalytic sites for the Diels–Alder reaction. (a) Diels–Alderase ribozyme and (b) retro-Diels–Alderase antibody 10F11 [21]. See Figure 16.11*

59. Z. Ma and J. Taylor, Nucleic acid-triggered catalytic drug release, *Proc. Natl. Acad. Sci. USA*, **97**, 11159–11163 (2000).
60. C. Drayton, K. du Plooy, K. Greenfield, R. Purchase and E. Smeaton (eds), *Science of Synthesis – Houben-Weyl Methods of Chemical Transformations*, Vols 1–35, Georg Thieme, Stuttgart, 2001.
61. C. T. Calderone, J. W. Puckett, Z. J. Gartner *et al.*, Directing otherwise incompatible reactions in a single solution by using DNA-templated organic synthesis, *Angew. Chem. Int. Ed.*, **41**, 4104–4108 (2002).
62. M. Oberhuber and G. F. Joyce, A DNA-templated aldol reaction as a model for the formation of pentose sugars in the RNA world, *Angew. Chem. Int. Ed.*, **44**, 7580–7583 (2005).
63. R. Breslow, On the mechanism of the formose reaction, *Tetrahedron Lett.*, **21**, 22–26 (1959).
64. M. Rueping, Aldol reactions within the RNA world, *Angew. Chem. Int. Ed.*, **45**, 1838–1840 (2006).
65. G. Illuminati and L. Mandolini, Ring closure reactions of bifunctional chain molecules, *Acc. Chem. Res.*, **14**, 95–102 (1981).
66. T. Snyder, B. Tse and D. Liu, Effects of template sequence and secondary structure on DNA-templated reactivity, *J. Am. Chem. Soc.*, **130**, 1392–1401 (2008).
67. S. Narayan, J. Muldoon, M. G. Finn *et al.*, 'On water': unique reactivity of organic compounds in aqueous suspension, *Angew. Chem. Int. Ed.*, **44**, 3275–3279 (2005).
68. G. Roelfes, DNA and RNA induced enantioselectivity in chemical synthesis, *Mo.l Biosyst.*, **3**, 126–135 (2007).
69. A. Jäschke, Artificial ribozymes and deoxyribozymes, *Curr. Opin. Struct. Biol.*, **11**, 321–326 (2001).
70. G. F. Joyce, G. M. Visser, C. A. A. Van Boeckel *et al.*, Chiral selection in poly(C)-directed synthesis of oligo(G), *Nature*, **310**, 602–604 (1984).
71. M. Bolli, R. Micura and A. Eschenmoser, Pyranosyl-RNA: chiroselective self-assembly of base sequences by ligative oligomerization of tetranucleotide-2′,3′-cyclophosphates (with a commentary concerning the origin of biomolecular homochirality), *Chem. Biol.*, **4**, 309–320 (1997).
72. I. Kozlov, L. Orgel and P. Nielsen, Remote enantioselection transmitted by an achiral peptide nucleic acid backbone, *Angew. Chem. Int. Ed.*, **39**, 4292–4295 (2000).
73. X. Y. Li and D. R. Liu, Stereoselectivity in DNA-templated organic synthesis and its origins, *J. Am. Chem. Soc.*, **125**, 10188–10189 (2003).
74. K. Luebke and P. Dervan, Nonenzymatic ligation of oligodeoxyribonucleotides on a duplex DNA template by triple-helix formation, *J. Am. Chem. Soc.*, **111**, 8733–8735 (1989).
75. T. Li and K. Nicolaou, Chemical self-replication of palindromic duplex DNA, *Nature*, **369**, 218–221 (1994).
76. Y. Chen and C. Mao, Reprogramming DNA-directed reactions on the basis of a DNA conformational change, *J. Am. Chem. Soc.*, **126**, 13240–13241 (2004).
77. A. T. Poulin-Kerstien and P. B. Dervan, DNA-templated dimerization of hairpin polyamides, *J. Am. Chem. Soc.*, **125**, 15811–15821 (2003).
78. H. Kolb, M. Finn and K. Sharpless, Click chemistry: diverse chemical function from a few good reactions, *Angew. Chem. Int. Ed.*, **40**, 2005–2021 (2001).
79. R. Rohatgi, D. P. Bartel and J. W. Szostak, Kinetic and mechanistic analysis of nonenzymic, template-directed oligoribonucleotide ligation, *J. Am. Chem. Soc*, **118**, 3332–3339 (1996).
80. R. Rohatgi, D. P. Bartel and J. W. Szostak, Nonenzymatic, template-directed ligation of oligoribonucleotides is highly regioselective for the formation of 3′–5′ phosphodiester bonds, *J. Am. Chem. Soc.*, **118**, 3340–3344 (1996).
81. N. Green, *Avidin, Adv. Protein Chem.*, **29**, 85–133 (1975).
82. Z. J. Gartner, M. W. Kanan and D. R. Liu, Multistep small-molecule synthesis programmed by DNA templates, *J. Am. Chem. Soc.*, **124**, 10304–10306 (2002).
83. X. Li, Z. J. Gartner, B. N. Tse *et al.*, Translation of DNA into synthetic *N*-acyloxazolidines, *J. Am. Chem. Soc.*, **126**, 5090–5092 (2004).
84. P. Brown and D. Botstein, Exploring the new world of the genome with DNA microarrays, *Nat. Genet.*, **21**, 33–37 (1999).
85. M. W. Kanan, M. M. Rozenman, K. Sakurai *et al.*, Reaction discovery enabled by DNA-templated synthesis and in vitro selection, *Nature*, **431**, 545–549 (2004).
86. K. Mullis, The polymerase chain reaction (Nobel Lecture), *Angew. Chem. Int. Ed.*, **33**, 1209–1213 (1994).
87. M. Kramer and D. Coen, Enzymatic amplification of DNA by PCR: standard procedures and optimization, in *Current Protocols in Molecular Biology*, ed. F. M. Ausubel, John Wiley & Sons, Inc., New York, 2000, pp. 15.11–15.13.

88. A. Katritzky and G. Musumarra, New insights into aliphatic nucleophilic substitution reactions from the use of pyridines as leaving groups, *Chem. Soc. Rev.*, **13**, 47–68 (1984).
89. M. Simon and K. Shokat, Engineering control over protein function using chemistry, in *Chemical Biology*, ed. S. Schreiber, T. Kapoor and G. Wess, Wiley-VCH Verlag GmbH, Weinheim, 2007, pp. 115–198.
90. D. S. Tan, Diversity-oriented Synthesis, in Chemical Biology, ed. S. Schreiber, T. Kapoor and G. Wess, Wiley-VCH Verlag GmbH, Weinheim, 2007, pp. 483–596.
91. S. Brenner and R. A. Lerner, Encoded combinatorial chemistry, *Proc. Natl. Acad. Sci. USA*, **89**, 5381–5383 (1992).
92. J. Nielsen, S. Brenner and K. D. Janda, Synthetic methods for the implementation of encoded combinatorial chemistry, *J. Am. Chem. Soc.*, **115**, 9812–9813 (1993).
93. Z. J. Gartner, R. Grubina, C. T. Calderone *et al.*, Two enabling architectures for DNA-templated organic synthesis, *Angew. Chem. Int. Ed.*, **42**, 1370–1375 (2003).
94. T. M. Snyder and D. R. Liu, Ordered multistep synthesis in a single solution directed by DNA templates, *Angew. Chem. Int. Ed.*, **44**, 7379–7382 (2005).
95. L. M. Smith, J. Z. Sanders, R. J. Kaiser *et al.*, Fluorescence detection in automated DNA sequence analysis, *Nature*, **321**, 674–679 (1986).
96. Z. J. Gartner, B. N. Tse, R. Grubina *et al.*, DNA-templated organic synthesis and selection of a library of macrocycles, *Science*, **305**, 1601–1605 (2004).
97. A. Jain, G. M. Whitesides, R. S. Alexander *et al.*, Identification of two hydrophobic patches in the active-site cavity of human carbonic anhydrase II by solution-phase and solid-state studies and their use in the development of tight-binding inhibitors, *J. Med. Chem.*, **37**, 2100–2105 (1994).
98. D. R. Halpin and P. B. Harbury, DNA display II. Genetic manipulation of combinatorial chemistry libraries for small-molecule evolution, *PLoS Biol.*, **2**, E174 (2004).
99. S. Melkko, J. Scheuermann, C. E. Dumelin *et al.*, Encoded self-assembling chemical libraries, *Nat. Biotechnol.*, **22**, 568–574 (2004).
100. J. B. Doyon, T. M. Snyder and D. R. Liu, Highly sensitive in vitro selections for DNA-linked synthetic small molecules with protein binding affinity and specificity, *J. Am. Chem. Soc.*, **125**, 12372–12373 (2003).
101. S. Klussman, *The Aptamer Handbook*, Wiley-VCH, 2006.
102. M. Kaiser, S. Wetzel, K. Kumar *et al.*, Biology-inspired synthesis of compound libraries, *Cell. Mol. Life Sci.*, **65**, 1186–1201 (2008).
103. K. Sakurai, T. M. Snyder and D. R. Liu, DNA-templated functional group transformations enable sequence-programmed synthesis using small-molecule reagents, *J. Am. Chem. Soc.*, **127**, 1660–1661 (2005).
104. J. Cai, X. Li, X. Yue *et al.*, Nucleic acid-triggered fluorescent probe activation by the Staudinger reaction, *J. Am. Chem. Soc.*, **126**, 16324–16325 (2004).
105. G. P. Smith and V. A. Petrenko, *Phage display, Chem. Rev.*, **97**, 391–410 (1997).
106. D. R. Halpin and P. B. Harbury, DNA display I. Sequence-encoded routing of DNA populations, *PLoS Biol.*, **2**, E173 (2004).
107. D. R. Halpin, J. A. Lee, S. J. Wrenn *et al.*, DNA display III. Solid-phase organic synthesis on unprotected DNA, *PLoS Biol.*, **2**, E175 (2004).
108. S. J. Wrenn, R. M. Weisinger, D. R. Halpin *et al.*, Synthetic ligands discovered by *in vitro* selection, *J. Am. Chem. Soc.*, **129**, 13137–13143 (2007).
109. M. Rozenman, M. Kanan and D. Liu, Development and initial application of a hybridization-independent, DNA-encoded reaction discovery system compatible with organic solvents, *J. Am. Chem. Soc.*, **129**, 14933–14938 (2007).
110. A. B. Beeler, S. Su, C. A. Singleton *et al.*, Discovery of chemical reactions through multidimensional screening, *J. Am. Chem. Soc.*, **129**, 1413–1419 (2007).
111. J. Kofoed, J. L. Reymond and T. Darbre, Prebiotic carbohydrate synthesis: zinc–proline catalyzes direct aqueous aldol reactions of α-hydroxy aldehydes and ketones, *Org. Biomol. Chem.*, **3**, 1850–1855 (2005).
112. Z. Tang and A. Marx, Proline-modified DNA as catalyst of the aldol reaction, *Angew. Chem. Int. Ed.*, **46**, 7297–7300 (2007).
113. J. Seayad and B. List, Asymmetric organocatalysis, *Org. Biomol. Chem.*, **3**, 719–724 (2005).
114. W. Tan, K. Wang and T. Drake, *Molecular beacons, Curr. Opin. Chem. Biol.*, **8**, 547–553 (2004).

115. J. Levsky and R. Singer, Fluorescence *in situ* hybridization: past, present and future, *J. Cell. Sci.*, **116**, 2833–2838 (2003).
116. C. Heid, J. Stevens, K. Livak *et al.*, Real time quantitative PCR, *Genome Res.*, **6**, 986–994 (1996).
117. R. R. Breaker, Natural and engineered nucleic acids as tools to explore biology, *Nature*, **432**, 838–845 (2004).
118. F. Barany, Genetic disease detection and DNA amplification using cloned thermostable ligase, *Proc. Natl. Acad. Sci. USA*, **88**, 189–193 (1991).
119. V. Demidov and M. Frank-Kamenetskii, Two sides of the coin: affinity and specificity of nucleic acid interactions, *Trends Biochem. Sci.*, **29**, 62–71 (2004).
120. P. Dawson, T. Muir, I. Clark-Lewis *et al.*, Synthesis of proteins by native chemical ligation, *Science*, **266**, 776–779 (1994).
121. S. Ficht, A. Mattes and O. Seitz, Single-nucleotide-specific PNA-peptide ligation on synthetic and PCR DNA templates, *J. Am. Chem. Soc.*, **126**, 9970–9981 (2004).
122. C. Dose and O. Seitz, Single nucleotide specific detection of DNA by native chemical ligation of fluorescence labeled PNA-probes, *Bioorg. Med. Chem.*, **16**, 65–77 (2008).
123. J. A. Prescher and C. R. Bertozzi, Chemistry in living systems, *Nat. Chem. Biol.*, **1**, 13–21 (2005).
124. K. Fujimoto, S. Matsuda, N. Takahashi *et al.*, Template-directed photoreversible ligation of deoxyoligonucleotides via 5-vinyldeoxyuridine, *J. Am. Chem. Soc.*, **122**, 5646–5647 (2000).
125. I. Saito, Y. Miyauchi, Y. Saito *et al.*, Template-directed photoreversible ligation of DNA via 7-carboxyvinyl7-deaza–2′-deoxyadenosine, *Tetrahedron Lett.*, **46**, 97–99 (2005).
126. Y. Yoshimura, Y. Noguchi, H. Saito *et al.*, Template-directed DNA photoligation in rapid and selective detection of RNA point mutations, *ChemBioChem*, **7**, 598–601 (2006).
127. T. Ihara, T. Fujii, M. Mukae *et al.*, Photochemical ligation of DNA conjugates through anthracene cyclodimer formation and its fidelity to the template sequences, *J. Am. Chem. Soc.*, **126**, 8880–8881 (2004).
128. M. Dobrikov, S. Gaidamakov, T. Gainutdinov *et al.*, Sensitized photomodification of single-stranded DNA by a binary system of oligonucleotide conjugates, *Antisense Nucleic Acid Drug Dev.*, **7**, 309–317 (1997).
129. M. Dobrikov, T. Gainutdinov and V. Vlassov, Visible light activatable binary system of oligonucleotide conjugates for nucleic acids modification, *Nucleosides Nucleotides*, **18**, 1517–1518 (1999).
130. S. Sando, H. Abe and E. Kool, Quenched auto-ligating DNAs: multicolor identification of nucleic acids at single nucleotide resolution, *J. Am. Chem. Soc.*, **126**, 1081–1087 (2004).
131. C. Ziegler and W. Göpel, Biosensor development, *Curr. Opin. Chem. Biol.*, **2**, 585–591 (1998).
132. N. Rosi and C. Mirkin, Nanostructures in biodiagnostics, *Chem. Rev.*, **105**, 1547–1562 (2005).
133. S. Melkko, C. Dumelin, J. Scheuermann *et al.*, Lead discovery by DNA-encoded chemical libraries, *Drug Discov. Today*, **12**, 465–471 (2007).
134. S. L. Schreiber, Small molecules: the missing link in the central dogma, *Nat. Chem. Biol.*, **1**, 64–66 (2005).
135. C. Lipinski, F. Lombardo, B. Dominy *et al.*, Experimental and computational approaches to estimate solubility and permeability in drug discovery and development settings, *Adv. Drug Deliv. Rev.*, **23**, 3–25 (1997).
136. M. M. Rozenman and D. R. Liu, DNA-templated synthesis in organic solvents, *ChemBioChem*, **7**, 253–256 (2006).
137. D. M. Rosenbaum and D. R. Liu, Efficient and sequence-specific DNA-templated polymerization of peptide nucleic acid aldehydes, *J. Am. Chem. Soc.*, **125**, 13924–13925 (2003).
138. L. H. Eckardt, K. Naumann, W. M. Pankau *et al.*, DNA nanotechnology: chemical copying of connectivity, *Nature*, **420**, 286 (2002).
139. K. Tanaka, G. H. Clever, Y. Takezawa *et al.*, Programmable self-assembly of metal ions inside artificial DNA duplexes, *Nat. Nanotechnol.*, **1**, 190–194 (2006).

5

Chemical Biology of Peptide Nucleic Acids (PNAs)

Peter E. Nielsen

5.1 Introduction

Peptide nucleic acids (PNAs), which were developed almost 20 years ago [1], are DNA analogs and mimics based on pseudopeptide backbones to which nucleobases are attached via linkers. The original, very simple (aeg) PNA contains an achiral, uncharged aminoethylglycine backbone and the nucleobases are connected via ethanoyl linkers (Figure 5.1) [1,2]. Although many derivatives and analogs of aegPNA have been synthesized and studied over recent years (e.g. [3]), the large majority of biological and application-related studies still use the original aegPNA due to the favorable DNA and RNA binding (hybridization) properties, high chemical and especially biological stability and easy accessibility of (modified) aegPNA oligomers. Thus, in addition to attracting wide attention in bioorganic chemistry, the properties and possibilities of PNA have also inspired researchers from as diverse fields as molecular biology, drug discovery/medicinal chemistry, genetic diagnostics, electrochemistry, combinatorial chemistry, nanoscience, cell biology and origin of life to include PNAs in their molecular repertoire. The present chapter will highlight the most recent advances in these areas.

5.2 Gene targeting

PNAs were originally conceived as a sequence-specific agent for gene targeting at the DNA level [1] and much of the subsequent biologically oriented studies with PNA have had the underlying theme of gene therapeutic drug discovery by DNA targeting and especially by mRNA (RNA interference/antisense) targeting. It became clear almost from the very beginning that cellular delivery and subsequent *in vivo*

The Chemical Biology of Nucleic Acids Edited by Günter Mayer

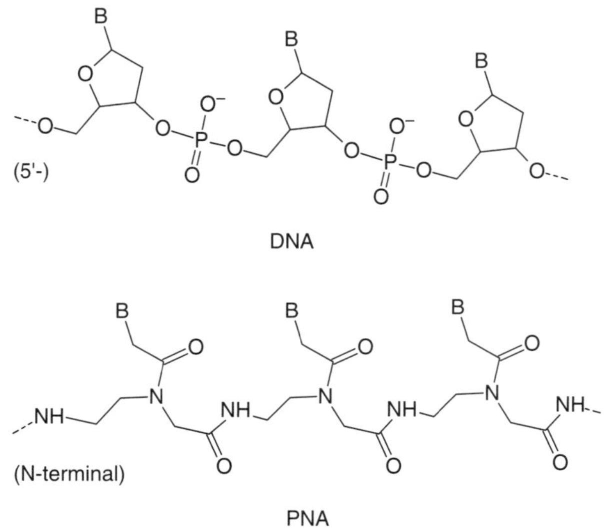

Figure 5.1 *Chemical structures of PNA compared with DNA*

bioavailability were at least as great a challenge for PNA drug discovery as it is for traditional oligonucleotide drug discovery, and the importance of addressing this issue is emphasized by the increased focus that this area has attracted over the past few years. Nonetheless, it is the broad spectrum of gene targeting principles that over the past years have been discovered and described for PNAs that justifies the increased focus on (*in vivo*) bioavailability and foster the hope that a range of effective gene therapeutic drugs may eventually be discovered and developed based on PNA technologies.

5.3 RNA interference

RNA interference in the broadest sense would include any sequence-directed and dependent interference with RNA function, whether the target is mRNA, microRNA or other RNAs (e.g. structural such as ribosomal RNAs) and using any kind of oligonucleotides or oligonucleotide derivatives or analogs as RNA interference agents. PNA is truly a non-biological molecule and therefore does not enter siRNA-type RISC pathways or RNase H-dependent antisense pathways. Rather, RNA interference by PNAs relies predominantly (or solely) on steric interference or blocking of the target RNA [4]. Consequently, potent RNA interference using PNAs has been focusing on translation interference through blockage of the ribosome assembly on the mRNA by targeting of the 5′UTR or around the translation initiation site. Alternatively, splicing interference and redirection by targeting intron–exon junctions have proven very potent and versatile (e.g. [5–10]) and, more recently, studies have been reported that show efficient inhibition of microRNA function [11]. From a drug discovery point of view, it is particularly encouraging that PNAs capable of inducing exon 23 skipping in the dystrophin mRNA have been identified and that the partly active, truncated dystrophin protein to a significant extent is able to restore muscle function in an mdx mouse model upon intramuscular administration of the PNA [12,13]. Thus RNA interfering PNA oligomers are finding their way into drug discovery.

5.4 Transcription interference

PNA may bind to duplex DNA targets by various mechanisms depending on the PNA and target sequence (Figure 5.2). Most surprisingly, homopyrimidine PNAs were found to bind complementary homopurine targets via triplex invasion in which an internal extremely stable PNA–DNA–PNA triplex is formed with the complementary purine DNA strand while the pyrimidine DNA strand is left as a single-stranded 'P-loop' [1,14]. The formation of such PNA triplex invasion complexes is driven by the extraordinarily high stability of PNA_2–DNA triplexes (10-mer complexes typically exhibit thermal stabilities around 70 °C) and, as a consequence, their formation is kinetically rather than thermodynamically controlled [15]. Furthermore, because of the invasion process, their rate of formation is inversely dependent on the concentration of cations such a Na^+, K^+ and to a greater extent Mg^{2+} and spermine/spermidine that stabilize the DNA double helix and thus inhibit DNA breathing and the invasion process. Duplex invasion may take place with PNA oligomers forming particularly stable PNA–DNA duplexes such as homopurine PNAs [16]. Alternatively, pseudo-complementary pcPNAs containing modified nucleobases that by steric blocking interfere with PNA–PNA duplex formation but do not destabilize PNA–DNA duplex formation can be exploited for double duplex invasion [17]. The diaminopurine–thiouracil base pair (Figure 5.3) allows efficient targeting of duplex DNA targets containing 50% or more AT base pairs. Furthermore, 12–15-mer homopyrimidine PNAs (composed of thymine and pseudoisocytosine nucleobases) were recently shown to form conventional parallel triplexes with complementary homopurine duplex DNA

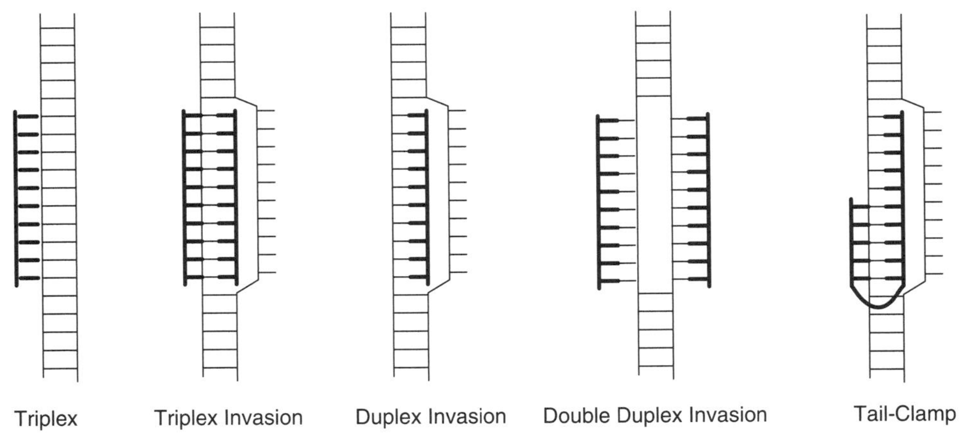

Figure 5.2 Schematic models of different PNA dsDNA complexes. Helix structures are represented as ladders and PNA oligomers are in bold

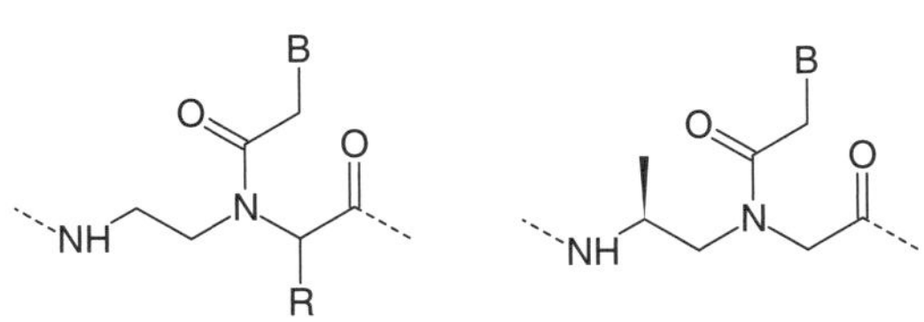

Figure 5.3 Chemical structures of backbone-modified aegPNA: lysinePNA [R = $-(CH_2)_4NH_2$] and γ-methyl aegPNA

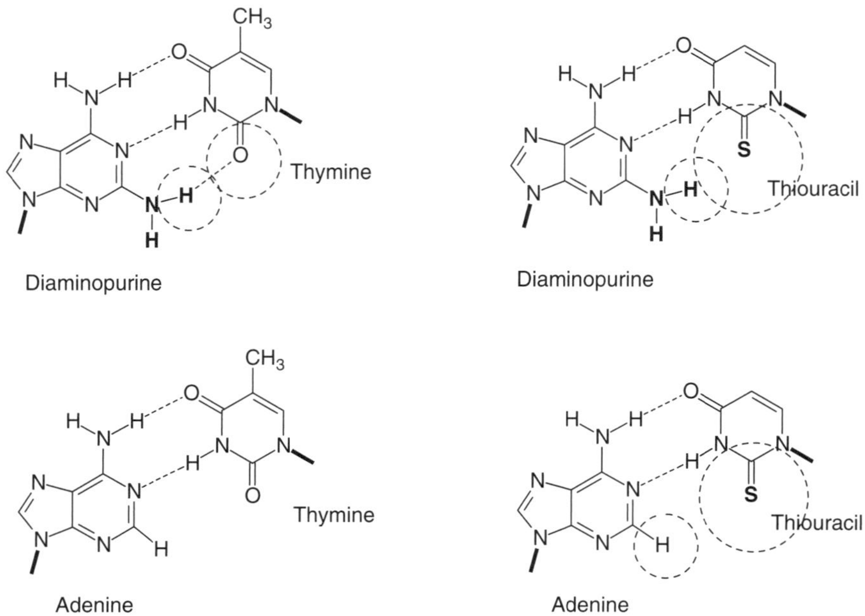

Figure 5.4 *Chemical structures of base pairs formed with adenine, thymine, diaminopurine and thiouracil*

targets with nanomolar binding constants [18]. Additionally, double duplex invasion complexes form more efficiently using backbone-modified, cationic PNAs in which glycine has been partly exchanged with lysine [19] (Figure 5.4) and by using a γ-methyl backbone modification (Figure 5.4) which stabilizes PNA–DNA duplexes, duplex invasion complexes can be formed with mixed-sequence PNAs at low ionic strength [20]. Finally, quadruplex-forming regions of the *c-myc* promoter have been targeted efficiently with sequence-homologous PNA oligomers, illustrating a very special binding mode, presumably via competing formation of PNA–DNA quadruplexes [21].

Originally, transcription interference with PNA was thought to have to rely on the very stable binding of triplex invading (bis)PNAs (or perhaps pseudo-complementary PNAs) to the target gene promoter causing inhibition of transcription initiation by blocking the access of transcription factors or the RNA polymerase itself. Alternatively, PNA binding downstream from the initiation inside the transcribed gene itself may cause transcription arrest of the elongating RNA polymerase [22,23]. However, very recently it was discovered that simple duplex-forming PNA oligomers may cause transcription interference if targeted to the transcription initiation region of the gene [24]. Most likely, the PNA is binding to the single-stranded DNA region of the RNA polymerase open complex and thereby 'jamming' and arresting the polymerase. In fact, fully analogous complexes were described for binding of RNA to the *Escherichia coli* RNA polymerase open complex almost 10 years ago [25]. If this mechanism can be generally exploited, novel possibilities open up for gene therapeutic drug discovery using PNAs by direct targeting of the transcription process.

5.5 Targeted gene repair

Targeted somatic gene repair as a therapeutic modality is coming closer to reality, mainly due to the finding that site-specific double-stranded genomic DNA breaks are very efficient inducers of gene-specific repair via (homologous) recombination with an exogenous DNA donor fragment [26]. Unfortunately, this technology requires cellular delivery (e.g. via genetic transfection) of a genetically engineered restriction endonuclease in addition to the DNA donor fragment [26,27]. Interestingly, it has been found that PNA triplex invasion complexes also stimulate gene-targeted repair via a DNA donor fragment, although at much lower efficiency, as recently demonstrated by the cellular repair of a point mutation responsible for sickle cell anemia [28]. Hence if the repair efficiency of such systems can be improved, sequence-specific targeting using PNA or other small- to medium-sized DNA recognition ligands may lead to therapeutically applicable somatic gene repair strategies.

5.6 Antibacterials

Because the bacterial genome differs dramatically from the human genome, it is not difficult to identify antisense sequence targets that are unique for essential genes in bacteria (or other microorganisms). Hence it should be possible to develop novel antibacterial drugs based on RNA interference. Indeed, it was demonstrated over 10 years ago that a PNA oligomer targeting the start codon on the *acpP* gene, an essential bacterial gene involved in fatty acid synthesis, is able to inhibit the growth of the *E. coli* strain AS19, a cell wall mutant that is more permeable to drugs, but not of wild-type *E. coli* [29]. However, when using the same PNA conjugated to the peptide KFFKFFKFFK, which was known to facilitate bacterial uptake and thus enhance antibiotic activity, antibacterial activity (MIC $\approx 1\ \mu$M) in wild-type *E. coli* was obtained [30], even in the presence of eukaryotic cells (HeLa cells) which were not affected by the PNA conjugate. Many subsequent studies have confirmed these findings [31–34] and extended them to other bacteria such as (multiresistant) *Klebsiella pneumoniae* [35], and also demonstrated antisense effects in Gram-positive bacteria, such as *Staphylococcus aureus* [36]. Most recently, simple mouse peritonitis model experiments were used to demonstrate that such antibacterial effects may be extended to an *in vivo* situation [37]. Unfortunately, toxicity studies in dogs (but not in mice) have indicated severe problems with this type of peptide carrier in terms of inflammatory reactions due to histamine release. Also, one should consider the development of bacterial resistance, by simple, silent point mutations within the PNA target. It can be argued that targets in stable RNAs such as ribosomal RNAs [38,39] would be better targets because functional regions which are very sensitive to mutations may be identified. Nonetheless, the results obtained so far do lay the ground for further developments which it is hoped could result in a novel class of antibacterial drugs, especially for the treatment of infections caused by the increasing number of pathogenic bacterial strains which have developed resistance towards present-day antibiotics.

5.7 Cellular and *in vivo* delivery

It is generally recognized that PNA oligomers are not readily taken up by either pro- or eukaryotic cells and that unmodified PNAs have very poor pharmacokinetic properties due to rapid excretion through the kidneys [40]. Therefore, chemical modification and/or formulation of PNA oligomers is required in order to obtain cellular or *in vivo* activity. A large variety of PNA conjugates mainly based on cationic cell-penetrating peptides (CPPs) (e.g. [41–45]) have been studied for improved cellular delivery, but unfortunately with only limited success so far in terms of obtaining high-potency gene-targeting PNA agents. In general, conjugates having IC_{50} values not less than micromolar have resulted. Indeed, it might be that this limitation in potency could be due to inherent properties of the PNA oligomers and their mode of biological action relying on steric interference rather than on enzymatic (RNase H/RISC) activation. However, by using phosphonate–peptide PNA conjugates [46], which due to their anionic character in analogy with antisense oligonucleotides and siRNA may be delivered by cationic lipids such as Lipofectamine, it was shown recently that in a luciferase splicing redirection system in which PNA–CPP conjugates are active in low micromolar concentrations, phosphonate–peptide PNA conjugates (like antisense oligonucleotides and siRNA in general) are active in the low nanomolar range. Therefore, the lower potency of the PNA–CPP conjugates must be due to lower availability for the mRNA target in the cells. In support of this conclusion, it has been known for some years that cellular uptake of cationic CPPs proceeds predominantly through an endosomal pathway and therefore efficient endosomal escape is paramount for good cytoplasmatic or nuclear delivery [47–49]. Recently, modified CPPs (e.g. [50]) and CPP conjugates (e.g. containing a lipophilic domain, a fatty acid) [51,52] with improved delivery properties have been identified, thereby providing hope that CPP derivatives combining efficient endosomal uptake and escape or perhaps even more effective CPPs that are taken up by endosomal-independent pathways may be discovered. Most interestingly, it was reported recently that the anthrax-protective antigen which has endosomal disruption/escape properties is able to deliver active PNAs to cells in culture [53].

5.8 Drug discovery/*in vivo* activity

Although drug discovery projects based on PNAs are hampered by the relatively poor *in vivo* bioavailability of PNA delivered i.p., i.v. or orally, several studies have been able to demonstrate *in vivo* activity of PNAs [12,37,43,50,54,55]. However, only a few pharmacokinetic (and preliminary toxicity) studies have been published so far and it is the general conclusion from these that unmodified PNAs administered i.v. or i.p. are very quickly excreted through the kidneys ($t_{1/2} < 1$ h) [40,56]. Peptide conjugation (tetralysine, penetratin, tat) may improve the *in vivo* bioavailability in terms of biological antisense effects [43,50,54] and for the $(KFF)_3K$ bacterial delivery peptide a fivefold increase in serum half-life in mice was measured [57]. However, in a recent pharmacokinetic study comparing the behavior in mice of unmodified and peptide (penetratin and tat)-conjugated PNAs only minor differences in tissue distributions and clearance half-lives were found upon i.p. or oral administration [58]. How these observations relate to the cellular behavior of such conjugates is not clear and more *in vivo* pharmacokinetic and efficacy studies of various chemically modified PNAs and PNA conjugates are clearly needed before the therapeutic potential of PNAs can be judged.

5.9 Origin of life

Peptide nucleic acids, in principle being ‘peptides’ with nucleic acid function, have also attracted attention in the discussion of various prebiotic scenarios in the origin of life preceding the ‘RNA world’ (e.g. [59–61]). These thoughts gained further momentum with the demonstration of possible prebiotic synthesis of central PNA building blocks, such as the aminoethylglycine backbone and pyrimidineacetic acids [62], and the detection and identification of diamino acids in the Murthison meteorite [63]. The idea of creating ‘living systems’ or even alternative life based on genetic information contained in PNA oligomers is also being experimentally entertained [64]. It is mainly the simple chemical structure of PNAs in combination with high chemical (and biological) stability while still retaining very efficient and specific nucleobase paired structures (not least duplexes) that makes this molecule relevant in these contexts. Whether PNA (-like) molecules played a role during the early origin of life (on Earth) is of course still an open question that may never be answered, and whether it will be possible to create (by design and/or selection) ‘living’, self-sustaining and-replication systems based an PNA oligomers is a challenge for the future.

5.10 Genetic diagnostics

PNAs have obvious potential applications as hybridization probes and tools in genetic diagnostics and especially in two areas this aspect has been successfully pursued. Many sensitive and specific protocols for *in situ* hybridization detection of especially infective agents and bacteria in environmental samples using PNA probes have been published during the past 15 years and this is still an active area of research (e.g. [65–67]).

PCR clamping [68] is another powerful and successful application. By exploiting the high sequence specificity and affinity of PNAs combined with the fact that PNAs are not a substrate for DNA polymerase (or for other enzymes for that matter), PCR templates differing by only one base can be selectively amplified by suppressing one of the templates with a PNA. In the most frequently used protocol, a PCR primer is designed to cover the mutated base (e.g. in a single base mutant oncogene) and a PNA is designed complementary to the wild-type, normal gene. By proper choice of PNA and primer target stabilities, PNA primer ratio and PCR annealing temperature, one can reach a situation where, the primer is able to bind only the mutant target as the PNA is blocking the wild-type target. Indeed, it has been possible using this strategy specifically to amplify a mutant target in the presence of over a 1000-fold excess of wild-type target. This is, of course, very useful for the diagnostic detection of, for example, a few cancer cells in a tissue. Most recently, it has been exploited for the development of a PCR assay for prenatal diagnosis of β-thalassemia in the fetus using a maternal blood sample [69] and for the detection of mutations in fibroblast growth factor 1 (TGFR1) in urine from patients with bladder cancer [70] and of mutations in epidermal growth factor receptor (EGFR) in non-small-cell lung cancer patients no longer responding to gefitinib treatment [71].

Finally, a light-directed synthesis of PNA oligomers on a solid support has been developed for the production of PNA arrays on chips [72]. These may become of interest for future diagnostic parallel hybridization formats in analogy with similar DNA array technology. The higher biological and chemical stability of PNAs may allow for extended reuse of the arrays.

5.11 PNA information tagging

A PNA can also be considered as a synthetic, organic information molecule, which may be exploited for identity tagging of individual components in combinatorial libraries or for sequence instructed combinatorial synthesis. For instance, active components of a PNA-tagged peptide library for profiling tyrosine kinases were studied through hybridization of the selected compounds to a DNA microarray in which the position defined the sequence of the PNA and thus of the peptide [73]. In another approach, a PNA-tagged small-molecule library was used to identify active protease inhibitors through hybridization to a DNA array [74–76].

5.12 Prospects

It is encouraging that almost 20 years after the discovery of PNAs, these molecules are still able to fascinate chemists and biologists, and scientists are still developing new applications and refining older ones. It is nonetheless somewhat discouraging that these molecules have not yet been able to enter drug discovery efficiently. The future will tell whether novel approaches to drug delivery or maybe novel drug targets or regimes will eventually allow this. In the meantime, I am confident that the simplicity of PNAs will still inspire chemists and biologists to develop the concept further and discover novel properties and applications within the broad field of molecular recognition and self-assembly.

Acknowledgments

The financial support of the European Commission through the Sixth Framework projects PACE (contract No. IST 002035) and SYNTHCELLS (contract No. NEST 043359) is gratefully acknowledged.

References

1. Nielsen PE, Egholm M, Berg RH, Buchardt O, Sequence-selective recognition of DNA by strand displacement with a thymine-substituted polyamide. *Science* 1991, **254**, 1497–1500.
2. Egholm M, Buchardt O, Nielsen PE, Berg RH, Peptide nucleic acids (PNA). Oligonucleotide analogs with an achiral peptide backbone. *Journal of the American Chemical Society* 1992, **114**, 1895–1897.
3. Ganesh KN, Nielsen PE, Peptide nucleic acids: analogs and derivatives. *Current Organic Chemistry* 2000, **4**, 931–943.
4. Nielsen E, RNA targeting using peptide nucleic acid. In *RNA Towards Medicine*, ed. Erdmann V, Brosius J, Barciszewski J, Springer, Berlin, 2006, pp. 395–403.
5. Sazani P, Kang SH, Maier MA, Wei C, Dillman J, Summerton J, Manoharan M, Kole R, Nuclear antisense effects of neutral, anionic and cationic oligonucleotide analogs. *Nucleic Acids Res* 2001, **29**, 3965–3974.
6. Karras JG, Maier MA, Lu T, Watt A, Manoharan M, Peptide nucleic acids are potent modulators of endogenous pre-mRNA splicing of the murine interleukin-5 receptor-alpha chain. *Biochemistry* 2001, **40**, 7853–7859.
7. Sazani P, Astriab-Fischer A, Kole R, Effects of base modifications on antisense properties of 2′-*O*-methoxyethyl and PNA oligonucleotides. *Antisense and Nucleic Acid Drug Development* 2003, **13**, 119–128.
8. Siwkowski AM, Malik L, Esau CC, Maier MA, Wancewicz EV, Albertshofer K, Monia BP, Bennett CF, Eldrup AB, Identification and functional validation of PNAs that inhibit murine CD40 expression by redirection of splicing. *Nucleic Acids Research* 2004, **32**, 2695–2706.
9. Wilusz JE, Devanney SC, Caputi M, Chimeric peptide nucleic acid compounds modulate splicing of the *bcl-x* gene *in vitro* and *in vivo*. *Nucleic Acids Research* 2005, **33**, 6547–6554.
10. Shiraishi T, Pankratova S, Nielsen PE, Calcium ions effectively enhance the effect of antisense peptide nucleic acids conjugated to cationic tat and oligoarginine peptides. *Chemistry and Biology* 2005, **12**, 923–929.
11. Fabani MM, Gait MJ, miR-122 targeting with LNA/2′-*O*-methyl oligonucleotide mixmers, peptide nucleic acids (PNA) and PNA–peptide conjugates. *RNA* 2008, **14**, 336–346.
12. Yin H, Lu Q, Wood M, Effective exon skipping and restoration of dystrophin expression by peptide nucleic acid antisense oligonucleotides in mdx mice. *Molecular Therapy* 2008, **16**, 38–45.
13. Wolf Y, Pritz S, Abes S, Bienert M, Lebleu B, Oehlke J, Structural requirements for cellular uptake and antisense activity of peptide nucleic acids conjugated with various peptides. *Biochemistry* 2006, **45**, 14944–14954.
14. Nielsen PE, Egholm M, Buchardt O, Evidence for (PNA)2/DNA triplex structure upon binding of PNA to dsDNA by strand displacement. *Journal of Molecular Recognition* 1994, **7**, 165–170.
15. Demidov VV, Yavnilovich MV, Belotserkovskii BP, Frank-Kamenetskii MD, Nielsen PE, Kinetics and mechanism of polyamide (‘peptide’) nucleic acid binding to duplex DNA. *Proceedings of the National Academy of Sciences of the United States of America* 1995, **92**, 2637–2641.
16. Nielsen PE, Christensen L, Strand displacement binding of a duplex-forming homopurine PNA to a homopyrimidine duplex DNA target. *Journal of the American Chemical Society* 1996, **118**, 2287–2288.
17. Lohse J, Dahl O, Nielsen PE, Double duplex invasion by peptide nucleic acid: a general principle for sequence-specific targeting of double-stranded DNA. *Proceedings of the National Academy of Sciences of the United States of America* 1999, **96**, 11804–11808.
18. Hansen ME, Bentin T, Nielsen PE, High-affinity triplex targeting of double stranded DNA Using chemically modified peptide nucleic acid oligomers. *Nucleic Acids Research* 2009, **37**, 4498–4507.
19. Ishizuka T, Yoshida J, Yamamoto Y, Sumaoka J, Tedeschi T, Corradini R, Sforza S, Komiyama M, Chiral introduction of positive charges to PNA for double-duplex invasion to versatile sequences. *Nucleic Acids Research* 2008, **36**, 1464–1471.

20. Rapireddy S, He G, Roy S, Armitage BA, Ly DH, Strand invasion of mixed-sequence B-DNA by acridine-linked, gamma-peptide nucleic acid (gamma-PNA). *Journal of the American Chemical Society* 2007, **129**, 15596–15600.
21. Roy S, Tanious FA, Wilson WD, Ly DH, Armitage BA, High-affinity homologous peptide nucleic acid probes for targeting a quadruplex-forming sequence from a MYC promoter element. *Biochemistry* 2007, **46**, 10433–10443.
22. Nielsen PE, Egholm M, Buchardt O, Sequence-specific transcription arrest by peptide nucleic acid bound to the DNA template strand. *Gene* 1994, **149**, 139–145.
23. Hanvey JC, Peffer NJ, Bisi JE, Thomson SA, Cadilla R, Josey JA, Ricca DJ, Hassman CF, Bonham MA, Antisense and antigen properties of peptide nucleic acids. *Science* 1992, **258**, 1481–1485.
24. Janowski BA, Kaihatsu K, Huffman KE, Schwartz JC, Ram R, Hardy D, Mendelson CR, Corey DR, Inhibiting transcription of chromosomal DNA with antigene peptide nucleic acids. *Nature Chemical Biology* 2005, **1**, 210–215.
25. Milne L, Xu Y, Perrin DM, Sigman DS, An approach to gene-specific transcription inhibition using oligonucleotides complementary to the template strand of the open complex. *Proceedings of the National Academy of Sciences of the United States of America* 2000, **97**, 3136–3141.
26. Urnov FD, Miller JC, Lee YL, Beausejour CM, Rock JM, Augustus S, Jamieson AC, Porteus MH, Gregory PD, Holmes MC. Highly efficient endogenous human gene correction using designed zinc-finger nucleases. *Nature* 2005, **435**, 646–651.
27. Lombardo A, Genovese P, Beausejour CM, Colleoni S, Lee YL, Kim KA, Ando D, Urnov FD, Galli C, Gregory PD, Holmes MC, Naldini L. Gene editing in human stem cells using zinc finger nucleases and integrase-defective lentiviral vector delivery. *Nature Biotechnology* 2007; **25**, 1298–1306.
28. Chin JY, Kuan JY, Lonkar PS, Krause DS, Seidman MM, Peterson KR, Nielsen PE, Kole R, Glazer PM, Correction of a splice-site mutation in the beta-globin gene stimulated by triplex-forming peptide nucleic acids. *Proceedings of the National Academy of Sciences of the United States of America* 2008, **105**, 13514–13519.
29. Good L, Nielsen PE, Antisense inhibition of gene expression in bacteria by PNA targeted to mRNA. *Nature Biotechnology* 1998, **16**, 355–358.
30. Good L, Awasthi SK, Dryselius R, Larsson O, Nielsen PE, Bactericidal antisense effects of peptide–PNA conjugates. *Nature Biotechnology* 2001, **19**, 360–364.
31. Dryselius R, Aswasti SK, Rajarao GK, Nielsen PE, Good L, The translation start codon region is sensitive to antisense PNA inhibition in *Escherichia coli*. *Oligonucleotides* 2003, **13**, 427–433.
32. Dryselius R, Nekhotiaeva N, Nielsen PE, Good L, Antibiotic-free bacterial strain selection using antisense peptide nucleic acid. *BioTechniques* 2003, **35**, 1060–1064.
33. Geller BL, Antibacterial antisense. *Current Opinion in Molecular Therapeutics* 2005, **7**, 109–113.
34. Nikravesh A, Dryselius R, Faridani OR, Goh S, Sadeghizadeh M, Behmanesh M, Ganyu A, Klok EJ, Zain R, Good L, Antisense PNA accumulates in *Escherichia coli* and mediates a long post-antibiotic effect. *Molecular Therapy* 2007, **15**, 1537–1542.
35. Kurupati P., Tan KS, Kumarasinghe G, Poh CL, Inhibition of gene expression and growth by antisense peptide nucleic acids in a multiresistant beta-lactamase-producing *Klebsiella pneumoniae* strain. *Antimicrobial Agents and Chemotherapy* 2007, **51**, 805–811.
36. Nekhotiaeva N, Awasthi SK, Nielsen PE, Good L, Inhibition of *Staphylococcus aureus* gene expression and growth using antisense peptide nucleic acids. *Molecular Therapy* 2004, **10**, 652–659.
37. Tan XX, Actor JK, Chen Y, Peptide nucleic acid antisense oligomer as a therapeutic strategy against bacterial infection: proof of principle using mouse intraperitoneal infection. *Antimicrobial Agents and Chemotherapy* 2005, **49**, 3203–3207.
38. Good L, Nielsen PE, Inhibition of translation and bacterial growth by peptide nucleic acid targeted to ribosomal RNA. *Proceedings of the National Academy of Sciences of the United States of America* 1998, **95**, 2073–2076.
39. Huang XW, Pan J, An XY, Zhuge HX, Inhibition of bacterial translation and growth by peptide nucleic acids targeted to domain II of 23S rRNA. *Journal of Peptide Science* 2007, **13**, 220–226.
40. Hamzavi R, Dolle F, Tavitian B, Dahl O, Nielsen PE, Modulation of the pharmacokinetic properties of PNA: preparation of galactosyl, mannosyl, fucosyl, *N*-acetylgalactosaminyl and *N*-acetylglucosaminyl derivatives of aminoethylglycine peptide nucleic acid monomers and their incorporation into PNA oligomers. *Bioconjugate Chemistry* 2003, **14**, 941–954.

41. Abes R, Arzumanov AA, Moulton HM, Abes S, Ivanova GD, Iversen PL, Gait MJ, Lebleu B, Cell-penetrating-peptide-based delivery of oligonucleotides: an overview. *Biochemical Society Transactions* 2007, **35**, 775–779.
42. Abes S, Turner JJ, Ivanova GD, Owen D, Williams D, Arzumanov A, Clair P, Gait MJ, Lebleu B, Efficient splicing correction by PNA conjugation to an R6-penetratin delivery peptide. *Nucleic Acids Research* 2007, **35**, 4495–4502.
43. Pooga M, Soomets U, Hällbrink M, Valkna A, Saar K, Rezaei K, Kahl U, Hao JX, Xu XJ, Wisenfeld-Hallin Z, Hökfelt T, Bartfai T, Langel Ü, Cell penetrating PNA constructs regulate galanin receptor levels and modify pain transmission *in vivo*. *Nature Biotechnology* 1998, **16**, 857–861.
44. Bendifallah N, Rasmussen FW, Zachar V, Ebbesen P, Nielsen PE, Koppelhus U, Evaluation of cell-penetrating peptides (CPPs) as vehicles for intracellular delivery of antisense peptide nucleic acid (PNA). *Bioconjugate Chemistry* 2006, **17**, 750–758.
45. El-Andaloussi S, Johansson HJ, Holm T, Langel Ü, A novel cell-penetrating peptide, M918, for efficient delivery of proteins and peptide nucleic acids. *Molecular Therapy* 2007, **15**, 1820–1826.
46. Shiraishi T, Hamzavi R, Nielsen PE, Subnanomolar antisense activity of phosphonate–peptide nucleic acid (PNA) conjugates delivered by cationic lipids to HeLa cells. *Nucleic Acids Research* 2008, **36**, 4424–4432.
47. Abes S, Williams D, Prevot P, Thierry A, Gait MJ, Lebleu B, Endosome trapping limits the efficiency of splicing correction by PNA–oligolysine conjugates. *Journal of Controlled Release* 2006, **110**, 595–604.
48. Shiraishi T, Nielsen PE, Photochemically enhanced cellular delivery of cell penetrating peptide–PNA conjugates. *FEBS Letters* 2006, **580**, 1451–1456.
49. Shiraishi T, Nielsen PE, Enhanced delivery of cell-pepetrating peptide–peptide nucleic acid conjugates by endosomal disruption. *Nature Protocols* 2006, **1**, 633–636.
50. Ivanova GD, Arzumanov A, Abes R, Yin H, Wood MJ, Lebleu B, Gait MJ, Improved cell-penetrating peptide–PNA conjugates for splicing redirection in HeLa cells and exon skipping in mdx mouse muscle. *Nucleic Acids Research* 2008, **36**, 6418–6428.
51. Koppelhus U, Shiraishi T, Zachar V, Pankratova S, Nielsen PE, Improved cellular activity of antisense peptide nucleic acids by conjugation to a cationic peptide–lipid (CatLip) domain. *Bioconjugate Chemistry* 2008, **19**, 1526–1534.
52. Hu J, Corey DR, Inhibiting gene expression with peptide nucleic acid (PNA)–peptide conjugates that target chromosomal DNA. *Biochemistry* 2007, **46**, 7581–7589.
53. Wright DG, Zhang Y, Murphy JR, Effective delivery of antisense peptide nucleic acid oligomers into cells by anthrax protective antigen. *Biochemical and Biophysical Research Communications* 2008, **376**, 200–205.
54. Sazani P, Gemignani F, Kang S-H, Maier MA, Manoharan M, Persmark M, Bortner D, Kole R, Systemically delivered antisense oligomers upregulate gene expression in mouse tissues. *Nature Biotechnology* 2002, **20**, 1228–1233.
55. Boffa LC, Cutrona G, Cilli M, Matis S, Damonte G, M Mariani. R, Millo E, Moroni M, Roncella S, Fedeli F, Ferrarini M, Inhibition of Burkitt's lymphoma cells growth in SCID mice by a PNA specific for a regulatory sequence of the translocated c-myc. *Cancer Gene Therapy* 2007, **14**, 220–226.
56. McMahon BM, Mays D, Lipsky J, Stewart JA, Fauq A, Richelson E, Pharmacokinetics and tissue distribution of a peptide nucleic acid after intravenous administration. *Antisense and Nucleic Acid Drug Development* 2002, **12**, 65–70.
57. Kristensen E, *In vitro* and *in vivo* studies on pharmacokinetics and metabolism of PNA constructs in rodents. In *Peptide Nucleic Acids: Methods and Protocols*, ed. Nielsen PE, Humana Press, Totowa, NJ, 2002, pp. 259–269.
58. Ganguly S, Chaubey B, Tripathi S, Upadhyay A, Neti PV, Howell RW, Pandey VN, Pharmacokinetic analysis of polyamide nucleic-acid-cell penetrating peptide conjugates targeted against HIV-1 transactivation response element. *Oligonucleotides* 2008, **18**, 277–286.
59. Nielsen PE, Peptide nucleic acid (PNA): a model structure for the primordial genetic material. *Origins of Life and Evolution of the Biosphere* 1993, **23**, 323–327.
60. Böhler C, Nielsen PE, Orgel LE, Template switching between PNA and RNA oligonucleotides. *Nature* 1995, **376**, 578–581.
61. Schmidt JG, Nielsen PE, Orgel LE, Information transfer from peptide nucleic acids to RNA by template-directed syntheses. *Nucleic Acids Research* 1997, **25**, 4797–4802.

62. Nelson KE, Levy M, Miller SL, Peptide nucleic acids rather than RNA may have been the first genetic molecule. *Proceedings of the National Academy of Sciences of the United States of America* 2000, **97**, 3868–3871.
63. Meierhenrich UJ, Caro GMM, Bredehöft JH, Jessberger EK, Thiemann WHP, Identification of diamino acids in the Murchison meteorite. *Proceedings of the National Academy of Sciences of the United States of America* 2004, **101**, 9182–9186.
64. Rasmussen S, Bedau MA, Chen L, Deamer D, Krakauer DC, Packard NH *et al.* (eds), *Protocells, Bridging Nonliving and Living Matter*, MIT Press, Cambridge, MA, 2008.
65. Braganra SM, Azevedo NF, Simoes LC, Keevil CW, Vieira MJ, Use of fluorescent *in situ* hybridization for the visualization of *Helicobacter pylori* in real drinking water biofilms. *Water Science and Technology* 2007, **55**, 387–393.
66. Reller ME, Mallonee AB, Kwiatkowski NP, Merz WG, Use of peptide nucleic acid-fluorescence *in situ* hybridization for definitive, rapid identification of five common *Candida* species. *Journal of Clinical Microbiology* 2007, **45**, 3802–3803.
67. Forrest GN, Roghmann MC, Toombs LS, Johnson JK, Weekes E, Lincalis DP, Venezia RA, PNA FISH for hospital acquired enterococcal bacteremia: delivering earlier effective antimicrobial therapy. *Antimicrobial Agents and Chemotherapy* 2008, **52**, 3558–3563.
68. Ørum H, Nielsen PE, Egholm M, Berg RH, Buchardt O, Stanley C, Single base pair mutation analysis by PNA directed PCR clamping. *Nucleic Acids Research* 1993, **21**, 5332–5336.
69. Galbiati S, Foglieni B, Travi M, Curcio C, Restagno G, Sbaiz L, Smid M, Pasi F, Ferrari A, Ferrari M, Cremonesi L, Peptide–nucleic acid-mediated enriched polymerase chain reaction as a key point for non-invasive prenatal diagnosis of beta-thalassemia. *Haematologica* 2008, **93**, 610–614.
70. Miyake M, Sugano K, Kawashima K, Ichikawa H, Hirabayashi K, Kodama T, Fujimoto H, Kakizoe T, Kanai Y, Fujimoto K, Hirao Y, Sensitive detection of FGFR3 mutations in bladder cancer and urine sediments by peptide nucleic acid-mediated real-time PCR clamping. *Biochemical and Biophysical Research Communications* 2007, **362**, 865–871.
71. Miyazawa H, Tanaka T, Nagai Y, Matsuoka M, Sutani A, Udagawa K, Zhang J, Hirama T, Murayama Y, Koyama N, Ikebuchi K, Nagata M, Kanazawa M, Nukiwa T, Takenoshita S, Kobayashi K, Hagiwara K, Peptide nucleic acid-locked nucleic acid polymerase chain reaction clamp-based detection test for gefitinib-refractory T790M epidermal growth factor receptor mutation. *Cancer Science* 2008, **99**, 595–600.
72. Liu ZC, Shin DS, Shokouhimehr M, Lee KN, Yoo BW, Kim YK, Lee YS, Light-directed synthesis of peptide nucleic acids (PNAs) chips. *Biosensors and Bioelectronics* 2007, **22**, 2891–2897.
73. Pouchain D, Diaz-Mochon JJ, Bialy L, Bradley M, A 10,000 member PNA-encoded peptide library for profiling tyrosine kinases. *ACS Chemical Biology* 2007, **2**, 810–818.
74. Winssinger N, Damoiseaux R, Tully DC, Geierstanger BH, Burdick K, Harris JL, PNA-encoded protease substrate microarrays. *Chemistry and Biology* 2004, **11**, 1351–1360.
75. Urbina HD, Debaene F, Jost B, Bole-Feysot C, Mason DE, Kuzmic P, Harris JL, Winssinger N, Self-assembled small-molecule microarrays for protease screening and profiling. *ChemBioChem* 2006, **7**, 1790–1797.
76. Debaene F, Da Silva JA, Pianowski Z, Duran FJ, Winssinger N, Expanding the scope of PNA-encoded libraries: divergent synthesis of libraries targeting cysteine, serine and metallo-proteases as well as tyrosine phosphatases. *Tetrahedron* 2007, **63**, 6577–6586.

6

The Interactions of Small Molecules with DNA and RNA

Yun Xie, Victor K. Tam and Yitzhak Tor

6.1 Introduction

Targeting DNA and RNA with small molecules represents an area of great academic and pharmaceutical interest; such interactions could potentially affect nucleic acid function and cellular response [1]. Examination of the central dogma of biology (Figure 6.1a) shows multiple points for potential intervention by small molecules that specifically target nucleic acids. DNA binders can interfere with DNA replication and impact cell proliferation or regulate transcription and ultimately halt gene expression. Similar effects can be brought about by RNA binders that meddle with translation by binding mRNA directly, as in the recently discovered ligand-mediated riboswitches (Figure 6.1b), but can also have a more universal impact on translation by binding ribosomal RNA. Nucleic acid binders can also potentially alter other nucleic acid interconversions, such as reverse transcription, an essential process for retroviral replication.

Many intricacies are involved in the binding of small molecules to nucleic acids; their structure, electrostatics and hydration must be understood. It is also essential to consider the modes of binding and their thermodynamics. Thus, it is necessary to first grasp the global biophysical properties that mediate ligand binding, affinity and selectivity before addressing current developments in targeting DNA and RNA. In this chapter, the fundamental features of nucleic acids and their interactions with small molecules are covered. While such important recognition events ultimately have to be individually deciphered, we attempt to provide certain generalizations, while referring the reader to key contributions and review articles. Both classical and contemporary representative examples will be presented along with a discussion of future prospects in this fascinating area of research at the boundaries of chemistry and biology.

The Chemical Biology of Nucleic Acids Edited by Günter Mayer

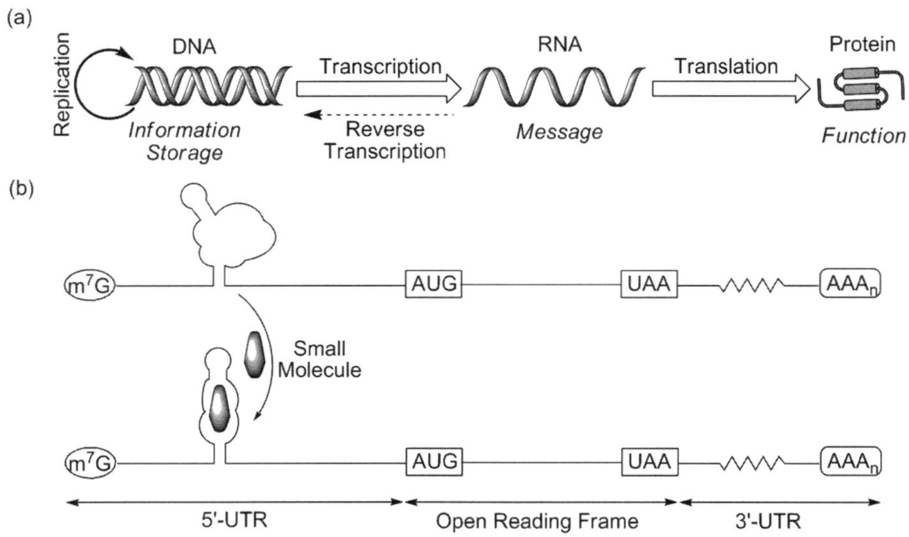

Figure 6.1 *(a) The central dogma of biology. (b) RNA binders could affect the translation of mRNA by binding at the 5′-untranslated region and interfering with ribosome scanning, a process preceding translation*

6.2 Nucleic acid building blocks

To understand the molecular features that govern nucleic acid–ligand interactions, one needs to appreciate the molecular information concealed within these biopolymers. The smallest repeating units of such biopolymers are nucleotides, phosphorylated nucleosides (Figure 6.2). In each nucleoside, nitrogenous heterocyclic bases are connected to D-ribose through a β-N-glycosidic bond [2]. Although both the charged

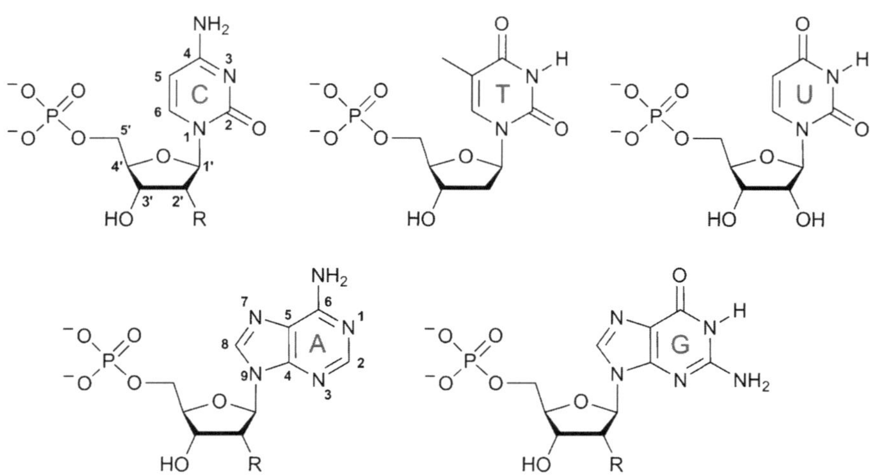

Figure 6.2 *Structures of common nucleotides found in nucleic acids. Top left to right, pyrimidines: cytidine (C) monophosphate, thymidine (T) monophosphate and uridine (U) monophosphate. Bottom left to right, purines: adenosine (A) monophosphate and guanosine (G) monophosphate. DNA: R=H. RNA: R=OH*

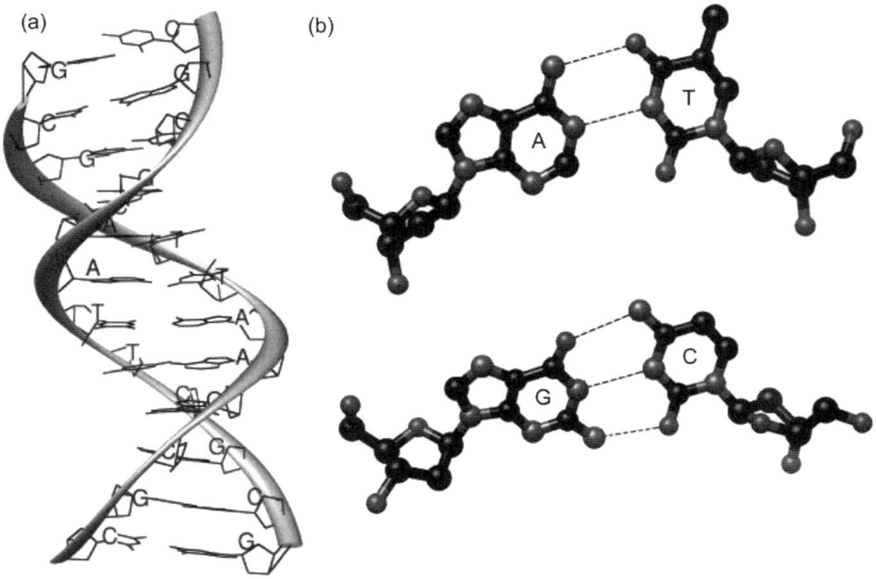

Figure 6.3 X-ray crystal structure of the Dickerson–Drew B-DNA dodecamer (a) and slices of the structure that represent Watson–Crick base pairs (b). The self-complementary sequence of the dodecamer is 5′-d(CGCGAATTCGCG)-3′. Hydrogen bonding is represented by dashed lines. PDB ID: 436D

phosphate and sugar moiety play important roles in numerous molecular recognition events, it is the nucleobases that contain a significant amount of specific molecular recognition information. These polarizable aromatic hydrocarbons, categorized into pyrimidines or purines, contain, in addition to a flat aromatic surface conducive to stacking and aromatic–aromatic interactions, a number of exocyclic functional groups, which are all capable of directional H-bonding interactions.

The ultimate manifestation of the recognition potential of nucleosides is seen in the formation of specific base pairs in the double helical structure of DNA as proposed by James Watson and Francis Crick in 1953 (Figure 6.3) [3]. The standard Watson–Crick base pairs are the dominant hydrogen-bonding patterns in

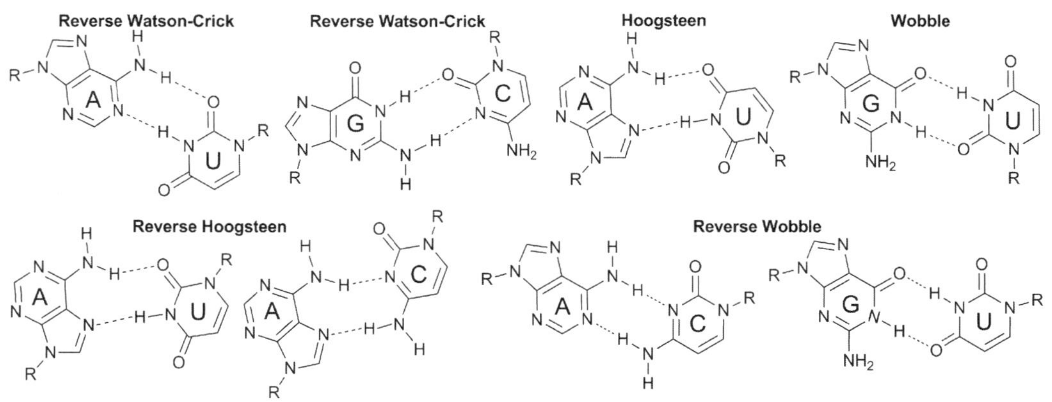

Figure 6.4 Non-canonical hydrogen bonding base pairs. DNA: R=deoxy-D-ribose. RNA: R=D-ribose

nucleic acids; however, other possibilities exist, taking advantage of additional hydrogen bonding sites on each nucleobase (Figure 6.4). Non-canonical interactions are essential for higher ordered structures, such as triple helical nucleic acids and quadruplexes.

6.3 Secondary structures of DNA and RNA

Once the nucleotides have oligomerized into long oligonucleotides, folding into higher structures takes place, governed by the information stored in the individual building blocks and conditions. Although a number of polymorphs exist, the right-handed B and A forms are the predominant secondary structure conformations of double-stranded DNA and RNA, respectively [4]. Pairing interactions consume many of the available noncovalent interactions. Clefts, grooves and surfaces defined by these folds present, however, specific functional groups for further molecular recognition events by exogenous ligands.

B-form DNA and A-form RNA differ in the width and depth of their grooves, which greatly impact the accommodation of small molecule ligands (Table 6.1). The major groove of DNA is wide and shallow compared to RNA, while DNA's minor groove is narrower and deeper than that of RNA (Figures 6.5 and 6.6). Thus, the proportions of the nucleic acid grooves influence the binding mode and affinity of ligands, depending on their shape and size. Other than dimensional disparities, DNA and RNA grooves also differ in their electrostatic potential, another important determinant for small ligand binding.

6.4 Electrostatics of the major and minor grooves of nucleic acids

Since nucleic acids are negatively charged polymers, electrostatics considerably influence their interactions with ions and ligands [5]. Within the major and minor grooves, differences in electrostatic potential are determined primarily by the orientation of individual base pairs [6,7]. In an A·T base pair, the N3 of adenine and the O2 of thymine both point towards the minor groove, resulting in an overall negative potential. In comparison, the minor groove of G·C is less densely charged, as the contribution from N3 in guanine and O2 carbonyl of cytosine is partially neutralized by the exocyclic N2 amino functionality [4,5]. In the major groove, there is less of a difference in electrostatic potential between A·T and G·C bases pairs. In A·T, a negative potential is contributed by O4 of thymine and N7 of adenine; however, positive electrostatic

Table 6.1 *Physical characteristics of B-form DNA and A-form RNA*

Characteristic	DNA (B-form)	RNA (A-form)
Base pairs per helical turn	10	11
Helical twist per base pair (°)	36	32.7
Distance between base pair (helical rise) (Å)	3.4	2.8
Pitch height (full helical turn) (Å)	34	32.7
Helix diameter (Å)	20	26
Major groove width (Å)	11.7	4.7
Major groove depth (Å)	8.8	12.9
Minor groove width (Å)	5.7	10.8
Minor groove depth (Å)	7.5	3.3

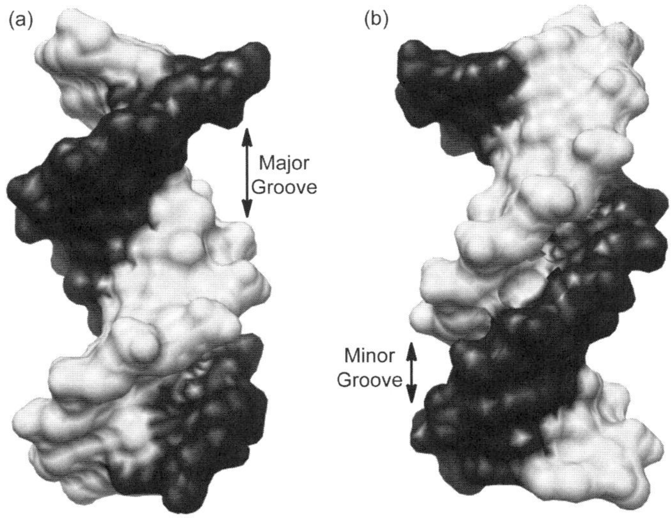

Figure 6.5 *X-ray crystal structure views of the (a) major groove and (b) minor groove of the Dickerson–Drew B-DNA dodecamer. Each strand is shaded separately, and each view is rotated approximately 180° from one another. Major and minor groove widths are the distances between inter-strand phosphates. PDB ID: 2BNA*

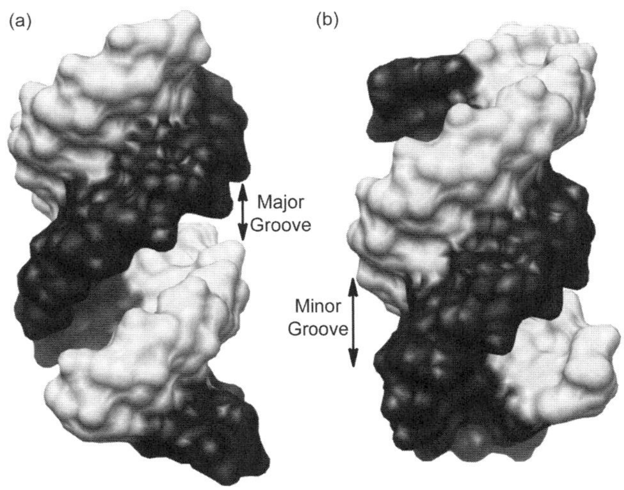

Figure 6.6 *Views of the (a) major groove and (b) minor groove of a representative A-form RNA duplex. The RNA duplex is an X-ray crystal structure of the self complementary 14-mer sequence 5′-U(UA)$_6$A-3′. Each strand is shaded differently, and each view is rotated approximately 180° from one another. Major and minor groove widths are the distances between inter-strand phosphates. PDB ID: 1RNA*

potential is provided by the exocyclic 6-amino functionality of adenine. Similarly, in G·C, positive potential from the exocyclic amine at N4 of cytosine partially diminishes the negative potential contributions from N7 and N6 of guanine. Theoretical calculations have supported the parity in groove electrostatics for DNA [8,9].

In the minor groove of DNA, the difference in electrostatics between A·T and G·C base pairs can affect sequence selectivity for ligand–nucleic acid binding. For instance, G·C base pairs are more electron rich, providing a preferable environment for intercalation of electron-deficient aromatic ligands. Additionally, the exocyclic 2-amino functionality of guanine in the minor groove can sterically deter binding to G·C tracts. Instead, A·T base pair–ligand interactions would be favored. Likewise, along the major groove, the steric bulk of the thymine methyl group can direct ligands to G·C base pairs. The methyl group can also facilitate van der Waals interactions with other hydrophobic residues such as the side-chains of leucine, isoleucine and alanine to promote protein binding [4].

In the A-polymorph, double-stranded RNA, the shallow and wide minor groove of RNA possesses a low electrostatic potential, whereas the deep major groove has a large electronegative potential. The major groove is, therefore, the main binding site for small molecules due to Coulombic attraction and hydrogen-bonding interactions. Proteins, on the other hand, generally prefer to bind in the shallow minor groove because of van der Waals contacts between the groove floor and the hydrophobic amino acid side-chains.

6.5 The impact of ions

The electrostatic potential of nucleic acids has a dramatic effect on molecules in the surrounding environment, particularly on metal cations. In the early 1970s, Manning proposed a quantitative model, known as the counterion condensation theory, to describe such interactions [10]. To reduce the effective negative charge of DNA strands and stabilize higher structures, simple cations aggregate within a few ångströms of the helix in no particular orientation, depending on the axial charge density. Record *et al.* later incorporated Debye–Hückel activity coefficients into Manning's model to suggest that it is necessary to consider the polarizable electron density of ions to fully quantify cation association [11].

Condensed ions energetically affect exogenous ligand binding as they compete for the same space [12]. Release of condensed ions upon ligand binding generally provides a positive entropic contribution to the overall free energy [10,11,13]. Since counterions provide structural stabilization, releasing condensed ions causes, however, an enthalpic penalty. Experimentally, ligand binding depends on salt concentration and buffer type with affinity typically increasing as bulk salt ion concentration decreases due to lowering of the enthalpic penalty for ion displacement.

6.6 The importance of hydration

Water plays a pivotal role in structural stabilization of nucleic acids and in ligand binding [12]. Being a highly polarizable molecule, water clusters around the negatively charged phosphate backbone and hydrogen bonds with donors and acceptors on the base pairs [14]. The water molecules involved in these interactions form the first hydration shell around double helices, estimated to consist of three water molecules per phosphate and approximately 20 water molecules throughout both grooves. A second, less-structured hydration shell exists, but it is short-lived and dependent on the orientation of the first [15]. Higher hydration shells were once thought to be stable; however, they have been determined to be extremely short-lived and similar to the bulk solvent [16].

When comparing A-DNA and A-RNA duplexes, a higher degree of hydration was found around RNA duplexes, most likely due to the additional 2′-OH group on the ribose moiety. The more compactly spaced phosphate backbone in A-form RNA duplexes allows water molecules to bridge the strands and enthalpically stabilize the helix. Due to the higher ordered hydration shell, ligand binding to RNA causes a higher entropic cost compared with DNA [17].

The displacement of water molecules upon ligand binding is difficult to quantify thermodynamically. Simplistically, the hydration shell and ions around DNA are expected to be ejected into the bulk solvent upon ligand binding, causing a positive change in entropy. Osmotic stress studies have, however, demonstrated uptake of water upon ligand binding and structural studies have illustrated cases of water-mediated ligand binding [18].

6.7 Modes and thermodynamics of binding

Ligand–nucleic acid interactions can be broadly categorized into three modes: (i) nonspecific electrostatic binding, (ii) intercalation and (iii) groove binding. Preference for one binding mode over another is largely influenced by ligand structure and the properties of the target nucleic acid. Ligands may bind to nucleic acids utilizing more than one mode.

Nonspecific electrostatic binding refers to the indiscriminate interactions between positively charged ions (e.g. Na^+, K^+ or Mg^{2+}) or polyamines (e.g. spermine, spermidine) that cluster around the negatively charged sugar–phosphate backbone. Although this binding mode governs the behavior of electrolytes and water in the nucleic acid's vicinity, it has little effect on ligands that occupy much larger surface area.

Intercalation describes the insertion of polyaromatic, planar molecules between consecutive base pairs [19,20]. Maximizing van der Waals contacts between the planar, aromatic system with Watson–Crick base pairs stabilizes the intercalated complex [21]. In addition, polarization of the stacked π–π systems minimizes repulsion within the complex.

Groove binding is typically governed by electrostatics and van der Waals forces. Ligands that bind the major or minor grooves of nucleic acids can hydrogen bond directly to the edges of base pairs. Generally, small ligands bind in the minor groove of DNA and major groove of RNA based on the buildup of electrostatic potential in those grooves. In contrast, proteins bind to the RNA minor groove and DNA major groove because those grooves have complementary hydrophobic environments.

Partitioning free energies involved with nucleic acid–ligand recognition can provide insight into how water and ions mediate binding. Chaires proposed to divide the overall free energy of ligand binding (ΔG_{obs}) into a sum of individual free energies associated with different molecular motions and interactions [13]:

$$\Delta G_{obs} = \Delta G_{DNA} + \Delta G_{r+t} + \Delta G_{hyd} + \Delta G_{pe} + \Delta G_{mol}\sqrt{a^2 + b^2}$$

where ΔG_{DNA} is the energetic cost of distorting the DNA (or RNA) structure upon ligand binding. This is observed mainly for intercalators due to base pair separation and unwinding that occur at the site of binding. In contrast, groove binders minimally perturb nucleic acid structure. Upon binding, degrees of freedom in translation or rotation are lost, which normally results in an entropic loss for complex formation (ΔG_{r+t}) [22]. One driving force for complex formation is the hydrophobic effect. Binding a hydrophobic ligand to a hydrophobic macromolecule (ΔG_{hyd}) contributes favorably to the overall free energy. The polyelectrolyte term (ΔG_{pe}) accounts for the energy from counterion and water release. Finally, weak forces (ΔG_{mol}), such as van der Waals, hydrogen bonding and dipole–dipole interactions, are also included. Experimentally, measuring the individual free energy contributions is difficult as they are all interrelated, but approximations, based on well-studied cases, are instructive.

Contrasting the binding of a groove binder and an intercalator to DNA is illustrative. A parsing of the binding free energies for Hoechst 33258, a groove binder, reveals no energetic cost from structural DNA distortion (ΔG_{DNA}), but a penalty of approximately 15 kcal mol^{-1} from a loss of molecular rotation and translation. Although the hydrophobic effect is the main driving force for binding ($\Delta G_{hyd} = -15$ kcal mol^{-1}), the release of condensed ions ($\Delta G_{pe} = -2$ kcal mol^{-1}) and weak intermolecular forces ($\Delta G_{mol} = 1$ kcal mol^{-1}) also contribute, generating approximately –12 kcal mol^{-1} for the overall binding free energy (ΔG_{obs}). In comparison, the binding of ethidium bromide, an intercalator, results in a small energetic penalty from the distortion of nearby base pairs ($\Delta G_{DNA} = +4$ kcal mol^{-1}) [23]. Similarly to Hoechst 33258, ethidium bromide causes a loss of rotational freedom ($\Delta G_{r+t} = +15$ kcal mol^{-1}) and release of condensed polyelectrolytes ($\Delta G_{pe} = -2$ kcal mol^{-1}). In contrast, hydrophobic interactions ($\Delta G_{hyd} = -12$ kcal mol^{-1}) and weak intermolecular forces ($\Delta G_{mol} = -12$ kcal mol^{-1}) play a larger role, resulting in –7 kcal mol^{-1} for the overall free energy of binding (ΔG_{obs}).

6.8 Targeting DNA

6.8.1 Intercalators

Early studies of small molecule–nucleic acid interactions were done on natural dyes [24]. Modified acridines, such as proflavine, were thought to only interact with the exterior of the DNA double helix [25]. Increased solution viscosity of acridine–DNA duplexes prompted Leonard Lerman to suggest his intercalation hypothesis, describing the insertion of planar aromatic molecules, specifically, proflavine, between the bases of DNA [19]. Although fluorescent dyes such as Acridine Orange were previously known to stain the nuclei of cells and were utilized as antiseptics and antimalarials by Paul Ehrlich (who laid out the foundation for chemotherapy in his Nobel Lecture 100 years ago), the publication of Lerman's hypothesis in 1961 is considered to be the beginning of selective targeting of nucleic acids with small organic molecules [26,27].

Lerman proposed two perturbations to DNA structure upon intercalation: (i) the DNA length would increase proportionally to acridine concentration, leading to increased viscosity, and (ii) DNA supercoiling would diminish due to helix extension and unwinding [19,26]. Since the latter is unique to intercalation, it can be independently observed to verify this mode of binding. Additionally, the distortion that intercalation imparts on adjacent base pairs prevents subsequent binding, known as the nearest-neighbor principle. Insertion of ethidium reduces the normal DNA helical twist from 36° to 10°. Acridines, such as proflavine, have been shown to unwind the double helix by 17°, whereas anthracyclines, such as daunorubicin and adriamycin, untwist the intercalation site by 11° [2]. In addition to structural effects, photophysical changes in the absorption and emission of the chromophoric ligands, as originally shown by Waring in 1965 [28] and Le Pecq and Paoletti in 1967 [29], respectively, are typically associated with intercalation.

Simple intercalators typically contain a single aromatic core that mediates binding by insertion between base pairs. Complex intercalators, such as daunomycin and nogalamycin, contain additional functional groups (e.g. peptides, saccharides) that facilitate groove binding. Higher affinity is typically displayed by ligands that contain multi-intercalator scaffolds, such as echinomycin [30], a naturally occurring bisintercalator. Synthetic derivatives such as WP631 (a daunorubicin-based bisintercalator) and bis(methidium)spermine (a phenanthridinium-based bisintercalator), also exhibit extremely high affinity for DNA [31,32]. Importantly, intercalators displaying selectivity towards higher structures such as triplexes [33] and quadruplexes [34,35] have been explored in recent years.

6.8.2 DNA groove binders

Hoechst 33258, berenil, netropsin and DAPI bind nucleic acid grooves (Figure 6.7). Instead of the fused aromatic ring architecture seen in intercalators, groove binders are typically positively charged 'crescent'-shaped linked aromatic rings. Van der Waals interactions and hydrophobic contacts overcome the unfavorable hydration displacement. Many groove binders preferentially bind A·T-rich duplexes (avoiding the bulky 2-aminoguanine moiety) and display selectivity for B- over A-form DNA or RNA [4,36].

Two well-known groove binders, neotropsin and distamycin, are naturally occurring antibiotics [37]. Structural data reveal that the DNA double helix is not extensively elongated or perturbed upon netropsin

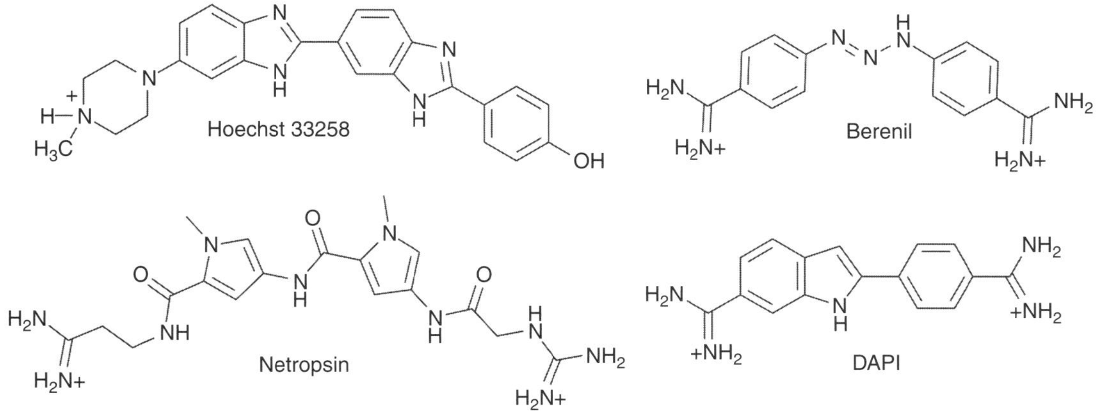

Figure 6.7 *Examples of groove binders. Most groove binders adopt a 'crescent'-shaped form*

binding (Figure 6.8). The netropsin amide protons hydrogen bond with adenine N3 and thymine O2, the same functional groups that are bridged by the spine of hydration, which is expelled upon binding. These structures reveal that the 2-amino group of guanine would disrupt binding and sterically clash with netropsin-like molecules. Replacing methylpyrrole with methylimidazole facilitates hydrogen bonding between G and the imidazole nitrogen (Figure 6.9) [38]. By replacing the pyrrole building blocks with imidazoles, these

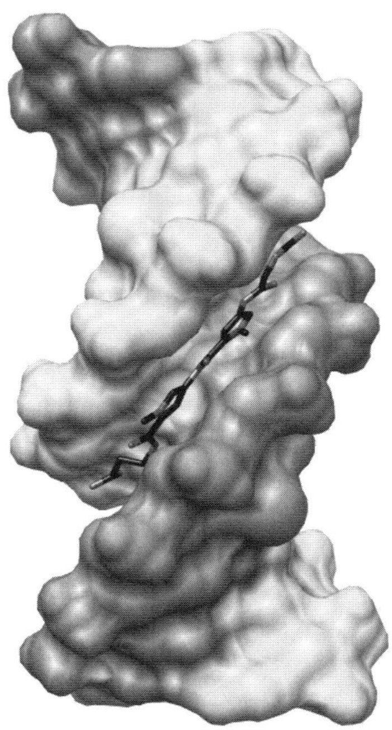

Figure 6.8 *X-ray crystal structure of a 1:1 netropsin–DNA complex. Netropsin resides in the minor groove of the DNA helix. PDB ID: 6BNA*

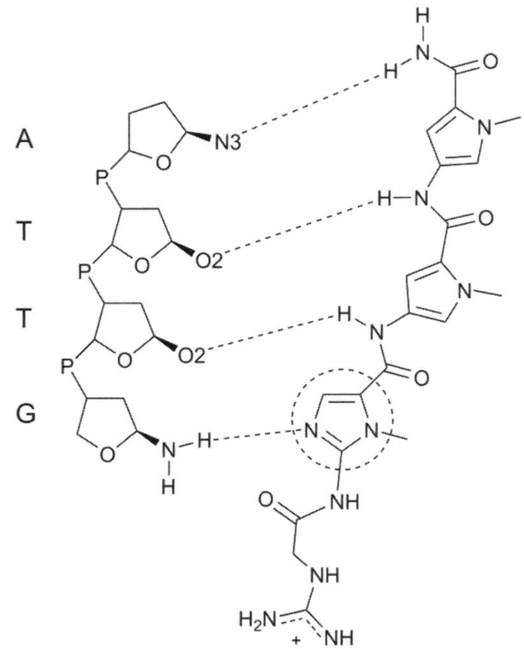

Figure 6.9 *Schematic representation of the hydrogen-bonding contacts between a PyPyIm 'lexitropsin' and ATTG bases in the minor groove of DNA. The additional imidazole (circled) nitrogen allows for hydrogen bonding with the N2 of guanosine, which normally clashes with a pyrrole hydrogen. The nucleic acid structure has been minimized with base identities indicated along the left column. The DNA sugar–phosphate backbone is represented by a reduced ribose and 'P' (for phosphorus). Only the purine or pyrimidine atoms involved in hydrogen bonding are indicated. Hydrogen bonding is represented by dashed lines*

'lexitropsins' can conceivably recognize A·T or G·C tracts; however, the G·C selectivity of lexitropsins has been limited at best [4].

6.8.3 Selective targeting of specific DNA sequences

A sought-after method for modulating gene expression is the selective targeting of regulatory, unique sequences of DNA. Netropsin and distamycin provided insight into how natural polyamide molecules selectively target specific base sequences. The observation that distamycin binds the DNA minor groove in a 2:1 ratio [39] inspired the design of sequence-specific binders (Figure 6.10). Dervan developed a heterocycle-based pairing system for targeting short DNA sequences using hairpin polyamides, with affinities reported in the nano- to picomolar range (Table 6.2) [36,40]. An imidazole–pyrrole (Im–Py) pair selectively targets G·C base pairs, while a pyrrole–imidazole (Py–Im) pairing selectively targets C·G base pairs. To resolve the indiscrimination of A·T or T·A base pairs by a pyrrole–pyrrole (Py–Py) pair, a hydroxypyrrole (Hp) was introduced [41]. This facilitated the selective targeting of T·A base pairs by hydroxypyrrole–pyrrole (Hp–Py) pairings and A·T base pairs by pyrrole–hydroxypyrrole (Py–Hp) pairings (Figure 6.11). Structural studies confirmed these key hydrogen-bonding contacts [42]. Examples of successful transcription regulation by such synthetic polyamides have been reported [43–46].

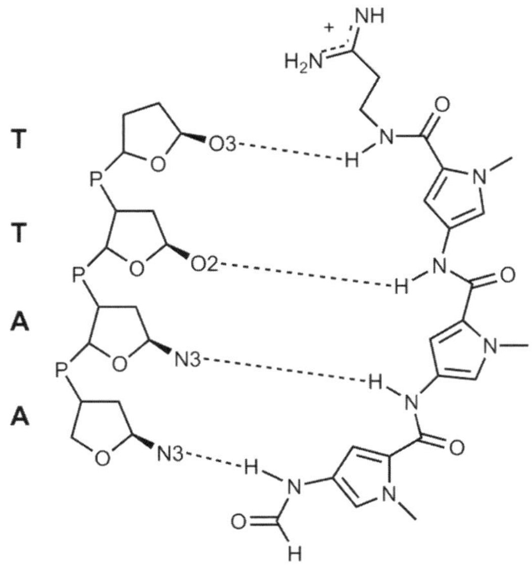

Figure 6.10 Schematic representation of the hydrogen-bonding contacts between distamycin and AATT bases in the minor groove of DNA. The nucleic acid structure has been minimized with base identities indicated along the left column. The DNA sugar–phosphate backbone is represented by a reduced ribose and 'P' (for phosphate). Only the purine or pyrimidine atoms involved in hydrogen bonding are indicated. Hydrogen bonding is represented by dashed lines

Table 6.2 Dervan's pairing rules for sequence selective targeting with polyamides

Pair[a]	G·C	C·G	T·A	A·T
Im/Py	√	×	×	×
Py/Im	×	√	×	×
Hp/Py	×	×	√	×
Py/Hp	×	×	×	√

[a] Im = imidazole; Py = pyrrole; Hy = hydroxypyrrole

6.8.4 Targeting DNA triple helices and quadruplexes

First observed in 1957 for a 2:1 mixture of poly r(U) and poly r(A), respectively, triple helices are formed when a single strand binds to the Hoogsteen face of base pairs in a double helix [47,48]. To advance 'anti-gene' strategies, short synthetic oligonucleotides have been developed to target stretches of DNA duplexes sequence-specifically and interfere with their expression [40,49]. Modified triplex-forming oligonucleotides have also been demonstrated to chemically modify or oxidatively cleave their DNA targets [50–53]. To enhance the affinity of triplex-forming oligonucleotides to their targets or to bind naturally occurring triple helical DNA (e.g. H-DNA, long tracts of repeating homopurines and homopyrimidines that are located near the promoter region of many genes), triplex-specific intercalators and binders have been developed [33,54–56] (Figure 6.12).

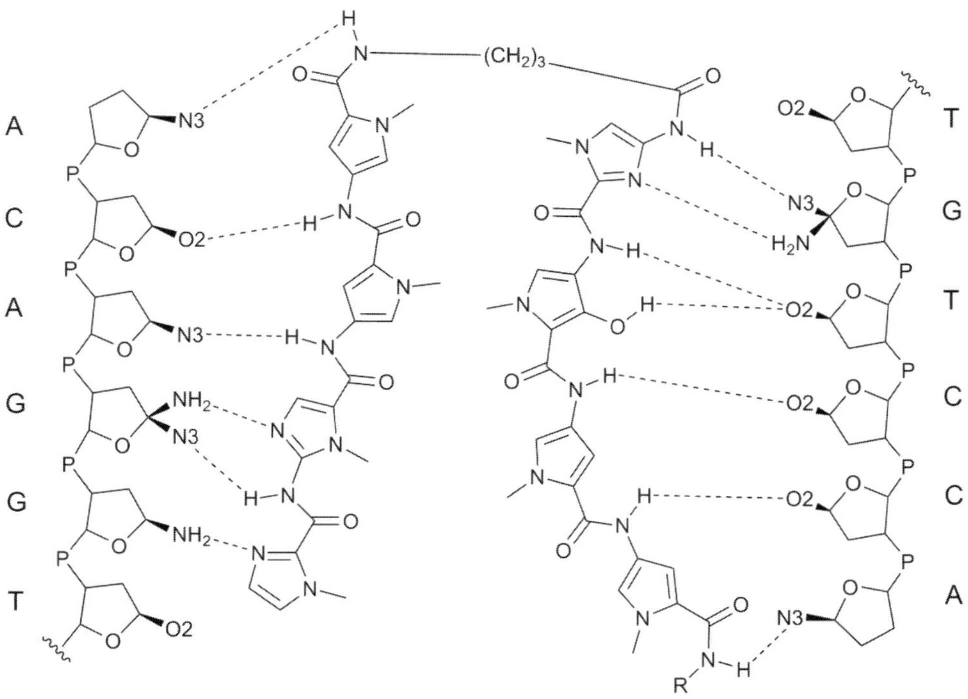

Figure 6.11 *Schematic representation of the hydrogen-bonding contacts between a sequence-selective hairpin polyamide in the minor groove of DNA. Note: the hydroxypyrrole allows for discrimination of an T·A base pair over A·T due to additional hydrogen-bonding interactions. The nucleic acid structure has been minimized with base identities indicated along the left and right columns. The DNA sugar–phosphate backbone is represented by a reduced ribose and 'P' (for phosphate). Only the purine or pyrimidine atoms involved in hydrogen bonding are indicated. Hydrogen bonding is represented by dashed lines*

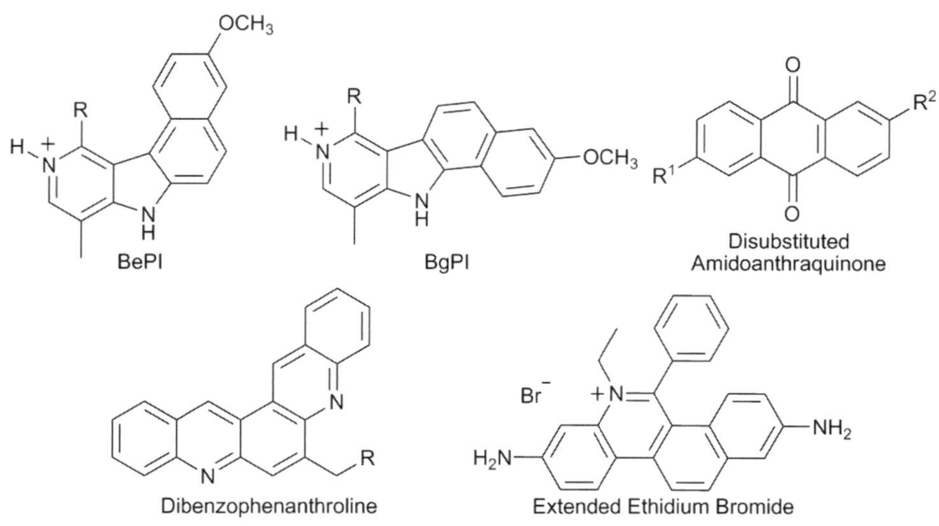

Figure 6.12 *Representative structures of triple helix binders*

Single-stranded DNA with long tracts of guanines could form G-quadruplexes through Watson–Crick and Hoogsteen base pairings on all four bases (Figure 6.13) [57]. Telomeres, specialized structures at the end of eukaryotic chromosomes, are thought to be rich in G-quadruplexes and are suspected to play a role in governing cell longevity by inhibiting telomerase activity (telomerase, an enzyme found mostly in cancer cells, normally extends telomeres, giving cancer cells the ability to proliferate indefinitely) [58–60]. Although every quadruplex contains G quartets, structural variations in the loop and groove regions exist. The development of quadruplex-selective and quadruplex-discriminating ligands is therefore an active area of exploration [61–64]. Several new heterocyclic ligands are shown in Figure 6.14.

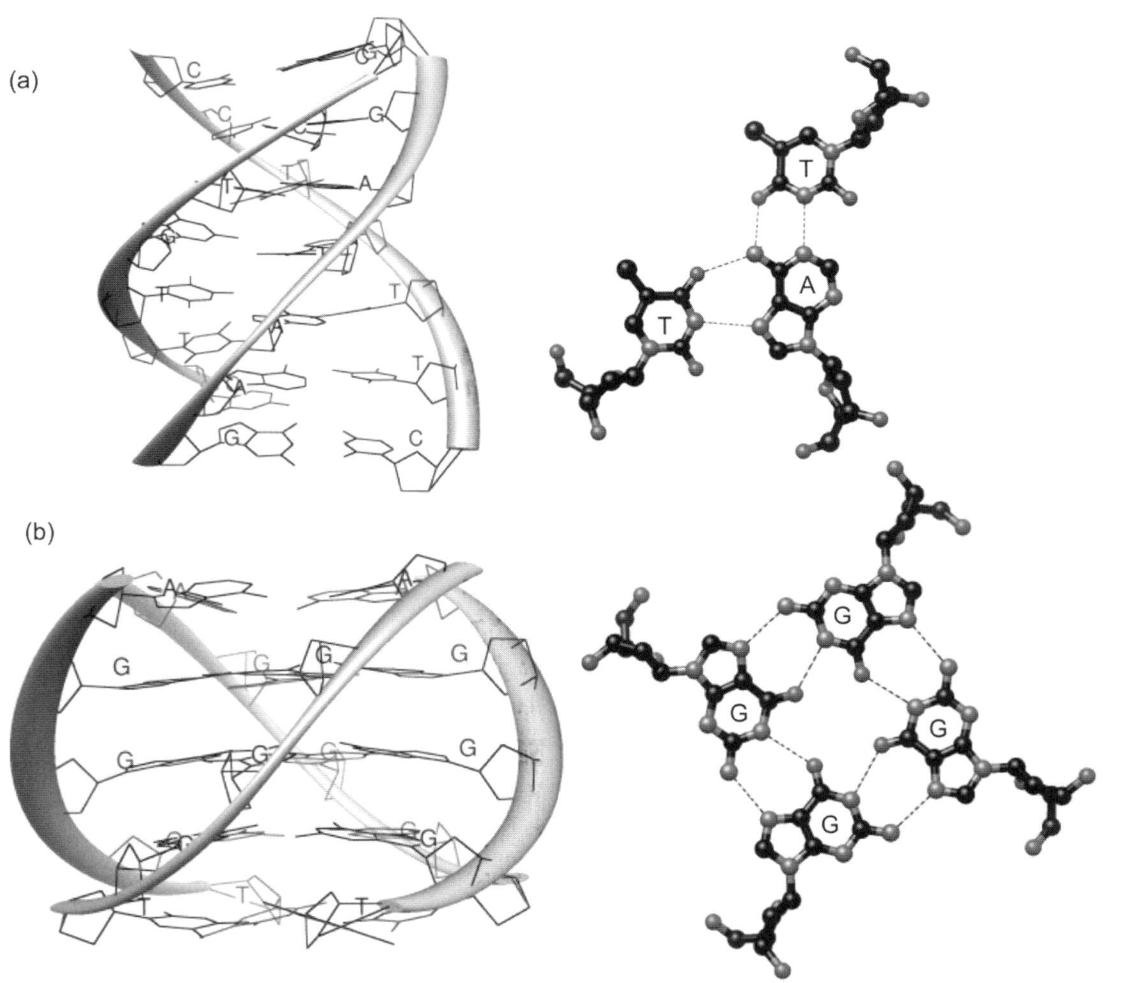

*__Figure 6.13__ (a) An NMR structure that is representative of a DNA triplex accompanied by a slice of the structure that shows the hydrogen-bonding pattern of a TA*T triplex. PDB ID: 149D. (b) An NMR structure of a truncated sequence of human telomeric DNA that is representative of quadruplex DNA accompanied by a slice of the structure that shows the hydrogen-bonding pattern of a G tetraplex. PDB ID: 1EVN. Hydrogen bonding is represented by dashed lines*

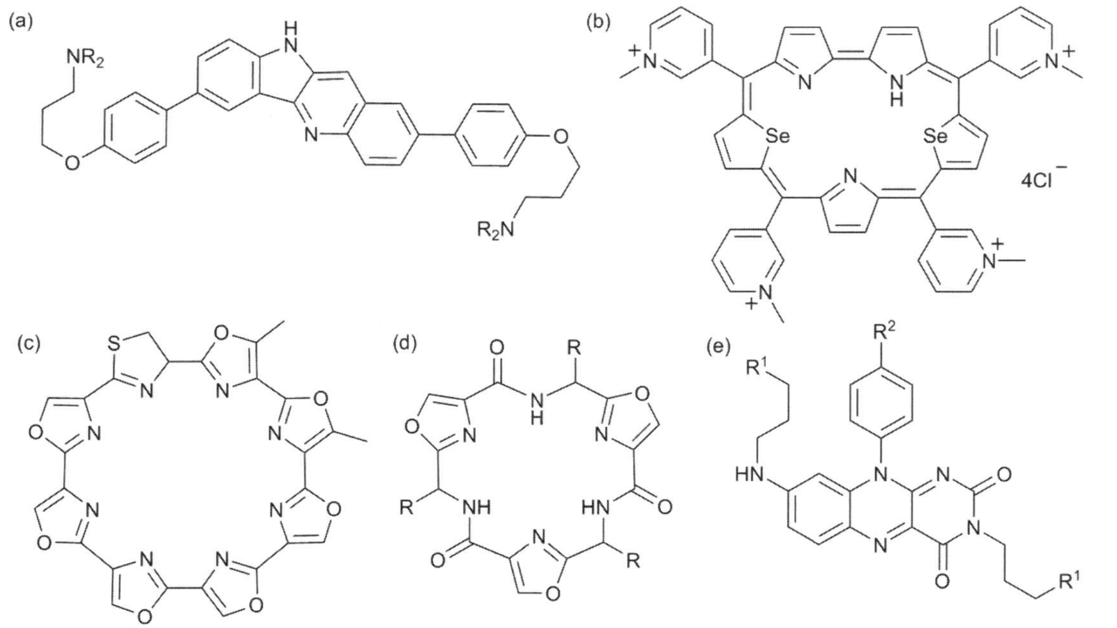

Figure 6.14 Scaffolds of G-quadruplex binders from the recent literature: (a) 2,7-disubstituted indoloquinolines [61]; (b) Se2SAP [62]; (c) telomstatin [62]; (d) oxazole-based peptide macrocycles [63]; (e) isoalloxazine ligands [64]

6.9 Targeting RNA

6.9.1 Importance and general principles

Until the mid-1990s, the exploration of small molecule–nucleic acid interactions was mostly centered on DNA, not RNA. Several reasons contributed to this trend, among them the accessibility of DNA oligonucleotides via efficient solid-phase synthesis protocols and their higher stability. Related challenges have also hindered high-resolution structural studies of RNA, further slowing structure-based design of specific RNA binders. The past two decades have seen, however, tremendous progress in our basic understanding of RNA structure and folding and also significant growth of our knowledge regarding RNA–ligand interactions [65,66]. Further motivation was provided by the urgent need to advance new approaches for antiviral and antibacterial therapy, due to the appearance of new and resistant pathogens.

Despite its significance, RNA recognition has yet to achieve the same level of understanding enjoyed by sequence-selective targeting of DNA. Unlike DNA, where Dervan's pairing rules allow one to design sequence-specific binders, few principles exist for targeting RNA with small molecules. Cellular RNA is intricately folded, generating internal and hairpin loops, bulges and stem junctions of various sizes and sequences [67]. While this structural diversity is critically important for RNA function, it hinders the decoding of recognition patterns and their exploitation for the judicious design and implementation of RNA-selective binders [68].

In a double-helical RNA, the major groove, although narrow, possesses the greatest negative electrostatic potential, but it is relatively inaccessible [69]. Bulges and loops, while distorting the RNA backbone and expanding the deep major groove, expose the discriminatory face of base pairs facilitating recognition by small ligands [70]. Selected examples of therapeutically relevant RNA targets illustrate the prevalence of such motifs (Figure 6.15). The widened major groove of the HIV-1 TAR accommodates the Tat peptide, its cognate biological ligand. The asymmetric bulge of the ribosomal prokaryotic A-site, and also the kissing hairpin loops of the HIV-1 DIS and the junction of the HIV-1 RRE, are also known to bind low molecular weight (MW) ligands [71–74]. It is important to appreciate that, while significant sequence and structural plasticity exist among hairpin loops, bulges and junctions, these domains can fold into well-defined structures via base stacking and also mismatched and non-canonical base pairing [2,68].

Two complementary models have been discussed in describing RNA–ligand binding. As a modified version of the prevalent 'induced-fit' model, the conformational capture theory hypothesizes that RNA, due to its dynamic nature, assumes numerous conformations, both biologically functional and nonfunctional [75–77]. Upon ligand binding, the RNA is locked into a given conformation that also governs its functional viability. The second model, focusing on electrostatic complementarity, suggests that the electrostatic potential generated by the RNA fold attracts ligands of complementary spatial charge density [66,78,79].

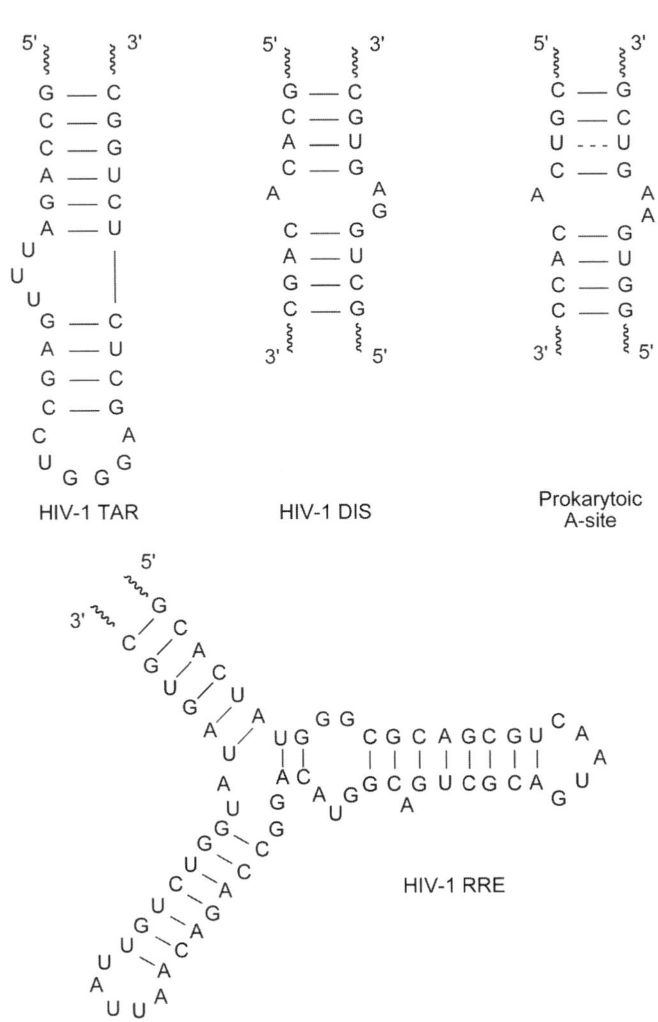

Figure 6.15 *RNA secondary structure of therapeutically targeted RNA constructs: the HIV-1 TAR (multi-base bulge), DIS (asymmetric internal loop), RRE (asymmetric internal loop) and the prokaryotic A-site (three-stem junction)*

This dictates that in many cases, ligands may compete with tightly bound cations that tend to condense at the sites of highest negative potential. Note, however, that these two models are not mutually exclusive and can represent different phases in the binding process.

The impact of ligand binding on RNA function can be complex. Small exogenous ligands may compete with an endogenous protein for the same binding pocket. Such competitive binding typically disfavors low MW ligands, as high affinity might be needed to compete with a tight cognate binder. Binding of a low MW ligand at a nearby site, concomitantly inducing a conformational change and locking the RNA into an unfavorable structure, can overcome such limitations. Such 'allosteric' inhibitory mechanisms have been suggested [80–82]. Another recognition mechanism that alleviates the need for extremely high affinity relies on RNA functioning as an on/off switch. To interfere with the biological function of the decoding site, for example, inhibitors with extreme affinities are not required (as competition with high-affinity endogenous ligands is not necessary; see below). Rather, ligands that display a combination of moderate affinity, high selectivity and appropriate molecular architecture are biologically active [68]. Note that quantifying the energetics associated with ligand-induced RNA structural changes is challenging. Isothermal calorimetry is often used for quantifying binding thermodynamics, but the overall calculated free energies (ΔG_{obs}) include contributions from multiple parameters, as described above [83–85].

6.9.2 Aminoglycoside antibiotics and the ribosome

The interactions between antibiotics and the ribosome have served as a paradigm for understanding RNA–ligand interactions and have inspired numerous contemporary research groups. Of particular significance is the established ability of aminoglycoside antibiotics to interfere with ribosomal protein biosynthesis [86–90]. Examples of these naturally occurring pseudo-oligosaccharides are shown in Figure 6.16. 2-Deoxystreptamine (2-DOS), a highly functionalized aminocyclitol, serves as the common core of most aminoglycosides. Glycosylation of the 2-DOS core with other amino sugars, usually at the 4- and 5- or 4- and 6-positions, yields most naturally occurring aminoglycosides. The abundance of protonatable amino groups on these structures suggests a highly cationic character that may facilitate interactions with negative electrostatic potential of RNA [91].

Several key discoveries zeroed in on the precise molecular target of aminoglycoside antibiotics. Moazed and Noller identified specific binding sites on ribosomal RNA using footprinting techniques [92]. Purohit and Stern demonstrated the direct binding of aminoglycosides to a short RNA construct representing the decoding site, suggesting that this site is an autonomous RNA domain capable of mimicking the function and recognition features of the entire 16S rRNA [93]. Observations by both Puglisi and co-workers [94,95] and Yokoyama and co-workers [96] illustrated that a minimal 27-mer A-site RNA construct is capable of binding aminoglycoside antibiotics of the neomycin family in a similar fashion to the decoding region. These discoveries facilitated the high-resolution structural characterization of short A-site constructs bound to aminoglycosides, by NMR and X-ray crystallography [71,95,97–99]. Importantly, crystal structures of aminoglycosides bound to the entire 30S ribosomal subunit revealed intermolecular contacts similar to the interactions observed in the structures determined with the short RNA constructs [100]. These important observations have validated the A-site as a *bona fide* drug target and paved the way for the development of useful biophysical assays for the discovery of A-site binders. It has also stimulated the design and synthesis of aminoglycoside analogs and mimetics as new RNA binders.

It is now accepted that aminoglycoside antibiotics exert their antimicrobial effect by tampering with protein biosynthesis by binding to the A-site and stabilizing an RNA conformation similar to that induced by the binding of the cognate acyl-tRNA to the corresponding mRNA (Figure 6.17) [101]. Distinguishing between cognate and noncognate tRNA–mRNA hybrids is hampered and the fidelity of translation is

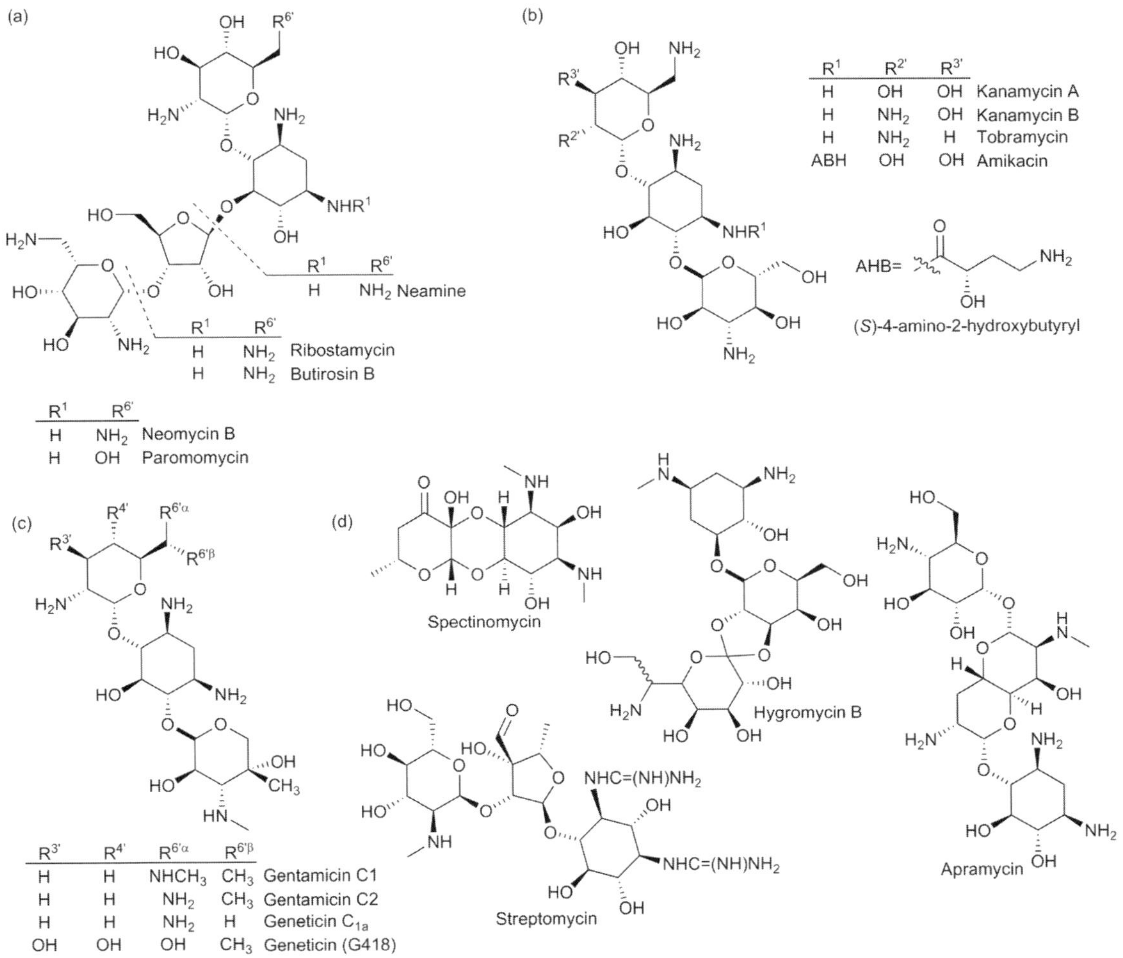

Figure 6.16 Structures of different classes of aminoglycosides: (a) neomycin family; (b) kanamycin and tobramycin family; (c) gentamicin family; (d) atypical aminoglycosides. Typical disubstituted 2-deoxystreptamine core is depicted along with related streptamine cores found in atypical aminoglycosides

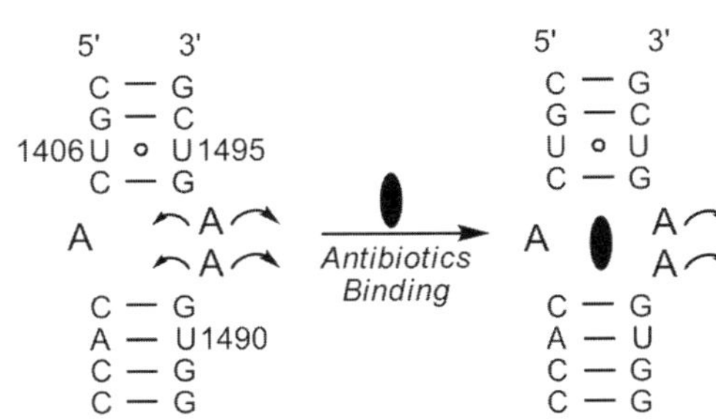

Figure 6.17 Upon antibiotic binding, the dynamics of the unpaired A1492 and A1493 residues are altered and they become extrahelical

compromised as near/noncognate charged tRNAs are not rejected by the ribosome. The A-site therefore operates as a small, yet critical, switch, as its conformational states impact the entire protein biosynthetic machinery. It is conceivable that the actual molecular mechanisms underlying this incredible riboswitch are more elaborate. In particular, recent observations suggest a complex interplay of structural and dynamic factors that are likely to govern the impact of these antibiotics on translation [102,103].

Three-dimensional structures of A-site–aminoglycoside complexes suggest that the 2-DOS moiety plays a key role in RNA recognition (Figure 6.18). Whereas aminoglycosides of the neomycin family bind the A-site in a

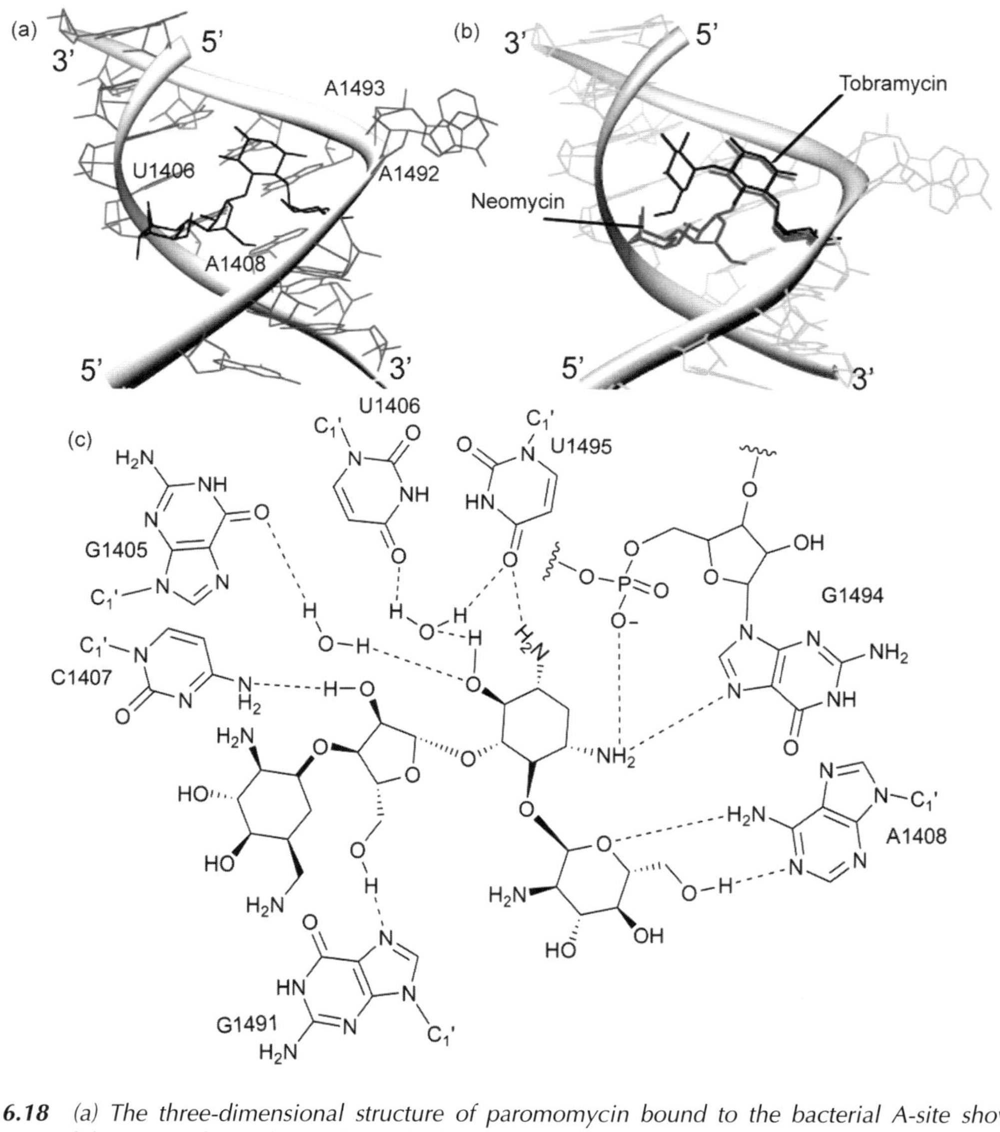

Figure 6.18 (a) The three-dimensional structure of paromomycin bound to the bacterial A-site shows key interactions of the aminoglycoside involving the 2-deoxystreptamine moiety (2-DOS) and the aminopyranose. PDB ID: 1J7T. (b) The overlap of the three-dimensional structures of tobramycin and neomycin bound to the bacterial A-site. PDB IDs: 1LC4 and 2ET4. (c) A more detailed look at specific base pair-like hydrogen bonding and water-mediated interactions between paromomycin and the bases of the bacterial A-site

different orientation compared with the kanamycin family, the 2-DOS maintains its location and orientation, suggesting that this *cis*-diaminocyclohexane moiety is the central pharmacophore in this recognition phenomenon. Additionally, a multitude of noncovalent and water-mediated interactions anchor the aminoglycosides within their binding site. Indeed, ITC studies of paromomycin binding to the A-site show a substantial energetic contribution from molecular interactions ($\Delta G_{mol} = -22.5$ kcal mol^{-1}) that are mediated by the charged ammonium groups [104].

6.9.3 Aminoglycoside antibiotics as universal RNA binders

The renaissance of aminoglycoside antibiotics, in recent years, can be attributed to Schroeder and co-workers' 1991 report, which demonstrated that these antibiotics could also inhibit splicing of group I introns [105], and to Green and co-workers' 1993 observations that aminoglycosides can also inhibit binding of the HIV-1 Rev protein to its RNA target, the Rev Response Element (RRE) [74]. Following these seminal reports, aminoglycosides and their derivatives were found to bind various RNA targets [106,107]. Small and large ribozymes [108–110], and also transfer RNAs [111,112], were found to bind aminoglycosides. Additional RNA targets, including HIV-1 TAR, DIS, the hepatitis delta virus ribozymes and RNase P, were also shown to interact with aminoglycoside antibiotics [113–115].

The underlying reasons for the promiscuity displayed by aminoglycosides are related to their highly charged nature and their conformational flexibility. Whereas aminoglycosides display inherent affinity to A-form duplexes [116,117], structural studies of aminoglycoside–RNA complexes reveal conformational plasticity, where related aminoglycosides assume drastically different conformations when bound to their corresponding RNA target. The inherent flexibility of aminoglycosides together with their electrostatically driven binding mode facilitate structural adaptation [118], so modestly stable complexes can form with diverse RNA targets. RNA–aminoglycoside interactions at these opportunistic binding sites tend to differ markedly from the contacts observed at the A-site, where a high-affinity complex of a compact bound structure is formed [119]. The A-site represents a unique target, since relatively few backbone conformational changes are associated with aminoglycoside binding. This predisposition towards aminoglycosides is not unforeseen, as these naturally occurring antibiotics have evolved specifically to target this Achilles heel of the ribosome [120].

6.9.4 Designing new RNA binders

Although the understanding of RNA–ligand recognition has advanced significantly [68], the discovery of new RNA-specific binders has nevertheless remained a taxing task. In particular, attempts to design ligands for RNA targets that do not share the unique traits of the A-site face the following challenges. (1) Deep solvent-excluded clefts are not abundant in most RNA folds. Numerous intermolecular contacts covering a significant surface area have to be established to secure affinity and selectivity. (2) A dominant contribution by electrostatic interactions creates a delicate interplay between overall charge, affinity and selectivity. (3) RNA–ligand binding frequently involves a mutual induced fit, where both the RNA host and the incoming ligand undergo conformational changes. Furthermore, the capability of flexible small molecules to 'remodel' structurally according to the RNA topography and electrostatic potential, and also the involvement of water-mediated contacts, further complicate ligand design. It is no surprise, then, that most new RNA binders have been fabricated by modifying the natural products or their fragments and by mimicking the specific building blocks of the natural antibiotics such as the 2-DOS ring (Figure 6.19) [121–127].

6.10 Conclusion and prospects

The evolution of this field, nowadays described as the chemical biology of nucleic acids, is threaded with discoveries that highlight small molecules as specific effectors of nucleic acid structure and function. While

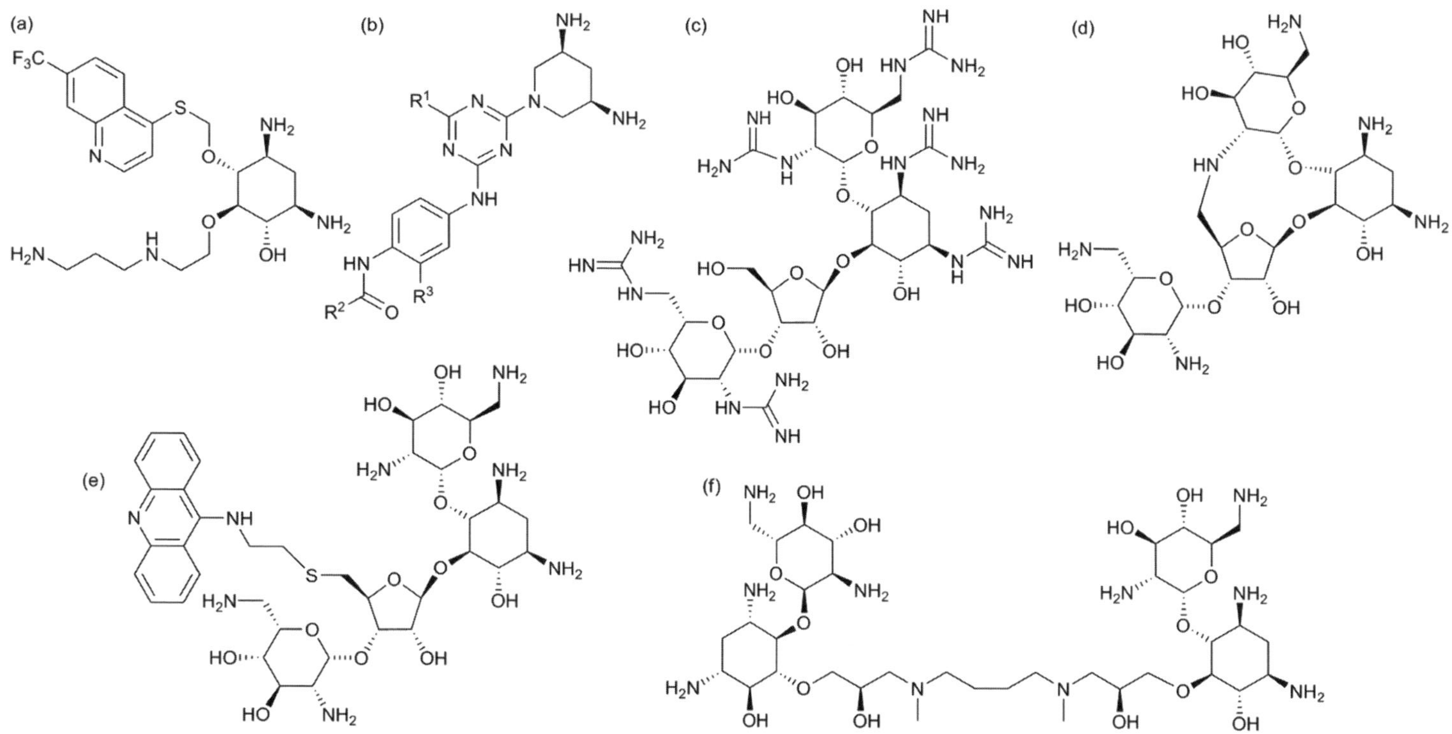

Figure 6.19 *Aminoglycoside derivatives and mimetics: (a) a carbohydrate-free mimetic derived from 2-DOS [121]; (b) DAPT antibacterials [122]; (c) guanidino-neomycin B [127]; (d) restricted neomycin derivative [123]; (e) aminoglycoside acridine conjugate [124]; (f) dimeric aminoglycoside [125]*

many of the early discoveries, such as specific nuclear staining by small organic dyes, go back to the days when little molecular understanding existed, contemporary explorations benefit from significant advances in the structural biology of nucleic acids. The biophysical exploration of nucleic acid–ligand interactions is currently a vibrant field, benefiting from advances in natural product chemistry, combinatorial synthesis and instrumentation, in addition to *in vitro* selection and directed evolution experiments.

Although small molecules that target nucleic acids have enormous therapeutic potential, one needs to appreciate that, in the cell, DNA is normally found in the nucleus, in a highly compacted form, and is extensively complexed to proteins. Similarly, RNA, being a single-stranded biopolymer, is typically folded into intricate structures, clearly distinct from the 'classical' forms observed in short oligonucleotides. Similarly to DNA, RNA is rarely found in 'free' form, dissociated from other cellular components. It is therefore important to appreciate that our knowledge regarding recognition events of such biopolymers is often gathered from *in vitro* biophysical experiments done with model, typically substantially shortened, oligonucleotides. Extrapolating ligand recognition properties into the cellular environment is, therefore, a complex exercise, full of caveats, which rarely succeeds. Only recently, researchers have started to explore such events systematically in a true biological milieu. In addition, compounds displaying promising *in vitro* recognition properties, in terms of affinity and selectivity, are not assured of success when translated into cell cultures. Numerous other parameters, including cell uptake and efflux, metabolism and off-target binding, tend to complicate *in vivo* behavior.

Despite these complications, there is compelling evidence that small molecules can regulate cellular processes by targeting nucleic acids. In addition to the proven therapeutic potency of DNA-targeting chemotherapeutic agents and rRNA-targeting aminoglycosides, recent reports suggest specific regulatory roles for small molecules at the RNA level and hence their ability to operate in a cellular environment [128–131]. Such ligand-mediated riboswitches can therefore alter cellular events that depend on the native folding or recognition properties of the targeted RNA. In addition, the discovery of new regulatory roles for RNA, such as that seen in RNA interference, is likely to fuel further efforts focusing on the design and synthesis of double-stranded RNA ligands as potential effectors of this important process [132,133]. As nucleic acids, and particularly RNA, are likely to continue to surprise us with new cellular functions [134], new targets for small molecule targeting are likely be emerge. These will provide synthetic and biophysical chemists with opportunities to utilize the enormous knowledge gathered over the past half century and exercise their creativity to develop novel, potent and selective nucleic acid ligands. The future is bright!

References

1. M. Demeunynck, C. Bailly and W. D. Wilson, *DNA and RNA Binders: From Small Molecules to Drugs*, Wiley-VCH Verlag GmbH, Weinheim, 2003.
2. G. M. Blackburn, M. J. Gait, D. Loakes and D. M. Williams, *Nucleic Acids in Chemistry and Biology*, Oxford University Press, Oxford, 2006.
3. J. D. Watson and F. H. C. Crick, Molecular structure of nucleic acids, *Nature*, **171**, 737–738 (1953).
4. S. Neidle, *Nucleic Acid Structure and Recognition*, Oxford University Press, New York, 2002.
5. W. Saenger, *Principles of Nucleic Acid Structure*, Springer, New York, 1984.
6. R. Larvey and B. Pullman, The molecular electrostatic potential of the B-DNA helix, *Theor. Chim. Acta*, **53**, 175–181 (1979).
7. R. Larvey and B. Pullman, The electrostatic field of DNA: the role of the nucleic acid conformation, *Nucleic Acids Res.*, **10**, 4383–4395 (1982).
8. K. Chin, K. A. Sharp, B. Honig and A. M. Pyle, Calculating the electrostatic properties of RNA provides new insights into molecular interactions and function, *Nat. Struct. Biol.*, **6**, 1055–1061 (1999).
9. D. Perahia and A. Pullman, The molecular electrostatic potential of the B-DNA helix, *Theor. Chim. Acta*, **50**, 351–354 (1979).

10. G. S. Manning, Counterion binding in polyelectrolyte theory, *Acc. Chem. Res.*, **12**, 443–449 (1979).
11. M. T. Record, C. F. Anderson and T. M. Lohman, Thermodynamic analysis of ion effects on the binding and conformational equilibria of proteins and nucleic acids: the roles of ion association or release, screening and ion effects on water activity, *Q. Rev. Biophys.*, **11**, 103–178 (1978).
12. C. J. Alden and S. H. Kim, Solvent-accessible surfaces of nucleic acids, *J. Mol. Biol.*, **132**, 411–434 (1979).
13. J. B. Chaires, Energetics of drug–DNA interactions, *Biopolymers*, **44**, 201–215 (1997).
14. T. V. Chalikian, A. P. Sarvazyan, G. E. Plum and K. J. Breslauer, Influence of base composition, base sequence and duplex structure on DNA hydration: apparent molar volumes and apparent molar adiabatic compressibilities of synthetic and natural DNA duplexes at 25°C, *Biochemistry*, **33**, 2394–2401 (1994).
15. M. L. Kopka, A. Fratini, H. R. Drew and R. E. Dickerson, Ordered water structure around a B-DNA dodecamer, *J. Mol. Biol.*, **163**, 129–146 (1983).
16. X. Shui, L. McFail-Isom, G. Hu and L. D. Williams, The B-DNA dodecamer at high resolution reveals a spine of water on sodium, *Biochemistry*, **37**, 8341–8355 (1998).
17. M. Egli, S. Portmann and N. Usman, RNA hydration: a detailed look, *Biochemistry*, **35**, 8489–8494 (1996).
18. J. R. Kiser, R. W. Monk, R. L. Smalls and J. T. Petty, Hydration changes in the association of Hoechst 33258 with DNA, *Biochemistry*, **44**, 16988–16997 (2005).
19. L. S. Lerman, Structural considerations in the interaction of DNA and acridines, *J. Mol. Biol.*, **3**, 18–30 (1961).
20. O. Persil and N. V. Hud, Harnessing DNA intercalation, *Trends Biotechnol.*, **25**, 433–436 (2007).
21. C. A. Hunter, K. R. Lawson, J. Perkins and C. J. Urch, Aromatic interactions, *J. Chem. Soc., Perkin Trans.* **2**, 651–669 (2001).
22. J. B. Chaires, A thermodynamic signature for drug–DNA binding mode, *Arch. Biochem. Biophys.*, **453**, 26–31 (2006).
23. J. Ren, T. C. Jenkins and J. B. Chaires, Energetics of DNA intercalation reactions, *Biochemistry*, **39**, 8439–8447 (2000).
24. H. Ihmels and D. Otto, Intercalation of organic dye molecules into double-stranded DNA – general principles and recent developments. In *Supermolecular Dye Chemistry*, ed. F. Würthner, *Topics in Current Chemistry* , Vol. **258**, Springer, Berlin, 2005.
25. A. R. Peacocke and J. N. H. Skerrett, The interaction of aminoacridines with nucleic acids, *Trans. Faraday Soc.*, **52**, 261–279 (1956).
26. L. S. Lerman, The structure of the DNA–acridine complex, *Proc. Natl. Acad. Sci. USA*, **49**, 94–102 (1963).
27. S. Altman, Masters of DNA, *J. Biol. Chem.*, **280**, 14361–14365 (2005).
28. M. Waring, Complex formation between ethidium bromide and nucleic acids, *J. Mol. Biol.*, **13**, 269–282 (1965).
29. J. B. Le Pecq and C. Paoletti, A fluorescent complex between ethidium bromide and nucleic acids. Physical–chemical characterization., *J. Mol. Biol.*, **27**, 87–106 (1967).
30. M. J. Waring and L. P. G. Wakelin, Echinomycin: a bifunctional intercalating antibiotic, *Nature*, **252**, 653–657 (1974).
31. P. B. Dervan and M. M. Becker, Molecular recognition of DNA by small molcules. Synthesis of bis(methidium)-spermine, a DNA polyintercalating molecule, *J. Am. Chem. Soc.*, **100**, 1968–1970 (1978).
32. F. Leng, W. Priebe and J. B. Chaires, Ultratight DNA binding of a new bisintercalating anthracycline antibiotic, *Biochemistry*, **37**, 1743–1753 (1998).
33. C. Escude, T. Garestier and J. S. Sun, Drug interaction with triple-helical nucleic acids, *Methods Enzymol.*, **340**, 340–357 (2001).
34. K. Shin-ya, K. Wierzba, K. Matsuo, T. Ohtani, Y. Yamada, K. Furihata, Y. Kayakawa and H. Seto, Telomestatin, a novel telomerase inhibitor from *Stretomyces anulatus*, *J. Am. Chem. Soc.*, **123**, 1262–1263 (2001).
35. A. Siddiqui-Jain, C. L. Grand, D. J. Bearss and L. H. Hurley, Direct evidence for a G-quadruplex in a promoter region and its targeting with a small molecule to repress c-*MYC* transcription, *Proc. Natl. Acad. Sci. USA*, **99**, 11593–11598 (2002).
36. P. B. Dervan, Design of sequence-specific DNA-binding molecules, *Science*, **232**, 464–471 (1986).
37. F. Arcamone, S. Penco, P. Orezzi, V. Nicolella and A. Pirelli, Structure and synthesis of distamycin A, *Nature*, **203**, 1064–1065 (1964).
38. M. L. Kopka, D. S. Goodsell, G. W. Han, T. K. Chiu, J. W. Lown and R. E. Dickerson, Defining GC-specificity in the minor grooove: side-by-side binding of the di-imidazole lexitropsin to C-A-T-G-G-C-C-A-T-G, *Structure*, **5**, 1033–1046 (1997).
39. J. F. Pelton and D. E. Wemmer, Structural characterization of a 2:1 distamycin A·d(CGCAAATTGGC) complex by two-dimensional NMR, *Proc. Natl. Acad. Sci. USA*, **86**, 5723–5727 (1989).
40. P. B. Dervan, Molecular recognition of DNA by small molecules, *Bioorg. Med. Chem.*, **9**, 2215–2235 (2001).

41. A. R. Urbach, J. W. Szewczyk, S. White, J. M. Turner, E. E. Baird and P. B. Dervan, Sequence selectivity of 3-hydroxypyrrole/pyrrole ring pairings in the DNA minor groove, *J. Am. Chem. Soc.*, **121**, 11621–11629 (1999).
42. M. Mrkish, W. S. Wade, T. J. Dwyer, B. H. Geierstanger, D. E. Wemmer and P. B. Dervan, Antiparallel side-by-side dimeric motif for sequence-specific recognition in the minor groove of DNA by the designed peptide 1-methylimidazole-2-carboxamide netropsin, *Proc. Natl. Acad. Sci. USA*, **89**, 7586–7590 (1992).
43. J. M. Gottesfeld, J. M. Belitsky, C. Melander, P. B. Dervan and K. Luger, Blocking transcription through a nucleosome with synthetic DNA ligands, *J. Mol. Biol.*, **321**, 249–263 (2002).
44. T. Bando and H. Sugiyama, Synthesis and biological properties of sequence-specific DNA-alkylating pyrrole–imidazole polyamides, *Acc. Chem. Res.*, **39**, 935–944 (2006).
45. P. B. Dervan, Regulation of her2/neu oncogene expression by synthetic polyamides, International Patent Application WO 0015242, 2000.
46. J. M. Turner and J. M. Gottesfeld, Regulation of gene expression with synthetic DNA-binding ligands, *Chemtracts*, **14**, 563–572 (2001).
47. G. Felsenfeld, D. R. Davies and A. Rich, Formation of a three-stranded polynucleotide molecule, *J. Am. Chem. Soc.*, **79**, 2023–2024 (1957).
48. A. Rich, Formation of a two- and three- stranded helical molecules by polyinosinic acid and polyadenylic acid, *Nature*, **181**, 521–525 (1958).
49. P. S. Sarkar and S. K. Brahmachari, Intramolecular triplex potential sequence within a gene down-regulates its expression *in vivo*, Sequence-specific double-strand alkylation and cleavage of DNA mediated by triple-helix formation, *Nucleic Acids Res.*, **20**, 5713–5718 (1992).
50. H. E. Moser and P. B. Dervan, Sequence-specific cleavage of double helical DNA by triple helix formation, *Science*, **238**, 645–650 (1987).
51. T. Le Doan, L. Perrouault, D. Praseuth, N. Habhoub, J.-L. Decout, N. T. Thoung, J. Lhomme and C. Hélène, Sequence-specific recognition, photocrosslinking and cleavage of the DNA double helix by an oligo-[α]-thymidylate covalently linked to an azidoproflavine derivative, *Nucleic Acids Res.*, **15**, 7749–7760 (1987).
52. T. J. Povsic, S. A. Strobel and P. B. Dewan, *J. Am. Chem. Soc.*, **114**, 5934–5941 (1992).
53. S. A. Strobel, L. A. Doucette-Stamm, L. Riba, D. E. Housman and P. B. Dervan, Site-specific cleavage of human chromosome 4 mediated by triple-helix formation, *Science*, **254**, 1639–1642 (1991).
54. J. L. Mergny, G. Duval-Valentin, C. H. Nguyen, L. Perrouault, B. Faucon, M. Rougée, T. Montenay-Garestier, E. Bisagni and C. Hélène, Triple helix-specific ligands, *Science*, **256**, 1681–1684 (1992).
55. A. Granzhan and H. Ihmels, Selective stabilization of triple-helical DNA by diazoniapolycyclic intercalators, *ChemBioChem*, **7**, 1031–1033 (2006).
56. V. K. Tam, Q. Liu and Y. Tor, Extended ethidium bromide analogue as a triple helix intercalator: synthesis, photophysical properties and nucleic acids binding, *Chem. Commun.*, 2684–2686 (2006).
57. M. Gellert, M. N. Lipsett and D. R. Davies, Helix formation by guanylic acid, *Proc. Natl. Acad. Sci. USA*, **48**, 2013–2018 (1962).
58. N. W. Kim, M. A. Piatyszek, K. R. Prowse, C. B. Harley, M. D. West, P. L. C. Ho, G. M. Coviello, W. E. Wright, S. L. Weinrich and J. W. Shay, Specific association of human telomerase activity with immortal cells and cancer, *Science*, **266**, 2011–2015 (1994).
59. A. M. Zahler, J. R. Williamson, T. R. Cech and D. M. Prescott, Inhibition of telomerase by G-quartet DNA structures, *Nature*, **350**, 718–720 (1991).
60. C. M. Counter, A. A. Avilion, C. E. LeFeuvre, N. G. Stewart, C. W. Greider, C. G. Harley and S. Bacchetti, Telomere shortening associated with chromosome instability is arrested in immortal cells which express telomerase activity, *EMBO J.*, **11**, 1921–1929 (1992).
61. C. L. Sann, J. Huddlestona and J. Mann, Synthesis and preliminary evaluation of novel analogues of quindolines as potential stabilisers of telomeric G-quadruplex DNA, *Tetrahedron*, **62**, 12903–12911 (2007).
62. E. M. Rezler, J. Seenisamy, S. Bashyam, M.-Y. Kim, E. White, W. D. Wilson and L. H. Hurley, Telomestatin and diseleno sapphyrin bind selectively to two different forms of the human telomeric G-quadruplex structure, *J. Am. Chem. Soc.*, **127**, 9439–9447 (2005).
63. K. Jantos, R. Rodriguez, S. Ladame, P. S. Shirude and S. Balasubramanian, Oxazole-based peptide macrocycles: a new class of G-quadruplex binding ligands, *J. Am. Chem. Soc.*, **128**, 13662–13663 (2006).

64. M. Bejugam, S. Sewitz, P. S. Shirude, R. Rodriguez, R. Shahid and S. Balasubramanian, Trisubstituted isoalloxazines as a new class of G-quadruplex binding ligands: small molecule regulation of c-kit oncogene expression, *J. Am. Chem. Soc.*, **129**, 12926–12927 (2007).
65. Y. Tor, Targeting RNA with small molecules, *ChemBioChem*, **4**, 998–1007 (2003).
66. T. Hermann and Y. Tor, RNA as a target for small-molecule therapeutics, *Expert Opin. Ther. Patents*, **15**, 49–62 (2005).
67. T. Hermann and D. J. Patel, RNA bulges as architectural and recognition motifs, *Structure*, **8**, R47–R54 (2000).
68. J. R. Thomas and P. J. Hergenrother, Targeting RNA with small molecules, *Chem. Rev.*, **104**, 1171–1224 (2008).
69. K. M. Weeks and D. M. Crothers, Major groove accessibility of RNA, *Science*, **261**, 1574–1577 (1993).
70. C. S. Chow and F. M. Bogdan, A structural basis for RNA–ligand interactions, *Chem. Rev.*, **97**, 1489–1514 (1997).
71. Q. Vicens and E. Westhof, Crystal structure of a complex between the aminoglycoside tobramycin and an oligonucleotide containing the ribosomal decoding A site, *Chem. Biol.*, **9**, 747–755 (2002).
72. A. Mujeeb, J. L. Clever, T. M. Billeci, T. L. James and T. G. Parslow, Structure of the dimer initiation complex of HIV-1 genomic RNA, *Nat. Struct. Biol.*, **5**, 432–436 (1998).
73. E. Ennifar, J. C. Paillart, R. Marquet, B. Ehresmann, C. Ehresmann, P. Dumas and P. Walter, HIV-1 RNA dimerization initiation site is structurally similar to the ribosomal A site and binds aminoglycoside antibiotics, *J. Biol. Chem.*, **278**, 2723–2730 (2003).
74. M. L. Zapp, S. Stern and M. R. Green, Small molecules that selectively block RNA binding of HIV-1 Rev protein inhibit Rev function and viral production, *Cell*, **74**, 969–978 (1993).
75. T. Hermann, Rational ligand design for RNA: the role of static structure and conformational flexibility in target recognition, *Biochimie*, **84**, 869–875 (2002).
76. N. Leulliot and G. Varani, Current topics in RNA–protein recognition: control of specificty and biological function through induted fit and conformational capture, *Biochemistry*, **40**, 7947–7956 (2001).
77. F. Pitici, D. L. Beveridge and A. M. Baranger, Molecular dynamics simulation studies of induced fit and conformational capture in U1A–RNA binding: do molecular substrates code for specificity?, *Biopolymers*, **65**, 424–435 (2002).
78. T. Hermann, P. Auffinder and E. Westhof, Molecular dynamics investigations of hammerhead ribozyme RNA, *Eur. Biophys. J.*, **27**, 153–165 (1998).
79. R. T. Batey, R. P. Rambo and J. A. Doudna, Tertiary motifs in RNA structure and folding, *Angew. Chem. Int. Ed.*, **38**, 2326–2343 (1999).
80. J. R. Williamson, Induced fit in RNA–protein recognition, *Nat. Struct. Biol.*, **7**, 834–837 (2000).
81. M. Froeyen and P. Herdewijn, RNA as a target for drug design, the example of Tat–TAR interaction, *Curr. Top. Med. Chem.*, **2**, 1123–1145 (2002).
82. T. E. Edwards and S. T. Sigurdsson, Electron paramagnetic resonance dynamic signatures of TAR RNA–small molecule complexes provide insight into RNA structure and recognition, *Biochemistry*, **41**, 14843–14847 (2002).
83. D. S. Pilch, M. Kaul, C. M. Barbieri and J. E. Kerrigan, Thermodynamics of aminoglycoside–rRNA recognition, *Biopolymers*, **70**, 58–79 (2003).
84. J. Suurkuusk, J. Alvarez, E. Freire and R. Biltonen, Calorimetric determination of the heat capacity changes associated with the conformational transitions of polyriboadenylic acid and polyribouridylic acid, *Biopolymers*, **16**, 2641–2652 (1977).
85. J. B. Chaires, Calorimetry and thermodynamics in drug design, *Annu. Rev. Biophys.*, **37**, 135–151 (2008).
86. A. Hutchin and G. Cortopassi, Proposed molecular and cellular mechanism for aminoglycoside ototoxicity, *Antimicrob. Agents Chemother.*, **38**, 2517–2520 (1994).
87. A. Schatz, E. Bugie and S. A. Waksman, Streptomycin, a substance exhibiting antibiotic activity against Gram-positive and Gram-negative bacteria, *Proc. Natl. Acad. Sci. USA*, **55**, 66–69 (1944).
88. H. Umezawa, K. Ueda and K. Maeda, Production and isolation of a new antibiotic, kanamycin, *J. Antibiot.*, **A10**, 187–189 (1957).
89. S. A. Waksman and H. A. Lichevalier, Neomycin, a new antibiotic active against streptomycin resistant bacteria, including tuberculosis organisms, *Science*, **109**, 305–307 (1949).
90. M. J. Weinstein, G. M. Leudemann and E. M. Oden, Gentamicin, a new broad spectrum antibiotic complex, *Antimicrob. Agents Chemother.*, **4**, 1–7 (1964).

91. C. M. Barbieri, M. Kaul, M. Bozza-Hingos, F. Zhao, Y. Tor, T. Hermann and D. S. Pilch, Defining the molecular forces that determine the impact of neomycin on bacterial protein synthesis: importance of the 2′-amino functionality, *Antimicrob. Agents Chemother.*, **51**, 1760–1769 (2007).
92. D. Moazed and H. F. Noller, Interaction of antibiotics with functional sites in 16S ribosomal RNA, *Nature*, **327**, 389–394 (1987).
93. P. Purohit and S. Stern, Interactions of a small RNA with antibiotic and RNA ligands of the 30S subunit, *Nature*, **370**, 659–662 (1994).
94. M. I. Recht, D. Fourmy, S. C. Blanchard, K. D. Dahlquist and J. D. Puglisi, RNA sequence determinants for aminoglycoside binding to an A-site rRNA model oligonucleotide, *J. Mol. Biol.*, **262**, 421–436 (1996).
95. D. Fourmy, M. Recht, S. Blanchard and J. Puglisi, Structure of the A site of *Escherichia coli* 16S ribosomal RNA complexed with an aminoglycoside antibiotic, *Science*, **274**, 1367–1371 (1996).
96. H. Miyaguchi, H. Narita, K. Sakamoto and S. Yokoyama, An antibiotic-binding motif of an RNA fragment derived from the A-site- related region of *Escherichia coli* 16S rRNA, *Nucleic Acids Res.*, **24**, 3700–3706 (1996).
97. Q. Han, Q. Zhao, S. Fish, K. B. Simonsen, D. Vourloumis, J. M. Froelich, D. Wall and T. Hermann, Molecular recognition by glycoside pseudo base pairs and triples in an apramycin–RNA complex, *Angew. Chem. Int. Ed.*, **44**, 2694–2700 (2005).
98. Q. Vicens and E. Westhof, Crystal structure of geneticin bound to a bacterial 16S ribosomal RNA A site oligonucleotide, *J. Mol. Biol.*, **326**, 1175–1188 (2003).
99. Q. Vicens and E. Westhof, Crystal structure of paromomycin docked into the eubacterial ribosomal decoding A site, *Structure*, **9**, 647–658 (2001).
100. A. P. Carter, W. M. Clemons, D. E. Broderson, R. J. Morgan-Warren, B. T. Wimberly and V. Ramakrishnan, Functional insights from the structure of the 30S ribosomal subunit and its interactions with antibiotics, *Nature*, **407**, 340–348 (2000).
101. J. Davies and B. D. Davis, Misreading of ribonucleic acid code words induced by aminoglycoside antibiotics, *J. Biol. Chem.*, **243**, 3312–3316 (1968).
102. C. M. Barbieri, M. Kaul and D. S. Pilch, Use of 2-aminopurine as a fluorescent tool for characterizing antibiotic recognition of the bacterial rRNA A-site, *Tetrahedron*, **63**, 3567–3574 (2007).
103. M. Kaul, C. M. Barbieri and D. S. Pilch, Fluorescence-based approach for detecting and characterizing antibiotic-induced conformational changes in ribosomal RNA: comparing aminoglycoside binding to prokaryotic and eukaryotic ribosomal RNA sequences, *J. Am. Chem. Soc.*, **126**, 3447–3453 (2004).
104. C. M. Barbieri and D. S. Pilch, Complete thermodynamic characterization of the multiple protonation equilibria of the aminoglycoside antibiotic paromomycin: a calorimetric and natural abundance ^{15}N NMR study, *Biophys J.*, **90**, 1338–1349 (2006).
105. U. v. Ahsen, J. Davies and R. Schroeder, Antibiotic inhibition of group I ribozyme function, *Nature*, **353**, 368–370 (1991).
106. T. Hermann, Aminoglycoside antibiotics: old drugs and new therapeutic approaches, *Cell. Mol. Life Sci.*, **64**, 1841–1852 (2007).
107. J. G. Silva and I. Carvalho, New insights into aminoglycoside antibiotics and derivatives, *Curr. Med. Chem.*, **14**, 1101–1119 (2007).
108. T. Hermann and E. Westhof, Aminoglycoside binding to the hammerhead ribozyme: a general model for the interaction of cationic antibiotics with RNA, *J. Mol. Biol.*, **276**, 903–912 (1998).
109. H. Wang and Y. Tor, RNA–aminoglycoside interactions: design, synthesis and binding of ‘amino-aminoglycosides’ to RNA, *Angew. Chem. Int. Ed.*, **37**, 109–111 (1998).
110. T. K. Stage, K. J. Hertel and O. C. Uhlenbeck, Inhibition of the hammerhead ribozyme by neomycin, *RNA*, **1**, 95–101 (1995).
111. S. R. Kirk and Y. Tor, $tRNA^{Phe}$ binds aminoglycoside antibiotics, *Bioorg. Med. Chem.*, **7**, 1979–1991 (1999).
112. N. E. Mikkelsen, K. Johansson, A. Virtanen and L. A. Kirsebom, Aminoglycoside binding displaces a divalent metal ion in a tRNA–neomycin B complex, *Nat. Struct. Biol.*, **8**, 510–514 (2001).
113. S. Freisz, K. Lang, R. Micura, P. Dumas and E. Ennifar, Binding of aminoglycoside antibiotics to the duplex form of the HIV-1 genomic RNA dimerization initiation site, *Angew. Chem. Int. Ed.*, **47**, 1–5 (2008).
114. D. P. Arya, *Aminoglycoside Antibiotics: From Chemical Biology to Drug Discovery*, Wiley-Interscience, Hoboken, NJ, 2007.

115. K. F. Blount and Y. Tor, A tale of two targets: differential RNA selectivity of nucleobase-aminoglycoside conjugates, *ChemBioChem*, **7**, 1612–1621 (2006).
116. H. Robinson, Y.-G. Gao, R. Sanishvili, A. Joachimiak and A. H.-J. Wang, Hexahydrated magnesium ions bind in the deep major groove and at the outer mouth of A-form nucleic acid duplexes, *Nucleic Acids Res.*, **28**, 1760–1766 (2000).
117. H. Robinson and A. H.-J. Wang, Neomycin, spermine and hexaamminecobalt(III) share common structural motifs in converting B- to A-DNA, *Nucleic Acids Res.*, **24**, 676–682 (1996).
118. Y. Tor, T. Hermann and E. Westhof, Deciphering RNA recognition: aminoglycoside binding to the hammerhead ribozyme, *Chem. Biol.*, **5**, R277–R283 (1998).
119. K. F. Blount, F. Zhao, T. Hermann and Y. Tor, Conformational constraint as a means for understanding RNA–aminoglycoside specificity, *J. Am. Chem. Soc.*, **127**, 9818–9829 (2005).
120. Y. Tor, The ribosomal A-site as an inspiration for the design of RNA binders, *Biochimie*, **88**, 1045–1051 (2006).
121. X. Wang, M. T. Migawa, K. A. Sannes-Lowery and E. E. Swayze, The synthesis and 16S A-site rRNA recognition of carbohydrate-free aminoglycosides, *Bioorg. Med. Chem. Lett.*, **15**, 4919–4922 (2005).
122. Y. Zhou, V. E. Gregor, Z. Sun, B. K. Ayida, Geoffrey C. Winters, D. Murphy, K. B. Simonsen, D. Vourloumis, S. Fish, J. M. Froelich, D. Wall and T. Hermann, Structure-guided discovery of novel aminoglycoside mimetics as antibacterial translation inhibitors, *Antimicrob. Agents Chemother.*, **49**, 4942–4949 (2005).
123. F. Zhao, Q. Zhao, K. F. Blount, Q. Han, Y. Tor and T. Hermann, Molecular recognition of RNA by neomycin and a restricted neomycin derivative, *Angew. Chem. Int. Ed.*, **44**, 5329–5334 (2005).
124. N. W. Luedtke, Q. Liu and Y. Tor, RNA–ligand interactions: affinity and specificity of aminoglycoside dimers and acridine conjugates to the HIV-1 Rev response element, *Biochemistry*, **42**, 11391–11403 (2003).
125. F. Agnelli, S. J. Sucheck, K. A. Marby, D. Rabuka, S.-L. Yao, P. S. Sears, F.-S. Liang and C.-H. Wong, Dimeric aminoglycosides as antibiotics, *Angew. Chem. Int. Ed.*, **43**, 1562–1566 (2004).
126. X. Liu, J. R. Thomas and P. J. Hergenrother, Deoxystreptamine dimers bind to RNA hairpin loops, *J. Am. Chem. Soc.*, **126**, 9196–9197 (2004).
127. N. W. Luedtke, T. J. Baker, M. Goodman and Y. Tor, Guanidinoglycosides: a novel family of RNA ligands, *J. Am. Chem. Soc.*, **122**, 12035–12036 (2000).
128. T. E. Edwards, D. J. Klein and A. R. Ferre-D'Amare, Riboswitches: small-molecule recognition by gene regulatory RNAs, *Curr. Opin. Struct. Biol.*, **17**, 273–279 (2007).
129. E. Nudler, Flipping riboswitches, *Cell*, **129**, 19–22 (2006).
130. M. Mandal, M. Lee, J. E. Barrick, Z. Weinberg, G. M. Emilsson, W. L. Ruzzo and R. R. Breaker, A glycine-dependent riboswitch that uses cooperative binding to control gene expression *Science*, **306**, 275–279 (2004).
131. J. E. Barrick, K. A. Corbino, W. C. Winkler, A. Nahvi, M. Mandal, J. Collins, M. Lee, A. Roth, N. Sudarsan, I. Jona, J. K. Wickiser and R. R. Breaker, New RNA motifs suggest an expanded scope for riboswitches in bacterial genetic control, *Proc. Natl. Acad. Sci. USA*, **101**, 6421–6426 (2004).
132. M. Krishnamurthy, K. Simon, A. M. Orendt and P. A. Beal, Macrocyclic helix-threading peptides for targeting RNA, *Angew. Chem. Int. Ed.*, **46**, 7044–7047 (2007).
133. B. D. Gooch, M. Krishnamurthy, M. Shadid and P. A. Beal, Binding of helix-threading peptides to *E. coli* 16S ribosomal RNA and inhibition of the S15–16S complex, *ChemBioChem*, **6**, 2247–2254 (2005).
134. A. Petherick, The production line, *Nature*, **454**, 1042–1045 (2008).

7

The Architectural Motifs of Folded RNAs

Valérie Fritsch and Eric Westhof

7.1 Introduction

Nucleic acids occur in two main families, DNA and RNA, depending on the presence of an oxygen atom on the 2′-position of the sugar ring. In RNA molecules, the presence of the 2′-hydroxyl group, with its donor and acceptor hydrogen bonding possibilities, leads to distinct and powerful chemical and structural properties compared with DNA molecules. The catalytic potential of natural and artificial RNAs relies primarily on the 2′-hydroxyl group. The presence of the 2′-hydroxyl group promotes clear energetic minima in the sugar conformations of the ribose compared with those of deoxyribose. Hence in RNA helices, the ribose sugars occupy the C3′-*endo* pucker with the occurrence of the C2′-*endo* pucker restricted to some single-stranded regions or terminal riboses. Furthermore, the hydrogen bonding capabilities of the 2′-hydroxyl group are used extensively in short- and long-range intramolecular contacts in addition to the base–base interactions. Such characteristics promote the self-assembling properties of RNA molecules and subtend the astonishing architectural diversity that folded RNAs reveal. Here we describe elements of RNA architecture together with some of the main building blocks of RNA architecture without addressing the biochemical and biophysical aspects of the folding process *per se*.

7.2 Definitions

Both purine and pyrimidine bases possess three edges for interaction through hydrogen bonds: the Watson–Crick edge (used in canonical pairing), the Hoogsteen edge [atoms N6 (or O6) and N7 in purines and atoms N4 (or O4) and C5 in pyrimidines] and the Sugar edge (Figure 7.1).

A given edge of one base can potentially interact in a plane with any one of the three edges of a second base and can do so in either the *cis* or *trans* orientation of the glycosidic bonds (Figure 7.2). To simplify, a triangle can be used to represent a nucleotide (Figure 7.3) [1]. The 12 possible, distinct

The Chemical Biology of Nucleic Acids Edited by Günter Mayer

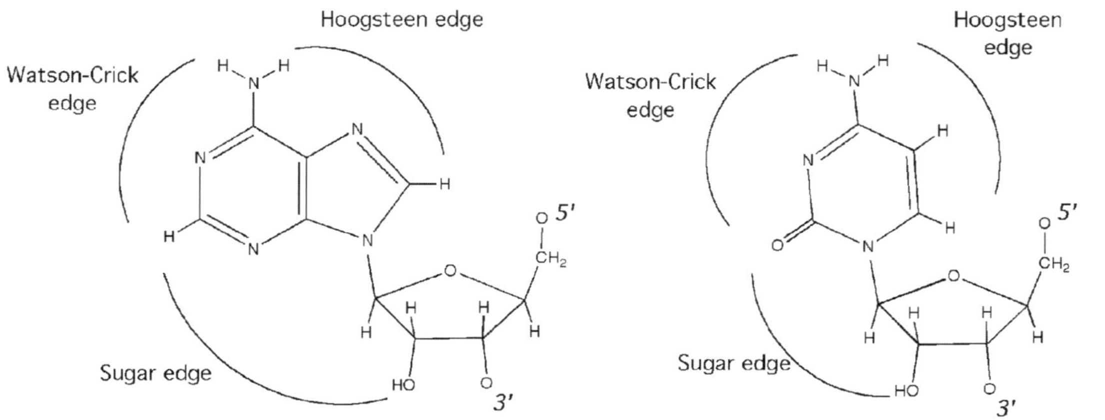

Figure 7.1 *The three edges on the nucleic acid bases that can be involved in edge-to-edge base–base interactions mediated by hydrogen bonding*

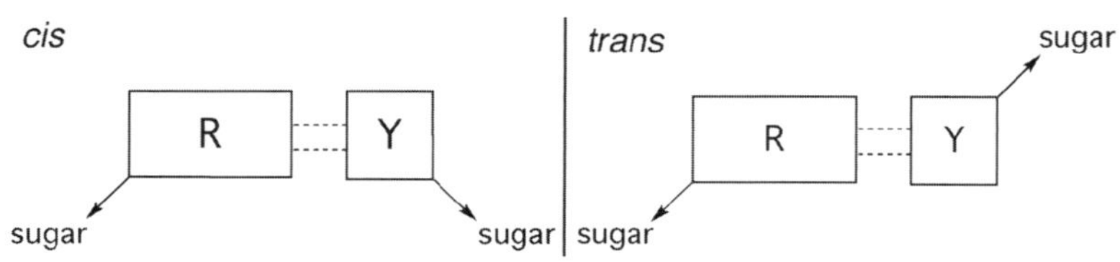

Figure 7.2 *Schematic views of* cis *and* trans *base pairs. Dotted lines symbolize hydrogen bonds. In a* cis *orientation, the nucleoside sugars are on the same side of a line drawn between the hydrogen bonds connecting the two bases and parallel to them. In a* trans *orientation, sugars are on either side of this line*

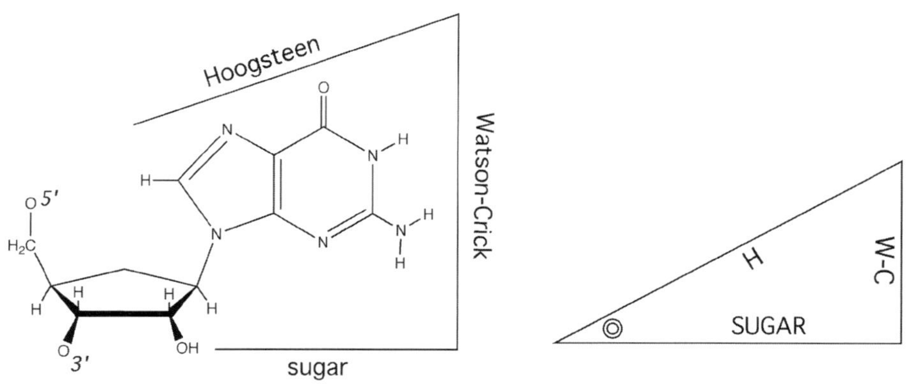

Figure 7.3 *Representation of an RNA nucleotide as a triangle. The symbol in the triangle corner where the Sugar and Hoogsteen edges meet indicates the orientation of the sugar–phosphate backbone relative to the plane of the pairing: a circle means that the 5′ → 3′ direction comes from back to front and a cross means that it goes from front to back. Two strands can be parallel (both strand 5′ → 3′ directions are the same way) or antiparallel (in the case of opposite directions)*

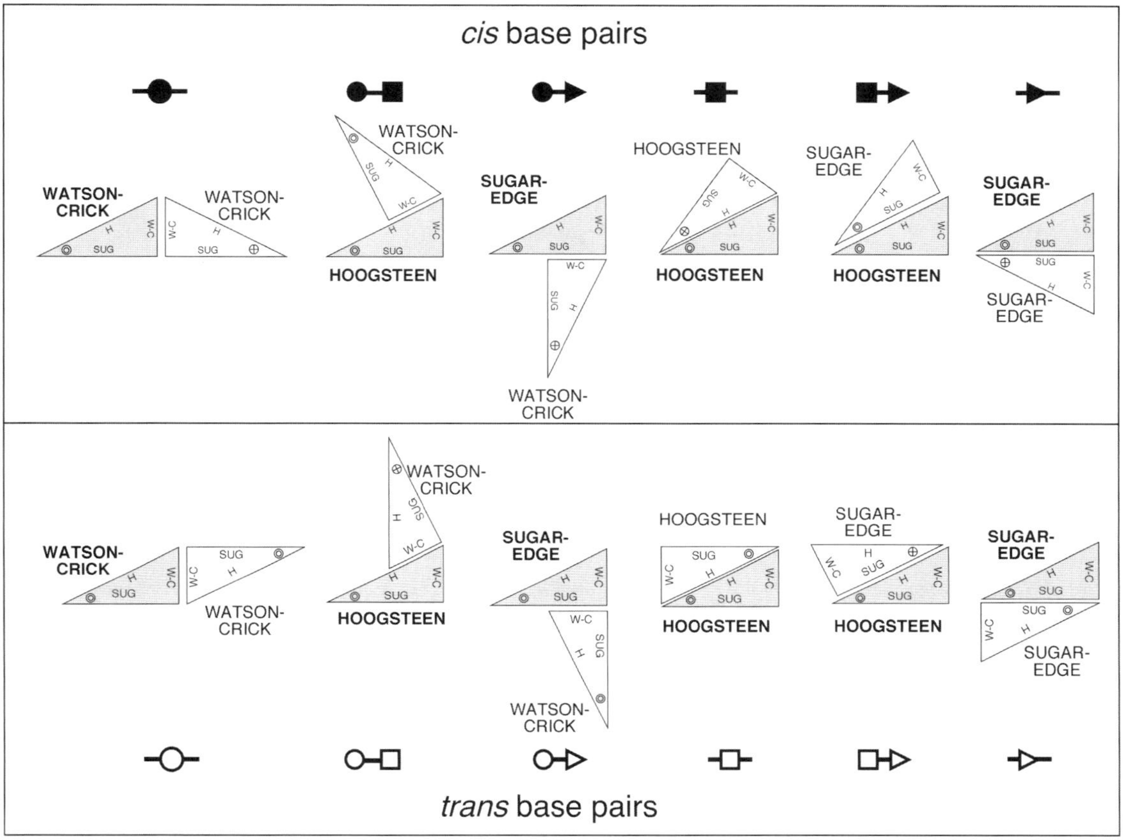

Figure 7.4 *The 12 possible base-pairing geometries. Each geometry is designated by stating the interacting edges of the two bases (Watson–Crick, Hoogsteen or Sugar edge) and the relative glycosidic bond orientation,* cis *or* trans. *Symbols used are described in Figure 7.5. After [1]*

edge-to-edge base-pairing geometries (or families) are illustrated in Figure 7.4. Symbols used are described in Figure 7.5. The 12 base pair geometries are listed in Table 7.1, with the local strand orientations in the default *anti* configurations of the bases with respect to the sugars. They cover all internucleotide interactions with at least two H-bonds, some of which can be described under other names such as the sheared base pair (*trans* Hoogsteen/Sugar edge), the adenosine platform (*cis* Hoogsteen/Sugar edge), the ribose zippers (*cis* Sugar edge/Sugar edge) and the A-minor contacts (*cis* and *trans* Sugar edge/Sugar edge).

7.3 The annotation of RNA structure

In order to facilitate the 2D representation of complex three-dimensional structures, conventions have been suggested for presenting the essential 3D features of RNA structures in a visually accessible and informative 2D diagram (Figure 7.5). These conventions are illustrated by examples of 2D annotations in Figure 7.6.

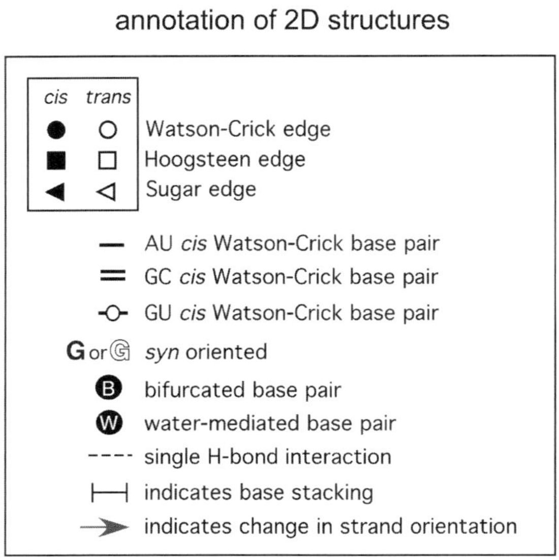

Figure 7.5 *Symbols used for indicating three-dimensional structural features in 2D representations of RNA structures. After [1]*

Table 7.1 *The 12 geometric families of nucleic acid base pairs with symbols for annotating secondary structure diagrams. The local strand orientation is given in the last column, assuming that all bases are in the default* anti *conformation; a* syn *orientation for one base would imply a reversal of orientation; for the global orientation, the stereochemistry at the phosphate groups has to be considered. In the very rare case that both bases are* syn*, the strand orientations revert to those given in the table*

No.	Glycosidic bond orientation	Interacting edges	Symbol	Default local strand orientation
1	*cis*	Watson–Crick/Watson–Crick		Antiparallel
2	*trans*	Watson–Crick/Watson–Crick		Parallel
3	*cis*	Watson–Crick/Hoogsteen		Parallel
4	*trans*	Watson–Crick/Hoogsteen		Antiparallel
5	*cis*	Watson–Crick/Sugar edge		Antiparallel
6	*trans*	Watson–Crick/Sugar edge		Parallel
7	*cis*	Hoogsteen/Hoogsteen		Antiparallel
8	*trans*	Hoogsteen/Hoogsteen		Parallel
9	*cis*	Hoogsteen/Sugar edge		Parallel
10	*trans*	Hoogsteen/Sugar edge		Antiparallel
11	*cis*	Sugar edge/Sugar edge		Antiparallel
12	*trans*	Sugar edge/Sugar edge		Parallel

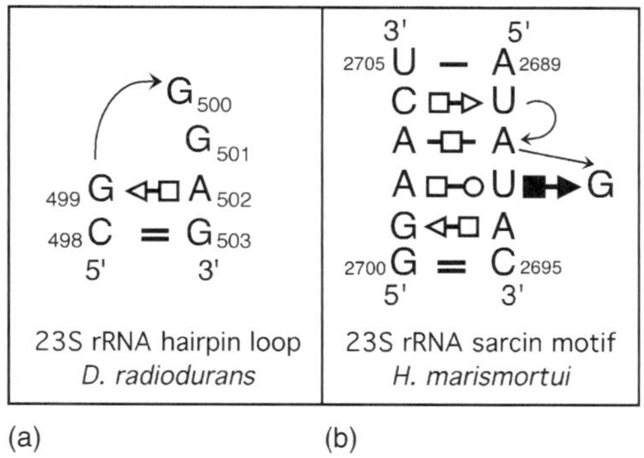

Figure 7.6 *Examples of annotations of RNA motifs. (a) A GNRA tetraloop and (b) a loop E or sarcin/ricin motif. After [1]*

The most important structural element in RNA is the A-form double helix. About 60–70% of RNA bases are base paired with canonical Watson–Crick and stacked together to form A-form double helices. In contrast to the B-DNA form (with its wide major groove and its narrow minor groove both with the same depth), the A-form has a narrow deep groove (corresponding to the major groove) and a wide shallow groove (corresponding to the minor groove).

7.4 Elements of RNA architecture

The set of all *cis* Watson–Crick/Watson–Crick base-paired helices defines the secondary structure of an RNA molecule. Nucleotides not involved in a double helix form single-stranded regions linking the helical segments and partitioned into internal loop, hairpin loop or junction. A hairpin loop is formed when an RNA strand folds back on itself, forming intrastrand base pairs and leaving some bases unpaired. An internal loop can be symmetric or asymmetric and does not usually contain canonical base pairs. Junctions are defined as regions where two or more double helical stems come together. Many types of junctions are possible depending on the number of stems (two-way junction, three-way junction, four-way junction). They play key roles in positioning helical domains at specific angles. It is now clear that the single-stranded regions are generally highly structured with stacking and non-Watson–Crick pairing interactions in the folded three-dimensional or tertiary structures. Thus, the non-Watson–Crick pairs are central to the formation of the three-dimensional architectures of RNA molecules. The recent upsurge in crystal structures demonstrated also that several of these structured 'single-stranded' regions were observed recurrently with a similar organization of non-Watson–Crick pairs in various functional RNAs. These recurrent folds are then called 'motifs'. Here, we will restrict 'motif' to conservation in sequence and use 'module' for conservation of three-dimensional fold. Thus, a 'module' is a conserved building block which stays unchanged (at least to some extent to be defined) wherever it is inserted; in other words, it maintains its intrinsic properties irrespective of what it is connected to. A 'module' is organized through an ensemble of ordered non-Watson–Crick pairs. The recent crystal structures have demonstrated that the modules are recurrent and limited in number.

7.4.1 Hairpin loops

7.4.1.1 GNRA and other tetraloops

The most common and studied hairpin loops are loops of four nucleotides called tetraloops. In these tetraloops, several types of sequences are well characterized and keep conserved structures: GNRA,

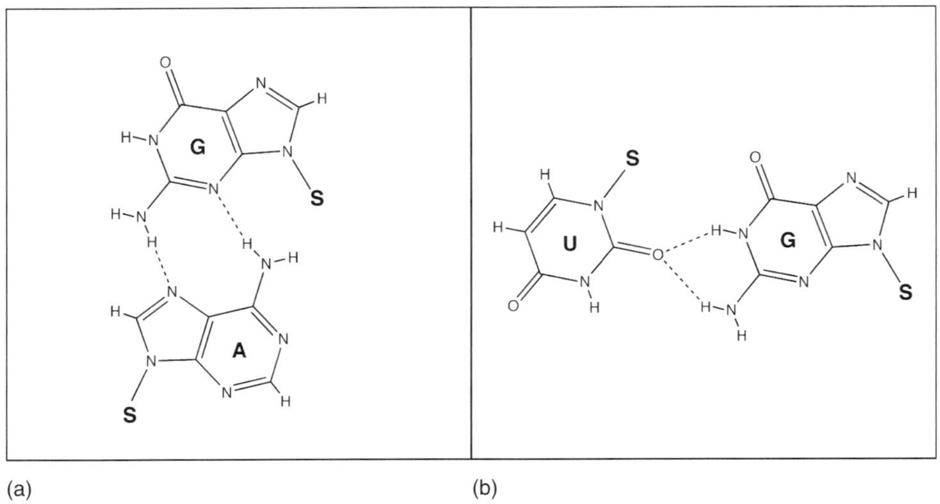

Figure 7.7 *(a) In GNRA tetraloops, the first and fourth nucleotides usually form a* trans *Sugar edge/Hoogsteen G•A base pair. (b) In the UUCG loop, bifurcated hydrogen bonds are observed between the first and fourth nucleotides with the G in the* syn *conformation*

UNCG, ANYA and (U/A)GNN (N designates any nucleotide, R corresponds to a purine and Y to a pyrimidine) [2–4]. In such tetraloops, the first and fourth nucleotides form a *trans* Sugar edge/Hoogsteen G • A base pair (also called sheared base pair) (Figure 7.7). All nucleotides have an *anti* conformation. In UNCG tetraloops, the first and fourth nucleotides usually interact by bifurcated hydrogen bonding (Figure 7.7) and the guanosine adopts a *syn* conformation [5]. In GNRA tetraloops, the backbone turn, after which the direction of the polynucleotide chain is reversed, occurs between the first and second nucleotides with the standard U-turn [6,7]. In AGNN and UGAA tetraloops, the phosphodiester backbone turn occurs between the second and third nucleotides (Figure 7.8) [8–10]. In these cases, the second nucleotide is observed in a *syn*

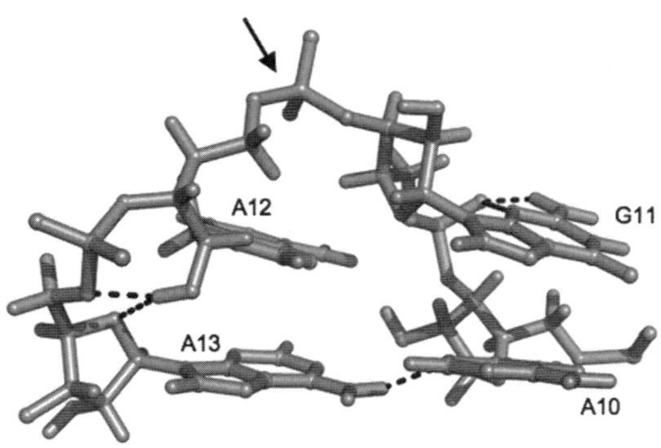

Figure 7.8 *Example of an AGAA tetraloop (from 1K6H.PDB). In this case, the reversal of the backbone chain direction occurs after the second nucleotide of the loop (illustrated by the arrow). Some direct intramolecular hydrogen bonds are observed: between N3(A10) and N6H(A13), N3(G11) and OP(G11), N2H(G11) and OP(G11), O2′H(A12) and O4′(A13), and O2′H(A12) and O5′(A13). Guanosine G11 adopts a* syn *conformation*

conformation. An unusual fold was observed for a tetraloop with methylated nucleotides [11]. Because the first and fourth loop nucleotides are methylated, the loop-closing base pair cannot be formed. The methyl group of the fourth residue also creates a steric clash, forcing the nucleotide to deviate from the plane by about 60°. The second loop residue must adopt the *syn* conformation to avoid a steric clash with the 3′-following nucleotide. The backbone turn occurs between the fourth nucleotide and the 3′-stem nucleotide (Figure 7.9a). The conformation of this loop is then radically different from that of a standard GNRA tetraloop. Another unusual fold is observed for UNCG tetraloops often found in ribosomal and other functional RNAs. In this family loop, the second loop nucleotide is extruded from the standard stacking arrangement. This loop has a very compact structure (Figure 7.9b). As mentioned previously, the last residue adopts a *syn* conformation and a bifurcated base pair is observed between 5′ and 3′ terminal loop nucleotides that contains both base–base and base–sugar hydrogen bonds [12]. The CUUG is a tetraloop also frequently found in ribosomal RNAs [13]. In this loop, the first and fourth residues form a canonical C=G Watson–Crick base pair. The second residue is folded into the minor groove and interacts with the Sugar edge sites of the loop closing C=G base pair and with the first stem base pair [6] (Figure 7.9c).

For several structures of hairpin loops containing five nucleotides, it was shown that they belong to the GNR(N)A pentaloop family [14–19]. In this structural family, the pentaloops GNR(N)A adopt a GNRA-like fold with the fourth nucleotide extruded from the loop [(N) corresponds to the bulged nucleotide] (Figure 7.10). Nucleotides 1, 2, 3 and 5 fold as a GNRA tetraloop with the formation of a G • A sheared base pair between the first and the last loop nucleotides with the change in direction of the sugar–phosphate backbone occurring between the first and the second loop nucleotides. This leads usually to sequential stacking of the second, third and fifth bases on the 3′ stem. The CUCAA pentaloop [20] adopts similar structural features to GNRA tetraloops despite the fact that, in this case, it is the second residue that is extruded and that none of the characteristic U-turn interactions are observed. The loop structure is asymmetric with the first C stacks on the 5′ stem and the last CAA forming a 3′ stack. Other structures of pentaloops were obtained and differ significantly from these GNRA-like structures by nucleotide stacking,

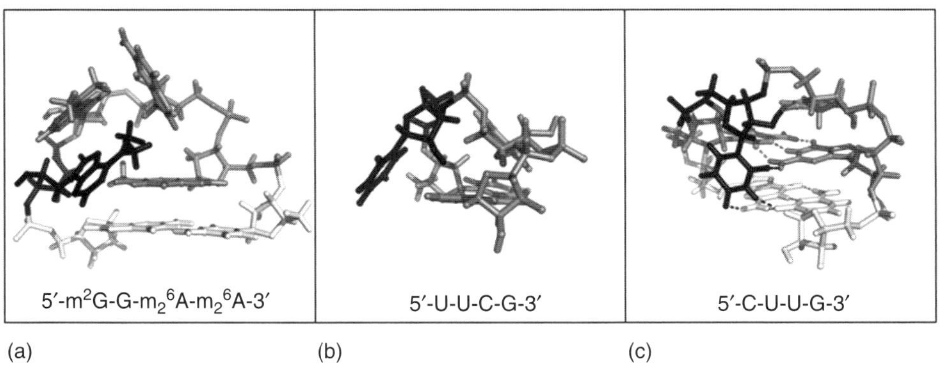

Figure 7.9 *Illustration of some particular tetraloop structures. (a) A tetraloop with three methylated residues. The orientation of the last loop residue, shown in black, illustrates the absence of stacking and the steric problem due to the methyl groups. A* cis *Watson–Crick C=G base pair (in white) constituted the first stem base pair (from 1WTS.PDB). (b) In this structure, the second residue (in black) is extruded from the loop. The loop is stabilized by a* trans *Watson–Crick/Sugar edge G·U base pair (from 1HLX.PDB). (c) In the CUUG tetraloop structure, a* cis *Watson–Crick C=G pair is formed between the first and fourth residues. The second loop residue (in black) is folded into the minor groove and makes hydrogen bonds with the Sugar edge sites of the loop closing C=G base pair (in gray) and with the first stem G=C base pair (in white) (from 1RNG.PDB)*

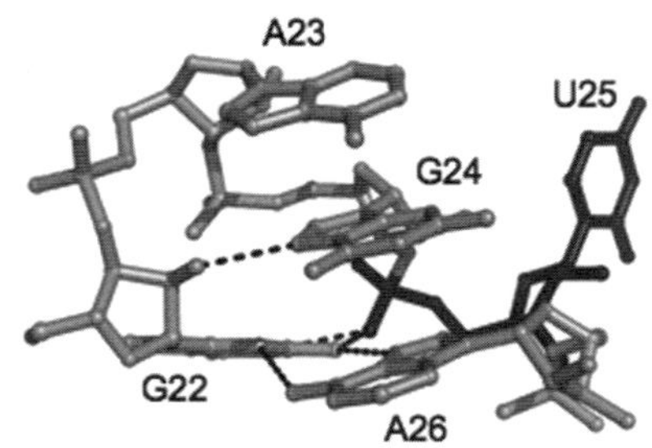

Figure 7.10 *The pentaloop GAGUA adopts a GNRA-like fold with the fourth uridine extruded (in black). A sheared G• A base pair is formed between the first and fifth residues. Additional hydrogen bonds contribute to the loop stability. A sequential stacking of the three purines A23, G24 and A26 is observed (from 1XJR.PDB)*

extrusion or not of a nucleotide or the shape of the sugar–phosphate backbone [21–24]. For all these pentaloops, the extruded nucleotide orientation makes it available for interactions with other RNA regions or proteins.

7.4.1.2 The lonepair triloop

This module is characterized by a hairpin comprising a loop of three unpaired nucleotides closed by a single base pair (Figure 7.11). The closing base pair of such a module belongs to several different geometric families and only a small fraction belongs to the Watson–Crick base pairs family. Lee *et al.* [25] classified the different lonepair triloops in three classes and five groups (R1–R5), based on the following criteria: the proximity to the 5′-adjacent helix, the presence of a U-turn in the triloop, the recruitment of a tertiary base from another region of the RNA molecule and the possibility for one or several nucleotides in the triloop to be involved in tertiary interactions with another RNA region. They proposed the following consensus sequence motifs for the five types of lonepair triloops: type R1 (U**GNR**A), type R2 (U**UYR**A), type R3 (N**RWA**N), type R4 (N**RYA**N) and type R5 (N**CNU**N). Lisi and Major [26] prefer to base the lonepair triloop classification on interaction graphs in which they specify the stack movement along the backbone, the intraloop base stacking formation and the intraloop base-pairing interactions. They also show that the flanking base pair is the most informative sequence element of

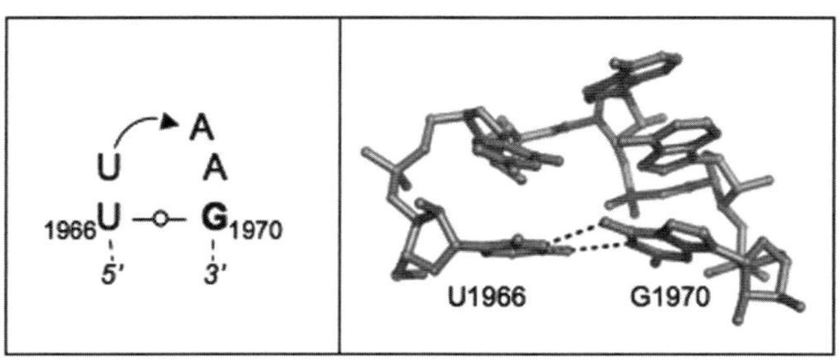

Figure 7.11 *Example of a lonepair triloop module observed in* H. marismortui *23S rRNA. In this module, the loop is closed by a* trans *Watson–Crick U·G base pair. G1970 is observed in a* syn *orientation (from 1JJ2.PDB)*

the triloop 3D structure. In general, the structural class of a specific triplet changes in function of its flanking base pair type.

7.4.1.3 The thymine loop or T-loop

The T-loop module was originally observed in the TΨC loop of yeast tRNA^{Phe} structure [27]. It is composed of a loop of seven nucleotides (sometimes fewer or more; see Krasilnikov and Mondragon [28] for more details) and contains a U-turn flanked by a non-Watson–Crick base pair between the first and the fifth residues of the loop, with the last two residues being extruded before joining the canonical base pair of the helix (Figure 7.12). Thus, the T-loop may be considered as a five-nucleotide loop including a U-turn and closed by a non-canonical base pair. This module has been found to exist not only in tRNAs but also in some plant viral RNAs [29], tmRNAs [30], local regions of ribosomal RNAs [28,31] and RNase P [32]. The consensus sequence for the first five residues seems to be U(G/U)NR(A/U) [4,31]. In their study, Krasilnikov and Mondragon [28] showed that the base pair closing the loop is almost always a *trans* Watson–Crick/Hoogsteen base pair but not necessarily a U·A pair as suggested previously [31]: C·A or A·A base pairs are also observed. The substitution of U·A by C·A base pairs is not surprising given that both pairs are isosteric [33]. It is more surprising to find a *trans* Watson–Crick/Hoogsteen A·A base pair that is not isosteric to a *trans* Watson–Crick/Hoogsteen U·A base pair. The T-loop appears flexible enough to accommodate different *trans* base pairs. In most cases, because of the large gap located between the fourth and fifth nucleotides, a T-loop module can accept the insertion of an external base (Figure 7.12). This base is a G in tRNAs and otherwise mostly an A [28]. Thus, the module is also used to bind selectively a small ligand such as the thiamine moiety, a flat but non-nucleic acid base, in the TPP riboswitch [34,35]. The presence of such an intercalated base slightly changes the geometry of the module. Because the T-loop module contains three unpaired residues closed by a non-Watson–Crick base pair, it is reminiscent of the construction of the lonepair triloop [36].

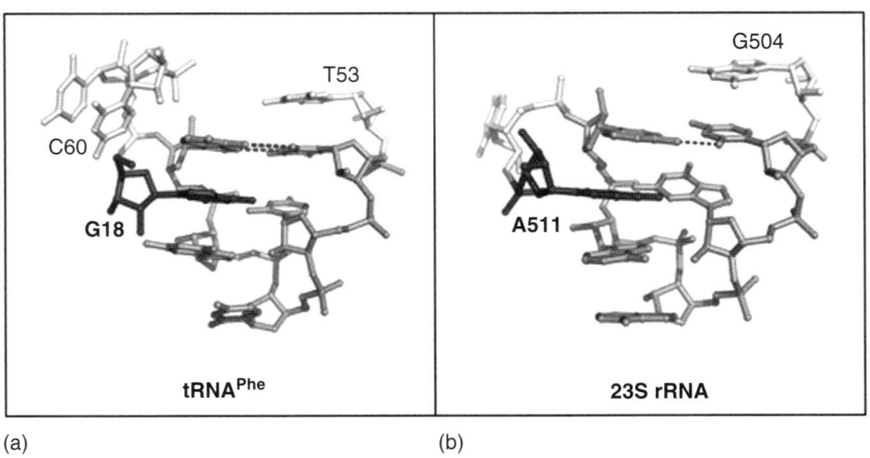

Figure 7.12 *Examples of T-loop modules. The five nucleotides composing the core of the T-loop module are shown in gray and nucleotides before and after the T-loop core are shown in white. The five nucleotide module is often followed by a sharp turn in the sugar–phosphate backbone. Because of the large gap located between the fourth and fifth core residues, T-loop modules accept an intercalated base (drawn in black). In the case of tRNA^{Phe}, the intercalating base is G18 of the D-loop [(a), from 1EHZ.PDB[. In the case of 23S rRNA, it is A511 located at the end of the T-loop which intercalates in the module [(b), from 1JJ2.PDB]. In both cases, only hydrogen bonds of the flanking base pairs are shown (for the sake of clarity)*

7.4.2 Internal loops

Internal loops contain nucleotides that do not form canonical Watson–Crick base pairs: they separate double helical RNA into segments. If the nucleotide number is the same on each strand, the internal loop is symmetric. An asymmetric loop is observed when a different number of nucleotides is inserted on the opposite strand. A bulge is a particular case of an internal loop and is observed when one or several nucleotides of a strand are unpaired whereas the nucleotides of the other strand are all base paired. Unpaired nucleotides can be positioned inside or outside the helix. The sizes of bulges can vary from a single nucleotide up to several nucleotides and frequently form flexible extrusions from paired segments [37]. Incorporation of unpaired bases into the duplex introduces distortion into a regular A-form helix. In solution, internal loops can be flexible but many of them have well-defined structures with several hydrogen bonding schemes, extending stacking interactions and non-Watson–Crick base pairs. Based on their interaction modes, several internal loop modules were defined.

7.4.2.1 The sarcin/ricin or eukaryotic loop E

The sarcin/ricin module contains a bulged-G module, in which a guanosine is extra-helical bulged, and was first observed in the eukaryotic loop E of 5S ribosomal RNA [38] and in the sarcin/ricin loop [39]. The G-bulged motif term was first proposed by Moore [2] and was also referred to as G-bulged cross-strand A stack [40,41], sarcin–ricin motif [42,43] or S-motif [44].

In such a module, the bulged guanosine generally forms a base triplet and is flanked on either side by non-Watson–Crick base pairs which are followed by Watson–Crick base pairs (Figures 7.13 and 7.14). It is composed of six conserved nucleotides which preserved the fold of the module. The remaining seven to nine nucleotides are variable and include two or three non-Watson–Crick base pairs which stack on the 5′-side of the bulged guanosine. The module is closed by an A·G *trans* Hoogsteen/Sugar edge base pair. The phosphodiester backbone forms an S-turn surrounding the bulged nucleotide: the backbone adjacent to the bulged guanosine is distorted by two bends, giving it a S shape (Figure 7.14) [40,41]. The first reverses the chain direction and involves the two nucleotides on the 5′-side of the bulged guanosine. The second restores the initial chain direction and involves the bulged nucleotide and the 3′-adjacent

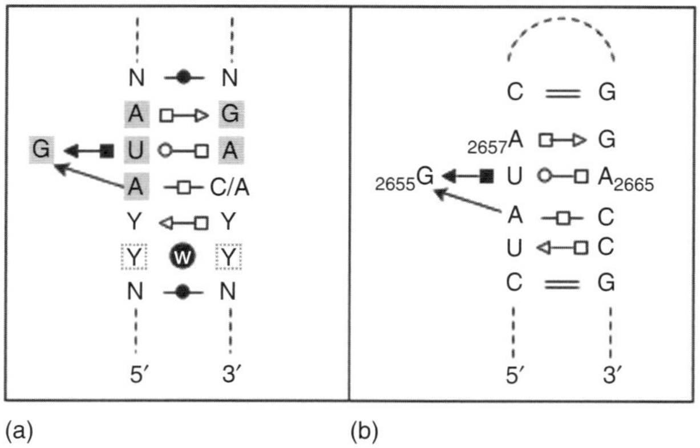

Figure 7.13 *(a) Consensus sequence of the sarcin/ricin module. The six conserved nucleotides are shown in gray boxes. Each terminal base pair form a Watson–Crick base pair. Conventional symbols are used to indicate unambiguously the nature of each base. Nucleotides in dashed boxes indicate nucleotides present only in certain variants. (b) Example of a sarcin/ricin module observed in the structure of the* E. coli *CG-SRL loop*

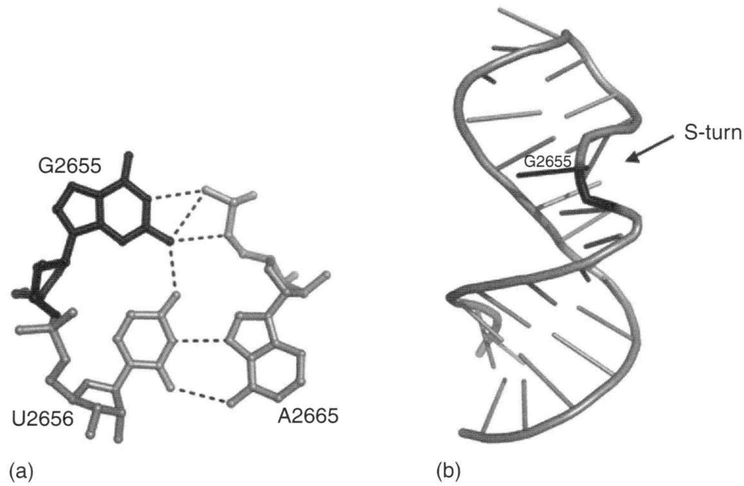

Figure 7.14 *(a) The base triplet observed in the sarcin/ricin module (from 1Q9A.PDB). U2656 and A2665 form a* trans *Watson–Crick/Hoogsteen base pair. The 3D proximity of Watson–Crick sites of G2655 and the phosphate-backbone atoms of A2665 allow the formation of hydrogen bonds which stabilize the module. (b) A cartoon view of the structure of the* E. coli *23S rRNA CG-sarcin/ricin loop with bases drawn as sticks and backbones as ribbons. The backbone adjacent to G2655 is distorted by two bends giving it an S appearance. This S-turn encompasses four nucleotides (U2653, A2654, G2655 and U2656). The 3D structure is represented using a cartoon view*

nucleotide [45]. Each bend reverses the chain direction through backbone torsion angle modifications and inverted sugar puckers [3,39,40]. The bulged guanosine and its 5′-nucleotide present their sugar 2′-hydroxyl groups in the deep (major) groove, in contrast to A-form RNA where the 2′-hydroxyl groups lies in the shallow (minor) groove. In this module, two adenosines are cross-strand stacked and present their free Watson–Crick and Sugar edges to the shallow groove, ideally positioned to interact with other molecules [45].

Corell *et al.* [45] determined the structures of three variants of 27-nucleotide mimics of the sarcin/ricin loop from *Escherichia coli* 23S ribosomal RNA. They compared the bulged-G structures and concluded that these modules have a common geometry and hydration scheme with 10 common hydration sites. The conserved solvent sites coincide with site-specific interactions made by proteins or RNAs. Metal ions are also observed in several structures but they do not share common binding sites. The G-bulged module seems to be a rigid RNA module, but junctions between the flanking structural elements and the G-bulged module are flexible.

7.4.2.2 The bacterial loop E

The 5S ribosomal RNA of *E. coli* comprises five regular Watson–Crick helices and five loop regions. It binds three proteins called L25, L18 and L5 and is an integral part of the 50S ribosome subunit. One of the internal loops, named loop E, is a symmetrical loop characterized by a large part of non-canonical base pairs. In the crystal structure of *E. coli* 5S rRNA [46], loop E forms a severely distorted double helix via its seven non-Watson–Crick base pairs. In this loop are found two isosteric submotifs related by 180° rotation [42,46]. Each submotif is referred to as ‘bacterial loop E’ and consists of three non-Watson–Crick base pairs (Figure 7.15).

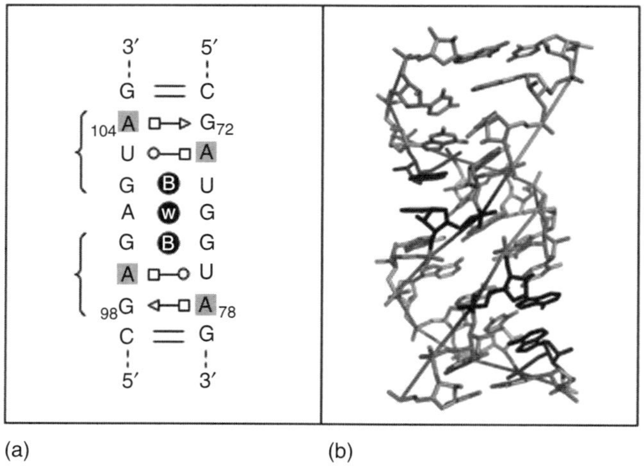

Figure 7.15 (a) E. coli *secondary structure of bacterial 5S rRNA loop E. Each 'bacterial loop E' module is constituted by three noncanonical base pairs. In each submotif are observed two cross-strand adenosines (in gray boxes). (b) Crystal structure of* E. coli *bacterial 5S rRNA loop E (from 364D.PDB). Cross-stranded adenosines are shown in black*

The module is the binding site of the L25 ribosomal protein. The consensus motif is constituted of the following base pairs:

- *trans* Hoogsteen/Sugar edge
- *trans* Watson–Crick/Hoogsteen or *trans* Sugar edge/Hoogsteen
- *cis* bifurcated or *trans* Sugar edge/Hoogsteen.

Such modules comprise a core of three base pairs which position three bases (mainly adenosines, but not exclusively) to face the shallow groove with their Watson–Crick and Sugar edges. The bases of these modules can interact with other molecules as RNA, protein or small molecules via their free edges. In each half, two cross-strand adenines are observed (Figure 7.15). Two guanines are also involved in another cross-strand base in the loop E. These three cross-strand purine stacks alter the shapes of both the shallow and the deep grooves. In eukaryal and archaeal 5S ribosomal RNA, modules observed in loops E are G-bulged. Thus, the eukaryal and archaeal versions of loop E are structurally different from the bacterial loop E [47].

7.4.2.3 *The hook-turn*

This module was initially characterized with the crystal structure of a 26-nucleotide RNA that contains the loop E sequence from *Chromatium minutissimum* [48]. The crystal structure consists of an A-form RNA helix that splits into two separated strands after a *trans* Sugar edge/Hoogsteen G•A base pair. Following this separation, the backbone of the strand containing the guanosine folds at almost 180° in the space of two nucleotides and interacts in the shallow groove of the helix from which it is coming (Figure 7.16). Such turns, with similar backbone folds, also occur in both 16S and 23S rRNAs. Despite local variations in the geometry and in the hydrogen bonding interactions, Szep *et al.* [48] identified a consensus sequence for these hook-turns (Figure 7.16). Almost all of the hook-turns identified start at the end of a regular double helix that terminates with a guanosine on the strand that is going to fold and an adenosine on the opposite strand. The last base pair in the helix, before the G•A base pair, is almost always a canonical C=G or G=C base pair.

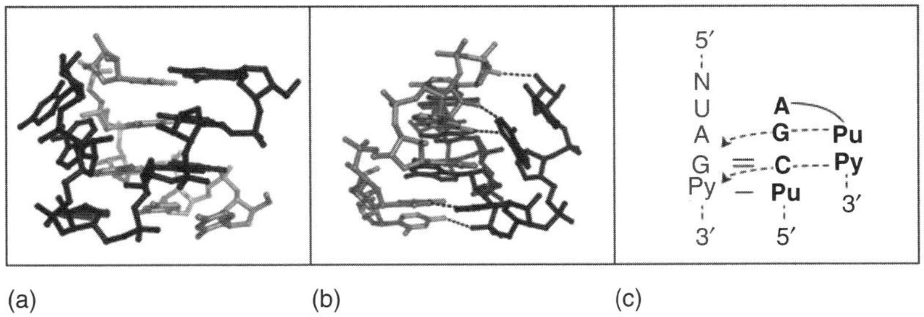

Figure 7.16 *(a, b) Two views of a hook-turn crystal structure. The strand that folds back is colored in black and the opposite one is in gray. Some hydrogen bonds are shown (from 1MHK.PDB). (c) The consensus secondary structure derived for the hook-turn module. Tertiary interactions (dashed lines) depend on the sequence. For example, they can be, in the case of the sequence from* Chromatium minutissimum *[48], a type I A-minor interaction between the purine and the guanosine or a sugar–sugar interaction between both pyrimidines*

7.4.2.4 The K-turn and reverse K-turn

The K-turn module (or kink turn) was first described in a human U4 snRNA fragment [49] and in *Haloarcula marismortui* rRNAs [50]. It is a helix–internal loop–helix motif that adopts an unusual conformation which produces a sharp bend (or kink) of the helical axis of almost 120° between two continuous RNA helices (Figure 7.17). This kink brings the shallow groove side of the two helices together. Klein *et al.* [50] proposed a consensus sequence (Figure 7.18): it is composed of a first helical stem called the 'canonical stem' which comprises only canonical Watson–Crick base pairs and usually ends with two canonical C=G base pairs before the internal loop. The second helical stem (named the 'non-canonical stem'), which follows the internal loop, starts with two non-Watson–Crick base pairs, usually *trans* Sugar edge/Hoogsteen G • A base pairs. The internal loop between the two helical stems is always asymmetric with usually three nucleotides on one strand

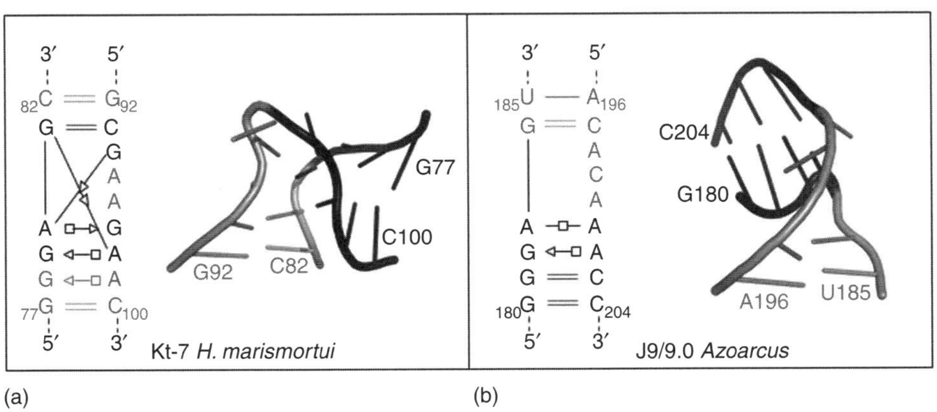

Figure 7.17 *(a) Secondary and tertiary structures of a kink turn found in the* Haloarcula marismortui *23S rRNA crystal structure. The canonical stem and the internal loop are shown in gray and the non-canonical stem in black. A tight bend (almost 120°) of the helical axis is observed between both stems (from 1JJ2.PDB). (b) Secondary and tertiary structures of the reverse kink turn found in the Azoarcus group I intron crystal structure. A 90° bend occurs between both helices in the opposite direction to that observed in the kink turn module (from 1U6B.PDB). This module kinks toward the deep groove of the flanking helices rather than to the shallow groove as in the case of kink turn*

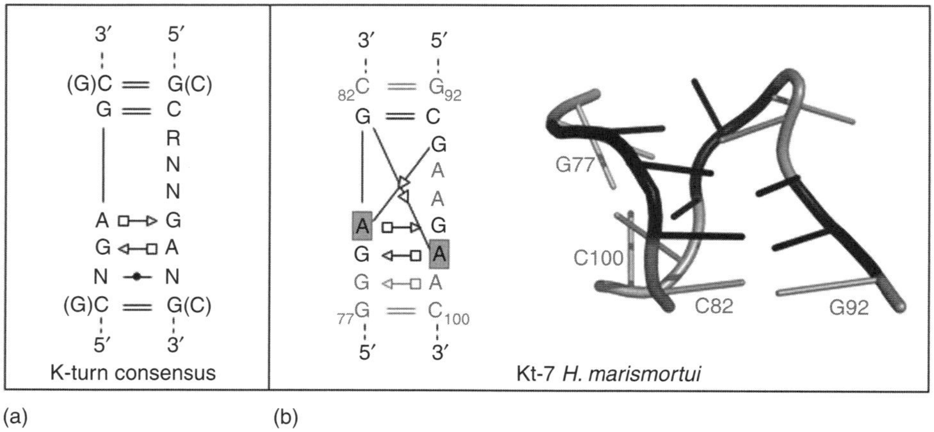

Figure 7.18 *(a) Secondary structure diagram for the consensus sequence of kink turns. (b) Five base pairs constitute the core of the kink turn module (in black) (from 1JJ2.PDB). In the Kt-7 kink turn found in* Haloarcula marismortui *23S rRNA, two adenines are cross-stranded stacked (gray boxes)*

and none on the other. The K-turn module comprises about 15 nucleotides and is characterized by five base pairs found in almost all K-turns (Figure 7.18). The core motif comprises two or three cross-stranded stacked adenosines [49,50]. Two of these adenosines are involved in *trans* Sugar edge/Hoogsteen G • A base pairs (Figure 7.18) [51]. Because they present their Sugar edge sites into the shallow groove of the canonical stem, they are also involved in type I A-minor motifs (see Section 7.5.4). Kink turns are universal motifs found in Archaea, Prokarya and Eukarya [49–51]. They are involved in protein recognition and intermolecular tertiary interactions.

In the reverse kink turn, which is also composed of a helix–internal loop–helix arrangement, the bend occurs in the opposite direction (Figure 7.17). This turn kinks towards the deep groove of the two surrounding helices instead of the shallow groove as observed in kink turn modules [52]. Such modules contain only a single invariant Sugar edge/Hoogsteen G • A base pair (which corresponds to the second G • A located after the internal loop) [51]. Goody *et al.* [53] examined the structural and dynamic properties of kink turns in free solution. They concluded that kink turn RNA elements exist in a dynamic equilibrium between a tightly kinked conformation and a more open conformation similar to a simple bulge bend. The transition between the two conformations involves major changes in both the bend angle and the bend direction and are induced by the noncooperative binding of metal ions. Under physiological conditions, they found that the K-turn is not fully folded into the form observed in the ribosome crystal structure. In ribosomes, K-turns are bound to proteins and these interactions could stabilize the kinked conformation. RNA tertiary interactions may also contribute to the folded structure stabilization. On the basis of this study, Strobel *et al.* [52] suggested that kink turns could be in a dynamic equilibrium between an unbent conformation and the two kinked states (K-turn and reverse K-turn) and that external stabilizing elements, such as protein binding or RNA tertiary interactions, may be necessary to lock the RNA fragment into one conformation or the other.

7.4.2.5 The C-loop

The C-loop module is an asymmetric internal loop with a longer strand and a shorter strand. It was observed in *Thermus thermophilus* 16S rRNA [54,55], in *Haloarcula marismortui* 23S rRNA [56] and in the *E. coli* mRNA

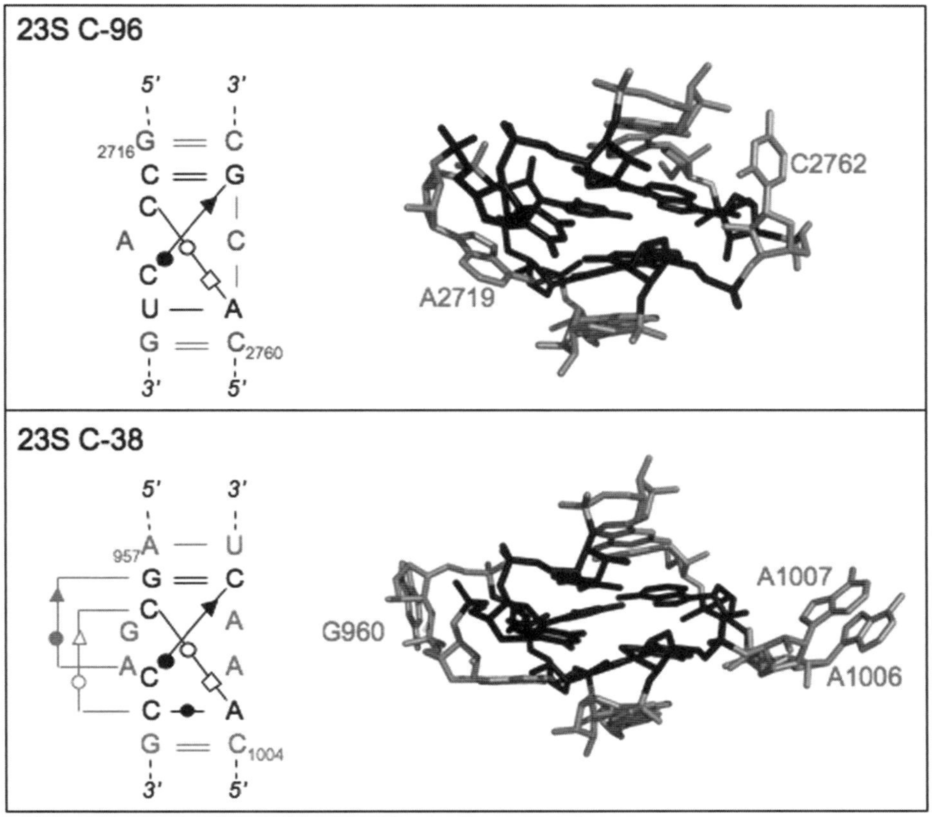

Figure 7.19 Secondary and tertiary structures for variant C-loop modules in Haloarcula marismortui *23S rRNA. The four characteristic interactions of this module are shown in black. In the 3D views, the numbers of the extruded nucleotides are indicated (from 1S72.PDB)*

of threonine synthetase [57]. The module usually contains four characteristic base pairs: two Watson–Crick base pairs and two non-Watson–Crick base pairs. The non-Watson–Crick base pairs involve loop bases of the longer strand and bases paired in the shorter strand. Then, at least two bases are involved in base triples (Figure 7.19).

The C-loop increases the stem helical twist between the two Watson–Crick base pairs flanking the module. The first base of the loop is usually a cytosine (hence the name C-loop). It stacks below the preceding flanking base pair and forms a *trans* Watson–Crick/Hoogsteen pair with the following Watson–Crick base pair of the C-loop. This interaction reinforces the large helical twist between the two flanking base pairs. The shorter strand usually has one or two unpaired nucleotides that are invariably extruded and lie accessible (Figure 7.19). In ribosome C-loops, these extruded bases are engaged in tertiary interactions. C-loop modules can differ in the number of extruded bases in the shorter strand and in the number of bases in the longer strand [51].

7.4.3 Junctions

RNA junctions are defined as regions where two or more interconnected double helical stems come together [58]. Different types of junctions are observed depending on the number of joined stems and the size of the

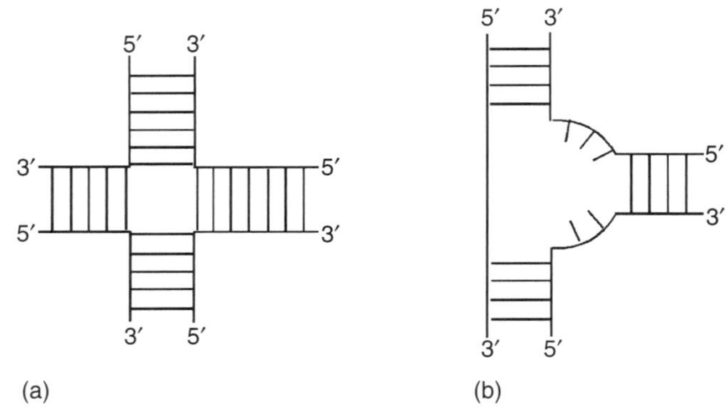

Figure 7.20 (a) A simple four-way junction, without nucleotides in branch regions. According to the nomenclature of junctions and branch points in nucleic acids, such a junction becomes HHHH, which could be shortened to 4H. (b) This three-way junction with two single-stranded sections could be shortened to $2HS_2HS_3$

branch regions. Junctions can be perfect several-way junctions or junctions with bulged bases located between the different stems (Figure 7.20). A systematic nomenclature is used to describe the different types of junctions [59]. With this nomenclature, a letter H is used for each helix and a letter S is used for a single-stranded section traversed. According to this system, a three-way junction becomes HHH or 3H and a junction between three helices with two bulge bases will be denoted 3HS2. Helical junctions in nucleic acids are observed in both DNA and RNA.

Multibranched junctions are stabilized by coaxial stacking of helical arms. A coaxial stack is formed when the terminal base pairs of two helices are in van der Waals contact leading to the formation of a quasi-continuous helical domain between 2D separated helical regions. Favorable interactions between the aromatic bases maximize base stacking interactions (see Section 7.5.1). Because of electrostatic repulsion between the negatively charged phosphate groups, metal ions play a critical role in such tight folding. The importance of metal-ion binding on the folding of such a structural module is demonstrated in the studies of the three-way junction at the catalytic center of the hammerhead ribozyme (Figure 7.21). In the absence of

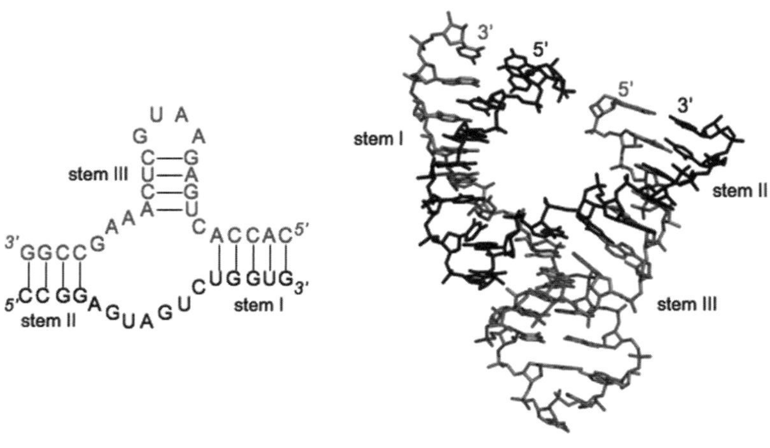

Figure 7.21 Secondary and three-dimensional structures of the hammerhead ribozyme (from 1MME.PDB)

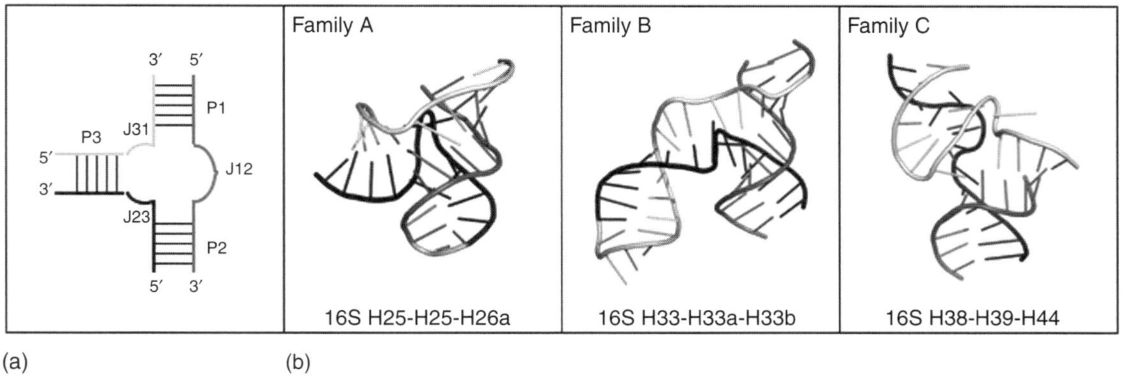

Figure 7.22 *(a) 2D nomenclature used for the three-way junctions (P = helix, J = junction segment). (b) Examples of family A, family B and family C junctions. In family A, helix P3 is roughly perpendicular to the P1/P2 coaxial stack. In family B, helix P3 bends toward helix P2 whereas it bends toward P1 in family C (from 1J5E.PDB)*

added cations, the pattern of electrophoretic mobilities suggest a structure with non interacting helical stems connected by extended single-stranded regions [60]. Under these conditions, electrostatic repulsions between phosphate groups prevent any folding into a more compact structure and the three-way junction forms an extended structure in which none of the helices are stacked (T-shape). Upon addition of a low concentration of magnesium ions (500 μM), the hammerhead folds into a structure with coaxial stacking between helices II and III while helix I forms an acute angle with respect to helix II involving a Y-shape. Helix I is oriented in the same direction as helix III. Over the range 1–15 mM magnesium ions, there is a further folding resulting in rotation of stem I towards stem II [60] as observed in crystal structures [61,62] (Figure 7.21). The ribozyme is a three-way junction where three helical stems are connected by single-stranded segments of 1, 7 and 3 nucleotides forming an HS1HS7HS3 junction. Lescoute and Westhof [63] showed that three-way junctions in folded RNAs can be divided into three main families depending on the lengths of the segments linking the three Watson–Crick helices (Figure 7.22). In family A, the number of nucleotides in junction J31 is smaller than that in junction J23. The helix P3 is roughly perpendicular to the P1/P2 coaxial stack and the junction contacts are the least extensive (Figure 7.22). In family B, the number of nucleotides in J31 and J23 are almost the same and the helix P3 bends toward helix P2. In family C, where the number of nucleotides in J31 is greater than that in J23, the helix P3 bends towards helix P1. Interactions at the junction are diverse but control the overall fold.

A well-characterized four-way junction is observed in transfer RNA structures. This junction consists of two pairs of coaxially stacked helices. The two extended helical regions are roughly perpendicular to each other and the molecule adopts an overall L-shape (see Section 7.5.1). The unpaired nucleotides at the junction are involved in tertiary interactions that stabilize this geometry. Generally in RNA, four-way junctions consist of two coaxial stacks but, in the absence of unpaired nucleotides in branch regions, the stems assume an X-shape [64,65].

7.5 RNA–RNA tertiary interaction protocols

Because the modules are recurrent and limited in number, the assembly of an RNA molecule can be viewed as modular and hierarchical. Modules should possess two important properties: (i) their interfaces should be easily identified and (ii) they should interact via defined protocols. The protocols define the rules of association between modules; they should be robust to sequence variations. Ideally, the modules should evolve independently of their environment and the protocols of interaction should promote evolution of the system. Following the description of the most common modules, we now describe some of the most important interaction protocols between modules. The influence of base sequence on the interaction protocols will be also described.

7.5.1 Coaxiality and interhelical stacking

Coaxial stacking of RNA helical regions is the most fundamental way by which RNA achieves higher order organization [66]. It corresponds to the end-to-end stacking of double helices and provides a way to form long column structures without continuous backbones on each side. Nucleotide bases from two separates helices (or stems) stack and align their helical axis to form a continuous coaxial helix. The importance of coaxial stacking to the RNA global fold was first observed in the crystal structure of yeast phenylalanine transfer RNA [67,68]. $tRNA^{Phe}$ folds into an overall L-shape, with two arms at a right-angle to each other. The arms consist of short co-axial stacked A-RNA helices. One arm is constituted by the stacking of the D short helix on the anticodon helix and the other by the stacking of the T helix on the longer acceptor helix (Figure 7.23). The predominance of coaxial stacking is also observed in the structures of the hepatitis delta ribozyme [69], in the *Tetrahymena* group I intron P4–P6 domain [70,71], in the hammerhead ribozyme [61,62] and in the L11-binding domain of 23S rRNA [72]. In some cases, these structures can be described as sets of coaxially stacked helices docked side-by-side (Figure 7.24). Tertiary RNA–RNA interactions (e.g. tetraloop-tetraloop receptor) and metal ion binding contribute further to the stabilization of the fold.

7.5.2 Mediated by Watson–Crick/Watson–Crick base pairing

7.5.2.1 Kissing hairpin loops

Kissing hairpin loops are formed by base pairing between the single-stranded residues of complementary hairpin loops. In such kissing complexes, the loop intermolecular interactions create a new double helix. The

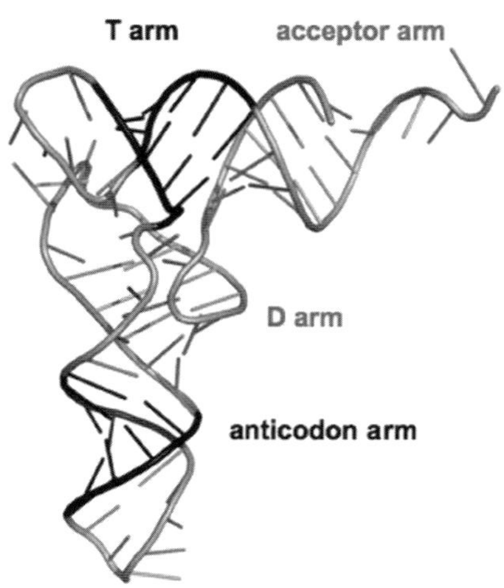

Figure 7.23 *Tertiary structure of yeast phenylalanine transfer RNA illustrating the coaxial stacking of separated 2D helices (from 1EVV.PDB)*

Figure 7.24 *Tertiary structure of* Tetrahymena *group I intron illustrating the docking of helices stabilized by, for example, GAAA/11 nt receptor interactions (in black) (from 1GID.PDB)*

duplex formed by the kissing loops stacks on both sides coaxially with each of the flanking hairpin helices. The overall structure of the kissing complex looks like a bent helix. Kissing hairpin loops can form either intramolecularly (between hairpins belonging to the same polynucleotide chain) or intermolecularly (between hairpins belonging to two polynucleotide chains). Intramolecular kissing hairpins formally constitute pseudoknots (see below).

The dimerization initiation site (DIS) of human immunodeficiency virus type 1 (HIV-1) constitutes the essential part of the dimer linkage structure (DLS) *in vitro* [73–76] and is crucial for efficient HIV-1 replication in cell culture [77]. The DIS adopts a hairpin structure with a hairpin loop composed of nine nucleotides with a six-nucleotide self-complementary sequence. *In vitro*, two RNA monomers form a kissing loop complex via intermolecular interactions of the six-nucleotide self-complementary sequences located in the DIS loops. The self-complementary sequence is flanked at the 5′ side by two conserved purines and at the 3′ side by a highly conserved adenine (Figure 7.25). These three purines are essential for stability of the kissing loop complex [78,79]. The overall structures of the kissing loop complexes are very similar, with the two hairpins interacting so that the canonical helices of each hairpin stem are in perfect coaxial stacking (Figure 7.26). The conserved pattern of the three unpaired purines is such that the 3′ conserved adenosine is stacked within the helix, whereas the two 5′ conserved purines are always extra-helical and stacked on one another [80]. An alternative structure, in which the two 5′ purines are bulged-in, has also been published [81].

Several other kissing loop complex structures have been identified, for example, kissing complexes between the RNA I and RNA II stem loops of the ColE1 plasmid of *E. coli* [82], between yeast

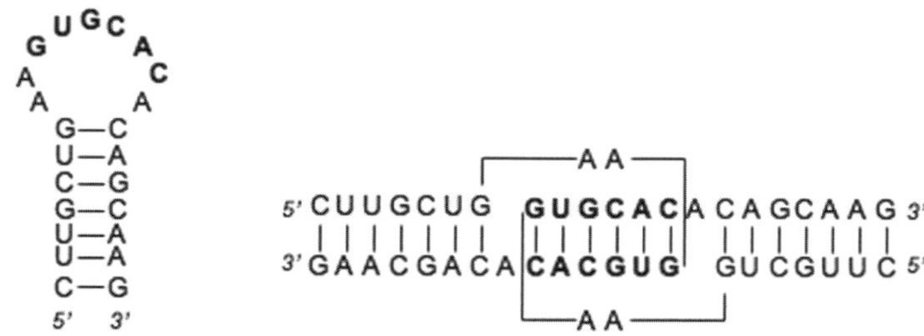

Figure 7.25 *Secondary and tertiary structures of the HIV-1 DIS (F subtype) kissing loops complex. Loop residues involved in the kissing loop interactions are drawn in bold. All the base pairs are canonical Watson–Crick base pairs*

phenylalanine transfer RNA and RNA aptamers [83], between the HIV trans-activating responsive (TAR) hairpin loop and its complement [84], between TAR and RNA hairpin aptamers [85], between TAR and DNA aptamers [86] and for the Moloney murine leukemia virus H3 stem–loop [87]. These complexes share many common features and they all display a quasi-continuous helix formed by the stacking of the two stems and the central loop–loop interactions. The main differences lie in the number of residues in the linker fragments between the two stems and the central duplex and in the number of residues in the central duplex.

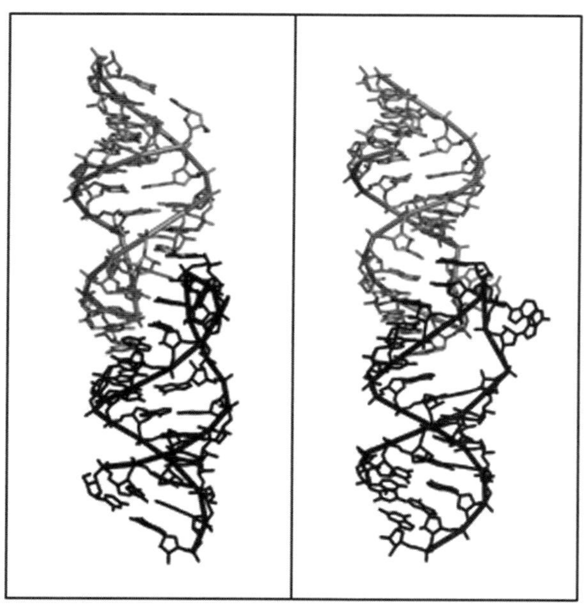

(a) (b)

Figure 7.26 *Two views of the HIV-1 RNA DIS kissing complex. Two colors are used to distinguish the two monomers. (a) The quasi-continuous helix is formed by the stacking of the two stems and the central loop–loop interactions. (b) One can observe the stacking of the extra-helical adenosines (from 1XP7.PDB)*

7.5.2.2 Pseudoknotting

A pseudoknot results from the formation of Watson–Crick base pairs between part of a hairpin loop and a complementary sequence outside the loop (Figure 7.27). Gultyaev *et al.* [88] gave a more precise definition of this module: a pseudoknot is a structure where for two paired nucleotide positions (i, k) there exists a nucleotide m, with $i < m < k$, that pairs with a nucleotide n such that $n < i$ or $n > k$. Within this definition, kissing hairpin loops formed intramolecularly constitute pseudoknots. The simplest type of pseudoknot is called the classical or H-type pseudoknot (H for hairpin loop). Two other types can be considered, depending on the secondary structure nature of the loop implied: I-type pseudoknot (for internal loop) and B-type pseudoknot (for bulge loop). H-type pseudoknots are the most abundant and the best characterized of all known pseudoknots.

RNA pseudoknots are found in all classes of RNA and play key roles in a variety of biological processes. There were first discovered as a part of the tRNA-like structure of the plant turnip yellow mosaic virus RNA [89]. Since then, pseudoknot modules have been found in all types of naturally occurring RNAs (for reviews, see [88,90]). They vary in size and complexity.

A pseudoknot is an RNA tertiary structure (the most common secondary structure prediction tools based on dynamic programming exclude them) that is composed at least of two double-stranded helices (named S1 and S2) connected by single-stranded regions (Figures 7.27 and 7.28). By definition, S1 is the first double helix encountered from the 5′-end. In principle, there will be three different loop regions in a classical pseudoknot [91]. For the classical pseudoknots, one of them does not exist (it contains zero nucleotides). In that case, the two double-stranded helices S1 and S2 are immediately adjacent and coaxial stacking could be expected. However, a perfect coaxial stacking of S1 and S2 is usually not found. A bend at the junction between S1 and S2 is observed for many pseudoknots, caused mainly by the presence of unpaired nucleotides intercalated between the two helices [92]. Concerning the loops, they bridge the double helix: one crosses the shallow groove of S1 (loop 2) and the other the deep groove of S2 (loop 1). The nucleotides of both loops are able to form triple interactions with helical base pairs. The crystal structure of the beet western yellow virus (BMYV) [93] reveals an RNA triplex formed by loop 2 in the shallow groove of stem 1 and quadruple interaction between loop 1 and the deep groove of stem 2 (Figure 7.29). Many interactions in the triplex involve 2′-OH groups of stem 1 base pairs. It is notable that the BMYV pseudoknot is stabilized by more hydrogen bonds involved in tertiary interactions than those engaged in Watson–Crick base pairs. Metal ion binding also stabilizes RNA pseudoknots against unfolding [94].

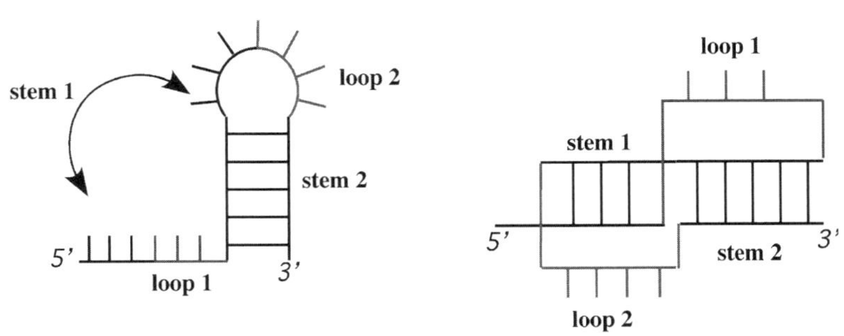

Figure 7.27 *The classical type of pseudoknot is formed through base-pairing interactions between nucleotides of a hairpin loop and those of an adjacent single-stranded region. The module results in two double helices stacked [stem 1 (or S1) and stem 2 (or S2)] and joined by two loops. The nucleotides in loops 1 and 2 can form further interactions of base pairs with the other elements*

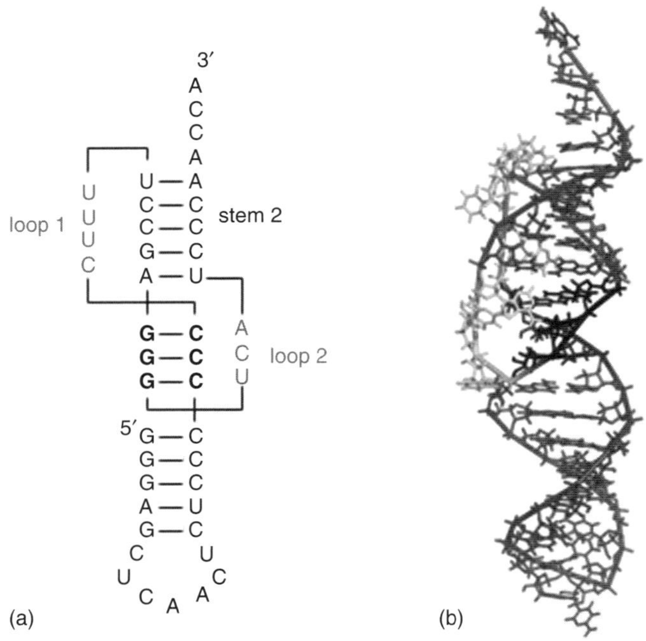

Figure 7.28 (a) Secondary structure of the pseudoknot found in the genomic RNA of the turnip yellow mosaic virus (TYMV). (b) Corresponding three-dimensional structure of the TYMV pseudoknot (from A160.PDB)

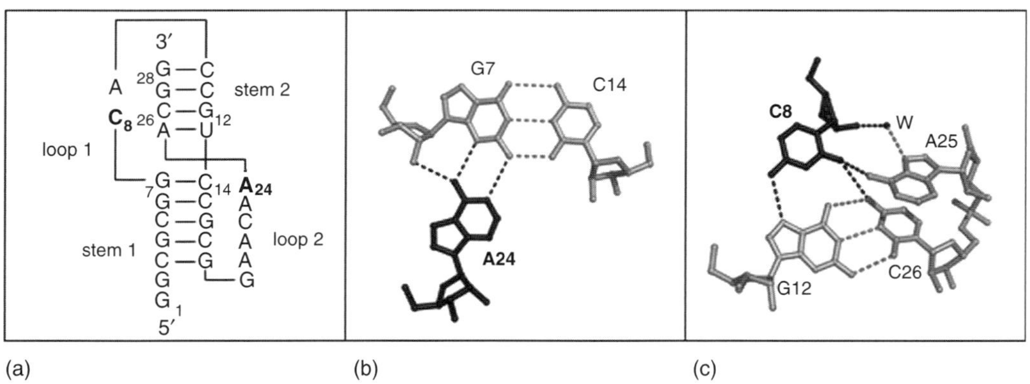

Figure 7.29 (a) Secondary structure of the beet western yellow virus (BWYV) pseudoknot. Examples of (b) triple interaction between a residue of loop 2 (A24) and H-bond sites located in the shallow groove of stem 1 and (c) quadruple interaction between a residue of loop 1 (C8) and H-bond sites located in the deep groove of stem 2 as observed in the crystal structure of BWYV. In the latter case, a water molecule (w) mediates interaction between C8 and A25 (from 437D.PDB)

7.5.3 Mediated by non-Watson–Crick base pairing

7.5.3.1 GNRA–helix

Because of the GNRA tetraloop conformations, the Sugar edges of the last two purine residues are exposed on one side of the hairpin and used for RNA–RNA interactions. The receptor of GNRA tetraloops is constituted of two consecutive regularly stacked *cis* Watson–Crick/Watson–Crick base pairs within a helical domain. The fourth A of the GNRA tetraloop interacts with any type of canonical base pairs; however, for the third R residue, the type of contacted base pair is dictated by the nature of the purine base (if R=A → A–U or G=C pairs and if R=G → A–U pairs only) [95]. The two purine residues form Sugar edge/Sugar edge base pairs *cis* and *trans* with the canonical base pairs.

7.5.3.2 GAAA/11 nucleotide (nt) receptor

The 11 nt receptor mediates RNA tertiary folding by providing a specific binding site for GAAA tetraloops. Such long-range tertiary interactions are found in large ribozymes (such as group I and group II introns [96–99] and RNase P [100–102]).

The most widely studied GNRA/receptor is the GAAA/11-nt receptor [71,99,103,104]. Costa and Michel [97] observed that GAAA tetraloops are often found in interaction with an 11-nt receptor. In this module, the RNA receptor, in interaction with a GAAA tetraloop, contains an internal loop with two consecutive adenosines forming a *cis* Hoogsteen/Sugar edge pair or adenosine platform (Figure 7.30) [71]. The GAAA tetraloop binds the receptor by hydrogen bonding into a complementary shallow groove pocket and stacking upon the adenosine platform (see Section 7.5.4). The tetraloop receptor is stabilized by a potassium ion coordinated immediately below the adenosine platform [105]. Metal ions are important for the tetraloop–receptor interactions. In several crystal and NMR structures, metal ions were found associated with the bound tetraloop–receptor complex [71,106–108]. Butcher *et al.* [103] showed that the isolated GAAA tetraloop

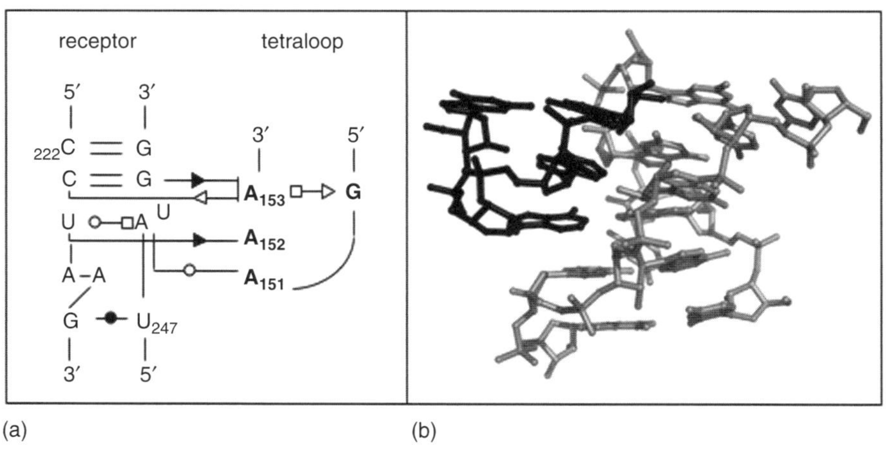

Figure 7.30 *(a) Secondary structure of the GAAA/11-nt receptor as observed in the* Tetrahymena thermophila *group I intron crystal structure. In this module, the three adenosines of the tetraloop interacts by hydrogen bonds with receptor nucleotides: A153 is involved in a type I A-minor interaction with C223 and G250, A152 is involved in a type II A-minor interaction with U224 and A151 forms a* trans *Watson–Crick/Watson–Crick base pair with A248 (see Section 7.5.4 for definition of A-minor interactions). (b) Illustration of the interactions between the GAAA tetraloop and the 11-nt tetraloop receptor (from 1GID.PDB)*

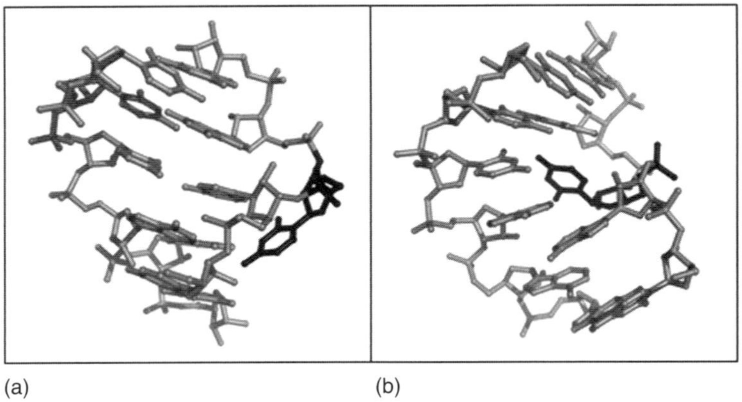

Figure 7.31 Structure of the tetraloop receptor when it is in interaction with the GAAA tetraloop [(a), from 1GID.PDB] and when it is isolated [(b), from 1TLR.PDB]. One can note the position of the internal loop uridine (in black, respectively U249 and U19), which is extra-helical in the bound form. The adenosine platform is not observed in the isolated form

receptor forms a structure different from that of the bound tetraloop receptor (Figure 7.31). Some structural elements are common to both structures including the G∘U wobble base pair at the bottom and the tandem C=G base pairs at the top of the internal loop. The major differences between the two forms are observed for the adenosine and the uridine locations. In the free form, the adenosine platform is not observed and uridines and adenosines are stacked together. The unpaired uridine is sometimes flipped out from the stem structure in the bound form (Figure 7.31). Thus, the 11-nt receptor undergoes a structural rearrangement upon binding of its GAAA target. Such conformational changes were also suggested by biochemical data [109]. In the structure of the bound GAAA/tetraloop receptor complex, Cate *et al.* [71] observed that the three adenosines of the GAAA loop stack on the 5′ side of the tetraloop receptor helix. In addition to stacking, each adenosine of the GAAA loop is involved in specific hydrogen bonds with tetraloop receptor nucleotides: the first adenosine is involved in a triple base interaction, the second one stacks on this triplet and is involved in ribose–ribose and base–ribose interactions, and the third one forms a triplet in addition to the *trans* Sugar edge/Hoogsteen G•A closing loop base pair to which it belongs (Figure 7.32). Many of the observed

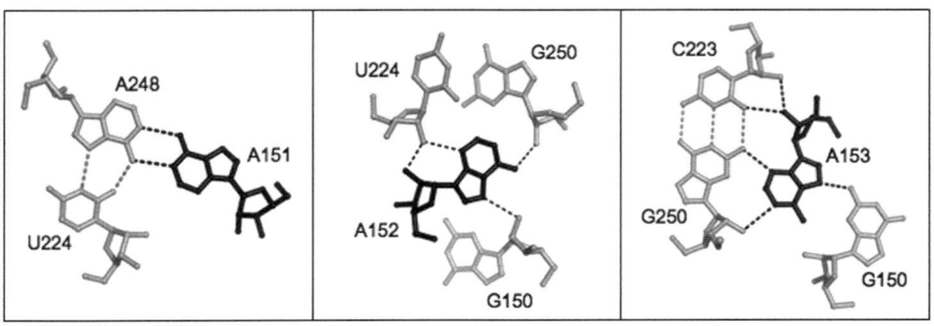

Figure 7.32 Details of some interactions observed between the GAAA tetraloop and the 11-nt receptor. As described in Figure 7.30, A151 forms a trans Watson–Crick/Watson–Crick base pair with A248 (which itself forms a trans Hoogsteen/Watson–Crick pair with U224), A152 is involved in a type II A-minor interaction with U224 and A153 is involved in a type I A-minor interaction with C223 and G250 (from 1GID.PDB)

interactions between the tetraloop and the receptor minor groove involve the 2′-sugar hydroxyl groups as was observed in the crystal structure of a hammerhead ribozyme [110]. The IC3 module is another class of GNRA receptor that distinguishes less stringently the sequences of GNRA loops than the 11-nt receptor [111]. This module contains 12 nucleotides and shares with the 11-nt receptor two canonical Watson–Crick C=G base pairs, a *trans* Watson–Crick/Hoogsteen U·A base pair and two unpaired adenosines. The IC3 module possesses an additional C=G base pair, lacks an unpaired uridine compared with the 11-nt module and ends with a C=G base pair (G∘U in the case of the 11-nt module). Characteristic base pairs and the uridine residue were identified as elements that determine the binding specificity by means of gel mobility shift essay [112], but no crystal or NMR structures are available to allow a structural description with more atomic details.

Geary *et al.* [113] reported the selection of novel receptors that bind tightly and specifically to GGAA tetraloops. Their selection scheme was directly based on RNA self-assembly properties, irrespective of catalytic activity. After several rounds of selection, they identified and characterized three classes of receptors: class-R1, class-R2 and class-R3. In the class-R3 group, the receptor bases form canonical Watson–Crick base pairs with GGAA bases. In contrast, class-R1 and class-R2 receptors are A-minor-dependent classes of receptors without obvious Watson–Crick base complementarity to the GAAA probe. Both class-R1 and class-R2 receptors present a conserved G=C base pair located 11 base pairs from the tetraloop.

7.5.4 A-minor interactions and ribose zippers

The modes of interactions between the GNRA tetraloops and their receptors (either helix or 11-nt receptor) constitute a particular instance of an extremely common type of RNA–RNA interactions, the A-minor interactions. In A-minor interactions, single-stranded adenosines insert into the shallow groove of an A-RNA helix (equivalent to the minor groove of a B-DNA helix), making hydrogen bonds and van der Waals contacts with canonical base pairs. As discussed above, such interactions can be described with the proposed nomenclature for base pairs. In the contacts described in this chapter, the ribose subunits are crucial and interactions between nucleotides often belong to the *cis* or *trans* Sugar edge/Sugar edge base pair families [36]. The isostericity matrices for the *cis* and *trans* Sugar edge/Sugar edge are shown in Figure 7.33. Such matrices indicate which pairs are geometrically equivalent and in this particular instance they show the partition between A-minor type of contacts and ribose zippers. Four types of A-minor interactions were

cis **Sugar Edge/Sugar Edge**

►	A	C	G	U
A	S	RZ	S	RZ
C	S	RZ	S	RZ
G	S	RZ		RZ
U	S	RZ	S	RZ

trans **Sugar Edge/Sugar Edge**

▷	A	C	G	U
A	S		S	
C	S		S	
G	S		S	
U	S		S	

Figure 7.33 *The isostericity matrices corresponding to the* cis *and* trans *Sugar edge/Sugar edge base pairs geometrically equivalent [130]. 'S' indicates base pairing with base–base H-bonds and 'RZ' indicates ribose zipper contacts which do not imply base–base H-bonds. An empty square indicates that in the relative orientations of the nucleotides, the chosen nucleotides do not form a pair*

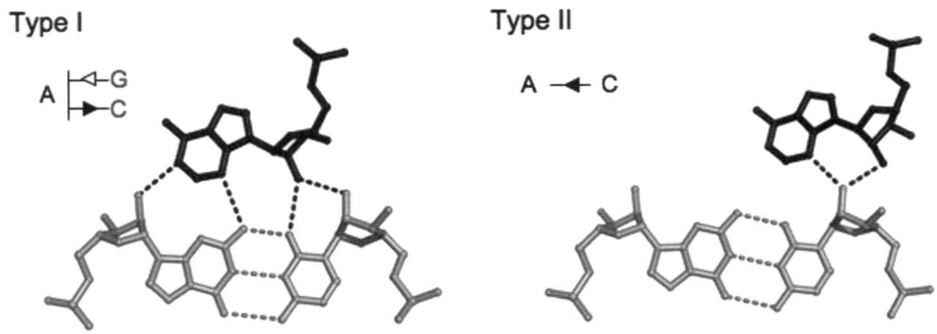

Figure 7.34 *Examples of type I and type II A-minor interactions found in* H. marismortui *23S ribosomal RNA (from 1FFK.PDB). In type I, the adenosine (in black) interacts by hydrogen bonds with both helix strands. In type II, only one strand is involved in interactions with adenosine. The triangle (in the 2D diagram), is oriented so that the arrow points from the HO2′ donor atom of a nucleotide to an acceptor atom of the other nucleotide and is filled in case of* cis *orientation and empty in case of* trans *orientation*

initially described, but only types I and II (Figure 7.34) are adenine dependent [114]. The other two types are not adenine specific: in type 0, it is the sugar of the adenosine which interacts with the Watson–Crick base pair (belongs to the ribose zipper family); in type IV, the Watson–Crick face of the adenosine hydrogen bonds with the ribose. Other nucleotides are able to make type 0 or type IV interactions with base pairs.

Adenines involved in A-minor interactions often cluster with other adenines stacked together and engaged in similar contacts. The number of adenines seldom exceeds three and is generally two. In the latter case, the 5′-adenine (A_n) makes a type II interaction with a Watson–Crick base pair and the 3′-adenine (A_{n+1}) is involved in a type I interaction (Figure 7.35). Thus, A_n interacts with one strand of the double helix whereas A_{n+1} interacts with both strands. It was shown that consecutive adenines have a high preference for canonical Watson–Crick base pairs rather than for wobble pairs [115,116]. Instances of A-minor contacts are very abundant in ribosomal RNAs and constitute the most frequent RNA–RNA recognition protocols [117]. They are also found in almost every ribozyme or structured RNA [36]. For more details about A-minor interactions, see [118]. It was also demonstrated that this frequent RNA–RNA interaction is essential for the recognition of codon–anticodon and that it is central to the fidelity of the decoding process by the ribosomal RNA 30S subunit

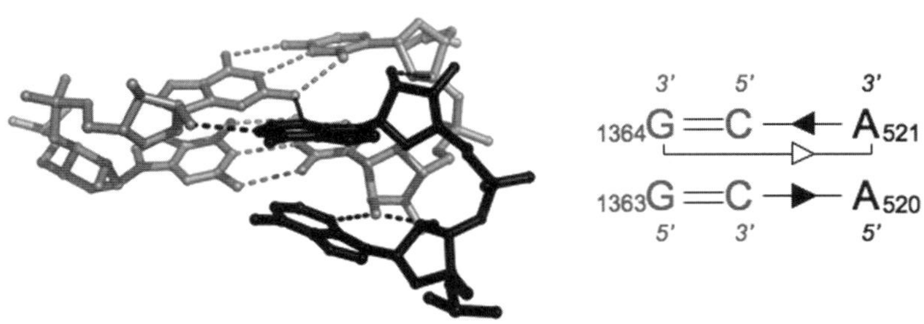

Figure 7.35 *Example of A-minor interactions formed between stacked adenines and the shallow (or minor) groove of an RNA helical arrangement as observed in 23S ribosomal RNA of* H. marismortui *(from 1FFK.PDB). In this cluster, the first A-minor interaction, between A520 and the canonical base pair G1363=C638, is of type II, whereas the next A-minor interaction, between A521 and the base pair G1364=C637, is of type I*

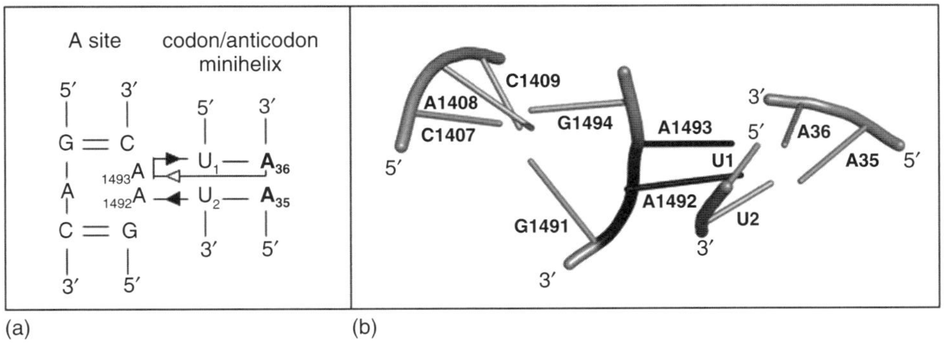

Figure 7.36 (a) In the 16S decoding A site, A1492 and A1493 are involved in A-minor interactions with two base pairs of the minihelix formed between the anticodon bases and the mRNA codon. As in A-minor interaction type I/type II cluster, the 3′ adenine (A1493) interacts with both the mRNA codon first base and the tRNA anticodon third base, whereas the 5′ adenosine (A1492) interacts only with the second base of mRNA codon (from 1IBM.PDB). (b) Interactions are represented using a cartoon view

[119–123]. It was shown that two consecutives adenosines (A1492 and A1493) of the 16S A site interact by forming A-minor interactions with the minihelix formed between the anticodon nucleotides and the RNA messenger codon triplet (Figure 7.36). As in the cluster of A-minor interactions described previously, the 5′-adenosine only interacts with one nucleotide of the messenger RNA codon whereas the 3′-adenosine interacts with both a nucleotide of the codon and one of the anticodon RNA. This conformation of the A site is highly stabilized in the presence of aminoglycoside antibiotics [124–126]. As discussed above, A-minor interactions are also observed in other RNA–RNA interactions: for example, in the binding of a GAAA tetraloop and the 11-nt receptor (as illustrated in Section 7.5.3.2) or in the ribose zipper interactions reported in the hammerhead ribozyme crystal [110] and in the crystal structure of the P4–P6 domain of the *Tetrahymena thermophila* intron [71]. As shown in Figure 7.37, A-minor contacts generally involve two consecutive adenines. The type I/type II A-minor assembly constitutes the most prevalent one, but there are instances of an assembly made of type I/type I A-minor interactions. In ribose zipper interactions, two consecutive nucleotides from one RNA segment contact two consecutive nucleotides from another RNA segment via their riboses: these interactions are intermolecular in the case of the hammerhead ribozyme structure whereas they are intramolecular in the

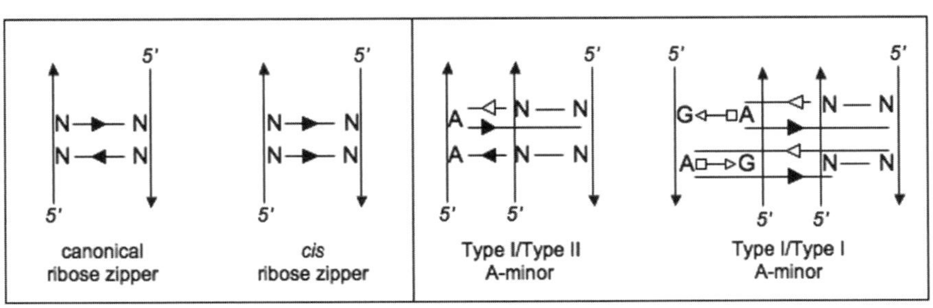

Figure 7.37 The ribose zipper and A-minor interactions occur generally between consecutive nucleotides. For each, two main arrangements have been observed. In the 'canonical' ribose zipper, the giving hydroxyl group alternates strands whereas in the 'cis' arrangement they belong to the same nucleotide strand. Similarly, but in a more complex fashion, for the type I/type II versus the type I/type I A-minor contacts

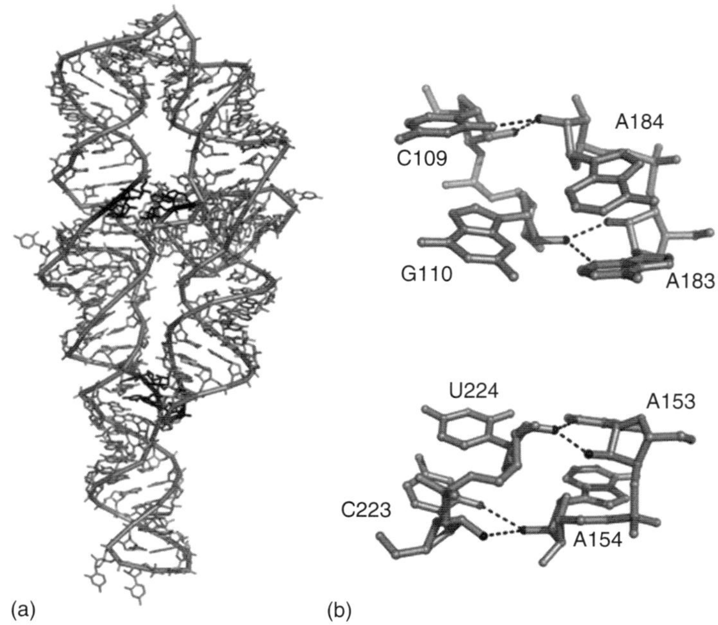

Figure 7.38 *(a) Structure of the P4–P6 group I intron domain and its ribose zipper interactions drawn in black (from 1GID.PDB). (b) More atomic details of the two ribose zipper interactions found in this structure: the first (upper) mediates the interactions between the A-rich bulge and the P4 stem. The second mediates the interactions between the tetraloop and the tetraloop receptor. Ribose O2′ atoms are colored in black*

structure of the P4–P6 domain. As for the A-minor contacts, two assemblies are mainly observed depending on the origins of the giving hydroxyl groups (illustrated in Figure 7.37).

In ribose zipper interactions, bifurcated hydrogen bonds are observed between the 2′-hydroxyl group of a ribose from one segment and the 2′-hydroxyl group and the purine N3 atom (or the pyrimidine O2 atom) from the other segment (Figure 7.38). Tamura and Holbrook [127] analyzed all the ribose zipper interactions present in the crystal structures of the large ribosomal subunit RNA of *Haloarcula marismortui* and *Deinococcus radiodurans* and the small ribosomal subunit RNA of *Thermus thermophilus*. They identified a total of 97 ribose zippers in these structures that they classified into seven classes based on the type and number of ribose–base interactions [127]. The common feature of these classes is that ribose zippers are formed by antiparallel chain interactions and are mainly constituted by interactions between two consecutive sugars. Base triplets are sometimes observed between the nucleotides of the ribose zippers and their Watson–Crick paired bases. The ribose zipper interactions bridge between ribosomal RNA domains for only one-third of the observed motifs while the rest form intradomains bridges. Finally, two-thirds of the ribosomal RNA ribose zippers interact with ribosomal protein by hydrogen bonding and charge neutralization. These protein–RNA interactions contribute to stabilizing ribose zipper tertiary interactions by bridging the backbones of different RNA segments.

Adenosines are also involved in the adenosine platform in which two consecutive adenosines located in a helical strand form a base pair within the helix: they stay side-by-side in the same plane interacting by *cis* Hoogsteen/Sugar edge hydrogen bonds between the 3′-adenosine and the 5′-adenosine (Figure 7.39). Such a contact was first observed in the crystal structure of the 160-nucleotide P4–P6 domain of the *Tetrahymena thermophila* self-splicing intron [128]. In this structure, the 3′-adenosine continues the stacking of the helix below it whereas the 5′-adenosine stacks on the opposite strand of the helix (Figure 7.39): the adenosine

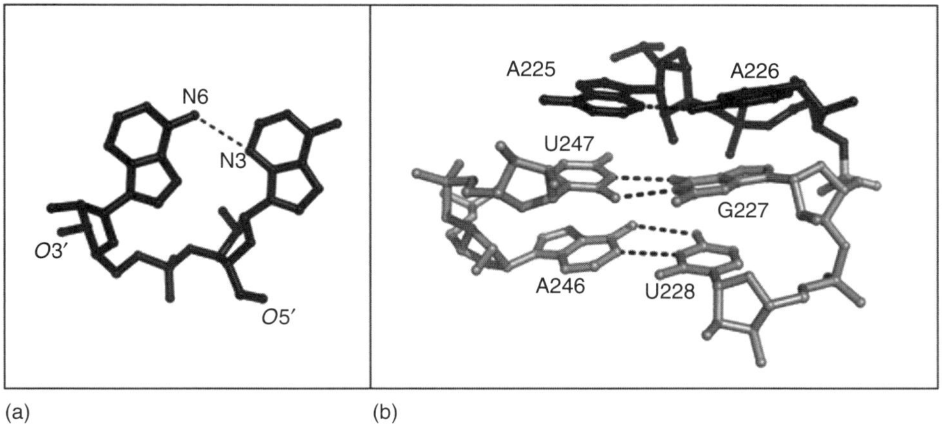

Figure 7.39 *(a) The* cis *Hoogsteen/Sugar edge A·A base pair observed in adenosine platform motif. (b) In the adenosine platform, the 3′-A (A226) extends the stacking of the helix below it whereas the 5′-A (A225) stacks on the opposite helix strand (from 1GID.PDB)*

platform is associated with cross-strand stacking and leaves one face of the 3′-adenosine available for stacking with bases of other regions. Side-by-side platforms are not restricted to 5′AA3′ dinucleotides but have also been observed for the 5′GU3′ sequence in the sarcin loop and the L11 complex [41,72].

7.6 Conclusion

It is surprising to observe how weakly the base sequence influences the interaction protocols, stressing the primordial role of the topological links on the final architecture. Thus, the coaxial stacking is very poorly dependent on the base pairs at the interface [129] and is rather generally dictated by the length of the nucleotide segment in the continuous strand of the stacked helices. The sequence dependence is the clearest in loop–loop contacts (either between two complementary hairpin loops as in kissing loops or between a hairpin loop and a single-stranded segment as in the commonest pseudoknots). However, the interaction protocols based on non-Watson–Crick pairs are dominated by Sugar edge/Sugar edge base contacts, which are known for displaying a prominent character of non-specificity, the least base-specific type being the ribose zipper. The topology of the junctions between helices (three-way or four-way junctions, etc.) is thus central to the folding of RNA, since only the proper choice of coaxial stack or relative orientations of helices will allow for correctly chosen and positioned RNA–RNA contacts.

References

1. N. Leontis and E. Westhof, Geometric nomenclature and classification of RNA base pairs, *RNA*, **7**, 499–512 (2001).
2. P.B. Moore, Structural motifs in RNA, *Annu Rev Biochem*, **68**, 287–300 (1999).
3. C. Correll and K. Swinger, Common and distinctive features of GNRA tetraloops based on a GUAA tetraloop structure at 1.4 Å resolution, *RNA*, **9**, 355–363 (2003).
4. D. Hendrix, S. Brenner and S. Holbrook, RNA structural motifs: building blocks of a modular biomolecule, *Q Rev Biophys*, **38**, 221–243 (2005).
5. G. Varani, C. Cheong and I.J. Tinoco, Structure of an unusually stable RNA hairpin, *Biochemistry*, **30**, 3280–3289 (1991).

6. F. Jucker and A. Pardi, Solution structure of the CUUG hairpin loop: a novel RNA tetraloop motif, *Biochemistry*, **34**, 14416–14427 (1995).
7. R.R. Gutell, J. Cannone, D. Konings and D. Gautheret, Predicting U-turns in ribosomal RNA with comparative sequence analysis, *J Mol Biol*, **300**, 791–803 (2000).
8. S. Butcher, T. Dieckmann and J. Feigon, Solution structure of the conserved 16 S-like ribosomal RNA UGAA tetraloop, *J Mol Biol*, **268**, 348–358 (1997).
9. I. Lebars, B. Lamontagne, S. Yoshizawa, S. Aboul-Elela and D. Fourmy, Solution structure of conserved AGNN tetraloops: insights into Rnt1p RNA processing, *EMBO J*, **20**, 7250–7258 (2001).
10. H. Wu, P.K. Yang, S. Butcher, S. Kang, G. Chanfreau and J. Feigon, A novel family of RNA tetraloop structure forms the recognition site for *Saccharomyces cerevisiae* RNase III, *EMBO J*, **20**, 7240–7249 (2001).
11. J.P. Rife and P.B. Moore, The structure of a methylated tetraloop in 16S ribosomal RNA, *Structure*, **6**, 747–756 (1998).
12. F. Allain and G. Varani, Structure of the P1 helix from group I self-splicing introns, *J Mol Biol*, **250**, 333–353 (1995).
13. C. Woese, S. Winker and R. Gutell, Architecture of ribosomal RNA: constraints on the sequence of 'tetra-loops', *Proc Natl Acad Sci USA*, **87**, 8467–8471 (1990).
14. D. Abramovitz and A. Pyle, Remarkable morphological variability of a common RNA folding motif: the GNRA tetraloop-receptor interaction, *J Mol Biol*, **266**, 493–506 (1997).
15. Z. Cai, A. Gorin, R. Frederick, X. Ye, W. Hu, A. Majumdar, A. Kettani and D. Patel, Solution structure of P22 transcriptional antitermination N peptide–boxB RNA complex, *Nat Struct Biol*, **5**, 203–212 (1998).
16. P. Legault, J. Li, J. Mogridge, L. Kay and J. Greenblatt, NMR structure of the bacteriophage lambda N peptide/boxB RNA complex: recognition of a GNRA fold by an arginine-rich motif, *Cell*, **93**, 289–299 (1998).
17. M. Scharpf, H. Sticht, K. Schweimer, M. Boehm, S. Hoffmann and P. Rosch, Antitermination in bacteriophage lambda. The structure of the N36 peptide–boxB RNA complex, *Eur J Biochem*, **267**, 2397–2408 (2000).
18. A. Huppler, L. Nikstad, A. Allmann, D. Brow and S. Butcher, Metal binding and base ionization in the U6 RNA intramolecular stem–loop structure, *Nat Struct Biol*, **9**, 431–435 (2002).
19. M. Robertson, H. Igel, R. Baertsch, D. Haussler, M.J. Ares and W. Scott, The structure of a rigorously conserved RNA element within the SARS virus genome, *PLoS Biol*, **3**, e5 (2005).
20. U. Nagaswamy, X. Gao, S. Martinis and G. Fox, NMR structure of a ribosomal RNA hairpin containing a conserved CUCAA pentaloop, *Nucleic Acids Res*, **29**, 5129–5139 (2001).
21. C. Theimer, L. Finger, L. Trantirek and J. Feigon, Mutations linked to dyskeratosis congenita cause changes in the structural equilibrium in telomerase RNA, *Proc Natl Acad Sci USA*, **100**, 449–454 (2003).
22. S. Baba, M. Kajikawa, N. Okada and G. Kawai, Solution structure of an RNA stem–loop derived from the 3′ conserved region of eel LINE UnaL2, *RNA*, **10**, 1380–1387 (2004).
23. R. Stefl and F. Allain, A novel RNA pentaloop fold involved in targeting ADAR2, *RNA*, **11**, 592–597 (2005).
24. F. Oberstrass, A. Lee, R. Stefl, M. Janis, G. Chanfreau and F. Allain, Shape-specific recognition in the structure of the Vts1p SAM domain with RNA, *Nat Struct Mol Biol*, **13**, 160–167 (2006).
25. J. Lee, J. Cannone and R. Gutell, The lonepair triloop: a new motif in RNA structure, *J Mol Biol*, **325**, 65–83 (2003).
26. V. Lisi and F. Major, A comparative analysis of the triloops in all high-resolution RNA structures reveals sequence structure relationships, *RNA*, **13**, 1537–1545 (2007).
27. G. Quigley and A. Rich, Structural domains of transfer RNA molecules, *Science*, **194**, 796–806 (1976).
28. A. Krasilnikov and A. Mondragon, On the occurrence of the T-loop RNA folding motif in large RNA molecules, *RNA*, **9**, 640–643 (2003).
29. P. Fechter, J. Rudinger-Thirion, C. Florentz and R. Giege, Novel features in the tRNA-like world of plant viral RNAs, *Cell Mol Life Sci*, **58**, 1547–1561 (2001).
30. S. Barends, K. Bjork, A. Gultyaev, M.H. De Smit, C. Pleij and B. Kraal, Functional evidence for D- and T-loop interactions in tmRNA, *FEBS Lett*, **514**, 78–83 (2002).
31. U. Nagaswamy and G. Fox, Frequent occurrence of the T-loop RNA folding motif in ribosomal RNAs, *RNA*, **8**, 1112–1119 (2002).
32. A. Krasilnikov, X. Yang, T. Pan and A. Mondragon, Crystal structure of the specificity domain of ribonuclease P, *Nature*, **421**, 760–764 (2003).

33. N. Leontis and E. Westhof, Conserved geometrical base-pairing patterns in RNA, *Q Rev Biophys*, **31**, 399–455 (1998).
34. A. Serganov, A. Polonskaia, A. Phan, R. Breaker and D. Patel, Structural basis for gene regulation by a thiamine pyrophosphate-sensing riboswitch, *Nature*, **441**, 1167–1171 (2006).
35. S. Thore, M. Leibundgut and N. Ban, Structure of the eukaryotic thiamine pyrophosphate riboswitch with its regulatory ligand, *Science*, **312**, 1208–1211 (2006).
36. N. Leontis and E. Westhof, Analysis of RNA motifs, *Curr Opin Struct Biol*, **13**, 300–308 (2003).
37. T. Hermann and D. Patel, RNA bulges as architectural and recognition motifs, *Structure*, **8**, R47–54 (2000).
38. B. Wimberly, G. Varani and I.J. Tinoco, The conformation of loop E of eukaryotic 5S ribosomal RNA, *Biochemistry*, **32**, 1078–1087 (1993).
39. A. Szewczak, P. Moore, Y. Chang and I. Wool, The conformation of the sarcin/ricin loop from 28S ribosomal RNA, *Proc Natl Acad Sci USA*, **90**, 9581–9585 (1993).
40. C. Correll, A. Munishkin, Y. Chan, Z. Ren, I. Wool and T. Steitz, Crystal structure of the ribosomal RNA domain essential for binding elongation factors, *Proc Natl Acad Sci USA*, **95**, 13436–13441 (1998).
41. C. Correll, I. Wool and A. Munishkin, The two faces of the *Escherichia coli* 23 S rRNA sarcin/ricin domain: the structure at 1.11 Å resolution, *J Mol Biol*, **292**, 275–287 (1999).
42. N. Leontis and E. Westhof, The 5S rRNA loop E: chemical probing and phylogenetic data versus crystal structure, *RNA*, **4**, 1134–1153 (1998).
43. N. Leontis and E. Westhof, A common motif organizes the structure of multi-helix loops in 16 S and 23 S ribosomal RNAs, *J Mol Biol*, **283**, 571–583 (1998).
44. C. Duarte, L. Wadley and A. Pyle, RNA structure comparison, motif search and discovery using a reduced representation of RNA conformational space, *Nucleic Acids Res*, **31**, 4755–4761 (2003).
45. C. Correll, J. Beneken, M. Plantinga, M. Lubbers and Y. Chan, The common and the distinctive features of the bulged-G motif based on a 1.04 Å resolution RNA structure, *Nucleic Acids Res*, **31**, 6806–6818 (2003).
46. C. Correll, B. Freeborn, P. Moore and T. Steitz, Metals, motifs and recognition in the crystal structure of a 5S rRNA domain, *Cell*, **91**, 705–712 (1997).
47. P. Romby, F. Baudin, C. Brunel, I. Leal De Stevenson, E. Westhof, P. Romaniuk, C. Ehresmann and B. Ehresmann, Ribosomal 5S RNA from *Xenopus laevis* oocytes: conformation and interaction with transcription factor IIIA, *Biochimie*, **72**, 437–452 (1990).
48. S. Szep, J. Wang and P.B. Moore, The crystal structure of a 26-nucleotide RNA containing a hook-turn, *RNA*, **9**, 44–51 (2003).
49. I. Vidovic, S. Nottrott, K. Hartmuth, R. Luhrmann and R. Ficner, Crystal structure of the spliceosomal 15.5 kD protein bound to a U4 snRNA fragment, *Mol Cell*, **6**, 1331–1342 (2000).
50. D. Klein, T. Schmeing, P. Moore and T. Steitz, The kink-turn: a new RNA secondary structure motif, *EMBO J*, **20**, 4214–4221 (2001).
51. A. Lescoute, N. Leontis, C. Massire and E. Westhof, Recurrent structural RNA motifs, isostericity matrices and sequence alignments, *Nucleic Acids Res*, **33**, 2395–2409 (2005).
52. S. Strobel, P.L. Adams, M. Stahley and J. Wang, RNA kink turns to the left and to the right, *RNA*, **10**, 1852–1854 (2004).
53. T. Goody, S. Melcher, D. Norman and D. Lilley, The kink-turn motif in RNA is dimorphic and metal ion-dependent, *RNA*, **10**, 254–264 (2004).
54. W.J. Clemons, D. Brodersen, J. Mccutcheon, J. May, A. Carter, R.J. Morgan-Warren, B. Wimberly and V. Ramakrishnan, Crystal structure of the 30 S ribosomal subunit from *Thermus thermophilus*: purification, crystallization and structure determination, *J Mol Biol*, **310**, 827–843 (2001).
55. B. Wimberly, D. Brodersen, W.J. Clemons, R.J. Morgan-Warren, A. Carter, C. Vonrhein, T. Hartsch and V. Ramakrishnan, Structure of the 30S ribosomal subunit, *Nature*, **407**, 327–339 (2000).
56. N. Ban, P. Nissen, J. Hansen, P. Moore and T. Steitz, The complete atomic structure of the large ribosomal subunit at 2.4 Å resolution, *Science*, **289**, 905–920 (2000).
57. A. Torres-Larios, A.C. Dock-Bregeon, P. Romby, B. Rees, R. Sankaranarayanan, J. Caillet, M. Springer, C. Ehresmann, B. Ehresmann and D. Moras, Structural basis of translational control by *Escherichia coli* threonyl tRNA synthetase, *Nat Struct Biol*, **9**, 343–347 (2002).

58. J. Nowakowski and I.J. Tinoco, RNA structure in solution. In *Oxford Handbook of Nucleic Acid Structure*, ed. S. Neidle, Oxford Science Publications, Oxford, 1998, pp. 567–602
59. D. Lilley, R.M. Clegg, S. Diekmann, N. Seeman, E. Von Kitzing and P. Hagerman, Nomenclature Committee of the International Union of Biochemistry and Molecular Biology (NC-IUBMB). A nomenclature of junctions and branchpoints in nucleic acids. Recommendations 1994, *Eur J Biochem*, **230**, 1–2 (1995).
60. G. Bassi, A. Murchie and D. Lilley, The ion-induced folding of the hammerhead ribozyme: core sequence changes that perturb folding into the active conformation, *RNA*, **2**, 756–768 (1996).
61. H. Pley, K. Flaherty and D. McKay, Three-dimensional structure of a hammerhead ribozyme, *Nature*, **372**, 68–74 (1994).
62. W. Scott, J.T. Finch and A. Klug, The crystal structure of an all-RNA hammerhead ribozyme: a proposed mechanism for RNA catalytic cleavage, *Cell*, **81**, 991–1002 (1995).
63. A. Lescoute and E. Westhof, Topology of three-way junctions in folded RNAs, *RNA*, **12**, 83–93 (2006).
64. A. Krol, E. Westhof, M. Bach, R. Luhrmann, J. Ebel and P. Carbon, Solution structure of human U1 snRNA. Derivation of a possible three-dimensional model, *Nucleic Acids Res*, **18**, 3803–3811 (1990).
65. D.A. Pomeranz Krummel, C. Oubridge, A. Leung, J. Li and K. Nagai, Crystal structure of human spliceosomal U1 snRNP at 5.5 Å resolution, *Nature*, **458**, 475–480 (2009).
66. R. Batey, R. Rambo and J. Doudna, Tertiary motifs in RNA structure and folding, *Angew Chem Int Ed*, **38**, 2326–2343 (1999).
67. S. Kim, F. Suddath, G. Quigley, A. Mcpherson, J. Sussman, A. Wang, N. Seeman and A. Rich, Three-dimensional tertiary structure of yeast phenylalanine transfer RNA, *Science*, **185**, 435–440 (1974).
68. J. Robertus, J. Ladner, J. Finch, D. Rhodes, R. Brown, B. Clark and A. Klug, Structure of yeast phenylalanine tRNA at 3 Å resolution, *Nature*, **250**, 546–551 (1974).
69. A.R. Ferre-D'Amare, K. Zhou and J. Doudna, Crystal structure of a hepatitis delta virus ribozyme, *Nature*, **395**, 567–574 (1998).
70. F. Murphy, Y. Wang, J. Griffith and T. Cech, Coaxially stacked RNA helices in the catalytic center of the *Tetrahymena* ribozyme, *Science*, **265**, 1709–1712 (1994).
71. J. Cate, A. Gooding, E. Podell, K. Zhou, B. Golden, C. Kundrot, T. Cech and J. Doudna, Crystal structure of a group I ribozyme domain: principles of RNA packing, *Science*, **273**, 1678–1685 (1996).
72. B. Wimberly, R. Guymon, J. McCutcheon, S. White and V. Ramakrishnan, A detailed view of a ribosomal active site: the structure of the L11–RNA complex, *Cell*, **97**, 491–502 (1999).
73. M. Laughrea and L. Jette, A 19-nucleotide sequence upstream of the 5′ major splice donor is part of the dimerization domain of human immunodeficiency virus 1 genomic RNA, *Biochemistry*, **33**, 13464–13474 (1994).
74. E. Skripkin, J. Paillart, R. Marquet, B. Ehresmann and C. Ehresmann, Identification of the primary site of the human immunodeficiency virus type 1 RNA dimerization *in vitro*, *Proc Natl Acad Sci USA*, **91**, 4945–4949 (1994).
75. D. Muriaux, P. Girard, B. Bonnet-Mathoniere and J. Paoletti, Dimerization of HIV-1Lai RNA at low ionic strength. An autocomplementary sequence in the 5′ leader region is evidenced by an antisense oligonucleotide, *J Biol Chem*, **270**, 8209–8216 (1995).
76. J. Paillart, E. Skripkin, B. Ehresmann, C. Ehresmann and R. Marquet, A loop–loop 'kissing' complex is the essential part of the dimer linkage of genomic HIV-1 RNA, *Proc Natl Acad Sci USA*, **93**, 5572–5577 (1996).
77. B. Berkhout and J.L. Van Wamel, Role of the DIS hairpin in replication of human immunodeficiency virus type 1, *J Virol*, **70**, 6723–6732 (1996).
78. J. Clever, M. Wong and T. Parslow, Requirements for kissing-loop-mediated dimerization of human immunodeficiency virus RNA, *J Virol*, **70**, 5902–5908 (1996).
79. J. Paillart, E. Westhof, C. Ehresmann, B. Ehresmann and R. Marquet, Non-canonical interactions in a kissing loop complex: the dimerization initiation site of HIV-1 genomic RNA, *J Mol Biol*, **270**, 36–49 (1997).
80. E. Ennifar and P. Dumas, Polymorphism of bulged-out residues in HIV-1 RNA DIS kissing complex and structure comparison with solution studies, *J Mol Biol*, **356**, 771–782 (2006).
81. S. Baba, K. Takahashi, S. Noguchi, H. Takaku, Y. Koyanagi, N. Yamamoto and G. Kawai, Solution RNA structures of the HIV-1 dimerization initiation site in the kissing-loop and extended-duplex dimers, *J Biochem*, **138**, 583–592 (2005).
82. J. Marino, R.J. Gregorian, G. Csankovszki and D. Crothers, Bent helix formation between RNA hairpins with complementary loops, *Science*, **268**, 1448–1454 (1995).

83. D. Scarabino, A. Crisari, S. Lorenzini, K. Williams and G.P. Tocchini-Valentini, tRNA prefers to kiss, *EMBO J*, **18**, 4571–4578 (1999).
84. K. Chang and I.J. Tinoco, The structure of an RNA 'kissing' hairpin complex of the HIV TAR hairpin loop and its complement, *J Mol Biol*, **269**, 52–66 (1997).
85. I. Lebars, T. Richard, C. Di Primo and J. Toulme, NMR structure of a kissing complex formed between the TAR RNA element of HIV-1 and a LNA-modified aptamer, *Nucleic Acids Res*, **35**, 6103–6114 (2007).
86. D. Collin, C. Van Heijenoort, C. Boiziau, J. Toulme and E. Guittet, NMR characterization of a kissing complex formed between the TAR RNA element of HIV-1 and a DNA aptamer, *Nucleic Acids Res*, **28**, 3386–3391 (2000).
87. C. Kim and I.J. Tinoco, A retroviral RNA kissing complex containing only two G·C base pairs, *Proc Natl Acad Sci USA*, **97**, 9396–9401 (2000).
88. A. Gultyaev, C. Pleij and E. Westhof, RNA structure: pseudoknots, *Encyclopedia Life Sci*, 1–7 (2005).
89. K. Rietveld, R. Van Poelgeest, C.W. Pleij, J.H. Van Boom and L. Bosch, The tRNA-like structure at the 3′ terminus of turnip yellow mosaic virus RNA. Differences and similarities with canonical tRNA, *Nucleic Acids Res*, **10**, 1929–1946 (1982).
90. T. Hermann and D. Patel, Stitching together RNA tertiary architectures, *J Mol Biol*, **294**, 829–849 (1999).
91. E. Westhof and L. Jaeger, RNA pseudoknots, *Curr Opin Struct Biol*, **2**, 327–333 (1992).
92. L. Shen and I.J. Tinoco, The structure of an RNA pseudoknot that causes efficient frameshifting in mouse mammary tumor virus, *J Mol Biol*, **247**, 963–978 (1995).
93. L. Su, L. Chen, M. Egli, J. Berger and A. Rich, Minor groove RNA triplex in the crystal structure of a ribosomal frameshifting viral pseudoknot, *Nat Struct Biol*, **6**, 285–292 (1999).
94. J. Wyatt, J. Puglisi and I.J. Tinoco, RNA pseudoknots. Stability and loop size requirements, *J Mol Biol*, **214**, 455–470 (1990).
95. F. Michel and E. Westhof, Modelling of the three-dimensional architecture of group I catalytic introns based on comparative sequence analysis, *J Mol Biol*, **216**, 585–610 (1990).
96. L. Jaeger, F. Michel and E. Westhof, Involvement of a GNRA tetraloop in long-range RNA tertiary interactions, *J Mol Biol*, **236**, 1271–1276 (1994).
97. M. Costa and F. Michel, Frequent use of the same tertiary motif by self-folding RNAs, *EMBO J*, **14**, 1276–1285 (1995).
98. M. Costa, E. Deme, A. Jacquier and F. Michel, Multiple tertiary interactions involving domain II of group II self-splicing introns, *J Mol Biol*, **267**, 520–536 (1997).
99. P. Adams, M. Stahley, M. Gill, A.B. Kosek, J. Wang and S. Strobel, Crystal structure of a group I intron splicing intermediate, *RNA*, **10**, 1867–1887 (2004).
100. M. Tanner and T. Cech, An important RNA tertiary interaction of group I and group II introns is implicated in Gram-positive RNase P RNAs, *RNA*, **1**, 349–350 (1995).
101. C. Massire, L. Jaeger and E. Westhof, Phylogenetic evidence for a new tertiary interaction in bacterial RNase P RNAs, *RNA*, **3**, 553–556 (1997).
102. A. Torres-Larios, K. Swinger, T. Pan and A. Mondragon, Structure of ribonuclease P – a universal ribozyme, *Curr Opin Struct Biol*, **16**, 327–335 (2006).
103. S. Butcher, T. Dieckmann and J. Feigon, Solution structure of a GAAA tetraloop receptor RNA, *EMBO J*, **16**, 7490–7499 (1997).
104. J.H. Davis, M. Tonelli, L.G. Scott, L. Jaeger, J. Williamson and S. Butcher, RNA helical packing in solution: NMR structure of a 30 kDa GAAA tetraloop–receptor complex, *J Mol Biol*, **351**, 371–382 (2005).
105. S. Basu, R. Rambo, J. Strauss-Soukup, J. Cate, A.R. Ferre-D'Amare, S. Strobel and J. Doudna, A specific monovalent metal ion integral to the AA platform of the RNA tetraloop receptor, *Nat Struct Biol*, **5**, 986–992 (1998).
106. K. Juneau, E. Podell, D. Harrington and T. Cech, Structural basis of the enhanced stability of a mutant ribozyme domain and a detailed view of RNA–solvent interactions, *Structure*, **9**, 221–231 (2001).
107. P. Adams, M. Stahley, A. Kosek, J. Wang and S. Strobel, Crystal structure of a self-splicing group I intron with both exons, *Nature*, **430**, 45–50 (2004).
108. J. Davis, T. Foster, M. Tonelli and S. Butcher, Role of metal ions in the tetraloop–receptor complex as analyzed by NMR, *RNA*, **13**, 76–86 (2007).
109. F. Murphy and T. Cech, GAAA tetraloop and conserved bulge stabilize tertiary structure of a group I intron domain, *J Mol Biol*, **236**, 49–63 (1994).

110. H. Pley, K. Flaherty and D.B. McKay, Model for an RNA tertiary interaction from the structure of an intermolecular complex between a GAAA tetraloop and an RNA helix, *Nature*, **372**, 111–113 (1994).
111. Y. Ikawa, D. Naito, N. Aono, H. Shiraishi and T. Inoue, A conserved motif in group IC3 introns is a new class of GNRA receptor, *Nucleic Acids Res*, **27**, 1859–1865 (1999).
112. Y. Ikawa, K. Nohmi, S. Atsumi, H. Shiraishi and T. Inoue, A comparative study on two GNRA-tetraloop receptors: 11-nt and IC3 motifs, *J Biochem*, **130**, 251–255 (2001).
113. C. Geary, S. Baudrey and L. Jaeger, Comprehensive features of natural and *in vitro* selected GNRA tetraloop-binding receptors, *Nucleic Acids Res*, **36**, 1138–1152 (2008).
114. P. Nissen, J. Ippolito, N. Ban, P. Moore and T. Steitz, RNA tertiary interactions in the large ribosomal subunit: the A-minor motif, *Proc Natl Acad Sci USA*, **98**, 4899–4903 (2001).
115. E. Doherty, R.T. Batey, B. Masquida and J. Doudna, A universal mode of helix packing in RNA, *Nat Struct Biol*, **8**, 339–343 (2001).
116. D. Battle and J. Doudna, Specificity of RNA–RNA helix recognition, *Proc Natl Acad Sci USA*, **99**, 11676–11681 (2002).
117. H. Noller, RNA structure: reading the ribosome, *Science*, **309**, 1508–1514 (2005).
118. A. Lescoute and E. Westhof, The A-minor motifs in the decoding recognition process, *Biochimie*, **88**, 993–999 (2006).
119. J.M. Ogle, D. Brodersen, W.J. Clemons, M. Tarry, A. Carter and V. Ramakrishnan, Recognition of cognate transfer RNA by the 30S ribosomal subunit, *Science*, **292**, 897–902 (2001).
120. J.M. Ogle, F. Murphy, M. Tarry and V. Ramakrishnan, Selection of tRNA by the ribosome requires a transition from an open to a closed form, *Cell*, **111**, 721–732 (2002).
121. V. Ramakrishnan, Ribosome structure and the mechanism of translation, *Cell*, **108**, 557–572 (2002).
122. J.M. Ogle, A. Carter and V. Ramakrishnan, Insights into the decoding mechanism from recent ribosome structures, *Trends Biochem Sci*, **28**, 259–266 (2003).
123. J.M. Ogle and V. Ramakrishnan, Structural insights into translational fidelity, *Annu Rev Biochem*, **74**, 129–177 (2005).
124. D. Brodersen, A. Carter, W.J. Clemons, R.J. Morgan-Warren, F.T. Murphy, J. Ogle, M. Tarry, B. Wimberly and V. Ramakrishnan, Atomic structures of the 30S subunit and its complexes with ligands and antibiotics, *Cold Spring Harbor Symp Quant Biol*, **66**, 17–32 (2001).
125. B. Francois, R. Russell, J. Murray, F. Aboul-Ela, B. Masquida, Q. Vicens and E. Westhof, Crystal structures of complexes between aminoglycosides and decoding A site oligonucleotides: role of the number of rings and positive charges in the specific binding leading to miscoding, *Nucleic Acids Res*, **33**, 5677–5690 (2005).
126. J. Kondo, A. Urzhumtsev and E. Westhof, Two conformational states in the crystal structure of the *Homo sapiens* cytoplasmic ribosomal decoding A site, *Nucleic Acids Res*, **34**, 676–685 (2006).
127. M. Tamura and S. Holbrook, Sequence and structural conservation in RNA ribose zippers, *J Mol Biol*, **320**, 455–474 (2002).
128. J. Cate, A. Gooding, E. Podell, K. Zhou, B. Golden, A. Szewczak, C. Kundrot, T. Cech and J. Doudna, RNA tertiary structure mediation by adenosine platforms, *Science*, **273**, 1696–1699 (1996).
129. R. Tyagi and D. Mathews, Predicting helical coaxial stacking in RNA multibranch loops, *RNA*, **13**, 939–951 (2007).
130. J. Stombaugh, C. Zirbel, E. Westhof and N. Leontis, Frequency and isostericity of RNA base pairs, *Nucleic Acids Res*, **37**, 2294–2312 (2009).

8

Genesis and Biological Applications of Locked Nucleic Acids (LNAs)

Harleen Kaur and Souvik Maiti

8.1 Introduction

The discovery of the Central Dogma of life propelled biological research into a new dimension and marked a beginning in understanding the regulation of gene expression. One particular mechanism of interest is regulation through molecules, complementary and encoded within the DNA itself. This is well illustrated by the contribution of small RNA molecules such as miRNA and siRNA in regulating gene expression. The challenge now lies in finding novel approaches that allow for the regulation and manipulation of these functions, to treat diseases characterized by the aberrant gene expression. As an alternative to conventional therapy, the immense potential of short synthetic oligonucleotides has been examined to improve the clinical outcome of patients suffering from genetic anomalies [1].

Modulation of gene expression can be achieved either by targeting the respective gene sequence at the level of DNA to inhibit its transcription (the antigene approach) or at the level of mRNA to inhibit translation (the antisense approach). Harvesting the potential of oligonucleotides not only yields breakthrough diagnostic and therapeutic strategies but also creates biotechnological advancements. While this entire set of applications rests upon the 'simple' theory of sequence-specific hybridization between the synthetic oligonucleotide and the target nucleic acid, the theory fails when put into practice. The commonly encountered constraints include (i) the inherent tendency of the target or the oligonucleotide to fold into reasonably stable secondary structures that prevent target–oligonucleotide hybridization, (ii) difficulty in predicting the dose of the oligonucleotide required to produce a given *in vivo* effect without mediating any other toxic effect and (iii) the specificity and the biological stability of the oligonucleotide. These practical limitations have led to the downfall of unmodified oligonucleotides. However, these have acted as triggers for the exploration and examination of the so-called nucleic acid 'analogs' or 'mimics'. These are the modifications of usual nucleic acids at either the nucleobase, the sugar ring or the phosphodiester backbone, conferring desirable traits on

The Chemical Biology of Nucleic Acids Edited by Günter Mayer

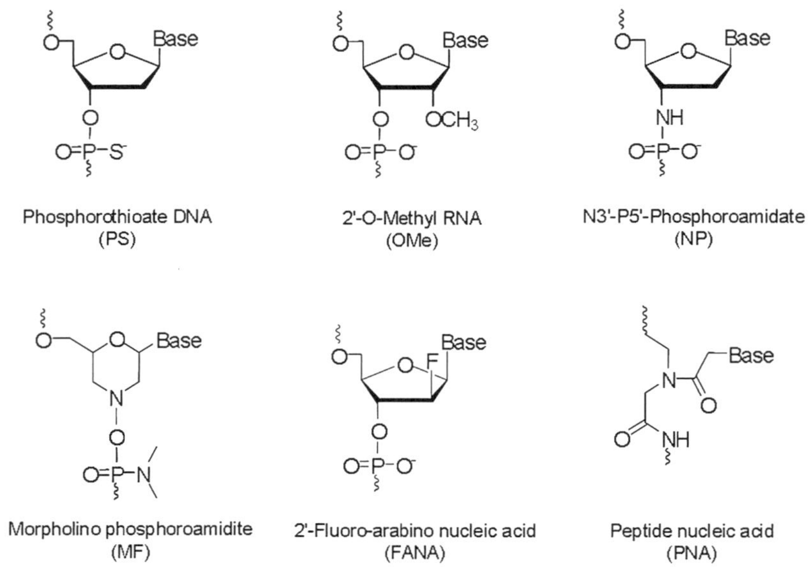

Figure 8.1 *Examples of synthetic nucleic acid analogs. From H. Kaur, B.R. Babu and S. Maiti, Perspectives on chemistry and therapeutic applications of locked nucleic acid (LNA),* Chem. Rev., ***107****, 4672–4697 (2007). Reprinted with permission. © 2007 American Chemical Society*

the oligonucleotide [2]. Figure 8.1 shows some of the well-characterized and widely used nucleic acid modifications.

One of the most promising candidates for chemically modified nucleotides developed in the last few years is locked nucleic acids (LNAs) [also known as bridged nucleic acids (BNAs)]. LNAs are ribonucleotide analogs containing a methylene linkage between the 2′-oxygen and 4′-carbon of the ribose ring (Figure 8.2). LNAs are known to confer dramatically enhanced binding affinity on complementary RNA sequences when incorporated into oligonucleotides. In addition, LNA drugs are resistant to degradation when given systemically, have long tissue half-lives, are taken up readily by many tissues and have improved therapeutic ratios over first- and second-generation antisense drugs. Furthermore, LNA oligonucleotides can be synthesized using conventional phosphoramidite chemistry, allowing automated synthesis of fully modified LNA and chimeric oligonucleotides such as LNA/DNA and LNA/RNA [3]. The following sections describe the numerous attributes of LNA that make it advantageous over other modified oligonucleotides. Its novel applications in the field of therapeutics and diagnostics are also discussed, along with its current status in clinical trials.

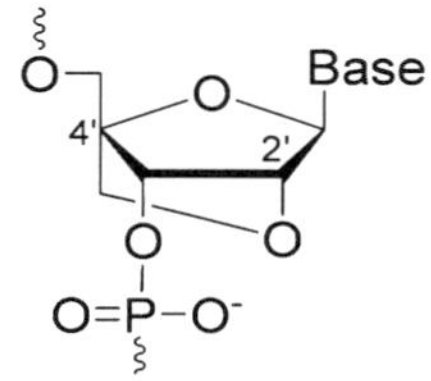

Figure 8.2 *Locked nucleic acid (β-D-LNA). From H. Kaur, B.R. Babu and S. Maiti, Perspectives on chemistry and therapeutic applications of locked nucleic acid (LNA),* Chem. Rev., ***107****, 4672–4697 (2007). Reprinted with permission. © 2007 American Chemical Society*

8.2 The locked nucleic acid: an overview

8.2.1 'Locked on to' the Target

The bridging of the 2′-oxygen of the ribose with the 4′-carbon in LNAs results in a locked 3′-*endo* (*N*-type) conformation. This constraint in the sugar moiety confers outstanding affinity on the complementary RNA and also with ssDNA, without the loss of sequence specificity. In fact, LNAs exhibit mismatch discrimination, equal or superior to native DNA [4]. The hybridization properties of LNAs containing oligonucleotides have been evaluated in different sequence contexts, ranging from six to 20 nucleotide long oligomers with varying levels of LNA content (such as fully modified LNA, LNA/DNA mixmers, LNA/RNA mixmers and LNA/PS-DNA mixmers). The unprecedented hybridization potential of LNAs with either RNA or DNA targets is reflected in the increased thermostability of the LNA-containing duplexes. Substitution by an LNA monomer leads to a rise in the ΔT_m values up to +1 to +8°C against DNA and an increase of +2 to +10°C against RNA [5,6]. This is possibly the largest increase in thermostability observed for a nucleic acid analog, which nevertheless saturates when the relative substitution by LNA monomer reaches to about 50% of the total residues in the LNA/DNA chimera.

8.2.2 LNA – an RNA mimic

NMR and circular dichroism (CD) spectroscopy and X-ray crystallography have been used to characterize structurally duplexes comprising of an LNA-modified strand and an unmodified complementary RNA or DNA strand. In general, these hybrids retain features common for the native nucleic acid duplexes, namely the usual Watson–Crick base pairing, nucleobases in an *anti* orientation, base stacking and a right-handed helical conformation. A noteworthy observation is the gradual increase towards the *A*-type helical geometry concomitant with the increase in the LNA content of the helix. In nature, such *A*-type helical geometry is a structural signature of an RNA which arises due to the N-type sugar pucker (3′-*endo* conformation), whereas *B*-type geometry associated with an S-type sugar pucker or a C2′-*endo* conformation defines a DNA structure. Normally, in a DNA·RNA (hybrid) there is an equilibrium between the N- and S-type sugar conformations (Figure 8.3). The incorporation of one or a few LNA substitutions in the DNA strand of the duplex produces a structural perturbation which drives the DNA nucleotides to attain an N-type sugar pucker.

The *B* conformation of the DNA duplex is thus compromised with a resultant acquisition of an *A* conformation. This increase in the population of N-type sugar puckers in the hybrid impels an *A*-like geometry that increases progressively with increase in LNA content [7,8]. However, this structural effect

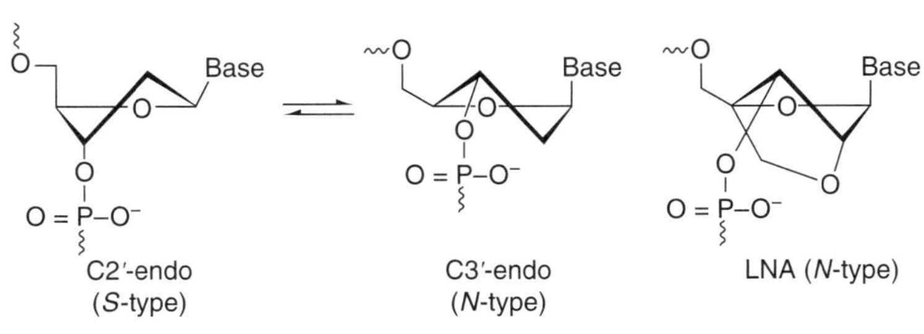

Figure 8.3 *(a) The C2′-*endo*–C3′-*endo *sugar ring equilibrium present in nucleic acids. (b) The molecular structure of locked nucleic acid (LNA), which shows the locked C3′-*endo *sugar conformation. From H. Kaur, B.R. Babu and S. Maiti, Perspectives on chemistry and therapeutic applications of locked nucleic acid (LNA),* Chem. Rev., ***107****, 4672–4697 (2007). Reprinted with permission. © 2007 American Chemical Society*

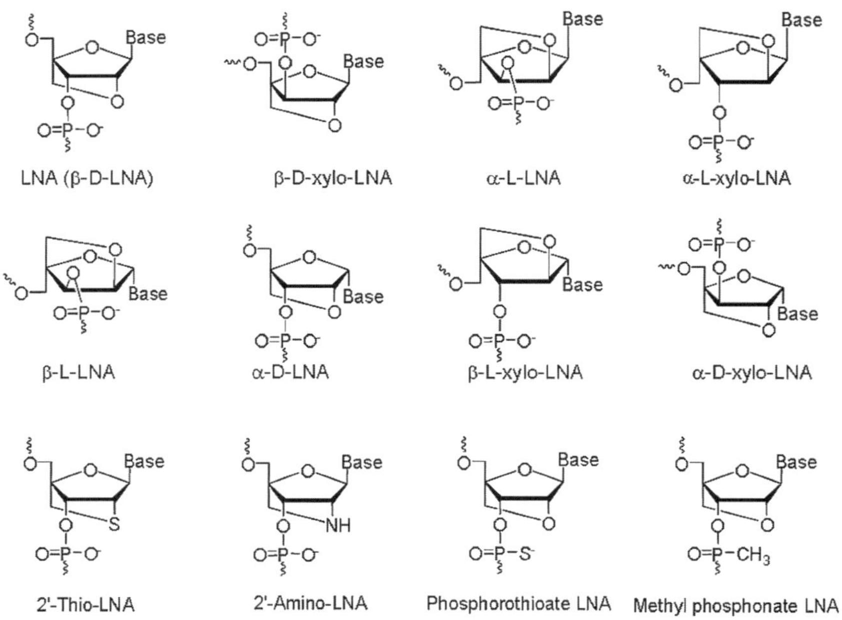

Figure 8.4 *The locked nucleic acid molecular family. From H. Kaur, B.R. Babu and S. Maiti, Perspectives on chemistry and therapeutic applications of locked nucleic acid (LNA),* Chem. Rev., ***107****, 4672–4697 (2007). Reprinted with permission. © 2007 American Chemical Society*

saturates when the LNA content of the strand reaches about 50% of the oligonucleotide length. This correlates well with the observed increase in helical thermostability, which reaches a maximum when the number of LNA monomers in the modified strands increases to 50% of the total residues [7].

8.2.3 The LNA family tree

The commonly used nomenclature 'LNA' structurally represents the β-D-LNA form – the parent molecule of the LNA family. However, it is not the only ribonucleotide mimic with a locked chemistry. The remarkable benefits of β-D-LNA have led to the extension of family to other structural analogs that have been developed and investigated for their ability to bind and knock down mRNA (Figure 8.4). Although most of the analogs exhibit high serum stability and high efficacy in mRNA knockdown, the efficacy of the parent LNA, namely the β-D-LNA, is known to be the highest of all diastereomeric forms. The next fascinating molecular analog is the α-L-LNA with RNA binding efficiency, comparable to that of the parent LNA. The remarkable binding affinity and specificity obtained for LNA and α-L-LNA, when fully and partially modified, have established these molecules as unique nucleic acids mimics [9].

8.3 LNA as a tool in therapeutics and diagnostics

The high potency and exquisite precision of RNA targeting by LNA oligonucleotide makes it an obvious choice for the development of new therapeutic strategies and efficient diagnostic tools. In addition, it also is a valuable tool for probing in various molecular applications such as detecting spatio-temporal expression of

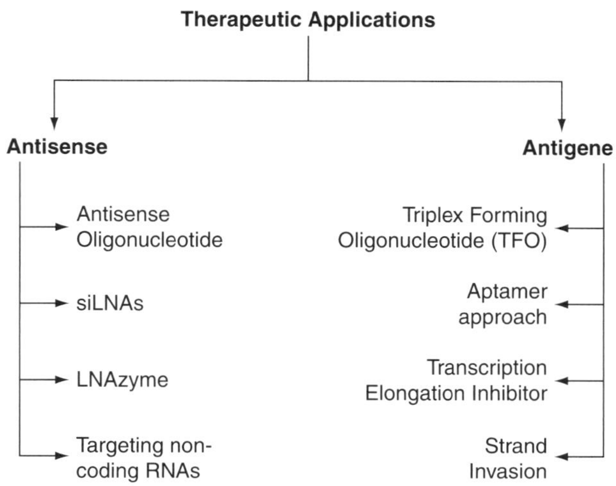

Scheme 8.1 *Different types of LNA-based therapeutic strategies*

an miRNA (Scheme 8.1). In the following, we discuss the attributes of this molecule which make it suitable over other modifications in an oligonucleotide.

High binding strength without loss of specificity

There is the usual trade-off between the binding strength and specificity of the modified oligonucleotide. This is a major challenge in *in vivo* applications, where fishing the target out of a soup of a huge number of molecules with partial sequence similarity with the target is required. Incorporating LNA into an oligonucleotide significantly and asymptotically increases its binding strength without compromising specificity. Thus efficient binding can be achieved even if the target region is highly structured or if the target is a short-length molecule such as an miRNA where specificity is the main concern. In diagnostic applications such as hybridization-based genotyping, which is limited by the presence of thermostable mismatches, the inclusion of LNA residues in probe oligonucleotides creates a significant difference between the T_m of the perfectly matched and mismatched targets. This broadening of the temperature window allows for greater flexibility in setting the optimal conditions for specific hybridization to a matched target.

High sensitivity

This attribute is particularly useful for diagnostic applications where incorporation of LNAs into the polymerase chain reaction (PCR) primers increases the amplification efficiency, thereby reducing the amount of template requirement by at least 10-fold. Such a decrease in template requirement is especially valuable in areas such as forensic profiling and archeological genetics, where often trace DNA samples, containing less than 100 pg of genomic DNA, are available for examination.

Antisense efficacy

The antisense efficacy of an oligonucleotide lies in its potential to alter gene expression by targeting the respective mRNA. This alteration may result from a decrease in the overall expression level or it may involve complete knockdown/blockage of expression. Depending on where it is targeted, the antisense oligonucleotide (AON) can use several mechanisms to interrupt mRNA translation (Figure 8.5). The three mechanisms depicted in the figure affect the profile of the representative proteins of the transcripts while the integrity of

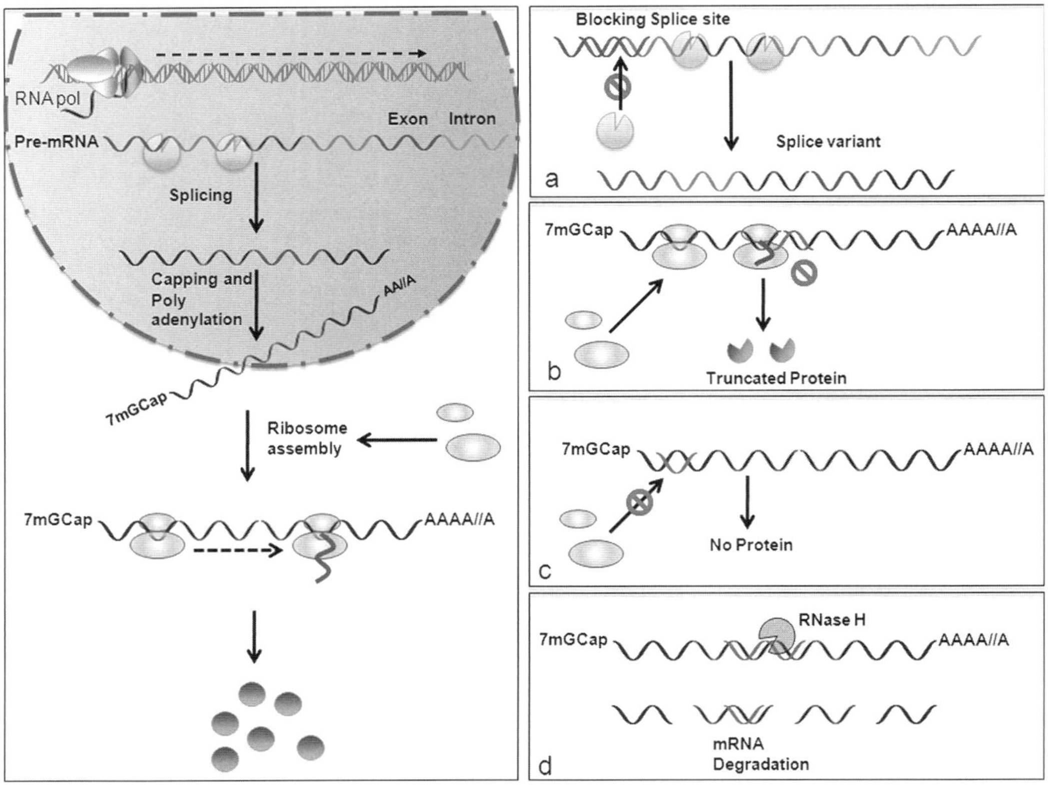

Figure 8.5 Different mRNA knockdown strategies: (a) blocking of the splice site by an AON leads to the production of a different splice variant [10]; (b) targeting the coding sequence of the mRNA leads to the formation of a truncated protein; (c) targeting of 5′-UTR prevents ribosomal binding; (d) hybridization of an RNase H active AON to mRNA at any site leads to mRNA knockdown

mRNA transcript remains intact. In addition to these, there exist mechanisms that result in complete degradation of the mRNA, thereby greatly amplifying the efficacy of the antisense approach. This occurs when the AON binds to the target strand at any site to produce a resultant duplex conformation of a DNA·RNA heteroduplex. This conformation serves as a recognition element for the recruitment of an enzyme RNase H that cleaves the RNA strand of the heteroduplex. Since this enzyme is ubiquitously expressed in all cell types, the ability of an AON to induce RNase H activation has an important role in its efficacy as a therapeutic agent. Amongst the various AONs that display promising affinities for RNA, only a handful (DNA, PS, FANA) form hybrids with RNA that are recognized and cleaved by RNase H. LNA-based oligonucleotides have excellent binding affinity to their respective complementary strand and, when used in a mixmer or chimeric design (LNA monomers interspersing DNA monomers), they are able to block sterically mRNA translation.

In fact, they also exhibit significant RNase H activation, accounting for the major part of their antisense efficacy. LNAs themselves are not good substrates for RNase H; however, when used in the form of a gapmer design (where LNA monomers flank a central core of DNA or phosphorothioate DNA), they work as efficient RNase H-recruiting AONs. RNase H activation is thus a feature restricted to LNA gapmers alone and not to the LNA mixmers designs [11]. This is attributed to the LNA-induced change in the helical geometry.

LNA-based oligonucleotides tend to adopt an *A*-type helix geometry, whereas RNase H requires a minor groove width intermediate between an *A*-type RNA and a *B*-type DNA helix, as its recognition element. When LNA monomers are incorporated into the antisense DNA strand, in a gapmer design, such that a central core of 6–8 DNA monomers is flanked by LNA monomers from both ends, the resulting (LNA/DNA/LNA)·RNA hybrid assumes a conformation intermediate between *A*- and *B*-type helixes, which is readily compatible with the enzyme. This allows highly efficient RNase H recruitment. On the other hand, substitution of LNA in the antisense strand in a mixmer design excessively propels the helical geometry to *A*-type, which thus fails to activate RNase H.

A noteworthy finding is the ability of α-L-LNA to mediate RNase H activation even in a non-gapmer design, thereby proving that a locked furanose conformation is not incompatible with the RNase H activity. In contrast to LNA, an α-L-LNA-substituted DNA·DNA helix adopts S-type sugar puckers and retain the features of a *B*-type helix. The same α-L-LNA/DNA mixmer when hybridized to RNA imparts a helical geometry intermediate between *A*- and *B*-type helixes, which is very much compliant with RNase H cleavage [12,13]. An α-L-LNA thus mimics a deoxyribonucleotide in DNA·RNA hybrids and dsDNA duplexes, leaving the global structure unperturbed yet conferring substantial elevation of the duplex stability.

Resistance against nucleases and serum stability

The *in vivo* efficiency of an AON is also greatly dependent on its biological stability. In this context, LNAs score over other modified nucleotides because of their high biological stability and low toxicity. End protection with LNA, as in the case of gapmer design, significantly increases the half-life of oligonucleotides in serum by rendering significant protection against exonucleases that cleave at the terminals [11]. LNA/DNA mixmers, on the other hand, are much more resistant to nuclease digestion. Furthermore, the stability in serum is sequence dependent. A significant resistance to exonucleolytic activity is observed on blocking the 3′-end of the oligonucleotide with two LNA monomers in tandem (LNA/DNA mixmers), whereas little or no protection is induced with one penultimate LNA nucleotide or with a single LNA nucleotide in the middle of the sequence [13,14]. In addition, biostability issues have also been addressed for double-stranded LNA-based oligonucleotides [15]. A rather counterintuitive finding in this context is that end-capping of double-stranded oligonucleotides with LNA could also confer significant protection from double-stranded specific endonucleases such as DNaseI. This is attributed to the LNA-induced change in the helical geometry such as groove width and flexibility, which are critical for recognition by the enzyme.

Limited in vivo toxicity

The intent behind the leveraging of the oligonucleotide platform to therapeutics is the development of safe, effective drugs. Safety is a vital concern before certification of an oligonucleotide for application *in vivo*. So far, most antisense applications have relied on the use of phosphorothioate DNA analogs for human antisense applications. Although markedly resistant to the action of serum nucleases, their polyanionic backbone is known to mediate nonspecific interactions with proteins, which contributes to the toxic profile of phosphorothioates (fever, hypotension, asthenia, complement activation, thrombocytopenia), thus limiting their application *in vivo* [3]. In contrast, LNAs are well tolerated in biological systems [16]. LNA-based AONs, by virtue of their high affinity and specificity, can be reduced to a length shorter than that required by the conventionally used DNA AONs. A short length of LNA oligomer can successfully mediate a desired biological effect while exhibiting minimal binding to proteins. An LNA oligomer, therefore, is devoid of many of the toxic manifestations that arise merely from the longer polyanionic backbone of the classical AONs. Toxicity of LNAs has been investigated in murine model systems [16–18]. Studies have shown that continuous treatment with LNA oligonucleotides is well tolerated and minor toxicity manifests only at doses well above the optimum level.

Short probe length

In addition to limiting the toxicity, the shorter length of LNA oligonucleotides also makes them suitable for designing probes for diagnostic applications where their enhanced thermostability permits the convenient use of short-length probes. These short probes are especially advantageous in addressing the classical AT- and GC-rich regions or in cases where genotyping of a genome, such as that of a virus, is limited by the low degree of conservation. The short LNA primers are also best suited for multiplex PCR because they are less prone to primer–primer interactions and allow running of LNA probes and primers in the same reaction.

Cellular delivery

Once an AON has passed through the checks of efficacy and toxicity, the next check is its ease of delivery into the cells. Delivery concerns are especially important when oligonucleotides with new chemistry are characterized and compared with others. Unlike PNAs and morpholinos, the charged phosphate backbone of LNA allows its convenient delivery into cells using conventional methods for oligonucleotide transfection. Delivery of LNA-based oligonucleotides has been studied in different cell lines, using different LNA designs and different transfection agents [19,20]. Efficient delivery of LNA oligonucleotides can easily be achieved for all mammalian cell lines and with almost all carrier agents, but their exact localization within the cell varies greatly with the cell type, sequence of oligonucleotide, the carrier chosen and the duration of transfection, etc.

8.3.1 Therapeutic benefits of LNA

8.3.1.1 LNA in antisense applications

Antisense oligonucleotides

The desirable attributes of LNA have made it an obvious candidate for antisense applications. High-affinity LNAs can be readily designed as fully modified LNA, LNA/DNA chimera and LNA/RNA chimera or in combination with other modifications, such as phosphorothioate linkages or 2′-*O*-Me RNA. This makes LNA compatible with different oligonucleotide chemistries and technologies. Further, the flexibility in the architecture of LNA oligonucleotides can be well tuned to achieve the desired level of gene expression, for example RNase H-mediated gene silencing by employing LNA gapmers or LNA chimera-mediated increase in protein expression by blockade of splice sites. Compared with α-L-LNA, β-D-LNA exhibits higher antisense efficacy, higher nuclease and thermal stability and comparable RNase H recruitment ability, except that more versatile design possibilities are displayed by α-L-LNA, eliciting RNase H recruitment even in the non-gapmer design [11]. Nevertheless, the β-D-LNA form is extensively used in cases where RNase H-mediated antisense effect is desired. The LNA-based gapmer designs have been used successfully for RNase H-mediated knockdown of expression of intracellular adhesion molecule-1 (ICAM-1) [21], δ-type opoid receptors (DORs) in the central nervous system of the rat [16], firefly luciferase mRNA [22], vallonoid receptor subtype 1 [23], and so on. Such an attractive set of properties including biostability, RNase H activation, lack of toxicity and potent biological activity prompted exploration of LNA as a prospective anticancer agent. The antisense efficacy of LNAs has been successfully used in MCF-7 cell line to down-regulate expression of two endogenous proteins, namely cyclin-dependent kinase inhibitor p21 and estrogen receptor alpha (ER alpha), both implicated in cancer development [19]. A well-known antisense ODN design of an LNA/DNA/LNA gapmer has also been successfully used in cancer cell cultures and in a nude mouse model bearing prostate tumor xenografts to inhibit tumor growth by targeting H-Ras mRNA [18]. High antisense efficacy was also exhibited by anti-H-Ras ODN containing α-L-LNA. LNA AONs have also been shown to mediate effective downregulation of anti-apoptotic proteins Bcl-2, Bcl-xL and survivin in different cancer cell lines. Since all known human tumors express either Bcl-2 or Bcl-xL, or both, an antisense strategy targeting Bcl-2 and Bcl-xL simultaneously has broad clinical applicability [24,25].

The increasing success in cell line-based experiments has paved the way to introduce LNA technology into clinical trials. An LNA-based antagonist of Bcl-2, referred to as SPC2996, was developed against chronic lymphocytic leukemia (CLL) and has reached the stage of multi-center Phase I/II clinical trials. SPC2996 mediated down-regulation of Bcl-2 expression provides an attractive means by which CLL and many other cancers can be resensitized to natural apoptotic stimuli and to chemotherapeutic agents that induce apoptosis. Two other LNA-based AONs undergoing clinical trials are SPC3042 and SPC 2968. Whereas SPC3042 is a potent inhibitor of survivin, which plays a vital regulatory role in both apoptosis and cell division, SPC2968 inhibits HIF-1α, a key sensor of cellular hypoxia in response to which it transcriptionally upregulates a host of genes that play an important role in cancer progression, including apoptosis, angiogenesis, cell migration and metastasis. Clinically, survivin and HIF-1α expression are associated with poor prognosis, increased cancer recurrence and resistance to radiotherapy and cytotoxic drugs. Thus, LNA-based antisense compounds are potential anticancer drugs in terms of both multiple modes of action and the wide variety of malignant solid tumors that can be targeted.

LNAzymes

Effective sequence-specific cleavage of RNA can also be achieved using LNA-modified DNAzymes, which are catalytically active DNA molecules capable of cleaving complementary RNA sequences in a site-specific manner. As shown in Figure 8.6, the two binding arms of the DNAzyme hybridize with the RNA substrate and hence control the specificity of DNAzyme while the 5-nucleotide catalytic core recognizes a pyrimidine·purine sequence at the cleavage site [26]. The chemical stability of these catalysts, their ease of synthesis and their ability to induce sequence-specific cleavage of RNA pave their way into therapeutics. However, in practice several factors limit their usage. For example, (i) stable hybridization of DNAzyme to its target RNA requires a high intracellular concentration, which is difficult to achieve. This problem can be easily circumvented by lengthening of the enzymatic arms, which can significantly increase the affinity between the two reactants, but this entails the synthesis of larger molecules that exhibit an increased tendency to form intermolecular structures. (ii) Hybridization is further challenged by the limited accessibility of the target sequence within the mRNA structure. It is estimated that 90% of the potential cleavage sites are buried deep in the RNA structure and, hence, not accessible to the conventionally used DNAzymes. An important strategy to overcome this problem relies on the incorporation of LNA residues in the binding arms of the enzyme. The so-formed LNAzymes exhibit high affinity, increased accessibility and enhanced endonuclease activity at concentrations much lower than that required by classical DNAzymes. Inclusion of LNA in the two arms

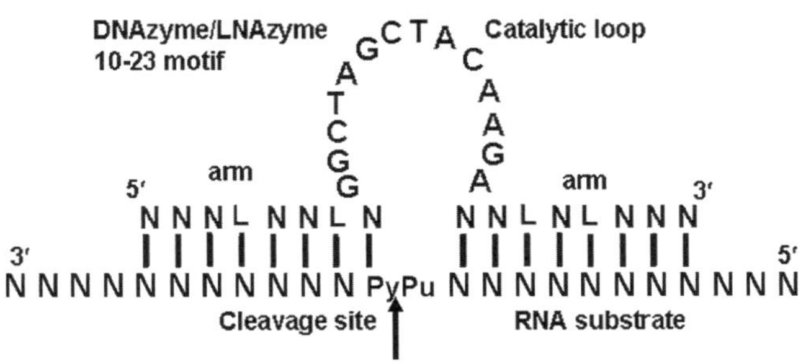

Figure 8.6 *Diagram illustrating the target recognition for 10–23 motif DNAzymes and LNAzymes. Lines represent Watson–Crick base pairs. The letter N stands for unmodified nucleotides and L represents LNA-substituted monomers. From H. Kaur, B.R. Babu and S. Maiti, Perspectives on chemistry and therapeutic applications of locked nucleic acid (LNA),* Chem. Rev., ***107****, 4672–4697 (2007). Reprinted with permission. © 2007 American Chemical Society*

of DNAzyme significantly improves the efficacy of this catalyst against small RNA substrates and more pronouncedly against the long substrates. Optimization of RNA cleavage by LNAzymes, however, requires careful adjustment of the arm length, sequence composition and number of LNA monomers [27].

A detailed analysis of underlying thermodynamics and kinetics behind the LNAzyme-mediated RNA cleavage reveals that (i) while inclusion of LNA in arms increases affinity to the substrate, it has a minimal effect on substrate cleavage, (ii) the increased efficiency of LNAzyme compared with DNAzyme stems from a reduction in the substrate dissociation rate (k–1), which suggests a stabilizing effect of LNA on the substrate–enzyme complex, and (iii) the effect on association depends on the type of substrate, the association rate (k1) of LNAzyme for short RNA substrates being nearly equivalent to DNAzymes, whereas for longer substrates the association is substantially faster for LNAzymes. Such faster annealing to the target can be explained by a model where LNA residues favor initial hybridization to the short stretches of unpaired residues of the structured target (nucleation), which subsequently leads to disruption of the local structure and completion of the binding [28].

The remarkable advantages offered by LNAzymes have encouraged their use in therapeutics. Recently, LNAzymes have been shown to inhibit human vascular smooth muscle cell growth by targeting early growth response-1 (EGR-1) transcripts. Even at lower concentrations, LNAzymes are more effective than DNAzymes with respect to *in vitro* substrate cleavage, protein expression, smooth muscle cell proliferation and regrowth after mechanical injury. Thus, introduction of LNAzymes is indeed a significant step towards the realization of oligonucleotide-based therapeutics with intrinsic endonucleolytic activity [29].

SiLNAs

siRNAs, short ~21 bp double-stranded RNA, cause nearly complete inhibition of gene expression by targeting the transcript of interest. When introduced exogenously into the cytoplasm, these siRNAs are incorporated into the RNA-induced silencing complex (RISC) that lies in the cytoplasm of the cell. Although initially both the strands of the siRNA duplex are incorporated into RISC machinery, one of these strands is cleaved by an endonuclease component of the RISC machinery, while the other strand in association with RISC selectively targets and cleaves a complementary cellular mRNA. RNA interference by siRNA is thus a promising therapeutic strategy to ensure complete knockdown of specific mRNA.

Recently, the use of LNA has been proposed in siRNA technology to develop the so-called siLNA technology that may improve the thermal stability, serum stability, cellular activity and pharmacokinetic properties of siRNAs. LNA is readily compatible with the siRNA machinery and substantially increases the biostability and thermal stability of siRNAs without compromising their efficiency [30]. A major concern with the use of siRNA as a genomic tool is the finding that cells might incorporate any of the strands into the RISC complex. Incorporation of the unwanted, non-target complementary, sense strand leads to the so-called 'off-target' effects that decrease the potency of siRNA by simply lowering the number of RISC complexes loaded on to the antisense strand. In this context, incorporation of LNA in siRNA is particularly useful if carefully optimized to achieve the desired level of its activity without mediating any off-target effects. Positioning of LNA substitution at the 5′ end of the 'antisense strand' (strand with sequence complementary to the target mRNA) of the siRNA duplex leads to a complete loss of activity; however, the 5′ end of the 'sense strand' (strand with sequence identical with the target mRNA) can be modified without loss of activity. Interestingly, activity lost by 5′ end modification of the antisense strand can be recovered by substituting the 3′ sense end and more so if modifications are made in the 5′ sense end [31]. This is attributed to the phenomenon called 'strand biasing' or the 'functional asymmetry' exhibited by the two strands of siRNA duplex, with only one strand being capable of triggering mRNA cleavage and the other destined to be destroyed. It is proposed that the strand displaying weaker binding energy at the 5′ base pair is preferentially incorporated in the RISC assembly, which subsequently mediates target mRNA cleavage [32,33]. Substitution of LNA at the 5′ antisense end increases the binding energy of its 5′ base pair and prevents it

```
5′-N N L N N N N N L N   N N L N N N N N N L L N-3′
   | | | | | | | | | |   | | | | | | | | | | | |
3′-N N N N N N N N N N — N N L N N N N N N L L N-5′
```

Figure 8.7 An example of sisiRNA design. The letter N stands for unmodified nucleotides and L represents LNA-substituted monomers added to increase the biostability of sisiRNA

from becoming assembled in RISC. Further, compensatory modifications at the 5′ and 3′ sense ends can successfully restore this effect. Apart from terminal modifications, certain internal substitutions in the antisense strand might also hamper siRNA activity. These internal positions are the ones that lie close to the site where mRNA cleavage occurs. It is suggested that the presence of LNA modification at these sites exerts a direct conformational or functional effect on the catalytic site, which subsequently affects siRNA activity.

In the light of the above findings, it can be inferred that the LNA substitution can effectively reduce the off-target effects of siRNA by two different mechanisms. Substitution at the 5′ sense position encourages strand loading of the antisense strand while additional internal substitutions in the sense strand might impair its ability to participate in target cleavage, once it is loaded in the RISC assembly. Recently, new siLNA design possibilities have been described [34]. This novel siRNA architecture comprises an intact continuous antisense strand complemented with two short segmented sense strands (Figure 8.7).

This three-stranded 'LNA-modified' construct referred to as the sisiRNA scored over the classical LNA-modified siRNA versions because:

1. The segmented nature of the design completely eliminates off-target effects produced by the sense strand as this design possibility allows only the antisense strand of the duplex to remain functional.
2. The sisiRNA design can readily support inclusion of chemical modifications at positions for which the standard siRNA is intolerant, which means that sisiRNA offers much more design flexibility for introducing modifications.
3. The new design has six terminal ends compared with four in the classical siRNA, which can conveniently be used to tether functional moieties to enhance effects such as cellular delivery.

   Together, these findings emphasize the potential of LNA in providing therapeutic benefits or enhancement of efficacy to siRNA technology.

Targeting non-coding RNA (antagomir approach)

A large part of the RNA pool is comprised of non-coding RNAs. MicroRNAs (miRNAs) are short, about 21 nucleotides in length, noncoding, regulatory RNA molecules representing a new layer in post-transcriptional regulation of gene expression. MicroRNAs regulate gene expression by base pairing to their target mRNAs and mediating mRNA cleavage or translational repression. miRNA expression is perturbed in many human cancers for which both tumor suppressor and oncogenic miRNAs have been identified. Dysregulated miRNA expression either may be a cause of cancer or it may simply be an effect of the less-differentiated state of cancerous cells. In addition, more than 50% of the human miRNA genes are located in cancer-associated genomic regions or at fragile sites, which substantiates a function for miRNAs in cancer. Elucidating the role of miRNA is hence important not only to study the wide regulatory network orchestrated by miRNAs but also to allow identification of miRNA-regulated genes that could serve as new potential therapeutic targets. Characterization of miRNAs can be easily achieved using the so-called 'antagomir' approach based on the use of LNA-modified antagomirs. These LNA-antagomirs are short LNA/RNA molecules that bind to miRNA and sequester it in a duplex form so that it is no longer available to bind to its target mRNA. Such loss-of-function studies allow characterization of specific functions of particular miRNAs and identification of miRNA-regulated genes that represent new potential therapeutic targets. By using various antagomir oligonucleotides comprising LNA modifications, several oncogenic miRNAs have been validated as potential therapeutic targets and delineated for their biological roles in cancers and in other pathophysiological conditions [35,36]. Although antagonizing the function of an miRNA using the LNA-based antagomir approach seems similar to antisense inhibition of mRNA transcripts, significant differences exist when targeting an miRNA versus an mRNA. Silencing the expression of an mRNA requires entry of the antisense oligonucleotide inside the nucleus to induce RNase H-mediated cleavage of the RNA target of the RNA·DNA heteroduplex. The antisense oligonucleotide used in this case is essentially a gapmer design and possesses the ability to enter the nuclease so as to activate the nuclear specific RNase H enzyme. An antagomir, on the other hand, is a non-RNase H activating species that sequesters the miRNA lying in the cytoplasm. In addition to LNA, other modifications such as morpholino [37] and the 2′-*O*-methyl- [38] and the 2′-*O*-methoxyethyl-modified [39] oligonucleotides have also proven useful as anti-miRNAs. However, the potency of LNA has been shown to be highest for both *in vivo* and *in vitro* applications. Furthermore, the desirable properties of LNA have eased the technically challenging task of detecting small-sized, low-expression miRNA in a cellular milieu. An account of the successful application of LNA-based probes in miRNA diagnostics and characterization is presented in detail the later sections.

8.3.1.2 LNA as an antigene agent

Antigene approaches to inhibit gene expression operate at the level of chromosomal DNA to block transcription. These may involve direct inhibition of transcription or interference with protein–DNA interactions. Some interesting strategies making use of antigene approach are outlined below.

The triplex approach

This involves the use of a short strand to bind to the major groove of double-stranded DNA in a sequence-specific manner. Binding of this third strand, referred to as the triplex-forming oligonucleotide (TFO), requires the presence of relatively long and uninterrupted homopurine:homopyrimidine tracts in DNA, to which it binds by forming Hoogsteen or reverse Hoogsteen hydrogen bonds (Figure 8.8). Studies in cell-free systems have shown that optimal conditions such as the presence of Mg^{2+}, low pH (to ensure protonation of cytosines) and limited concentrations of monovalent cations favor the binding of TFOs to homopurine sequences with dissociation constants (K_d) in the nanomolar range. The low K_d values and high stability allow TFOs to compete effectively with proteins such as transcription regulatory factors for binding to DNA and thus inhibit transcription in cell-free systems. However, various factors may limit the activity of TFOs in cells. Intracellular degradation of the oligonucleotide, sub-optimal ionic and pH conditions, insufficient nuclear accumulation and limited accessibility of the target site can prevent triplex formation. Once the binding has occurred, changes in DNA, chromatin dynamics and the intracellular environment may lead to dissociation of the triple helical complex. DNA unwinding associated with replication, transcription and DNA repair may displace the TFO bound to DNA. Therefore, in addition to its intrinsic triplex-forming ability, the residence time of a TFO on the target is a critical factor determining the extent and duration of its biological effects. Accordingly, any approach designed to increase the half-life of the triplex DNA in physiological conditions is likely to increase the biological activity of a TFO.

Systematic studies with different LNA constructs as TFOs have shown that incorporation of one LNA monomer centrally in TFO promotes triplex formation at physiological pH and leads to a significant increase in stability ($\Delta T_m > 10°C$). An exceptionally high stability is observed for triplexes containing monomers of N2′-glycyl-functionalized 2′-amino-LNA [40,41]. This increased stability due to increased rigidity of LNA-modified TFO in the free state permits a significant increase in the binding constant by ~20-fold at neutral pH without loss of sequence specificity. The LNA-based TFOs have been shown to inhibit gene

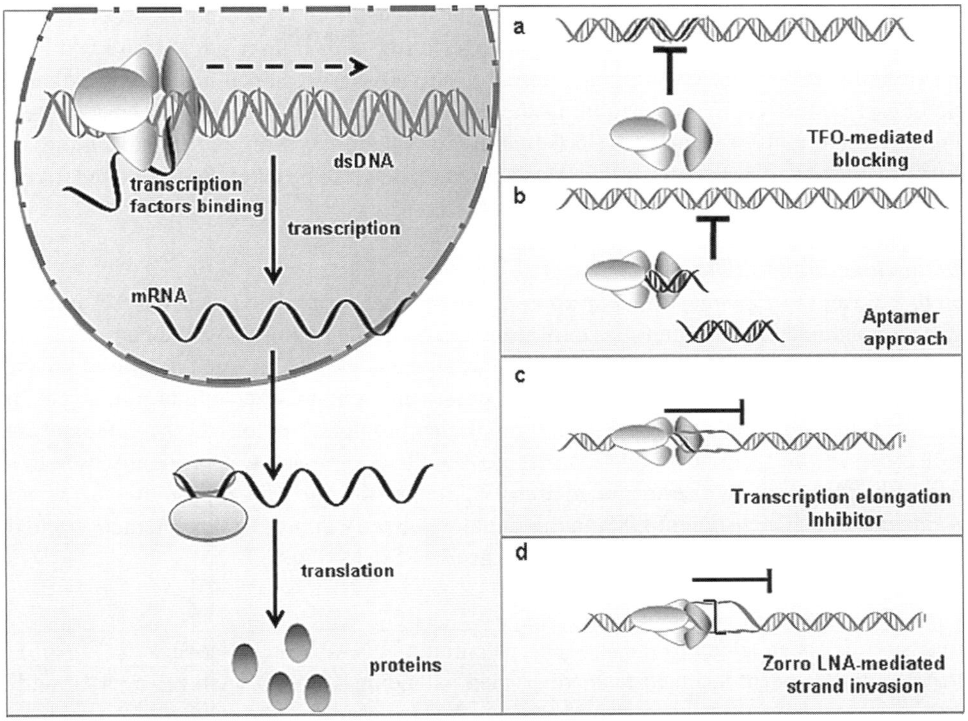

Figure 8.8 *Schematic representation of different antigene strategies: (a) presence of a third strand TFO prevents binding of transcription factor; (b) presence of short dsDNA sequestering transcription factors; (c) short single-stranded oligonucleotide invading the transcription complex; (d) strand invasion using 'Z-shaped' Zorro-LNA*

expression even in cases where the phosphodiester TFOs failed to act. *In vivo* studies have shown that while LNAs increase the binding affinity, stability and, therefore, the gene-silencing property of the TFOs, the expression of the mutated gene remains unaffected. This suggests that an LNA-modified oligonucleotide retains the specificity towards the native (unmutated) sequence and is at the same time highly effective in its silencing activity [42].

The decoy/aptamer approach

Another widely used technique to inhibit gene expression at the level of transcription relies on the use of the decoy approach that involves the use of short double-stranded DNA bearing the consensus binding sequence for a specific transcription factor. Such sequences when transfected into the cell compete with the native genomic sequence to interact with the target factor and prevent its binding to the respective native promoter, causing a marked reduction in transcriptional activation (Figure 8.8). This method can be used to antagonize the function of transcription-activating proteins involved in the pathophysiology of human diseases caused by aberrant gene activation or expression. The presence of LNAs in these dsDNA molecules increases the functional efficiency of the approach by protecting these dsDNAs against nucleases. However, when designing such LNA-based decoys, the number of LNA substitutions and their relative positioning should be carefully fine-tuned, as a slight excess or incorrect positioning might lead to changes in the helical geometry of the decoy such that it no longer binds to the transcription factor. In practice, this approach has been successfully used to interrupt the binding of NF-κB to its native genomic target. In this case, inclusion of one or two terminal LNA monomers in both the strands, outside the κB binding sequence, appreciably increased the protection against nucleases without interfering with transcription factor binding, whereas positioning of additional LNA substitutions internal to the NF-κB sequence in most cases badly affected NF-κB binding [15]. Interestingly, positioning the diastereoisomeric form of LNA (α-L-LNA) internally reversed the adverse effects of LNA on NF-κB binding. It is therefore inferred that positioning of substitutions, whether in one strand or in both strands of the aptamer, needs to be carefully optimized to reconcile the biostability with affinity to the target transcription factor.

Transcription elongation inhibition

In addition to TFO and the aptamer antigene approaches, another strategy to attack dsDNA employs a short, high-affinity oligonucleotide that binds to DNA at the transcription complex. At the beginning of transcription, RNA polymerase unwinds the dsDNA promoter region and exposes a small stretch of ssDNA. During this stage, a short, high-affinity oligoribonucleotide possessing a non-extendable terminal 3′-deoxyribonucleotide can hybridize to the sequence upstream of the promoter region of the template and inhibit transcription by an *in situ* block to elongation (Figure 8.9). This approach was originally used to inhibit the *Escherichia coli* RNA polymerase-mediated transcription at the *lac* UV-5 promoter using a 2′-*O*-Me-ribonucleotide pentamer. Inclusion of LNA in this pentamer greatly improved this interaction and drastically reduced the efficiency of transcription efficiency by nearly 95% [43].

Strand invasion by 'Zorro LNA'

A new dimension has been added to the antigene potential of LNA by the discovery of 'Zorro' LNA. This Z-shaped molecule has been designed to mediate gene silencing at the DNA level by a 'strand invasion' mechanism. Amongst the various nucleic acid analogs developed so far, the capacity of strand invasion has been best described for dsPNAs. Strand invasion is most effective when pyrimidine PNAs are connected by a flexible linker to form bisPNA in which one strand hybridizes to DNA via Watson–Crick base pairing and other binds to the 'same' strand via the Hoogsteen mode. Formation of such a complex at the dsDNA interrupts transcription through RNA polymerase, thereby producing effective gene downregulation. Since bisPNAs recognize only the homopurine sequences of duplex DNA, this limits their usage when the target comprises a mixed sequence. This limitation of bisPNAs has been successfully overcome by the discovery of a novel sequence-specific antigene molecule – the Zorro LNAs, in which a 14-mer LNA oligonucleotide binds to the coding strand, while a connected 16-mer LNA binds to the template strand, with the two arms connected by a seven base pair bridge (Figure 8.8). The simultaneous hybridization of Zorro LNA to both strands of the duplex has been shown to inhibit RNA polymerase-II and U6-driven transcription [44,45].

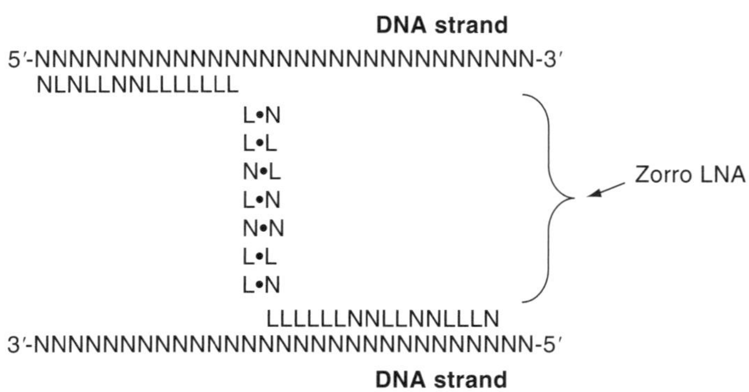

Figure 8.9 *Schematic representation of the LNA-modified 'Zorro LNA'*

8.3.1.3 Guidelines for designing LNA-based therapeutic probes

1. For most applications, the LNA content of the modified oligonucleotide may be restricted to less than 50% of the total length.
2. Since LNA·LNA base pairing is very strong, the self-annealing capacity of LNA must be taken into account when designing fully modified LNA or LNA mixmers with large numbers of LNA substitutions.
3. Fully modified and LNA mixmers are useful for steric block approaches, including the noncoding RNAs. LNA gapmers should be used where complete knockdown of expression is required by the RNase-mediated gene silencing approach.
4. LNAs should be introduced at positions where specificity and discrimination are most required, especially when using LNA as a diagnostic probe.
5. Thermodynamic data show that compared with LNA purines, LNA pyrimidines confer more thermostability, with the order being $A^L << G^L < T^L < C^L$ in terms of $\Delta\Delta G°_{37}$, although the contextual sequence may alter binding.

8.3.3 LNA as a diagnostic tool

In addition to creating avenues for gene-based therapy, LNAs find wide application in diagnostics as probes for *in vivo* and *in vitro* assays and as primers in PCR. When used for *in vivo* applications such as miRNA detection or fluorescence *in situ* hybridization, considerations such as cellular toxicity, biological stability, high selectivity and hybridization, as discussed above, become vital. However, when used in *in vitro* PCR-based assay, the following qualifications provide an edge to LNA-based probes over other nucleic acid chemistries. These desired features include the following.

Increased binding strength and high specificity

Since hybridization-based genotyping methodologies are limited by the presence of thermally stable mismatches, the inclusion of LNA residues in probe oligonucleotides significantly improves their thermostability and specificity. Compared with a DNA probe, an LNA probe increases the T_m difference between the perfectly matched and mismatched targets. This broadening of the temperature window allows for greater flexibility in setting the optimal conditions for specific hybridization to a matched target. Such high-specificity LNA probes have considerably eased the genotyping of complex disorders such as HLA-DQBI alleles associated with susceptibility to Type I diabetes. The genotyping in this case is complex as there is no single nucleotide polymorphism (SNP) that is distinct from other alleles. Despite this, the LNA probes used, in a one-step assay for asymmetric PCR, were very specific and produced no cross-reactivity. It is remarkable that where DNA probes failed to give readable results due to the presence of secondary structure in the PCR product, LNA probes allowed good separation of genotypes [46].

High sensitivity

The greater binding strength and the enhanced mismatch discrimination observed for LNAs allow significant enhancement of the amplification efficiency when LNAs are incorporated into the PCR primers, thereby reducing the amount of template required by at least 10-fold. Such a decrease in template requirement is especially valuable in areas such as forensic profiling and archeological genetics, where often trace DNA samples, containing less than 100 pg of genomic DNA, are available for examination.

Short probe length

The enhanced hybridization characteristics of LNAs with a high T_m contribution have allowed the convenient use of 'short' length probes. These short probes are especially advantageous in addressing the classical

AT- and GC-rich region or in cases where genotyping of a genome, such as that of a virus, is limited by a low degree of conservation. The short LNA primers are also best suited for multiplex PCR because they are less prone to primer–primer interactions and allow running of LNA probes and primers in the same reaction.

8.3.3.1 LNA-based probes and primers in PCR

The heightened ability of LNA to discriminate between matched and mismatched target nucleic acids has attracted the attention of the genotyping community. LNA-substituted oligonucleotides are routinely used as probes in a number of hybridization-based assays such as expression profiling, DNA sequencing and SNP genotyping. Applications of LNA probes and primers have been extended to allele-specific PCR, TaqMan and molecular beacons, real-time PCR, etc. Assays based on the ability of allele-specific LNA-based probes to discriminate between matched and mismatched targets with enhanced signal-to-noise ratio have been described. These enzyme-independent genotyping methods have used colorimetry, direct fluorescence and fluorescence polarization to signal the hybridization event and can be run in parallel or multiplexed formats.

Solid-phase SNP genotyping

The advantage of LNA probes can be exploited for enzyme-linked immunosorbent assay (ELISA)-like assays where allele-specific LNA capture probes covalently immobilized on a microtiter plate can be used to detect SNPs in PCR-amplicons from human genomic DNA. The technique has been used to screen for SNPs in a number of genes where biotinylated PCR-amplicons were hybridized to immobilized LNA probes and scored colorimetrically using a horseradish peroxidase–anti-biotin Fab conjugate and tetramethylbenzidine, a chromogenic substrate for horseradish peroxidase. The presence of LNA probes in this case allowed efficient and specific interrogation of SNPs in which the hybridization signal from the perfectly matched targets was 10 times higher than that from the single nucleotide mismatched targets [47].

Homogeneous SNP genotyping

In addition to 'solid-phase' SNP genotyping, LNA probes have also been used in real-time homogeneous (liquid) genotyping assays. These probe-based assays allow quantification of a real-time target using the relative change in the probe signal, e.g. fluorescence polarization or fluorescence intensity. In the former case, the fluorescence polarization values of a fluorescent dye (such as rhodamine or hexachlorofluore) attached to the LNA probe can be used to score a match versus a mismatch. Whereas fluorescence polarization increases significantly upon hybridization of the probe to the target DNA molecule, the presence of single mismatch results in only a small or even no change in this parameter. Based on its high sensitivity and relative simplicity, this assay is suitable for high-throughput screening of SNPs. Furthermore, multiplexing in the assay could be achieved using differently labeled wild-type and mutant-specific probes in the same solution to allow simultaneous detection of several SNPs [48].

In addition to fluorescence polarization, homogeneous genotyping might rely on estimation of the relative change in the fluorescence intensity of the fluorophore attached to the probe. Such fluorophore-labeled probes have been used in different experimental designs. Some of the most commonly used experimental strategies are the following:

1. The 5′-nuclease assay where dual-labeled fluorogenic LNA probes based on TaqMan chemistry hybridize within the DNA target sequence, bound by PCR primers. In the hybridized state, the reporter florescence is suppressed, as it lies in close proximity to the quencher. However, during primer extension the 5′-nuclease activity of the Taq polymerase cleaves the probe, resulting in increased fluorescence of the reporter probe. The assay pointed to the enhanced discriminatory power of LNA oligonucleotides, thus outperforming the comparable DNA probes [49].

2. Similarly to a 5′-nuclease assay, real-time monitoring can also be achieved using an LNA-based probe tagged to a terminal self-quenching fluorophore. During the annealing stage, the quenching moiety would change its properties when stacked with the neighboring base during binding, lowering quenching and, hence, increasing the fluorescence signal [50].
3. A rather advanced strategy based on the principle of 5′-nuclease assays is the use of molecular beacons (MBs). These single-stranded fluorescent probes form a stem-and-loop structure that do not fluoresce when they are free in solution. However, when hybridized to a target nucleic acid, they undergo a conformational change that enables them to fluoresce brightly. The LNA-MBs have proven to be far more superior than the analogous DNA-MBs, readily allowing detection of even one nucleotide differences. The high thermostability of LNA-MBs prevented their opening even at temperatures as high as 95°C. The presence of the target strand, however, led to their efficient opening and subsequent hybridization even at room temperature without loss of higher selectivity, nuclease stability and signal-to-background ratio in the intracellular environment. The longer lifetime with extremely low background of LNA-MBs makes them an excellent probe for gene expression in cellular studies. The high sensitivity of detection exhibited by LNA-based MBs coupled with LNA-modified primers has been used successfully to detect hepatitis C viral load in plasma and serum samples. The detection and accurate quantification of this and many other such viruses is complicated by a low degree of conservation and the presence of extensive repetitive sequences in the genotype. Accurate detection therefore calls for the selection of a primer and probe set that generates relatively short amplicons and detects the different genotypes with the same sensitivity [51].
4. In a few other assays, a non-extendable LNA-containing oligomer has been used as a 'clamp' to facilitate the screening of minority mutations in a tissue sample containing excess wild-type DNA (Figure 8.10). The clamping oligomer binds preferentially to the wild-type DNA sequence, thus suppressing its PCR amplification and changing the PCR bias towards the mutated target. During the melting analysis, the LNA oligomer competes for binding with the detection probe, enhancing the detection of the mutant variant many-fold compared with the conventional real-time PCR [52–54].

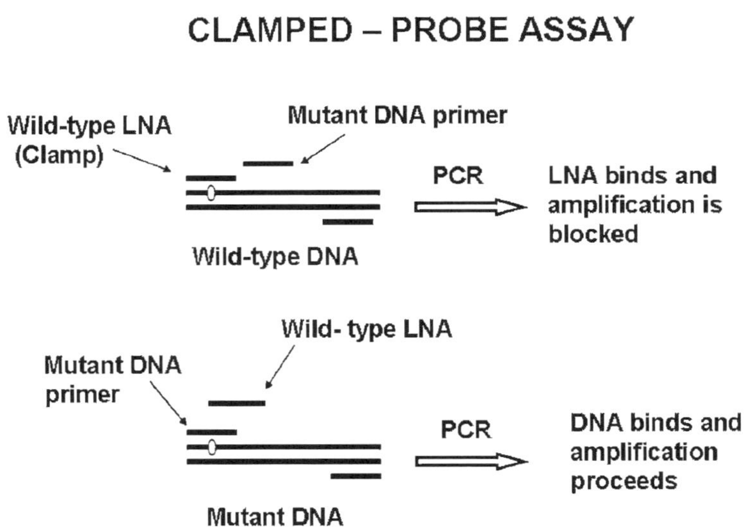

Figure 8.10 *Schematic representation of clamp-probe assay. From H. Kaur, B.R. Babu and S. Maiti, Perspectives on chemistry and therapeutic applications of locked nucleic acid (LNA),* Chem. Rev., ***107****, 4672–4697 (2007). Reprinted with permission. © 2007 American Chemical Society*

8.3.3.2 Some novel applications in diagnostics

The above strategies illustrate the compatibility and the versatility offered by LNAs, allowing them to work under different experimental conditions and with different probe chemistries. Such flexibility in design of LNA-based oligonucleotides has improved the performance of probes and primers in approaches such as DNA fingerprinting, capturing of specific RNA against a background of total cellular RNA and fluorescence *in situ* hybridization. Some of the useful applications where LNA-based probes or primers have greatly improved the experimental outcomes of these techniques are described in the following.

DNA fingerprinting

DNA fingerprinting refers to the use of techniques based on PCR to reveal the specific DNA profile for a particular organism that is as unique as a fingerprint. An advanced method for fingerprinting comprises the exploitation of SNPs that flag individual alleles of conventional markers. Detection of SNPs using allele-specific PCR based on LNA probes allows the highly sensitive detection of polymorphism in close relatives with a narrow genetic background. The sensitivity and specificity of 3′-locked LNA probes has been used to improve the hybridization performance in fingerprinting assays, allowing reliable clustering of DNA sequences that reflected the underlying homology between the sequences [55]. LNA-modified probes can therefore be used to unravel the sequence similarity of DNA sequences and characterize the unknown.

Quantifying the methylation level of a genome

Aberrant DNA methylation of tumor suppressor genes indicates an early event in tumor progression, therefore accurate detection and quantification of methylated alleles is particularly important in diagnosis and risk assessment for cancer. Quantification of the extent of methylation in a complex genome has been performed using allele-specific PCR – a technique relying on the positioning of the LNA residue at the 3′-position of the primer to a specific nucleotide. Incorporation of an LNA residue at this position allows for specific recognition and accurate extension of the correctly matched primer under stringent conditions, thereby allowing for accurate and sensitive discrimination of the methylated versus the unmethylated cytosines, even when the methylation level is as low as 1% of the overall population. LNA-based primers have also been successfully used in real-time quantitative methylation-specific polymerase chain reaction (QMSP) to enhance the analytical specificity for methylated alleles and to eliminate the formation of nonspecific products because of mispriming from unmethylated alleles [56].

Fluorescence in situ hybridization

In addition to the PCR-based *in vitro* assays, the fluorogenic LNA probes have been successfully applied to FISH (fluorescence *in situ* hybridization) where labeled LNA/DNA mixmers have been used to probe different tandem repeats sequence elements, such as the satellite-2 repeat in heterochromatic regions, the alpha-satellite repeats in centromeres and the telomeric repeats. Compared with DNA probes, LNA probes produced a strong signal, with a short time of hybridization and detect regions as small as 1–3 kb [57,58].

LNA as an mRNA capture probe

The heightened ability of LNAs to allow specific and sensitive detection of nucleic acids has been extended to the *in situ* and *in vitro* detection of RNA. Unlike DNA, the detection of a specific mRNA against a large background of RNA pool is particularly complicated by low signal-to-background ratios. Further, the propensity of RNA to form considerably stable secondary structure poses another challenge to the accessibility by the probe. Nevertheless, successful detection of RNA has been achieved using LNA-based probes [59]. *In vitro* detection of low expressed RNA has also been made possible by the use of real-time (RT)-PCR. However, when using this technique for amplification of a low copy number mRNA in biological samples, a stochastic off-targeted amplification occurs, which merely results from the low specific to non-specific ratio.

This drawback decreases the assay reliability, but can be readily overcome by the incorporation of an LNA residue at the 5′-position of the primer which significantly strengthens the 5′-hybridization. Another serious problem encountered while amplifying a low copy number mRNA in biological samples is the interference due to DNA contamination, leading to false-positive results. In such cases, by using an oligonucleotide complementary to the intron of a gene, genomic DNA amplification could be eliminated without affecting the amplification of reverse-transcribed spliced mRNA [60].

miRNA detection tool

The emerging evidence that miRNAs play a pivotal role in human cancers has given an additional impetus to miRNA research and characterization. miRNAs have been increasingly realized as valuable biomarkers for cancer prognostics and diagnostics. As discussed in earlier sections, the detection of miRNAs is technically demanding and calls for robust and improved technologies. The high affinity and enhanced mismatch discrimination offered by LNA probes have greatly assisted in understanding the complete biology of miRNA. LNA probes have been used successfully (i) for probing miRNA in Northern blots, where the presence of an LNA-based probe significantly increased the detection of low-abundance miRNAs by at least one order of magnitude [61], (ii) as capture probes in microarray and (iii) as probes for *in sit*u detection of miRNA in developing embryos and tissue sections. A prerequisite for the development of a microarray platform is to optimize the design of T_m-normalized probes so as to establish uniform hybridization conditions for the expression profiling of a genome-wide set of miRNAs. The flexibility in the LNA/DNA design possibilities has allowed the design of an miRNA microarray (miChip), which permitted maintenance of stringent hybridization, while maintaining discrimination of miRNAs differing by a single nucleotide mismatch. Yet another advantage of LNA alleviates the need for purification or amplification of miRNAs. MicroRNA profiling of normal versus malignant tissue provides substantial evidence for the specific miRNAs that are aberrantly activated or downregulated. The so-classified miRNAs can be assessed by further studies as putative therapeutic targets and may serve as biomarkers for cancer prognostics and diagnostics. Compared with mRNA profiles, the miRNA-based classification of poorly differentiated tumors has proved to be more accurate. Although miRNA arrays allow highly accurate screening of miRNAs, they are less preferred in routine clinical assessment of tissue samples, where *in situ* detection of miRNA accumulation in tumor biopsies serves as an approach for diagnostic and prognostic evaluation. *In situ* detection of miRNAs is an effective tool to characterize the temporal and spatial expression of individual miRNAs and has been used successfully in developing embryos of zebrafish, mouse and chicken and different tissue types such as dissected tissues or tissue sections [62–66].

8.3.3.4 General guidelines for LNA primer design

Although increasing the binding strength of the primers improves the amplification efficiency considerably in PCR reactions, it may prove to be detrimental, leading to decreased amplification of the target, with an increase in primer–dimers and other artifacts. Therefore, designing LNA primers needs to be carefully optimized as too many LNAs can decrease amplification, as can positioning of the LNA at an incorrect location. Each LNA base increases T_m by approximately 2–4°C, so the following generalizations should be considered when designing LNA-based primers:

1. LNAs should be introduced at the positions where specificity and discrimination are needed (e.g. the 3′ end in allele-specific PCR and in the SNP position in allele-specific hybridization probes).
2. Stretches of more than four LNA residues should be avoided as LNA hybridizes very tightly when several consecutive residues are substituted with LNA bases.

3. Poorly performing primers can be rescued most efficiently by positioning an LNA substitution at the 5′ end of the primer. This allows the 5′ end of the primer to anneal at high T_m, avoiding random priming by unspecific annealing of the 3′ end.
4. LNA self-complementarity and complementarity to other LNA-containing oligonucleotides in the assay should be avoided.

8.4 Conclusion

LNAs constitute a versatile tool for application in biotechnology. The exceptionally high stability of LNAs without a loss of specificity has allowed it to score over and even replace other conventionally used nucleotide analogs. In addition to the high potency, the high nuclease stability and adequate cellular uptake and biodistribution of LNA have successfully addressed the safety and efficacy concerns raised by many other analogs. LNAs thus possess various physiochemical properties that are mandatory for the development of effective and safe drugs for therapeutics, as further confirmed by the promising results shown by LNA-based drugs in clinical trials. In addition to revolutionizing the field of medicine, the high specificity and enhanced mismatch discrimination offered by LNAs have allowed the successful application of LNA-based primers and probes in diagnostics. As discussed in detail, the ease of designing LNA-based oligonucleotides can be exploited to introduce stringency under given experimental conditions. Such desirable attributes of LNAs have not only led to the improvisation of the current technology platform of therapeutics and diagnostics, but have also permitted the implementation of newer technologies. A major boom has been witnessed in the field of miRNA, where the sole application of 'locked' nucleic acids has unlocked the secrets of miRNA biology. LNAs, in a true sense, have been a boon in all spheres of biotechnology and have opened up newer avenues for further scientific exploration and development.

References

1. C. Marwick, First 'antisense' drug will treat CMV retinitis, *J. Am. Med. Assoc.*, **280**, 871 (1998).
2. J. Kurreck, Antisense technologies. Improvement through novel chemical modifications, *Eur. J. Biochem.*, **270**, 1628–1644 (2003).
3. H. Kaur, B.R. Babu and S. Maiti, Perspectives on chemistry and therapeutic applications of locked nucleic acid (LNA), *Chem. Rev.*, **107**, 4672–4697 (2007).
4. J. Wengel, Synthesis of 3′-C- and 4-C-branched oligodeoxynucleotides and the development of locked nucleic acids (LNA), *Acc. Chem. Res.*, **32**, 301–310 (1999).
5. S. Obika, D. Nanbu, Y. Hari, J. Andoh, K. Morio, T. Doi and T. Imanishi, Stability and structural features of the duplexes containing nucleoside analogues with a fixed N-type conformation, 2′-O,4′-C-methylene ribonucleosides, *Tetrahedron Lett.*, **39**, 5401–5404 (1998).
6. A.A. Koshkin, P. Nielson, M. Meldgaard, V.K. Rajwanshi, Singh, S.K. and J. Wengel, An RNA mimic forming exceedingly stable LNA:LNA duplexes, *J. Am Chem. Soc.*, **120**, 13252–13253 (1998).
7. M. Petersen, K. Bondensgaard, J. Wengel and J.P. Jacobsen, Locked nucleic acid (LNA) recognition of RNA: NMR solution structures of LNA:RNA hybrids, *J. Am. Chem. Soc.*, **124**, 5974–5982 (2002).
8. K. Bondensgaard, M. Peterson, S.K. Singh, V.K. Rajwanshi, R. Kumar, J. Wengel and J.P. Jacobsen, Structural studies of LNA:RNA duplexes by NMR: conformations and implications for RNase H activity. *Chemistry*, **6**, 2687–2695 (2000).
9. V.K. Rajwanshi, A.E. Hakansson, M.D. Sorensen, S. Pitsch, S.K. Singh, R. Kumar, P. Nielsen and J. Wengel, The eight stereoisomers of LNA (locked nucleic acid): a remarkable family of strong RNA binding molecules, *Angew. Chem. Int. Ed.*, **39**, 1656–1659 (2000).
10. S.T. Crooke, Progress in antisense technology, *Annu. Rev. Med.*, **55**, 61–95 (2004).

11. J. Kurreck, E. Wyszko, C. Gillen and V.A. Erdmann, Design of antisense oligonucleotides stabilized by locked nucleic acids. *Nucleic Acids Res.*, **30**, 1911–1918 (2002).
12. J.T. Nielsen, P.C. Stein, M. Petersen, J.T. Nielsen, P.C. Stein and M. Petersen, NMR structure of an alpha-L-LNA:RNA hybrid: structural implications for RNase H recognition. *Nucleic Acids Res.*, **31**, 5858–5867 (2003).
13. M. Frieden, S.M. Christensen, N.D. Mikkelsen, C. Rosenbohm, C.A. Thrue, M. Westergaard, H.F. Hansen, H. Orum and T. Koch, Expanding the design horizon of antisense oligonucleotides with alpha-L-LNA. *Nucleic Acids Res.*, **31**, 6365–3672 (2003).
14. K. Morita, C. Hasegawa, M. Kaneko, S. Tsutsumi, J. Sone, T. Ishikawa, T. Imanishi and M. Koizumi, 2′-O,4′-C-ethylene-bridged nucleic acids (ENA): highly nuclease-resistant and thermodynamically stable oligonucleotides for antisense drug, *Bioorg. Med. Chem. Lett.*, **12**, 73–76 (2002).
15. R. Crinelli, M. Bianchi, L. Gentilini and M. Magnani, Design and characterization of decoy oligonucleotides containing locked nucleic acids, *Nucleic Acids Res.*, **30**, 2435–2443 (2002).
16. C. Wahlestedt, P. Salmi, L. Good, J. Kela, T. Johnsson, T. Hokfelt, C. Broberger, F. Porreca, J. Lai, K. Ren, M. Ossipov, A. Koshkin, N. Jakobsen, J. Skouv, H. Orum, M.H. Jacobsen and J. Wengel, Potent and nontoxic antisense oligonucleotides containing locked nucleic acids. *Proc. Natl. Acad. Sci. USA*, **97**, 5633–5636 (2000).
17. A. Arzumanov, A.P. Walsh, V.K. Rajwanshi, R. Kumar, J. Wengel and M.J. Gait, Inhibition of HIV-1 Tat-dependent *trans* activation by steric block chimeric 2′-O-methyl/LNA oligoribonucleotides. *Biochemistry*, **40**, 14645–14654 (2001).
18. K. Fluiter, A.L. ten Asbroek, M.B. De Wissel, M.E. Jakobs, M. Wissenbach, H. Olsson, O. Olsen, H. Oerum and F. Baas, *In vivo* tumor growth inhibition and biodistribution studies of locked nucleic acid (LNA) antisense oligonucleotides. *Nucleic Acids Res.*, **31**, 953–962 (2003).
19. J.S. Jepsen, H.M. Pfundheller and A.E. Lykkesfeldt, Downregulation of p21(WAF1/CIP1) and estrogen receptor alpha in MCF-7 cells by antisense oligonucleotides containing locked nucleic acid (LNA). *Oligonucleotides*, **14**, 147–156 (2004).
20. A.N. Elayadi, D.A. Braasch and D.R. Corey, Implications of high-affinity hybridization by locked nucleic acid oligomers for inhibition of human telomerase. *Biochemistry*, **41**, 9973–9981 (2002).
21. S. Obika, R. Hemamayi, T. Masuda, T. Sugimoto, S. Nakagawa, T. Mayumi and T. Imanishi, Inhibition of ICAM-1 gene expression by antisense 2′,4′-BNA oligonucleotides. *Nucleic Acids Res. Suppl.*, **1**, 145–146 (2001).
22. D.A. Braasch, Y. Liu and D.R. Corey, Antisense inhibition of gene expression in cells by oligonucleotides incorporating locked nucleic acids: effect of mRNA target sequence and chimera design. *Nucleic Acids Res.*, **30**, 5160–5167 (2002).
23. A. Grunweller, E. Wyszko, B. Bieber, R. Jahnel, V.A. Erdmann and J. Kurreck, Comparison of different antisense strategies in mammalian cells using locked nucleic acids, 2′-O-methyl RNA, phosphorothioates and small interfering RNA. *Nucleic Acids Res.*, **31**, 3185 (2003).
24. J.B. Hansen, M. Westergaard, C.A. Thrue, B. Giwercman and H. Orum, Antisense knockdown of PKC-alpha using LNA-oligos. *Nucleosides Nucleotides Nucleic Acids*, **22**, 1607–1609 (2003).
25. A.P. Simoes-Wust, S. Hopkins-Donaldson, B. Sigrist, L. Belyanskaya, R.A. Stahel and U. Zangemeister-Wittke, A functionally improved locked nucleic acid antisense oligonucleotide inhibits Bcl-2 and Bcl-xL expression and facilitates tumor cell apoptosis. *Oligonucleotides*, **14**, 199–209 (2004).
26. S.W. Santoro and G.F. Joyce, A general purpose RNA-cleaving DNA enzyme, *Proc. Natl. Acad. Sci. USA*, **94**, 4262–4266 (1997).
27. S. Schubert, D.C. Gul, H.P. Grunert, H. Zeichhardt, V.A. Erdmann and J. Kurreck, RNA cleaving '10–23' DNAzymes with enhanced stability and activity. *Nucleic Acids Res.*, **31**, 5982–5992 (2003).
28. S. Donini, M. Clerici, J. Wengel, B. Vester and A. Peracchi, The advantages of being locked. Assessing the cleavage of short and long RNAs by locked nucleic acid-containing 8–17 deoxyribozymes *J. Biol. Chem.*, **282**, 35510–35518 (2008).
29. R.G. Fahmy and L.M. Khachigian, Locked nucleic acid modified DNA enzymes targeting early growth response-1 inhibit human vascular smooth muscle cell growth. *Nucleic Acids Res.*, **32**, 2281 (2004).
30. D.A. Braasch, S. Jensen, Y. Liu, K. Kaur, K. Arar, M.A. White and D.R. Corey, RNA interference in mammalian cells by chemically-modified RNA. *Biochemistry*, 7967–7975 (2003).

31. J. Elmen; H. Thonberg, K. Ljungberg, M. Frieden, M. Westergaard, Y. Xu, B. Wahren, Z. Liang, H. Orum, T. Koch and C. Wahlestedt, Locked nucleic acid (LNA) mediated improvements in siRNA stability and functionality. *Nucleic Acids Res.*, **33**, 439–447 (2005).
32. D.S. Schwarz, G. Hutvagner, T. Du, Z. Xu, N. Aronin and P.D. Zamore, Asymmetry in the assembly of the RNAi enzyme complex, *Cell*, **115**, 199–208 (2003).
33. A. Khvorova, A. Reynolds and S.D. Jayasena, Functional siRNAs and miRNAs exhibit strand bias, *Cell*, **115**, 209 (2003).
34. J.B. Bramsen, M.B. Laursen, C.K. Damgaard, S.W. Lena, B.R. Babu, J. Wengel and J. Kjems, Improved silencing properties using small internally segmented interfering RNAs, *Nucleic Acids Res.*, **35**, 5886–5897 (2007).
35. J. Stenvang, A.N. Silahtaroglu, M. Lindowb, J. Elmenb and S. Kauppinen, The utility of LNA in microRNA-based cancer diagnostics and therapeutics, *Semin. Cancer Biol.*, **18**, 89–102 (2008).
36. A.S. Flynt, N. Li, E.J. Thatcher, L. Solnica-Krezel and J.G. Patton, Zebrafish miR-214 modulates Hedgehog signaling to specify muscle cell fate. *Nat. Genet.*, **39**, 259–263 (2007).
37. B. Yang, H. Lin, J. Xiao, Y. Lu, X. Luo, B. Li, Y. Zhang, C. Xu, Y. Bai, H. Wang, G. Chen and Z. Wang, The muscle-specific microRNAmiR-1 regulates cardiac arrhythmogenic potential by targeting GJA1 and KCNJ2. *Nat. Med.*, **13**, 486–491 (2007).
38. C. Esau, S. Davis, S.F. Murray, X.X. Yu, S.K. Pandey, M. Pear *et al.*, miR-122 regulation of lipid metabolism revealed by *in vivo* antisense targeting. *Cell Metab.*, **6**, 87–98 (2006).
39. I. Naguibneva, M. Meyar-Zazoua, N. Nonne, A. Polesskaya, S. Ait-Si-Ali, R. Groisman, M. Souidi, L.L. Pritchard and A. Harel-Bellan, An LNA-based loss-of-function assay for micro-RNAs, *Biomed. Pharmacother.*, **60**, 633–638 (2006).
40. H. Torben; B.R. Babu, B. Torsten and J. Wengel, Triplex-forming ability of modified oligonucleotides, *Nucleosides Nucleotides Nucleic Acids*, **26**, 1411–1414 (2007).
41. T. Højland, S. Kumar, B.R. Babu, T. Umemoto, N. Albæk, P.K. Sharma, P. Nielsen and J. Wengel, LNA (locked nucleic acid) and analogs as triplex-forming oligonucleotides, *Org. Biomol. Chem.*, **5**, 2375–2379 (2007).
42. E. Brunet, M. Corgnali, L. Perrouault, V. Roig, U. Asseline, M.D. Sorensen, B.R. Babu, J. Wengel and C. Giovannangeli, Intercalator conjugates of pyrimidine locked nucleic acid-modified triplex-forming oligonucleotides: improving DNA binding properties and reaching cellular activities, *Nucleic Acids Res.*, **33**, 4223–4234 (2005).
43. N. Jacobsen, J. Bentzen, M. Meldgaard, M.H. Jakobsen, M. Fenger, S. Kauppinen and J. Skouv, Enhanced inhibition of transcription start by targeting with 2′-OMe pentaribonucleotides comprising locked nucleic acids and intercalating nucleic acids, *Nucleic Acids Res.*, **30**, e100 (2002).
44. R. Ge, J.E. Heinonen, M.G. Svahn, A.J. Mohamed, K.E. Lundin and C.I. Smith. Zorro locked nucleic acid induces sequence-specific gene silencing. *FASEB J.*, **8**, 1902–1914 (2007).
45. R. Ge, M.G. Svahn, O.E. Simonson, A.J. Mohamed, K.E. Lundin and C.I. Smith, Sequence-specific inhibition of RNA polymerase III-dependent transcription using Zorro locked nucleic acid (LNA) *J. Gene Med.*, **10**, 101–109 (2008).
46. M. Kiviniemi, J. Nurmi, T. Lovgren and J. Ilonen, Locked nucleic acid (LNA) probes in high-throughput genetic analysis:application to an assay for type 1 diabetes-related HLA-DQB1 alleles. *Clin. Biochem.*, **38**, 1015–1022 (2005).
47. P. Mouritzen, A.T. Nielsen, H.M. Pfundheller, Y. Choleva, L. Kongsbak and S. Moller, Single nucleotide polymorphism genotyping using locked nucleic acid (LNA), *Expert. Rev. Mol. Diagn.*, **3**, 27–38 (2003).
48. A. Simeonov, A. and T.T. Nikiforov, Single nucleotide polymorphism genotyping using short, fluorescently labeled locked nucleic acid (LNA) probes and fluorescence polarization detection, *Nucleic Acids Res.*, **30**, e91 (2002).
49. L.A. Ugozzoli, D. Latorra, R. Puckett, K. Arar and K. Hamby, Real-time genotyping with oligonucleotide probes containing locked nucleic acids, *Anal. Biochem.*, **324**, 143–152 (2004).
50. S. Grannemann, O. Landt, S. Breuer and B. Blomeke, LightTyper assay with locked-nucleic-acid-modified oligomers for genotyping of the toll-like receptor 4 polymorphisms A896G and C1196T, *Clin. Chem.*, **51**, 1523–1525 (2005).
51. L. Morandi, D. Ferrari, C. Lombardob, A. Pession and G. Tallini, Monitoring HCV RNA viral load by locked nucleic acid molecular beacons real time PCR, *J. Virol. Methods*, **140**, 148–154 (2007).

52. A. Senescau, A. Berry, F. Benoit-Vical, O. Landt, R. Fabre, J. Lelievre, S. Cassaing and J.F. Magnaval, Use of a locked-nucleic-acid oligomer in the clamped-probe assay for detection of a minority Pfcrt K76T mutant population of *Plasmodium falciparum*. *J. Clin. Microbiol.*, **43**, 3304–3308 (2005).
53. P.L. Dominguez and M.S. Kolodney, Wild-type blocking polymerase chain reaction for detection of single nucleotide minority mutations from clinical specimens, *Oncogene*, **24**, 6830–6834 (2005).
54. Y. Nagai, H. Miyazawa, H.; Huqun, T. Tanaka, K. Udagawa, M. Kato, S. Fukuyama, A. Yokote, K. Kobayashi, M. Kanazawa and K. Hagiwara, Genetic heterogeneity of the epidermal growth factor receptor in non-small cell lung cancer cell lines revealed by a rapid and sensitive detection system, the peptide nucleic acid-locked nucleic acid PCR clamp. *Cancer Res.*, **65**, 7276–7282 (2005).
55. J. Nakitandwe, F. Trognitz and B. Trognitz, Reliable allele detection using SNP-based PCR primers containing locked nucleic acid: application in genetic mapping, *Plant Methods*, **3**, R2 (2007).
56. K.S. Gustafson, Locked nucleic acids can enhance the analytical performance of quantitative methylation-specific polymerase chain reaction, *J. Mol. Diagn.*, **10**, 33–42 (2008).
57. A.N. Silahtaroglu, N. Tommerup and H. Vissing, FISHing with locked nucleic acids (LNA): evaluation of different LNA/DNA mixmers, *Mol. Cell Probes*, **17**, 165–169 (2003).
58. A. Silahtaroglu, H. Pfundheller, A. Koshkin, N. Tommerup and S. Kauppinen, LNA-modified oligonucleotides are highly efficient as FISH probes, *Cytogenet. Genome Res.*, **107**, 32–37 (2004).
59. N. Jacobsen, P.S. Nielsen, D.C. Jeffares, J. Eriksen, H. Ohlsson, P. Arctander and S. Kauppinen, Direct isolation of poly(A)+ RNA from 4 M guanidine thiocyanate-lysed cell extracts using locked nucleic acid-oligo(T) capture. *Nucleic Acids Res.*, **32**, e64 (2004).
60. L. Hummelshoj, L.P. Ryder, H.O. Madsen and L.K. Poulsen, Locked nucleic acid inhibits amplification of contaminating DNA in real-time PCR, *Biotechniques*, **38**, 605–610 (2005).
61. A. Valoczi, C. Hornyik, N. Varga, J. Burgyan, S. Kauppinen and Z. Havelda, Sensitive and specific detection of microRNAs by northern blot analysis using LNA-modified oligonucleotide probes, *Nucleic Acids Res.*, **32**, e175 (2004).
62. W.P. Kloosterman, E. Wienholds, E. de Bruijn, S. Kauppinen and R.H. Plasterk, *In situ* detection of miRNAs in animal embryos using LNA-modified oligonucleotide probes, *Nat. Methods*, **3**, 27–29 (2006).
63. E. Wienholds and R.H. Plasterk, MicroRNA function in animal development, *FEBS Lett.*, **579**, 5911–5922 (2007).
64. D.K. Darnell, S. Kaur, S. Stanislaw, J.H. Konieczka, T.A. Yatskievych and P.B. Antin, MicroRNA expression during chick embryo development. *Dev. Dyn.* **235**, 3156–3165 (2006).
65. G. Wheeler, S. Ntounia-Fousara, B. Granda, T. Rathjen and T. Dalmay, Identification of new central nervous system specific mouse microRNAs, *FEBS Lett.*, **580**, 2195–2200 (2006).
66. P.T. Nelson, D.A. Baldwin, W.P. Kloosterman, S. Kauppinen, R.H. Plasterk and Z. Mourelatos, RAKE and LNA-ISH reveal microRNA expression and localization in archival human brain, *RNA*, **12**, 187–191 (2006).

9

Small Non-coding RNA in Bacteria

Sabine Brantl

9.1 Introduction

Small RNAs (sRNAs) that act as regulators of gene expression have been identified in all kingdoms of life. Only since 2001 have a series of systematic computational approaches revealed that bacteria encode an amazing number of sRNAs. Meanwhile (in 2008) in *Escherichia coli* more than 100 sRNAs are known. To date, only about 25 sRNAs have been allocated a biological function, indicating that it is still a challenging issue both to identify targets of these sRNAs and to elucidate their mechanisms of action.

sRNAs can be divided into two major groups. The first group comprises so-called bona fide antisense-RNAs that regulate gene expression by a base-pairing mechanism. Thereby, we can distinguish between *cis*- and *trans*-encoded RNAs (Figure 9.1). Whereas *cis*-encoded RNAs are fully complementary to their target RNAs and, thus, can form complete duplexes with them, *trans*-encoded RNAs are only partially complementary to their targets and can form only partial duplexes with their – often multiple – target RNAs. In both cases, the interaction between antisense RNA and target mRNA results in post-transcriptional inhibition or activation of target RNA function. The second group of sRNAs encompasses RNAs that act by binding of small proteins (see Figure 9.5).

In this context, so-called sensory RNAs should also be mentioned, which directly sense environmental conditions of either physical (RNA thermometers) or biochemical (riboswitches) origin. Additionally, other small RNAs, such as tmRNA, RNase P RNA and 4.5 S RNA, with diverse functions have been discovered and investigated over the past 30 years.

The Chemical Biology of Nucleic Acids Edited by Günter Mayer

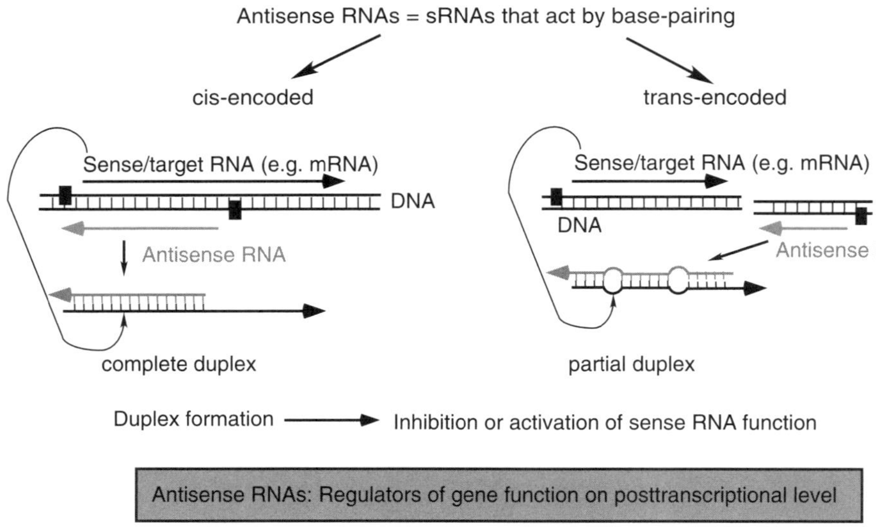

Figure 9.1 *Overview of* cis- *and* trans-*encoded antisense RNAs. Antisense RNAs are drawn in gray and sense RNAs in black. Black rectangles denote promoters*

9.2 Small regulatory RNAs

9.2.1 *cis*-Encoded sRNAs

The majority of *cis*-encoded antisense RNAs were found in so-called accessory genetic elements, that is, plasmids, phages and transposons. The first of these RNAs were discovered in the *E. coli* plasmids ColE1 [1] and R1 [2], where they regulate replication and, hence, control copy numbers. Subsequently, in a wide variety of plasmids, phages and transposons, *cis*-encoded antisense RNAs were found and intensively characterized (reviewed in [3]). In plasmids, antisense RNAs regulate replication, maintenance and segregational stability. In phages, they have a fine-tuning function in the decision between lysis and lysogeny. In transposons, they control transposition frequency. Meanwhile, a few chromosomally encoded *cis*-acting sRNAs have also been found. Among them are an sRNA from the *Bacillus subtilis* genome, RatA, that regulates a toxin TxpA [4] (Figure 9.2a) and IsrR from the cyanobacterium *Synechococcus* that regulates the amount of the photosynthesis component IsiA [5]. The other two examples are from *E. coli*: GadY involved in the acid stress response regulates the synthesis of GadX and GadW ([6], Figure 9.2a) and recently, SymR (formerly RyjC) was found to regulate a toxin, the SOS-induced endonuclease SymE [7]. Two systematic searches have been performed in the *E. coli* chromosome, one for very small [35 nucleotide (nt)] RNAs [8] and the other for sRNAs complementary to known ORFs (R. Schroeder, personal communication); however, no characterization has been performed on the majority of these RNAs. Furthermore, a systematic search for sRNAs in the pathogenic bacterium *Salmonella typhimurium* [9] revealed 19 novel sRNAs associated with pathogenicity islands, many of which seem to be *cis*-encoded. Table 9.1 provides an overview of all currently known *cis*-encoded antisense RNAs for which targets have been identified.

9.2.1.1 Mechanisms of action

In the majority of cases, antisense RNA action entails post-transcriptional inhibition of target RNA function, but, in a few cases, activating mechanisms have been found. The currently known regulatory mechanisms employed by *cis*-encoded antisense RNAs are discussed below and summarized schematically in Figure 9.2a.

Transcription attenuation

This mechanism was first discovered for the replication control of staphylococcal plasmid pT181 [10] and later for the streptococcal plasmids pIP501 and pAMβ1 [11]. Transcription attenuation as a replication control mechanism has not been detected in Gram-negative bacteria, although there is experimental proof that it principally functions in *E. coli* [12]. The nascent *rep* mRNA can adopt two mutually exclusive conformations: Upon binding of the antisense RNA, a terminator stem–loop is induced in the nascent *rep* mRNA, and, consequently, transcription is terminated prematurely upstream of the *rep* SD sequence

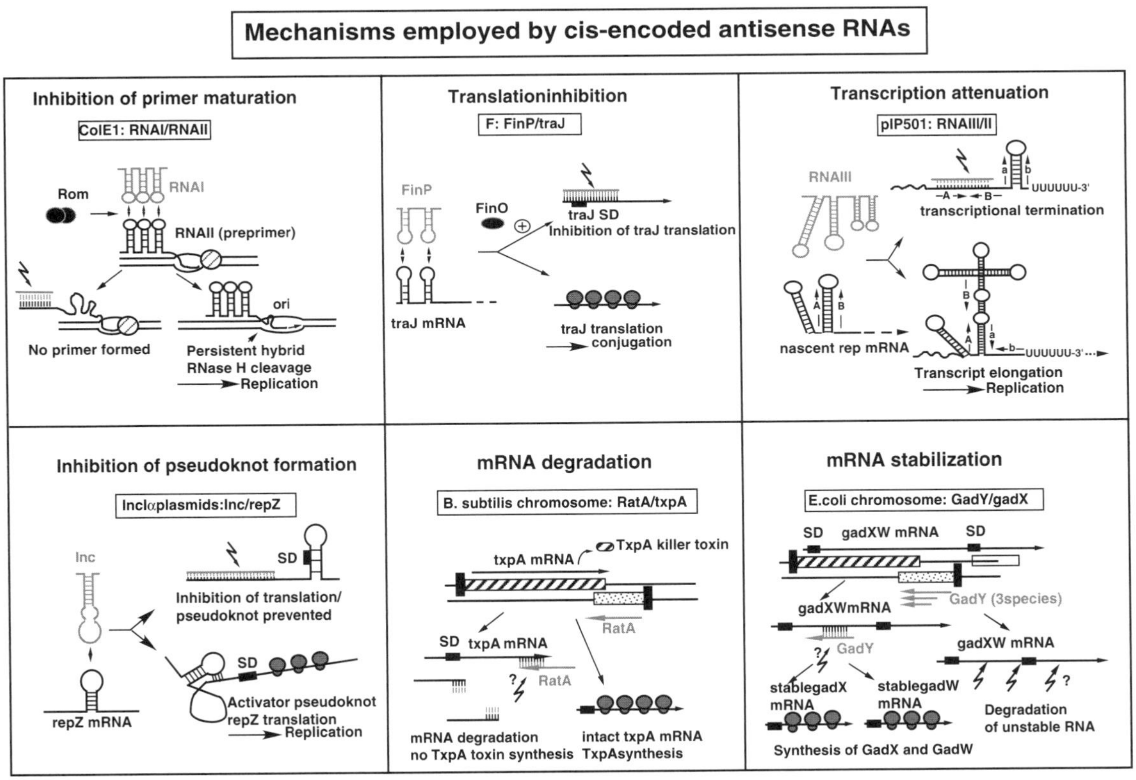

(a)

Figure 9.2 *Overview of regulatory mechanisms employed by* cis-encoded *antisense RNAs. (a) Antisense RNAs are drawn in gray and sense RNAs in black. Black rectangles denote promoters, hatched and stippled boxes sense and antisense RNA genes, respectively. Hatched circle, DNA polymerase I; gray symbols, ribosomes. Thunderbolt-like arrows indicate action of RNase III, black arrows the putative action of other RNases. Details are described in the text. One example for each mechanism is given. ?, Not yet experimentally confirmed. The upper part of the figure is based on [22]. (b) Two principle binding pathways of antisense/sense RNA pairs. Upper part: multistep binding pathway illustrated with the CopA/CopT interaction from plasmid R1 as example. Based on [20]. Lower part: the one-step binding pathway of the sense/antisense RNA system RNA-IN/RNA-OUT regulating transposition of IS10. The initial contact occurs between the first 3Gs or Cs at the 5′ end of RNA-IN and the complementary bases of RNA-OUT followed by an extension of stable pairing through the loop domain and the stem domain of RNA-OUT. Based on [20]*

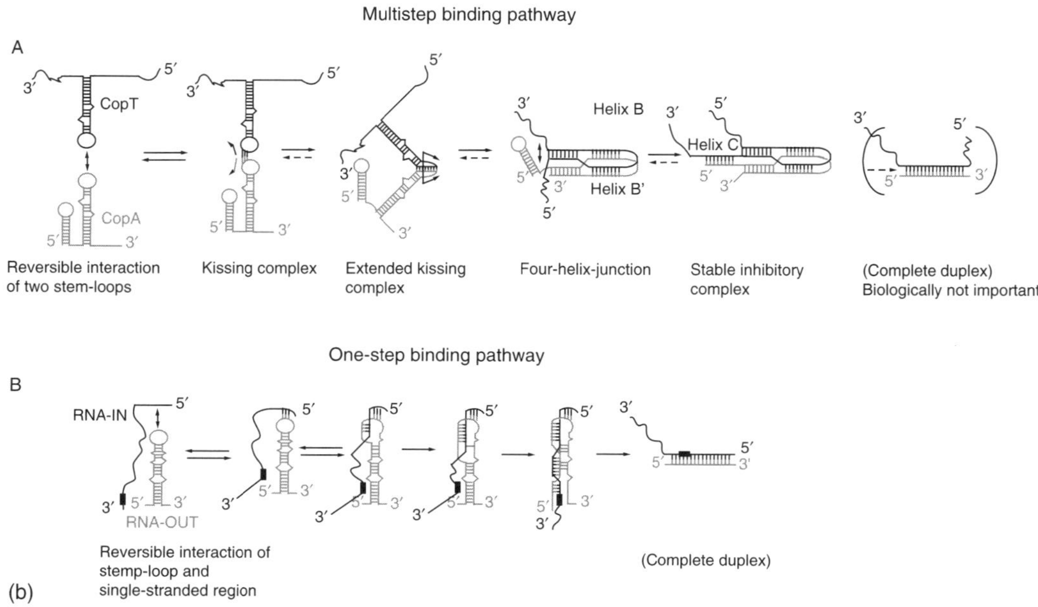

Figure 9.2 *(Continued)*

preventing Rep protein synthesis, and, hence, replication. If the nascent *rep* RNA escapes antisense RNA binding, it can refold by complementary base pairing between two alternative sequences preventing terminator formation and allowing transcriptional read-through. This results in Rep protein synthesis, and, consequently, replication (see Figure 9.2a). The antisense RNA binds and exerts its inhibitory effect only during a short time window [13] and without a helper protein. RNAIII of plasmid pIP501 is composed of four stem–loops, L1 to L4, of which the first two are dispensable both *in vivo* and *in vitro*. Both loops L3 and L4 must interact simultaneously with the complementary loops L1 and L2 of RNAII (*repR* mRNA) to yield efficient inhibition [14]. The inhibition rate constant of RNAIII has been determined to be $1 \times 10^6\ M^{-1}\ s^{-1}$, whereas the pairing rate constant was 10-fold lower, indicating that inhibition occurs faster than stable duplex formation between RNAIII and its target, RNAII [13]. Structure probing of the complex between RNAII and RNAIII of pIP501 showed that it is not a complete duplex [14]. The intracellular concentrations of both RNAII and RNAIII have been determined to be 1–2 μM and 50 nM, respectively, ensuring a 10-fold excess of the regulator over its target [15]. Normally, antisense RNAs that regulate plasmid copy numbers are short-lived with half-lives of 1–2 min to correct quickly fortuitous copy number fluctuations. However, RNAIII is unusually long-lived with a half-life of ~30 min [15]. Therefore, CopR (10.6 kDa), a second control element with a dual function, is needed for proper control [16].

Both the pIP501 and the pT181 systems work similarly and have been studied biochemically in terms of secondary structures of sense and antisense RNAs, binding kinetics and sequence requirements for efficient inhibition ([17]; also see below).

Translation inhibition

This conceptionally simplest mechanism, inhibition of translation of the sense RNA by direct blocking of the ribosome binding site, has been found, for example, in control of plasmid replication and maintenance (reviewed in [3] and [18]. In plasmid pLS1, RNAII complementary to the *repB*-RBS directly inhibits

Table 9.1 *Overview of cis-encoded antisense RNAs*[a]

Antisense RNA/ target RNA	Length of antisense RNA (nt)	Location	Biological function	Mechanism of action	Peculiarity
Plasmid encoded RNAs		*Plasmid*			
RNAI/RNAII	108	ColE1	Replication control	Inhibition of primer maturation	Rom protein promotes kissing
RNAI/rep	115	ColE2 and relatives	Replication control	Translation inhibition	
CopA/CopT	90	R1 and other IncFII relatives	Replication control	Translation inhibition	Leader peptide translat. coupling
RNAIII/RNAII	136	pIP501 and *Inc18* relatives	Replication control	Transcription attenuation	Unusually stable antisense RNA
RNAI, II/repC	84, 141	pT181	Replication control	Transcription attenuation	2 antisense RNAs
Inc/repZ	71	Inc1α/IncB relatives	Replication control	Inhibition of translation and pseudoknot formation	Leader peptide
RNAII/rep	50	pLS1	Replication control	Translation inhibition	
RNAI/repA	82	pSK41	Replication control	Translation inhibition	
Incα/repC	54–57	pTiR10 relatives	Replication control	Transcription attenuation[b]	
Incα/repC	67	pRmeGR4	Replication control	mRNA stability[b] Transcriptional interference[b]	
Sok/hok	64	R1	Segregational stability	Translation inhibition	Activity in absence of gene locus
RNAII/RNAI	64	pAD1	Segregational stability	Translation inhibition	2 complementary regions involved in complex
FinP/traJ	79	F, R1 and relatives	Conjugation control	Translation inhibition	FinO protein stabilizes duplex
Qa, mD/prgX		pCF10, pAD1	Conjugation control	Transcription attenuation[b]	Pheromone represses Qa/mD
RNAα/fatA,B	650	pJM1	Iron transport regulation	mRNA stability[b]	Regulation by chromosomal Fur protein; iron
Transposon encoded RNAs		*Transposon*			
RNA-OUT/RNA-IN		IS10/Tn10	Transposition	Translation inhibition	
RNA-C		IS30	Transposition	Inhibition of translation elongation[b]	

(*continued overleaf*)

Table 9.1 (*Continued*)

Antisense RNA/ target RNA	Length of antisense RNA (nt)	Location	Biological function	Mechanism of action	Peculiarity
Phage encoded RNAs		*Phage*			
OOP/cII	77	λ	Switch lysis/lysogeny	mRNA stability	
C4, CI/icd-ant	77	P1, P4	Switch lysis/lysogeny	Transcription termination	
Sar/arc-ant	68	P22	Switch lysis/lysogeny	Translation inhibition	
Sas/sieB-esc	105	P22	Superinfection override	Switch of translation start site	
$T_{an}t$/T1	151	ΦH	Switch lysis/lysogeny	RNA processing	Archaebacterial
Chromosomally encoded RNAs		*Species*			
P3 RNA/glnA	43	*C. acetobutylicum*	Glutamine synthetase	Translation inhibition[b]	
Sof/gef		*E. coli*	Toxin/antitoxin system	Translation inhibition[b]	
Isf/sulA	350	*E. coli*	SOS response?	?	
RdlD/ldrD	66	*E. coli*/relatives	Toxin/antitoxin system	mRNA stability[b]	
GadY/gadXW	105/90/59	*E. coli*	Acid response regulation	mRNA stabilization	3 GadY species different length
RatA/txpA	222	*B. subtilis*	Antitoxin/toxin	mRNA degradation	75 nt overlap
IsrR/isiA	176	*Synechocystis* sp.	Photosynthesis component	mRNA degradation[b]	Regulation by iron stress
SymR/symE	77	*E. coli*	Antitoxin/toxin	Translation inhibition	SOS induced mRNA degradation
RyjB/sgcA	90	*E. coli*	Phosphotransferase II comp.	?	
RyeA, B/pphA	100, 275	*E. coli*	Phosphatase I	?	
IsrJ	74	*S. typhimurium*	Virulence-associated effector	?	Low $O_2 + Mg$ ind.
IsrE	98	*S. typhimurium*	Virulence		RyhB homolog, but different function

[a] For plasmid encoded RNAs, the antisense RNA of the best studied example is given.
[b] Mechanism proposed but not experimentally substantiated.?, no mechanism proposed.

ribosome loading. The same mechanism is used by the FinP antisense RNA that blocks the RBS of *traJ*, an activator of the plasmid F and R1 conjugal transfer operons. The FinO protein promotes the *traJ*/FinP duplex and prolongs FinP half-life by protecting it against RNase E. Additionally, Sok and RNA I antisense RNAs (plasmids R1, pAD1) and RNA-OUT (IS10) use translational inhibition. In the first two cases, the synthesis of the killer toxin that kills plasmid free cells is inhibited; in the last, the expression of transposase mRNA is regulated. In the replication control systems of plasmid R1 and the IncB/Inc$I\alpha$ plasmids, the antisense RNAs CopA and Inc inhibit translation of a leader peptide that itself is – via translational coupling – required for efficient Rep translation.

SymR, an sRNA from the *E. coli* chromosome, shows that the mechanism of translational inhibition – in this case coupled to a decrease in target mRNA (*symE*) stability – is not confined to plasmid systems [7].

Inhibition of primer maturation

This mechanism has been only found for ColE1 and its related plasmids (reviewed in [19]) that require a plasmid-encoded replication primer that is synthesized as a 550 nt pre-primer (RNAII). ColE1 and relatives do not encode their own replication initiator protein, but instead rely solely on host proteins. For the formation of a persistent RNAII/DNA hybrid within the origin, RNAII must acquire specific secondary and tertiary structures which form during RNAII synthesis in a well-characterized series of events. Subsequently, the mature primer which can be extended by DNA polymerase I is generated by RNase H cleavage of the RNA strand of the RNAII/DNA hybrid. Binding of the antisense RNA (RNAI) that must occur within a short time window induces a change in the nascent primer thereby preventing primer maturation. The kissing complex between RNAI and RNAII is stabilized by the plasmid-encoded Rom protein. However, inhibition functions also in the absence of Rom.

Prevention of formation of an activator RNA pseudoknot

Rep expression in IncB, Inc$I\alpha$, IncK and IncL/M plasmids involves a long-distance activator RNA pseudoknot. As in R1, a leader peptide ORF, *repY*, must be translated to allow RepZ synthesis to disrupt an inhibitory stem–loop at the *rep* RBS (reviewed in [18]). This permits the formation of a short helix between the target loop and disrupted stem, located 100 nt apart. This long-distance pseudoknot activates *repZ* translation. The corresponding antisense RNAs have a dual function: they block leader peptide translation and pseudoknot formation.

Promotion of mRNA degradation or processing

A few antisense RNAs are known to influence mRNA stability, among them the λ OOP RNA that facilitates RNase III-dependent decay of the *cII* mRNA (reviewed in [18] and [20]) and RNAα expressed from plasmid pJM1 that affects the stability of both *fatA* and *fatB*-mRNA in *Vibrio anguillarum* (reviewed in [20]). Interestingly, nearly all *cis*-encoded antisense RNAs from the bacterial chromosome known to date seem to affect either translation or mRNA stability. The RdlD antitoxin RNA of *E. coli* was reported to exert its function post-transcriptionally and regulate the expression of the *ldrD* toxin gene. It is still unclear whether RdlD affects translation or stability of *ldrD* mRNA by base pairing [21]. Similarly, for the IsrR/*isiA* system involved in photosynthesis in *Synechocystis* sp., IsrR-mediated *isiA*-mRNA degradation has been suggested [5]. By contrast, the experimental data for the RatA/TxpA antitoxin/toxin system from *B. subtilis* support antisense RNA-mediated mRNA degradation, although an involvement of RNase III is still elusive [4]. In the recently published SymR/SymE system, so far, repression of translation is suggested as the putative mechanism of action of the unusually long-lived SymR antitoxin antisense RNA (half-life 60 min). It was shown that RNase III and Hfq are not required for inhibition [7].

An exception is the *E. coli* GadY/*gadX* system, for which an mRNA stabilizing effect by the antisense RNA has been proposed [6]. The most recent data suggest that base pairing between GadY and the *gadX* mRNA stimulates cleavage of a longer *gadXW* mRNA, resulting in two products that are more stable than the full-length transcript (G. Storz, personal communication) (Figure 9.2a).

9.2.1.2 Binding kinetics, binding pathway and requirement of RNase III

For many systems, antisense/sense RNA binding pathways have been studied in detail and binding kinetics measured (reviewed in [20] and [22]). Analysis of pairing rate constants usually yielded values of $\sim10^6\ M^{-1}\ s^{-1}$. The initial contact between antisense and sense RNA that often form complementary structures can occur either between two complementary loops (many replication control systems) or between a loop and a single-stranded region (e.g. RNA-IN/RNA-OUT of IS10, hok/Sok of plasmid R1). In the first case, simple helix progression in both directions is topologically impossible due to accumulating torsional stress. Therefore, loop–loop initiating systems require a subsequent interaction at a distal site to circumvent this limitation. Irrespective of a one-step or multi-step pathway (see Figure 9.2b), the final result of the interaction is a complete duplex that is often degraded by the double strand-specific RNase III. However, despite the fact that *cis*-encoded antisense RNAs are fully complementary to their targets, formation of complete duplexes is too slow to account for the observed biological effects. Instead, many antisense RNAs mediate inhibition by forming complexes that involve limited numbers of base pairs with their targets. In many instances, full duplex formation is not required for control (reviewed in [23]). In the case of R1 and IS10, RNase III cleavage was not found to be essential for control (reviewed in [20]). For replication control of R1, a partially paired binding intermediate is sufficient for inhibition *in vivo* [24]. Similarly, *in vitro* analyses of pT181 and pIP501 from Gram-positive bacteria also yielded only partial duplexes [14,17]. In pIP501, complexes between the complementary loop pairs of sense and antisense RNA form and progress into the stems, but the spacer between the two stem–loops remains single-stranded and is unimportant for inhibition [14].

For the replication control system of plasmid R1, the binding pathway between antisense RNA CopA and sense RNA CopT has been elucidated in detail [25–27] (Figure 9.2b). Binding initiates with an unstable loop–loop interaction (kissing complex) that is converted into an extended kissing complex. Later, a single-stranded region is required to overcome the torsional stress created upon the unidirectional progression of this loop–loop interaction. Next, a binding intermediate is formed which contains a four-helical junction. This intermediate is transformed into a stable inhibitory complex which is only a partial duplex and is only slowly converted into a stable duplex, which is cleaved by RNAse III. Apparently, this stepwise binding pathway is conserved and the four-helix junction is not restricted to R1 and its relatives, but is also a binding intermediate in the Inclα and related plasmids [28].

In contrast to R1 and IS10, in λ OOP RNA, RNase III cleavage was found to be necessary for control [29,30], most probably because the mechanism exerted by OOP RNA is mRNA degradation, whereas in the other systems steps preceding degradation, i.e. translation initiation, are inhibited.

9.2.2 *trans*-Encoded sRNAs

The first *trans*-encoded RNA from the bacterial chromosome, MicF, was detected in 1984 [31] and its biological function was elucidated 5 years later [32]. This 93 nt RNA forms a 20 bp imperfect RNA duplex with the translation-initiation region of the *ompF* mRNA that encodes a porin of the outer *E. coli* membrane, thereby inhibiting translation of this mRNA [33]. Until the middle of the 1990s, only a handful of *trans*-encoded sRNAs were known [34]. Only since 2001 systematic genome searches in intergenic regions have been performed and revealed that bacteria encode a tremendous number of sRNAs (e.g. [35–37]). Many of the *E. coli* sRNA genes are conserved in closely related pathogens. Interestingly, the majority of the sRNAs with known functions regulate porins of the outer *E. coli* membrane [38]. Many other sRNAs still await the identification of their targets and their characterization. Recently, two systematic searches have been performed in *Pseudomonas aeruginosa* [39,40]. Whereas the first search identified 17 RNAs of unknown function, the latter detected 11 sRNAs, among them two novel Hfq-binding RNAs, designated PhrS and PhrD. In Gram-positive bacteria, only a few systematic searches have been carried out, among them in *B. subtilis* [41,42], *Staphylococcus aureus* [43] and *Listeria monocytogenes* [44,45]. Table 9.2 provides an overview of all currently known *trans*-encoded sRNAs.

9.2.2.1 Biological functions

trans-Encoded sRNAs have been implicated in a wide variety of biological functions. They are often involved in the fine-tuning of metabolic processes, which is reflected by the lack of severe phenotypes upon deletion or overexpression of such RNAs. Examples include iron transport and storage (RyhB [46]), membrane composition {MicA, MicC, MicF, RseX, OmrA/B, RybB (reviewed in [38]), InvR [47], CyaR [48]}, sugar utilization (Spot 42 [49]), phosphosugar stress (SgrS, [50,51]) stationary phase regulation (OxyS [52], DsrA [53], RprA [54]), quorum sensing (Qrr1–4 [55]) and SOS response (IstR-1 [56]). In several cases, one sRNA regulates a set of mRNAs implicated in the same metabolic pathway. For example, RyhB RNA of *E. coli* is involved in the regulation of at least seven target mRNAs, *sodB*, *sdhD*, *acnA*, *fumA*, *bfr*, *ftn* and *shiA*, six of which encode proteins required for iron transport and storage [46]. In bacterial envelope stress response, two σ^E-controlled sRNAs, MicA and RybB RNA of *Salmonella*, were found to be involved [57–60]. Whereas MicA facilitates selectively the decay of *ompA* mRNA [57,58], RybB was found to accelerate the decay of the mRNA for at least eight major *omp* mRNAs encoding porins of the outer membrane [60]. Another example is the *E. coli* sRNA GcvB that regulates seven ABC transporter mRNAs by targeting C/A-rich elements inside and upstream of ribosome-binding sites using its G/U-rich single-stranded central region [61]. In other cases, one mRNA is the target of several sRNAs under different environmental conditions. For example, under oxidative stress, RpoS levels are downregulated by OxyS [52], whereas at low temperatures or under osmotic shock, DsrA or RprA activate translation of *rpoS* mRNA [53,54].

Interestingly, some of the *trans*-encoded sRNAs contain small open reading frames that are in some cases translated. Examples of translated ORFs include the δ-hemolysin ORF of the 514 nt *S. aureus* RNAIII [62], the ORF in *Streptococcus* pel RNA [63] and the recently identified 43 codon SgrT ORF on *E. coli* SgrS RNA that inhibits glucose transporter activity [51]. By contrast, the function of the 39 codon ORF on *B. subtilis* SR1 RNA and the ORF on PhrS of *Pseudomonas aeruginosa* are still elusive [42,40].

9.2.2.2 Mechanisms employed by trans-encoded antisense RNAs

The most important mechanism is inhibition of translation. Recent publications demonstrated that this inhibition can occur in three different ways: (i) by direct blocking of the ribosome binding site (RBS), (ii) by induction of structural alterations downstream from the RBS or (iii) by blocking of a ribosome standby site required for efficient translation (Figure 9.3). Furthermore, a few *trans*-encoded sRNAs are known to activate translation of their target mRNAs. In some cases, translational inhibition is accompanied by mRNA degradation, whereas in others, no mRNA degradation has been observed. The currently known mechanisms employed by *trans*-encoded sRNAs are discussed below and summarized in Figure 9.3.

Translation inhibition

(i) Translation inhibition by direct blocking of the RBS Translation inhibition by direct blocking of the RBS is the most widespread mechanism, shared by *cis*- and *trans*-encoded antisense RNAs, for example OxyS, Spot42, MicA, MicC, MicF, RyhB and SgrS. In all cases, the complementary regions of the sRNAs and their

Table 9.2 *Overview of* trans-*encoded antisense RNAs*[a]

sRNA (length)	Target RNA(s)	Biological function	Mechanism of action	Control of expression[b]
Escherichia coli				
MicA (72 nt)	ompA	Membrane composition	Translation inhibition	σ^E, stationary phase
MicC (109 nt)	ompC	Membrane composition	Translation inhibition and mRNA degradation	Low temperature
MicF (93 nt)	ompF	Membrane composition	Translation inhibition	High temperature, salt, HU, H-NS, Lrp, OmpR, SoxS, MarA, Rob
RseX (∼90 nt)	ompA, ompC	Membrane composition		?
OmrA/OmrB (88/82 nt)	cirA, fecA, fepA	Iron transport		High osmolarity, OmpR
	ompT	Protease		
IpeX (167 nt, on prophage)	ompC, ompF	Membrane composition	mRNA degradation?	Phage-encoded porin gene
RybB (78 nt)	ompC, ompW	Membrane composition		σ^E
RyhB (90 nt)	sodB, sdhD, acnA, fumA, bfr, ftn	Iron metabolism and mRNA degradation	Translation inhibition	Fur, iron
	shiA	Shikimate permease	Translation activation	
Spot42 (109 nt)	galK	Galactose utilization	Translation inhibition	cAMP-CAP, glucose
SgrS (227 nt)	ptsG	Glucose transport	Translation inhibition (mRNA degradation)	SgrR, phosphosugar stress
DsrA (87 nt)	rpoS	Stationary phase	Translation activation	Low temperature, osmotic shock, LeuO?
	hn-S	Stationary phase	Translation inhibition	
RprA (106 nt)	rpoS	Stationary phase	Translation activation	Osmotic shock, cell surface stress, RcsB
OxyS (109 nt)	rpoS	Stationary phase	Sequestration of Hfq	Oxidative stress
	fhlA	Formate metabolism	Translation inhibition	
GcvB (130 nt)	oppA, dppA, gltI	Peptide transport	Translation inhibition	GcvA, GcvR, glycine excess
	livK, livJ, argT	ABC transporter		
	STM4351			
IstR-1 (75 nt)	tisB	Antitoxin	Translation inhibition by competing with standby ribosomes	
GlmZ	glmS	Amino-sugar metabolism	Translation inhibition	GlmY prevents processing

Salmonella typhimurium				
MicA (72 nt)	ompA	Membrane composition	Translation inhibition?	σ^E
RybB (78 nt)	ompC, D, F, N, S	Membrane composition	Translational inhibition? + mRNA degradation?	σ^E
InvR (80 nt)	ompD	Membrane composition	Translation inhibition	HilD
CyaR (86 nt)	ompX	Membrane composition	Translation inhibition	CRP
Pseudomonas aeruginosa				
PrrF1/PrrF2	sodB, sdhD, bfr	Iron metabolism	Translation inhibition and mRNA degradation	
Chlamydia trachomatis				
IhtA (120 nt)	hctA	histon homolog	Translation inhibition	?
Bacillus subtilis				
SR1 (205 nt)	ahrC	arginine catabolism	Translation inhibition by inducing structural changes downstream of RBS	CcpN, glucose
Staphylococcus aureus				
RNAIII (514 nt)	hla	hemolysine synthesis	Translation activation	AgrC, AgrA, stationary phase
	spa, rot, sa1000, sa2353, coa	host–pathogen interaction	Translation inhibition and mRNA degradation by RNaseIII	
SprA (202 nt)	SA2216-ORF	ABC transporter?	Post-translational?	Strain-specific
Streptococcus pyogenes				
pel RNA (459 nt)	speB, emm, sic, nga	cysteine protease	Post-transcriptional transcriptional control	Stationary phase, conditioned media
		M- and related proteins		
Vibrio cholerae/harveyi				
Qrr1–4 (96, 108, 107, 107 nt)	hapR	quorum sensing	mRNA destabilization	LuxOP, σ^{54}

[a] Only sRNAs are summarized for which target genes have been identified, allowing the conclusion that they are *trans*-encoded.
[b] Control of expression: all proteins (transcription factors, sigma factors, etc.) and growth conditions known to regulate sRNA expression are listed. It is not indicated whether these factors promote or inhibit sRNA expression.

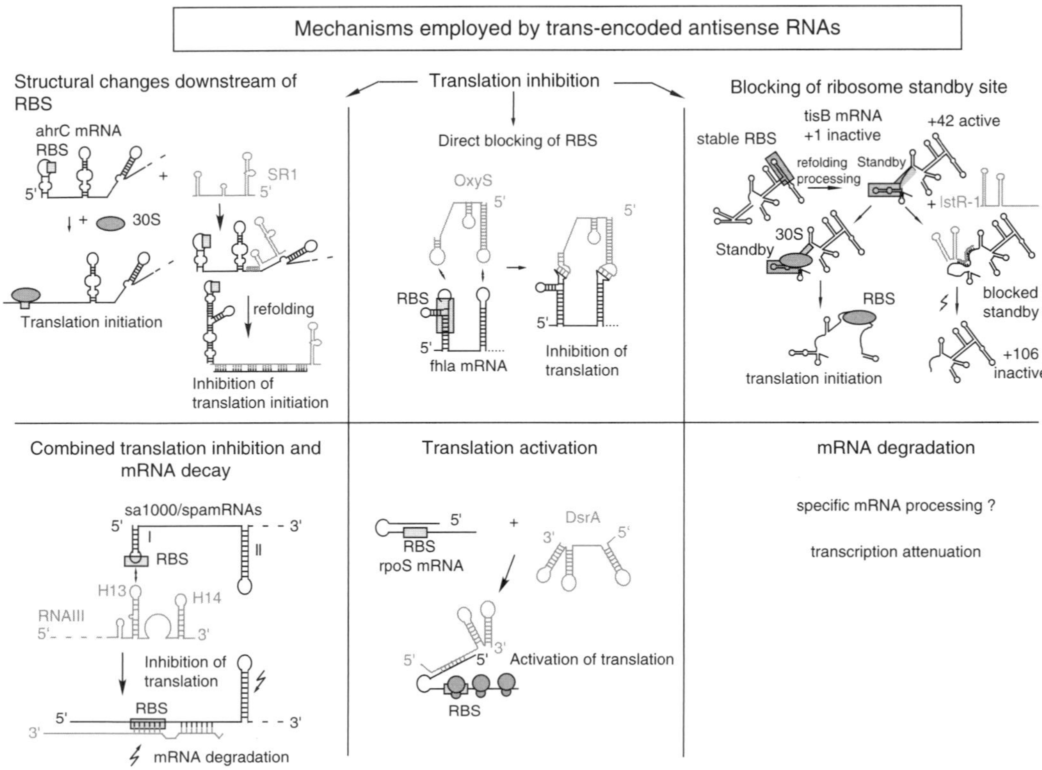

Figure 9.3 *Overview of regulatory mechanisms employed by* trans-*encoded antisense RNAs. Antisense RNAs are drawn in gray and sense RNAs in black. Light gray rectangles, ribosome binding sites. Thunderbolt-like arrows denote RNase III action. Light gray circular symbols indicate ribosomes*

target mRNAs overlap the RBS or the RBS and/or the immediate adjacent 5′ or 3′ regions. For MicA and MicF, base pairing interactions include the *ompA* and *ompF* RBS, respectively, whereas for MicC, the two complementary regions with *ompC* comprise 6 and 16 continuous base pairs, respectively, immediately upstream of the RBS (summarized in [38]). In the case of OxyS/*fhla*, two interacting regions of 7 and 9 bp were found overlapping the RBS and about 25 nt downstream from the AUG start codon, respectively [64]. For SgrS, 23 complementary base pairs were found to overlap the RBS and the AUG of *ptsG* mRNA; however, mutational analyses revealed that only 6 bp around the RBS are crucial [65].

(ii) Translation inhibition by induction of structural changes downstream from the RBS To date, the only known example for this mechanism is SR1 from *B. subtilis*, which interacts with *ahrC* mRNA encoding a transcriptional activator of the arginine catabolic operons [66]. Both RNAs share seven complementary regions A–G in the 3′ half of SR1 and the central and 3′ part of *ahrC* mRNA. Region G is located ~100 nt downstream from the RBS of *ahrC* mRNA. Binding of SR1 induces structural alterations not only in all complementary regions, but also immediately downstream from the *ahrC* RBS and upstream of region G, which results in inhibition of translation initiation, as shown by toeprinting analysis [67]. This means that a base pairing interaction far downstream from the RBS can nevertheless prevent binding of the 30S ribosomal subunit at the RBS.

(iii) Translation inhibition by blocking of a ribosome standby site To date, the only example of this mechanism is IstR-1/*tisAB* of *E. coli* [56]. Here, ~100 nt upstream of the *tisB* RBS a standby site for ribosomes has been found, which is required for efficient translation of the TisB toxin from this highly structured RBS. The antisense RNA, IstR-1, is complementary to this site and competes with standby ribosomes for binding. The IstR-1/*mRNA* interaction generates a cleavage site for RNase III which in turn results in a 5′ truncated *tisB* mRNA which cannot be translated any longer.

(iv) Combined translation inhibition and mRNA degradation Frequently, translation inhibition by a *trans*-encoded sRNA was found to be accompanied by degradation of the target mRNA by RNase E or RNase III. In some cases, degradation by RNase III was demonstrated to be necessary for inhibition, as in the case of RNAIII of *S. aureus* [68]. RNAIII (514 nt) is composed of 14 stem–loops and two long-distance interactions that define independent structural domains [69], among which three hairpins–loops carry C-rich sequences that could potentially base pair with SD sequences. RNAIII is of particular interest as it is both a regulatory antisense RNA and a protein-encoding mRNA carrying an ORF for δ-hemolysin in its 5′ domain [62]. As an antisense RNA, it not only activates translation of the hemolysin α mRNA [62] (see below), but also inhibits translation of a variety of targets such as *spa* encoding the main surface adhesin protein [68] (see Figure 9.3), SA1000 mRNA encoding a novel fibrinogen-binding protein and the *rot* mRNA encoding the pleiotropic transcriptional factor Rot [70] (repressor of toxins). The last one was identified as a general antagonist of the *agr* regulation. For the action of RNAIII as base pairing antisense RNA, the 3′ domain carrying two redundant hairpin–loop motifs that target *rot* mRNA is decisive. One loop–loop contact occurs with the *rot*–SD sequence, the other with another stem–loop. In all systems, the formation of RNAIII–mRNA duplexes results in inhibition of ribosome binding and favors specific recognition by RNase III (see Figure 9.3). Specificity of RNAIII action is obtained by either propagating the first loop–loop contact into the stem regions (*sa1000* mRNA, *sa2353* mRNA) or by the addition of a second loop–loop interaction as in *rot* mRNA. Interestingly, the 3′ domain of RNAIII itself is sufficient to promote the synthesis of several exoproteases and exotoxins, which is most likely the consequence of RNAIII-mediated *rot* translation inhibition.

Generally, it is suggested that in *E. coli* and other Gram-negative bacteria, degradation of translationally repressed mRNAs is probably a consequence of ribosome exclusion rather than the primary event, because translation inhibition can occur in the absence of mRNA degradation. This was shown for SgrS/*ptsG* [71] and RyhB/*sodB* of *E. coli* [72]. Likewise, for the IstR-1/*tisAB* system, RNase III degradation of both RNAs is not required for inhibition [56]. In *S. aureus*, however, RNAIII-dependent inhibition of the ribosome is coupled to RNase III-dependent cleavage to render the arrest of translation irreversible [68]. Thereby, RNA III was shown to guide the RNase III to the repressed mRNAs *in vivo*.

(v) Translation activation In some cases, RBSs are located in double-stranded structures which prevent the access of ribosomes. Melting of these regions is promoted by binding of the antisense RNA to one strand, thus liberating the complementary strand containing the RBS and hence activating translation. Four examples are known so far, three being RNAIII from *S. aureus*, which activates translation of *hla* mRNA encoding α hemolysin [62], and DsrA [53] and RprA [73] from *E. coli*, both of which promote translation of *rpoS* mRNA encoding the stationary phase σ factor. The fourth example is RyhB RNA from *E. coli* that activates translation of *shiA* mRNA encoding shikimate permease [74]. Whereas for DsrA and RyhB action, Hfq is required, for the action of RNAIII, Hfq does not play a role (see below).

(vi) Other mechanisms? Although other mechanisms of action of *trans*-encoded sRNAs are conceivable, e.g. mere promotion of mRNA degradation, induction of a specific target RNA processing event that could either result in more stable or unstable messages or transcriptional attenuation, such mechanisms have not

been discovered so far. At least the examples of RNAIII, DsrA and RyhB show that the same *trans*-encoded sRNA can use different mechanisms on different targets. However, since research on the functions and mechanisms of *trans*-encoded sRNAs is still in its infancy, the detection and elucidation of new mechanisms can be expected in the near future.

(vii) Identification of novel RNAs and novel targets for known trans-encoded RNAs With different experimental approaches, novel sRNAs are being detected. Such approaches comprise a combination of bioinformatics and Northern blotting, chemical or enzymatic sequencing, generation of specialized cDNA libraries, microarray analysis using arrays encompassing intergenic regions and genomic SELEX (using the association of an sRNA with a specific protein as Hfq), and have been described and discussed recently with regard to possible advantages and disadvantages [75]. The identification of targets for novel *trans*-encoded sRNAs is still a challenging issue, which is reflected in the relatively small number of sRNAs for which targets have been found. Methods for target identification include bioinformatics-aided approaches, biochemical fishing experiments, microarray-based target screening and proteomics, which have been reviewed recently [76].

9.2.2.3 Role of Hfq

One important hallmark of many *trans*-endcoded antisense RNAs from *E. coli* is their ability to bind the RNA chaperone Hfq (reviewed in [77]). Hfq was identified in *E. coli* as host factor for the replication of bacteriophage Qβ [78]. It is present in half of all sequenced bacterial species [79–81]. Several bacteria such as *Bacillus anthracis* encode even two Hfq proteins. At least one archaeon, *Methanococcus jannaschii*, contains a protein that is related to Hfq. Hfq comprises between 70 and 110 amino acids and forms homohexamers. It is an abundant protein, which is present in up to 60 000 monomers/*E. coli* cells in the stationary phase. The major fraction is associated with the ribosomes, whereas a minor fraction appears to be associated with the nucleoid. Hfq was shown to bind to AU-rich sequences in single-stranded regions generally flanked by one or two stem–loops. Hfq is a homohexamer that is very similar to the eukaryotic Sm and Sm-like proteins involved in splicing and to the archaeal Lsm proteins (Figure 9.4a and b). It forms a toroid with an outer diameter of ~70 Å and a thickness of 25 Å. The central pore is 8–12 Å wide. The N-terminal α-helix is followed by five β-strands that form a tightly bent sheet. A recent key finding was that Hfq has two RNA binding sites, a proximal and a distal [77]: the proximal site binds sRNA and mRNA and the distal site binds polyA tails. The crystal structure of *S. aureus* Hfq with an AU_5G oligo shows that the RNA expands and fills the central, basic pore on the 'proximal side' of the hexamer and binds in a circular manner. The repetition of identical binding pockets on the Hfq hexamer suggests that the binding surface can accommodate more than just a single RNA target. This would allow simultaneous binding of two RNA strands and could promote the interaction between these strands, which is particularly important in sRNA–target mRNA interactions.

In *E. coli*, inactivation of the *hfq* gene causes pleiotropic effects, such as a reduction in growth rate and UV sensitivity or an increase in cell length. Hfq is involved in mRNA stability, mRNA polyadenylation and translation. The importance of Hfq is further underscored by its multiple roles in virulence, bacteriocin production and nitrogen fixation. It has been reported to have a relatively weak ATPase activity.

Interestingly, Hfq was shown to interact with the 30S ribosomal subunit, RNase E, PNP, PAPI and the ribosomal protein S1. The C-terminal scaffold domain of RNase E, which was shown to participate in RyhB-mediated degradation of target mRNAs, acts as an Hfq binding site. Hfq interacts with the RNA polymerase in an S1-dependent manner. Surprisingly, it was found that Hfq interacts with itself and can form well-ordered fibers with 36 Hfq subunits per helical pitch, the physiological relevance of which is still unclear.

Many *trans*-encoded sRNAs, for example DsrA, Spot42 and RyhB, require Hfq for their stability [81]. In other cases, for example OxyS/*fhlA*, Spot42/*galK*, RyhB/*sodB* and MicA/*ompA*, Hfq was shown to promote the interaction between sRNAs and their targets [82]. A FRET study on the intermolecular base pairing

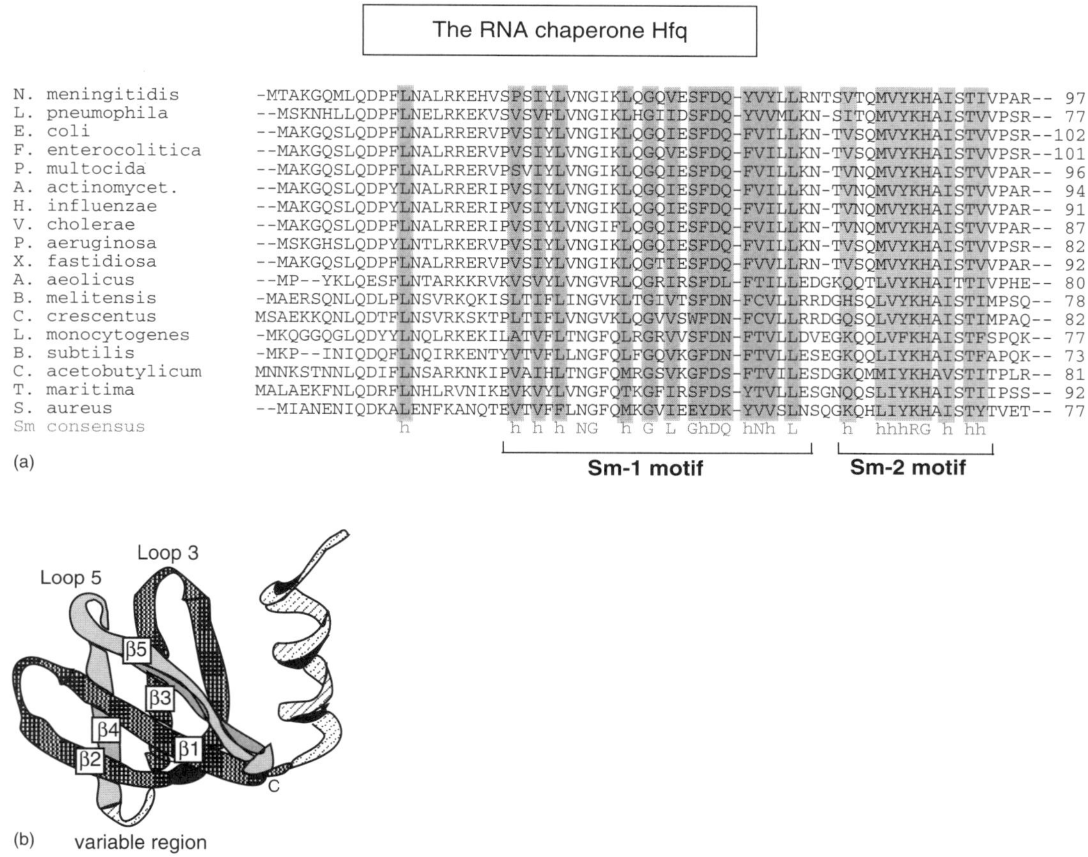

Figure 9.4 *The RNA chaperone Hfq. (a) Amino acid sequence alignment of a subset of bacterial Hfq proteins. The Sm consensus based on 80 Sm and Sm-related proteins is shown in gray at the bottom. The Sm1- and Sm2-consensus positions are highlighted in light gray; h, Bulky hydrophobic residue (L, I, M, V, F, Y or W). Based on [80]. (b) Ribbon diagram of an Hfq subunit. The Sm1 motif is dark stippled and the Sm2 motif is light gray. Regions outside the two motifs, i.e. the N-terminal α-helix and the variable region, are shown as hatched areas*

between DsrA and *rpoS* mRNA demonstrated that Hfq accelerates strand exchange and subsequent annealing between sRNA and *rpoS* mRNA, which results in exposure of the *rpoS* RBS (see also Figure 9.3). For *E. coli* Hfq, it has been shown recently that the C-terminal domain is the hitherto unrecognized RNA interaction surface with specificity for mRNAs [83]. The high Hfq concentration under standard conditions suggests that additional regulators might be required to trigger Hfq-dependent regulation.

For sRNAs from Gram-positive bacteria, the putative function of Hfq is still elusive. Some of the identified sRNAs bind Hfq (RNAIII of *S. aureus*, LhrA and LhrC from *Listeria monocytogenes* [44], SR1 of *B. subtilis* [67]), but at least in two cases, SR1/*ahrC* from *B. subtilis* and RNAIII/*spa* or other target mRNAs from *S. aureus*, no influence of Hfq has been found on the interaction with their targets [67,84]. This might be due to the lack of the C-terminal domain found to be required for RNA annealing activity [83]. Interestingly, Hfq is not highly expressed in *S. aureus*. It cannot be excluded that in Gram-positive bacteria Hfq is needed for other, still unidentified functions. Alternatively, it is conceivable that other RNA binding proteins might fulfill the role of Hfq in these bacteria.

9.2.2.4 *Differences between cis- and trans-encoded antisense RNAs*

cis-Encoded RNAs are complementary to their targets over a large nucleotide stretch and can, therefore, form stable duplexes with their target RNAs. Although in two cases, ColE1 and R1/F, plasmid-encoded RNA binding proteins (Rom and FinO, respectively) were shown to have an effect [85,86], *cis*-encoded RNAs usually do not require an additional protein to facilitate complex formation with their targets. Rom promotes RNAI/RNAII pairing only fivefold, since the inhibition rate is primarily determined by the binding rate constant and not the binding affinity between the loop–loop complexes. The FinO protein of F plasmid acts by promoting strand exchange between *traJ* and FinP, but its key function is to protect FinP against RNase E degradation [87]. By contrast, for R1 (CopA) and ColE1 (RNAI), replication control *in vivo* was found to be functional in an *E. coli* strain lacking the RNA chaperone Hfq, although these two antisense RNAs bound Hfq (G. Wagner, personal communication). In RNAII/III of pIP501, Hfq is not expected to play a role *in vivo*, since the original host of this plasmid, *Streptococcus agalactiae*, does not encode *hfq*. For F plasmid, it was recently demonstrated that Hfq even destabilizes the sense RNA *traJ* by binding 5′ of the stem–loops interacting with FinP, but does not bind FinP specifically [88].

In contrast, a large number of *trans*-encoded antisense RNAs from Gram-negative bacteria need Hfq either for stabilization or for complex formation, most probably to facilitate the interaction with their only partially complementary target RNA(s). In other cases, the role of Hfq binding to the regulatory RNA is still elusive (e.g. SR1 from *B. subtilis* [67]) and it is not clear whether in Gram-positive bacteria another protein fulfills the role of Hfq.

Another interesting finding was the existence of U-turn motifs [5′ YUNR) and the formation of such U-turn structures which has, so far, only been observed for many *cis*-encoded antisense RNAs and their targets. Here, one loop in either the antisense or the sense RNA forms a U-turn structure [89], which provides a scaffold for the rapid interaction with the complementary RNA. For Both the hok/Sok system of plasmid R1 [89] and the RNAIII/RNAII system of plasmid pIP501 [90] it has been demonstrated experimentally that the U-turn structure – present in the sense RNA loop – is important for an efficient interaction with the antisense RNA.

Each *cis*-encoded antisense RNA uses on its single target one defined mechanism of action. In contrast, *trans*-encoded sRNAs can employ different mechanisms on different target mRNAs. For instance, RNAIII, DsrA and RyhB inhibit translation of one or a few targets, but activate translation initiation of others (see above). It cannot be excluded that a *trans*-encoded sRNA might use three or four different mechanisms to regulate its several targets.

Recently, a *trans*-encoded antisense-RNA was found that is under the control of another small RNA: the GlmY/GlmZ sRNA pair regulates hierarchically GlmS synthesis in *E. coli*. Thereby, GlmY inhibits processing of GlmZ that in turn regulates translation of *glmS* mRNA involved in amino sugar metabolism [91,92]. It remains to be seen whether such an RNA-controlled antisense RNA is an exception or a fairly widespread control mechanism which might be also found for *cis*-encoded sRNAs.

9.2.3 *trans*-Encoded sRNAs that act via protein binding

Unlike the *trans*-encoded antisense RNAs that act by base pairing with their target mRNAs, another class of *trans*-encoded sRNA employs protein binding as its mechanism of action. Thereby, we can differentiate between two groups of protein binding sRNAs: 6S RNA that binds RNA polymerase and sRNAs that sequester small proteins that inhibit target mRNA translation. A third group, with so far only one member, Rcd of *E. coli* plasmid ColE1, comprise sRNAs that bind and modulate the activity of enzymes. Table 9.3 provides an overview of all currently known *trans*-encoded sRNAs that act via protein binding.

Table 9.3 *Overview of* trans-*encoded sRNAs acting via protein binding*

sRNA (length)	Interacting protein	Biological function	Mechanism	Control
Escherichia coli				
CsrB (366 nt)	CsrA (61 aa)	Glycogen biosynthesis	Protein sequestration	BarA/UvrY
CsrC (245 nt)		Biofilm production		
6S RNA (200 nt)	σ^{70}-RNAP, σ^{S}-RNAP	Stationary phase regulation	Protein sequestration	Stationary phase
Rcd (70 nt)	Tryptophanase (TnaA)	Stable plasmid maintenance	Activation of enzyme	Plasmid multimers
Vibrio cholerae				
CsrB/CsrC/CsrD (416/366/351 nt)	CsrA	Virulence	Protein sequestration	GacS/GacA
Erwinia carotovora				
RsmB (479 nt)	RsmA (61 aa)	Proteases, pectinases, etc.	Protein sequestration	GacS/GacA
Pseudomonas fluorescens				
RsmX/RsmY (119/118 nt)	RsmE (61 aa)	Secondary metabolism	Protein sequestration	GacS/GacA
RsmX (127 nt)				
Pseudomonas aeruginosa				
RsmY/RsmZ (126/117 nt)	RsmE	Secondary metabolism	Protein sequestration	GacS/GacA
Salmonella enterica				
CsrB/CsrC (363/240 nt)	CsrA	Epithelial cell invasion	Protein sequestration	
Yersinia pseudotuberculosis				
CsrB/CsrC (~310 nt)	CsrA	Virulence	Protein sequestration	UvrY (csrC)

9.2.3.1 6S RNA

6S RNA was discovered in the late 1960s [93], but only in 2000 was it found that it comigrates with RNA polymerase in glycerol gradient gels and might interact with this enzyme [94]. Now it is known to be ubiquitous among bacteria. To date, more than 100 putative 6S RNA homologs have been identified by bioinformatic procedures and many of them have been confirmed experimentally (for a review on 6S RNA, see [95]). In *Bacillus subtilis*, even two 6S RNA species exist. 6S RNA has a characteristic secondary structure consisting of a central region with a largely single-stranded internal loop, which is flanked by two long, irregular, double-stranded stem regions that are interrupted by small bulges. This secondary structure resembles a partially single-stranded DNA bubble and has led to a hypothesis for the potential function of 6S RNA (Figure 9.5). 6S RNA forms a stable complex with RNA polymerase and, therefore, it was assumed that it acts as an open promoter DNA mimic that interferes with the formation of transcription initiation complexes. Its levels increase ~10-fold to about 10 000 molecules during stationary phase. 6S RNA interacts preferentially with RNA polymerase containing the exponential phase sigma factor σ^{70}. Therefore, it was proposed that 6S RNA participates in shifting global gene expression from exponential to stationary phase. However, a recent demonstration of weak contacts between 6S RNA and stationary phase

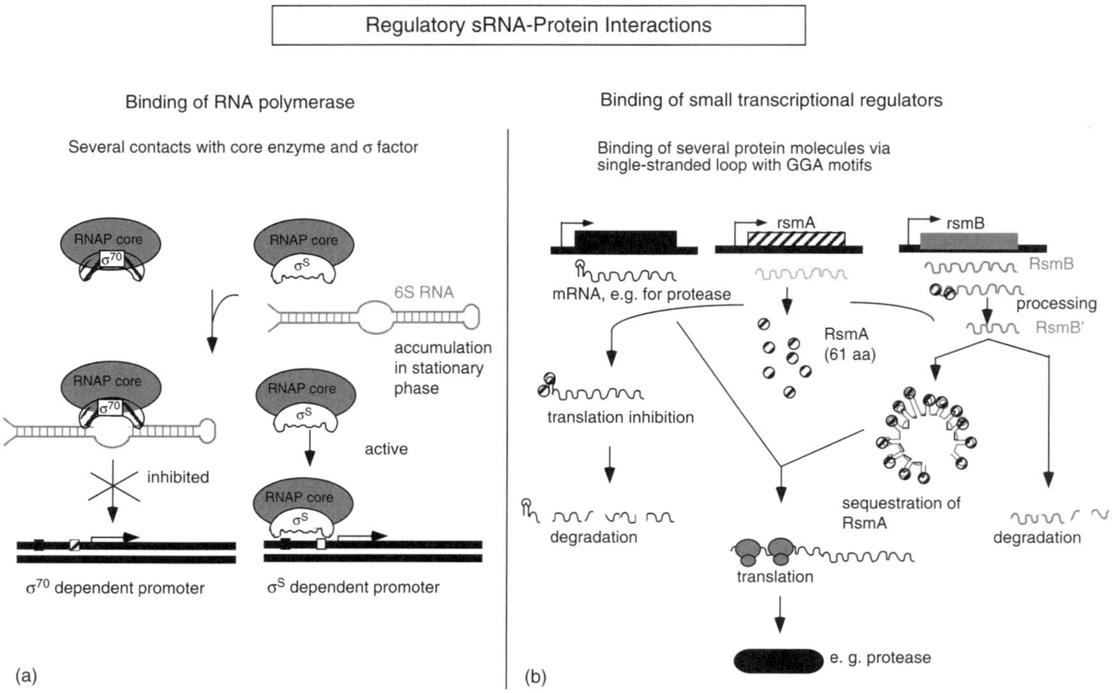

Figure 9.5 *Overview of sRNAs acting by protein binding. Gray, 6S RNA; gray ellipse, RNA polymerase core. Hatched and white symbols denote σ^{70} and σ^{S}, respectively. Black and white rectangles indicate –35 and –10 boxes, respectively. (a) Mechanism of 6S RNA action. (b) Mechanism of action of sRNAs binding translational regulators. The RsmA–RsmB regulatory system of* Erwinia carotovora *ssp.* carotovora. *RsmB primary RNA is subsequently processed yielding a 259 nt long functional RNA (RsmB′). Free RsmA protein induces degradation of target mRNA. Binding of RsmA protein to RsmB′ depletes the pool of free RsmA, thereby inhibiting target mRNA degradation. The binding of RsmA to RsmB′ also protects the RsmB′ transcript from degradation*

sigma factor σ^{38} and the fact that 6S can inhibit both RNA polymerase holoenzymes [96] shows that the molecular details for the mechanism of action of 6S RNA have still to be elucidated. Surprisingly, it was found that 6S RNA can act as a template for the production of small *de novo* transcripts: 14–22 nt transcripts (pRNA) were observed when 6S concentration was high [96,97], whereas 170 nt transcripts (whose role is still unclear) were observed when RNA polymerase (RNAP) concentrations were high [96]. Synthesis of the pRNA destabilizes the 6S–RNAP complexes and leads to release of the pRNA–6S RNA hybrid [97]. *In vivo*, 6S RNA-directed RNA synthesis occurs during outgrowth from stationary phase and likely is responsible for liberating RNAP from 6S RNA in response to nutrient availability [97]. The short transcripts show that 6S RNA is functionally engaged in the active site of RNAP. Recently, it was shown that 6S-sensitive promoters have weak –35 boxes or extended –10 boxes or both, whereas transcription from promoters with strong –35 boxes is not inhibited by 6S RNA. Based on identified contacts between 6S RNA and the 4.2 region of σ^{70}, a model of competition between 6S RNA and promoters for σ^{70} binding was proposed [98]. Recently, it was shown that 6S RNA transcription depends on a network of global regulators and is inhibited by H-NS and LRP and, to a lesser extent, by StpA, whereas FIS seems to act as a dual regulator [99].

9.2.3.2 Small RNAs that act by sequestration of translational regulators

In 1997, it was shown that a small untranslated RNA, CsrB from *E. coli*, does not act via base pairing on an mRNA target, but exerts its function by binding of the small protein CsrA, a repressor of stationary phase processes such as gluconeogenesis and biofilm production and activator of glycolysis, motility and acetate metabolism [100]. CsrA binds to the untranslated leader of the *glgCAP* transcript, where it blocks translation and causes rapid mRNA degradation. In contrast, positive control of gene expression by CsrA involves binding to the untranslated leader and stabilization of the corresponding mRNAs; however, the detailed mechanism responsible for activation is still elusive. Highly conserved CsrA homologs with monomer sizes of about 7 kDa are found in diverse eubacteria and regulate virulence factors of animal and plant pathogens (reviewed in [101]). The structures of three CsrA homologs have been reported. They act as dimers and encompass five β-strands and a C-terminal α-helix. The CsrA dimer contains two symmetrical surfaces that function in RNA recognition.

The CsrA/CsrB ribonucleoprotein complex comprises 18 CsrA subunits and a single CsrB molecule. Binding of CsrA to CsrB or the SD sequence and untranslated leader region of, for example, the *glgC* transcript involves the same imperfect repeat sequence CAGGAUG that is located primarily in the loops of the CsrB hairpins. In 2003, a second RNA, CsrC, was found that acts by binding nine CsrB molecules [102]. Null mutations of either CsrB or CsrC cause a modest increase in each other's levels. The binding sequence of CsrA could be narrowed down to the minimal motif 5′ GGA present on both sRNAs (CsrB and CsrC) and the corresponding mRNAs. However, the SELEX-derived consensus sequence is RU*AC*ARGGAU*GU* and the conserved AC and GU residues shown in italics were always base paired to one another (reviewed in [101]). Recently, CsrD was discovered to target CsrB and CsrC in *E. coli* for degradation [103]. Two CsrA homologs (RsmA and RsmE) and three redundant sRNAs (RsmX, RsmY and RsmZ) that function as antagonists of RsmA and RsmE have been identified in *Pseudomonas fluorescens*, whereas only RsmY and RsmZ were found in *Pseudomonas aeruginosa* (reviewed in [104]). The size of the CsrB type sRNAs in the different bacteria varies between 100 and 479 nt, and all share multiple unpaired GGA motifs. All sRNAs are under control of the two-component GacS/GacA system. Furthermore, the sRNAs feedback inhibit the transcription of their own genes by interfering with the function of the GacS/GacA system in an unknown manner. It is not clear whether the sRNA families have common ancestors or whether they are functional homologs. The three sRNAs from *Pseudomonas fluorescens* regulate secondary metabolism and biocontrol traits (e.g. antifungal metabolites and extracellular enzymes that protect plant roots from pathogenic fungi). An *rsmY*/*rsmZ* double mutant is strongly impaired in the synthesis of extracellular enzymes, whereas single mutants do not show significant effects. A similar pathway is present in *Erwinia carotovora* with the largest known sRNA – RsmB – and homologous elements are found in *Salmonella typhimurium*, where they control the expression of genes related to invasiveness. Recently, a Csr-type regulatory system has been found in *Yersinia pseudotuberculosis* to regulate virulence [105].

9.2.3.3 A small RNA that acts by binding tryptophanase

The *cer* site is required for the resolution of ColE1 plasmid dimers by the site-specific Xer-cer recombination system composed of four proteins, XerC, XerD, ArgR and PepA. The resolution of plasmid multimers must be achieved before the cell divides, because this is the only time when a plasmid may be lost. Upon induction by plasmid multimers, the untranslated Rcd RNA is transcribed from a promoter within the *cer* site. It has been found that Rcd acts by binding the enzyme tryptophanase (TnpA), increasing its affinity for tryptophan and stimulating indole production by rapidly dividing cells (low-density cultures). By a still unknown mechanism, indole seems to act as an intracellular signaling molecule to delay cell division, giving the plasmid time to convert multimers into monomers [106].

References

1. J. Tomizawa, T. Itoh, G. Selzer, T. Som, Inhibition of ColE1 RNA primer formation by a plasmid-specified small RNA, *Proc. Natl. Acad. Sci. USA*, **78**, 1421–1425 (1981).
2. P. Stougaard, S. Molin, K. Nordström, RNAs involved in copy-number control and incompatibility of plasmid R1, *Proc. Natl. Acad. Sci. USA*, **78**, 6008–6012 (1981).
3. E. G. H. Wagner, S. Altuvia, P. Romby, Antisense RNAs in bacteria and their genetic elements, *Adv. Genet.*, **46**, 361–398 (2004).
4. J. M. Silvaggi, J. B. Perkins, R. Losick, Small untranslated RNA antitoxin in *Bacillus subtilis*, *J. Bacteriol.*, **187**, 6641–6650 (2005).
5. U. Dühring, I. M. Axmann, W. R. Hess, A. Wilde, An internal antisense RNA regulates expression of the photosynthesis gene *isiA*, *Proc. Natl. Acad. Sci. USA*, **103**, 7054–7058 (2006).
6. J. A. Opdyke, J. G. Kang, G. Storz, GadY, a small-RNA regulator of acid response genes in *Escherichia coli*, *J. Bacteriol.*, **186**, 6698–6705 (2004).
7. M. Kawano, L. Aravind, G. Storz, An antisense RNA controls synthesis of an SOS-induced toxin evolved from an antitoxin, *Mol. Microbiol.*, **64**, 738–754 (2007).
8. M. Kawano, A. A. Reynolds, J. Miranda-Rios, G. Storz, Detection of 5′- and 3′-UTR-derived small RNAs and *cis*-encoded antisense RNAs in *Escherichia coli*, *Nucleic Acids Res.*, **33**, 1040–1050 (2005).
9. G. Padalon-Brauch, R. Hershberg, M. Elgrably-Weiss, K. Baruch, I. Rosenshine, H. Margalit, S. Altuvia, Small RNAs encoded within genetic islands of *Salmonella typhimurium* show host-induced expression and role in virulence, *Nucleic Acids Res.*, **36**, 1913–1927 (2008).
10. R. P. Novick, S. Iordanescu, S. J. Projan, J. Kornblum, I. Edelman, pT181 plasmid replication is regulated by a countertranscript-driven transcriptional attenuator, *Cell*, **59**, 395–404 (1989).
11. S. Brantl, E. Birch-Hirschfeld, D. Behnke, RepR protein expression on plasmid pIP501 is controlled by an antisense RNA-mediated transcription attenuation mechanism, *J. Bacteriol.*, **175**, 4052–4061 (1993).
12. S. Brantl, E. G. H. Wagner, An antisense RNA-mediated transcriptional attenuation mechanism functions in *Escherichia coli*, *J. Bacteriol.*, **184**, 2740–2747 (2002).
13. S. Brantl, E. G. H. Wagner, Antisense RNA-mediated transcriptional attenuation occurs faster than stable antisense/target RNA pairing: an *in vitro* study of plasmid pIP501, *EMBO J.*, **13**, 3599–3607 (1994).
14. N. Heidrich, S. Brantl, Antisense RNA-mediated transcriptional attenuation in plasmid pIP501: the simultaneous interaction between two complementary loop pairs is required for efficient inhibition by the antisense RNA, *Microbiology*, **153**, 420–427 (2007).
15. S. Brantl, E. G. H. Wagner, An unusually long-lived antisense RNA in plasmid copy number control: *in vivo* RNAs encoded by the streptococcal plasmid pIP501, *J. Mol. Biol.*, **255**, 275–288 (1996).
16. S. Brantl, E. G. H. Wagner, Dual function of the *copR* gene product of plasmid pIP501, *J. Bacteriol.*, **179**, 7016–7024 (1997).
17. S. Brantl, E. G. H. Wagner, Antisense RNA-mediated transcriptional attenuation: an *in vitro* study of plasmid pT181, *Mol. Microbiol.*, **35**, 1469–1482 (2000).
18. S. Brantl, Plasmid replication controlled by antisense RNAs. In *The Biology of Plasmids*, ed. B. Funnell, G. Phillips, ASM Press, Washington, DC, 2004, pp. 47–62.
19. Y. Eguchi, T. Itoh, J. Tomizawa, Antisense RNA, *Annu. Rev. Biochem.*, **60**, 631–652 (1991).
20. S. Brantl, Regulatory mechanisms employed by *cis*-encoded antisense RNAs, *Curr. Opin. Microbiol.*, **10**, 102–109 (2007).
21. M. Kawano, T. Oshima, H. Kasai, H. Mori, Molecular characterization of long direct repeat (LDR) sequences expressing a stable mRNA encoding for a 35-amino-acid cell-killing peptide and a *cis*-encoded small antisense RNA in *Escherichia coli*, *Mol. Microbiol.*, **45**, 333–349 (2002).
22. S. Brantl, Antisense-RNA regulation and RNA interference, *Biochim. Biophys. Acta*, **1575**, 15–25 (2002).
23. E. G. H. Wagner, S. Brantl, Kissing and RNA stability in antisense control of plasmid replication, *Trends Biochem. Sci.*, **23**, 451–454 (1998).
24. C. Malmgren, E. G. H. Wagner, C. Ehresmann, B. Ehresmann, P. Romby, Antisense RNA control of plasmid R1 replication. The dominant product of the antisense RNA–mRNA binding is not a full RNA duplex, *J. Biol. Chem.*, **272**, 12508–12512 (1997).

25. F. A. Kolb, C. Malmgren, E. Westhof, C. Ehresmann, B. Ehresman, E. G. H. Wagner, P. Romby, An unusual structure formed by antisense-target RNA binding involves an extended kissing complex with a four-way junction and a side-by-side helical alignment. *RNA*, **6**, 311–324 (2000).
26. F. A. Kolb, H. M. Engdahl, J. B. Slagter-Jäger, B. Ehresmann, C. Ehresmann, E. Westhof, E. G. H. Wagner, P. Romby, Progression of a loop–loop complex to a four-way junction is crucial for the activity of a regulatory antisense RNA. *EMBO J.*, **19**, 5905–5915 (2000).
27. F. A. Kolb, E. Westhof, C. Ehresmann, B. Ehresmann, E. G. H. Wagner, P. Romby, Bulged residues promote the progression of a loop–loop interaction to a stable and inhibitory antisense-target RNA complex, *Nucleic Acids Res.*, **29**, 3145–3153 (2001).
28. F. A. Kolb, E. Westhof, C. Ehresmann, B. Ehresmann, E. G. H. Wagner, P. Romby, Four-way junctions in antisense RNA–mRNA complexes involved in plasmid replication control: a common theme? *J. Mol. Biol.*, **209**, 6005–614 (2001).
29. L. Krinke, D. L. Wulff, OOP RNA, produced from multicopy plasmids, inhibits *cII* gene expression through an RNase III-dependent mechanism, *Genes Dev.*, **1**, 1005–1012 (1987).
30. L. Krinke, D. L. Wulff, RNase III-dependent hydrolysis of *cII-O* gene mRNA by OOP antisense RNA, *Genes Dev.*, **4**, 2223–2233 (1990).
31. T. Mizuno, M. Y. Chou, M. Inouye, A unique mechanism regulating gene expression: translational inhibition by a complementary RNA transcript (micRNA) *Proc. Natl. Acad. Sci USA*, **81**, 1966–1970 (1984).
32. J. Andersen, S. A. Forst, K. Zhao, M. Inouye, N. Delihas, The function of micF RNA. micF RNA is a major factor in the thermal regulation of OmpF protein in *Escherichia coli*, *J. Biol. Chem.*, **264**, 17961–17979 (1989).
33. J. Andersen, N. Delihas, micF RNA binds to the 5′ end of *ompF* mRNA and to a protein from *Escherichia coli*, *Biochemistry*, **29**, 9249–9256 (1990).
34. K. M. Wassarman, A. Zhang, G. Storz, Small RNAs in *Escherichia coli*, *Trends Microbiol.*, **7**, 37–45 (1999).
35. L. Argamann, R. Hershberg, J. Vogel, G. Bejerano, E. G. H. Wagner, H. Margalit, S. Altuvia, Novel small RNA-encoding genes in the intergenic regions of *Escherichia coli*, *Curr. Biol.*, **11**, 941–950 (2001).
36. K. M. Wassarman, F. Repoila, C. Rosenow, G. Storz, S. Gottesman, Identification of novel small RNAs using comparative genomics and microarrays, *Genes Dev.*, **15**, 1637–1651 (2001).
37. J. Vogel, V. Bartels, T. H. Tang, G. Churakov, J. G. Slagter-Jäger, A. Hüttenhofer, E. G. H. Wagner, RNomics in *Escherichia coli* detects new sRNA species and indicates parallel transcriptional output in bacteria, *Nucleic Acids Res.*, **31**, 6435–6443 (2003).
38. J. Vogel, K. Papenfort, Small noncoding RNAs and the bacterial outer membrane, *Curr. Opin. Microbiol.*, **9**, 605–611 (2006).
39. J. Livny, A. Brencic, S. Lory, M. K. Waldor, Identification of 17 *Pseudomonas aeruginosa* sRNAs and prediction of sRNA-encoding genes in 10 diverse pathogens using the bioinformatic tool sRNAPredict2, *Nucleic Acids Res.*, **34**, 3484–3493 (2006).
40. E. Sonnleitner, T. Sorger-Domenigg, M. J. Madej, S. Findeiss, J. Hackermüller, A. Hüttenhofer, P. Stadler, U. Bläsi, I. Moll, Detection of Hfq-binding small RNAs by RNomics and by structure-based bioinformatic tools in *Pseudomonas aeruginosa*, *Microbiology*, **154**: 3175–3187 (2008).
41. M. Lee, S. Zhang, S. Saha, S. Santa Anna, C. Jiang, J. Perkins, RNA expression analysis using an antisense *Bacillus subtilis* genome array, *J. Bacteriol.*, **183**, 7371–7380 (2001).
42. A. Licht, S. Preis, S. Brantl, Implication of CcpN in the regulation of a novel untranslated RNA (SR1) in *B. subtilis*, *Mol. Microbiol.*, **58**, 189–206 (2005).
43. C. Pichon, B. Felden, Small RNA genes expressed from *Staphylococcus aureus* genomic and pathogenicity islands with specific expression among pathogenic strains, *Proc. Natl. Acad. Sci. USA*, **102**, 14249–14254 (2005).
44. J. K. Christiansen, J. S. Nielsen, T. Ebersbach, P. Valentin-Hansen, L. Søgaard-Andersen, B. H. Kallipolitis, Identification of small Hfq-binding RNAs in *Listeria monocytogenes*, *RNA*, **12**, 1–14 (2006).
45. P. Mandin, F. Repoila, M. Vergassola, T. Geissmann, P. Cossart, Identification of new noncoding RNAs in *Listeria monocytogenes* and prediction of mRNA targets, *Nucleic Acids Res.*, **35**, 962–974 (2007).
46. E. Massé, S. Gottesman, A small RNA regulates the expression of genes involved in iron metabolism in *Escherichia coli*, *Proc. Natl. Acad. Sci. USA*, **99**, 4620–4625 (2002).

47. V. Pfeiffer, A. Sittka, R. Tomer, K. Tedin, V. Brinkmann, J. Vogel, A small non-coding RNA of the invasion gene island (SPI-1) represses outer membrane protein synthesis from the *Salmonella* core genome, *Mol. Microbiol.*, **66**, 1174–1191 (2007).
48. K. Papenfort, V. Pfeiffer, S. Lucchini, A. Sonawane, J. C. Hinton, J. Vogel, Systematic deletion of *Salmonella* small RNA genes identifies CyaR, a conserved CRP-dependent riboregulator of OmpX synthesis, *Mol. Microbiol.*, **68**, 890–906 (2008).
49. T. Møller, T. Franch, C. Udesen, K. Gerdes, P. Valentin-Hansen, Spot 42 RNA mediates discoordinate expression of the *E. coli* galactose operon, *Genes Dev.*, **16**, 1696–1706 (2002).
50. C. K. Vanderpool, S. Gottesman, Involvement of a novel transcriptional activator and small RNA in post-transcriptional regulation of the glucose phosphoenolpyruvate phosphotransferase system, *Mol. Microbiol.*, **54**, 1076–1089 (2004).
51. C. S. Wadler, C. K. Vanderpool, A dual function for a bacterial small RNA: SgrS performs base-pairing dependent regulation and encodes a functional polypeptide, *Proc. Natl. Acad. Sci. USA*, **104**, 20454–20459 (2007).
52. S. Altuvia, D. Weinstein-Fischer, A. Zhang, L. Postow, G. Storz, A small, stable RNA induced by oxidative stress: role as a pleiotropic regulator and antimutator, *Cell*, **90**, 43–53 (1997).
53. D. D. Sledjeski, A. Gupta, S. Gottesman, The small RNA, DsrA, is essential for the low temperature expression of RpoS during exponential growth in *Escherichia coli*, *EMBO J.*, **15**, 3993–4000 (1996).
54. N. Majdalani, S. Chen, J. Murrow, K. St John, S. Gottesman, Regulation of RpoS by a novel small RNA: the characterization of RprA, *Mol. Microbiol.*, **39**, 1382–1394 (2001).
55. K.C. Tu, B. L. Bassler, Multiple small RNAs act additively to integrate sensory information and control quorum sensing in *Vibrio harveyi*, *Genes Dev.*, **21**, 221–233 (2007).
56. F. Darfeuille, C. Unoson, J. Vogel, E. G. H. Wagner, An antisense RNA inhibits translation by competing with standby ribosomes, *Mol. Cell*, **26**, 381–392 (2007).
57. K. I. Udekwu, F. Darfeuille, J. Vogel, J. Reimegård, E. Holmqvist, E. G. H. Wagner, Hfq-dependent regulation of OmpA synthesis is mediated by an antisense RNA, *Genes Dev.*, **19**, 2355–2366 (2005).
58. A. A. Rasmussen, M. Eriksen, K. Gilany, C. Udesen, T. Franch, C. Petersen, P. Valentin-Hansen, Regulation of *ompA* mRNA stability: the role of a small regulatory RNA in growth phase-dependent control, *Mol. Microbiol.*, **58**, 1421–1429 (2005).
59. J. Johansen, A. A. Rasmussen, M. Overgaard, P. Valentin-Hansen, Conserved small non-coding RNAs that belong to the sigmaE regulon: role in down-regulation of outer membrane proteins, *J. Mol. Biol.*, **364**, 1–8 (2006).
60. K. Papenfort, V. Pfeiffer, F. Mika, S. Lucchini, J. C. Hinton, J. Vogel, SigmaE-dependent small RNAs of *Salmonella* respond to membrane stress by accelerating global *omp* mRNA decay, *Mol. Microbiol.*, **62**, 1674–1688 (2006).
61. C. M. Sharma, F. Darfeuille, T. H. Plantinga, J. Vogel, A small RNA regulates multiple ABC transporter mRNAs by targeting C/A-rich elements inside and upstream of ribosome-binding sites, *Genes Dev.*, **21**, 2804–2817 (2007).
62. E. Morfeldt, D. Taylor, A. von Gabain, S. Arvidson, Activation of alpha-toxin translation in *Staphylococcus aureus* by the *trans*-encoded antisense RNA, RNAIII, *EMBO J.*, **14**, 4569–4577 (1995).
63. M. Mangold, M. Siller, B. Roppenser, B. J. Vlaminckx, T. A. Penfound, R. Klein, R. Novak, R. P. Novick, E. Charpentier, Synthesis of group A streptococcal virulence factors is controlled by a regulatory RNA molecule, *Mol. Microbiol.*, **53**, 1515–1527 (2004).
64. S. Altuvia, A. Zhang, L. Argaman, A. Tiwari, G. Storz, The *Escherichia coli* OxyS regulatory RNA represses *fhlA* translation by blocking ribosome binding, *EMBO J.*, **17**, 6069–6075 (1998).
65. H. Kawamoto, Y. Koide, T. Morita, H. Aiba, Base-pairing requirement for RNA silencing by a bacterial small RNA and acceleration of duplex formation by Hfq, *Mol. Microbiol.*, **61**, 1013–1022 (2006).
66. N. Heidrich, A. Chinali, U. Gerth, S. Brantl, The small untranslated RNA SR1 from the *B. subtilis* genome is involved in the regulation of arginine catabolism, *Mol. Microbiol.*, **62**, 520–536 (2006).
67. N. Heidrich, I. Moll, S. Brantl, *In vitro* analysis of the interaction between the small RNA SR1 and its primary target *ahrC* mRNA, *Nucl. Acids Res.*, **35**, 4331–4346 (2007).
68. E. Huntzinger, S. Boisset, C. Saveanu, Y. Benito, T. Geissmann, A. Namane, G. Lina, J. Etienne, B. Ehresmann, C. Ehresmann, A. Jacquier, F. Vandenesch, P. Romby, *Staphylococcus aures* RNA III and endoribonuclease III coordinately regulate *spa* gene expression, *EMBO J.*, **24**, 824–835 (2005).

69. Y. Benito, F. A. Kolb, P. Romby, G. Lina, J. Etienne, F. Vandenesch, Probing the structure of RNAIII, the *Staphylococcus aureus agr* regulatory RNA and identification of the RNA domain involved in repression of protein A expression, *RNA*, **5**, 6668–6679 (2000).
70. S. Boisset, T. Geissmann, E. Huntzinger, P. Fechter, N. Bendridi, M. Possedko, C. Chevalier, A. C. Helfer, Y. Benito, A. Jacquier, C. Gaspin, F. Vandenesch, P. Romby, *Staphylococcus aureus* RNAIII coordinately represses synthesis of virulence factors and the transcription regulator Rot by an antisense mechanism, *Genes Dev.*, **21**, 1353–1366 (2007).
71. T. Morita, Y. Mochizuki, H. Aiba, Translational repression is sufficient for gene silencing by bacterial small noncoding RNAs in the absence of mRNA destruction, *Proc. Natl. Acad. Sci. USA*, **103**, 4858–4863 (2006).
72. E. Massé, F. E. Escorcia, S. Gottesman, Coupled degradation of a small regulatory RNA and its mRNA targets in *Escherichia coli*, *Genes Dev.*, **17**, 2374–2383 (2003).
73. N. Majdalani, D. Hernandez, S. Gottesman, Regulation and mode of action of the second small RNA activator of RpoS translation, RprA, *Mol. Microbiol.*, **46**, 813–826 (2002).
74. K. Prevost, H. Salvail, G. Desnoyers, J. F. Jacques, E. Phaneuf, E. Massé, The small RNA RyhB activates the translation of *shiA* mRNA encoding a permease of shikimate, a compound involved in siderophore synthesis, *Mol. Microbiol.*, **64**, 1260–1273 (2007).
75. A. Hüttenhofer J. Vogel, Experimental approaches to identify non-coding RNAs, *Nucleic Acids Res.*, **34**, 635–646 (2006).
76. J. Vogel, E. G. H. Wagner, Target identification of small noncoding RNAs in bacteria, *Curr. Opin. Microbiol.*, **10**, 262–270 (2007).
77. R. G. Brennan, T. M. Link, Hfq structure, function and ligand binding, *Curr. Opin. Microbiol.*, **10**, 12–133 (2007).
78. M. T. Franze de Fernandez, L. Eoyang, J. T. August, Factor fraction required for the synthesis of bacteriophage Qβ-RNA, *Nature*, **219**, 588–590 (1968).
79. X. Sun, I. Zhulin, R. M. Wartell, Predicted structure and phyletic distribution of the RNA-binding protein Hfq, *Nucleic Acids Res.*, **30**, 3662–3671 (2002).
80. T. Møller, T. Franch, P. Hojrup, R. K. Douglas, H. P. Bächinger, R. G. Brennan, P. Valentin-Hansen, Hfq: A bacterial Sm-like protein that mediates RNA-RNA interaction, *Mol. Cell*, **9**, 23–30 (2002).
81. P. Valentin-Hansen, M. Eriksen, C. Udesen, The bacterial Sm-like protein Hfq: a key player in RNA transactions, *Mol. Microbiol.*, **51**, 1525–1533 (2004).
82. A. Zhang, K. M. Wassarman, C. Rosenow, B. C. Tjaden, G. Storz, S. Gottesman, Global analysis of small RNA and mRNA targets of Hfq, *Mol. Microbiol.*, **50**, 1111–1124 (2003).
83. B. Vecerek, L. Rajkowitsch, E. Sonnleitner, R. Schroeder, U. Bläsi, The C-terminal domain of *Escherichia coli* Hfq is required for regulation, *Nucleic Acids Res.*, **36**, 133–143 (2007).
84. C. Bohn, C. Rigoulay, P. Bouloc, No detectable effect of RNA-binding protein Hfq absence in *Staphylococcus aureus*, *BMC Microbiol.*, **7**, 10 (2007).
85. J. Tomizawa, T. Som, Control of ColE1 plasmid replication: enhancement of binding of RNAI to the primer transcript by the Rom protein, *Cell*, **38**, 871–878 (1984).
86. D. C. Arthur, A. F. Ghetu, M. J. Gubbins, R. A. Edwards, L. S. Frost, J. N. Glover, FinO is an RNA chaperone that facilitates sense-antisense RNA interactions. *EMBO J.*, **22**, 6346–6355 (2003).
87. L. J. Jerome, T. van Biesen, L. S. Frost, Degradation of FinP antisense RNA from F-like plasmids: the RNA-binding protein, FinO, protects FinP from ribonuclease E, *J. Mol. Biol.* **285**, 1457–1473 (1999).
88. W. R. Will, L. S. Frost, Hfq is a regulator of F-plasmid TraJ and TraM synthesis in *Escherichia coli*, *J. Bacteriol.*, **188**, 124–131 (2006).
89. T. Franch, M. Petersen, E. G. H. Wagner, J. P. Jacobsen, K. Gerdes, Antisense RNA regulation in prokaryotes: rapid RNA/RNA interaction facilitated by a general U-turn-loop structure, *J. Mol. Biol.*, **294**, 1115–1125 (1999).
90. N. Heidrich, S. Brantl, Antisense-RNA mediated transcriptional attenuation: importance of a U-turn loop structure in the target RNA of plasmid pIP501 for efficient inhibition by the antisense-RNA, *J. Mol. Biol.*, **333**, 917–929 (2003).
91. J. H. Urban, J. Vogel, Two seemingly homologous noncoding RNAs act hierarchically to activate *glmS* mRNA translation, *PloS Biol.*, **6**, e64 (2008).
92. B. Reichenbach, A. Maes, F. Kalamorz, E. Hajnsdorf, B. Görke, The small RNA GlmY acts upstream of the sRNA GlmZ in the activation of *glmS* expression and is subject to regulation by polyadenylation in *Escherichia coli*, *Nucleic Acids Res.*, **36**, 2570–2580 (2008).

93. J. Hindley, Fractionation of 32P-labeled ribonucleic acids on polyacrylamide gels and their characterization by fingerprinting, *J. Mol Biol.*, **30**, 125–136 (1967).
94. K. M. Wassarman, G. Storz, 6S RNA regulates *E. coli* RNA polymerase activity, *Cell*, **101**, 613–623 (2000).
95. K. M. Wassarman, 6S RNA, a small RNA regulator of transcription, *Curr. Opin. Microbiol.*, **10**, 164–168 (2007).
96. N. Gildehaus, T. Neusser, R. Wurm, R. Wagner, Studies on the function of the riboregulator 6S RNA from *E. coli*: RNA polymerase binding, inhibition of *in vitro* transcription and synthesis of RNA-directed *de novo* transcripts, *Nucleic Acids Res.*, **35**, 1885–1896 (2007).
97. K. M. Wassarman, R. M. Saecker, Synthesis-mediated release of a small RNA inhibitor of RNA polymerase, *Science*, **314**, 1601–1603 (2006).
98. A. T. Cavanagh, A. D. Klocko, X. Liu, K. M. Wassarman, Promoter specifictiy for 6S RNA regulation of transcription is determined by core promoter sequences and competition for region 4.2 of σ^{70}, *Mol. Microbiol.*, **67**, 242–256 (2008).
99. T. Neusser, N. Gildehaus, R. Wurm, R. Wagner, Studies on the expression of 6S RNA from *E. coli*: involvement of regulators important for stress and growth adaptation, *Biol. Chem.*, **389**, 285–297 (2008).
100. M. Y. Liu, G. Gui, B. Wei, J. F. 3rd Preston, L. Oakford, U. Yüksel, D. P. Giedroc, T. Romeo, The RNA molecule CsrB binds to the global regulatory protein CsrA and antagonizes its activity in *Escherichia coli*, *J. Biol. Chem.*, **272**, 17502–17510 (1997).
101. P. Babitzke, T. Romeo, CsrB sRNA family: sequestration of RNA-binding regulatory proteins, *Curr. Opin. Microbiol.*, **10**, 156–163 (2007).
102. T. Weilbacher, K. Suzuki, A. K. Dubey, X. Wang, S. Gudapaty, I. Morozov, C. S. Baker, D. Georgelis, P. Babitzke, T. Romeo, A novel sRNA component of the carbon storage regulatory system of *Escherichia coli*, *Mol. Microbiol.*, **48**, 657–670 (2003).
103. K. Suzuki, P. Babitzke, S. R. Kushner, T. Romeo, Identification of a novel regulatory protein (CsrD) that targets the global regulatory RNAs CsrB and CsrC for degradation by RNase E, *Genes Dev.*, **20**, 2605–25617 (2006).
104. K. Lapouge, M. Schubert, F. H-T. Allain, D. Haas, Gac/Rsm signal transduction pathway of γ-proteobacteria: from RNA recognition to regulation of social behaviour, *Mol. Microbiol.*, **67**, 241–253 (2008).
105. A. K. Heroven, K. Böhme, M. Rohde, P. Dersch, A Csr-type regulatory system, including small non-coding RNAs, regulates the global virulence regulator RovA of *Yersinia pseudotuberculosis* through RovM, *Mol. Microbiol.*, **68**, 1179–1195 (2008).
106. E. L. Chant, D. K. Summers, Indole signalling contributes to the stable maintenance of *Escherichia coli* multicopy plasmids, *Mol. Microbiol.*, **63**, 35–43 (2007).

10

MicroRNA-guided Gene Silencing

Gunter Meister

10.1 Introduction

MicroRNAs (miRNAs) are small non-coding RNAs that play fundamental roles in diverse biological processes. After the genetic discovery in *Caenorhabditis elegans* by the groups of Ambros and Ruvkun [1,2], miRNAs have now been identified in many different organisms ranging from single-cellular algae to mammals. Even some DNA viruses express specific sets of miRNAs. MiRNAs have inhibitory functions on the level of mRNA translation and/or stability and regulate the expression of distinct genes. To date, more than 1000 miRNAs have been identified in humans and it has been estimated that more than 30% of all human genes are regulated by miRNAs. Based on these estimations, miRNAs can be viewed as key-regulators of gene expression.

10.2 miRNA biogenesis

In contrast to other small non-coding RNAs, miRNAs are endogenously expressed from distinct genomic loci. MiRNA genes can be located within introns or untranslated regions (UTRs) of other genes or within intergenic regions. Furthermore, several miRNA genes are very often clustered and expressed as a single transcript (Figure 10.1). Such transcripts are referred to as primary miRNA transcripts (pri-miRNAs) and are like mRNAs capped at their 5′ ends and polyadenylated at their 3′ ends [3,4]. In the nucleus, primary miRNA transcripts are processed to miRNA precursors or pre-miRNAs by the RNase III enzyme Drosha as part of a multi-protein complex termed microprocessor [5]. Pre-miRNAs are stem–loop structured and contain 5′ phosphate groups in addition to two nucleotide 3′ overhangs typical for RNase III processing. Pre-miRNAs are subsequently transported to the cytoplasm by employing the export receptor exportin-5, where the RNase III enzyme Dicer cleaves off the loop of the hairpin and generates a double-stranded (ds) small RNA intermediate [6]. Such ds, 20–25 nucleotides (nt) long, small RNA molecules are subsequently unwound and one strand gives rise to the mature miRNA. The other strand, which is referred to as miRNA* (miRNA

The Chemical Biology of Nucleic Acids Edited by Günter Mayer

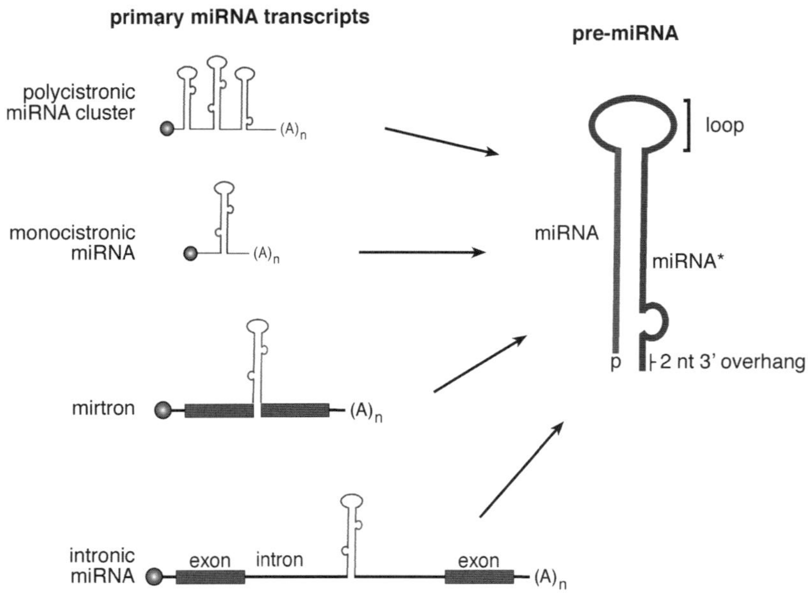

Figure 10.1 Architecture of primary miRNA transcripts and miRNA precursors (pre-miRNAs). MiRNAs can be transcribed from polycistronic miRNA clusters that contain multiple miRNA genes or from a single miRNA gene (monocistronic). Many miRNA genes, however, are found in introns of protein coding genes. Mirtrons form an entire intron without any extra intronic sequences (for details, see text). Pre-miRNAs are characterized by a stem–loop structure. One strand of the stem gives rise to the mature miRNA. The opposite strand forms the miRNA* sequence, which is unstable. Pre-miRNAs are further characterized by 5′ phosphates (p) and 2 nt 3′ overhangs. $(A)_n$, polyA tail. See color plate section

star), is rapidly degraded (Figure 10.2). Selection of the right miRNA strand is very important for gene regulation and is therefore highly coordinated and regulated. Generally, miRNA strand selection follows so called 'asymmetry rules' and the miRNA strand with the less stably paired 5′ end is preferentially selected and incorporated into functional miRNA protein complexes often referred to as micro-ribonucleoprotein particles or miRNPs [7,8]. The asymmetry rules are also applicable for siRNA strand selection and are now the basis for most siRNA design algorithms.

Alternative miRNA biogenesis pathways have been reported in vertebrates and invertebrates (Figures 10.1 and 10.2). Unusual short introns about 70 nt in length with conserved 5′ and 3′ splice sites have been found. Like any other intron, such short introns are spliced by the spliceosome but the released introns fold back into hairpin structures and are not further degraded. Instead, the miRNA machinery recognizes the spliced out intron as pre-miRNA and transports it into the cytoplasm where Dicer further processes it to mature miRNAs. Processing of such miRNAs is independent of Drosha and, due to their intronic origin, such miRNAs have been named 'mirtrons' [9,10].

10.3 Argonaute proteins

Argonaute (Ago) proteins were originally identified in the model plant *Arabidopsis thaliana* and named after the squid-like phenotype of plants lacking functional Ago proteins. Extensive research in recent

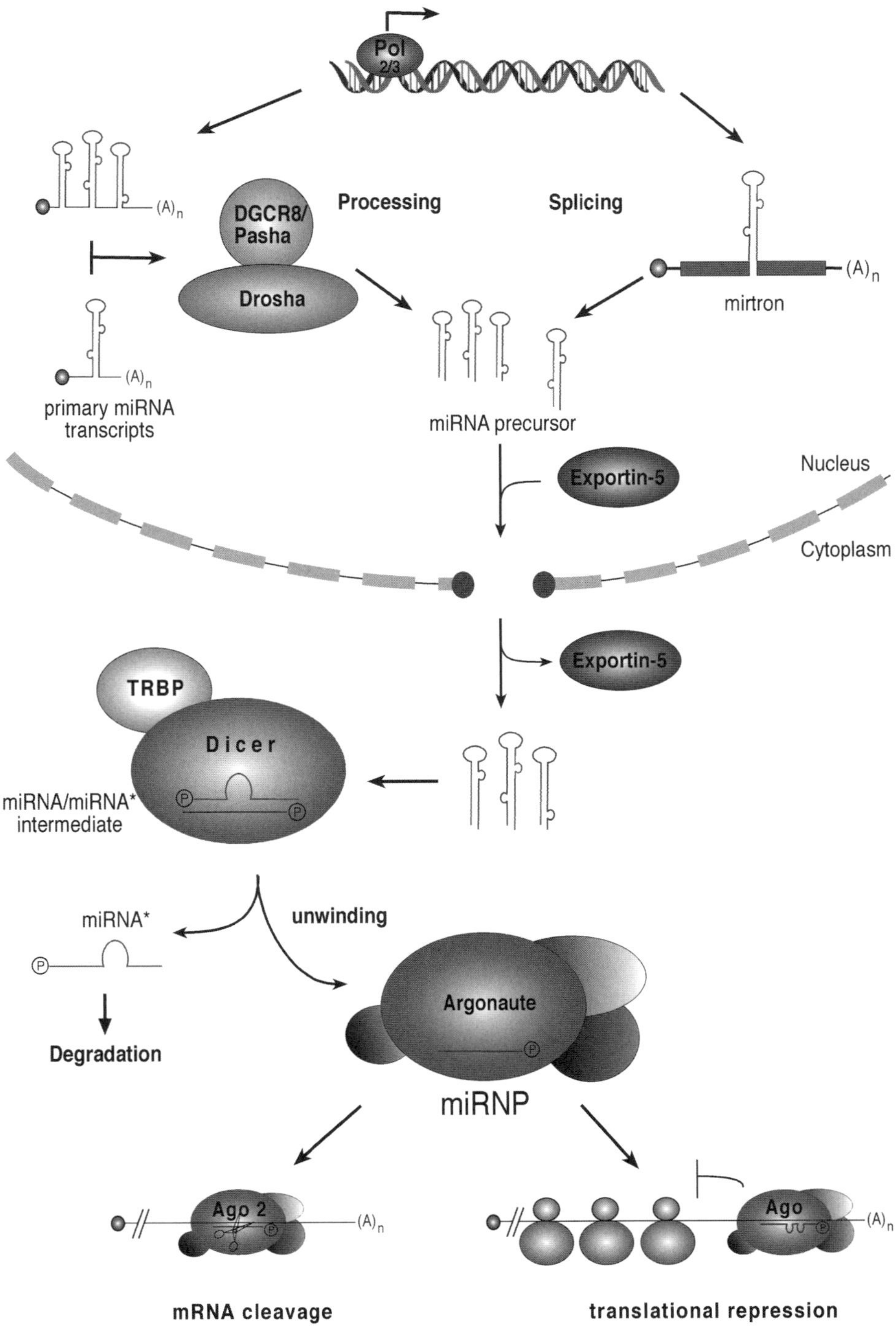

Figure 10.2 *Schematic representation of miRNA biogenesis and function. MiRNA genes are transcribed as primary miRNA transcripts by RNA polymerases II and III and are further processed to pre-miRNAs by the RNase III enzyme Drosha and its co-factor DGCR8/Pasha. The export receptor exportin-5 transports pre-miRNAs to the*

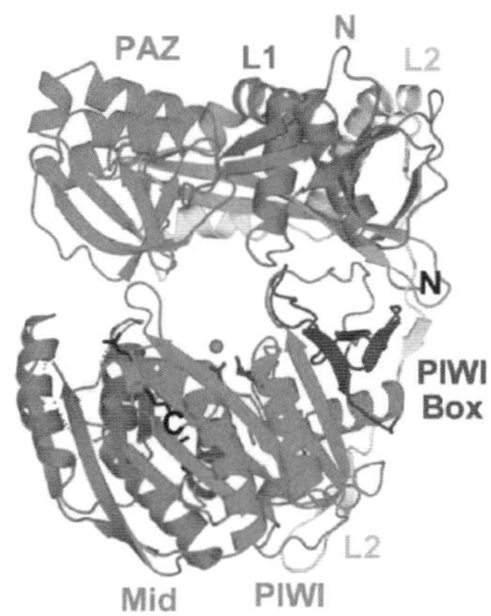

Figure 10.3 Architecture of the Aquifex aeolicus Argonaute protein. An N-terminal domain with so far unknown functions is linked to the PAZ domain by linker L1. The PAZ domain interacts with the 3′ end of miRNAs. Linker L2 connects the PAZ domain to the MID domain, which anchors the 5′ end of the miRNA. The MID domain is followed by the RNase H-like PIWI domain. See color plate section

years revealed that small RNAs are just guides that target Ago proteins to distinct RNAs. Ago proteins are believed to be the actual mediators of miRNA function and are therefore key factors in small RNA-guided gene silencing pathways [11,12]. Ago proteins are characterized by PAZ [P-element-induced wimpy testes (PIWI)–Argonaute–Zwille)] and PIWI domains and are found in a variety of different species ranging from Archaea to humans. Ago proteins specifically bind single-stranded small RNAs including miRNAs by anchoring the 5′ end into a specific binding pocket within the MID domain located adjacent to the PIWI domain. The PAZ domain tightly binds the 3′ end of the miRNA and may play a key role in miRNA loading by recognizing the 2 nt 3′ overhang structure of ds miRNA intermediates. The PIWI domain of Ago proteins is structurally highly similar to RNase H, a nuclease that typically cleaves RNA in DNA–RNA hybrids (Figure 10.3). Indeed, biochemical studies have shown that Ago proteins cleave target RNAs that are fully complementary to bound miRNAs and Ago proteins with nuclease activity have therefore been named 'slicers' [13–15]. Most animal miRNAs are only partially complementary to their targets and slicer-mediated cleavage may play a minor role for miRNA-guided gene silencing.

***Figure 10.2** (Continued)* cytoplasm, where the RNase III enzyme Dicer cleaves off the loop and generates a double-stranded short RNA. Such short intermediate RNA duplexes are rapidly unwound and one strand is incorporated into protein complexes (miRNPs) as the mature miRNA. Key components of such miRNA 'effector complexes' are members of the Argonaute protein family. In plants, and very rarely in animals, miRNAs hybridize to highly complementary target sequences and guide its sequence-specific cleavage in a process identical with RNA interference (RNAi). Many miRNAs, however, bind to only partially complementary target sites typically located in the 3′ untranslated region of distinct target mRNAs. Such imperfect pairing leads either to translational repression or de-adenylation followed by degradation of the mRNA. See color plate section

10.4 Mechanisms of miRNA-guided gene silencing

In plants, many miRNAs hybridize to nearly perfectly complementary target sites within target mRNAs and induce their sequence-specific cleavage. In animal organisms, however, miRNAs bind to partially complementary target sites, which are preferentially located in the 3′ UTR of target mRNAs [3]. Early functional genetics approaches in *C. elegans* revealed that the miRNA lin-4 is required for accurate developmental timing by targeting the 3′ UTR of the lin-14 mRNA. The intriguing finding, however, was that lin-14 mRNA levels remained unchanged, whereas protein levels dropped dramatically when lin-4 was present [1,2]. Based on these studies, it rapidly became apparent that miRNAs hybridize to partially complementary binding sites on target mRNAs and represses their translation. The molecular mechanism underlying miRNA-guided translational repression has been studied in detail and it has been suggested that miRNAs might employ multiple mechanisms to facilitate gene silencing [16]. In addition to cleavage of perfectly complementary target mRNAs, which is a rare event in animals but very common in plants, miRNAs can induce unspecific degradation of target mRNAs by recruiting protein components of the deadenylation and also the decapping machinery (Figure 10.2).

The molecular mechanisms of miRNA-guided translational repression are not fully understood. Initially, it was demonstrated that miRNAs co-sediment with polyribosomes in sucrose density gradients, suggesting that miRNAs inhibit translation after initiation of translation. Later, it was shown for other targets that miRNAs can also inhibit translational initiation by preventing ribosome assembly on the target mRNA [16]. It is reasonable that miRNAs can function differently depending on the mRNA target, which might even be influenced by RNA binding proteins that are associated with distinct mRNA targets.

It is now well established that miRNAs form large regulatory networks with probably more than 100 mRNA targets for a single miRNA [17]. It has been suggested that miRNAs fine-tune expression levels of target genes because miRNAs and at least some of their targets are frequently co-expressed. However, it has also been shown that miRNAs can function as regulatory switches during development. In *C. elegans*, for example, miRNAs are important for correct developmental timing and mutations in miRNA genes result in severe developmental defects. Individual functions of miRNAs are therefore diverse and are influenced by developmental processes in addition to cellular conditions and environments.

10.5 Subcellular localization of miRNAs

Intracellular localization studies revealed that Ago proteins in addition to miRNAs localize to the diffuse cytoplasm but are also enriched in cytoplasmic foci termed cytoplasmic processing bodies or P-bodies (Figure 10.4) [18]. P-bodies exist in different sizes and are most like protein–RNA aggregates that are involved in RNA turnover or regulation of gene expression [19]. Their precise molecular functions, however, are still elusive. Biochemical and genetic studies in different organisms have shown that Ago proteins interact with GW182, a protein that also localizes to P-bodies. In humans, three paralogs of GW182, namely trinucleotide repeats containing 6A–C (TNRC6A–C), exist and it has been shown that TNRC6A and -B are involved in miRNA function [12]. Experimental studies revealed that GW182 family proteins interact directly with Ago proteins and represent another family of key proteins in miRNA-guided gene silencing with functions in mRNA destabilization and translational repression. Deadenylation and decapping processes in miRNA-guided gene silencing take place may at least partially in P-bodies.

In liver cells, miR-122 targets the CAT-1 mRNA to P-bodies. The CAT-1 protein is required for stress response in liver cells and expressed upon cellular stress. It has been demonstrated that the CAT-1 mRNA is not degraded in P-bodies but stored and translationally repressed by miR-122 under stress-free conditions. However, upon cellular stress the miRNA inhibition is released and the CAT-1 mRNA relocalizes to the

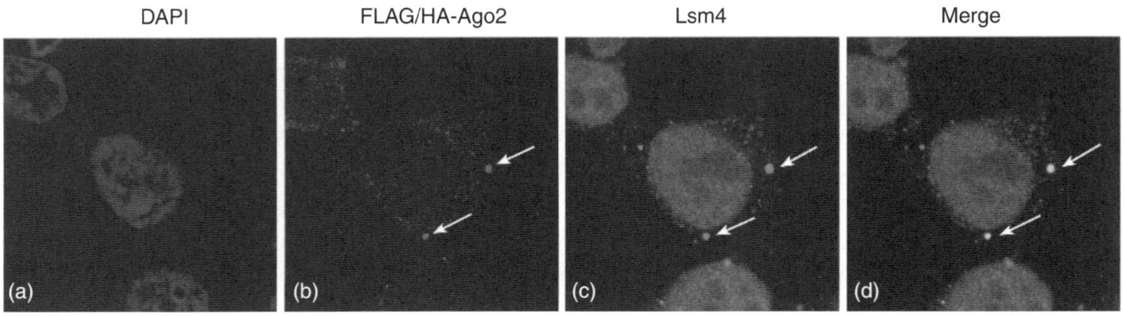

Figure 10.4 Localization of Argonaute proteins in human cells revealed by indirect immunofluorescence. Cells were treated with chromophore-conjugated antibodies directed against human Ago2 (b) or LSm4 (c), a marker protein that localizes to cytoplasmic processing bodies (P-bodies). The dot-like cytoplasmic structure represents P-bodies. The nuclei of the cells are stained with DAPI (a, blue), a dye that intercalates with DNA. Panel (d) shows merged panels (b) and (c) and demonstrates that Ago2 is present in P-bodies. See color plate section

diffuse cytoplasm where it is actively translated [20]. Therefore, P-bodies can also function as storage pools for mRNAs where miRNA hold them in translationally repressed states.

10.6 Viral miRNAs

Soon after the discovery of mammalian miRNAs, viruses were also analyzed for miRNA expression [21]. Indeed, DNA viruses of the herpes virus family including Epstein–Barr virus (EBV), Kaposi's sarcoma virus (KSHV) and human cytomegalovirus (HCMV) [22,23] encode and express miRNAs that are easily detectable in infected cells. The individual functions of the majority of the viral miRNAs are still unknown. However, some viral miRNAs have been functionally characterized in more detail. For example, HCMV miR-UL112 targets genes of the host immune system leading to reduced killing activity of natural killer cells [24]. Moreover, simian virus 40 (SV40) expresses a miRNA that regulates viral gene expression to reduce susceptibility to cytotoxic T-cells [25]. It has also been reported that KSHV miR-K12-11 functions as an ortholog of the cellular miR-155 and may therefore exploit pre-existing pathways in B-cells [26].

10.7 miRNAs and cancer

MiRNAs have been implicated in a variety of diseases, including cancer. Initially, it was reported that miR-15 and miR-16 expression is frequently downregulated or even deleted in B-cell chronic lymphocytic leukemia (B-CLL) [27]. To date, a variety of different types of cancer have been analyzed for miRNA expression [28,29]. In many cases, distinct miRNA profiles have been reported. Moreover, oncogenes such as RAS or BCL2 are under the control of the miRNA pathway and alterations in miRNA expression result in upregulation of these oncogenes [30,31]. MiRNAs therefore very often function as tumor suppressors. Reciprocally, miRNAs have also been reported to function as oncogenes by targeting tumor suppressors [28,29,32].

The miR-17–92 cluster, a miRNA gene cluster containing six individual miRNAs, has oncogenic activity and is frequently amplified in several types of lymphoma and solid tumors. Expression profiling studies that measured the expression of miRNA genes in different tumor types revealed that the miR-17–92 cluster is overexpressed in hematopoietic malignancies (tumors of the blood system) or solid tumors derived from

Table 10.1 *Selected miRNAs that have been associated with cancer*

miRNA	Type of cancer	Reference
Let-7 family	Regulates the ras oncogenes in lung cancer	[30]
miR-15 and miR-16	B cell chronic lymphocytic leukemia (B-CLL)	[27]
miR-21	Glioblastoma	[37]
miR-155	Burkitt and Hodgkin lymphomas, B cell lymphomas	[28,29]
miR-372 and miR-373	Testicular germ cell tumors	[32]
miR-10b	Metastasis, breast cancer	[38]
miR-17–92 cluster	Hepatocellular carcinomas, B cell lymphomas	[39,40]
miR-373, miR-520c, miR-200 family	Tumor migration, metastasis	[41,42]

breast, colon, lung, pancreas, prostate and stomach (well-characterized cancer-associated miRNAs are summarized in Table 10.1) [33].

Although the existence of mammalian miRNAs has just been reported, it is becoming more and more apparent that they play important roles in cancer. This discovery, however, opens new doors for cancer therapies. A variety of strategies for efficient miRNA inhibition or administration *in vivo* are being tested and the first results are very encouraging.

10.8 Strategies for miRNA inhibition

The identification of viral miRNAs and also the discovery that miRNAs have distinct roles in cancer pathogenesis led to the development of inhibition strategies that aim at interfering with miRNA function. Sequence-specific antisense inhibitors proved to be advantageous since individual miRNAs can be targeted specifically. Major obstacles for such single-stranded antisense oligonucleotides are stability within the cell or organisms and the binding affinity to the endogenous miRNA molecule. To increase the stability of the inhibitors and also the binding affinity to the miRNA, 2′-O-methylated (2′-OMe) nucleotides have been used for miRNA inhibition (Figure 10.5) [34]. Such inhibitors can be transfected similarly to siRNAs into transfectable cell lines and efficiently block the function of the targeted miRNA. 2′-OMe inhibitors have been further modified and sulfur groups have been introduced into the phosphate backbone of the inhibitor molecule. Furthermore, a cholesterol group has been added to the 3′ end in order to ensure cellular uptake. Such inhibitors have been termed ‘antagomirs’ and injection into the blood system of mice resulted in efficient sequence-specific inhibition of a distinct miRNA in various tissues [35].

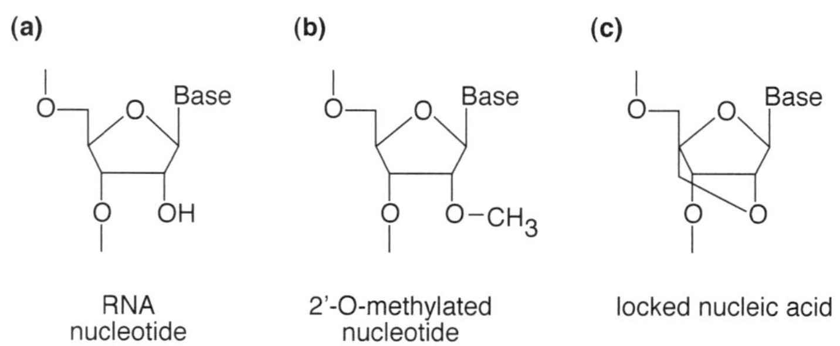

Figure 10.5 *Nucleic acid modifications that are most commonly used for miRNA inhibition. (a) An RNA nucleotide; (b) a 2′-O-methylated RNA nucleotide; (c) a locked nucleic acid*

Another nucleotide modification that is frequently used for miRNA inhibition is locked nucleic acids (LNAs) (Figure 10.5). LNA-modified oligonucleotides show a very high affinity to RNA molecules and form highly stable duplexes. *In vivo* applications of LNA-modified anti-miRNA molecules have been very promising. Injection of an LNA-modified oligonucleotide complementary to the liver-specific miR-122 efficiently inhibited the miRNA even without additional modifications such as cholesterol residues [36].

Based on such studies, it is becoming more and more apparent that inhibition of miRNAs *in vivo* is possible and we will see the advent of new modified oligonucleotide-based drugs against various diseases in the future.

References

1. Lee, R.C., Feinbaum, R.L. and Ambros, V. The *C. elegans* heterochronic gene lin-4 encodes small RNAs with antisense complementarity to lin-14. *Cell* **75**, 843–54 (1993).
2. Wightman, B., Ha, I. and Ruvkun, G. Posttranscriptional regulation of the heterochronic gene *lin-14* by *lin-4* mediates temporal pattern formation in *C. elegans*. *Cell* **75**, 855–62 (1993).
3. Bushati, N. and Cohen, S.M. microRNA functions. *Annu Rev Cell Dev Biol* **23**, 175–205 (2007).
4. Chen, P.Y. and Meister, G. microRNA-guided posttranscriptional gene regulation. *Biol Chem* **386**, 1205–18 (2005).
5. Tomari, Y. and Zamore, P.D. Perspective: machines for RNAi. *Genes Dev* **19**, 517–29 (2005).
6. Meister, G. and Tuschl, T. Mechanisms of gene silencing by double-stranded RNA. *Nature* **431**, 343–9 (2004).
7. Khvorova, A., Reynolds, A. and Jayasena, S.D. Functional siRNAs and miRNAs exhibit strand bias. *Cell* **115**, 209–16 (2003).
8. Schwarz, D.S. *et al.* Asymmetry in the assembly of the RNAi enzyme complex. *Cell* **115**, 199–208 (2003).
9. Okamura, K., Hagen, J.W., Duan, H., Tyler, D.M. and Lai, E.C. The mirtron pathway generates microRNA-class regulatory RNAs in *Drosophila*. *Cell* **130**, 89–100 (2007).
10. Ruby, J.G., Jan, C.H. and Bartel, D.P. Intronic microRNA precursors that bypass Drosha processing. *Nature* **448**, 83–6 (2007).
11. Hutvagner, G. and Simard, M.J. Argonaute proteins: key players in RNA silencing. *Nat Rev Mol Cell Biol* **9**, 22–32 (2008).
12. Peters, L. and Meister, G. Argonaute proteins: mediators of RNA silencing. *Mol Cell* **26**, 611–23 (2007).
13. Parker, J.S. and Barford, D. Argonaute: a scaffold for the function of short regulatory RNAs. *Trends Biochem Sci* **31**, 622–30 (2006).
14. Patel, D.J. *et al.* Structural biology of RNA silencing and its functional implications. *Cold Spring Harbor Symp Quant Biol* **71**, 81–93 (2006).
15. Tolia, N.H. and Joshua-Tor, L. Slicer and the argonautes. *Nat Chem Biol* **3**, 36–43 (2007).
16. Pillai, R.S., Bhattacharyya, S.N. and Filipowicz, W. Repression of protein synthesis by miRNAs: how many mechanisms? *Trends Cell Biol* **17**, 118–26 (2007).
17. Chen, K. and Rajewsky, N. The evolution of gene regulation by transcription factors and microRNAs. *Nat Rev Genet* **8**, 93–103 (2007).
18. Sen, G.L. and Blau, H.M. Argonaute 2/RISC resides in sites of mammalian mRNA decay known as cytoplasmic bodies. *Nat Cell Biol* **7**, 633–636 (2005).
19. Eulalio, A., Behm-Ansmant, I. and Izaurralde, E. P bodies: at the crossroads of post-transcriptional pathways. *Nat Rev Mol Cell Biol* **8**, 9–22 (2007).
20. Bhattacharyya, S.N., Habermacher, R., Martine, U., Closs, E.I. and Filipowicz, W. Relief of microRNA-mediated translational repression in human cells subjected to stress. *Cell* **125**, 1111–24 (2006).
21. Pfeffer, S. *et al.* Identification of virus-encoded microRNAs. *Science* **304**, 734–6 (2004).
22. Cullen, B.R. Viruses and microRNAs. *Nat Genet* **38** Suppl, S25–30 (2006).
23. Pfeffer, S. and Voinnet, O. Viruses, microRNAs and cancer. *Oncogene* **25**, 6211–9 (2006).
24. Stern-Ginossar, N. *et al.* Host immune system gene targeting by a viral miRNA. *Science* **317**, 376–81 (2007).
25. Sullivan, C.S., Grundhoff, A.T., Tevethia, S., Pipas, J.M. and Ganem, D. SV40-encoded microRNAs regulate viral gene expression and reduce susceptibility to cytotoxic T cells. *Nature* **435**, 682–6 (2005).

26. Gottwein, E. *et al.* A viral microRNA functions as an orthologue of cellular miR-155. *Nature* **450**, 1096–9 (2007).
27. Calin, G.A. *et al.* Frequent deletions and down-regulation of microRNA genes miR-15 and miR-16 at 13q14 in chronic lymphocytic leukemia. *Proc Natl Acad Sci USA* **99**, 15524–9 (2002).
28. Calin, G.A. and Croce, C.M. MicroRNA signatures in human cancers. *Nat Rev Cancer* **6**, 857–66 (2006).
29. Esquela-Kerscher, A. and Slack, F.J. Oncomirs – microRNAs with a role in cancer. *Nat Rev Cancer* **6**, 259–69 (2006).
30. Johnson, S.M. *et al.* RAS is regulated by the let-7 microRNA family. *Cell* **120**, 635–47 (2005).
31. Cimmino, A. *et al.* miR-15 and miR-16 induce apoptosis by targeting BCL2. *Proc Natl Acad Sci USA* **102**, 13944–9 (2005).
32. Voorhoeve, P.M. *et al.* A genetic screen implicates miRNA-372 and miRNA-373 as oncogenes in testicular germ cell tumors. *Cell* **124**, 1169–81 (2006).
33. Mendell, J.T. miRiad roles for the miR-17–92 cluster in development and disease. *Cell* **133**, 217–22 (2008).
34. Meister, G., Landthaler, M., Dorsett, Y. and Tuschl, T. Sequence-specific inhibition of microRNA- and siRNA-induced RNA silencing. *RNA* **10**, 544–50 (2004).
35. Krutzfeldt, J. *et al.* Silencing of microRNAs *in vivo* with 'antagomirs'. *Nature* **438**, 685–9 (2005).
36. Elmen, J. *et al.* LNA-mediated microRNA silencing in non-human primates. *Nature* **452**, 896–9 (2008).
37. Chan, J.A., Krichevsky, A.M. and Kosik, K.S. MicroRNA-21 is an antiapoptotic factor in human glioblastoma cells. *Cancer Res* **65**, 6029–33 (2005).
38. Ma, L., Teruya-Feldstein, J. and Weinberg, R.A. Tumour invasion and metastasis initiated by microRNA-10b in breast cancer. *Nature* **449**, 682–8 (2007).
39. He, L. *et al.* A microRNA polycistron as a potential human oncogene. *Nature* **435**, 828–33 (2005).
40. O'Donnell, K.A., Wentzel, E.A., Zeller, K.I., Dang, C.V. and Mendell, J.T. c-Myc-regulated microRNAs modulate E2F1 expression. *Nature* **435**, 839–43 (2005).
41. Gregory, P.A. *et al.* The miR-200 family and miR-205 regulate epithelial to mesenchymal transition by targeting ZEB1 and SIP1. *Nat Cell Biol* **10**, 593–601 (2008).
42. Huang, Q. *et al.* The microRNAs miR-373 and miR-520c promote tumour invasion and metastasis. *Nat Cell Biol* **10**, 202–10 (2008).

11

Nucleic Acid-based Therapies

Britta Hoehn and John J. Rossi

11.1 Introduction

The completion of the sequence of the whole human genome has provided an estimated number of 20 000–25 000 genes and has led to an enhanced understanding of gene function and disease pathogenesis. In the last few years, high-throughput screens and microarray analyses have provided an accumulation of huge amounts of information about the relationship between genes and proteins. Moreover, these analyses have given new insights into the complex networks of gene interactions and mechanisms through which uncoupling of these interactions can lead to diseases such as cancer. Hence the role of mRNAs as targets for therapeutic interventions has gained a new significance.

The past decades have seen the rapid evolution of oligonucleotide-based gene-silencing strategies. Several molecules have been examined in the setting of clinical trials, resulting recently in the successful transition from the bench to the bedside. Oligonucleotides present themselves as powerful tools as they target specifically the complementary mRNA sequence of the gene of interest, thereby interfering with the flow of the genetic information. Different mechanisms of action, which can be modulated by the chosen target sequence and by chemical modifications of the oligonucleotide, present a variety of ways to inhibit gene expression by interfering with expression of the genetic information encoded within the mRNAs. Strategies targeting mRNA include antisense oligonucleotides, ribozymes, DNAzymes and, more recently, RNA interference.

In addition to blocking mRNA function, oligonucleotide-based therapies encompass a wide range of properties beyond the interaction with mRNA. Aptamers or decoys are selected *in vitro* for high affinity and specificity of interactions with their target molecules. These oligonucleotides are able to bind to proteins and inhibit their function. Aptamers have also been shown to form potent delivery tools for other oligonucleotide therapeutics such as siRNAs. Furthermore, a new class of immunostimulatory oligonucleotides, which interact with Toll-like receptors to induce an immune response, have been utilized as adjuvants for vaccines in the treatment of infectious diseases and cancer. Oligonucleotides themselves can additionally be employed as vaccines. Plasmid DNA or messenger RNA coding for antigens can be expressed *in vivo* after injection,

The Chemical Biology of Nucleic Acids Edited by Günter Mayer

triggering the development of a humoral and cellular immune response specific for the antigen. Moreover, DNA strategies have been developed to repair specific mutations in the genome and also to correct faulty splicing. Most recently, miRNAs have become potential drug candidates and are showing their capability as biomarkers or targets for disease therapy.

Oligonucleotides have emerged as a promising class of biopharmaceuticals and the field has grown rapidly with new and exciting discoveries every year. The relatively short time frames for screening of effective oligonucleotides compared with conventional drugs and the fast progression of these molecules from bench to therapeutic applications have created optimism for these nucleic acid-based drugs for treating a wide range of diseases. Understanding of the basic mechanisms of oligonucleotide-based therapeutics has resulted in the development of effector molecules with improved potency and efficacy. The first US Food and Drug Administration (FDA)-approved oligonucleotide, an antisense oligodeoxynucleotide and an aptamer have paved the way for oligonucleotides as therapeutic agents. Nevertheless, in spite of the large number of oligonucleotides which are under investigation for potential clinical application, there are some drawbacks, which have to be overcome to meet the promise of rapid development for therapeutics. Key challenges such as the stability of the nucleic acid reagent *in vivo*, efficient and specific delivery to certain tissue or cell types, off-target effects and toxicity hampered the successful outcome of clinical trials in many cases and need to be solved in the future. However, solutions to these problems are quickly being developed and will be incorporated in upcoming designs and strategies.

Table 11.1 summarizes the different classes of oligonucleotides that present themselves as promising candidates for therapeutics. Because of the complexity of the various approaches, only a few of these oligonucleotides will be examined in detail. We will explain the different mechanisms of action and address the advantages and disadvantages of oligonucleotide-based therapeutics, leading into a discussion of the challenges for their successful use as therapeutic compounds. Finally, the application of oligonucleotides in therapy and the outcome of clinical trials will be briefly reviewed (Table 11.2).

11.2 Antisense oligonucleotides

Antisense oligonucleotides (AONs) are single-stranded DNA molecules of 15–25 nucleotides in length that are complementary to their target mRNA and form RNA–DNA hybrids that either sterically block translation or trigger RNase H cleavage [26,27] (Figure 11.1). The mechanism of action is dependent on the target sequence site and the backbone modification of the AON. AON-based post-transcriptional gene silencing has had the longest time frame for clinical development. During the late 1970s, Paterson *et al.* were the first to demonstrate antisense mediated gene inhibition via the use of DNA oligonucleotides [28]. Shortly thereafter, Zamecnik and Stephenson demonstrated the potential of small, single-stranded DNA molecules to act as antisense agents by inhibition of the replication of the Rous sarcoma virus in cell culture [29,30]. Since then, antisense technology has been widely used in target validation and has been tested as a therapeutic modality.

Several considerations have to be taken into account in the design of AONs for therapeutic applications. First, accessible target sites have to be identified in the mRNA of question to avoid complex secondary and tertiary structures or bound protein interaction sites. Experimental- and computer-based approaches have been developed in the past to support the choice of the best available sequence site. Second, certain sequence motifs should be avoided that are known to trigger Toll-like receptor innate immune responses such as cytosine–phosphate–guanine (CpG) motifs. Third, stretches of guanosine residues should be excluded in the oligonucleotide designs because they can form G-quartets via Hoogsteen base-pair formation and reduce the affinity of the AONs for the mRNA target. The formation of other secondary motifs in the sequence or intermolecular base pairing should also be avoided. Finally, database searches for significant homology with

Table 11.1 *Overview of oligonucleotide-based therapeutics*

Oligonucleotide	Type	Mechanism	Advantages	Challenge
Antisense oligonucleotide (AON)	ssDNA	Complementary sequence binds to target RNA in nucleus, degradation of target RNA by RNase H, blocking of translation	Specific silencing of gene expression	High levels of antisense RNA necessary, specific delivery
Ribozyme, DNAzyme	ssRNA, ssDNA	*trans*-Cleavage of target RNA leading to translational inhibition, *trans*-Splicing leading to repair of mutant RNA	Specific, autocatalytic	Efficient specific delivery, long-term inhibition was difficult, target specificity was concern for *trans*-splicing
siRNA	dsRNA, 21-mer or Dicer substrates	Complementary sequence in RISC complex binds to target mRNA leading to degradation of target RNA	Specific silencing of gene expression, using natural cell machinery, very potent	Off-target effects, toxicity, cell-specific delivery
Aptamer, decoy	ssRNA, ssDNA	RNA or DNA aptamers bind to proteins and inhibit their activity or act as antagonists	High specificity and affinity, antidote, not immunogenic	Systemic cell-specific delivery
Immunostimulatory oligonucleotides	dsRNA, ssRNA, CpG motifs	Activate the immune response through TLR9	Vaccine or vaccine adjuvant, T-cell activation	Toxicity, potential induction of autoimmune diseases
Triplex-forming oligonucleotides (TFOs) [1,2]	ssDNA	Binding to the major groove of DNA, transcription inhibition, guided homologous recombination, mutation repair	Maintenance of the corrected gene, correct gain-of-function mutations	Low correction frequency, purine- or pyrimidine-rich sequences

(*continued overleaf*)

Table 11.1 *(Continued)*

Oligonucleotide	Type	Mechanism	Advantages	Challenge
Small DNA fragment (SDF) [3,4]	ssDNA, dsDNA	Small fragment homologous replacement (SFHR)	Maintenance of the corrected gene, correct gain-of-function mutations	Low correction frequency, delivery *in vivo*
Single-stranded oligonucleotides (SSOs) [5,6]	ssDNA	Mismatch repair by homologous recombination	Maintenance of the corrected gene, correct gain-of-function mutations	Low correction frequency, delivery *in vivo*
Chimeric RNA–DNA oligonucleotides (RDOs) [7]	RNA–DNA double hairpin	Mismatch repair by homologous recombination	Maintenance of the corrected gene, correct gain-of-function mutations	Low correction frequency, delivery *in vivo*
Pre-*trans*-splicing molecule (PTM) [8,9]	ssDNA	Spliceosome-mediated RNA *trans*-splicing (SMaRT)	Use of the cellular machinery, delivery of molecules	Toxicity, long-term effect? delivery *in vivo*
Anti-miRNA AON	ssDNA	Complementary sequence binds to miRNA and inhibits function	Targeting subset of proteins	Cell-specific delivery, toxicity
miRNA-mimic	dsRNA	Restores miRNA function	Targeting subset of proteins	Cell-specific delivery, toxicity
Antigen-coding mRNA [10,11]	ssRNA	Tumor vaccination, T-cell-mediated cancer immunotherapy	Coding of the whole antigen, no gene therapy	Local injection, fast degradation or adaptive cell transfer

Table 11.2 *Selected examples for oligonucleotide-based therapeutics in clinical trials*

Compound	Category	Target/disease	Company	Status	Ref.
Vitravene (fomivirsen)	AON	CMV IE2/CMV retinitis	ISIS Pharmaceuticals	Approved as drug	[12,13]
Affinitac/aprinocarsen (ISIS 3521)	AON	PKC /solid tumors	ISIS Pharmaceuticals	Phase III	[14]
Alicaforsen (ISIS 2302)	AON	ICAM-1/psoriasis, Crohn's disease, ulcerative colitis	ISIS Pharmaceuticals	Phase II/III	http://www.isispharm.com
Genasense (G3139)	AON	Bcl-2/solid tumors	Genta	Phase II/III	http://www.genta.com
LY-2181308	AON	Survivin/cancer	ISIS with Lilly	Phase I	http://www.isispharm.com
ASM8	AON	Cellular receptors/asthma	Topigen	Phase I	http://www.topigen.com
Resten-MP (AVI-4126)	AON	c-myc/cardiovascular restenosis	AVI Biopharma	Phase II	http://www.avibio.com
AV-4557	AON	Cytochrome P450/metabolism	AVI Biopharma	Phase I	http://www.avibio.com
Resten-NG (AVI-4126)	AON	c-myc/restenosis, cancer, polycystic kidney disease	AVI Biopharma	Phase II/III	http://www.avibio.com
GTI 2040	AON	Ribonucleotide reductase R2/AML	Lorus Therapeutics	Phase II	http://www.lorusthera.com/rd_anti.asp
GTI 2051	AON	Ribonucleotide reductase R1/AML	Lorus Therapeutics	Phase II	http://www.lorusthera.com/rd_anti.asp
GRN163L	AON	Telomerase/lymphocytic leukemia	Geron	Phase I	http://www.geron.com
ANGIOZYME	Ribozyme	VEGF receptor/metastatic colorectal cancer	SIRNA	Phase II	http://www.sirna.com
HERZYME	Ribozyme	HER-2, breast and ovarian cancer	SIRNA	Phase I	http://www.sirna.com
RSV01	siRNA	Respiratory syncytical virus	Alnylam	Phase II	http://www.alnylam.com [15,16]
SIRNA-027 (AGN-211745)	siRNA	VEGF-R/AMD	SIRNA/Allergan	Phase II	http://www.sirna.com
Bevasiranib (Cand5)	siRNA	VEGF/AMD	Acuity Pharmaceuticals/ Opko Health	Phase II/III	[17]
sh/TAR/CCR5	shRNA	HIV	Benitec/City of Hope	Phase I	http://clinicaltrials.coh.org
ICS-283	siRNA	Solid tumor cancer	Intradigm	Preclinical	http://www.intradigm.com

(*continued overleaf*)

Table 11.2 (*Continued*)

Compound	Category	Target/disease	Company	Status	Ref.
Macugen/pegaptanib	Aptamer	VEGF/AMD	EyeTech/Pfizer	Approved as drug	[18]
REG1 (RB006, RB007)	Aptamer–antidote	Factor IXa/anticoagulation	Regardo Biosciences	Phase I/II	[19]
ARC1779	Aptamer	vWF/TTP	Archemix	Phase III	http://www.archemix.com
ARC183	Aptamer	Thrombin/anticoagulation	Archemix	Phase I	http://www.archemix.com
NU172 (ARC2172)	Aptamer	Thrombin/anticoagulation	Nuvelo/Archemix	Preclinical	http://www.archemix.com
AS-1411	Aptamer	Nucleolin/cancer	Antisoma	Phase I/II	http://www.antisoma.com
HIV-1 Rev response element decoy	Natural decoy	HIV-1	Childrens Hospital Los Angeles	Phase I	[20]
Edifoligide	Decoy	E2F/CABG surgery	Anesiva	Phase III	http://www.anesiva.com
AVT-01	Decoy	STAT1/asthma	Avontec	Phase II	http://avontec.com
AVT-02	Decoy	STAT1/psoriasis	Avontec	Phase II	http://avontec.com
Avrina	Decoy	NF-B/eczema	Anesiva	Phase III	http://www.anesiva.com
CPG-7909/PF-3512676	TLR9 agonist	Hepatitis B, influenza, cancer and others	Pfizer/Coley	Phase I/II	[21,22] http://www.coleypharma.com
CPG-10101	TLR9 agonist	Hepatitis B	Coley	Phase II	[23]
ISS 1018	TLR9 agonist	Cancer, asthma	Dynavax	Phase I/II	http://www.dynavax.com
AVE-7279	TLR9 agonist	Asthma	Sanofi-Aventis/Coley	Phase I	http://www.coleypharma.com
Autologous tumor mRNA library	mRNA	Colorectal cancer	Wakefield Gastroenterology Centre	Phase I	[24]
Autologous tumor mRNA library	mRNA	Melanoma	CureVac	Phase I/II	[25]

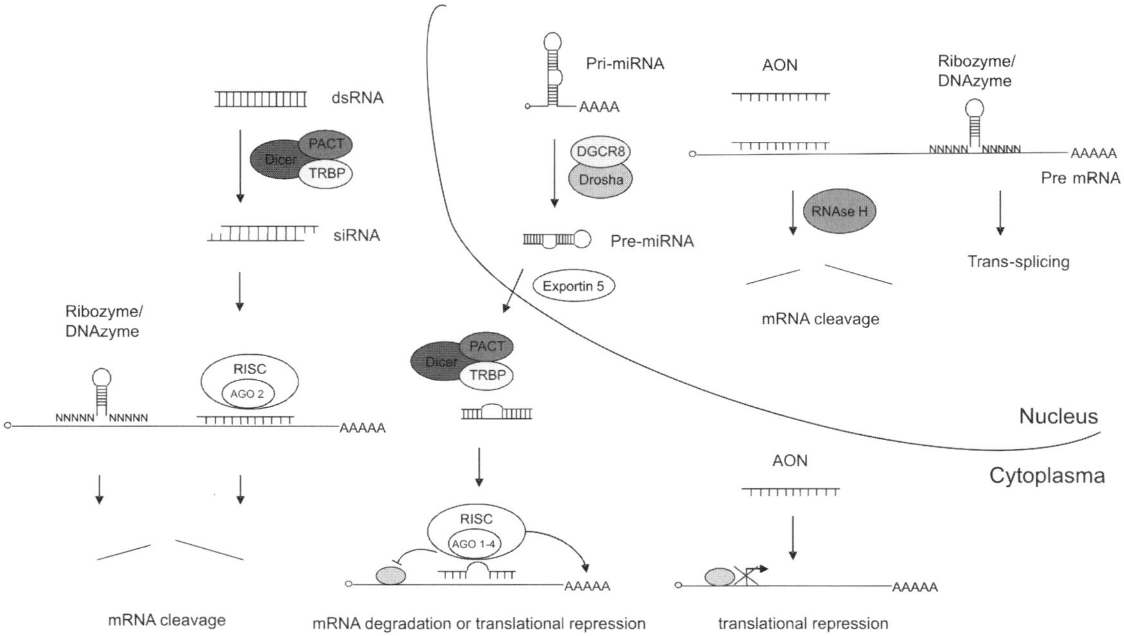

Figure 11.1 *Different mechanisms of action of AONs, ribozymes/DNAzymes, siRNAs and miRNAs*

other RNAs should be included in the design of nucleic acid sequences to abrogate so-called off-target effects, the nonspecific binding of oligonucleotides to non-targeted RNAs.

For *in vivo* applications, phosphodiester ssDNA molecules are not stable. In the past two decades, a vast number of chemical modifications have been identified on the sugar phosphate backbone and also the bases which significantly improve the stability of AONs while simultaneously increasing their efficacy (reviewed in [31,32]). The first generation of modified AONs consisted of phosphorothioate (PS) backbones in which one of the nonbridging oxygen atoms of the phosphodiester bond is replaced by sulfur [33]. Phosphorothioate oligonucleotides are still the most widely used AONs to date [34]. This modification results in some major advantages, such as resistance to nuclease degradation, a strong negative charge, good pharmacokinetics and the activation of RNase H cleavage. Most of the AONs in ongoing clinical trials belong to this group of first-generation modification. Nevertheless, these molecules elicit a number of side-effects and can induce cellular toxicity [35]. Consequently, a second generation of modifications has been developed with substitutions at the 2′-position of the ribose, including 2′-*O*-methyl (2′-*O*-Me) and 2′-*O*-methoxyethyl (2′-*O*-MOE) nucleotide analogs [36]. These AONs are less toxic than the PS-modified AONs and show an increased target affinity due to their ability to bind the mRNA in an A-type duplex [37]. On the downside, the same property that promotes A-type structure inhibits the cleavage by RNase H, which requires availability of the 2′-OH group of the RNA for efficient activity [38]. In the absence of RNase H, the antisense effect is most likely caused by blocking the translational machinery, but this is less potent in silencing efficiency [39]. Thus, a compromise has been made in the form of so-called gapmers [40,41]. Here, the 2′-ribose modifications are only incorporated at the ends of the oligonucleotide and flank a center of DNA or PS DNA sufficient in length for the activation of RNase H. Modifications at the 3′- and 5′-ends prevent nuclease degradation, allowing systemic delivery. Another advantage of gapmer technology is the reduction of aberrant cleavages of non-targeted mRNAs because the sequence that induces RNase H cleavage is reduced to 6–8 nucleotides [42]. More recently, the development of new chemical modifications has led to improved target affinity, nuclease resistance, good pharmacokinetics and reduction of toxicities along with some cost minimization. Prominent representatives of this third generation of AON modifications include locked nucleic acids (LNAs) [43,44], 2′-fluoro-arabino nucleic acids (FANAs) [45], 2′-*O*,4′-*C*-ethylene-bridged nucleic acids (ENAs) [46], peptide nucleic acids (PNAs) [47] and morpholinophosphoroamidates (MFs) [48]. These nucleic acid analogs have proven to be efficient in terms of stability and hybridization affinity, but they show poor activation of RNase H cleavage. Therefore, they are also best incorporated in gapmer designs when used in the classical antisense methodology. For example, the LNA–DNA–LNA gapmers against the oncogene H-Ras have shown high efficacy of inhibition of tumor growth in xenograft models [49] and chimeric FANA-DNA AONs were shown to inhibit specific gene expression via RNase H activation with a 30-fold lower IC_{50} than the corresponding phosphorothioate DNA [50].

Although the majority of third-generation modifications are not able to activate RNase H, they can be effective silencers of gene expression via blocking of translation. Morpholino oligonucleotides directed against the translation start site prevent ribosome initiation. Several clinical trials of morpholino-modified AONs have been conducted to address their potential in translational inhibition in different diseases. AVI-4126, for example, a phosphorodiamidate–morpholino hybrid oligomer against c-myc has been investigated in Phase I studies for prostate and breast cancer [51]. N3′–P5′ phosphoroamidates (NPs) are another example of AON modifications that can block translation [52]. The related analog N3′–P5′ thiophosphoramidate designated GRN163L was designed to be complementary to the RNA template of telomerase, thus functioning as an enzyme inhibitor. This AON has reached Phase II clinical trials for anticancer therapy in breast cancer [53].

AONs exhibiting a completely different mechanism of action have been developed using backbone modifications that do not recruit RNase H to the hybrid. One of these is the correction or alteration of splicing patterns for the purpose of blocking splice sites caused by mutations or skipping exons that harbor deleterious mutations. Some of the AON modifications that are used for this purpose are morpholino, LNA, PNA and 2′-*O*-Me analogs. The genetic blood disorder β-thalassemia can be caused by a mutation in intron 2 resulting in an aberrant splicing pattern. A phosphorothioate 2′-*O*-Me oligonucleotide blocking this mutant splice site was shown to restore the correct splicing pattern of the β-globin transcript [54]. AONs have also been used to exclude mutated exons from the mature mRNA in Duchenne muscular dystrophy (DMD) models, resulting in the production of a shorter, functional protein reminiscent of the less serious Becker's dystrophy [55].

Vitravene (fomivirsen) is to date the only FDA-approved antisense drug [12]. This is a phosphorothioate DNA approved by the FDA in 1998 for the treatment of cytomegalovirus (CMV) retinitis in patients with acquired immune deficiency syndrome (AIDS). Isis Pharmaceuticals (Carlsbad, CA, USA) developed the drug and licensed the worldwide commercial rights to Novartis (Basel, Switzerland) in 2001. Although Vitravene's approval demonstrated that antisense drugs are effective in the treatment of local disease and could be manufactured for commercial use, the field has suffered from setbacks when some AONs failed in Phase III trials. Alicaforsen (ISIS 2302) was designed against intercellular adhesion molecule 1 (ICAM-1) and tested for the treatment of Crohn's disease. In 1999, the oligonucleotide drug did not show significant efficacy in a Phase III study [56,57] and is now being investigated for the treatment of active ulcerative colitis and pouchitis, an inflammatory bowel disease of the colon [58]. Genta (Berkeley Heights, NJ, USA) has developed the anti-bcl-2 oligonucleotide Genasense (oblimersen), which failed to show a survival benefit using an add-on design of Genasense plus dacarbazine (DTIC) versus DTIC alone in metastatic melanoma [59]. Nevertheless, this antisense molecule is currently in Phase II/III for a variety of other cancers, including chronic lymphocytic leukemia (CLL), acute myelocytic leukemia and non-Hodgkin's lymphoma, and has been administered via intravenous and subcutaneous routes [60–62]. Surprisingly, some of the Genasense off-target effects may represent useful therapeutic principles for new anticancer mechanisms [63]. Recently, Affinitac (aprinocarsen sodium) (ISIS 3521) targeting protein kinase C-α (PKC-α) was also unsuccessful in improving patient outcomes in a standard chemotherapeutic regimen for non-small-cell lung carcinoma [14]. Further antiviral or anticancer AONs in clinical trials exemplify more promising results. The antisense compounds GTI-2501 and GTI-2040 directed against the ribonucleotide reductase subunits R1 and R2, respectively, have successfully completed Phase I trials leading to the initiation of a Phase II for GTI-2501 in hormone-refractory prostate cancer in combination with docetaxel and for GTI-2040 in renal cell carcinoma and other types of cancer (http://www.lorusthera.com/rd_anti.asp). Mipomersen (ISIS 301012) reduces the production of apolipoprotein B-100 and has been developed as a new treatment for high cholesterol. The drug is now in Phase II clinical trials [64].

Despite the initial enthusiasm of the outcome of early clinical trials and the increasing number of tested AONs, the successful development of antisense-based therapeutics remains low. The field is still challenged by some major problems of stability, bioavailability and delivery. Advances in medicinal chemistry and basic research may improve the design of specific AON compounds. Newly developed modifications for second- and third-generation AONs are expected to exhibit more favorable pharmacokinetic properties and reduced toxicity and have the potential for oral administration [65]. Therefore, the outcome of future clinical studies with new designs of AONs is eagerly anticipated by many investigators in the field of antisense therapy.

11.3 Ribozymes and DNAzymes

In 1981, the self-splicing activity of the group I intron of *Tetrahymena thermophilia* was discovered by Cech and co-workers [66,67], which paved the way for a whole new class of catalytic RNAs, which were termed ribozyme, a fusion of ribonucleic acid and enzyme. The first *trans*-acting RNA enzyme which was described is the RNA component of the pre-tRNA processing enzyme RNase P [68]. Subsequent to these ground-breaking findings, a variety of ribozymes have been found in lower eukaryotes, viruses and bacteria with different mechanisms of actions [69] (Figure 11.1). In general, the hydrolysis and transesterification of nucleotide bonds are catalyzed by a pH- or metal-dependent mechanism. Ribozymes appeared to represent promising therapeutic tools for the silencing of deleterious gene transcripts. The clear advantage of RNA enzymes is the specificity of the cleavage reaction. As true RNA enzymes, they are also capable of releasing the cleaved products and multiple turnovers. Beyond the cleavage of RNA substrates, ribozymes have extended their therapeutic potential to RNA repair applications. *trans*-Splicing ribozymes have been engineered to catalyze the *trans*-cleavage and ligation of target mRNA yielding the correction of mutant mRNAs [9,70,71].

The most widely studied natural RNA enzyme is the hammerhead ribozyme, which was initially isolated from plant viroid RNA [72,73]. The original *cis*-cleaving molecule was modified and minimized into a target-specific *trans*-cleaving ribozyme that can be employed against any accessible target site. This designed hammerhead ribozyme is ~40 nucleotides in length and consists of two binding arms that can specifically recognize the RNA substrate via an intermolecular helix and harbors a conserved catalytic domain that cleaves the target RNA at a specific position.

In addition to naturally occurring ribozymes, several catalytic RNA or DNA molecules have been generated by *in vitro* selection using combinatorial libraries. The *in vitro* environment allows the finding of accessible target sites [74], selection for improved activity [75], cleavage sequences [76] and biostability [77] and the development of catalysis of new chemical reactions [78]. A highly active ribozyme, called Zinzyme, against a K-*ras* target sequence was isolated from a 2′-fluoro- and 2′-amino-modified RNA library and was further modified to yield a half-life of >100 h in human serum [76].

For clinical applications, the same problems that confront all oligonucleotide biopharmaceuticals exist, including stability, delivery and biosafety. Modified nucleotides have to be carefully introduced in the sequence of ribozymes so as not to disrupt the catalytic activity. A comprehensive study of optimized stabilized hammerhead ribozymes has been reported [79,80]. For other ribozymes, modifications including 2′-*C*-allyluridine, 2′-*O*-Me, phosphorothioate or 3′–3′ inverted thymidine can be introduced outside the catalytic center. The viral genome of human immunodeficiency virus (HIV-1) has been the target of a number of ribozyme-mediated gene regulation studies. Retroviral vectors were used as delivery vehicles for hammerhead ribozymes directed against sequences in the HIV-1 genome. Peripheral $CD4^+$ lymphocytes were transduced *ex vivo* with the ribozyme construct and infusion of these transformed cells into the HIV-positive patients resulted in increased T-lymphocyte survival even though the effect was transient [81–86]. Hammerhead ribozymes were also developed against the highly conserved 5′-untranslated region of hepatitis C virus (HCV). This ribozyme was denoted HEPTAZYME, resulted in 90% inhibition of viral replication in cell culture [87] and showed promising results in Phase I and II trials. However, the clinical studies were suspended because of toxicity concerns [88]. HERZYME is the name given a ribozyme directed against the human epidermal growth factor-2 (HER-2), which is overexpressed in breast and ovarian cell carcinomas [76]. A Phase I trial was conducted to collect information about the safety and pharmacokinetics of this ribozyme. Another hammerhead ribozyme developed for cancer therapy was denoted ANGIOZYME and targeted the vascular endothelial growth factor receptor to reduce tumor growth via the inhibition of the formation of new blood vessels [89]. ANGIOZYME has been tested clinically in Phase I and II studies but has not progressed into an approved therapeutic.

DNAzymes are autocatalytic RNA-cleaving deoxyribonucleic acids that are derived by *in vitro* evolution and selection. The well-characterized DNAzyme 10–23 contains a cation-dependent catalytic core of 15 deoxyribonucleotides that cleaves the target RNA through a de-esterification reaction [90]. A number of structural modifications have been made to enhance the stability and to improve target accessibility including 3′–3′ inverted nucleotides, 2′-*O*-Me nucleotides or LNAs [91,92]. Catalytic DNA molecules have been widely examined as tools for silencing gene expression *in vitro* and *in vivo* [93–95]. DNAzymes against the transcription factor Egr-1 showed promising results in experimental models of restenosis via inhibition of smooth muscle cell hyperplasia and also in rat models of myocardial infarction [95–97]. Zhang *et al.* employed DNAzymes targeting vascular endothelial growth factor (VEGF) receptor 2 in rats and demonstrated a potent reduction in tumor growth with an associated reduction in tumor angiogenesis [98]. Although *in vivo* studies have given hope for the potential of DNAzymes as therapeutic drugs, none of the molecules tested has reached the level of clinical trials thus far. Recent studies suggest that the biological activities of several published DNAzymes were not mediated through the catalytic degradation of target mRNAs but rather via off-target cytotoxic effects, particularly for DNA molecules with G-rich sequences in the 5′-end [99,100].

11.4 siRNAs and miRNAs

The field of oligonucleotide-based therapy experienced a revival with the discovery of RNA interference (RNAi) in 1998 [101]. RNAi is a conserved endogenous mechanism, which is triggered by double-stranded (ds) RNAs leading to target-specific inhibition of gene expression by promoting mRNA degradation or translational repression. There are two RNAi pathways that are guided either by small interfering RNAs (siRNAs), which are perfectly complementary to the mRNA or by microRNAs (miRNAs), which bind imperfectly to their target mRNA [102] (Figure 11.1). SiRNAs can also induce direct transcriptional gene silencing (TGS) in the nucleus, although the mechanisms underlying this are not well understood in mammalian systems [103,104]. In lower eukaryotes, such as nematodes, insects and plants, RNAi represents an antiviral defense mechanism, in which viral dsRNA molecules are processed by the RNase III enzyme Dicer into small 19–23 nucleotide long double-stranded siRNAs, which contain 3′-end overhangs and 5′-end phosphate groups. These siRNAs are then incorporated into the RNA-induced silencing complex (RISC), where the passenger or sense strand of the siRNA duplex is degraded and the guide or antisense strand leads with full complementarity to the targeted mRNA sequence. The endonuclease Argonaute 2 (Ago2), a member of the RISC complex, cleaves the guide strand–mRNA duplex leading to subsequent degradation of the mRNA. Once activated, RISC can mediate multiple rounds of mRNA cleavage causing a potent knockdown effect. A breakthrough in the field of siRNA therapeutic agents was achieved in 2001 by Elbashir *et al.*, who demonstrated that synthetic, exogenously applied dsRNAs 21 nucleotides in length can induce silencing in mammalian cells [105]. In addition to the siRNA design of a 21-mer duplex with 3′-overhangs at both sides, Dicer-substrate formats such as 27-mers or short hairpin (sh) RNAs have been developed that elicit a more potent gene-silencing effect at lower concentrations than conventional 21-mer siRNAs [106–108].

It is remarkable how quickly after its discovery RNAi has been established as the method of choice for targeted inhibition of gene expression in mammalian systems. Because RNAi uses a natural pathway for gene silencing, it generally results in a greater potency of knockdown than AONs or ribozymes. Preclinical results have confirmed the effectiveness of RNAi and have generated serious optimism about the potential for siRNA drugs. As with the other oligonucleotide-based approaches, the applications of siRNAs as therapeutic agents face most of the above-mentioned challenges. Some of these challenges, however, have already been addressed in the course of AON and ribozyme development.

Two siRNA-specific limitations will be discussed in more detail: off-target effects and immunogenic toxicity. Off-target effects can be most likely attributed to the function of siRNAs as microRNAs, if they contain seed sequences that match 3′-untranslated regions in non-targeted mRNAs. Careful selection of the sequence can minimize off-target effects and chemical modifications at the second position of the seed sequence in the guide strand can even further reduce the unwanted transcript silencing [109,110]. Caution, is needed, however, when siRNAs enter preclinical and clinical trials and unspecific silencing should be precisely evaluated. Another concern with siRNA-mediated therapy is the possible interactions of siRNAs with Toll-like receptors or internal dsRNA receptors such as retinoic acid-inducible protein I (RIG-I) and dsRNA-dependent protein kinase (PKR). SiRNAs have been shown to interact with extracellular receptors, including Toll-like receptor (TLR) 3, TLR7 and TLR8, which trigger the activation of nuclear factor-κB (NF-κB), mitogen-activated protein kinase (MAPK), interferon regulatory factor (IRF) and subsequent interferon and cytokine production [111]. TLRs recognize specific immunostimulatory sequence motifs that are often associated with viral RNAs such as U- and G- rich sequences, which should be avoided in siRNA designs [112,113]. Selective 2′-*O*-Me or LNA modifications of sense or antisense strands can abolish the induction of the interferon response by evading TLR detection [112,114–117]. Application of siRNAs at the lowest effective concentrations could also help in preventing innate immune responses, and this could be accomplished by the design of potent siRNA triggers. In addition to the TLRs, cytoplasmic receptors such as RIG-I, melanoma differentiation-associated gene 5 (Mda-5), IRF and dsRNA-dependent protein kinase (PKR) respond to small dsRNAs and activate intracellular pathways for the interferon induction [111]. It is noteworthy that some of the above-mentioned effects were only seen with liposome carriers, whereas naked siRNAs conjugated to cholesterol or in a complex with atelocollagen did not induce the immune system [118–120].

The interferon response can also be avoided by the use of intracellularly expressed shRNAs that follow more closely the endogenous RNAi pathway as opposed to exogenously supplied synthetic siRNAs [121]. A limitation of the shRNA approach is the potential for toxic effects *in vivo* when shRNAs are overexpressed and saturate or compete for access to the endogenous microRNA interacting components [122]. Careful attention to levels of expression is hence needed in therapeutic settings.

In a different scenario, the siRNA-mediated activation of the interferon response may be a useful attribute for enhancing the therapeutic benefits. For this purpose, common stimulatory motifs for TLRs, RIG-I or Mda-5 are currently being evaluated as stimulatory agents of innate immune responses [123]. These oligonucleotides are denoted immunostimulatory RNAs (isRNAs) and are being investigated in the treatment of viral infection and cancer.

Many of the standard stabilizing oligonucleotide modifications that have already been explored for antisense strategies were employed in siRNA designs. SiRNA properties can be beneficially improved by the introduction of certain chemical modifications at distinct positions in the sequence, including thermal stability of the duplex, resistance against degradation, specificity for the target mRNA, reduction of off-target effects, biodistribution and cellular uptake [124]. In a systematic study, Jackson and co-workers reported that many individual nucleotides in the antisense strand may be modified with 2′-*O*-Me groups without loss of the silencing potential. A similar study has been performed with 2′-fluoro (2′-F) and 2′-*O*-MOE [125]. An additional advantage of using 2′-*O*-Me nucleotides is a reduction in off-target effects [126] and also avoidance of the interferon responses [116]. The strategic placement of these modifications is crucial. Modifications at the 5′-end of the guide strand can inhibit the silencing effect [127], while modifications at the 5′-end of the passenger strand can improve stability and also guide strand selection and targeting specificity [128,129]. Incorporation of 3′-*S*-phosphorothiolate [130], boranophosphates [131], 4′-thioriboses [132] and LNAs [133,134] has also been reported to enhance target-binding affinity and increase silencing potency.

As discussed above, cell-specific delivery *in vivo* is one of the crucial steps in the development of oligonucleotide-based therapeutic agents and remains a key challenge of siRNA therapeutic application. Systemic delivery of siRNAs to specific cells via cell-surface receptors would provide the maximum therapeutic benefit by decreasing the amount of drug and avoiding non-specific silencing or toxicity in healthy cells.

In the past, siRNAs have been delivered in animal models by intravenous or high-pressure tail vein injection, resulting in sufficient knockdown of primarily liver-specific gene expression. Such an approach suffers from obvious technical and practical limitations in a therapeutic setting [135]. Soutschek *et al.* conjugated a cholesterol group to the 3′-end of a siRNA passenger strand directed against apolipoprotein B (ApoB) and intravenously administered the complex in mice [118]. The siRNA was successfully delivered into the liver and jejunum through receptor-mediated endocytosis, where ApoB mRNA was reduced by 50% and 70%, respectively. Although this report demonstrates the efficiency of systemic delivery, this nonselective targeting approach is only appropriate for certain tissue types such as liver and jejunum.

The next steps towards cell-specific delivery involved the development of specially formulated liposomes and nanoparticles as carriers for siRNAs. These complexes fulfill several important criteria. They are not toxic *in vivo*, protect siRNAs that are often encapsulated in the particle, enhance efficient delivery and can be coupled to cell-penetrating peptides or poly(ethylene glycol) (PEG). Morrissey and co-workers encapsulated stabilized siRNAs targeting the hepatitis B virus in lipid bilayers, known as stable nucleic acid–lipid particles (SNALPs), with PEG at the outer surface and improved the bioavailability of the siRNAs for an effective silencing response in both mice and non-human primates [116,136]. Systemic delivery of cationic cardiolipid liposomes containing an siRNA targeting murine leukemia viral oncogene homolog 1 (Raf-1) silenced the expression of Raf-1 and inhibited tumor growth in a xenograft model of human prostate cancer [137]. A liposome-based complex that included an anti-transferrin receptor single-chain antibody fragment as the targeting moiety specifically and efficiently delivered siRNAs to primary and metastatic tumors when administered systemically [138]. Moreover, the surface of nanoparticles can be coated with cell-type specific ligands that might expand their properties towards specific delivery. In a Ewing sarcoma tumor mouse model, the self-assembled nanoparticles composed of cyclodextrin-containing polycations (CDPs), transferrin ligands, PEG and siRNAs were able to bind to the transferrin receptor and carry siRNAs into the tumor cells to inhibit tumor formation [139]. Another cell-specific delivery strategy includes the use of cell-type specific ligands such as antibodies or aptamers. Recent studies suggest that siRNA–ligand complexes enter the targeted cell through receptor endocytosis and are subsequently released to the cytosol to specifically silence target gene expression [140,141]. Heavy-chain antibody fragments (Fabs) specific for the HIV-1 envelope glycoprotein gp120 were fused to protamine, a nucleic-acid-binding protein, to bind the siRNAs [142]. Injection of this complex into mice only targeted HIV envelope-expressing B16 melanoma cells and inhibited tumor growth. Similarly, antibody–protamine fusion proteins targeting the human integrin lymphocyte function-associated antigen-1 efficiently delivered siRNAs in primary lymphocytes, monocytes and dendritic cells [143].

Two studies made use of an aptamer selected against the prostate-specific membrane antigen (PSMA), a cell surface receptor that is overexpressed in prostate cancer cells and vascular endothelium. McNamara *et al.* coupled the siRNA covalently to the aptamer via a nucleic acid linker [144] and Chu *et al.* used biotinylated 27-mer siRNAs and biotinylated anti-PSMA aptamers that were noncovalently bound together via a modular streptavidin bridge [145]. The aptamer–siRNA chimeras recognized specifically LNCaP, a prostate cancer cell line expressing PSMA and induced cell death; in contrast, no effect was determined in PSMA-negative cell lines. Another strategy used an anti-HIV envelope (gp120) aptamer fused to an anti-HIV 27-mer siRNA targeting HIV tat/rev common exon. Only cells infected with HIV were targeted with this chimeric construct and a several-fold reduction in viral p24 antigen production was observed over a 1 week period of treatment. In this setting, both the aptamer and siRNAs have inhibitory functions, making them a dual inhibitory system [146].

RNAi triggers can also be expressed as shRNAs or miRNA mimics from viral vectors [such as lentiviral, retroviral, adeno-associated virus (AAV) and adenovirus] and can be delivered in a gene therapy setting. A possible advantage of lentiviral vectors is the stable, long-term expression of shRNAs following vector integration into the host cell genome. A risk factor is the possibility of insertional mutagenesis by the vector when integrating vectors such as lentivirus or retroviruses are employed [147,148]. Adenoviruses or AAVs are non-integrating vectors that largely remain episomal, but possible drawbacks are limited loading capacity and their immunogenic potential, which preclude them from repeated administrations. Nevertheless, recent studies illustrate the potential of viral vector-delivered shRNAs in therapeutic settings [149,150].

Preclinical studies have demonstrated the safe use and the potential for therapeutic benefit of RNAi-mediated gene silencing [151,152]. SiRNAs are in early-stage clinical trials for the treatment of viral infections, cancer and ocular diseases. Phase I studies are planned for numerous other diseases, including neurodegenerative diseases, asthma/allergies and inflammatory diseases [153]. The most advanced stage testing for an siRNA-based drug is for the treatment of viral infection and was developed by Alnylam Pharmaceuticals (Cambridge, MA, USA). The siRNA ALN-RSV01 was designed against the respiratory syncytial virus (RSV), which causes severe respiratory illness, primarily in infants [15]. The unmodified siRNAs, administered by inhalation, showed significant viral reduction in experimentally infected adult volunteers compared with the placebo group in a Phase II GEMINI study, and is now being tested in RSV patients with naturally acquired infection. Other examples of antiviral applications have been proposed for severe acute respiratory syndrome (SARS) [154], herpes simplex virus 2 [155] and HIV-1 [156,157]. Serious concerns about the rapid development of drug-resistant HIV variants make the use of multiple drug combinations inevitable. Recently, a pilot study of the safety and feasibility of stem cell therapy for AIDS lymphoma patients was initiated using a lentivirus vector encoding three anti-HIV RNAs [150]. The combinatorial approach involves an shRNA targeting tat/rev, an RNA TAR decoy and an anti-chemokine receptor 5 (CCR5) ribozyme.

The lead product of Intradigm (Palo Alto, CA, USA) targets angiogenesis (http://www.intradigm.com) by an RNAi Nanoplex particle ICS-283, comprised of a nanoparticle and two siRNAs, one against vascular endothelial growth factor (VEGF) and the other against the VEGF's main receptor (VEGFR2). Two ongoing clinical trials also aim at angiogenesis in age-related macular deficiency (AMD). Bevasiranib (previously known as Cand5) was developed against VEGF and AGN 211745 (previously known as Sirna- 027) against its receptor (VEGFR1). Early clinical studies showed that the therapeutic reagents were well tolerated and could prevent neovascularization in the eye after intravitreal injection. AGN 211745 is being investigated in a Phase II study in combination with ranibizumab and patients are currently being enrolled in a Phase III study to evaluate the safety and effectiveness of bevasiranib. Controversially, a report was recently published suggesting that the suppression of neovascularization in two animal models is a generic property of siRNAs through TLR3 activation, independent of the sequence [158]. This example clearly demonstrates that preclinical studies need to be carefully conducted to prove safety and a specific siRNA-mediated silencing effect. Encouraged by earlier achievements of oligonucleotide-based therapeutics, some RNAi strategies may have been rushed into clinical trials. It is crucial to understand the basic mechanism of RNAi and its diverse related effectors to avoid toxic side-effects and to develop rationally designed biopharmaceuticals.

MiRNAs (see miRNA database: miRBase [159–161]) are endogenous, often highly conserved RNA molecules ~22 nucleotides in length that are involved in the regulation of major events in the cell, including differentiation, proliferation and apoptosis [162]. Estimates suggest that about 30% of the human genome may be under the control of miRNAs which regulate expression of multiple gene targets [163]. When incorporated into a silencing machinery similar to the siRNA complex, mature miRNAs bind within the 3′-untranslated region (3′-UTR) of the target mRNA, forming mismatched duplexes, and repress translation through different mechanisms [164].

In recent years, it has been discovered that altered expression of specific miRNA genes contributes to an increasing number of human diseases such as cancer, neurological diseases, metabolic disorders and cardiac diseases (reviewed in [165–167]). In cancer, miRNAs that can function as tumor suppressors or as oncogenes (referred to as oncomirs) have been identified by high-throughput and cell-based assays [168–170]. The reduction or deletion of an miRNA that normally regulates tumor suppressors leads to tumor formation and proliferation. On the other hand, the amplification or overexpression of miRNAs that have oncogenic function can also promote tumor formation. The role of miRNAs in cancer is further supported by the fact that about half of the annotated human miRNAs are encoded within fragile regions of chromosomes, which are associated with various human cancers [171]. Some of the first examples of miRNAs in cancers include the downregulation of miR-15a and miR-16- 1 in CLL and miR-143 and miR-145 in colorectal cancer [172,173]. The cluster miR-17–92 is upregulated in 65% of B-cell lymphoma samples that were tested and was associated with the MYC oncogene [174]. Examples of altered miRNA expression that are related to non-tumor diseases include the liver-specific miR-122 in hypercholesterolemia and hepatitis C infection, miR-9 and miR-128A in Alzheimer's disease and miR-1–2 in cardiac morphogenesis [175–177]. Analysis of SNP databases for humans and mice revealed that mutations in miRNA sequences or their target site might be important in diseases, in addition to suggesting interesting new therapeutic targets [178]. A role of miRNA regulation was also found in virus infection. However, the interaction is not well understood [179]. Pathogenic viruses of the herpes virus family encode multiple pre-miRNAs that are efficiently processed into mature miRNAs [167]. Targeting of cellular host mRNAs have to be experimentally confirmed. Another report demonstrated that the HCV replication in liver cells is dependent upon miR-122, but the mechanism remains obscure [180].

Expression profiling of miRNAs has revealed that miRNAs are promising diagnostic and prognostic biomarkers also for the classification, staging and progression of disease [181]. The widespread role in diseases makes miRNAs potentially interesting targets for therapeutic intervention. Two strategies for therapeutic treatment via miRNA regulation involve the delivery of miRNA sequences to restore proper function in tissues that show reduced miRNA expression or the inhibition of miRNA function when they are illicitly or overexpressed. Experience from gene therapy approaches and antisense and siRNA methodologies such as backbone modifications, delivery and pharmacokinetics has laid the groundwork for therapeutic strategies involving modulation of miRNA levels. For instance, the restoration of miRNA function was successfully achieved by the introduction of a synthetic miR-181a miRNA mimic in monocytes and macrophages [182]. Reintroduction of mature miRNAs expressed from short hairpin RNA encoding vectors was also achieved by viral delivery similar to a gene therapy approach [183,184].

Inhibition of miRNA function was successfully accomplished by an antisense approach via the use of anti-miRNA AONs in cell culture and *in vivo* (summarized in [165]). Modified antisense oligonucleotides were named, for example, antagomirs, miRNA ASOs or antimiRs when applied against miRNA targets. 2′-*O*-Me, 2′-*O*-MOE, morpholino and LNA modifications were demonstrated to be most efficient in terms of binding affinity and silencing efficacy. Antagomirs with 2′-*O*-Me- and PS-modified oligonucleotides and a cholesterol tag were shown to be efficient and specific silencers of endogenous miRNAs in mice after intravenous administration [185,186]. Most recently, a study was published using unconjugated LNA-AONs targeting miR–122 in nonhuman primates [187]. A simple systemic delivery could effectively antagonize liver-expressed miR-122 and lower plasma cholesterol levels, suggesting a potential new class of therapeutics for hypercholesterolemia.

Another approach for inhibiting miRNA function has been termed miRNA masking. This approach is to use AONs to block the mRNA binding site [178,188,189]. Moreover, alternative strategies have been developed to intervene with the regular miRNA pathways. MiRNA sponges use transcripts that have multiple miRNA binding sites to 'soak up' mature miRNAs [190,191]. Targeting components of the miRNA biogenesis pathway such as Drosha or Dicer resulting in a reduced level of mature miRNAs is another possible approach for inhibition of miRNA function [192,193], although his approach may interfere with other important pathways in cell biology.

Progress in the functional use of miRNAs as biomarkers in the diagnosis and prognosis of diseases and also exploiting potential therapeutic strategies against miRNAs is remarkable. Nevertheless, this field is still in its infancy and major hurdles remain the same as for siRNA and AON technology. Targeting miRNAs as a therapeutic strategy is also complicated by the fact that on the one hand one miRNA can regulate several different mRNAs and on the other the 3′-UTRs of mRNAs contain multiple binding sites and can be modulated by diverse miRNAs. The biology of miRNAs and the complex regulation network is far from understood and future investigations remain to provide new insights in this promising field.

11.5 Aptamers and decoys

Aptamers are single-stranded oligonucleotides (25–50 nucleotides), which fold into well-defined three-dimensional structures and thereby they are able to bind with high affinity and specificity ($10^{-12} < K_d > 10^{-9}$) to a variety of different target molecules. These capabilities make aptamers unique in comparison with the conventional oligonucleotide therapies that target the post-transcriptional machinery of the cell by destroying the mRNA. The name aptamer is derived from the latin word *aptus* that means 'to fit' [194]. Typically, aptamers are generated by *in vitro* selection from complex combinatorial libraries of nucleic acid sequences under defined conditions, but natural occurring aptamers have also been reported [195–197].

RNA ligands are attractive as possible therapeutics because they can directly bind and inhibit the activity of clinically relevant proteins. Modulating protein function can be accomplished in different ways. First, the function of the protein can be inhibited by binding of the aptamer to the active site; second, the binding site for a ligand can be blocked by the aptamer; or third, the aptamer can interfere with a specific domain in a multi-subunit protein. Sullenger *et al.* in 1990 first demonstrated the concept of therapeutic aptamers by over-expressing the *trans*-activation response (TAR) RNA of HIV-1 as decoy for the viral Tat and cellular cyclin T1 proteins in $CD4^+$T cells that could inhibit protein function and virus replication [198].

In the same year, two groundbreaking studies were published which made RNA ligands broadly accessible to a variety of targets and applications by developing the Systematic Evolution of Ligands by Exponential Enrichment (SELEX) technology [199]. Here, partial altered or complete synthetic oligonucleotide libraries were generated for the first time, which enlarges the diversity of the original library and allows the targeting of virtually any class of molecules. Tuerk and Gold partially randomized an RNA that binds to T4-DNA polymerase and selected again against the T4-DNA polymerase [199]. Besides the wild-type sequence, they were able to identify other sequences with the same binding affinity as the wild-type sequence. Ellington and Szostak showed that it is possible to generate RNA aptamers which can specifically target and bind to small chemical molecules [194].

The SELEX process comprises iterative cycles of *in vitro* selections of oligonucleotides binding to the target, partitioning and amplification. The initial library of oligonucleotides consists of single-stranded DNA molecules with typically 20–50 randomized nucleotides flanked by known primer sequences for amplification. The random region allows a high diversity with generally up to 10^{15} individual molecules, although in large libraries, each possible binding sequence exists as multiple copies. After several rounds of selection, the original pool is eventually enriched for specific binders and single sequences can be identified by molecular cloning. The isolation of binding motifs by affinity assays is most important for the potential use as biopharmaceuticals to reduce the best binders to their minimal sequence without loss of specificity.

For *in vivo* applications, resistance to nuclease degradation for aptamers can be achieved using a modified pre-SELEX or post-SELEX modifications of different chemical nucleotide analogs. Pre-SELEX modifications that are incorporated into the initial RNA libraries by *in vitro* transcription typically include 2′-fluoro- or 2′-aminopyrimidines [200,201]. 2′-*O*-Me nucleotides were also reported to be directly incorporated in the T7 transcripts [202]. Phosphorothioates, 2′-*O*-Me nucleotides, LNAs or capping at the 3′-end are post-SELEX modifications and promote high nuclease resistance [203,204]. The risk of substitutions of nucleotide analogs after the selection process is the loss of affinity of the aptamer for the target molecule due to the disruption of the specific three-dimensional RNA structure. Therefore, the position and the type of the modification have to be carefully engineered and optimization often requires large-scale screening approaches. An alternative method is the Spiegelmer technology [205,206]. Unnatural L-isomer form nucleotides have been incorporated into aptamers that are not recognized any longer as substrates by nucleases and exhibit extreme stability *in vivo*. Modified nucleotides are also very useful in manipulating bioavailability, renal clearance and cellular uptake. To improve pharmacokinetics further, aptamers have been coupled to PEG or other hydrophobic groups or attached to a liposome surface that extends the half-life to several hours [207,208]. A way to control aptamer activity independent of biological clearance was demonstrated by the design of an antidote [209].

Another important criterion for aptamers as therapeutics is the systemic delivery *in vivo*. Aptamers primarily act on extracellular targets and local administration is required by injection [210]. In recent years, the development of aptamers against intracellular targets has been pursued, evoking the necessity for specific delivery into the targeted tissue or cells, which still remains a challenge. So-called intramers (intracellular aptamers) have been developed in expression cassettes that need to be delivered via viral vectors [211,212]. For use in therapeutic settings, the same safety concerns have to be overcome that apply to other gene therapy approaches. An alternative route is delivery by conjugated lipid molecules, similarly to the siRNA approach.

Aptamers are often referred to as ‘chemical antibodies’ because the specificity and affinity for their target molecule can be compared to that of antibodies. For certain cases, aptamers may exhibit advantages over antibodies because of their nucleic acid character. Thus in some cases aptamers may be an attractive alternative to monoclonal antibodies or conventional drugs.

Nucleic acids have the property of combining genotype and phenotype, thereby allowing them to contain their own genetic information and also carry out certain functions as a result of their capacity to form elaborate three-dimensional structures. For aptamers, the entire selection process is usually carried out *in vitro* and can therefore target theoretically any possible molecule, even toxic proteins under a variety of conditions. Certain properties can be changed on demand just by employing additional rounds of selection. Furthermore, aptamers can be chemically synthesized and a variety of modifications or tags can be included in the design for diverse functions. Aptamers are also very stable compared with antibodies, with no loss of activity under a wide range of buffer conditions or extreme heat. Hence these molecules can be considered as regenerative, reliable and long-term storable potential drugs. Additionally, aptamers have so far shown evidence that they are immunogenic.

The potential of aptamers as therapeutic agents has been explored for a variety of diseases, including viral infection, cancer, inflammation and autoimmune disease (recently reviewed in, e.g., [213–215]). Macugen (pegaptanib sodium) is the first aptamer that received FDA approval (December 2004) [18]. This drug was initially generated as 2′-fluoropyrimidine, 2′-ribopurine RNA against a VEGF-165 isoform and was further minimized and modified with 2′-*O*-Me nucleotides and also a PEG moiety at the 5′-terminus [216,217]. VEGF is critical in angiogenesis and overproduction is associated with ocular neovascularization such as in wet age-related macular degeneration (AMD), causing adult blindness. In Phase III clinical trials with a total of 1186 AMD patients, Macugen, administered by local intravitreous injection into the eye every 6 weeks, was able to reduce visual loss significantly after 54 weeks [18]. Macugen is currently being investigated in clinical trials for anti-angiogenic therapies against diabetic macular edema, retinal vein occlusion and cancer [218].

Aptamers have also been selected against numerous coagulation factors to prevent clotting of blood vessels. The PEGylated aptamer ARC1779 was developed to inhibit the function of a protein called von Willebrand factor (vWF), which when activated is responsible for the adhesion, activation and aggregation of platelets in thrombotic thrombocytopenic purpura (TTP). This aptamer is currently being tested in Phase III trials. NU172 and ARC183 are aptamers designed to inhibit directly thrombin's ability to stimulate blood clot formation and are currently in early stages of clinical development for coronary artery bypass grafting (CABG) surgery (http://www.archemix.com).

Most interesting is the generation of antidote-controlled aptamers established in an elegant study by Rusconi and co-workers [19,209]. The aptamer is directed against Factor IXa, a key component of the coagulation cascade, and was proven to inhibit blot clots efficiently. The antidote is a second oligonucleotide complementary to a part of the aptamer sequence that binds strongly to alter the aptamer structure and thereby prevent it from binding to its target. Anticoagulation activity of the aptamer was found to be neutralized by more than 95% within 10 min after addition of the antidote [19]. Given the long half-life of the aptamer in blood, the relatively short-half of the antidote and the fast neutralization reaction, fine-tuning of the drug dose is feasible and reliable and renders this concept as an attractive enrichment in the field of oligonucleotide therapeutics. The aptamer–antidote pair has now entered Phase II clinical studies under the name REG1 by Regado Biosciences (Durham, NC, USA) (http://www.regadobiosciences.com).

Another example is the unmodified guanosine-rich aptamer AS1411 (also known as AGRO 100), which represents the first aptamer to be tested for the treatment of cancer [219]. The mechanism of action includes binding to the cell surface protein nucleolin, internalization and inhibition of NF-κB, which results eventually in antiproliferation and anticancer effects [220]. In a Phase I trial, more than half of patients suffering from advanced renal or non-small cell lung cancer (NSCLC) exhibited a stable disease profile 2 months after treatment started [221]. Antisoma (London, UK) is now preparing a Phase II trial for renal cancer and acute myeloid leukemia patients (http://www.antisoma.com/asm/products/as1411/). In addition to their therapeutic applications, aptamers present themselves as ideal tools for target validation, drug discovery, diagnosis and, recently, specific delivery of siRNA (see Section 11.4) due to the combination of their high specificity and binding affinity for their target.

The concept of decoys is very similar to that of aptamers. Decoys bind with high specificity and affinity to specific proteins and can decoy proteins from their endogenous targets such as genomic DNA binding sites and thus alter transcription of genes. Naturally occurring decoys were first described in HIV-1 [198] but in the meantime, diverse decoys have been designed or selected to target human transcription factors. For instance, edifoligide was developed against the mammalian transcription factor E2F that is known to play a key role in the upregulation of genes involved in neointimal hyperplasia, which leads to atherosclerosis and thrombosis. The decoy includes the consensus binding site sequence for E2F and showed sustained suppression of neointima formation in a rat model [222]. Unfortunately, edifoligide could not prove its potential in Phase III clinical investigation including 1508 patients who were undergoing CABG surgery. One year after the treatment, the decoy did not show any therapeutic benefit compared with the placebo group (http://www.a-nesiva.com). The reason might be the existence of at least eight isoforms in the E2F transcription factor family that could possibly rescue the inhibition of the targeted isoform. Avontec's (Munich, Germany) pipeline contains among others a short, double-stranded oligonucleotide decoy AVT-01 that effectively inhibits STAT-1, a transcription factor shown to be strongly involved in the regulation of chronic inflammation. Recently, AVT-01 entered a multiple dosing Phase II clinical study in asthmatic patients (http://avontec.com).

11.6 Toll-like receptor 9 agonists

Toll-like receptors (TLRs) belong to the innate immune system of vertebrates and interact with extracellular and intracellular pathogens, followed by the activation of the interferon pathway to stimulate the adaptive immune response. The 10 known human TLRs are expressed in a distinct subset in immune cells and recognize different types of pathogens [223]. TLR9 is the best studied among the Toll-like receptors and several synthetic oligodeoxynucleotide agonists are currently under investigation in clinical trials for the treatment of infectious disease, cancer and asthma/allergy [224]. TLR9 detects unmethylated cytosine–phosphate–guanine (CpG) dinucleotides, which are relatively common in bacterial and viral DNA but are methylated in vertebrates [225]. B-cells and plasmacytoid dendritic cells constitutively express TLR9, but after cellular activation, TLR9 expression is also induced in additional cell types even in some non-immune cells, including pulmonary epithelial cells, intestinal epithelium cells and keratinocytes [223,226–229]. Comprehensive studies have revealed the species-specific optimal CpG motif for mice (GACGTT) and for humans (GTCGTT) [230–233]. Clinical CpG oligonucleotides (CpG ODNs) usually contain two to four hexamer CpG motifs separated by at least two bases, preferably thymine residues as spacers. For better stability in serum, CpG ODNs are typically modified, at least partially, with phosphorothioate nucleotides leading to a half-life of about 48 h [234,235]. Three distinct classes of ODNs have been described that, depending on their sequences and modification pattern, result in distinct properties and trigger different phenotypes [224].

TLR9 agonists have been widely studied as standalone vaccines or as vaccine adjuvants since they can be strong inducers of $CD4^+$ and $CD8^+$ T-cell responses, leading to rapid production of antigen-specific antibodies. For instance, the addition of CPG 7909 to a peptide MART-1 vaccine in patients with human leukocyte antigen-$A2^+$ melanoma caused an approximately 10-fold increase of antigen-specific $CD8^+$ T-cells [236,237]. The same CpG ODN was used as an adjuvant for hepatitis B vaccination in humans. The production of hepatitis B surface antigen-specific antibodies appeared earlier and had higher titers in all recipients who received the combination of the vaccine and CPG 7909 as compared with control recipients [21,22]. Additional, CpG ODNs were investigated as therapeutic pharmaceuticals against virus diseases, cancer or asthma. Coley Pharmaceutical Group (Welleley, MA, USA) has developed CPG 10101 as a monotherapy agent against HCV. A Phase I trial involving 60 HCV-infected patients demonstrated decreased viral RNA levels in blood in a dose-dependent manner after CPG 10101 treatment [23]. It was reasoned that the activation of the immune system by TLR9 generated a strong T-cell response against the virus allowing control of the disease and viral clearance [238]. Comparably, the rationale for the use of synthetic CpG motifs in cancer therapy is based on the hypothesis that TLR9 promotes an anti-tumor T-cell response which directly influences the growth and survival of the tumor. TLR9 agonists targeting hematological malignancies, skin cancer, lung cancer and other tumors are under evaluation either in monotherapy or in combination with conventional therapies [239]. PF-3512676 (also known as CPG 7909) has reached Phase II in clinical trials against various types of cancers [239]. It was shown that the CpG ODN drug was generally well tolerated in patients and was associated with anti-tumor activity. Conversely, two Phase III trials of PF-3512676 in combination with chemotherapy for NSCLC patients have been discontinued because of the lack of incremental efficacy with the addition of the ODN [239].

As with other therapeutic immunotherapies, the concern of induction of autoimmune diseases raises concerns over chronic TLR9 activation. Clinical experience to date indicates that CpG ODN treatment is not associated with significant autoimmune diseases. However, data from long-term treatments of more than 6 months are still lacking and no definitive conclusions can be drawn about chronic use of TLR agonists. Moreover, the phosphorothioate backbone of ODNs can induce some sequence-independent effects such as the inhibition of coagulation or the activation of leukocytes [240,241]. Fortunately, this toxicity does not occur below the threshold of the drug concentration in blood [241]. Overall, CpG ODNs are well tolerated in patients and minor local injection-site reactions are transient [224]. Although clinical trials are still ongoing, examining the safety and the efficacy of TLR agonists in therapy, evidence indicates the potential clinical benefit of CpG ODNs either alone or in combination with other therapies.

11.7 Outlook

Although only one antisense molecule and one aptamer have found their way to becoming approved drugs, the progress in understanding and development of oligonucleotides for therapeutic interventions is impressive. The increasing interest in nucleic acids in therapy is also substantiated by the number of biopharmaceutical companies that are investigating at least one oligonucleotide-based drug in their pipeline.

Oligonucleotide-based therapeutics encompasses a wide range of properties, in terms of their mechanism of action and their medical applications. The field has evolved from the traditional strategies of targeting specific mRNA degradation by antisense and ribozyme technology to the more recent approaches of siRNA-mediated target degradation, aptamer-mediated inhibition of protein function and immunostimulation through binding to Toll-like receptors. As with all technologies that move from the bench to the bedside, major obstacles litter the long way to clinical applications. Early setbacks included unspecific properties such as off-target effects or toxic stimulation of the immune system. However, every new challenge was addressed and novel ideas and solutions were quickly developed to bypass them. Further discoveries will give novel opportunities in the design of rational therapeutic oligonucleotides.

References

1. M.P. Knauert and P.M. Glazer, Triplex forming oligonucleotides: sequence-specific tools for gene targeting, *Hum. Mol. Genet.*, **10**, 2243–2251 (2001).
2. K.M. Vasquez and P.M. Glazer, Triplex-forming oligonucleotides: principles and applications, *Q. Rev. Biophys.*, **35**, 89–107 (2002).
3. E. Bruscia *et al.*, Isolation of CF cell lines corrected at DeltaF508-CFTR locus by SFHR-mediated targeting, *Gene Ther.*, **9**, 683–685 (2002).
4. D.C. Gruenert *et al.*, Sequence-specific modification of genomic DNA by small DNA fragments, *J. Clin. Invest*, **112**, 637–641 (2003).
5. O. Igoucheva, V. Alexeev and K. Yoon, Targeted gene correction by small single-stranded oligonucleotides in mammalian cells, *Gene Ther.*, **8**, 391–399 (2001).
6. W. Yin, B.T. Kren and C.J. Steer, Site-specific base changes in the coding or promoter region of the human beta- and gamma-globin genes by single-stranded oligonucleotides, *Biochem. J.*, **390**, 253–261 (2005).
7. O. Igoucheva and K. Yoon, Targeted single-base correction by RNA–DNA oligonucleotides, *Hum. Gene Ther.*, **11**, 2307–2312 (2000).
8. Y. Yang and C.E. Walsh, Spliceosome-mediated RNA trans-splicing, *Mol. Ther.*, **12**, 1006–1012 (2005).
9. L.G. Mitchell and G.J. McGarrity, Gene therapy progress and prospects: reprograming gene expression by trans-splicing, *Gene Ther.*, **12**, 1477–1485 (2005).
10. B. Weide, C. Garbe, H.G. Rammensee and S. Pascolo, Plasmid DNA- and messenger RNA-based anti-cancer vaccination, *Immunol. Lett.*, **115**, 33–42 (2008).
11. S. Pascolo, Messenger RNA-based vaccines, *Expert Opin. Biol. Ther.*, **4**, 1285–1294 (2004).

12. C. Marwick, First 'antisense' drug will treat CMV retinitis, *JAMA*, **280**, 871 (1998).
13. D.A. Jabs and P.D. Griffiths, Fomivirsen for the treatment of cytomegalovirus retinitis, *Am. J. Ophthalmol.*, **133**, 552–556 (2002).
14. L. Paz-Ares *et al.*, Phase III study of gemcitabine and cisplatin with or without aprinocarsen, a protein kinase C-alpha antisense oligonucleotide, in patients with advanced-stage non-small-cell lung cancer, *J. Clin. Oncol.*, **24**, 1428–1434 (2006).
15. V. Bitko, A. Musiyenko, O. Shulyayeva and S. Barik, Inhibition of respiratory viruses by nasally administered siRNA, *Nat. Med.*, **11**, 50–55 (2005).
16. W. Zhang *et al.*, Inhibition of respiratory syncytial virus infection with intranasal siRNA nanoparticles targeting the viral NS1 gene, *Nat. Med.*, **11**, 56–62 (2005).
17. A.V. Chappelow and P.K. Kaiser, Neovascular age-related macular degeneration: potential therapies, *Drugs*, **68**, 1029–1036 (2008).
18. E.S. Gragoudas, A.P. Adamis, E.T. Cunningham Jr, M. Feinsod and D.R. Guyer, Pegaptanib for neovascular age-related macular degeneration, *N. Engl. J. Med.*, **351**, 2805–2816 (2004).
19. C.P. Rusconi *et al.*, RNA aptamers as reversible antagonists of coagulation factor IXa, *Nature*, **419**, 90–94 (2002).
20. D.B. Kohn *et al.*, A clinical trial of retroviral-mediated transfer of a rev-responsive element decoy gene into CD34(+) cells from the bone marrow of human immunodeficiency virus-1-infected children, *Blood*, **94**, 368–371 (1999).
21. S.A. Halperin *et al.*, A phase I study of the safety and immunogenicity of recombinant hepatitis B surface antigen co-administered with an immunostimulatory phosphorothioate oligonucleotide adjuvant, *Vaccine*, **21**, 2461–2467 (2003).
22. C.L. Cooper *et al.*, CPG 7909, an immunostimulatory TLR9 agonist oligodeoxynucleotide, as adjuvant to Engerix-B HBV vaccine in healthy adults: a double-blind phase I/II study, *J. Clin. Immunol.*, **24**, 693–701 (2004).
23. J.G. McHutchison *et al.*, Phase 1B, randomized, double-blind, dose-escalation trial of CPG 10101 in patients with chronic hepatitis C virus, *Hepatology*, **46**, 1341–1349 (2007).
24. N. Rains, R.J. Cannan, W. Chen and R.S. Stubbs, Development of a dendritic cell (DC)-based vaccine for patients with advanced colorectal cancer, *Hepatogastroenterology*, **48**, 347–351 (2001).
25. B. Weide *et al.*, Results of the first phase I/II clinical vaccination trial with direct injection of mRNA, *J. Immunother.*, **31**, 180–188 (2008).
26. S. Agrawal, S.H. Mayrand, P.C. Zamecnik and T. Pederson, Site-specific excision from RNA by RNase H and mixed-phosphate-backbone oligodeoxynucleotides, *Proc. Natl. Acad. Sci. USA*, **87**, 1401–1405 (1990).
27. P.J. Furdon, Z. Dominski and R. Kole, RNase H cleavage of RNA hybridized to oligonucleotides containing methylphosphonate, phosphorothioate and phosphodiester bonds, *Nucleic Acids Res.*, **17**, 9193–9204 (1989).
28. B.M. Paterson, B.E. Roberts and E.L. Kuff, Structural gene identification and mapping by DNA-mRNA hybrid-arrested cell-free translation, *Proc. Natl. Acad. Sci. USA*, **74**, 4370–4374 (1977).
29. P.C. Zamecnik and M.L. Stephenson, Inhibition of Rous sarcoma virus replication and cell transformation by a specific oligodeoxynucleotide, *Proc. Natl. Acad. Sci. USA*, **75**, 280–284 (1978).
30. M.L. Stephenson and P.C. Zamecnik, Inhibition of Rous sarcoma viral RNA translation by a specific oligodeoxyribonucleotide, *Proc. Natl. Acad. Sci. USA*, **75**, 285–288 (1978).
31. J. Kurreck, Antisense technologies. Improvement through novel chemical modifications, *Eur. J. Biochem.*, **270**, 1628–1644 (2003).
32. C. Wilson and A.D. Keefe, Building oligonucleotide therapeutics using non-natural chemistries, *Curr. Opin. Chem. Biol.*, **10**, 607–614 (2006).
33. C.E. De, F. Eckstein, H. Sternbach and T.C. Merigan, Interferon induction by and ribonuclease sensitivity of thiophosphate-substituted polyribonucleotides, *Antimicrob. Agents Chemother.*, **9**, 187–191 (1969).
34. F. Eckstein, Phosphorothioate oligodeoxynucleotides: what is their origin and what is unique about them?, *Antisense Nucleic Acid Drug Dev.*, **10**, 117–121 (2000).
35. M.A. Guvakova, L.A. Yakubov, I. Vlodavsky, J.L. Tonkinson and C.A. Stein, Phosphorothioate oligodeoxynucleotides bind to basic fibroblast growth factor, inhibit its binding to cell surface receptors and remove it from low affinity binding sites on extracellular matrix, *J. Biol. Chem.*, **270**, 2620–2627 (1995).
36. M. Manoharan, 2′-Carbohydrate modifications in antisense oligonucleotide therapy: importance of conformation, configuration and conjugation, *Biochim. Biophys. Acta*, **1489**, 117–130 (1999).
37. S.T. Crooke *et al.*, Kinetic characteristics of *Escherichia coli* RNase H1: cleavage of various antisense oligonucleotide–RNA duplexes, *Biochem. J.*, **312** (Pt 2), 599–608 (1995).

38. E. Zamaratski, P.I. Pradeepkumar and J. Chattopadhyaya, A critical survey of the structure–function of the antisense oligo/RNA heteroduplex as substrate for RNase H, *J. Biochem. Biophys. Methods*, **48**, 189–208 (2001).
39. B.F. Baker *et al.*, 2′-*O*-(2-Methoxy)ethyl-modified anti-intercellular adhesion molecule 1 (ICAM-1) oligonucleotides selectively increase the ICAM-1 mRNA level and inhibit formation of the ICAM-1 translation initiation complex in human umbilical vein endothelial cells, *J. Biol. Chem.*, **272**, 11994–12000 (1997).
40. B.P. Monia *et al.*, Evaluation of 2′-modified oligonucleotides containing 2′-deoxy gaps as antisense inhibitors of gene expression, *J. Biol. Chem.*, **268**, 14514–14522 (1993).
41. H. Wu, W.F. Lima and S.T. Crooke, Properties of cloned and expressed human RNase H1, *J. Biol. Chem.*, **274**, 28270–28278 (1999).
42. I. Lebedeva and C.A. Stein, Antisense oligonucleotides: promise and reality, *Annu. Rev. Pharmacol. Toxicol.*, **41**, 403–419 (2001).
43. A.A. Koshkin and J. Wengel, Synthesis of novel 2′,3′-linked bicyclic thymine ribonucleosides, *J. Org. Chem.*, **63**, 2778–2781 (1998).
44. R. Kumar *et al.*, The first analogues of LNA (locked nucleic acids): phosphorothioate-LNA and 2′-thio-LNA, *Bioorg. Med. Chem. Lett.*, **8**, 2219–2222 (1998).
45. M.J. Dhama *et al.*, Hybrids of RNA and arabinonucleic acids (ANA and FANA) are substrates of ribonuclease H, *J. Am. Chem. Soc.*, **120**, 12976–12977 (1989).
46. M. Koizumi *et al.*, Direct comparison of *in vivo* antisense activity of ENA oligonucleotides targeting PTP1B mRNA with that of 2′-*O*-(2-methoxy)ethyl-modified oligonucleotides, *Oligonucleotides*, **16**, 253–262 (2006).
47. P.E. Nielsen, M. Egholm, R.H. Berg and O. Buchardt, Sequence-selective recognition of DNA by strand displacement with a thymine-substituted polyamide, *Science*, **254**, 1497–1500 (1991).
48. J. Summerton, Morpholino antisense oligomers: the case for an RNase H-independent structural type, *Biochim. Biophys. Acta*, **1489**, 141–158 (1999).
49. K. Fluiter *et al.*, On the *in vitro* and *in vivo* properties of four locked nucleic acid nucleotides incorporated into an anti-H-Ras antisense oligonucleotide, *ChemBioChem*, **6**, 1104–1109 (2005).
50. C.N. Lok *et al.*, Potent gene-specific inhibitory properties of mixed-backbone antisense oligonucleotides comprised of 2′-deoxy-2′-fluoro-D-arabinose and 2′-deoxyribose nucleotides, *Biochemistry*, **41**, 3457–3467 (2002).
51. G.R. Devi *et al.*, *In vivo* bioavailability and pharmacokinetics of a c-MYC antisense phosphorodiamidate morpholino oligomer, AVI-4126, in solid tumors, *Clin. Cancer Res.*, **11**, 3930–3938 (2005).
52. O. Heidenreich, S. Gryaznov and M. Nerenberg, RNase H-independent antisense activity of oligonucleotide N3′→P5′ phosphoramidates, *Nucleic Acids Res.*, **25**, 776–780 (1997).
53. A.E. Hochreiter *et al.*, Telomerase template antagonist GRN163L disrupts telomere maintenance, tumor growth and metastasis of breast cancer, *Clin. Cancer Res.*, **12**, 3184–3192 (2006).
54. H. Sierakowska, M.J. Sambade, S. Agrawal and R. Kole, Repair of thalassemic human beta-globin mRNA in mammalian cells by antisense oligonucleotides, *Proc. Natl. Acad. Sci. USA*, **93**, 12840–12844 (1996).
55. G. McClorey, H.M. Moulton, P.L. Iversen, S. Fletcher and S.D. Wilton, Antisense oligonucleotide-induced exon skipping restores dystrophin expression *in vitro* in a canine model of DMD, *Gene Ther.*, **13**, 1373–1381 (2006).
56. A. Dove, Isis and antisense face crucial test without Novartis, *Nat. Biotechnol.*, **18**, 19 (2000).
57. B. Yacyshyn *et al.*, A randomized, double-masked, placebo-controlled study of alicaforsen, an antisense inhibitor of intercellular adhesion molecule 1, for the treatment of subjects with active Crohn's disease, *Clin. Gastroenterol. Hepatol.*, **5**, 215–220 (2007).
58. S.J. van Deventer *et al.*, A phase II dose ranging, double-blind, placebo-controlled study of alicaforsen enema in subjects with acute exacerbation of mild to moderate left-sided ulcerative colitis, *Aliment. Pharmacol. Ther.*, **23**, 1415–1425 (2006).
59. K. Garber, New apoptosis drugs face critical test, *Nat. Biotechnol.*, **23**, 409–411 (2005).
60. K.F. Pirollo, A. Rait, L.S. Sleer and E.H. Chang, Antisense therapeutics: from theory to clinical practice, *Pharmacol. Ther.*, **99**, 55–77 (2003).
61. J.N. Moreira, A. Santos and S. Simoes, Bcl-2-targeted antisense therapy (oblimersen sodium): towards clinical reality, *Rev. Recent Clin. Trials*, **1**, 217–235 (2006).
62. B. Pro *et al.*, Phase II multicenter study of oblimersen sodium, a Bcl-2 antisense oligonucleotide, in combination with rituximab in patients with recurrent B-cell non-Hodgkin lymphoma, *Br. J. Haematol.* **143**, 355–360 (2008).

63. B.T. Gjertsen, T. Bredholt, N. Anensen and O.K. Vintermyr, Bcl-2 antisense in the treatment of human malignancies: a delusion in targeted therapy, *Curr. Pharm. Biotechnol.*, **8**, 373–381 (2007).
64. M. Bayes and X. Rabasseda, Gateways to clinical trials, *Methods Findings Exp. Clin. Pharmacol.*, **30**, 67–99 (2008).
65. J. Roberts *et al.*, Efficient and persistent splice switching by systemically delivered LNA oligonucleotides in mice, *Mol. Ther.*, **14**, 471–475 (2006).
66. T.R. Cech, A.J. Zaug and P.J. Grabowski, *In vitro* splicing of the ribosomal RNA precursor of *Tetrahymena*: involvement of a guanosine nucleotide in the excision of the intervening sequence, *Cell*, **27**, 487–496 (1981).
67. K. Kruger *et al.*, Self-splicing RNA: autoexcision and autocyclization of the ribosomal RNA intervening sequence of *Tetrahymena*, *Cell*, **31**, 147–157 (1982).
68. C. Guerrier-Takada, K. Gardiner, T. Marsh, N. Pace and S. Altman, The RNA moiety of ribonuclease P is the catalytic subunit of the enzyme, *Cell*, **35**, 849–857 (1983).
69. J.A. Doudna and T.R. Cech, The chemical repertoire of natural ribozymes, *Nature*, **418**, 222–228 (2002).
70. B.A. Sullenger and T.R. Cech, Ribozyme-mediated repair of defective mRNA by targeted, trans-splicing, *Nature*, **371**, 619–622 (1994).
71. S.H. Hong *et al.*, *In vivo* reprogramming of hTERT by trans-splicing ribozyme to target tumor cells, *Mol. Ther.*, **16**, 74–80 (2008).
72. O.C. Uhlenbeck, A small catalytic oligoribonucleotide, *Nature*, **328**, 596–600 (1987).
73. J. Haseloff and W.L. Gerlach, Simple RNA enzymes with new and highly specific endoribonuclease activities, *Nature*, **334**, 585–591 (1988).
74. S.P. Ohuchi, Y. Ikawa and Y. Nakamura, Selection of a novel class of RNA–RNA interaction motifs based on the ligase ribozyme with defined modular architecture, *Nucleic Acids Res.*, **36**, 3600–3607 (2008).
75. J. Conaty, P. Hendry and T. Lockett, Selected classes of minimised hammerhead ribozyme have very high cleavage rates at low Mg^{2+} concentration, *Nucleic Acids Res.*, **27**, 2400–2407 (1999).
76. S.P. Zinnen *et al.*, Selection, design and characterization of a new potentially therapeutic ribozyme, *RNA*, **8**, 214–228 (2002).
77. F. Eckstein, A.R. Kore and K.L. Nakamaye, *In vitro* selection of hammerhead ribozyme sequence variants, *ChemBioChem.*, **2**, 629–635 (2001).
78. B. Seelig and A. Jaschke, A small catalytic RNA motif with Diels–Alderase activity, *Chem. Biol.*, **6**, 167–176 (1999).
79. L. Beigelman *et al.*, Chemical modification of hammerhead ribozymes. Catalytic activity and nuclease resistance, *J. Biol. Chem.*, **270**, 25702–25708 (1995).
80. L. Beigelman *et al.*, Synthesis of 2′-modified nucleotides and their incorporation into hammerhead ribozymes, *Nucleic Acids Res.*, **23**, 4434–4442 (1995).
81. V. Brower *et al.*, All clear for HIV-targeting ribozyme in phase II, *Nat. Biotechnol.*, **16**, 123 (1998).
82. P.M. Rowe, Ribozymes enter clinical trials for HIV-1 treatment, *Lancet*, **348**, 1302 (1996).
83. F. Wong-Staal, E.M. Poeschla and D.J. Looney, A controlled, Phase 1 clinical trial to evaluate the safety and effects in HIV-1 infected humans of autologous lymphocytes transduced with a ribozyme that cleaves HIV-1 RNA, *Hum. Gene Ther.*, **9**, 2407–2425 (1998).
84. R.G. Amado *et al.*, A phase I trial of autologous $CD34^+$ hematopoietic progenitor cells transduced with an anti-HIV ribozyme, *Hum. Gene Ther.*, **10**, 2255–2270 (1999).
85. N. Sarver *et al.*, Ribozymes as potential anti-HIV-1 therapeutic agents, *Science*, **247**, 1222–1225 (1990).
86. J.L. Macpherson *et al.*, Long-term survival and concomitant gene expression of ribozyme-transduced $CD4^+$ T-lymphocytes in HIV-infected patients, *J. Gene Med.*, **7**, 552–564 (2005).
87. D.G. Macejak *et al.*, Inhibition of hepatitis C virus (HCV)-RNA-dependent translation and replication of a chimeric HCV poliovirus using synthetic stabilized ribozymes, *Hepatology*, **31**, 769–776 (2000).
88. A. Peracchi, Prospects for antiviral ribozymes and deoxyribozymes, *Rev. Med. Virol.*, **14**, 47–64 (2004).
89. D.E. Weng *et al.*, A Phase I clinical trial of a ribozyme-based angiogenesis inhibitor targeting vascular endothelial growth factor receptor-1 for patients with refractory solid tumors, *Mol. Cancer Ther.*, **4**, 948–955 (2005).
90. S.W. Santoro and G.F. Joyce, A general purpose RNA-cleaving DNA enzyme, *Proc. Natl. Acad. Sci. USA*, **94**, 4262–4266 (1997).
91. C.R. Dass, E.G. Saravolac, Y. Li and L.Q. Sun, Cellular uptake, distribution and stability of 10–23 deoxyribozymes, *Antisense Nucleic Acid Drug Dev.*, **12**, 289–299 (2002).

92. S. Schubert *et al.*, RNA cleaving '10–23' DNAzymes with enhanced stability and activity, *Nucleic Acids Res.*, **31**, 5982–5992 (2003).
93. J.C. Achenbach, W. Chiuman, R.P. Cruz and Y. Li, DNAzymes: from creation *in vitro* to application *in vivo*, *Curr. Pharm. Biotechnol.*, **5**, 321–336 (2004).
94. S.D. Patil, D.G. Rhodes and D.J. Burgess, DNA-based therapeutics and DNA delivery systems: a comprehensive review, *AAPS J.*, **7**, E61–E77 (2005).
95. R. Bhindi *et al.*, Brothers in arms: DNA enzymes, short interfering RNA and the emerging wave of small-molecule nucleic acid-based gene-silencing strategies, *Am. J. Pathol.*, **171**, 1079–1088 (2007).
96. L.M. Khachigian, Early growth response-1 in cardiovascular pathobiology, *Circ. Res.*, **98**, 186–191 (2006).
97. L.M. Khachigian, Catalytic DNAs as potential therapeutic agents and sequence-specific molecular tools to dissect biological function, *J. Clin. Invest*, **106**, 1189–1195 (2000).
98. L. Zhang *et al.*, Angiogenic inhibition mediated by a DNAzyme that targets vascular endothelial growth factor receptor 2, *Cancer Res.*, **62**, 5463–5469 (2002).
99. A. Goodchild *et al.*, Cytotoxic G-rich oligodeoxynucleotides: putative protein targets and required sequence motif, *Nucleic Acids Res.*, **35**, 4562–4572 (2007).
100. L. Rivory *et al.*, The DNAzymes Rs6, Dz13 and DzF have potent biologic effects independent of catalytic activity, *Oligonucleotides*, **16**, 297–312 (2006).
101. A. Fire *et al.*, Potent and specific genetic interference by double-stranded RNA in *Caenorhabditis elegans*, *Nature*, **391**, 806–811 (1998).
102. T.M. Rana, Illuminating the silence: understanding the structure and function of small RNAs, *Nat. Rev. Mol. Cell Biol.*, **8**, 23–36 (2007).
103. M.A. Matzke and J.A. Birchler, RNAi-mediated pathways in the nucleus, *Nat. Rev. Genet.*, **6**, 24–35 (2005).
104. M. Wassenegger, The role of the RNAi machinery in heterochromatin formation, *Cell*, **122**, 13–16 (2005).
105. S.M. Elbashir *et al.*, Duplexes of 21-nucleotide RNAs mediate RNA interference in cultured mammalian cells, *Nature*, **411**, 494–498 (2001).
106. S.D. Rose *et al.*, Functional polarity is introduced by Dicer processing of short substrate RNAs, *Nucleic Acids Res.*, **33**, 4140–4156 (2005).
107. D.H. Kim *et al.*, Synthetic dsRNA Dicer substrates enhance RNAi potency and efficacy, *Nat. Biotechnol.*, **23**, 222–226 (2005).
108. D. Siolas *et al.*, Synthetic shRNAs as potent RNAi triggers, *Nat. Biotechnol.*, **23**, 227–231 (2005).
109. A.L. Jackson *et al.*, Widespread siRNA 'off-target' transcript silencing mediated by seed region sequence complementarity, *RNA*, **12**, 1179–1187 (2006).
110. A.L. Jackson *et al.*, Position-specific chemical modification of siRNAs reduces 'off-target' transcript silencing, *RNA*, **12**, 1197–1205 (2006).
111. J.T. Marques and B.R. Williams, Activation of the mammalian immune system by siRNAs, *Nat. Biotechnol.*, **23**, 1399–1405 (2005).
112. V. Hornung *et al.*, Sequence-specific potent induction of IFN-alpha by short interfering RNA in plasmacytoid dendritic cells through TLR7, *Nat. Med.*, **11**, 263–270 (2005).
113. A.D. Judge, G. Bola, A.C. Lee and I. MacLachlan, Design of noninflammatory synthetic siRNA mediating potent gene silencing *in vivo*, *Mol. Ther.*, **13**, 494–505 (2006).
114. M. Sioud, RNA interference below the immune radar, *Nat. Biotechnol.*, **24**, 521–522 (2006).
115. D.V. Morrissey *et al.*, Activity of stabilized short interfering RNA in a mouse model of hepatitis B virus replication, *Hepatology*, **41**, 1349–1356 (2005).
116. D.V. Morrissey *et al.*, Potent and persistent *in vivo* anti-HBV activity of chemically modified siRNAs, *Nat. Biotechnol.*, **23**, 1002–1007 (2005).
117. M. Robbins *et al.*, 2′-*O*-Methyl-modified RNAs act as TLR7 antagonists, *Mol. Ther.*, **15**, 1663–1669 (2007).
118. J. Soutschek *et al.*, Therapeutic silencing of an endogenous gene by systemic administration of modified siRNAs, *Nature*, **432**, 173–178 (2004).
119. J.D. Heidel, S. Hu, X.F. Liu, T.J. Triche and M.E. Davis, Lack of interferon response in animals to naked siRNAs, *Nat. Biotechnol.*, **22**, 1579–1582 (2004).

120. Y. Minakuchi *et al.*, Atelocollagen-mediated synthetic small interfering RNA delivery for effective gene silencing *in vitro* and *in vivo*, *Nucleic Acids Res.*, **32**, e109 (2004).
121. M.A. Robbins *et al.*, Stable expression of shRNAs in human CD34(+) progenitor cells can avoid induction of interferon responses to siRNAs *in vitro*, *Nat. Biotechnol.*, **24**, 566–571 (2006).
122. D. Grimm *et al.*, Fatality in mice due to oversaturation of cellular microRNA/short hairpin RNA pathways, *Nature*, **441**, 537–541 (2006).
123. M. Schlee, V. Hornung and G. Hartmann, siRNA and isRNA: two edges of one sword, *Mol. Ther.*, **14**, 463–470 (2006).
124. D.R. Corey, Chemical modification: the key to clinical application of RNA interference?, *J. Clin. Invest*, **117**, 3615–3622 (2007).
125. T.P. Prakash *et al.*, Positional effect of chemical modifications on short interference RNA activity in mammalian cells, *J. Med. Chem.*, **48**, 4247–4253 (2005).
126. Y. Fedorov *et al.*, Off-target effects by siRNA can induce toxic phenotype, *RNA*, **12**, 1188–1196 (2006).
127. Y.L. Chiu and T.M. Rana, siRNA function in RNAi: a chemical modification analysis, *RNA*, **9**, 1034–1048 (2003).
128. P.Y. Chen *et al.*, Strand-specific 5′-*O*-methylation of siRNA duplexes controls guide strand selection and targeting specificity, *RNA*, **14**, 263–274 (2008).
129. T. Kubo, Z. Zhelev, H. Ohba and R. Bakalova, Modified 27-nt dsRNAs with dramatically enhanced stability in serum and long-term RNAi activity, *Oligonucleotides*, **17**, 445–464 (2007).
130. J.W. Gaynor, J. Brazier and R. Cosstick, Synthesis of 3′-*S*-phosphorothiolate oligonucleotides for their potential use in RNA interference, *Nucleosides Nucleotides Nucleic Acids*, **26**, 709–712 (2007).
131. A.H. Hall, J. Wan, E.E. Shaughnessy, S.B. Ramsay and K.A. Alexander, RNA interference using boranophosphate siRNAs: structure–activity relationships, *Nucleic Acids Res.*, **32**, 5991–6000 (2004).
132. S. Hoshika, N. Minakawa and A. Matsuda, RNA interference induced by siRNAs modified with 4′-thioribonucleosides, *Nucleic Acids Symp. Ser.*, 77–78 (2005).
133. O.R. Mook, F. Baas, M.B. De Wissel and K. Fluiter, Evaluation of locked nucleic acid-modified small interfering RNA *in vitro* and *in vivo*, *Mol. Cancer Ther.*, **6**, 833–843 (2007).
134. J. Elmen *et al.*, Locked nucleic acid (LNA) mediated improvements in siRNA stability and functionality, *Nucleic Acids Res.*, **33**, 439–447 (2005).
135. A. Aigner, Gene silencing through RNA interference (RNAi) *in vivo*: strategies based on the direct application of siRNAs, *J. Biotechnol.*, **124**, 12–25 (2006).
136. T.S. Zimmermann *et al.*, RNAi-mediated gene silencing in non-human primates, *Nature*, **441**, 111–114 (2006).
137. A. Pal *et al.*, Systemic delivery of RafsiRNA using cationic cardiolipin liposomes silences Raf-1 expression and inhibits tumor growth in xenograft model of human prostate cancer, *Int. J. Oncol.*, **26**, 1087–1091 (2005).
138. K.F. Pirollo *et al.*, Tumor-targeting nanoimmunoliposome complex for short interfering RNA delivery, *Hum. Gene Ther.*, **17**, 117–124 (2006).
139. S. Hu-Lieskovan, J.D. Heidel, D.W. Bartlett, M.E. Davis and T.J. Triche, Sequence-specific knockdown of EWS-FLI1 by targeted, nonviral delivery of small interfering RNA inhibits tumor growth in a murine model of metastatic Ewing's sarcoma, *Cancer Res.*, **65**, 8984–8992 (2005).
140. B. Liu, Exploring cell type-specific internalizing antibodies for targeted delivery of siRNA, *Brief. Funct. Genomics Proteomics*, **6**, 112–119 (2007).
141. J.J. Rossi, Partnering aptamer and RNAi technologies, *Mol. Ther.*, **14**, 461–462 (2006).
142. E. Song *et al.*, Antibody mediated *in vivo* delivery of small interfering RNAs via cell-surface receptors, *Nat. Biotechnol.*, **23**, 709–717 (2005).
143. D. Peer, P. Zhu, C.V. Carman, J. Lieberman and M. Shimaoka, Selective gene silencing in activated leukocytes by targeting siRNAs to the integrin lymphocyte function-associated antigen-1, *Proc. Natl. Acad. Sci. USA*, **104**, 4095–4100 (2007).
144. J.O. McNamara *et al.*, Cell type-specific delivery of siRNAs with aptamer-siRNA chimeras, *Nat. Biotechnol.*, **24**, 1005–1015 (2006).
145. T.C. Chu, K.Y. Twu, A.D. Ellington and M. Levy, Aptamer mediated siRNA delivery, *Nucleic Acids Res.*, **34**, e73 (2006).
146. J. Zhou, H. Li, S. Li, J. Zaia and J.J. Rossi, Novel dual inhibitory function aptamer-siRNA delivery system for HIV-1 therapy, *Mol. Ther.* **16**, 1481–1489 (2008).

147. C. Baum *et al.*, Chance or necessity? Insertional mutagenesis in gene therapy and its consequences, *Mol. Ther.*, **9**, 5–13 (2004).
148. E. Marshall, Gene therapy. What to do when clear success comes with an unclear risk?, *Science*, **298**, 510–511 (2002).
149. H. Sumimoto and Y. Kawakami, Lentiviral vector-mediated RNAi and its use for cancer research, *Future Oncol.*, **3**, 655–664 (2007).
150. J.J. Rossi, C.H. June and D.B. Kohn, Genetic therapies against HIV, *Nat. Biotechnol.*, **25**, 1444–1454 (2007).
151. F.A. de, H.P. Vornlocher, J. Maraganore and J. Lieberman, Interfering with disease: a progress report on siRNA-based therapeutics, *Nat. Rev. Drug Discov.*, **6**, 443–453 (2007).
152. M.A. Behlke, Progress towards *in vivo* use of siRNAs, *Mol. Ther.*, **13**, 644–670 (2006).
153. D.H. Kim and J.J. Rossi, Strategies for silencing human disease using RNA interference, *Nat. Rev. Genet.*, **8**, 173–184 (2007).
154. B.J. Li *et al.*, Using siRNA in prophylactic and therapeutic regimens against SARS coronavirus in Rhesus macaque, *Nat. Med.*, **11**, 944–951 (2005).
155. D. Palliser *et al.*, An siRNA-based microbicide protects mice from lethal herpes simplex virus 2 infection, *Nature*, **439**, 89–94 (2006).
156. J.M. Jacque, K. Triques and M. Stevenson, Modulation of HIV-1 replication by RNA interference, *Nature*, **418**, 435–438 (2002).
157. G.A. Coburn and B.R. Cullen, Potent and specific inhibition of human immunodeficiency virus type 1 replication by RNA interference, *J. Virol.*, **76**, 9225–9231 (2002).
158. M.E. Kleinman *et al.*, Sequence- and target-independent angiogenesis suppression by siRNA via TLR3, *Nature*, **452**, 591–597 (2008).
159. S. Griffiths-Jones, R.J. Grocock, S. van Dongen, A. Bateman and A.J. Enright, miRBase: microRNA sequences, targets and gene nomenclature, *Nucleic Acids Res.*, **34**, D140–D144 (2006).
160. S. Griffiths-Jones, miRBase: the microRNA sequence database, *Methods Mol. Biol.*, **342**, 129–138 (2006).
161. S. Griffiths-Jones, H.K. Saini, S. van Dongen and A.J. Enright, miRBase: tools for microRNA genomics, *Nucleic Acids Res.*, **36**, D154–D158 (2008).
162. J. Kim *et al.*, microRNA-directed cleavage of ATHB15 mRNA regulates vascular development in Arabidopsis inflorescence stems, *Plant J.*, **42**, 84–94 (2005).
163. B.P. Lewis, C.B. Burge and D.P. Bartel, Conserved seed pairing, often flanked by adenosines, indicates that thousands of human genes are microRNA targets, *Cell*, **120**, 15–20 (2005).
164. W. Filipowicz, S.N. Bhattacharyya and N. Sonenberg, Mechanisms of post-transcriptional regulation by microRNAs: are the answers in sight?, *Nat. Rev. Genet.*, **9**, 102–114 (2008).
165. J. Stenvang and S. Kauppinen, MicroRNAs as targets for antisense-based therapeutics, *Expert Opin. Biol. Ther.*, **8**, 59–81 (2008).
166. B. Zhang and M.A. Farwell, microRNAs: a new emerging class of players for disease diagnostics and gene therapy, *J. Cell Mol. Med.*, **12**, 3–21 (2008).
167. H.S. Soifer, J.J. Rossi and P. Saetrom, MicroRNAs in disease and potential therapeutic applications, *Mol. Ther.*, **15**, 2070–2079 (2007).
168. A. Esquela-Kerscher and F.J. Slack, Oncomirs – microRNAs with a role in cancer, *Nat. Rev. Cancer*, **6**, 259–269 (2006).
169. C.H. Lawrie, MicroRNA expression in lymphoma, *Expert Opin. Biol. Ther.*, **7**, 1363–1374 (2007).
170. E. Tili, J.J. Michaille and G.A. Calin, Expression and function of micro-RNAs in immune cells during normal or disease state, *Int. J. Med. Sci.*, **5**, 73–79 (2008).
171. G.A. Calin *et al.*, Human microRNA genes are frequently located at fragile sites and genomic regions involved in cancers, *Proc. Natl. Acad. Sci. USA*, **101**, 2999–3004 (2004).
172. G.A. Calin *et al.*, Frequent deletions and down-regulation of micro-RNA genes miR15 and miR16 at 13q14 in chronic lymphocytic leukemia, *Proc. Natl. Acad. Sci. USA*, **99**, 15524–15529 (2002).
173. M.Z. Michael, S.M. O'Connor, N.G. van Holst Pellekaan, G.P. Young and R.J. James, Reduced accumulation of specific microRNAs in colorectal neoplasia, *Mol. Cancer Res.*, **1**, 882–891 (2003).
174. L. He *et al.*, A microRNA polycistron as a potential human oncogene, *Nature*, **435**, 828–833 (2005).

175. C. Esau *et al.*, miR-122 regulation of lipid metabolism revealed by *in vivo* antisense targeting, *Cell Metab*, **3**, 87–98 (2006).
176. W.J. Lukiw, Micro-RNA speciation in fetal, adult and Alzheimer's disease hippocampus, *Neuroreport*, **18**, 297–300 (2007).
177. Y. Zhao *et al.*, Dysregulation of cardiogenesis, cardiac conduction and cell cycle in mice lacking miRNA-1–2, *Cell*, **129**, 303–317 (2007).
178. A. Clop *et al.*, A mutation creating a potential illegitimate microRNA target site in the myostatin gene affects muscularity in sheep, *Nat. Genet.*, **38**, 813–818 (2006).
179. B.R. Cullen, Viruses and microRNAs, *Nat. Genet.*, **38** Suppl., S25–S30 (2006).
180. C.L. Jopling, M. Yi, A.M. Lancaster, S.M. Lemon and P. Sarnow, Modulation of hepatitis C virus RNA abundance by a liver-specific microRNA, *Science*, **309**, 1577–1581 (2005).
181. G.A. Calin and C.M. Croce, MicroRNA signatures in human cancers, *Nat. Rev. Cancer*, **6**, 857–866 (2006).
182. C. Guimaraes-Sternberg, A. Meerson, I. Shaked and H. Soreq, MicroRNA modulation of megakaryoblast fate involves cholinergic signaling, *Leuk. Res.*, **30**, 583–595 (2006).
183. T.R. Brummelkamp, R. Bernards and R. Agami, Stable suppression of tumorigenicity by virus-mediated RNA interference, *Cancer Cell*, **2**, 243–247 (2002).
184. P.J. Paddison, A.A. Caudy, E. Bernstein, G.J. Hannon and D.S. Conklin, Short hairpin RNAs (shRNAs) induce sequence-specific silencing in mammalian cells, *Genes Dev.*, **16**, 948–958 (2002).
185. J. Krutzfeldt and M. Stoffel, MicroRNAs: a new class of regulatory genes affecting metabolism, *Cell Metab*, **4**, 9–12 (2006).
186. J. Krutzfeldt *et al.*, Silencing of microRNAs *in vivo* with 'antagomirs', *Nature*, **438**, 685–689 (2005).
187. J. Elmen *et al.*, LNA-mediated microRNA silencing in non-human primates, *Nature*, **452**, 896–899 (2008).
188. J. Xiao *et al.*, Novel approaches for gene-specific interference via manipulating actions of microRNAs: examination on the pacemaker channel genes HCN2 and HCN4, *J. Cell Physiol*, **212**, 285–292 (2007).
189. A.D. George and S.A. Tenenbaum, MicroRNA modulation of RNA-binding protein regulatory elements, *RNA Biol.*, **3**, 57–59 (2006).
190. M.S. Ebert, J.R. Neilson and P.A. Sharp, MicroRNA sponges: competitive inhibitors of small RNAs in mammalian cells, *Nat. Methods*, **4**, 721–726 (2007).
191. J.M. Franco-Zorrilla *et al.*, Target mimicry provides a new mechanism for regulation of microRNA activity, *Nat. Genet.*, **39**, 1033–1037 (2007).
192. Y. Karube *et al.*, Reduced expression of Dicer associated with poor prognosis in lung cancer patients, *Cancer Sci.*, **96**, 111–115 (2005).
193. J.M. Thomson *et al.*, Extensive post-transcriptional regulation of microRNAs and its implications for cancer, *Genes Dev.*, **20**, 2202–2207 (2006).
194. A.D. Ellington and J.W. Szostak, *In vitro* selection of RNA molecules that bind specific ligands, *Nature*, **346**, 818–822 (1990).
195. R.R. Breaker, Natural and engineered nucleic acids as tools to explore biology, *Nature*, **432**, 838–845 (2004).
196. B.R. Cullen and W.C. Greene, Regulatory pathways governing HIV-1 replication, *Cell*, **58**, 423–426 (1989).
197. R.A. Marciniak, M.A. Garcia-Blanco and P.A. Sharp, Identification and characterization of a HeLa nuclear protein that specifically binds to the trans-activation-response (TAR) element of human immunodeficiency virus, *Proc. Natl. Acad. Sci. USA*, **87**, 3624–3628 (1990).
198. B.A. Sullenger, H.F. Gallardo, G.E. Ungers and E. Gilboa, Overexpression of TAR sequences renders cells resistant to human immunodeficiency virus replication, *Cell*, **63**, 601–608 (1990).
199. C. Tuerk and L. Gold, Systematic evolution of ligands by exponential enrichment: RNA ligands to bacteriophage T4 DNA polymerase, *Science*, **249**, 505–510 (1990).
200. W.A. Pieken *et al.*, Structure–function relationship of hammerhead ribozymes as probed by 2′-modifications, *Nucleic Acids Symp. Ser.*, 51–53 (1991).
201. J. Ruckman *et al.*, 2′-Fluoropyrimidine RNA-based aptamers to the 165-amino acid form of vascular endothelial growth factor (VEGF165). Inhibition of receptor binding and VEGF-induced vascular permeability through interactions requiring the exon 7-encoded domain, *J. Biol. Chem.*, **273**, 20556–20567 (1998).
202. P.E. Burmeister *et al.*, Direct *in vitro* selection of a 2′-*O*-methyl aptamer to VEGF, *Chem. Biol.*, **12**, 25–33 (2005).

203. K.S. Schmidt *et al.*, Application of locked nucleic acids to improve aptamer *in vivo* stability and targeting function, *Nucleic Acids Res.*, **32**, 5757–5765 (2004).
204. F. Darfeuille, J.B. Hansen, H. Orum, P.C. Di and J.J. Toulme, LNA/DNA chimeric oligomers mimic RNA aptamers targeted to the TAR RNA element of HIV-1, *Nucleic Acids Res.*, **32**, 3101–3107 (2004).
205. S. Klussmann, A. Nolte, R. Bald, V.A. Erdmann and J.P. Furste, Mirror-image RNA that binds D-adenosine, *Nat. Biotechnol.*, **14**, 1112–1115 (1996).
206. A. Nolte, S. Klussmann, R. Bald, V.A. Erdmann and J.P. Furste, Mirror-design of L-oligonucleotide ligands binding to L-arginine, *Nat. Biotechnol.*, **14**, 1116–1119 (1996).
207. C.E. Tucker *et al.*, Detection and plasma pharmacokinetics of an anti-vascular endothelial growth factor oligonucleotide-aptamer (NX1838) in rhesus monkeys, *J. Chromatogr. B Biomed. Sci. Appl.*, **732**, 203–212 (1999).
208. M.C. Willis *et al.*, Liposome-anchored vascular endothelial growth factor aptamers, *Bioconj. Chem.*, **9**, 573–582 (1998).
209. C.P. Rusconi *et al.*, Antidote-mediated control of an anticoagulant aptamer *in vivo*, *Nat. Biotechnol.*, **22**, 1423–1428 (2004).
210. C. Pestourie, B. Tavitian and F. Duconge, Aptamers against extracellular targets for *in vivo* applications, *Biochimie*, **87**, 921–930 (2005).
211. M. Famulok and G. Mayer, Intramers and aptamers: applications in protein-function analyses and potential for drug screening, *ChemBioChem.*, **6**, 19–26 (2005).
212. J. Mi *et al.*, H1 RNA polymerase III promoter-driven expression of an RNA aptamer leads to high-level inhibition of intracellular protein activity, *Nucleic Acids Res.*, **34**, 3577–3584 (2006).
213. G. Kaur and I. Roy, Therapeutic applications of aptamers, *Expert Opin. Investig. Drugs*, **17**, 43–60 (2008).
214. N.S. Que-Gewirth and B.A. Sullenger, Gene therapy progress and prospects: RNA aptamers, *Gene Ther.*, **14**, 283–291 (2007).
215. S.M. Nimjee, C.P. Rusconi and B.A. Sullenger, Aptamers: an emerging class of therapeutics, *Annu. Rev. Med.*, **56**, 555–583 (2005).
216. S.A. Vinores, Pegaptanib in the treatment of wet, age-related macular degeneration, *Int. J. Nanomed.*, **1**, 263–268 (2006).
217. S.A. Vinores, Technology evaluation: pegaptanib, Eyetech/Pfizer, *Curr. Opin. Mol. Ther.*, **5**, 673–679 (2003).
218. J. Huang *et al.*, Highly specific antiangiogenic therapy is effective in suppressing growth of experimental Wilms tumors, *J. Pediatr. Surg.*, **36**, 357–361 (2001).
219. C.R. Ireson and L.R. Kelland, Discovery and development of anticancer aptamers, *Mol. Cancer Ther.*, **5**, 2957–2962 (2006).
220. A.C. Girvan *et al.*, AGRO100 inhibits activation of nuclear factor-kappaB (NF-kappaB) by forming a complex with NF-kappaB essential modulator (NEMO) and nucleolin, *Mol. Cancer Ther.*, **5**, 1790–1799 (2006).
221. D.M. Miller *et al.*, Extended phase I study of AS1411 in renal and non-small cell lung cancers, *Ann. Oncol.*, **17**, 147–148 (2006).
222. R. Morishita *et al.*, A gene therapy strategy using a transcription factor decoy of the E2F binding site inhibits smooth muscle proliferation *in vivo*, *Proc. Natl. Acad. Sci. USA*, **92**, 5855–5859 (1995).
223. A. Iwasaki and R. Medzhitov, Toll-like receptor control of the adaptive immune responses, *Nat. Immunol.*, **5**, 987–995 (2004).
224. A.M. Krieg, Therapeutic potential of Toll-like receptor 9 activation, *Nat. Rev. Drug Discov.*, **5**, 471–484 (2006).
225. A.M. Krieg, Antitumor applications of stimulating Toll-like receptor 9 with CpG oligodeoxynucleotides, *Curr. Oncol. Rep.*, **6**, 88–95 (2004).
226. F. Hayashi, T.K. Means and A.D. Luster, Toll-like receptors stimulate human neutrophil function, *Blood*, **102**, 2660–2669 (2003).
227. J. Li *et al.*, CpG DNA-mediated immune response in pulmonary endothelial cells, *Am. J. Physiol. Lung Cell Mol. Physiol.*, **287**, L552–L558 (2004).
228. M.C. Lebre *et al.*, Human keratinocytes express functional Toll-like receptor 3, 4, 5 and 9, *J. Invest Dermatol.*, **127**, 331–341 (2007).
229. G. Pedersen, L. Andresen, M.W. Matthiessen, J. Rask-Madsen and J. Brynskov, Expression of Toll-like receptor 9 and response to bacterial CpG oligodeoxynucleotides in human intestinal epithelium, *Clin. Exp. Immunol.*, **141**, 298–306 (2005).

230. R. Rankin *et al.*, CpG motif identification for veterinary and laboratory species demonstrates that sequence recognition is highly conserved, *Antisense Nucleic Acid Drug Dev.*, **11**, 333–340 (2001).
231. A.M. Krieg, G. Hartmann and A.K. Yi, Mechanism of action of CpG DNA, *Curr. Top. Microbiol. Immunol.*, **247**, 1–21 (2000).
232. G. Hartmann and A.M. Krieg, Mechanism and function of a newly identified CpG DNA motif in human primary B cells, *J. Immunol.*, **164**, 944–953 (2000).
233. A.K. Yi, M. Chang, D.W. Peckham, A.M. Krieg and R.F. Ashman, CpG oligodeoxyribonucleotides rescue mature spleen B cells from spontaneous apoptosis and promote cell cycle entry, *J. Immunol.*, **160**, 5898–5906 (1998).
234. G. Hartmann *et al.*, Delineation of a CpG phosphorothioate oligodeoxynucleotide for activating primate immune responses *in vitro* and *in vivo*, *J. Immunol.*, **164**, 1617–1624 (2000).
235. T.L. Roberts, M.J. Sweet, D.A. Hume and K.J. Stacey, Cutting edge: species-specific TLR9-mediated recognition of CpG and non-CpG phosphorothioate-modified oligonucleotides, *J. Immunol.*, **174**, 605–608 (2005).
236. D.E. Speiser *et al.*, Rapid and strong human $CD8^+$ T cell responses to vaccination with peptide, IFA and CpG oligodeoxynucleotide 7909, *J. Clin. Invest*, **115**, 739–746 (2005).
237. V. Appay *et al.*, New generation vaccine induces effective melanoma-specific $CD8^+$ T cells in the circulation but not in the tumor site, *J. Immunol.*, **177**, 1670–1678 (2006).
238. B. Rehermann and M. Nascimbeni, Immunology of hepatitis B virus and hepatitis C virus infection, *Nat. Rev. Immunol.*, **5**, 215–229 (2005).
239. A.M. Krieg, Toll-like receptor 9 (TLR9) agonists in the treatment of cancer, *Oncogene*, **27**, 161–167 (2008).
240. J.P. Sheehan and H.C. Lan, Phosphorothioate oligonucleotides inhibit the intrinsic tenase complex, *Blood*, **92**, 1617–1625 (1998).
241. S.P. Henry *et al.*, Complement activation is responsible for acute toxicities in rhesus monkeys treated with a phosphorothioate oligodeoxynucleotide, *Int. Immunopharmacol.*, **2**, 1657–1666 (2002).

12

Innate Immune Recognition of Nucleic Acids

Stefan Bauer

12.1 Introduction

Deoxyribonucleic acid (DNA) and ribonucleic acid (RNA) are indispensable components in living organisms. DNA stores genetic information whereas RNA serves as genome for some viruses, is a prerequisite for protein translation and has important regulatory functions.

Evidence has accumulated over the past few decades that nucleic acids, when released from pathogens or host cells, can activate the innate immune system and lead to enhanced anti-tumor activity and type I interferon production [1–4].

The immunostimulatory capacity of viral and host cell-derived RNA has been known for several decades. For example, synthetic RNA mimics such as single-stranded poly(inosinic acid) or double-stranded polyinosine–poly(cytidylic acid) [poly(I:C)] were identified as immunostimulatory agents 40 years ago [5–7].

Studies in 1984 by Tokunaga *et al.* first demonstrated that bacterial DNA itself was the component of Bacillus Calmette–Guérin (BCG), which promoted immunostimulatory and anti-tumor effects [4]. Stimulation led to inhibition of virus replication and promoted immunostimulatory and anti-tumor effects. The stimulatory effect of bacterial DNA is due to the presence of unmethylated CpG dinucleotides in a particular base context named CpG motif [8]. In contrast vertebrate DNA is not or less stimulatory due to methylation of CpG dinucleotides, their low frequency (CpG suppression) and the presence of possibly inhibitory sequences [9,10]. However, eukaryotic DNA such as calf thymus DNA or DNA from dying cells can be immunostimulatory, although to a lesser content than bacterial DNA. Under certain circumstances, such as enhanced DNA uptake, eukaryotic DNA leads to cytokine production and inhibition of viral replication [1,11].

The last 10 years of intense research activities have brought tremendous insight into the various receptor systems that detect nucleic acids. This chapter focuses on nucleic acid receptors of the innate immune system that are located in the endosome/lysosome and the cytoplasm where they detect nucleic acid from pathogens, respectively.

The Chemical Biology of Nucleic Acids Edited by Günter Mayer

12.2 Innate immune receptors for DNA

Several receptors in the endosome and cytoplasm have been reported recently that recognize DNA and stimulate the innate immune system for initiation of an anti-pathogen immune response (Table 12.1 and Figure 12.1).

12.2.1 Endosomal detection of DNA by TLR9

Toll receptors are type I transmembrane proteins which are evolutionarily conserved between insects and vertebrates [12]. In *Drosophila*, Toll was first identified as an essential molecule for dorso-ventral patterning of the embryo and subsequently as a key molecule for the antifungal immune response in the adult animal [13,14]. A homologous family of Toll receptors, termed Toll-like receptors (TLRs), exists in vertebrates [12].

Table 12.1 *Overview of DNA-sensing pattern recognition receptors and their ligands*

DNA-sensing pattern recognition receptor	Ligand
Toll-like receptor 9 (TLR9)	CpG motif-containing DNA [18,54,133] and CpG-independent DNA recognition [32,33]
DAI (DNA-dependent activator of IFN-regulatory factors) (DLM-1/ZBP1)	Cytoplasmic double-stranded B-form DNA such as poly(dA–dT)·poly(dT–dA), plasmids, genomic or bacterial DNA and synthetic double-stranded oligonucleotides [59,60,62]
ADAR1, adenosine deaminase acting on dsRNA 1	Negative regulator of DAI [63]
NALP3 (NACHT-, LRR- and PYD-containing protein 3), cryopyrin, CIAS1	Cytoplasmic bacterial and viral RNA [129–132], cytoplasmic adenoviral DNA [70]
AIM2 (absent in melanoma 2)	Negative regulator of DAI [77–80]
p202	Negative regulator of AIM2 [80].

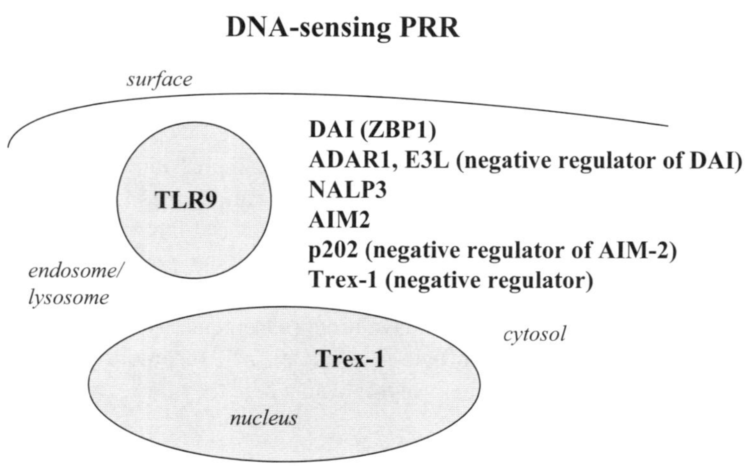

Figure 12.1 *Overview of DNA-sensing pattern recognition receptors in the endosome/lysosome and the cytoplasm*

In vertebrates, 13 members (TLR1–13) have been reported so far which are fundamental in recognition of PAMPs [15,16]. The family of TLRs recognizes various PAMPS from different pathogenic origins such as bacteria, viruses, fungi and protozoan parasites [17]. One subfamily of TLRs that consists of TLR1, -2, -4, -5, -6 and -11 is expressed on the surface of cells, recognizes a plurality of different structures and can be phagocytosed. The other subfamily is formed by TLR-3, -7, -8 and -9 that are localized inside the cell in vesicles where these receptors recognize nucleic acids [17].

TRL9 recognizes DNA with a core DNA sequence consisting of a hexamer motif with a central CpG dinucleotide flanked by 5′-purines and 3′-pyrimidines [8,18]. In activated cells, TLR9 is expressed in endosomal/lysosomal compartments but in non-activated immune cells TLR9 is found in the endoplasmic reticulum (ER) [19,20]. Upon cellular activation, TLR9 traffics to endosomal and lysosomal compartments where it interacts with DNA at acidic pH, a condition that is thought to be necessary for DNA recognition. The exact molecular basis for the retention of TLR9 in the endoplasmic reticulum (ER) in quiescent cells and the subsequent trafficking to the endosome upon cellular stimulation are not well understood. However, the recently described ER resident protein unc93b is involved in TLR9 trafficking since a dominant negative mutant of unc93b leads to non-responsiveness of TLR9 (and other endosomally expressed TLRs) accompanied by the disruption of TLR–unc93b interaction [21–23].

The immunostimulatory effects of DNA can be mimicked by synthetic oligodeoxynucleotides (ODNs) containing a hexamer motif consisting of a central CpG dinucleotide with 5′-purines and 3′-pyrimidines [8]. Meanwhile, three major classes of structurally and phenotypically different CpG ODN have been described [24–26]. The A-class CpG ODNs (also referred to as type D) are potent inducers of interferon-alpha (IFNα) secretion from plasmacytoid dendritic cells (pDCs), but poor inducers of B cell stimulation. B-class ODN (also referred to as type K) have a completely phosphorothioate (PTO) backbone and are strong B-cell stimulators but only weakly induce IFNα secretion. However, if B-class CpG ODNs are artificially forced into higher ordered structures with cationic lipid transfection or complexation to polymixin B, they show the same immune profile as the A-class CpG ODN [27,28]. The C-class CpG ODN exerts immune properties that are intermediate between those of the A and B classes, inducing both B-cell activation and IFNα secretion. The unique structure of these ODNs with a 5′ CpG motif and a 3′ palindrome may allow duplex formation within the endosomal environment leading to the characteristics of cytokine production [29–31].

Recently, the strict CpG-dependent recognition of phosphodiester (PD) DNA by TLR9 has been questioned, since non-CpG ODNs stimulated FLT3L (FMS-like tyrosine kinase 3 ligand)-generated DCs in a sequence-independent but TLR9-dependent manner. This stimulative capacity was dependent on complexation to cationic lipid transfection reagents such as *N*-[1-(2,3-dioleoyloxy)propyl]-*N*,*N*,*N*-trimethylammonium methylsulfate (DOTAP) or on addition of polyG tails at the 3′ end promoting aggregate formation [32,33]. Haas *et al.* further demonstrated that the sugar backbone determines TLR9-mediated DNA recognition since homopolymeric, base-free PD 2′-deoxyribose acted as a TLR9 agonist albeit at lower efficiency than base-containing DNA [33].

Furthermore, the activity of TLR9 depends on receptor cleavage in the endolysosome as previously reported. The involvement of certain cathepsins such as cathepsin K or cathepsin B is still controversial [34–37].

The signaling cascade leading to TLR-mediated cytokine production is well understood (Figure 12.2). The engagement of Toll-like receptors by their cognate ligands leads to the activation of a signaling cascade with subsequent induction of genes that are involved in the immune response against pathogens. In general, three major pathways are activated; the first culminates in the activation of the transcription factor NF-κB, which acts as a master switch for inflammation. The second leads to activation of the MAP kinases p38 and Jun amino-terminal kinase (JNK), which also participate in increased transcription, and the third pathway leads to type I IFN production via IFN regulatory factors (IRFs) [38].

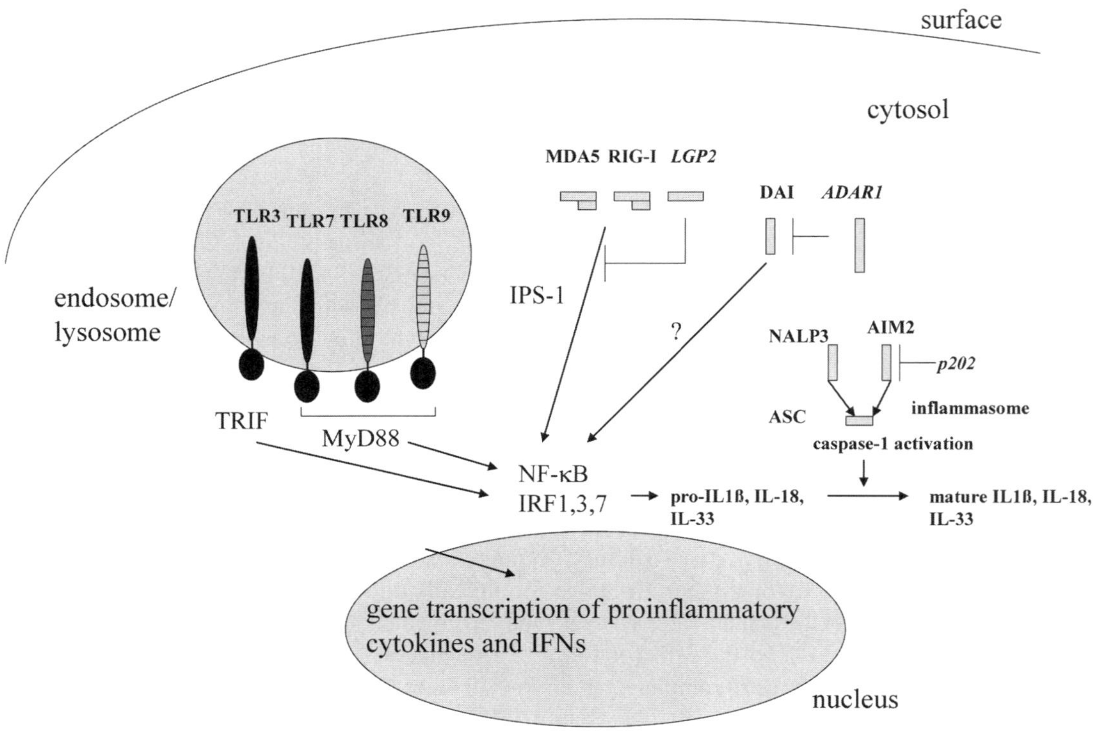

Figure 12.2 *Overview of RNA- and DNA-sensing receptors and signaling pathways. For details, see in the text*

In general, four different adapter molecules have been identified for TLRs that are recruited after ligand binding: MyD88 (myeloid differentiation protein 88) TIRAP (TIR-associated protein)/MAL (MyD88-adaptor like), TRIF (TIR-domain containing adaptor protein-inducing IFNβ/TICAM1 (TIR domain containing molecule 1) and TRIF-related adaptor molecule (TRAM) [38]. TLR9 (and TLR7/8) utilizes solely MyD88 to induce cytokines (Figure 12.2) [18,39,40]. Usually, recruitment of MyD88 is followed by engagement of IL-1 receptor-associated kinase 4 (IRAK-4) and IL-1 receptor associated kinase 1 (IRAK-1), which is phosphorylated by IRAK-4 and associates with TNF receptor-associated factor-6 (TRAF6) or/and TNF receptor-associated factor-3 (TRAF3) [41–43]. TRAF3 is recruited along with TRAF6 and is essential for the induction of type I IFN, but it is dispensable for expression of pro-inflammatory cytokines [44,45]. IRF7 is the key component in IFN-α induction which is recruited to a complex consisting of MyD88, IRAK-4, IRAK-1 and TRAF-3 [46,47]. Oligomerization of TRAF6 and subsequent ubiquitination and degradation events lead to activation of the transcription NF-κB followed by the production of inflammatory cytokines and costimulatory molecules [43].

One role of TLR9 *in vivo* is the recognition of DNA viruses such as murine cytomegalovirus (MCMV), herpes simplex virus 1 and 2 (HSV-1/HSV-2) which induces production of inflammatory cytokines and type I IFN [48–51]. The TLR9-mediated IFNα response to HSV-1 and HSV-2 is limited to a subtype of dendritic cells, called plasmacytoid dendritic cells (pDCs) or natural interferon-α producing cells (NIPCs) that secrete high amounts of IFNα in response to viral infection [48–52]. Cellular activation does not require viral infection since live, heat or UV-inactivated HSV-1/HSV-2 produce high levels of IFNα. Recently, a fundamental role of TLR9 in the in the initiation of a protective immune response against poxvirus has been reported [53]. Accordingly, recognition of ectromelia virus (ECTV), the causative agent of mousepox, was strictly TLR9 dependent and mice lacking TLR9 showed a drastically increased susceptibility to ECTV infection [53].

TLR9 also recognizes various bacterial genomic DNAs and activates immune cells [54,55]. Surprisingly, the role of TLR9 in fighting bacterial infection has not been elucidated in detail. Interestingly, one study showed that TLR9 activation is critical for the production of IFNγ during infection with *Propionibacterium acnes* (formerly *Corynebacterium parvum*) that is part of the human flora and associated with several human pathologies. TLR9-dependent activation via *P. acnes* primes IFNγ-dependent enhanced resistance to murine typhoid fever which is abolished in TLR9-deficient mice [56]. Further insight into the *in vivo* role of TLR9 comes from a study with TLR-deficient mice using *Neisseria meningitidis* as a model for meningococcal sepsis. Interestingly, TLR9-deficient mice showed reduced survival upon infection compared with wild-type mice, demonstrating an important role for TLR9 in host defense against *N. meningitidis* [57].

12.2.2 Cytoplasmic recognition of DNA and interferon production

Alternative DNA-sensing receptors apart from TLR9 have been proposed from experiments showing that IFNα was secreted by macrophages upon HSV infection in a TLR9-independent manner, suggesting that pDC- and TLR9-independent mechanisms exist that induce an effective immune response against DNA viruses [49,58]. Further reports have demonstrated the cytoplasmic DNA in its right-handed B-form [such as poly(dA–dT)·poly(dT–dA)] triggers type I interferon production via a TBK-1 (TANK-binding kinase-1, a pivotal kinase of interferon regulatory factors) and interferon regulatory factor-3 (IRF-3). DNA with a left-handed Z-conformation [such as poly(dC–dG)·poly(dG–dC)] was not stimulatory [59–61].

A protein named DNA-dependent activator of IFN-regulatory factor (DAI) or Z-DNA binding protein 1 (ZBP1) has been reported as a candidate cytoplasmic DNA sensor in the murine fibroblast cell line L929. DAI contains two Z-DNA binding domains and a putative signaling function (Figure 12.3).

It is expressed in the cytoplasm and overexpression increases type I IFN production upon stimulation with B-form DNA in a TBK-1 (TANK-binding kinase-1, a pivotal kinase of interferon regulatory factors) and IRF-3-dependent manner. RNA interference of DAI led to a decrease in B-form DNA-mediated type I interferon production [62]. Since DAI contains Z-DNA binding domains, the domain for B-form DNA recognition was not defined. Wang *et al.* partly solved this issue by the identification of a new B-form binding domain termed D3 [63]. However, since DAI-deficient mouse embryonic fibroblasts or dendritic cells show no defect in recognition of cytoplasmic B-DNA, the role and importance of DAI are still controversial [64].

Although the interferon-inducing cytoplasmic DNA sensor has not yet been convincingly identified, the immunostimulatory capacity of B-DNA is well documented. Therefore, it is envisioned that some pathogens may interfere with DNA recognition to evade efficient immune recognition. Accordingly, E3L, a protein produced by vaccinia virus that is necessary for mortality, efficiently blocks B-DNA-induced IFNβ production [63,65]. Since E3L contains a Z-DNA and double-stranded RNA-binding domain, the mechanism of action is currently unknown.

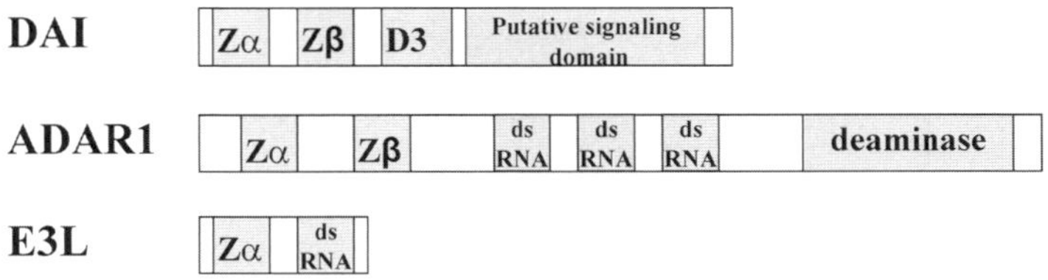

Figure 12.3 Schematic structure of the cytoplasmic DNA-sensing pattern recognition receptors inducing or modulating type I interferon production

Since excessive immune stimulation by DNA is potentially self-destructive or may lead to autoimmunity, negative regulators of B-DNA sensing have been postulated. Apart from DNases that limit the occurrence of excess DNA, further proteins have been identified that are involved in regulation of immune activation. For example, another member of Z-DNA-binding proteins termed adenosine deaminase acting on RNA 1 (ADAR1) is probably involved in modulating B-DNA-driven immune stimulation. Accordingly, ADAR1-deficient cells show an increased induction of IFNβ upon infection with DNA viruses or stimulation with B-DNA [63,66]. The role of three double-stranded RNA binding domains within ADAR1 is currently unknown, but it is hypothesized that they may modulate cytoplasmic RNA recognition. 3′ repair exonuclease (Trex-1) is also involved in modulating DNA-mediated immune activation since Trex-1-deficient mice accumulate DNA followed by immune activation, type I interferon production and development of autoimmunity [67].

12.2.3 Inflammasome-dependent cytoplasmic DNA detection by AIM2 and NALP3

Apart from type I interferon production, DNA stimulation also leads to the production of inflammatory cytokines such as interleukin-1 beta (IL-1β) and to cell pyroptosis, a process of programmed cell death distinct from apoptosis [68,69]. In general, production and secretion of mature IL-1β (and IL-18, IL-33) depends on caspase-1 activity. In more detail, caspase-1 activity is regulated by the inflammasome, a multi-protein complex. Upon sensing of cytoplasmic ligands by proteins of the NOD-like receptor (NLR) family, the adaptor proteins Cardinal and ASC are recruited, the latter by homotypic pyrin domain (PYD) interactions [17]. ASC clustering then leads to pro-caspase-1 recruitment via CARD–CARD (caspase recruitment domains) interactions and caspase-1 activation [68,69].

Recently, the NALP3 inflammasome pathway was identified for sensing cytoplasmic adenoviral DNA leading to IL-1β secretion [70]. Interestingly, NALP3 senses in addition other microbial molecules such as lipopolysaccharide, bacterial and viral RNA, the double-stranded RNA analog poly(I:C) and bacterial toxins. In addition, NLRP3 has been implicated in the activation of caspase-1 in response to non-microbial signals including uric acid crystals, the fibrillar peptide amyloid-beta, asbestos and silica [71–76]. However, a mechanistic insight into these molecular interactions is currently not available, but it is believed that the interaction is indirect, possibly mediated by yet unknown receptors. In experiments carried out by Muruve *et al.*, transfected cytoplasmic plasmid DNA was sensed independently of NALP3, but in an ASC-dependent manner, arguing for an additional unknown DNA sensor within the inflammasome [70].

Recently, the hematopoietic interferon-inducible nuclear proteins with a 200 amino acid repeat (HIN-200) family member AIM2 was identified as the sensor for cytoplasmic double-stranded (ds) DNA [77–80]. In contrast, other members of the HIN200 family such as IFIX, IFI16 and MNDA are expressed in the nucleus and are not involved in dsDNA recognition (Figures 12.2 and 12.4) [77–80]. These results further indicate

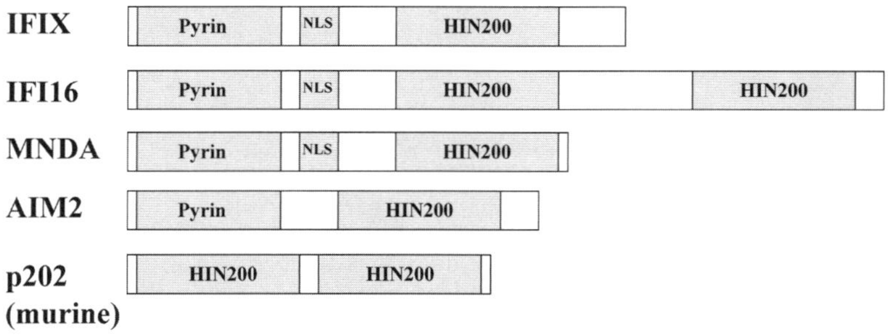

Figure 12.4 Schematic structure of HIN-200 domain containing proteins. Three members contain nuclear localization signal (NLS) and are therefore expressed in the nucleus

that AIM2 binds directly to cytoplasmic DNA in a length-dependent manner, oligomerizes and triggers the assembly of an AIM2 inflammasome, which results in caspase-1 activation and the maturation of IL-1β. The source and sequence of cytoplasmic dsDNA appear to be of no importance, because viral, bacterial, mammalian and synthetic dsDNA led to caspase-1 activation [77–80].

Roberts *et al.* further reported that the murine HIN-200 protein p202 negatively regulates the AIM2 inflammasome [80]. P202 contains two HIN domains but no pyrin domain and binds dsDNA similarly to AIM2 in a sequence-independent but length-dependent manner. The lack of the pyrin domain will therefore not support ASC binding and subsequent inflammasome formation. In contrast, it probably antagonizes the AIM2 inflammasome by sequestering dsDNA. Since currently no human homolog of p202 has been identified, the relevance of this mechanism to humans is currently unclear [80]. However, AIM2 and p202 fall within susceptibility loci for the autoimmune disease systemic lupus erythematosus (SLE) and therefore these studies clearly imply a role for cytoplasmic dsDNA-mediated immune activation in autoimmune diseases [80].

12.3 Innate immune receptors for RNA

In a similar manner to that depicted for DNA recognition, several receptors in the endosome and cytoplasm have been identified that recognize RNA and lead to immune activation (Table 12.2 and Figure 12.5).

12.3.1 TL3, TLR7 and TLR8

TLR3 recognizes dsRNA derived from virus or its mimic poly(I:C) and induces type I IFN [81]. The dsRNA can be generated as an intermediate during the replication cycle of single-stranded (ss) RNA or DNA viruses. The structure of TLR3 and how TLR3 binds dsRNA has recently been elucidated by X-ray crystallography [82–84]. TLR3 ectodomains (ECDs) dimerize on dsRNA at least 40–50 base pairs in length. They bind dsRNA at two sites located at opposite ends of the TLR3 horseshoe, which stabilizes the dimer and initiates signaling [82].

Table 12.2 *Overview of RNA-sensing pattern recognition receptors and their ligands*

RNA-sensing pattern recognition receptor	Ligand
TLR3	Poly(I:C), dsRNA [81]
TLR7	ssRNA [40,91,104], siRNA [92], nucleoside analogs such as imidazoquinolines [93,134] and loxoribine [19,100]
TLR8	ssRNA [40], nucleoside analogs such as imidazoquinolines [134]
TLR9	CpG motif-containing DNA [18,54,133] and CpG-independent DNA recognition [32,33]
RIG-I, retinoic acid-inducible gene I, DEAD (Asp–Glu–Ala–Asp) box polypeptide 58, Ddx58	Viral RNA [111], 5′-triphosphate RNA [114,115], dsRNA [112]
MDA5, melanoma differentiation-associated gene 5, Ifih1	Viral RNA from picornaviruses and poly(I:C) [110–112]
LGP2, laboratory of genetics and physiology 2, DEXH (Asp–Glu–X–His) box polypeptide 58, Dhx58	dsRNA [125,126], negative regulator of RIG-I [127] or MDA5 [128]

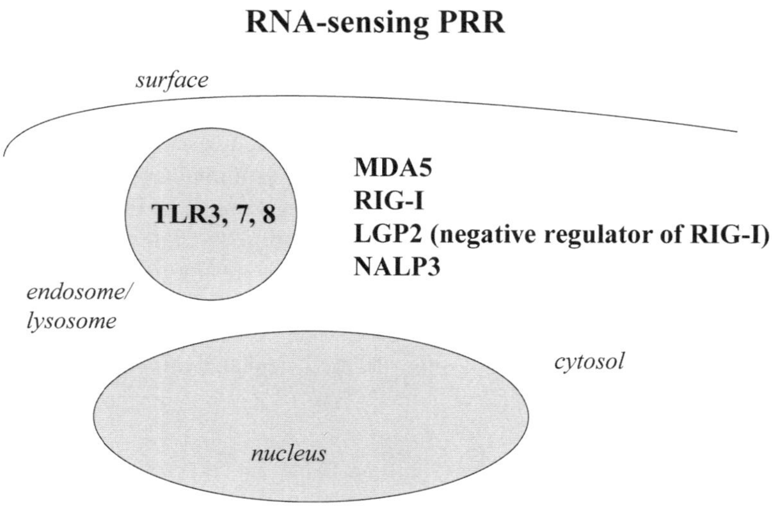

Figure 12.5 *Overview of RNA-sensing pattern recognition receptors in the endosome/lysosome and the cytoplasm*

Although MyD88 is the master adapter protein for most of the TLRs, TLR3 activates a signaling cascade independent of MyD88. TLR3 (and TLR4) utilize TRIF to produce IFNβ and IFN inducible genes (Figure 12.2) [85]. TRIF interacts with receptor-interacting protein-1 (RIP1) and TANK binding kinase 1 (TBK1) to initiate NF-κB activation and IRF-3 and IRF-7 phosphorylation, respectively [38]. Phosphorylated IRF-3 and IRF-7 form homodimers, translocate into the nucleus and regulate the expression of IFN-inducible genes by binding to interferon-stimulated response element (ISRE) [27,86].

Since dsRNA seems to be a universal viral PAMP, TLR3 was believed to be the key receptor in an antiviral immune response. However, early viral infection experiments with lymphocytic choriomeningitis virus (LCMV), vesicular stomatitis virus (VSV), murine cytomegalovirus (MCMV) and reovirus in wild-type and TLR3$^{-/-}$ mice revealed that TLR3 is not required for the antiviral response [87]. In contrast, utilizing an influenza A virus (IAV) infection model analyzing animal survival, respiratory suffering, viral clearance, leukocyte recruitment and secretion of inflammatory cytokines, mice deficient in TLR3 had a survival advantage compared with wild-type mice [88]. Surprisingly, West Nile virus (WNV), an ssRNA flavivirus that can cause neuronal injury in humans, utilizes TLR3-mediated proinflammatory cytokine production such as TNF-α. Since TNF-α leads to the disruption of the blood–brain barrier, virus-induced TLR3 activation facilitates the entry into the brain. Accordingly, TLR3$^{-/-}$ mice were resistant to peripheral WNV infection compared with other wild-type littermates [89]. However, these observations have been questioned by Daffis *et al.*, who demonstrated that the absence of TLR3 enhances the viral burden in the brain of mice and increases WNV mortality [90].

TLR7 and TLR8 have recently been shown to recognize single-stranded viral or synthetic RNA and double-stranded short-interfering RNA (siRNA) [40,91,92].

Initially, imidazoquinoline derivatives including imiquimod (R-837, S-26308) and resiquimod (R-848) were identified as TLR7 ligands in mice [93]. These derivatives are small synthetic compounds that induce type I IFN and IL-12 production in immune cells and show potent antiviral activity [94–97].

Recently, further variants of R848 have been identified (Table 12.3). 3M-001 is specific for TLR7, 3M–002 for TLR8 and 3M–003 activates both receptors [98] (Table 12.3). Of note, other base analogs such as adenosine, guanosine and pyrimidine derivatives, have been shown to activate TLR7 and/or TLR8.

Table 12.3 *Synthetic nucleoside analogs as TLR7/8 ligands*

Compound	Structure	TLR activity	Compound	Structure	TLR activity
Imiquimod (R-837, S-26308) 1-(2-methylpropyl)-1*H*-imidazo[4,5-*c*] quinolin-4-amine	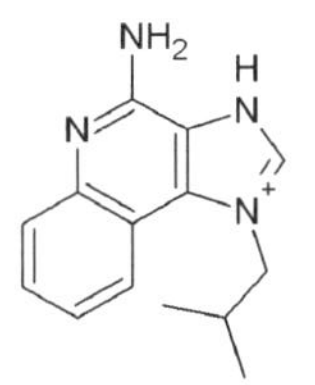	mTLR7, hTLR7 [93,100]	Resiquimod (R-848, S-28463) 4-amino-2-ethoxymethyl-α,α-dimethyl-1*H*-imidazo[4,5-*c*]quinoline-1-ethanol	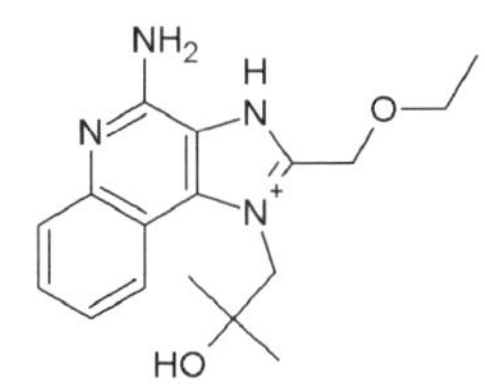	mTLR7, hTLR7, hTLR8 [93,134]
3M-001 *N*-[4-(4-amino-2-ethyl-1*H*-imidazo[4,5-*c*]quinolin-1-yl)butyl]methanesulfonamide	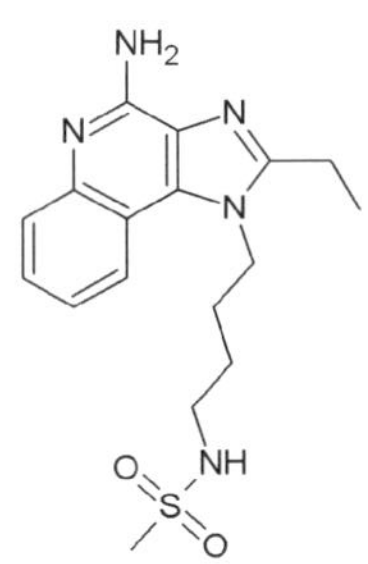	hTLR7 [98]	3M-002 2-propylthiazolo [4,5-*c*]quinolin-4-amine	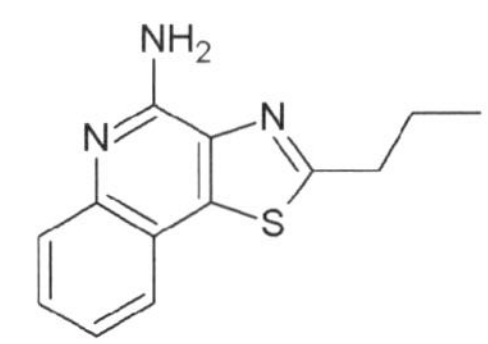	hTLR8, mTLR8 [if poly(dT)$_{17}$ added] ([98,135] and this chapter)
3M-003 4-amino-2-(ethoxymethyl)-α, -dimethyl-6,7,8,9-tetrahydro-1*H*-imidazo[4,5-*c*]quinoline-1-ethanol	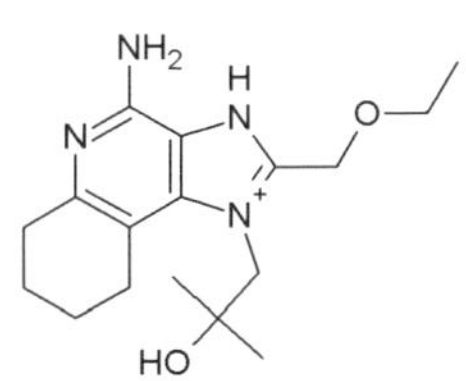	hTLR7, hTLR8 [98]	CL097	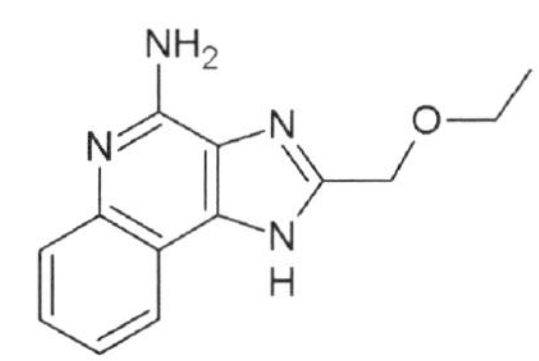	TLR7, TLR8 [136]

(*continued overleaf*)

Table 12.3 (*Continued*)

Compound	Structure	TLR activity	Compound	Structure	TLR activity
Brompirimine (2-amino-5-bromo-6-phenyl-4(3*H*)-pyrimidinone)	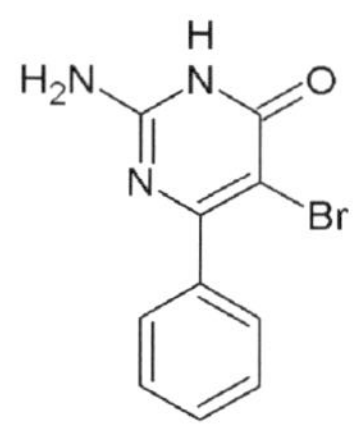	TLR7[99]	Loxoribine 7-allyl-8-oxoguanosine	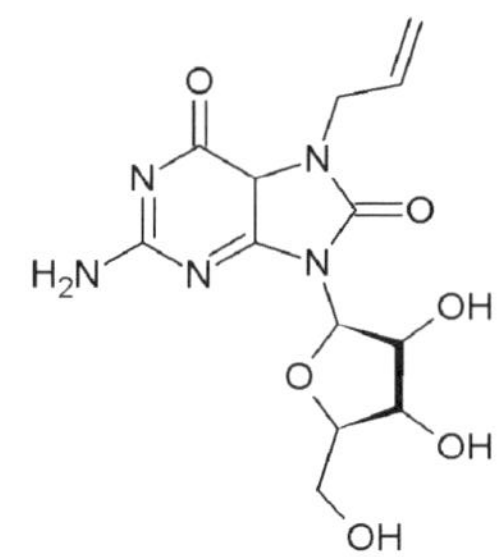	mTLR7, hTLR7 [19,100] and hTLR8 [if poly(dT)$_{17}$ added] [137]

Accordingly, the guanosine analog loxoribine (7-allyl-8-oxoguanosine) and the pyrimidine analog brompirimine activate human and murine TLR7 [19,99,100]. Loxoribine shows anti-viral and anti-tumor activity in murine animal models since it activates NK cells and B cells and induces various cytokines [101,102]. Brompirimine is an orally active immunomodulator that induces IFNα and is used for immunotherapy of carcinoma *in situ* of the bladder and upper urinary tract [103]. Overall, the synthetic nucleoside analogs prove to be universal compounds for activating TLR7 or TLR8 and modulating immune responses to fight viral infection or tumors.

The structural similarities of these synthetic compounds fueled the search for the natural ligand of TLR7 and TLR8, which turned out to be ssRNA [40,91,104]. TLR7 senses synthetic RNA oligonucleotides derived from the U5 region of human immunodeficiency virus (HIV), polyU RNA and RNA from influenza virus, Newcastle disease virus (NDV) and vesicular stomatitis virus (VSV) [40,91,104,105]. Human TLR8 recognizes RNA rich in guanosine and uridine and is involved in the recognition of coxsackie B virus and human parechovirus 1 [40,106,107]. The sequence dependence of RNA recognition by TLR7 and TLR8 is not well defined, although uridine is important for TLR7/8 activation [40,108]. Guanosines and uridines seem necessary for TLR7 mediated IFNα production, whereas RNA containing adenosines and uridines activates proinflammatory cytokines such as TNF-α and IL-12 in a TLR7- and TLR8-dependent manner [40].

The *in vivo* role of TLR7 and TLR8 in an antiviral response has not been thoroughly analyzed and, owing to redundancy in pattern recognition receptors recognizing viral nucleic acid in the cytoplasm and endosome/lysosome, a strong phenotype of TLR7-deficient mice is not likely to be identified. However, recently an important *in vivo* role for TLR7 has been described in bacterial infection with group B streptococcus. Whereas wild-type mice survived infection, TLR7-deficient mice showed enhanced lethality after infection with group B *Streptococcus* [109].

12.3.2 RNA helicases (RIG-I, MDA-5, LPG2)

In response to viral infections, RNA and DNA are recognized by TLRs and RIG-I-like helicases (RLHs). The family of RLHs contains three members, RIG-I (retinoic acid-inducible gene I), MDA5 (melanoma differentiation-associated gene 5) and LGP2 (laboratory of genetics and physiology 2) (Figure 12.6). Rig-I and MDA5 are composed of two N-terminal caspase recruitment domains (CARD) for signaling, a DExD/H RNA helicase domain followed by C-terminal repressor domain (CTD). In contrast, LPG2 carries an RNA helicase and a repressor domain and may function as negative regulator of RIG-I/MDA5 signaling (see below).

Initially, dsRNA such as poly(I:C) was reported as a ligand for RIG-I and MDA-5. However, analysis of RIG-I- and MDA5-deficient mice demonstrated that poly(IC) stimulated MDA-5 but not RIG-I [110,111]. In addition, both helicases also differentially recognize RNA viruses. Influenza virus, vesicular stomatitis virus and paramyxovirus did not induce type I interferon in RIG-I deficient cells, whereas MDA5-deficient cells

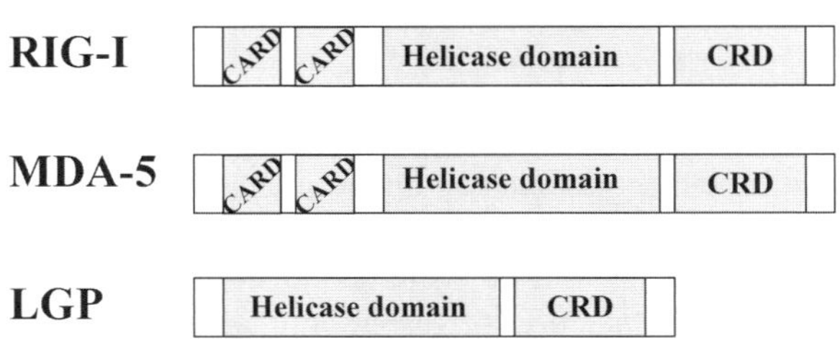

Figure 12.6 *Schematic structure of the RIG-I-like helicase*

did not respond to infection with picornaviruses such as EMCV (encephalomyocarditis virus) [110,111]. Interestingly, the differential recognition of dsRNA by Rig-I and MDA5 is dependent on RNA length. Accordingly, the dsRNA mimic poly(I:C) acts as an MDA-5 ligand; however, shortening the poly(I:C) length converted the RNA to a RIG-I ligand. Similarly, viral dsRNA also differentially activated RIG-I and MDA5 depending on the length [112].

Some viruses such as influenza virus do not produce (detectable) dsRNA intermediates during replication and therefore alternative ligands for RIG-I have been postulated [113]. In two seminal studies, 5′-triphosphate ssRNA or dsRNA was identified as a ligand for RIG-I that binds the C-terminal domain [114,115]. Host cell RNA such as rRNA or mRNA is usually not recognized due to 5′-triphosphate removal or modification by a 7-methylguanosine cap, respectively. Interestingly, some viruses such as Hantaan virus, Crimean–Congo hemorrhagic fever virus and Borna disease virus evade immune recognition by RIG-I, utilizing a viral-induced phosphatase activity that removes the 5′-triphosphate [116].

Sequence dependence of RIG-I mediated RNA recognition was reported by Saito *et al.*, who demonstrated that the polyuridine motif of the HCV genome 3′ non-translated region and its replication intermediate act as substrate for RIG-I [117]. In their study, 5′-terminal triphosphate was necessary but not sufficient for RIG-I binding, which was strongly dependent on the polyuridine composition of the RNA.

Collectively, 5′-triphosphate and dsRNA are two molecular patterns that enable RIG-I to discriminate pathogenic RNA from self-RNA. Interestingly, 5′-triphosphate binding mediates the dsRNA translocation activity of RIG-I, which is important for activity. In the absence of 5′-triphosphate-RNA, translocation on dsRNA is suppressed by the CARD domains [118].

In summary, 5′-triphosphate and dsRNA are ligands for RIG-I, although some controversy still exists on the importance of each molecular pattern.

Although the exact nature of the ligands for RIG-I and MDA5 is still a matter of debate or unknown, respectively, the signaling cascade leading to type I interferon production is well understood. In more detail, the CARDs of RIG-I and MDA5 initiate the signaling cascade by associating with the CARD domain containing adaptor molecule CARDIF (IPS-1, MAVS or VISA) [119–122]. Interestingly, CARDIF is located in the outer mitochondrial membrane, suggesting a role for mitochondria in innate immune responses. CARDIF associates with TRAF3 that recruits and activates TANK binding kinase 1 (TBK1) and inducible IκB kinase, which phosphorylates and activates the transcription factors IRF-3 and IRF-7 [123]. This phosphorylation event leads to their translocation into the nucleus and binding to the promoters of interferon-stimulated response elements (ISRE), leading to the transcription of IFN and IFN-inducible genes [124].

LGP2 lacks a CARD domain and therefore could act as a negative regulator. Supporting this view, LGP2 binds to the termini of dsRNA and overexpression of LGP2 inhibits immune activation by Sendai virus and Newcastle disease virus [125–127]. However, LGP2-deficient mice show a more complex phenotype. Stimulation with poly(I:C) or VSV leads to increased type I interferon production, whereas the immune response to EMCV infection is partly impaired. It is therefore suggested that LGP2 negatively regulates RIG-I but not MDA5 [128].

12.3.3 NALP3

NALP3 senses various microbial molecules such as lipopolysaccharides, bacterial and viral RNA and the dsRNA poly(I:C) with subsequent activation of the inflammasome and IL-1 secretion (see earlier) [129,130] However, an *in vivo* role of NALP3 in viral or bacterial infections has not been investigated in detail. Recently, a critical role for NALP3 induction of the inflammasome and subsequent protection against influenza virus infection has been reported, underscoring the importance of the inflammasome in coping with viral infections [131,132].

12.4 Conclusion

Overall, recent research on nucleic acid receptors has illuminated how the innate immune system senses RNA and DNA and how the corresponding cells are activated and induce cytokine production (Figure 12.2). Intense research on TLRs, RNA helicases and components of the inflammasome has strongly enhanced our knowledge on how the innate immune system utilizes the recognition of nucleic acid from pathogens to fight infections. Furthermore, this knowledge will help in the development of new therapeutics for the treatment of infectious diseases and cancer. Further efforts may also focus on the identification of so far unidentified nucleic acid-sensing receptors on the cell surface and within the nucleus.

12.5 Abbreviations

ADAR1	adenosine deaminase acting on double-stranded RNA 1
ADAR1	adenosine deaminase acting on RNA 1
AIM2	absent in melanoma 2
ASC	apoptosis-associated speck-like protein
CARD	caspase recruitment domains
DAI	DNA-dependent activator of IFN regulatory factor
DAI	DNA-dependent activator of IFN regulatory factors
FLT3L	FMS-like tyrosine kinase 3 ligand
HIN-200	the hematopoietic interferon-inducible nuclear proteins with a 200 amino acid repeat
IRF	interferon regulatory factor-3
LGP2	laboratory of genetics and physiology 2
MDA5	melanoma differentiation-associated gene 5
NALP3	NACHT-, LRR- and PYD-containing protein 3
ODN	oligodeoxynucleotide
PD	phosphodiester
poly(I:C)	polyinosine–poly(cytidylic acid)
PTO	phosphorothioate
PYD	pyrin domain
RIG-I	retinoic acid inducible gene I
TBK-1	TANK-binding kinase-1
Trex-1	3′ repair exonuclease 1

References

1. A. Isaacs, R.A. Cox and Z. Rotem, *Lancet* **ii**, 113–116 (1963).
2. K.E. Jensen, A.L. Neal, R.E. Owens and J. Warren, *Nature* **200**, 433–434 (1963).
3. Z. Rotem, R.A. Cox and A. Isaacs, *Nature* **197**, 564–566 (1963).
4. T. Tokunaga, H. Yamamoto, S. Shimada, H. Abe, T. Fukuda, Y. Fujisawa, Y. Furutani, O. Yano, T. Kataoka, T. Sudo *et al.*, *J Natl Cancer Inst* **72**, 955–962 (1984).
5. J.Y. Richmond and L.D. Hamilton, *Proc Natl Acad Sci USA* **64**, 81–86 (1969).
6. S. Baron, N.N. Bogomolova, A. Billiau, H.B. Levy, C.E. Buckler, R. Stern and R. Naylor, *Proc Natl Acad Sci USA* **64**, 67–74 (1969).
7. R.H. Adamson, S. Fabro, E.R. Homan, R.W. O'Gara and R.P. Zendzian, *Antimicrob Agents Chemother* **9**, 148–152 (1969).

8. A.M. Krieg, A.K. Yi, S. Matson, T.J. Waldschmidt, G.A. Bishop, R. Teasdale, G.A. Koretzky and D.M. Klinman, *Nature* **374**, 546–549 (1995).
9. A.P. Bird, *Nucleic Acids Res* **8**, 1499–1504 (1980).
10. A.M. Krieg, T. Wu, R. Weeratna, S.M. Efler, L. Love-Homan, L. Yang, A.K. Yi, D. Short and H.L. Davis, *Proc Natl Acad Sci USA* **95**, 12631–12636 (1998).
11. K.J. Ishii, K. Suzuki, C. Coban, F. Takeshita, Y. Itoh, H. Matoba, L.D. Kohn and D.M. Klinman, *J Immunol* **167**, 2602–2607 (2001).
12. F.L. Rock, G. Hardiman, J.C. Timans, R.A. Kastelein and J.F. Bazan, *Proc Natl Acad Sci USA* **95**, 588–593 (1998).
13. K.V. Anderson, L. Bokla and C. Nusslein-Volhard, *Cell* **42**, 791–798 (1985).
14. B. Lemaitre, E. Nicolas, L. Michaut, J.M. Reichhart and J.A. Hoffmann, *Cell* **86**, 973–983 (1996).
15. K. Takeda and S. Akira, *Int Immunol* **17**, 1–14 (2005).
16. K. Tabeta, P. Georgel, E. Janssen, X. Du, K. Hoebe, K. Crozat, S. Mudd, L. Shamel, S. Sovath, J. Goode, L. Alexopoulou, R.A. Flavell and B. Beutler, *Proc Natl Acad Sci USA* **101**, 3516–3521 (2004).
17. H. Kumar, T. Kawai and S. Akira, *Biochem J* **420**, 1–16 (2009).
18. H. Hemmi, O. Takeuchi, T. Kawai, T. Kaisho, S. Sato, H. Sanjo, M. Matsumoto, K. Hoshino, H. Wagner, K. Takeda and S. Akira, *Nature* **408**, 740–745 (2000).
19. F. Heil, P. Ahmad-Nejad, H. Hemmi, H. Hochrein, F. Ampenberger, T. Gellert, H. Dietrich, G. Lipford, K. Takeda, S. Akira, H. Wagner and S. Bauer, *Eur J Immunol* **33**, 2987–2997 (2003).
20. E. Latz, A. Schoenemeyer, A. Visintin, K.A. Fitzgerald, B.G. Monks, C.F. Knetter, E. Lien, N.J. Nilsen, T. Espevik and D.T. Golenbock, *Nat Immunol* **5**, 190–198 (2004).
21. K. Tabeta, K. Hoebe, E.M. Janssen, X. Du, P. Georgel, K. Crozat, S. Mudd, N. Mann, S. Sovath, J. Goode, L. Shamel, A.A. Herskovits, D.A. Portnoy, M. Cooke, L.M. Tarantino, T. Wiltshire, B.E. Steinberg, S. Grinstein and B. Beutler, *Nat Immunol* **7**, 156–164 (2006).
22. M.M. Brinkmann, E. Spooner, K. Hoebe, B. Beutler, H.L. Ploegh and Y.M. Kim, *J Cell Biol* **177**, 265–275 (2007).
23. Y.M. Kim, M.M. Brinkmann, M.E. Paquet and H.L. Ploegh, *Nature* **452**, 234–238 (2008).
24. G. Hartmann and A.M. Krieg, *J Immunol* **164**, 944–953 (2000).
25. D. Verthelyi, K.J. Ishii, M. Gursel, F. Takeshita and D.M. Klinman, *J Immunol* **166**, 2372–2377 (2001).
26. A. Krug, S. Rothenfusser, V. Hornung, B. Jahrsdorfer, S. Blackwell, Z.K. Ballas, S. Endres, A.M. Krieg and G. Hartmann, *Eur J Immunol* **31**, 2154–2163 (2001).
27. K. Honda, Y. Ohba, H. Yanai, H. Negishi, T. Mizutani, A. Takaoka, C. Taya and T. Taniguchi, *Nature* **434**, 1035–1040 (2005).
28. C. Guiducci, G. Ott, J.H. Chan, E. Damon, C. Calacsan, T. Matray, K.D. Lee, R.L. Coffman and F.J. Barrat, *J Exp Med* **203**, 1999–2008 (2006).
29. G. Hartmann, J. Battiany, H. Poeck, M. Wagner, M. Kerkmann, N. Lubenow, S. Rothenfusser and S. Endres, *Eur J Immunol* **33**, 1633–1641 (2003).
30. J.D. Marshall, K. Fearon, C. Abbate, S. Subramanian, P. Yee, J. Gregorio, R.L. Coffman and G. Van Nest, *J Leukoc Biol* **73**, 781–792 (2003).
31. J. Vollmer, R. Weeratna, P. Payette, M. Jurk, C. Schetter, M. Laucht, T. Wader, S. Tluk, M. Liu, H.L. Davis and A.M. Krieg, *Eur J Immunol* **34**, 251–262 (2004).
32. K. Yasuda, M. Rutz, B. Schlatter, J. Metzger, P.B. Luppa, F. Schmitz, T. Haas, A. Heit, S. Bauer and H. Wagner, *Eur J Immunol* **36**, 431–436 (2006).
33. T. Haas, J. Metzger, F. Schmitz, A. Heit, T. Muller, E. Latz and H. Wagner, *Immunity* **28**, 315–323 (2008).
34. B. Park, M.M. Brinkmann, E. Spooner, C.C. Lee, Y.M. Kim and H.L. Ploegh, *Nat Immunol* **9**, 1407–1414 (2008).
35. M. Asagiri, T. Hirai, T. Kunigami, S. Kamano, H.J. Gober, K. Okamoto, K. Nishikawa, E. Latz, D.T. Golenbock, K. Aoki, K. Ohya, Y. Imai, Y. Morishita, K. Miyazono, S. Kato, P. Saftig and H. Takayanagi, *Science* **319**, 624–627 (2008).
36. F. Matsumoto, S. Saitoh, R. Fukui, T. Kobayashi, N. Tanimura, K. Konno, Y. Kusumoto, S. Akashi-Takamura and K. Miyake, *Biochem Biophys Res Commun* **367**, 693–699 (2008).
37. S.E. Ewald, B.L. Lee, L. Lau, K.E. Wickliffe, G.P. Shi, H.A. Chapman and G.M. Barton, *Nature* **456**, 658–662 (2008).
38. L.A. O'Neill, *Immunol Rev* **226**, 10–18 (2008).

39. H. Hacker, R.M. Vabulas, O. Takeuchi, K. Hoshino, S. Akira and H. Wagner, *J Exp Med* **192**, 595–600 (2000).
40. F. Heil, H. Hemmi, H. Hochrein, F. Ampenberger, C. Kirschning, S. Akira, G. Lipford, H. Wagner and S. Bauer, *Science* **303**, 1526–1529 (2004).
41. Z. Cao, J. Xiong, M. Takeuchi, T. Kurama and D.V. Goeddel, *Nature* **383**, 443–446 (1996).
42. M. Muzio, G. Natoli, S. Saccani, M. Levrero and A. Mantovani, *J Exp Med* **187**, 2097–2101 (1998).
43. S. Li, A. Strelow, E.J. Fontana and H. Wesche, *Proc Natl Acad Sci USA* **99**, 5567–5572 (2002).
44. H. Hacker, V. Redecke, B. Blagoev, I. Kratchmarova, L.C. Hsu, G.G. Wang, M.P. Kamps, E. Raz, H. Wagner, G. Hacker, M. Mann and M. Karin, *Nature* **439**, 204–207 (2006).
45. G. Oganesyan, S.K. Saha, B. Guo, J.Q. He, A. Shahangian, B. Zarnegar, A. Perry and G. Cheng, *Nature* **439**, 208–211 (2006).
46. K. Honda, H. Yanai, T. Mizutani, H. Negishi, N. Shimada, N. Suzuki, Y. Ohba, A. Takaoka, W.C. Yeh and T. Taniguchi, *Proc Natl Acad Sci USA* **101**, 15416–15421 (2004).
47. T. Kawai, S. Sato, K.J. Ishii, C. Coban, H. Hemmi, M. Yamamoto, K. Terai, M. Matsuda, J. Inoue, S. Uematsu, O. Takeuchi and S. Akira, *Nat Immunol* **5**, 1061–1068 (2004).
48. J. Lund, A. Sato, S. Akira, R. Medzhitov and A. Iwasaki, *J Exp Med* **198**, 513–520 (2003).
49. H. Hochrein, B. Schlatter, M. O'Keeffe, C. Wagner, F. Schmitz, M. Schiemann, S. Bauer, M. Suter and H. Wagner, *Proc Natl Acad Sci USA* **101**, 11416–11421 (2004).
50. A. Krug, A.R. French, W. Barchet, J.A. Fischer, A. Dzionek, J.T. Pingel, M.M. Orihuela, S. Akira, W.M. Yokoyama and M. Colonna, *Immunity* **21**, 107–119 (2004).
51. A. Krug, G.D. Luker, W. Barchet, D.A. Leib, S. Akira and M. Colonna, *Blood* **103**, 1433–1437 (2004).
52. F.P. Siegal, N. Kadowaki, M. Shodell, P.A. Fitzgerald-Bocarsly, K. Shah, S. Ho, S. Antonenko and Y.J. Liu, *Science* **284**, 1835–1837 (1999).
53. C. Samuelsson, J. Hausmann, H. Lauterbach, M. Schmidt, S. Akira, H. Wagner, P. Chaplin, M. Suter, M. O'Keeffe and H. Hochrein, *J Clin Invest* **118**, 1776–1784 (2008).
54. S. Bauer, C.J. Kirschning, H. Hacker, V. Redecke, S. Hausmann, S. Akira, H. Wagner and G.B. Lipford, *Proc Natl Acad Sci USA* **98**, 9237–9242 (2001).
55. A. Dalpke, J. Frank, M. Peter and K. Heeg, *Infect Immun* **74**, 940–946 (2006).
56. C. Kalis, M. Gumenscheimer, N. Freudenberg, S. Tchaptchet, G. Fejer, A. Heit, S. Akira, C. Galanos and M.A. Freudenberg, *J Immunol* **174**, 4295–4300 (2005).
57. H. Sjolinder, T.H. Mogensen, M. Kilian, A.B. Jonsson and S.R. Paludan, *Infect Immun* **76**, 5421–5428 (2008).
58. L. Malmgaard, J. Melchjorsen, A.G. Bowie, S.C. Mogensen and S.R. Paludan, *J Immunol* **173**, 6890–6898 (2004).
59. K.J. Ishii, C. Coban, H. Kato, K. Takahashi, Y. Torii, F. Takeshita, H. Ludwig, G. Sutter, K. Suzuki, H. Hemmi, S. Sato, M. Yamamoto, S. Uematsu, T. Kawai, O. Takeuchi and S. Akira, *Nat Immunol* **7**, 40–48 (2006).
60. D.B. Stetson and R. Medzhitov, *Immunity* **24**, 93–103 (2006).
61. A.T. Phan, V. Kuryavyi and D.J. Patel, *Curr Opin Struct Biol* **16**, 288–298 (2006).
62. A. Takaoka, Z. Wang, M.K. Choi, H. Yanai, H. Negishi, T. Ban, Y. Lu, M. Miyagishi, T. Kodama, K. Honda, Y. Ohba and T. Taniguchi, *Nature* **448**, 501–505 (2007).
63. Z. Wang, M.K. Choi, T. Ban, H. Yanai, H. Negishi, Y. Lu, T. Tamura, A. Takaoka, K. Nishikura and T. Taniguchi, *Proc Natl Acad Sci USA* **105**, 5477–5482 (2008).
64. K.J. Ishii, T. Kawagoe, S. Koyama, K. Matsui, H. Kumar, T. Kawai, S. Uematsu, O. Takeuchi, F. Takeshita, C. Coban and S. Akira, *Nature* **451**, 725–729 (2008).
65. Y.G. Kim, M. Muralinath, T. Brandt, M. Pearcy, K. Hauns, K. Lowenhaupt, B.L. Jacobs and A. Rich, *Proc Natl Acad Sci USA* **100**, 6974–6979 (2003).
66. A.M. Toth, P. Zhang, S. Das, C.X. George and C.E. Samuel, *Prog Nucleic Acid Res Mol Biol* **81**, 369–434 (2006).
67. D.B. Stetson, J.S. Ko, T. Heidmann and R. Medzhitov, *Cell* **134**, 587–598 (2008).
68. F. Martinon, K. Burns and J. Tschopp, *Mol Cell* **10**, 417–426 (2002).
69. S.L. Fink and B.T. Cookson, *Infect Immun* **73**, 1907–1916 (2005).
70. D.A. Muruve, V. Petrilli, A.K. Zaiss, L.R. White, S.A. Clark, P.J. Ross, R.J. Parks and J. Tschopp, *Nature* **452**, 103–107 (2008).
71. A. Halle, V. Hornung, G.C. Petzold, C.R. Stewart, B.G. Monks, T. Reinheckel, K.A. Fitzgerald, E. Latz, K.J. Moore and D.T. Golenbock, *Nat Immunol* **9**, 857–865 (2008).

72. V. Hornung, F. Bauernfeind, A. Halle, E.O. Samstad, H. Kono, K.L. Rock, K.A. Fitzgerald and E. Latz, *Nat Immunol* **9**, 847–856 (2008).
73. C. Dostert, V. Petrilli, R. Van Bruggen, C. Steele, B.T. Mossman and J. Tschopp, *Science* **320**, 674–677 (2008).
74. S.L. Cassel, S.C. Eisenbarth, S.S. Iyer, J.J. Sadler, O.R. Colegio, L.A. Tephly, A.B. Carter, P.B. Rothman, R.A. Flavell and F.S. Sutterwala, *Proc Natl Acad Sci USA* **105**, 9035–9040 (2008).
75. F. Martinon, V. Petrilli, A. Mayor, A. Tardivel and J. Tschopp, *Nature* **440**, 237–241 (2006).
76. S. Mariathasan, D.S. Weiss, K. Newton, J. McBride, K. O'Rourke, M. Roose-Girma, W.P. Lee, Y. Weinrauch, D.M. Monack and V.M. Dixit, *Nature* **440**, 228–232 (2006).
77. T. Burckstummer, C. Baumann, S. Bluml, E. Dixit, G. Durnberger, H. Jahn, M. Planyavsky, M. Bilban, J. Colinge, K.L. Bennett and G. Superti-Furga, *Nat Immunol* **10**, 266–272 (2009).
78. T. Fernandes-Alnemri, J.W. Yu, P. Datta, J. Wu and E.S. Alnemri, *Nature* **458**, 509–513 (2009).
79. V. Hornung, A. Ablasser, M. Charrel-Dennis, F. Bauernfeind, G. Horvath, D.R. Caffrey, E. Latz and K.A. Fitzgerald, *Nature* **458**, 514–518 (2009).
80. T.L. Roberts, A. Idris, J.A. Dunn, G.M. Kelly, C.M. Burnton, S. Hodgson, L.L. Hardy, V. Garceau, M.J. Sweet, I.L. Ross, D.A. Hume and K.J. Stacey, *Science* **323**, 1057–1060 (2009).
81. L. Alexopoulou, A.C. Holt, R. Medzhitov and R.A. Flavell, *Nature* **413**, 732–738 (2001).
82. L. Liu, I. Botos, Y. Wang, J.N. Leonard, J. Shiloach, D.M. Segal and D.R. Davies, *Science* **320**, 379–381 (2008).
83. J.K. Bell, I. Botos, P.R. Hall, J. Askins, J. Shiloach, D.M. Segal and D.R. Davies, *Proc Natl Acad Sci USA* **102**, 10976–10980 (2005).
84. J. Choe, M.S. Kelker and I.A. Wilson, *Science* **309**, 581–585 (2005).
85. K. Hoebe, X. Du, P. Georgel, E. Janssen, K. Tabeta, S.O. Kim, J. Goode, P. Lin, N. Mann, S. Mudd, K. Crozat, S. Sovath, J. Han and B. Beutler, *Nature* **424**, 743–748 (2003).
86. S. Sakaguchi, H. Negishi, M. Asagiri, C. Nakajima, T. Mizutani, A. Takaoka, K. Honda and T. Taniguchi, *Biochem Biophys Res Commun* **306**, 860–866 (2003).
87. K.H. Edelmann, S. Richardson-Burns, L. Alexopoulou, K.L. Tyler, R.A. Flavell and M.B. Oldstone, *Virology* **322**, 231–238 (2004).
88. R. Le Goffic, V. Balloy, M. Lagranderie, L. Alexopoulou, N. Escriou, R. Flavell, M. Chignard and M. Si-Tahar, *PLoS Pathog* **2**, e53 (2006).
89. T. Wang, T. Town, L. Alexopoulou, J.F. Anderson, E. Fikrig and R.A. Flavell, *Nat Med* **10**, 1366–1373 (2004).
90. S. Daffis, M.A. Samuel, M.S. Suthar, M. Gale Jr and M.S. Diamond, *J Virol* **82**, 10349–10358 (2008).
91. S.S. Diebold, T. Kaisho, H. Hemmi, S. Akira and C. Reis e Sousa, *Science* **303**, 1529–1531 (2004).
92. V. Hornung, M. Guenthner-Biller, C. Bourquin, A. Ablasser, M. Schlee, S. Uematsu, A. Noronha, M. Manoharan, S. Akira, A. de Fougerolles, S. Endres and G. Hartmann, *Nat Med* **11**, 263–270 (2005).
93. H. Hemmi, T. Kaisho, O. Takeuchi, S. Sato, H. Sanjo, K. Hoshino, T. Horiuchi, H. Tomizawa, K. Takeda and S. Akira, *Nat Immunol* **3**, 196–200 (2002).
94. C.J. Harrison, L. Jenski, T. Voychehovski and D.I. Bernstein, *Antiviral Res* **10**, 209–223 (1988).
95. M.A. Tomai, S.J. Gibson, L.M. Imbertson, R.L. Miller, P.E. Myhre, M.J. Reiter, T.L. Wagner, C.B. Tamulinas, J.M. Beaurline, J.F. Gerster *et al.*, *Antiviral Res* **28**, 253–264 (1995).
96. T.L. Testerman, J.F. Gerster, L.M. Imbertson, M.J. Reiter, R.L. Miller, S.J. Gibson, T.L. Wagner and M.A. Tomai, *J Leukoc Biol* **58**, 365–372 (1995).
97. S.L. Spruance, S.K. Tyring, M.H. Smith and T.C. Meng, *J Infect Dis* **184**, 196–200 (2001).
98. K.B. Gorden, K.S. Gorski, S.J. Gibson, R.M. Kedl, W.C. Kieper, X. Qiu, M.A. Tomai, S.S. Alkan and J.P. Vasilakos, *J Immunol* **174**, 1259–1268 (2005).
99. S. Akira and H. Hemmi, *Immunol Lett* **85**, 85–95 (2003).
100. J. Lee, T.H. Chuang, V. Redecke, L. She, P.M. Pitha, D.A. Carson, E. Raz and H.B. Cottam, *Proc Natl Acad Sci USA* **100**, 6646–6651 (2003).
101. D.F. Smee, H.A. Alaghamandan, A. Jin, B.S. Sharma and W.B. Jolley, *Antiviral Res* **13**, 91–102 (1990).
102. B.L. Pope, J. Sigindere, E. Chourmouzis, P. MacIntyre and M.G. Goodman, *Cancer Immunol Immunother* **38**, 83–91 (1994).
103. M.F. Sarosdy, *Eur Urol* **31** Suppl 1, 20–26 (1997).

104. J.M. Lund, L. Alexopoulou, A. Sato, M. Karow, N.C. Adams, N.W. Gale, A. Iwasaki and R.A. Flavell, *Proc Natl Acad Sci USA* **101**, 5598–5603 (2004).
105. H. Kato, S. Sato, M. Yoneyama, M. Yamamoto, S. Uematsu, K. Matsui, T. Tsujimura, K. Takeda, T. Fujita, O. Takeuchi and S. Akira, *Immunity* **23**, 19–28 (2005).
106. K. Triantafilou, E. Vakakis, G. Orthopoulos, M.A. Ahmed, C. Schumann, P.M. Lepper and M. Triantafilou, *Eur J Immunol* **35**, 2416–2423 (2005).
107. K. Triantafilou, G. Orthopoulos, E. Vakakis, M.A. Ahmed, D.T. Golenbock, P.M. Lepper and M. Triantafilou, *Cell Microbiol* **7**, 1117–1126 (2005).
108. M. Sioud, *Eur J Immunol* **36**, 1222–1230 (2006).
109. G. Mancuso, M. Gambuzza, A. Midiri, C. Biondo, S. Papasergi, S. Akira, G. Teti and C. Beninati, *Nat Immunol* **10**, 587–594 (2009).
110. L. Gitlin, W. Barchet, S. Gilfillan, M. Cella, B. Beutler, R.A. Flavell, M.S. Diamond and M. Colonna, *Proc Natl Acad Sci USA* **103**, 8459–8464 (2006).
111. H. Kato, O. Takeuchi, S. Sato, M. Yoneyama, M. Yamamoto, K. Matsui, S. Uematsu, A. Jung, T. Kawai, K.J. Ishii, O. Yamaguchi, K. Otsu, T. Tsujimura, C.S. Koh, C. Reis e Sousa, Y. Matsuura, T. Fujita and S. Akira, *Nature* **441**, 101–105 (2006).
112. H. Kato, O. Takeuchi, E. Mikamo-Satoh, R. Hirai, T. Kawai, K. Matsushita, A. Hiiragi, T.S. Dermody, T. Fujita and S. Akira, *J Exp Med* **205**, 1601–1610 (2008).
113. F. Weber, V. Wagner, S.B. Rasmussen, R. Hartmann and S.R. Paludan, *J Virol* **80**, 5059–5064 (2006).
114. A. Pichlmair, O. Schulz, C.P. Tan, T.I. Naslund, P. Liljestrom, F. Weber and C. Reis e Sousa, *Science* **314**, 997–1001 (2006).
115. V. Hornung, J. Ellegast, S. Kim, K. Brzozka, A. Jung, H. Kato, H. Poeck, S. Akira, K.K. Conzelmann, M. Schlee, S. Endres and G. Hartmann, *Science* **314**, 994–997 (2006).
116. M. Habjan, I. Andersson, J. Klingstrom, M. Schumann, A. Martin, P. Zimmermann, V. Wagner, A. Pichlmair, U. Schneider, E. Muhlberger, A. Mirazimi and F. Weber, *PLoS ONE* **3**, e2032 (2008).
117. T. Saito, D.M. Owen, F. Jiang, J. Marcotrigiano and M. Gale Jr, *Nature* **454**, 523–527 (2008).
118. S. Myong, S. Cui, P.V. Cornish, A. Kirchhofer, M.U. Gack, J.U. Jung, K.P. Hopfner and T. Ha, *Science* **323**, 1070–1074 (2009).
119. R.B. Seth, L. Sun, C.K. Ea and Z.J. Chen, *Cell* **122**, 669–682 (2005).
120. E. Meylan, J. Curran, K. Hofmann, D. Moradpour, M. Binder, R. Bartenschlager and J. Tschopp, *Nature* **437**, 1167–1172 (2005).
121. T. Kawai, K. Takahashi, S. Sato, C. Coban, H. Kumar, H. Kato, K.J. Ishii, O. Takeuchi and S. Akira, *Nat Immunol* **6**, 981–988 (2005).
122. L.G. Xu, Y.Y. Wang, K.J. Han, L.Y. Li, Z. Zhai and H.B. Shu, *Mol Cell* **19**, 727–740 (2005).
123. B. Guo and G. Cheng, *J Biol Chem* **282**, 11817–11826 (2007).
124. K. Honda, H. Yanai, H. Negishi, M. Asagiri, M. Sato, T. Mizutani, N. Shimada, Y. Ohba, A. Takaoka, N. Yoshida and T. Taniguchi, *Nature* **434**, 772–777 (2005).
125. D.A. Pippig, J.C. Hellmuth, S. Cui, A. Kirchhofer, K. Lammens, A. Lammens, A. Schmidt, S. Rothenfusser and K.P. Hopfner, *Nucleic Acids Res* **37**, 2014–2025 (2009).
126. X. Li, C.T. Ranjith-Kumar, M.T. Brooks, S. Dharmaiah, A.B. Herr, C. Kao and P. Li, *J Biol Chem* **284**, 13881–13891 (2009).
127. S. Rothenfusser, N. Goutagny, G. DiPerna, M. Gong, B.G. Monks, A. Schoenemeyer, M. Yamamoto, S. Akira and K.A. Fitzgerald, *J Immunol* **175**, 5260–5268 (2005).
128. T. Venkataraman, M. Valdes, R. Elsby, S. Kakuta, G. Caceres, S. Saijo, Y. Iwakura and G.N. Barber, *J Immunol* **178**, 6444–6455 (2007).
129. T.D. Kanneganti, N. Ozoren, M. Body-Malapel, A. Amer, J.H. Park, L. Franchi, J. Whitfield, W. Barchet, M. Colonna, P. Vandenabeele, J. Bertin, A. Coyle, E.P. Grant, S. Akira and G. Nunez, *Nature* **440**, 233–236 (2006).
130. T.D. Kanneganti, M. Body-Malapel, A. Amer, J.H. Park, J. Whitfield, L. Franchi, Z.F. Taraporewala, D. Miller, J.T. Patton, N. Inohara and G. Nunez, *J Biol Chem* **281**, 36560–36568 (2006).
131. P.G. Thomas, P. Dash, J.R. Aldridge Jr, A.H. Ellebedy, C. Reynolds, A.J. Funk, W.J. Martin, M. Lamkanfi, R.J. Webby, K.L. Boyd, P.C. Doherty and T.D. Kanneganti, *Immunity* **30**, 566–575 (2009).

132. I.C. Allen, M.A. Scull, C.B. Moore, E.K. Holl, E. McElvania-Tekippe, D.J. Taxman, E.H. Guthrie, R.J. Pickles and J.P. Ting, *Immunity* **30**, 556–565 (2009).
133. F. Takeshita, C.A. Leifer, I. Gursel, K.J. Ishii, S. Takeshita, M. Gursel and D.M. Klinman, *J Immunol* **167**, 3555–3558 (2001).
134. M. Jurk, F. Heil, J. Vollmer, C. Schetter, A.M. Krieg, H. Wagner, G. Lipford and S. Bauer, *Nat Immunol* **3**, 499 (2002).
135. K.K. Gorden, X.X. Qiu, C.C. Binsfeld, J.P. Vasilakos and S.S. Alkan, *J Immunol* **177**, 6584–6587 (2006).
136. M. Salio, A.O. Speak, D. Shepherd, P. Polzella, P.A. Illarionov, N. Veerapen, G.S. Besra, F.M. Platt and V. Cerundolo, *Proc Natl Acad Sci USA* **104**, 20490–20495 (2007).
137. M. Jurk, A. Kritzler, B. Schulte, S. Tluk, C. Schetter, A.M. Krieg and J. Vollmer, *Eur J Immunol* **36**, 1815–1826 (2006).

13

Light-responsive Nucleic Acids for the Spatiotemporal Control of Biological Processes

Alexander Heckel and Günter Mayer

13.1 Introduction

Light represents an excellent trigger signal. It does not interfere with conventional read-out formats of biological experiments and, given the use of the correct wavelength and short irradiation durations light, is harmless to cells and organisms [1]. The technological advances of the past decades facilitated access to laser and even two-photon laser microscopy devices, which are no longer restricted to limited use by experienced experts in the field but can now also be applied by skilful researchers in different disciplines [2]. However, using light to control biological processes requires the interdisciplinary collaboration of researchers in different fields, such as physics, chemistry, biology and medicine. Starting with the synthesis of light-responsive molecules by chemists, the application requires at least some physical knowhow and/or experience to bring light to the right position. Data interpretation, above all the phenotypic characterization of living organisms, makes the trained eye of a biologist in some way indispensable. Thus, the approach to investigating biological phenomena with light as an exogenous trigger signal is certainly located at a threefold junction where physics, chemistry and biology overlap. Several substance classes and light-responsive derivatives thereof, ranging from small molecules to proteins, have been employed for the investigation of biological processes. The first compounds to be prepared in a light-activatable fashion were ATP and cAMP [3,4]. Both were prepared about 30 years ago and have been studied intensively since then. Only in the last decade have some research groups started to investigate the synthesis of nucleic acids and nucleic acid derivatives and their application to the analysis of biological functions under the control of light irradiation [5–7]. In this chapter, we summarize the latest progress in this research area and review the application of photochemically controlled nucleic acids to biological and biochemical issues.

The Chemical Biology of Nucleic Acids Edited by Günter Mayer

13.2 Irreversible light control of transcription, translation and replication

13.2.1 Light-controlled transcription

The first to study the possibility of light-activated transcription with modified nucleic acids were Haselton and co-workers [8]. They used plasmids for luciferase and GFP and treated them with compound **2** (Figure 13.1a). The latter can be obtained by the oxidation of compound **1** with MnO_2 and is prepared *in situ*. The anticipated reaction is the modification of the backbone phosphate groups with a caging group. Of course, this process is happening statistically. Unfortunately, when it comes to using photolabile groups as caging groups, there is a confusing diversity of abbreviations and their usage (and especially use in combination with the caging group precursors) often does not conform to the IUPAC rules. We have tried to solve this dilemma in the most reasonable way and also tried to be as chemically correct as possible. In our current example, one such plasmid was modified with ~270 DMNPE groups (from 4,5-dimethoxy-2-nitrophenyl-1-ethyl). The caged luciferase plasmid was then transfected into rat skin by particle bombardment and the caged GFP plasmid was transfected into HeLa cells using liposomes. In the rat skin there was no expression of luciferase until irradiation with a 355 nm laser. However, the recovery rate was very limited. Even with laser intensities as high as 20 J cm^{-2} only a $17 \pm 6\%$ recovery was achieved. Interestingly, exposure of the caged plasmid to 1.6 J cm^{-2} of light before the transfection into rat skin also led to a low recovery of only ~40%. Similar results were obtained with the caged GFP plasmid, in which case a maximum recovery rate of 25–50% was achieved. While the cited paper [8] has to be considered seminal, it also shows the room for improvement. An equally seminal paper was published by Okamoto and co-workers slightly later [9]. For the caging they used a coumarin-derived group that is often abbreviated as Bhc (from 6-bromo-7-hydroxycoumarin). It was introduced similarly as before using the *in situ* prepared precursor **3** but this time GFP mRNA was caged (at a rate of ~30 Bhc groups per 1 kb of

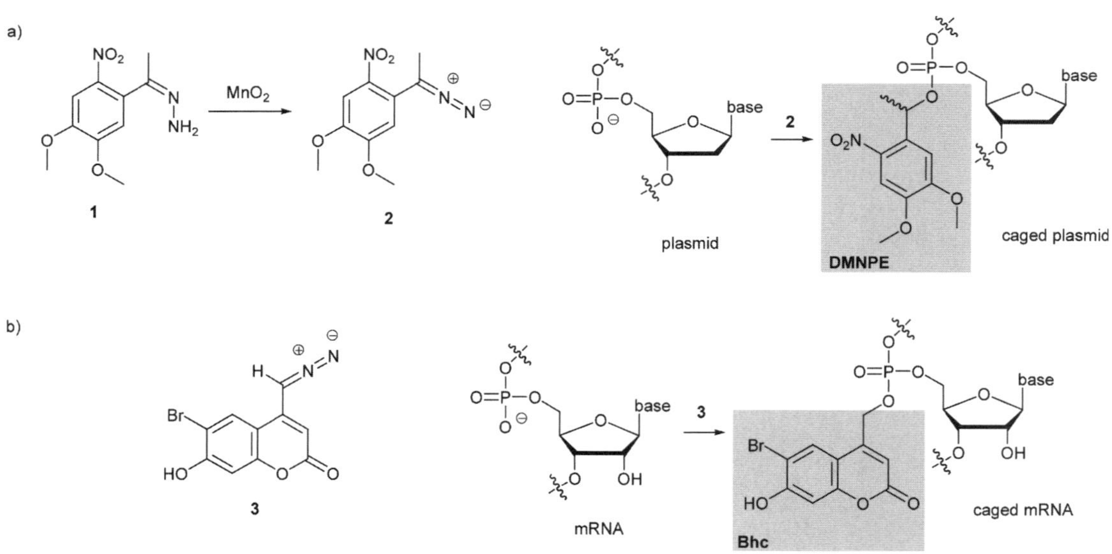

Figure 13.1 *Strategies for the 'statistical caging' of DNA and mRNA. (a) The precursor **1** can be oxidized* in situ *generating the reactive diazo compound **2**, which can be used for the reaction with DNA and mRNA. Haselton and co-workers used this strategy for the caging of plasmid DNA with DMNPE groups [8]. (b) In a similar way, Okamoto and co-workers generated the reactive compound **3** and modified mRNA with Bhc caging groups [9]*

RNA). Through *in vitro* translation assays, 4% translation was found before and 23% after light-activation. The caged mRNA was then introduced into zebrafish embryos through injection in the one-cell stage. This caused an almost ubiquitous distribution in the whole embryonic body. Before irradiation, GFP production was severely reduced but still clearly visible. Through irradiation of the whole embryos (350–365 nm, 0.1 J cm^{-2}), the GFP production could be partly restored. Partial irradiation of embryos led to only partial expression of GFP. Interestingly, the caged mRNA turned out to be very stable as even after 17 h of translation could still be induced with light. In later papers, the authors reported on the elaboration of the technical details [10] and use their approach to study the role of the two homeodomain proteins Lhx2 and Six3s in zebrafish forebrain growth [11].

A problem in both studies mentioned is that the introduction of the caging groups into RNA or DNA is a convenient one-step reaction, but it happens in a statistical fashion regarding the sites that are modified. Consequently, in order to hit 'vulnerable spots', a significant amount of caging groups has to be introduced that are later difficult to get rid of. Therefore, we started our own approach in which we wanted to introduce as few modifications as possible – preferably only one or two – and only in identified key spots responsible for the activity in the respective system. First we decided not to cage the backbone phosphate groups but rather the nucleobases. One of the first such residues that we prepared is shown as T^{NPP} in Figure 13.2 [12]. The NPP caging group severely interferes with Watson–Crick base pair formation as could be shown in melting temperature studies. For example, three T^{NPP} residues led to a reduction of 21.5 °C in melting temperature in a homothymidine 25-mer. Upon irradiation, the caging groups could be quantitatively removed, as proven by high-performance liquid chromatography (HPLC). To demonstrate the usefulness of these defined modifications, they were introduced into a T7 promoter. Due to the local structural perturbation in the promoter, the T7–RNA–polymerase was no longer able to recognize it and consequently no transcript formation was detected. Upon irradiation, all modifications could be cleanly removed restoring the transcript formation to the same level of the unmodified positive control reaction. Interestingly it turned out that a single T^{NPP} already prevented transcription completely.

a)

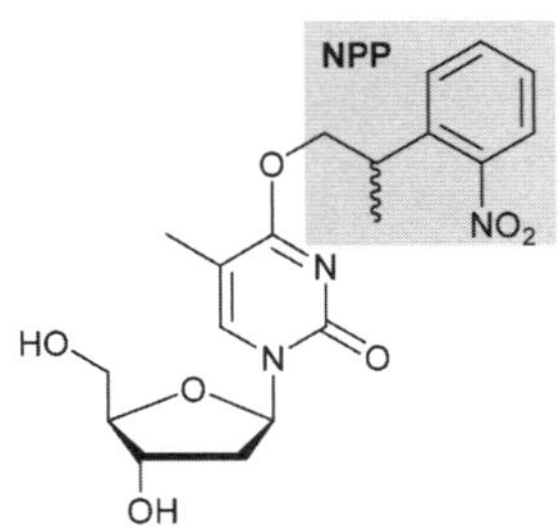

T^{NPP}

b)

sequence 1	sequence 2	melting temperature	
T_{25}	A_{25}	53.8 °C	
$T_{12}T^{NPP}T_{12}$	A_{25}	44.5 °C	before irradiation
		52.7 °C	after irradiation
$T_8T^{NPP}TTTT^{NPP}TTTT^{NPP}T_8$	A_{25}	32.3 °C	before irradiation
		53.2 °C	after irradiation

c) Light-induced transcription using T7 RNA polymerase

```
non-template strand 5'-TAAT(NPP)ACGACTCACTATA-3'
    template strand 3'-ATT A TGCTGAGTGATAT...-5'
                        ↑      ↑          ↑
                       -17    -10        -1
                       ——— T7 promoter ———
```

Figure 13.2 *Light-induced transcription through integration of temporary mismatches in a T7 promoter [12]. (a) The residue T^{NPP} which was used in these studies. (b) Melting temperature studies to demonstrate the influence of such a temporary mismatch. (c) The design of a T7 promoter with transient perturbation through a nucleobase-caged residue*

13.2.2 Light-induced RNA interference

One of the most promising tools for gene regulation today is RNA interference. Consequently, the discovery has been honored by the award of the Nobel Prize in Physiology or Medicine in 2006 to Fire and Mello [13,14]. Arbitrary choice of the region, the timing and the amount of the effect would make this technology even more powerful. The first to report on light-activatable RNA interference were Friedman and co-workers [15]. They again used a 'statistical approach' similar to those described above and the caging group used was the DMNPE group (Figure 13.3a). The double-stranded siRNA they used targeted GFP and was modified with an average of 1.4 cages per duplex. HeLa cells were transfected with both the GFP-plasmid and the (modified) siRNAs. A light-dependent effect was indeed found but the caged siRNAs were not completely inactive. Increasing the amount of caging groups led to siRNAs, which were more inactive, but they could not be fully activated any longer. In a later study, Friedman's group followed a different approach to try to block siRNA activity completely until irradiation (Figure 13.3b) [16]. They modified the 5′-phosphate of the antisense strand since it had been shown previously that modifications at this position are not tolerated by the RNAi mechanism. In their study they used a number of 5′-phosphate-modified antisense strands – including

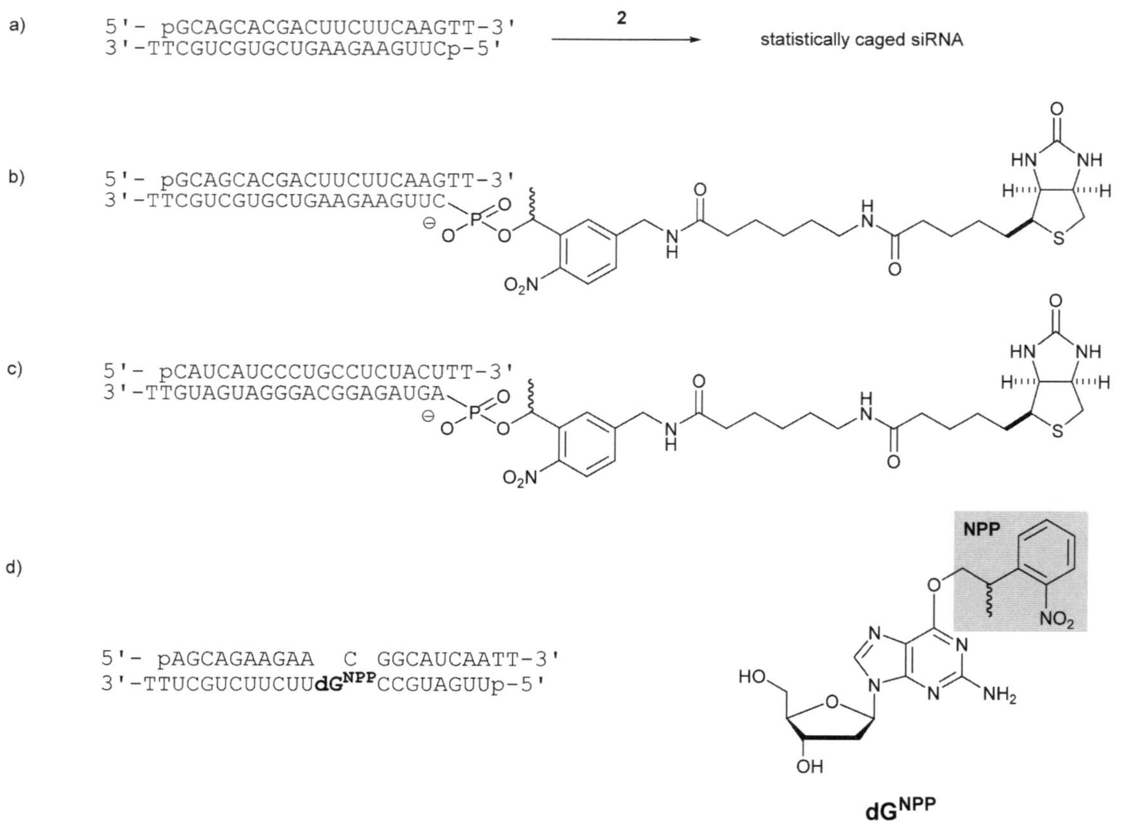

Figure 13.3 Several approaches to make RNA interference light dependent. (a) Friedman and co-workers first used 'statistical caging' on siRNAs [15]. (b) In a later study. the same group modified the 5′-phosphate in the guide strand [16]. (c) McMasters and co-workers used the same design on a different siRNA [17]. (d) Our approach to modify the guide strand at a specific central position which is opposite to the position where the mRNA is cleaved in the RISC complex [18]

the one shown in Figure 13.3b. To their surprise, they found that in contrast to their expectations from the aforementioned studies, the RNAi effect was not completely abrogated and this was not due to the presence of impurities of unintentional uncaging. A very similar approach was used by McMasters and co-workers slightly before the second Friedman study [17]. They used the same modification on an siRNA which was this time targeting GAPDH (Figure 13.3c) (GAPDH = glyceraldehyde-3-phosphate dehydrogenase). As in the Friedman study, it was difficult to obtain a clean ON–OFF behavior. The effects of impurities – especially ($n - 1$) products from the solid-phase synthesis which are still fully functioning siRNAs but do not have the caging group – was investigated. In our own contribution to the field, we again used our approach of nucleobase-caging to modify the antisense strand of siRNAs since these become incorporated as 'guide strands' into the RISC complex (Figure 13.3d) [18]. Even though we incorporated nucleobase-caged nucleotides in various positions (also in the sense strand for control purposes), we anticipated from a literature study analyzing the effect of chemical modifications in various positions [19] in siRNAs that caging groups in position 9 or 10 in the guide strand produce inactive siRNAs. This is plausible because it is exactly the opposite of these positions where the mRNA is cleaved in the RISC complex. On the other hand, the same literature studies showed that deoxynucleotides are readily tolerated there. This is an advantage because it allows the use of caged deoxynucleotides at these positions which are much easier to synthesize. Indeed, the caged siRNA shown in Figure 13.3d was completely inactive within error limits before irradiation but fully active after light activation. Cages in positions further away or on the sense strand had in most cases some influence on siRNA activity, but none of the tested sequences was really inactive before irradiation.

The idea of using caged siRNAs involves the generation of a stable depot of inactive compounds and one potential problem is not only accidental cage hydrolysis but also the cellular degradation of the siRNA. Monroe and co-workers addressed this by using fully 2′-fluorinated nucleic acids that they subjected again to their 'statistical caging approach' [20]. These modified siRNAs were tested *in vitro* and in zebrafish embryos. Caging again had an additional influence on the degradation stability. The caged fluorinated siRNAs were inactivated to a significant extent. The best results were obtained when the antisense ('guide') strand was caged and the activity could be partly but not fully restored upon irradiation with light.

13.2.3 Light-induced antisense approaches

Another way for the regulation of gene expression is the use of antisense oligonucleotides. This approach is much older than RNA interference and has already been reviewed many times [21,22]. In the classical picture, oligonucleotide analogs complementary to mRNAs are used that either block the accessibility of the mRNA or induce its degradation via the ubiquitous RNase H that degrades the RNA strand of an RNA–DNA duplex. These analogs can be, for example, phosphorothioate, peptide, morpholino or locked nucleic acids. Tang and Dmochowski were the first to come up with a light-activatable antisense strategy [23]. Their solution was similar to that which had been published by Asanuma and co-workers in which they used several reversibly photoswitchable azobenzenes to trigger duplex formation (see below) [24]. In Tang and Dmochowski's study, a thiolated antisense oligonucleotide was coupled via a photocleavable heterobifunctional linker to a blocking partially complementary oligonucleotide (Figure 13.4a). After photocleavage of the crosslinker, the duplex showed a reduction in melting temperature from 80 to 51 °C. After this light-initiated 'liberation' of the sequestered antisense oligonucleotide, it was available for the formation of an mRNA heteroduplex which was then cleaved by RNase H. In a further study, Dmochowski's group used negatively charged peptide nucleic acids as the antisense component and 2′-methylated RNA for blocking in an *in vivo* study [25]. The photocleavable linker was again the same as in the previous study. The targets in these studies were the Kozak sequence and start codon of *chordin* mRNA in zebrafish (*chordin* plays an important role in zebrafish embryonic development, especially in the dorsal-ventral patterning) and the *bozozok* gene. The constructs were delivered via injection into the embryos. The population was then divided and one part was irradiated for 8 min. It was found that 85% of the nonirradiated embryos developed normally, whereas in the irradiated group 81% of this population showed the expected aberrant phenotype. Similar efficiencies were obtained for both genes. Importantly, the zebrafish embryos were not harmed by the light irradiation, as was shown in control experiments. Another study by Dmochowski and co-workers investigated the light-controlled degradation of the mRNA of *c-myb* – a hematopoietic transcription factor [26]. Several combinations of sides for the photocleavable linker and introduction of mismatches were tested and evaluated. In a recent investigation by the same group, the architecture was slightly changed and the photocleavable linker was introduced in the middle of a 2′-methylated antisense oligonucleotide (Figure 13.4b) [27]. This construct was used to block the action of the ribosome on an mRNA temporarily and hence turn translation ON after irradiation (whereas in the previous approaches translation was turned OFF with light due to the degradation of the mRNA). This strategy was termed 'RNA bandages' in the

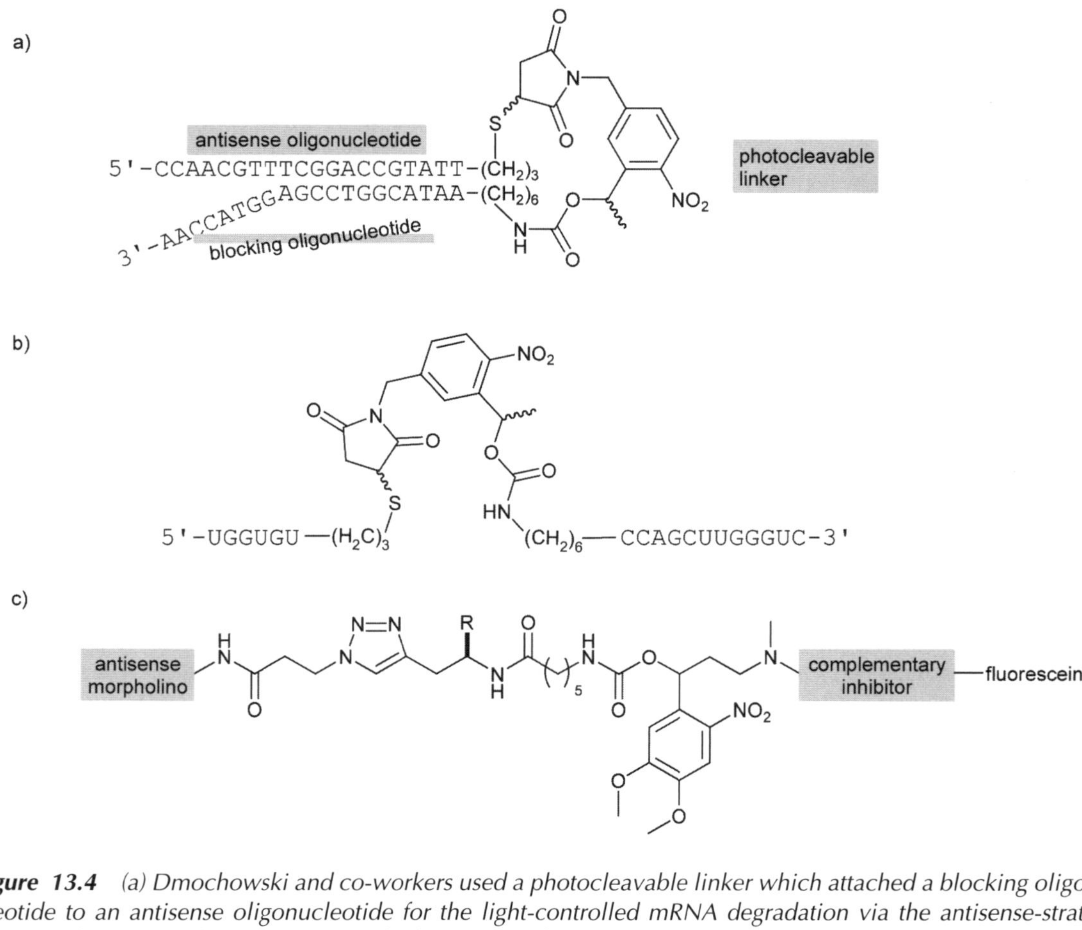

Figure 13.4 (a) Dmochowski and co-workers used a photocleavable linker which attached a blocking oligonucleotide to an antisense oligonucleotide for the light-controlled mRNA degradation via the antisense-strategy [23,25,26]. (b) In an alternative approach, they prepared an 'RNA-bandage' consisting of two 2′-methylated RNA strands connected with a photocleavable linker to turn translation ON [27]. (c) This temporarily blocked morpholino construct with photocleavable linker was used by Chen and co-workers. The substituent R is not further specified in the paper [28]. (d) Nucleobase-caged antisense phosphorothioates were presented by Deiters and co-workers [29]. (e) A crosslinker that can be cleaved by singlet oxygen generated by irradiation with red light in the presence of a photosensitizer [30]

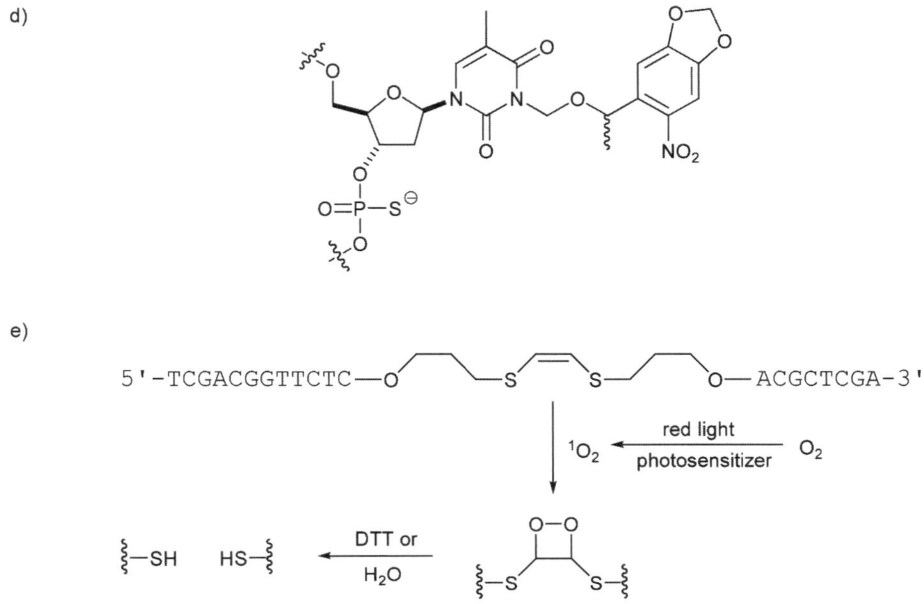

Figure 13.4 *(Continued)*

investigation. After optimization of the design they arrived at a downregulation of translation to 30%. Upon irradiation, the translational activity could be restored completely within error limits.

Three more groups have joined the field of light-activatable antisense-agents recently, one of them by Chen and co-workers [28]. They used morpholino antisense oligonucleotides that were again temporarily blocked by an inhibitor which was attached via a photocleavable linker (Figure 13.4c), but a more complex linker was applied that used the power of click chemistry. The target gene was the *no tail* gene in zebrafish. Delivery was again achieved via injection in the one-cell stage. Unirradiated embryos developed normally whereas irradiated embryos showed predominantly the expected phenotype. Deiters and co-workers introduced caging groups on the nucleobases of phosphorothioate antisense oligonucleotides (Figure 13.4d) [29]. An excellent ON/OFF ratio was obtained upon irradiation with these in a renilla luciferase reporter assay. Mokhir and co-workers presented another photolinker for antisense phosphorothioates that is cleavable with red light (Figure 13.4e) [30]. The red light is absorbed by a photosensitizer creating singlet oxygen, which reacts in a [2 + 2] addition with the crosslinker. The resulting four-membered ring decomposes with dithiothreitol or in water to liberate the two antisense parts.

13.2.4 DNA amplification and light control

Two more studies will be reported here that used caged nucleic acids for light control in DNA amplification: In the first publication on nucleobase-caged nucleic acids by Dmochowski's group, they used a combination of a DNA template strand and a DNA primer together with the Klenow fragment of *Escherichia coli* DNA polymerase I [31]. The template strand was modified twice: It contained both a fluorophore modification and a photocleavable quencher attached to a nucleobase (Figure 13.5a). Before irradiation, the polymerase could not extend the primer opposite the bulky residue CD, whereas after irradiation, the primer extension was possible. The advantage of this system is that uncaging can be followed directly via the emergence of a

a)

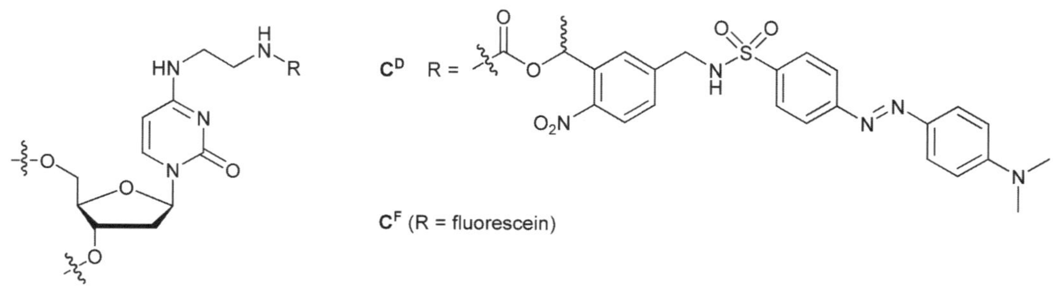

primer 5'-GGATGATGGAGG-3'
template strand 3'-CCTACTACCTCC^FC^DCGACGACACCTA-5'

a)

STOP for primer extension
primer 5'-CTCGATCCTCTACGCCG-3'
template strand 3'-GAGCTAGGAGATGCGGCCTGT^{NPP}CGA-5'

Figure 13.5 *Light control in DNA amplification. (a) Dmochowski and co-workers used the caged residue CD with a DABSYL quencher to not only induce primer extension but also to control the uncaging process via the increase of a fluorescence signal [31,32]. (b) Komiyama and co-workers used the residue TNPP to stop primer extension in a LACE-PCR reaction (light-assisted cohesive-ending PCR), which can thus generate amplified DNA with unmodified sticky ends when the caging groups are removed after completing the PCR reaction [33,34]*

fluorescence signal. The disadvantage, however, is – at least in the realization of this study – that the fluorophore is attached in a way that interferes with duplex formation and the uncaging is not traceless. This was also apparent in melting temperature studies [32]. Komiyama and co-workers used the previously presented residue T^{NPP} from our own group to stop primer extension at a defined site [33,34]. For the application they coined the term 'LACE-PCR' for light-assisted cohesive-ending poymerase chain reaction since their method allows the generation of amplified DNA with sticky ends which is not possible in a normal PCR reaction. Via photolysis, the caging groups can be removed and the uncaged sticky ends can then be used for subsequent ligations.

13.3 Nucleic acid folding

Photolabile protecting groups have been employed for the elucidation of the kinetics of RNA folding pathways. In this way, caging moieties were introduced at certain positions of an RNA molecule, thereby favoring the folding of an RNA structure in one out of two secondary motifs. After light-dependent release of the cage, the refolding kinetics could be resolved by NMR spectroscopy. In pioneering work, Pitsch, Schwalbe and co-workers applied this approach to investigate the refolding of short RNA hairpins [35]. They used an *o*-nitrophenylethyl (NPE)-protected guanine residue, incorporated into a 20-nucleotide RNA molecule that basically folds into two different hairpin structures. Upon incorporation of the NPE group, one hairpin was favored, with the NPE-guanine embedded in the loop moiety. Irradiation removed the cage whereas refolding was induced now allowing

Watson–Crick base pairing of the previously caged guanine being part of a stem structure rather than a loop motif. The refolding was monitored by NMR spectroscopy. In a subsequent study, this concept was extended to caged uridine and a similar RNA hairpin structure [36] and for the elucidation of time-resolved conformational dynamics of bistable RNA molecules [37]. Recently, this approach was also applied for the time-resolved study of the hammerhead ribozyme and to investigate ligand-induced conformational changes with the aptamer domain of naturally occurring riboswitches [38,39]. In the latter case, it is noteworthy that the cage was not implemented within a certain position of the RNA itself. In fact, the caging strategy was used to mask the ligand of the riboswitch, namely hypoxanthine that had previously been shown to bind to the guanine riboswitch with high affinity [40]. Based on this approach, three defined kinetic steps were revealed, starting with the primary complex formation and ultimately leading to a stabilization of the complex by distant long-range loop–loop interactions. These data impressively demonstrate how the caging approach in combination with time-resolved spectroscopic measurements can contribute to the understanding of the conformation dynamics of biological relevant nucleic acid macromolecules.

13.4 RNA and DNA enzymes

Cages can be used for the photocontrol of ribozymes either by modification of the 2′-OH position adjacent to the scissile phosphate diester bond or by employing ligands of hammerhead-derived aptazymes such as theophylline (Figure 13.6) [38,41–43].

While 2′-modifications of hammerhead ribozymes are useful tools to investigate time-resolved cleavage rates and associated conformational changes, the photocontrol of aptazymes offers a route for the spatiotemporal control of gene expression. In this regard, aptazymes can be used as gene control elements, when embedded in the 5′- or 3′-untranslated region (UTR) of mRNA molecules encoding for certain proteins. Aptazymes are RNA molecules that comprise an aptamer domain which binds to a defined small molecule, thereby triggering conformational changes within the adjacent ribozyme domain and thus regulating cleavage activity. When embedded in mRNA molecules, this can lead to a degradation of the mRNA and consequently to inhibited protein expression. In a first example, Young and Deiters demonstrated that caged variants of theophylline are capable to accomplish photocontrol of ribozyme activity (Figure 13.7) [43].

Lee *et al.* expanded this concept to a reversible control of hammerhead ribozyme activity [44]. The employed a pyrene derivative (10-carboxy-2,7-di-*tert*-butyl-*trans*-12c,12d-dimethyl-12c,12d-dihydrobenzo[*e*]pyrene – BDHP-COOH) that can be photo-isomerized at certain wavelengths (Figure 13.8).

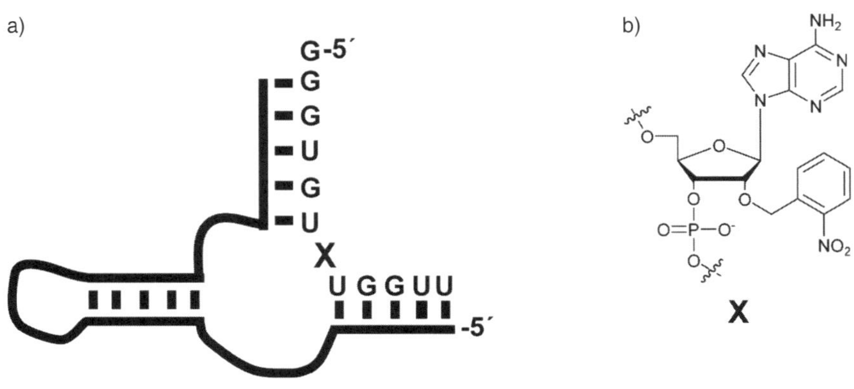

Figure 13.6 *(a) Schematic representation of the minimal hammerhead ribozyme bound to its substrate. (b) X = 2′-caged nucleotide for light induced RNA cleavage [41,42]*

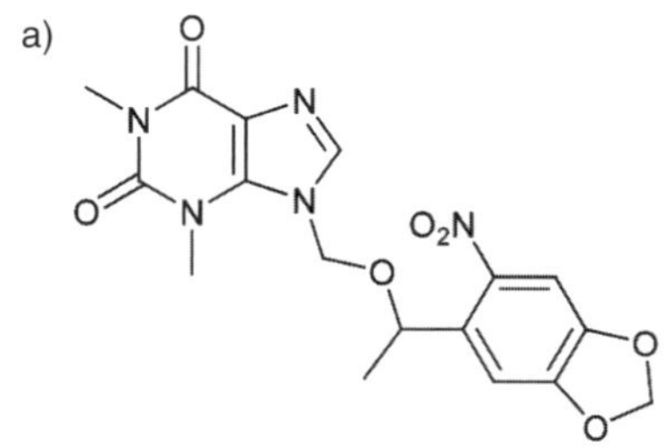

Figure 13.7 Caged theophylline (a) and theophylline (b) for the light control of theophylline-dependent hammerhead-related aptazymes [43]

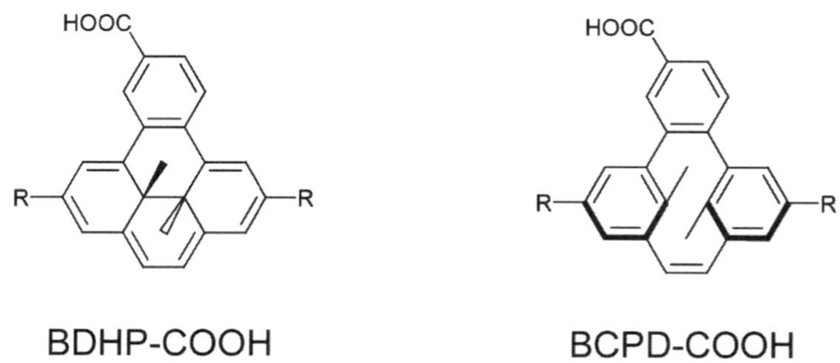

Figure 13.8 10-Carboxy-2,7-di-tert-butyl-trans-12c,12d-dimethyl–12c,12d-dihydrobenzo[e]pyrene (BDHP-COOH) and its photoisomerized counterpart (BCPD-COOH). R = tert-butyl [44]

Aptamers that were selected to bind to one isomer of this compound were shown to be useful as regulatory elements to construct hammerhead-based aptazymes capable of being regulated by light in a reversible manner.

An alternative approach to regulate nucleic acid enzymes reversibly by light was applied by Keiper and Vyle [45]. They introduced azobenzene-modified uridylates at a strategic position of a DNAzyme capable of cleaving RNA. The azobenzene moiety was introduced on the 2′-position of the 2′-deoxyribose. The azobenzene was attached either *ortho-*, *meta-* or *para-* with respect to the ribose and, interestingly, different cleavage and light-responsiveness could be determined (Figure 13.9).

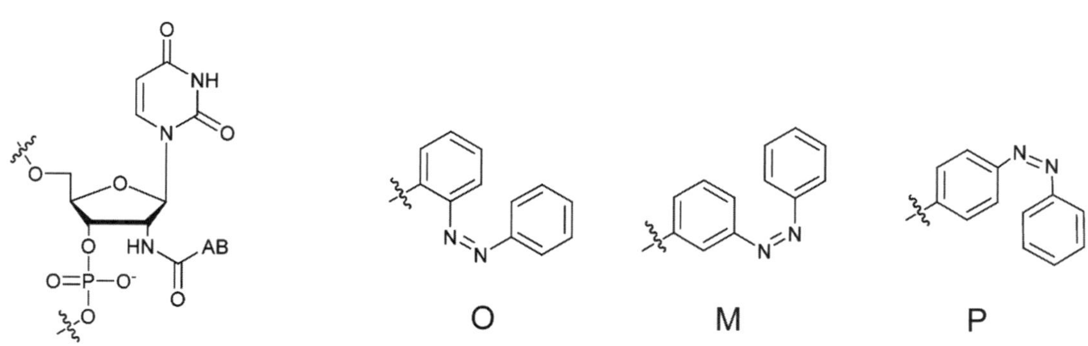

Figure 13.9 Ortho-, meta- *and* para-*substituted azobenzene moieties (O, M and P) incorporated into DNA at the 2′-position of a DNAzyme shown on the left [45]*

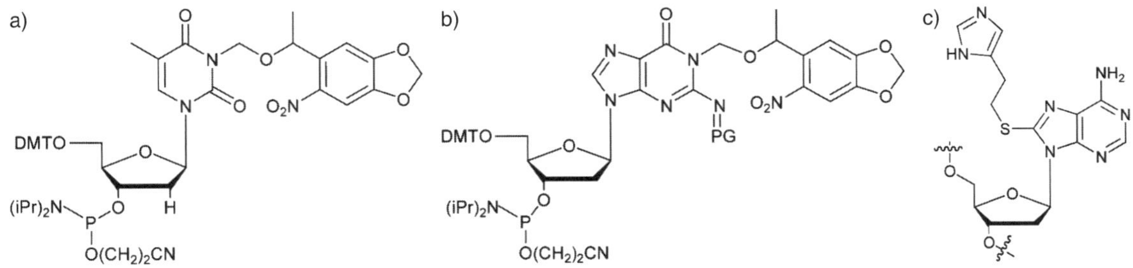

Figure 13.10 *Phosphoramidites of 6-nitropiperonyloxymethyl (NPOM)-caged thymidine (a) and guanine (b) nucleobases for the site-specific incorporation into DNA [46,47]. C8-linked thioether imidazole adenine for incorporation into a DNAzyme and its light-dependent regulation (c) [48]*

Whereas *ortho*- and *meta*-derivatized azobenzene were shown to be at least as active as the wild-type DNAzyme, the *para*-substituted molecule displayed only half of the activity of the wild-type DNAzyme.

Deiters and co-workers used the same DNAzyme and gained photocontrol of RNA cleavage by introducing N^3 nucleobase-caged thymidine residues at various positions, which either were located within the catalytic core of the DNAzyme or necessary to build an extended base-paired region for substrate recognition (Figure 13.10a and b) [46]. A similar approach was described by the same group now employing modified guanine nucleobases located at strategic positions of a G-quadruplex forming DNAzyme with peroxidase activity [47]. Perrin and co-workers used a different approach and introduced a C8-linked thioether imidazole adenine nucleoside into a DNAzyme which has RNA cleavage activity (Figure 13.10c) [48]. In this way, a light-sensitive DNAzyme was achieved by photochemical homolysis of the S–C bond.

The most comprehensive collection of caged nucleosides was introduced by Höbartner and Silverman [49]. Employing a synthesis-ligation strategy, they were able to synthesize the 160 nucleotides comprising the P4–P6 domain of the group I intron with cages at different positions. Using these derivatives, it was possible to investigate the folding of caged and subsequently light-induced uncaged P4–P6 RNAs by biochemical assays, such as RNA cleavage. This approach might reveal a further route for the time-resolved folding of larger RNA molecules.

13.5 Caged aptamers

In addition to ribozymes and DNAzymes, caging groups were incorporated into DNA aptamers, thus generating light-responsive protein inhibitors. We started to investigate the compatibility of the caging technology with nucleic acid aptamers in 2005 and were able to demonstrate that it is possible to create not only aptamers that can be activated but also those that can be inactivated by light [50,51]. We used an anti-thrombin aptamer that is built up of 15 nucleotides and forms a G-quadruplex structure [52]. Starting from the crystal structure of the aptamer-thrombin complex we identified crucial nucleotide positions, which were assumed to either contribute directly to the binding of the aptamer to thrombin or those residues which were necessary to allow the formation of the G-quadruplex structure (Figure 13.11) [53]. In both cases we obtained caged aptamer variants that were inactive as long as the cage was present and which were efficiently reactivated by irradiation with light. Moreover, and due to the nucleic acid nature of aptamers, we were also able to design aptamer variants that were active in the presence of a cage but could be essentially inactivated upon light irradiation (Figure 13.11).

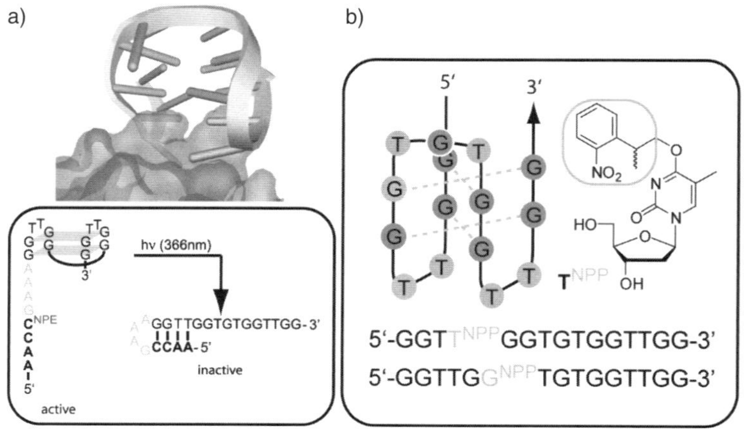

Figure 13.11 *Caged aptamers targeting the exosite I of human α-thrombin. The 15 nucleotide-containing aptamer was synthesized in order to obtain aptamer variants that can be deactivated (a) or activated (b) upon irradiation with light (λ = 365 nm). Light control was achieved by incorporation of* o*-nitrophenylethyl (NPE)-modified cytidine or* o*-nitrophenylpropyl (NPP)-modified thymidine and guanine nucleobases [50,51,53]. See color plate section*

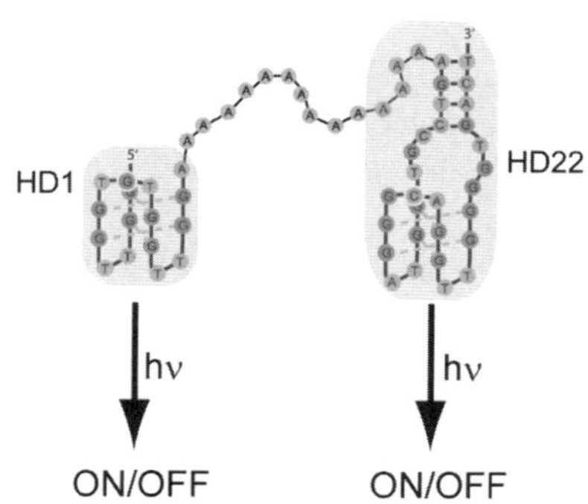

Figure 13.12 *Caged bivalent aptamer for differential protein subdomain regulation. A bivalent aptamer comprising the HD1 domain (targeting exosite I of thrombin) and the HD22 domain (targeting exosite II of thrombin) – both shaded gray – can be employed for being light controllable by the incorporation of caged nucleobases at strategic positions [54,55]. See color plate section*

We even went a step further and constructed bivalent aptamers targeting both exosites of thrombin [54,55]. By caging either sub-domain of the bivalent aptamer, we were able to regulate thrombin activity differentially in a spatial manner by light (Figure 13.12).

13.6 Miscellaneous

13.6.1 Strand breaks and abasic sites

A substantial body of literature exists on the development and application of light-responsive nucleic acid strand breaks. Dussy *et al.* described the synthesis and incorporation of *o*-nitrobenzyl esters into the backbone

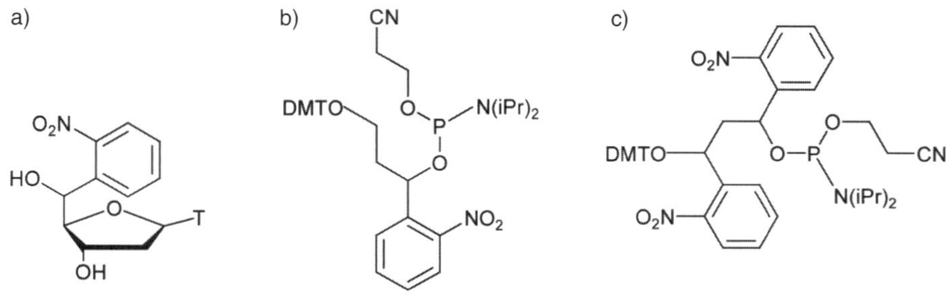

Figure 13.13 *C5′-caged nucleoside (a) and phosphoramidites for the synthesis of caged strand-breaks [(b) and (c)] [56,57–60]*

of DNA molecules (Figure 13.13) [56]. Ordoukhanian and Taylor demonstrated in 1995 the synthesis of a phosphoramidite *o*-nitrobenzyl derivative that can be incorporated in DNA [57]. Using this molecule, they demonstrated that this approach is useful for introducing single and double strand breaks. These kinds of artificial lesions might have an impact in studies of DNA repair mechanisms (Figure 13.13) [58]. In a subsequent study, the same group described a caged strand break employing two *o*-nitrobenzyl groups (Figure 13.13) [58]. They synthesized a small circular dsDNA molecule, a so-called plasmid mimic, comprising these caged strand breaks. Upon light irradiation, a site-specific strand break was induced and, noteworthily, this DNA scission could be repaired by enzymes of the DNA repair machinery. These data are in accord with a progenitor study, in which it was demonstrated that a light-induced strand break can be re-ligated by ligases in the presence of ATP as energy resource [59,60].

Sheppard and co-workers investigated the possibility of introducing abasic sites into DNA and RNA oligonucleotides (Figure 13.14) [61,62]. These studies facilitate the mimicking of depurinated nucleic acids and offer a straightforward approach to the study of DNA repair mechanisms. Zheng and Sheppard reported a method for photochemical generation and access to oxidized abasic sites in DNA. This strategy resembles the oxidative damage of DNA and therefore permits a synthetic route to exogenously controlled DNA damage and repair studies [63]. Aso, Suemune and co-workers prepared caged nucleosides with *o*-nitrobenzyl moieties located either at the 1′-carbon (base) or at the 4′-position. Double-stranded oligodeoxynucleotides that have these modified compounds incorporated bear the cages either in the major groove (4′-position) or within the base stacking area (1′-position). Uncaging experiments revealed that uncaging of differently located nitrobenzyl groups is different in double- compared with single-stranded oligodeoxynucleotides (Figure 13.14) [64].

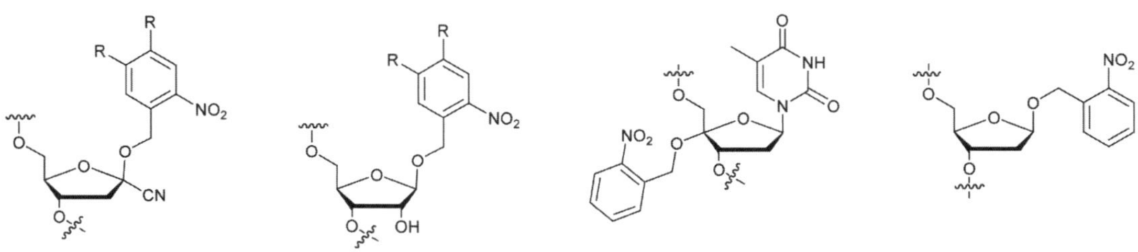

Figure 13.14 *Caged abasic sites for the light control of depurinated DNA strands [61–64]*

13.6.2 Nucleic acid ligands

Baigl and co-workers described the synthesis of a trimethylammonium bromide azobenzene derivative, AzoTAB, which reacts with DNA in a reversible manner. In its *trans* conformation it shows a high affinity for DNA, whereas upon light-induced *trans–cis* isomerization its DNA binding activities were strongly reduced. Using this molecule, the conformation of DNA could be essentially, and noteworthily in a reversible manner, controlled and thereby associated functions of the DNA might also be regulated [65].

Caged magnesium ions have been used to regulate the activity of Mg^{2+} ion-dependent enzymes. In this way, Burns and co-workers investigated the spatial regulation of the DNA endonuclease Sma1 [66]. They demonstrated that only regions of the DNA that were in close proximity of the irradiation beam were hydrolyzed. This approach offers a way to investigate site-specific enzyme activities, especially when a substrate offers more than one possible reactive site. More recently, Rotaru and Mokhir introduced a generic model for the investigation of nucleic acid interactors that react to light of different wavelengths [67]. They constructed an indirect operating system, employing photosensitizers that produce reactive oxygen species upon irradiation with light. This species can subsequently induce the strand break of complementary oligodeoxynucleotides connected via an electron-rich ethylene–thioether bridge. They demonstrated that two different photosensitizers, reacting to green or red light, were useful and irradiation finally led to a strand cleavage resulting in leverage of the two strands.

Monroe and co-workers described the use of 1-(4,5-dimethoxy-2-nitrophenyl)ethyl ester (DMNPE) to gain exogenous light control over DNA–DNA hybridization [68]. They used a molecular beacon to monitor complementary base-pair formation. Tanabe, Nishimoto and co-workers constructed a nucleic acid hairpin-based system, which can be useful for the light-triggered release of any compound interlinked with the nucleic acid via a 2-nitrobenzyl group [69]. The hairpin structure in this approach was mainly useful for monitoring compound release by FRET analysis. A similar approach was described by Saito and co-workers 2 years earlier [70].

A bivalent caged O^6-benzylguanine nucleobase derivative was synthesized by Johnsson and co-workers, revealing an *o*-nitroethyl group at position N^7 [71]. This molecule was shown to be useful for the irreversible, but light-dependent, labeling of O^6-alkylguanine alkyltransferases, so-called SNAP-tag. The bivalent molecule allows access to a synthetic dimerizer with exogenous control by light.

13.7 Reversibly photoswitchable DNA and RNA systems

13.7.1 Introduction

As already pointed out in the Introduction, caging is a technology that had its origin about 30 years ago in the late 1970s (see above). Its attractiveness comes from the fact that if one can identify 'hot spots' in biologically active molecules – spots that are sensitive to derivatization or can even be identified as being responsible for the activity – derivatives can potentially be synthesized that are completely inactive. Since after the photorelease the original molecule is recovered, its activity is completely restored – provided that there are no side reactions. Thus a clean ON–OFF behavior can often be easily achieved with the caging technology. However, upon photolysis, stoichiometric amounts of byproducts are formed (resulting from the cleavage of the caging group) and the activation process is of course irreversible. Therefore, approaches that allow a reversible photoswitching of activity would be very attractive and indeed the first attempts to realize this are even older than the caging approach. Early attempts tried to achieve reversible photoswitching of protein function and the results were modest [72]. Instead of photolabile groups, the key players are now bistable photoswitches.

The system which has been used by far the most often is the azobenzene system. The *trans* conformer is thermodynamically more stable and it is flat and pi-extended. Upon irradiation into the absorption band, typically around 350 nm, the system undergoes a photoisomerization to the *cis* form, which is helical because of the steric hindrance of the *ortho*-hydrogen atoms. The *cis* isomer now has an absorption band, for example at around 450 nm, and irradiation with visible light leads to the formation of the *trans* isomer again. It is important to note that this transition can usually be performed many times. A very elegant study by Trauner and co-workers used an azobenzene to put a glutamate reversibly into the binding pocket of an ionotropic glutamate receptor, and thus reversibly open and close a ligand-gated ion channel with light as input signal [73]. A critical issue is always how much of the population can be isomerized to the *cis* state. This conversion has sometimes been reported to be very high (see below) with the azobenzene system, but is usually rather in the range 70–80% [74]. Another problem with azobenzene is that it can be reduced in a cellular context and thus loses its switching ability [75]. An alternative system is based on a spiropyran skeleton. Upon irradiation with light in the near-UV region, the spiropyran opens to form a merocyanin isomer. The latter is flat, pi-extended and more polar than the former and can again be closed to the spiropyran by irradiation with visible light. This change in polarity has been harnessed, for example, in an elegant study by Feringa and co-workers in which they modified the mechanosensitive channel of high conductance in *E. coli* with a spiropyran and were able to open and close reversibly a 3 nm pore with light [76].

13.7.2 Photoswitchable derivatives of oligonucleotides

The first examples of reversibly photoswitchable oligonucleotides date back to the late 1990s and were performed by Yamana and co-workers [77–81]. Slightly later, Komiyama's group presented a system in which the azobenzene unit was not incorporated into the backbone but rather in the side-chain – replacing half of the ribose and the nucleobase. Figure 13.15a shows the first residue that they used (drawn here and in the following examples as a phosphoramidite building block to explain how the residue was incorporated and to show the 5′/3′-orientation) [82]. They hypothesized that such a residue $\mathbf{X^a}$ would lead to less perturbation in the nucleic acid backbone and the planar *trans*-azobenzene should fit well into the base pair stack – as opposed to the helical *cis* conformer. The residue $\mathbf{X^a}$ contains a stereogenic center and in the first studies an enantiomeric mixture of $\mathbf{X^a}$ was used for the solid-phase synthesis and the resulting diastereomeric oligodeoxynucleotides were later separated by reversed-phase HPLC. The configuration in the diastereomers was not elucidated until later [83], but both diastereomeric oligodeoxynucleotides could be switched between the *cis* and *trans* forms several times without any difference in the spectra obtained. Consequently, it was possible to control the formation of DNA duplexes with light [84]. As shown in Figure 13.15b, the melting temperature between the homoadenosine and homothymidine is only slightly different from diastereomer 1 of the hybrid containing the $\mathbf{X^a}$ residue. This shows how well this residue can indeed replace a nucleoside. On the other hand, a reduction of 8.9 °C in the melting temperature is observed upon photoisomerization to the *cis* state – albeit in a very short oligonucleotide containing only A=T base pairs. The *trans* conformer in diastereomer 2 did not fit as nicely into a DNA duplex. At this point, it was not yet known to which absolute configuration in the $\mathbf{X^a}$ residue this referred. In a second study, the regulation of a DNA duplex was studied more thoroughly [85]. Photoregulation of triplex formation was also possible [86]. Therefore, they used a DNA duplex with a central $(A{=}T)_{14}$ polypurine tract (ds1 in Figure 13.15b) and added a T_{13} homothymidine sequence with an $\mathbf{X^a}$ residue on the 5′-end. The melting temperature for the triplex formation of this hybrid was higher than that of T_{13} and even higher than that of a 14-mer homothymidine. Upon photoswitching, the melting temperature decreased by 13.3 °C.

After showing that duplex formation can be regulated, Komiyama and co-workers used this approach for the photoregulation of a DNA polymerase reaction [87]. For that purpose, they used a setup in which not only a 54 base pair template and an 18-mer primer but also a modulator was present (Figure 13.15c). The modulator was a short oligodeoxynucleotide of 12 nucleotides with a 3′-propanol cap (to prevent it from being extended by the polymerase) and bound downstream of the primer to the template. Upon addition of T7

a)

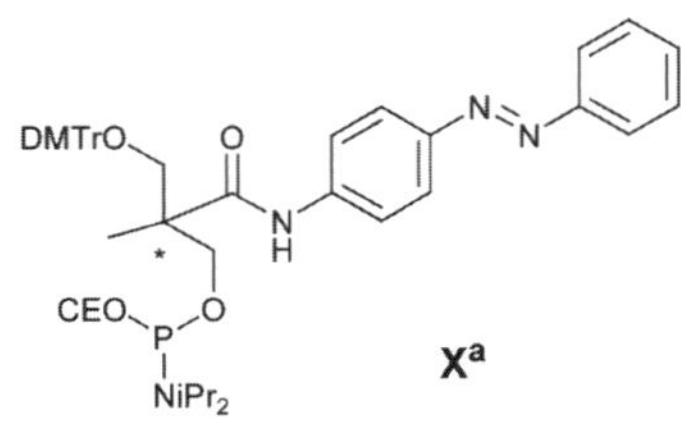

X^a

b)

sequence 1	sequence 2	melting temperature	
AAAAAAAA	TTTTTTTT	23.7 °C	
trans AAA**X**ᵃAAAA	TTTTTTTT	24.8 °C	diastereomer 1
cis AAA**X**ᵃAAAA	TTTTTTTT	15.9 °C	
trans AAA**X**ᵃAAAA	TTTTTTTT	19.9 °C	diastereomer 2
cis AAA**X**ᵃAAAA	TTTTTTTT	14.7 °C	
TTTTTTTTTTTTT	ds1	18 °C	triple helix formation
trans **X**ᵃTTTTTTTTTTTTT	ds1	26 °C	
cis **X**ᵃTTTTTTTTTTTTT	ds1	12.7 °C	
TTTTTTTTTTTTTT	ds1	22.5 °C	

c) Photoswitching of primer extension:

```
primer                              modulator
5'-CTAATACGACTCACTATA-3'            5'-GXaGGAATTCGGA-cap-3'
3'-GATTATGCTGAGTGATATAAAAAAAAAAAAAAAAACC CCTTAAGCCTAAAAAAAA-5'
                         template
```

Figure 13.15 *(a) The structure of the two enantiomers* X^a *with an azobenzene replacing the nucleobase and half of the ribose of a normal nucleoside [82]. (b) Data from melting temperature studies using this residue [84, 86]; ds1 denotes a 32 base pair DNA duplex with a central* $(A{=}T)_{14}$ *region and two flanking nine base pair regions of mixed sequence [86]. (c) Setup for the photoregulation of a DNA polymerase reaction (cap = propanol) [87]. Here and in the following figures all oligonucleotides are noted in the 5′–3′ direction unless noted otherwise and thick arrows represent nucleic acids in the 5′–3′ direction*

DNA polymerase, the primer was extended but only and cleanly up to the 5′-end of the modulator. After prior irradiation, the polymerase produced only the full-length product (under the irradiation conditions the polymerase activity was not impaired). The authors explained this by the fact that the polymerase was able to peel off the modulator only if the base pairing on the 5′-end was weakened by the *cis*-azobenzene. Interestingly, both the *trans* and the *cis* forms of the modulator had melting temperatures well beyond the reaction temperature of the polymerase reaction and were therefore always base paired to the template before the addition of the polymerase.

Around the time of these studies, Komiyama and co-workers also tested another way to include an azobenzene: The idea was to keep the nucleic acid even more intact by leaving the ribose and even the nucleobase in place. Therefore, they used a uridine derivative $\mathbf{X}^{\mathbf{b}}$ with an azobenzene attached to the 2′-OH group (Figure 13.16a and b) [88]. It turned out that a 5′-terminal introduction of $\mathbf{X}^{\mathbf{b}}$ into a homothymidine oligomer lead to a significant increase in duplex stability. The residue could again be switched with light and the conversion to the *cis* state was complete to 90%, affording a relative reduction of the melting temperature by 8.8 °C. It was possible to switch the system back to the *trans* conformation by irradiating with visible light, fully recovering the starting state. The residue $\mathbf{X}^{\mathbf{b}}$ could also be used as a switchable internal modification. However, melting temperature studies showed that it does not incorporate itself as nicely into the duplex as the previous residue $\mathbf{X}^{\mathbf{a}}$ because even in the **trans** state the duplex was less stable than the unmodified form by 10.3 °C. Hence the residue $\mathbf{X}^{\mathbf{b}}$ is an interesting alternative but probably useful only for terminal modification providing extra stabilization if it is in the *trans* state. Around the same time, Komiyama and co-workers also studied the 5′-cap residues shown in Figure 13.16c – one with an azobenzene and the other with a spiropyran [89]. However, the effect of photoswitching these loosely attached residues was rather small and they were not further investigated.

To further improve the usefulness of the residue $\mathbf{X}^{\mathbf{a}}$, it would be desirable to study the effect of multiple residues of this sort in one oligomer. However, the method of separation of the diastereomers after solid-phase

a)

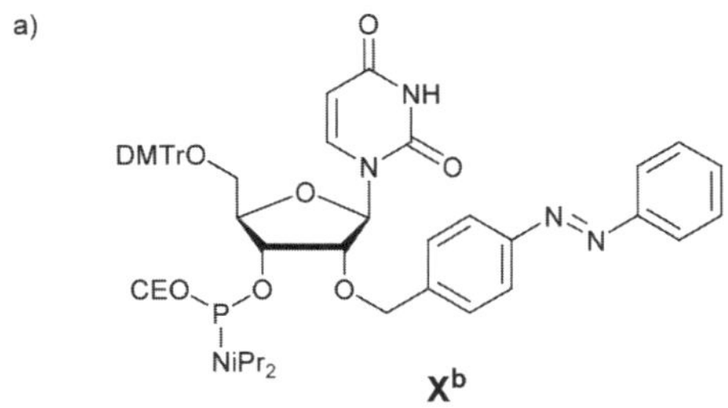

b)

sequence 1	sequence 2	melting temperature
TTTTTTTT	AAAAAAAA	23.7 °C
trans $\mathbf{X^b}$TTTTTTT	AAAAAAAA	30.2 °C
cis $\mathbf{X^b}$TTTTTTT	AAAAAAAA	21.4 °C
trans TTT$\mathbf{X^b}$TTTT	AAAAAAAA	13.4 °C
cis TTT$\mathbf{X^b}$TTTT	AAAAAAAA	8.4 °C

c)

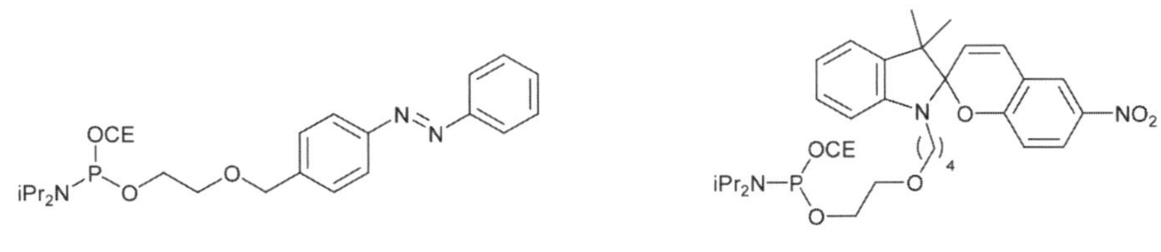

Figure 13.16 *(a) The structure of the residue* $\boldsymbol{X^b}$ *introduced by Komiyama and co-workers which leaves the ribose and the nucleobase of a nucleoside intact [88]. (b) Data from melting temperature studies using this residue [88]. (c) Two photoswitchable 5′-cap units introduced by Komiyama and co-workers [89]*

synthesis that Komiyama and coworkers had used up to that time would afford complicated diastereomeric mixtures. Therefore, they decided to use the residues $\mathbf{X^c}$ and $\mathbf{X^d}$ instead (Figure 13.17a) [90] (n this chapter, we decided to refrain from using the usual stereochemical notations such as *rac*-$\mathbf{X^a}$ or *ent*-$\mathbf{X^c}$ for the sake of readability when using these residues in the one-letter formalism in which oligonucleotides are written). They are both derived from L- or D-hreoninol which is available via the pool of chiral compounds and differ from the

a)

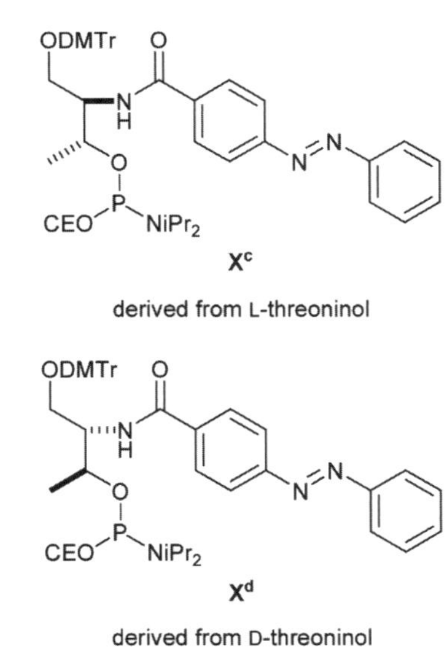

b)

sequence 1	sequence 2	melting temperature	
GCGAGTCG	CGACTCGC	46.6 °C	
trans GCGA$\mathbf{X^c}$GTCG	CGACTCGC	45.1 °C	
cis GCGA$\mathbf{X^c}$GTCG	CGACTCGC	40.8 °C	
trans GCGA$\mathbf{X^d}$GTCG	CGACTCGC	50.9 °C	
cis GCGA$\mathbf{X^d}$GTCG	CGACTCGC	36.6 °C	
trans GC$\mathbf{X^c}$GAGT$\mathbf{X^c}$CG	CGACTCGC	25.4 °C	
cis GC$\mathbf{X^c}$GAGT$\mathbf{X^c}$CG	CGACTCGC	25.5 °C	
trans GC$\mathbf{X^d}$GAGT$\mathbf{X^d}$CG	CGACTCGC	43.9 °C	
cis GC$\mathbf{X^d}$GAGT$\mathbf{X^d}$CG	CGACTCGC	22.4 °C	
TTTTTTTTTTTTT	ds2	22.5 °C	triple helix formation
trans $\mathbf{X^c}$TTTTTTTTTTTTT	ds2	29.7 °C	
cis $\mathbf{X^c}$TTTTTTTTTTTTT	ds2	12.6 °C	
trans $\mathbf{X^d}$TTTTTTTTTTTTT	ds2	17.7 °C	
cis $\mathbf{X^d}$TTTTTTTTTTTTT	ds2	12.6 °C	

Figure 13.17 *(a) The structures of the residues* $\boldsymbol{X^c}$ *and* $\boldsymbol{X^d}$ *introduced by Komiyama and co-workers [90]. (b) Data from melting temperature studies using these residues and investigating duplex [90] or triplex formation [92]. ds2 denotes a 31 base pair DNA duplex with a central* $(A{=}T)_{13}$ *region and two flanking nine base pair regions of mixed sequence [92]*

previous residue $\mathbf{X^a}$ by the presence of one extra methyl group and the orientation of the amide bond. Figure 13.17b shows again the results of melting temperature studies. It is important to note that at this point Komiyama and co-workers finally used mixed sequences in which the modified residue is not only at a rather special position (as in ref. [87]) and more importantly that the azobenzene residue is now used not as a nucleoside replacement for the duplex formation but rather as an extra residue with respect to the counter strand! The latter issue is not discussed in the paper. However, the results are astonishing: both residues were readily photoswitchable but the introduction of $\mathbf{X^c}$ rather lead to an overall destabilization with regard to the unmodified duplex (Figure 13.17b), whereas with one residue $\mathbf{X^d}$ the duplex was stabilized if the unit was in the *trans* conformation and destabilized if the unit was in the *cis* conformation with a T_m difference of 14.3 °C with one residue and 21.5 °C with two residues. Hence $\mathbf{X^d}$ – derived from D-threoninol – is the residue of choice to photoswitch duplex formation and the authors have explained this by the fact that in $\mathbf{X^d}$ the azobenzene residue should point towards the narrow minor groove where a structural change makes more difference compared with $\mathbf{X^c}$ where the azobenzene would point to the wide major groove. The authors also showed that by including nine $\mathbf{X^d}$ residues in a 20-mer they could obtain a melting temperature amplitude of a DNA duplex of ~40 °C! [91]. In a similar fashion, the formation of a triple helix was also photoswitchable – but here the residue $\mathbf{X^c}$ turned out to be the preferred one (Figure 13.17b) [92].

The residue $\mathbf{X^d}$ was then used for a series of studies to regulate transcription (Figure 13.18a). For example, a 39-nucleotide template strand was used and another complementary 20-mer containing the residue $\mathbf{X^d}$ at different locations (again as an extra residue) was added to form the double-stranded T7 promoter [93]. When $\mathbf{X^d}$ was in the *trans* state it slowed the transcription rate to 10% relative to the control reaction without any modification in the nontemplate strand. Irradiation with UV light did not harm the transcription reaction but generated the *cis* conformer, which led to an increase in the transcription rate. The effect depended on the position of the $\mathbf{X^d}$ residue and was optimal when it was located between the 10th and 11th nucleotides upstream of the GGG, in which case a rate recovery to a level of 60% was achieved. The authors rationalized this in a footnote with the explanation that the *trans* conformer of $\mathbf{X^d}$ can intercalate and thus widens and unwinds the T7 promoter, whereas in the *cis* conformer the azobenzene is 'positioned in the groove, where it does not disturb the binding of RNAP' [93]. In a follow-up study, Komiyama and co-workers investigated this system more closely [94]. The T7 promoter has two functional regions known as the 'loop binding region' of the RNA polymerase and the 'unwinding region' around the TATA box where the polymerase opens the double strand (Figure 13.18a). For this study, residues $\mathbf{X^d}$ were introduced in a series of locations both in the template and nontemplate strands and in both regions. It is important to note that again the residues $\mathbf{X^d}$ were used as insertions and not as nucleoside replacements and that in such a duplex they worked via a change of local structure and not by a reduction in the melting temperature below the reaction temperature. These subtle changes turned out to be sufficient. Whereas $\mathbf{X^d}$ residues in the loop binding region affected primarily K_m of the polymerase, $\mathbf{X^d}$ residues in the unwinding regions modulated the k_{cat} value. The best result was obtained with one residue $\mathbf{X^d}$ in each of the regions (such as in the example in Figure 13.18a), which afforded a 7.6-fold increase in reaction rate after irradiation with UV light. The authors suggested that the *trans* conformer of $\mathbf{X^d}$ in the unwinding region stabilizes this region and thus prevents duplex opening at the TATA box, whereas in the case of the loop binding region the *trans* form leads to a local perturbation in the structure but not the *cis* form, which flips out of the DNA duplex. In a similar study, Komiyama and co-workers investigated the photoregulation of the SP6 promoter (Figure 13.18b) [95]. The greatest effect was again found when two $\mathbf{X^d}$ residues were incorporated and the difference in rate before and after irradiation was threefold in this case.

As pointed out in the Section 13.2.3, one of the strategies to control gene expression is the antisense strategy in which the mRNA is hybridized to antisense DNA, thus activating the degradation of the RNA by RNase H (or an interruption of the translation at the ribosome). To control the function of RNase H with light, Komiyama and co-workers constructed a model system in which a 20-mer sense RNA strand could hybridize

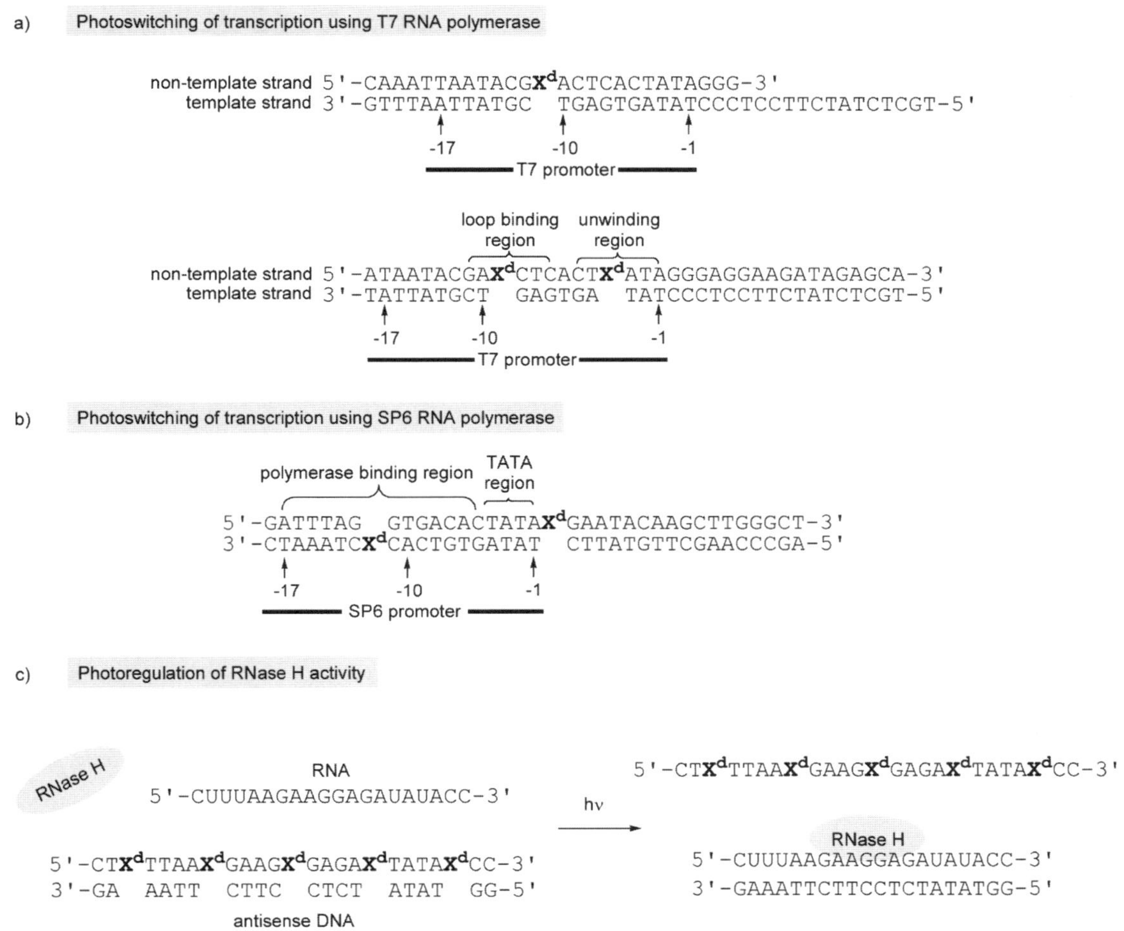

Figure 13.18 *Setups used for the photoswitching of transcription using T7 RNA polymerase [93,94] (a) or SP6 RNA polymerase [95] (b) and also for the photoregulation of RNase H activity (c) [96]*

to a 20-mer antisense DNA strand (Figure 13.18c) [96]. However, the antisense strand was first hybridized with another DNA sense strand containing five $\mathbf{X^d}$ residues. Thus, in the beginning all antisense DNA would be sequestered in a DNA duplex from which it could be released upon irradiation with UV light. Once free, the antisense strand would hybridize to the RNA strand, which would then be digested by RNase H. Before irradiation, a background activity of 25% digestion was observed, whereas afterwards the activity was restored to about 100%. The melting temperature of the modified DNA duplex was determined to be 60.8 °C in the *trans* form and 42.6 °C in the *cis* form and the RNase H digestion was run at 37 °C.

In an attempt to improve further the efficiency of the azobenzene residue, other substitution patterns were also investigated. For example, the *meta*-substituted residue $\mathbf{X^e}$ was synthesized by Komiyama and co-workers [97] and included in a homoadenosine sequence (Figure 13.19). It turned out that this residue had an interesting quality: the *trans* conformer is always the more thermodynamically stable one and therefore the rate of the reaction from *cis* to *trans* is always a question. It was known that *para* substituents contribute electronically to the rate. In the case of an oligomer with the *meta*-connected residue $\mathbf{X^e}$ a half-life of 13.2 h

a)

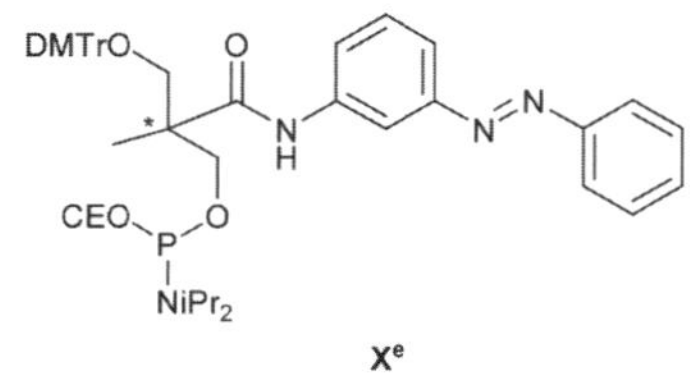

b)

sequence 1	sequence 2	melting temperature	
AAAAAAAA	TTTTTTTT	23.7 °C	
trans AAAXAAAA	TTTTTTTT	24.3 °C	
cis AAAX^{b}AAAA	TTTTTTTT	16.7 °C	
trans AAAX^{a}AAAA	TTTTTTTT	24.8 °C	
cis AAAX^{a}AAAA	TTTTTTTT	15.9 °C	
TTTTTTTTTTTTT	ds^2	40.0 °C	triple helix formation
trans **X**^{e}TTTTTTTTTTTTT	ds^2	46.3 °C	
cis **X**^{e}TTTTTTTTTTTTT	ds^2	26.2 °C	

c)

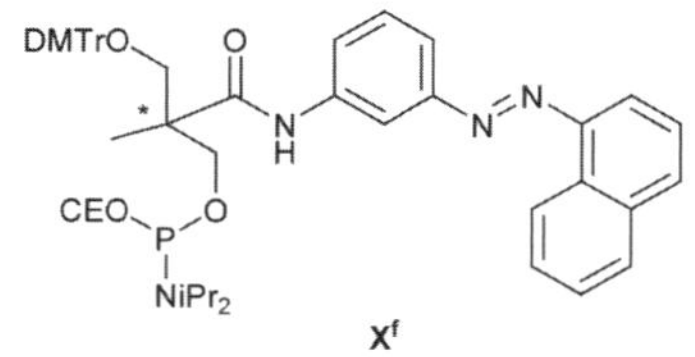

d)

sequence 1	sequence 2	melting temperature	
TTTTTTTTTTTTT	ds^2	22.5 °C	
trans **X**^{a}TTTTTTTTTTTTT	ds^2	29.5 °C	triple helix formation
cis **X**^{a}TTTTTTTTTTTTT	ds^2	13.5 °C	
trans **X**^{e}TTTTTTTTTTTTT	ds^2	30.2 °C	
cis **X**^{e}TTTTTTTTTTTTT	ds^2	10.7 °C	
trans **X**^{f}TTTTTTTTTTTTT	ds^2	45.5 °C	
cis **X**^{f}TTTTTTTTTTTTT	ds^2	5.0 °C	

Figure 13.19 *The structure of the* meta*-substituted residue* $\boldsymbol{X}^e$ *[97] (a) and data from melting temperature studies and the use of* $\boldsymbol{X}^e$ *in the photoregulation of DNA triplex formation (b) [98]. Even though the same DNA duplex ds2 as in Figure 13.21 was used, the melting temperature data cannot be directly compared with these studies because the buffer conditions were different. (c) The structure of the phenylazonaphthalene derivative* $\boldsymbol{X}^f$ *[99] and a comparison of its efficiency in the photoregulation of triplex formation. (d) Results of melting temperature studies in which again different conditions were used*

was determined at 50 °C, whereas for the oligomer with the *para*-substituted residue $\mathbf{X}^a$ it was 20 min. At the physiologically more relevant 37 °C, the half-life of the $\mathbf{X}^e$-containing oligomer was even 64 h. A very comprehensive study on the formation of triplex structures showed, for example, that the residue $\mathbf{X}^e$ – if used as the 5′-terminus – can lead to differences in triplex melting temperature of around 20 °C [98]. Even greater was the effect of the residue $\mathbf{X}^f$ with which – if used as the 5′-terminus of a triplex-forming oligonucleotide – differences in melting temperature upon photoswitching of 40.5 °C were achieved [99]. It must be noted that due to the significant influence of experimental conditions (salt concentration, etc.) on the melting temperature, it is difficult to compare values from different experiments – in this case, Komiyama and co-workers used a number of different conditions over the years.

Figure 13.20 gives an overview of more reversibly switchable residues: There is, for example, the residue $\mathbf{X}^g$ which was introduced by Asanuma and co-workers [24] (Asanuma had previously been publishing with Komiyama's group) in the search for a residue that shows a better photoswitching behavior. Indeed, the addition of the two methyl groups increased the melting temperature difference by a factor of three (14.6 °C compared with 5.7 °C with $\mathbf{X}^d$ in a 12-mer of mixed sequence) in DNA duplex formation and again this modification had a stabilizing effect on the thermal *cis–trans* back-reaction (half-life 25 h). Asanuma and co-workers further introduced the residues $\mathbf{X}^h$, $\mathbf{X}^i$ and $\mathbf{X}^j$ and found that they show reversed photoswitching behavior [100]. Now the *trans* conformer destabilizes the duplex formation and the *cis* form does not, or not much. Differences in melting temperature of up to 13.3 °C were obtained. Another recent study by Asanuma and co-workers opened a new door for the application of photoswitchable residues: they used the most promising residue $\mathbf{X}^d$ to power a DNA nanomachine by light irradiation [101]. The machine was based on

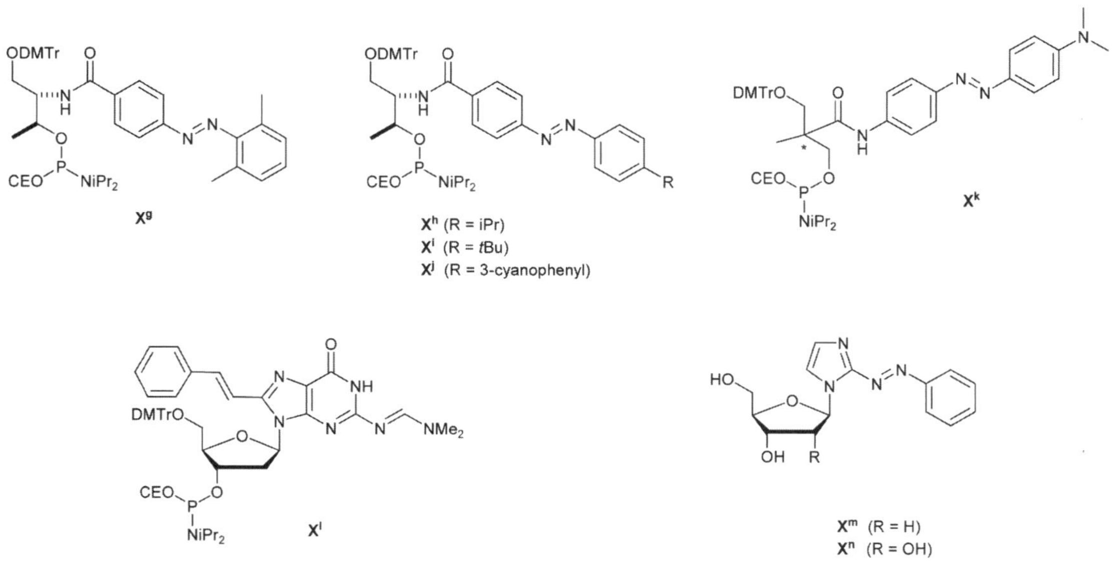

Figure 13.20 *The structures of diverse reversibly photoswitchable residues with optimized properties. Residues* $\mathbf{X^g}$ *and* $\mathbf{X^h}$, $\mathbf{X^i}$, $\mathbf{X^j}$ *were introduced by Asanuma and co-workers [24,99], the residue* $\mathbf{X^k}$ *by Uyeda and co-workers [102], the residue* $\mathbf{X^l}$ *by Ogasawara and Maeda [103] and the nucleosides* $\mathbf{X^m}$ *and* $\mathbf{X^n}$ *by Majima and co-workers [104]*

DNA-fueled tweezers that had been presented before. Upon photoswitching to the *cis* form, a helper strand containing 12 $\mathbf{X^d}$ residues dissociates off and opens the tweezer. With visible light the process could be reversed and the tweezers closed again. This process could be repeated 10 times.

The residue $\mathbf{X^k}$ (Figure 13.20) was introduced by Uyeda and co-workers [102] and is mentioned here rather for the sake of completeness because its switching behavior was studied in dimethyl sulfoxide. In aqueous solutions, the back-reaction from *cis* to *trans* is reported to be very rapid so that it was only detectable by laser flash photolysis. Much more interesting is the residue $\mathbf{X^l}$ (Figure 13.20) introduced by Ogasawara and Maeda [103]. Attaching a styryl group to the 8-position of guanine, they obtained not only a reversibly photoswitchable residue of a significantly different kind than those that had been presented earlier, but also a residue whose conformational state could be followed by monitoring the fluorescence. The *trans* conformer shows green–blue fluorescence whereas in the *cis* state the fluorescence is reduced to one-sixth. This time with light of 350–450 nm, the double bond can be isomerized to the *cis* conformation with a conversion between 63% and 86% and the back-reaction to the *trans* state occurs upon irradiation at 320 nm. No thermal conversion between the two states was observed even at 80 °C! Two more residues ($\mathbf{X^m}$ and $\mathbf{X^n}$, Figure 13.20) were presented by Majima and co-workers [104]. They have an intact (deoxy)ribose unit but a phenylazoimidazole as switchable principle and nucleobase replacement. This residue is readily switchable with wavelengths of 360 and 455 nm with reported quantitative conversion. However, these studies were performed on the nucleoside alone and no follow-up study could be found in which this nucleoside was built into an oligonucleotide.

13.7.3 Reversible photoswitching of oligonucleotides with small molecules

Instead of including the photoswitch in the oligonucleotide, another approach is to have a small molecule interact with the oligonucleotide and make this small molecule photoswitchable. The examples of this

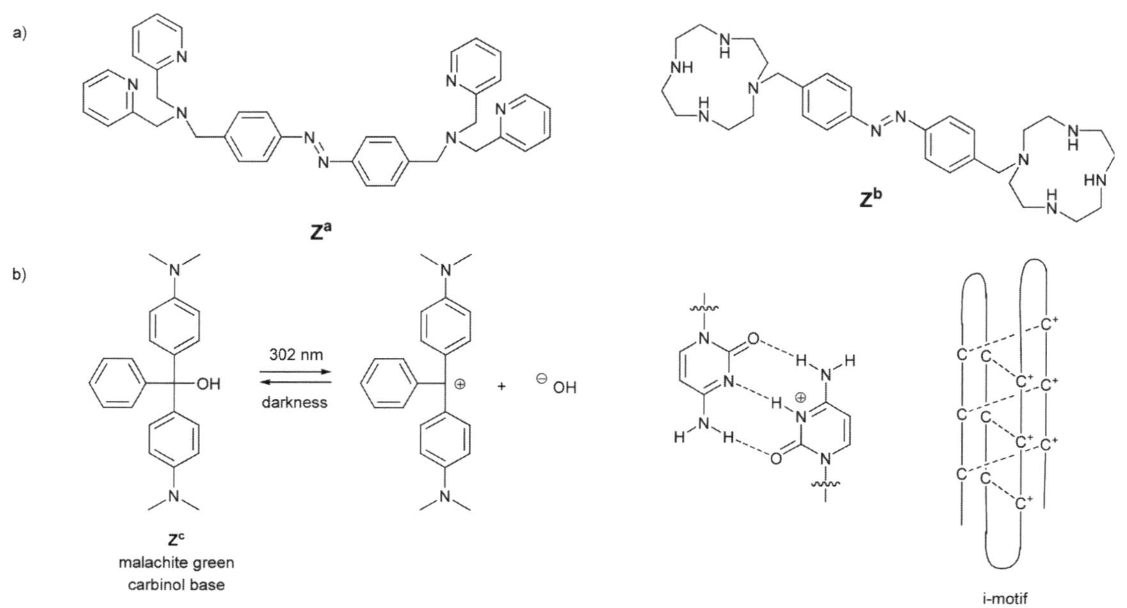

Figure 13.21 (a) The structures of the residues Z^a and Z^b used for the photocontrol of RNA hydrolysis [105] or conformation, respectively [106]. (b) An approach using the photocontrollable base source Z^c to regulate the formation of i-motif DNA [107]

approach are not as numerous those of the former approach and again Komiyama's group was one of the first to step into this field: One of their studies dealt with the photoregulation of RNA hydrolysis by the dizinc complex of azobenzene derivative $\mathbf{Z^a}$ (Figure 13.21a), but the effect was rather moderate and the study was not followed up [105]. Instead, 7 years later, Yamana and co-workers introduced the compound $\mathbf{Z^b}$, which can also form a complex with two Zn^{2+} ions and be used as a photoswitchable crosslinking agent for DNA [106]. Each Zn^{2+} complex can selectively bind to N3 of a thymidine after breaking an A=T base pair. Using the dizinc complex of $\mathbf{Z^b}$, the authors could show by circular dichroism (CD) spectroscopy that the conformation of an appropriate DNA duplex could be changed reversibly with light but the thermal back-reaction was very fast. In a very unconventional study, Liu *et al.* used the fact that the dye Malachite Green carbinol base ($\mathbf{Z^c}$, Figure 13.21b) can be reversibly dissociated into a stable cation and a hydroxide ion upon irradiation with UV light [107]. The fact that this compound dissolves poorly in water was solved via the use of a surfactant. Thus the molecule was ready to provide pH shifts photoreversibly. On the other hand, the so-called i-motif is a DNA structure in which a protonated cytidine forms a base pair with an unprotonated cytidine and in which two duplexes of this sort are interlocked by intercalation (see Figure 13.21b) [108]. The formation of this structure is obviously very pH dependent and indeed, using $\mathbf{Z^c}$, the authors showed via CD-spectroscopy that the i-motif can be reversibly formed and broken again using light.

An entirely different approach came from Nakatani and co-workers [109,110]. They used the azobenzene derivative $\mathbf{Z^d}$ (Figure 13.22) modified with positive amino acid side-chains to generate an aptamer (see above). Arginines were used to provide residues that could interact through coulomb interactions and through hydrogen bonding with RNA, whereas the lysine was used for technical immobilization purposes for the aptamer selection and the subsequent interaction studies. In their first investigation, they showed that $\mathbf{Z^d}$ can be readily photoswitched and that the *cis* form is stable for 10 h at room temperature. With UV light (360 nm), a population with 95% *cis* state could be reached. However, upon irradiation with visible light (430 nm)

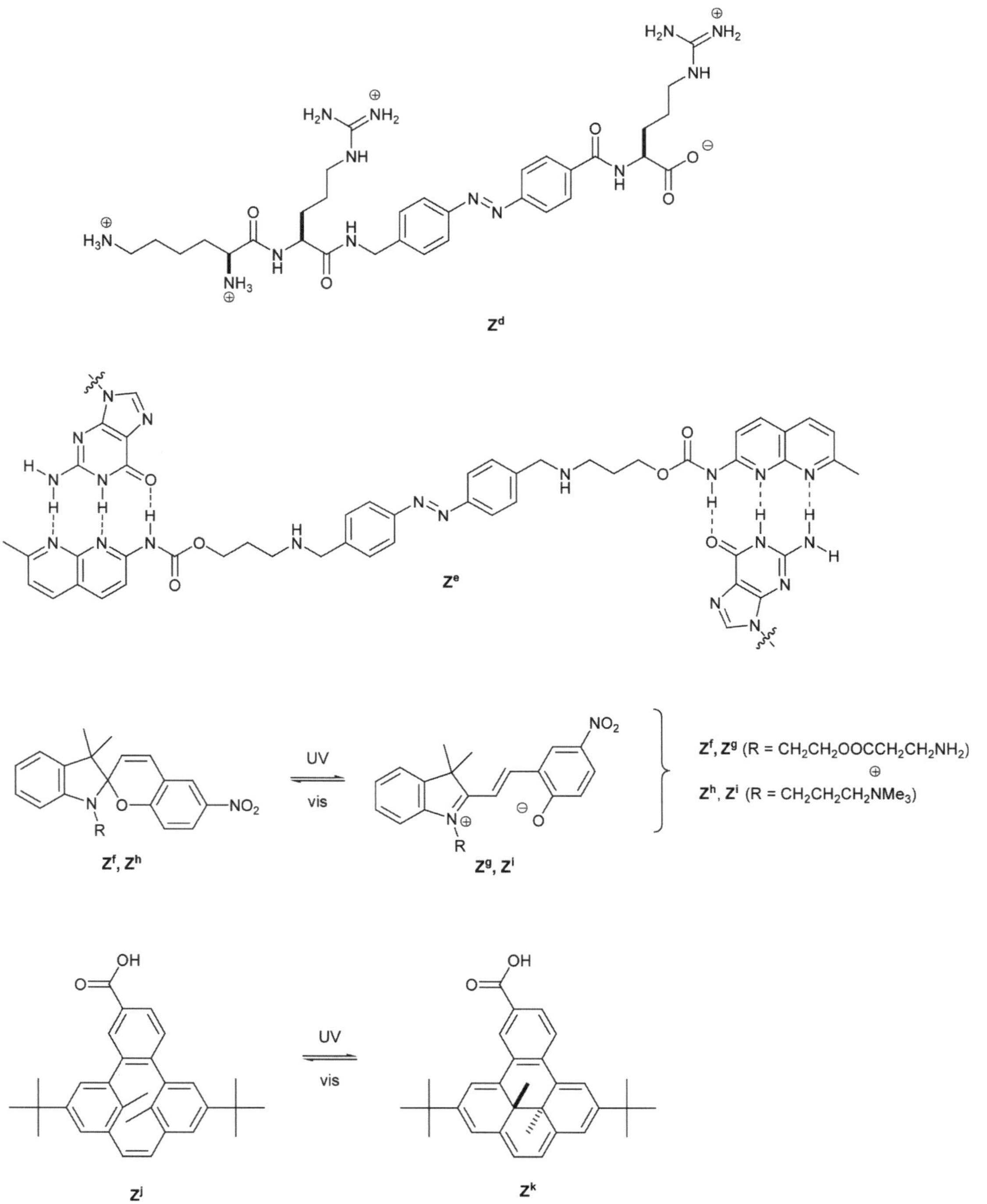

Figure 13.22 *The structures of the compounds* Z^d, Z^e, Z^f, Z^g *and* Z^i *that were used by Nakatani and co-workers [109–111], Young and Deiters [112] and Andréasson* et al. *[113] for the light-controlled interaction with nucleic acids. In the case of compound* Z^e, *the proposed mechanism of interaction with guanine nucleobases is drawn*

20% *cis* isomer remained. Several sequences were found and the K_d values in the bound state were around 1 μM. Nakatani and co-workers also presented the compound $\mathbf{Z^e}$ consisting of an azobenzene core unit and naphthyridine carbamate substituents which they called 'photoswitchable molecular glue for DNA' [111]. The naphthyridine carbamates were designed in such a way that they could interact with guanine nucleobases (see Figure 13.22) and this system worked best with a G–G mismatch in the DNA. Using this compound, the authors were able to photoswitch DNA hybridization both in solution and on a gold surface (surface plasmon resonance).

In a similar approach, Young and Deiters selected an RNA aptamer against the spiropyran/merocyanin system $\mathbf{Z^f}$/$\mathbf{Z^g}$ [112]. The affinity for $\mathbf{Z^f}$ was again determined by surface plasmon resonance and was 19 μM whereas $\mathbf{Z^g}$ was bound less efficiently. One problem with the spiropyran system and its recognition by aptamers is that the spiro form has a stereogenic center. This aspect is briefly discussed in their paper. Andréasson *et al.* used the comparable system $\mathbf{Z^h}$/$\mathbf{Z^i}$ [113]. By comparing the UV spectra of $\mathbf{Z^h}$ and $\mathbf{Z^i}$ with and without calf thymus DNA, they showed that only $\mathbf{Z^i}$ interacts with DNA and they approximated the dissociation constant to be ~50 μM. As already noted previously, Sen and co-workers also developed an aptamer against a photoswitchable small molecule (Figure 13.8) [44]. In their case, a dihydropyrene-derived system was used (Figure 13.8) that can undergo an electrocyclic ring closure and opening. RNA aptamers were selected that bind one form and not the other and used for the allosteric regulation of a hammerhead ribozyme. In the presence of micromolar concentrations of the effector, the ribozyme exhibited a >900-fold difference in catalytic rates between the two states.

References

1. J. Forman, M. Diertich and W. T. Monroe, Photobiological and thermal effects of photoactivating UVA light doses on cell cultures, *Photochem. Photobiol. Sci.*, **6**, 649–658 (2007).
2. W. R. Zipfel, R.M. Williams and W.W. Webb, Nonlinear magic: multiphoton microscopy in the biosciences, *Nat. Biotechnol.*, **21**, 1369–1377 (2003).
3. J. H. Kaplan, B. Forbush and J. F. Hoffmann, Rapid photolytic release of adenosine 5′-triphosphate from a protected analogue: utilization by the Na:K pump of human red blood cell ghosts, *Biochemistry*, **17**, 1929–1935 (1978).
4. J. Engels and E.-J. Schlaeger, Synthesis, structure and reactivity of adenosine cyclic 3′,5′-phosphate benzyl triesters, *J. Med. Chem.*, **20**, 907–911 (1977).
5. G. Mayer and A. Heckel, Biologically active molecules with a 'light switch', *Angew. Chem. Int. Ed.*, **45**, 4900–4921 (2006).
6. X. J. Tang and I. J. Dmochowski, Regulating gene expression with light activated oligonucleotides, *Mol. Biosyst.*, **3**, 100–110 (2007).
7. D. D. Young and A. Deiters, Photochemical control of biological processes, *Org. Biomol. Chem.*, **5**, 999–1005 (2007).
8. W. T. Monroe, M. M. McQuain, M. S. Chang, J. S. Alexander and F. R. Haselton, Targeting expression with light using caged DNA, *J. Biol. Chem.*, **274**, 20895–20900 (1999).
9. H. Ando, T. Furuta, R. Y. Tsien and H. Okamoto, Photo-mediated gene activation using caged RNA/DNA in zebrafish embryos, *Nat. Genet.*, **28**, 317–325 (2001).
10. H. Ando and H. Okamoto, Practical procedures for ectopic induction of gene expression in zebrafish embryos using Bhc-diazo-caged mRNA, *Methods Cell Sci.*, **25**, 25–31 (2003).
11. H. Ando, M. Kobayashi, T. Tsubokawa, K. Uyemura, T. Furuta and H. Okamoto, Lhx2 mediates the activity of Six3 in zebrafish forebrain growth, *Dev. Biol.*, **287**, 456–468 (2005).
12. L. Kröck and A. Heckel, Photoinduced transcription by using temporarily mismatched caged oligonucleotides, *Angew. Chem.*, **117**, 475–477 (2005); *Angew. Chem. Int. Ed.*, **44**, 471–473 (2005).
13. A. Z. Fire, Gene silencing by double-stranded RNA (Nobel Lecture), *Angew. Chem.*, **119**, 7094–7113 (2007); *Angew. Chem. Int. Ed.*, **46**, 6966–6984 (2007).

14. C. C. Mello, Return to the RNAi-world: rethinking gene expression and evolution (Nobel Lecture), *Angew. Chem.*, **119**, 7114–7124 (2007); *Angew. Chem. Int. Ed.*, **46**, 6985–6994 (2007).
15. S. Shah, S. Rangarajan and S. H. Friedman, Light-activated RNA interference, *Angew. Chem.* **117**, 1352–1356 (2005); *Angew. Chem. Int. Ed.*, **44**, 1328–1332 (2005).
16. S. Shah and S. H. Friedman, Tolerance of RNA interference toward modifications of the 5′ antisense phosphate of small interfering RNA, *Oligonucleotides*, **17**, 35–43 (2007).
17. Q. N. Nguyen, R. V. Chavli, J. T. Marques, P. G. Conrad II, D. Wang, W. He, B. E. Belisle, A. Zhang, L. M. Pastor, F. R. Witney, M. Morris, F. Heitz, G. Divita, B. R. G. Williams and G. K. McMasters, Light controllable siRNAs regulate gene suppression and phenotypes in cells, *Biochim. Biophys. Acta*, **1758**, 394–403 (2006).
18. V. Mikat and A. Heckel, Light-dependent RNA interference with nucleobase-caged siRNAS, *RNA*, **13**, 2341–2347 (2007).
19. Y. L. Chiu and T. M. Rana, siRNA function in RNAi: a chemical modification analysis, *RNA*, **9**, 1034–1048 (2003).
20. R. A. Bildner, K. R. Svoboda, R. P. Hammer and W. T. Monroe, Photoinduced RNA interference using DMNPE-caged 2′-deoxy-2′-fluoro substituted nucleic acids *in vitro* and *in vivo*, *Mol. Biosyst.*, **4**, 431–440 (2008).
21. E. Uhlmann and A. Peyman, Antisense oligonucleotides: a new therapeutic principle, *Chem. Rev.*, **4**, 544–584 (1990).
22. S. T. Crooke, Antisense strategies, *Curr. Mol. Med.*, **4**, 465–487 (2004).
23. X. Tang and I. J. Dmochowski, Controlling RNA digestion by RNase H with a light-activated DNA hairpin, *Angew. Chem.*, **118**, 3603–3606 (2006); *Angew. Chem. Int. Ed.*, **45**, 3523–3526 (2006).
24. H. Nishioka, X. Liang, H. Kashida and H. Asanuma, 2′,6′-Dimethylazobenzene as an efficient and thermo-stable photo-regulator for the photoregulation of DNA hybridization, *Chem. Commun.*, 4353–4356 (2007).
25. X. Tang, S. Maegawa, E. S. Weinberg and I. J. Dmochowski, Regulating gene expression in zebrafish embryos using light-activated negatively charged peptide nucleic acids, *J. Am. Chem. Soc.*, **129**, 11000–11001 (2007).
26. X. Tang, J. Swaminathan, A. M. Gewirtz and I. J. Dmochowski, Regulating gene expression in human leukemia cells using light-activated oligonucleotides, *Nucleic Acids Res.*, **36**, 559–569 (2008).
27. J. L. Richards, X. Tang, A. Turetsky, I. J. Dmochowski, RNA bandages for photoregulating *in vitro* protein synthesis, *Bioorg. Med. Chem. Lett.*, **18**, 6255–6258 (2008).
28. I. A. Shestopalov, S. Sinha and J. K. Chen, Light-controlled gene silencing in zebrafish embryos, *Nat. Chem. Biol.*, **3**, 650–651 (2007).
29. D. D. Young, H. Lusic, M. O. Liveley, J. A. Yoder and A. Deiters, Gene silencing in mammalian cells with light-activated antisense agents, *ChemBioChem*, **9**, 2937–2940 (2008).
30. A. Rotaru, J. Kovács and A. Mokhir, Red light-activated phosphorothioate oligodeoxyribonucleotides. *Bioorg. Med. Chem. Lett.*, **18**, 4336–4338 (2008).
31. X. Tang, J. L. Richards, A. E. Peritz and I. J. Dmochowski, Photoregulation of DNA polymerase I (Klenow) with caged fluorescent oligodeoxynucleotides, *Bioorg. Med. Chem. Lett.*, **15**, 5303–5306 (2005).
32. X. Tang and I. J. Dmochowski, Phototriggering of caged fluorescent oligodeoxynucleotides, *Org. Lett.*, **7**, 279–282 (2005).
33. K. Tanaka, A. Kuzuya and M. Komiyama, Site-selective termination of DNA replication by using a caged template, *Chem. Lett.*, **37**, 584–585 (2008).
34. K. Tanaka, H. Katada, N. Shigi, A. Kuzuya and M. Komiyama, Site-selective blocking of PCR by a caged nucleotide leading to direct creation of desired sticky ends in the products, *ChemBioChem*, **9**, 2120–2126 (2008).
35. P. Wenter, B. Fürtig, A. Hainard, H. Schwalbe and S. Pitsch, Kinetics of RNA refolding by real-time NMR spectrocopy, *Angew. Chem. Int. Ed.*, **44**, 2600–2603 (2005).
36. P. Wenter, B. Fürtig, A. Hainard, H. Schwalbe and S. Pitsch, A caged uridine for the selective preparation of an RNA fold and determination of its refolding kinetics by real-time NMR, *ChemBioChem*, **7**, 417–420 (2006).
37. B. Fürtig, P. Wenter, L. Reymond, C. Richter, S. Pitsch and H. Schwalbe, Conformational dynamics of bistable RNAs studied by time-resolved NMR spectroscopy, *J. Am. Chem. Soc.*, **129**, 16222–16229 (2007).
38. B. Fürtig, C. Richter, P. Schell, P. Wenter, S. Pitsch and H. Schwalbe, NMR-spectroscopic characterisation of phosphodiester bond cleavage catalysed by the minimal hammerhead ribozyme, *RNA Biol.*, **5**, 1–8 (2008).
39. J. Buck, B. Fürtig, J. Noeske, J. Wöhnert and H. Schwalbe, Time-resolved NMR methods resolving ligand-induced RNA folding at atomic resolution, *Proc. Natl. Acad. Sci. USA*, **104**, 15699–15705 (2007).

40. R. T. Batey, S. D. Gilbert and R. K. Montange, Structure of a natural guanine responsive riboswitch complexed with the metabolite hypoxanthine, *Nature*, **432**, 411–415 (2004).
41. S. Pitsch, P. A. Weiss, X. Wu, D. Ackermann and T. Honegger, Fast and reliable automated synthesis of RNA and partially 2′-*O*-protected precursors ('caged RNA') based on two novel, orthogonal 2′-*O*-protecting groups, *Helv. Chim. Acta*, **82**, 1753–1761 (1999).
42. S. G. Chaulk and A. M. MacMillan, Caged RNA: photo-control of a ribozyme reaction, *Nucleic Acids Res.*, **26**, 3173–3178 (1998).
43. D. D. Young and A. Deiters, Photochemical hammerhead ribozyme activation, *Bioorg. Med. Chem. Lett.*, **16**, 2658–2661 (2006).
44. H.-W. Lee, S. G. Robinson, S. Bandyopadhyay, R. H. Mitchell and D. Sen, Reversible photo-regulation of a hammerhead ribozyme using a diffusible effector, *J. Mol. Biol.*, **371**, 1163–1173 (2007).
45. S. Keiper and J. S. Vyle, Reversible photocontrol of deoxyribozyme-catalyzed RNA cleavage under multiple turnover conditions, *Angew. Chem. Int. Ed.*, **118**, 3384–3387 (2006).
46. H. Lusic, D. D. Young, M. O. Lively and A. Deiters, Photochemical DNA activation, *Org. Lett.*, **9**, 1903–1906 (2007).
47. H. Lusic, M. O. Lively and A. Deiters, Light-activated deoxyguanosine: photochemical regulation of peroxidase activity, *Mol. Biosystems*, **4**, 508–511 (2008).
48. R. Ting, L. Lerner and D. M. Perrin, Triggering DNAzymes with light: a photoactive C8 thioether-linked adenosine, *J. Am. Chem. Soc.*, **126**, 12720–12721 (2004).
49. C. Höbartner and S. K. Silverman, Modulation of RNA tertiary folding by incorporation of caged nucleotides, *Angew. Chem. Int. Ed.*, **117**, 1–6 (2005).
50. A. Heckel and G. Mayer, Light regulation of aptamer activity: an anti-thrombin aptamer with caged thymidin nucleobases, *J. Am. Chem. Soc.*, **127**, 822–823 (2005).
51. A. Heckel, M. C. R. Buff, M.-S. L. Raddatz, J. Müller, B. Pötzsch and G. Mayer, An anticoagulant with light triggered antidote activity, *Angew. Chem. Int. Ed.*, **45**, 6748–6750 (2006).
52. L. C. Bock, L. C. Griffin, J. A. Latham, E. H. Vermaas and J. J. Toole, Selection of single-stranded DNA molecules that bind and inhibit human thrombin, *Nature*, **355**, 564 (1992).
53. G. Mayer, L. Kröck, V. Mikat, M. Engeser and A. Heckel, Light-induced formation of G-quadruplex DNA secondary structures, *ChemBioChem*, **6**, 1966–1970 (2005).
54. J. Müller, B. Wulffen, B. Pötzsch and G. Mayer, Multidomain targeting generates a high affinity thrombin-inhibiting bivalent aptamer, *ChemBioChem*, **8**, 2223–2226 (2007).
55. G. Mayer, J. Müller, T. Mack, D. F. Freitag, T. Höver, B. Pötzsch and A. Heckel Differential regulation of protein sub-domain activity with caged bivalent ligands, *ChemBioChem*, **10**, 654–657 (2009).
56. A. Dussy, C. Meyer, E. Quennet, T. A. Bickle, B. Giese and A. Marx, New light-sensitive nucleosides for caged DNA strand breaks, *ChemBioChem*, **3**, 54–60 (2002).
57. P. Ordoukhanian and J.-S. Taylor, Design and synthesis of a versatile photocleavable DNA building block. Application to phototriggered hybridization, *J. Am. Chem. Soc.*, **117**, 9570–9571 (1995).
58. P. Ordoukhanian and J.-S. Taylor, Caged single and double strand breaks, *Bioconj. Chem.*, **11**, 94–103 (2000).
59. K. Zhang and J.-S. Taylor, Phototriggered formation and repair of DNA containing a site-specific single strand-break of the type produced by ionizing radiation or AP lyase activity, *Biochemistry*, **40**, 153–159 (2001).
60. K. Zhang and J.-S. Taylor, A caged ligatable DNA strand break, *J. Am. Chem. Soc.*, **121**, 11579–11580 (1999).
61. H. J. Lenox, C. P. McCoy and T. L. Sheppard, Site-specific generation of deoxyribonolactone lesions in DNA oligonucleotides, *Org. Lett.*, **3**, 2415–2418 (2001).
62. J. D. Trzupek and T. L. Sheppard, Photochemical generation of ribose abasic sites in RNA oligonucleotides, *Org. Lett.*, **7**, 1493–1496 (2005).
63. Y. Zheng and T. L. Sheppard, Half-life and DNA strand scission products of 2-deoxyribonolactone oxidative DNA damage lesions, *Chem. Res. Toxicol.*, **17**, 197–207 (2004).
64. K. Usui, M. Aso, M. Fukuda and H. Suemune, Photochemical generation of oligodeoxynucleotide containing a C4′-oxidized abasic site and its efficient amine modification: dependence on structure and microenvironment, *J. Org. Chem.*, **73**, 241–248 (2008).
65. M. Sollogoub, S. Guieu, M. Geoffroy, A. Yamada, A. Estevez-Torres, K. Yoshikawa and D. Baigl, Photocontrol of single chain DNA conformation in cell-mimicking microcompartments, *ChemBioChem*, **9**, 1201–6 (2008).

66. V. Namasivayam, R. G. Larson, D. T. Burke and M. A. Burns, Light-induced molecular cutting: localized reaction on a single DNA molecule, *Anal. Chem.*, **75**, 4188–4194 (2003).
67. A. Rotaru and A. Mokhir, Nucleic acid binders activated by light of selectable wavelength, *Angew. Chem. Int. Ed.*, **46**, 6180–3 (2007).
68. B. Ghosn, F. R. Haselton, K. R. Gee and W. T. Monroe, Control of DNA hybridization with photocleavable adducts, *Photochem. Photobiol.*, **81**, 953–959 (2005).
69. K. Tanabe, H. Nakata, S. Mukai and S. Nishimoto, Modulated drug-release from the stem-and-loop structured oligodeoxynucleotide upon UV-A irradiation in the presence of target DNA, *Org. Biomol. Chem.*, **3**, 3893–3897 (2005).
70. A. Okamoto, K. Tanabe, T. Inasaki and I. Saito, Phototriggered drug release from functionalized oligonucleotides by a molecular beacon strategy, *Angew. Chem.*, **115**, 2606–2608 (2003).
71. S. Banala, A. Arnold and K. Johnsson, Caged substrates for protein labeling and immobilization, *ChemBioChem*, **9**, 38–41 (2008).
72. M. Aizawa, K. Namba and S. Suzuki, Photo control of enzyme activity of α-amylase, *Arch. Biochem. Biophys.*, **180**, 41–48 (1977).
73. M. Volgraf, P. Gorostiza, R. Numano, R. H. Kramer, E. Y. Isacoff and D. Trauner, Allosteric control of an ionotropic glutamate receptor with an optical switch, *Nat. Chem. Biol.*, **2**, 47–52 (2006).
74. M. Volgraf, P. Gorostiza, S. Szobota, M. R. Helix, E. Y. Isacoff and D. Trauner, Reversibly caged glutamate: a photochromic agonist of ionotropic glutamate receptors, *J. Am. Chem. Soc.*, **129**, 260–261 (2007).
75. C. Boulègue, M. Löweneck, C. Renner and L. Moroder, Redox potential of azobenzene as amino acid residue in peptides, *ChemBioChem*, **8**, 591–594 (2007).
76. A. Koçer, M. Walko, W. Weijnberg and B. L. Feringa, A light-actuated nanovalve derived from a channel protein, *Science*, **309**, 755–758 (2005).
77. K. Yamana, A. Yoshikawa and H. Nakano, Synthesis of a new photoizomerizable linker for connecting two oligonucleotide segments, *Tetrahedron Lett.*, **37**, 637–640 (1996).
78. K. Yamana, A. Yoshikawa, R. Noda and H. Nakano, Synthesis and binding properties of oligonucleotides containing an azobenzene linker, *Nucleosides Nucleotides Nucleic Acids*, **17**, 233–242 (1998).
79. K. Yamana, K. Kan and H. Nakano, Synthesis of oligonucleotides containing a new azobenzene fragment with efficient photoizomerizability, *Bioorg. Med. Chem. Lett.*, **7**, 2977–2983 (1999).
80. M. Matsumoto, D. Miyazaki, M. Tanaka, R. Azumi, E. Manda, Y. Kondo, N. Yoshino and H. Tachibana, Reversible light-induced morphological change in Langmuir–Blodgett films, *J. Am. Chem. Soc.*, **120**, 1479–1484 (1998).
81. M. Fujimaki, Y. Matsuzawa, Y. Hayashi and K. Ichimura, Monolayers of calix [4]resorcinarenes with azobenzene residues exhibiting efficient photoizomerizability, *Chem. Lett.*, **27**, 165–166 (1998).
82. H. Asanuma, T. Ito and M. Komiyama, Photo-responsive oligonucleotides carrying azobenzene in the side-chains, *Tetrahedron Lett.*, **39**, 9015–9018 (1998).
83. X. Liang, H. Asanuma, H. Kashida, A. Takasu, T. Sakamoto, G. Kawai and M. Komiyama, NMR study on the photoresponsive DNA tethering an azobenzene. Assignment of the absolute configuration of two diastereomers and structure determination of their duplexes in the *trans*-form, *J. Am. Chem. Soc.*, **125**, 16408–16415 (2003).
84. H. Asanuma, T. Ito, T. Yoshida, X. Liang and M. Komiyama, Photoregulation of the formation and dissociation of a DNA duplex by using the *cis–trans* isomerization of azobenzene, *Angew. Chem.*, **111**, 2547–2549 (1999); *Angew. Chem. Int. Ed.*, **38**, 2393–2395 (1999).
85. H. Asanuma, X. Liang, T. Yoshida and M. Komiyama, Photocontrol of DNA duplex formation by using azobenzene-bearing oligonucleotides, *ChemBioChem*, **2**, 39–44 (2001).
86. H. Asanuma, X. Liang, T. Yoshida, A. Yamazawa and M. Komiyama, Photocontrol of triple-helix formation by azobenzene-bearing oligo(thymidine), *Angew. Chem.*, **112**, 1372–1374 (2000); *Angew. Chem. Int. Ed.*, **39**, 1316–1318 (2000).
87. A. Yamazawa, X. Liang, H. Asanuma and M. Komiyama, Photoregulation of the DNA polymerase reaction by oligonucleotides bearing an azobenzene, *Angew. Chem.*, **112**, 2446–2447 (2000); *Angew. Chem. Int. Ed.*, **39**, 2356–2357 (2000).
88. H. Asanuma, T. Yoshida, T. Ito and M. Komiyama, Photo-responsive oligonucleotides carrying azobenzene at the 2′-position of uridine, *Tetrahedron Lett.*, **40**, 7995–7998 (1999).

89. H. Asanuma, K. Shirasuka, T. Yoshida, T. Takarada, X. Liang and M. Komiyama, Spiropyran as a regulator of DNA hybridization with reversed switching mode to that of azobenzene, *Chem. Lett.*, **30**, 108–109 (2001).
90. H. Asanuma, T. Takarada, T. Yoshida, D. Tamaru, X. Liang and M. Komiyama, Enantioselective incorporation of azobenzenes into oligodeoxyribonucleotide for effective photoregulation of duplex formation, *Angew. Chem.*, **113**, 2743–2745; *Angew. Chem. Int. Ed.*, **40**, 2671–2673 (2001).
91. H. Asanuma, D. Matsunaga and M. Komiyama, *Nucl. Acids Symp. Ser.*, **49**, 35–36 (2005).
92. T. Takarada, D. Tamaru, X. Liang, H. Asanuma and M. Komiyama, L-Threoninol as a chiral linker of azobenzene for the effective photo-regulation of DNA triplex formation, *Chem. Lett.*, **30**, 732–733 (2001).
93. H. Asanuma, D. Tamaru, A. Yamazawa, M. Liu and M. Komiyama, Photoregulation of the transcription reaction of T7 RNA polymerase by tethering an azobenzene to the promoter, *ChemBioChem*, **3**, 786–789 (2002).
94. M. Liu, H. Asanuma and M. Komiyama, Azobenzene-tethered T7 promoter for efficient photoregulation of transcription, *J. Am. Chem. Soc.*, **128**, 1009–1015 (2006).
95. M. Liu, D. Tamaru, H. Asanuma and M. Komiyama, Synergistic effect of the two azobenzenes in the promoter on the photo-regulation of transcription reaction with SP6 RNA polymerase, *Chem. Lett.*, **32**, 1174–1175 (2003).
96. D. Matsunaga, H. Asanuma and M. Komiyama, Photoregulation of RNA digestion by RNase H with azobenzene-tethered DNA, *J. Am. Chem. Soc.*, **126**, 11452–11453 (2004).
97. H. Asanuma, X. Liang and M. Komiyama, *meta*-Aminoazobenzene as thermo-insensitive photo-regulator of DNA-duplex formation, *Tetrahedron Lett.*, **41**, 1055–1058 (2000).
98. X. Liang, H. Asanuma and M. Komiyama, Photoregulation of DNA triplex formation by azobenzene, *J. Am. Chem. Soc.*, **124**, 1877–1883 (2002).
99. X. Liang, H. Asanuma and M. Komiyama, Phenylazonaphthalene as a superb photo-regulator for DNA-triplex formation, *Tetrahedron Lett.*, **42**, 6723–6725 (2001).
100. X. Liang, N. Takenaka, H. Nishioka and H. Asanuma, Molecular design for reversing the photoswitching mode of turning on and off of DNA hybridization, *Chem. Asian J.*, **3**, 533–560 (2008).
101. X. Liang, H. Nishioka, N. Takenaka and H. Asanuma, A DNA nanomachine powered by light irradiation, *ChemBioChem*, **9**, 702–705 (2008).
102. T. Kamei, M. Kudo, H. Akiyama, M. Wada, J. Nagasawa, M. Funahashi, N. Tamaoki and T. Q. P. Uyeda, Visible-light photoresponsitivity of a 4-(dimethylamino)azobenzene unit incorporated into single strand DNA: demonstration of a large spectral change accompanying isomerization in DMSO and detection of rapid (*Z*)-to-(*E*) isomerization in aqueous solution, *Eur. J. Org. Chem.*, 1846–1853 (2007).
103. S. Ogasawara and M. Maeda, Straightforward and reversible photoregulation of hybridization by using a photochromic nucleoside, *Angew. Chem.*, **120**, 8971–8974 (2008); *Angew. Chem. Int. Ed.*, **47**, 8839–8842 (2008).
104. M. Endo, K. Nakayama, Y. Kaida and T. Majima, Photoisomerization of 2′-deoxyribofuranosyl and ribofuranosyl 2-phenylazoimidazole, *Tetrahedron Lett.*, **44**, 6903–6906 (2003).
105. J. Sumaoka, K. Kawata and M. Komiyama, Photo-regulation of RNA hydrolysis by the zinc(II) complex carrying azobenzene, *Chem. Lett.*, **28**, 439–440 (1999).
106. K. Maie, M. Nakamura and K. Yamana, Conformational changes of DNA by photoirradiation of DNA–bis(Zn(II)-cyclen)-azobenzene complex, *Nucleosides Nucleotides Nucleic Acids*, **25**, 453–462 (2006).
107. H. Liu, Y. Xu, F. Li, Y. Yang, W. Wang, Y. Song and D. Liu, Light-driven conformational switch of i-motif DNA, *Angew. Chem.*, **119**, 2567–2569 (2007); *Angew. Chem. Int. Ed.*, **46**, 2515–2517 (2007).
108. J. L. Leroy, M. Guéron, J. L. Mergny and C. Hélène, Intramolecular folding of a fragment of the cytosine-rich strand of telomeric DNA into an i-motif, *Nucleic Acids Res.*, **22**, 1600–1606 (1994).
109. G. Hayashi, M. Hagihara, C. Dohno and K. Nakatani, Photoregulation of a peptide–RNA interaction on a gold surface, *J. Am. Chem. Soc.*, **129**, 8678–8679 (2007).
110. G. Hayashi, M. Hagihara and K. Nakatani, RNA aptamers that reversibly bind photoresponsive azobenzene-containing peptides, *Chem. Eur. J.*, **15**, 424–432 (2009).
111. C. Dohno, S. N. Uno and K. Nakatani, Photoswitchable molecular glue for DNA, *J. Am. Chem. Soc.*, **129**, 11898–11899 (2007).
112. D. D. Young and A. Deiters, Light-regulated RNA–small molecule interactions, *ChemBioChem*, **9**, 1225–1228 (2008).
113. J. Andréasson, S. Li, P. Lincoln and J. Andréasson, Photoswitched DNA-binding of a photochromic spiropyran, *J. Am. Chem. Soc.*, **130**, 11836–11837 (2008).

14
DNA Methylation

Albert Jeltsch and Renata Z. Jurkowska

14.1 Introduction

DNA methylation was first discovered in calf thymus DNA by Hotchkiss in 1948 (Hotchkiss, 1948). It occurs at the N6-position of adenine residues and the N4- and C5-positions of cytosine residues, only the last type being observed in higher eukaryotes including mammals (Figure 14.1). These methylated bases are natural components of the DNA, which distinguishes them from a large variety of chemically modified bases formed during DNA damage. Physiological DNA methylation positions the methyl group in the major groove of the DNA, where they do not interfere with the Watson–Crick base-pairing capacities of the nucleotides (Figure 14.1). Thereby, methylation adds extra information to the DNA that is not encoded in the DNA sequence. It can be compared with the small modification of characters in the German Umlaute[1]. The presence of the methyl group can be detected by proteins interacting with the DNA and, thereby cause biological effects.

DNA methylation is introduced into DNA by a group of enzymes called DNA methyltransferases (MTases or Dnmts[2]). These enzymes use *S*-adenosyl-L-methionine (AdoMet) as donor for an activated methyl group and modify the DNA in a sequence-specific manner, often at palindromic sites. In 1964, Gold and Hurwitz identified the first DNA methyltransferase (MTase) in *Escherichia coli* (Gold and Hurwitz, 1964). DNA methylation since then has been discovered in nearly all the organisms investigated, ranging from bacteria to humans. However, its role and abundance vary considerably among different genomes, from totally unmethylated DNA of the yeast *Saccharomyces cerevisiae* and nematode *Caenorhabditis elegans* to heavily methylated mammalian and plant genomes.

The first mammalian DNA MTase activity was discovered by Razin's group (Gruenbaum *et al.*, 1982). The enzyme responsible for this activity today is called Dnmt1; its murine homolog was the first mammalian DNA methyltransferase to be cloned and expressed as recombinant protein (Bestor, 1988; Pradhan *et al.*, 1997). Since then, more members of the mammalian Dnmt enzyme family have been discovered and cloned

[1] To give an example: the German word 'Schaden' is translated as damage. 'Schäden' represents the plural form, a small but significant difference.
[2] The abbreviation Dnmt is derived from DNA methyltransferase.

The Chemical Biology of Nucleic Acids Edited by Günter Mayer

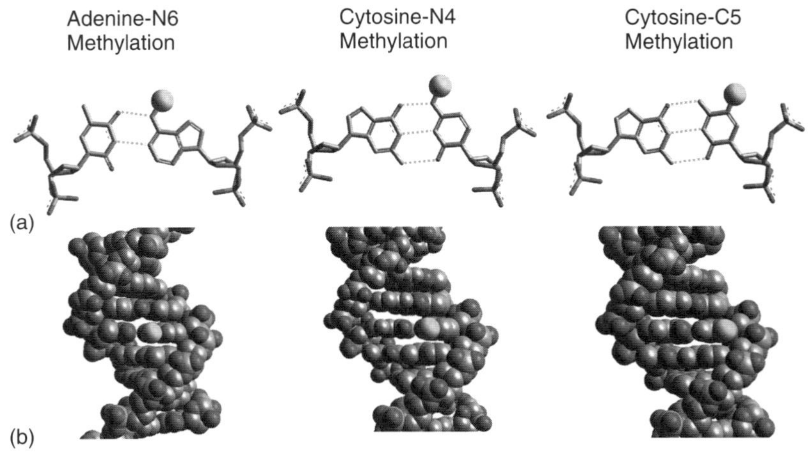

Figure 14.1 Types of methylated bases observed in DNA. (a) TA and GC base pairs are shown in stick representation. The positions of methyl groups in N6-methyladenine, N4-methylcytosine and C5-methylcytosine are indicated by green balls. The methylation does not interfere with Watson–Crick base pairing in any of these cases. (b) B-DNA structures are shown in space fill representation. The methyl groups positioned in the major grove of the DNA are highlighted in green. See color plate section

(Okano *et al.*, 1998) and biochemical, genetic and molecular biology approaches have been systematically applied to expand further our understanding of this important class of enzymes.

The process of DNA methylation is tightly connected with DNA replication, which inherently destroys DNA methylation, because the newly synthesized DNA strand does not carry any methylation. The methylation pattern, therefore, needs to be copied to the new strand after each round of DNA replication.

14.2 The role of DNA methylation in prokaryotes

In prokaryotic DNA, all three types of methylated bases described above are observed. Here, DNA methylation has four major biological roles (Figure 14.2) (for a reviews, see Wion and Casadesus, 2006):

- distinction of self and non-self DNA in restriction-modification systems
- marking of the parental DNA strand in post-replicative mismatch repair
- control of DNA replication and cell cycle
- regulation of gene expression.

Restriction/modification systems (RM systems) (for reviews, see Pingoud and Jeltsch, 2001; Pingoud *et al.*, 2005) function as a defense mechanism of bacteria against infection by bacteriophages and are the source of the overwhelming majority of DNA methyltransferases found in prokaryotes. In addition to the methyltransferase, RM systems contain a restriction endonuclease that specifically cleaves DNA invading the cell at defined recognition sites. The cellular DNA is protected against cleavage by methylation within the same recognition site, as the restriction endonuclease is not able to cleave the methylated DNA. Hence the methylation state of the DNA is detected by the restriction enzymes. Structural analyses showed that these enzymes form very close contacts to their target sites, which are prevented sterically if the DNA is methylated. Consequently, in the RM

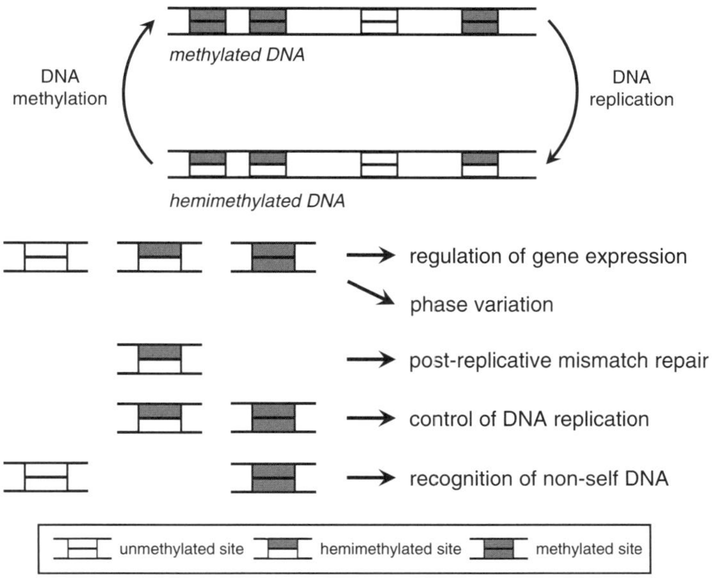

Figure 14.2 *Dynamics and roles of DNA methylation in prokaryotes*

systems, the sequence context of the DNA methylation carries the information that allows the cell to discriminate between self and non-self DNA. If a phage DNA is not cleaved and the phage successfully infects the bacterial cell, the progeny phages will carry the same methylation pattern as the host cell and all other bacteria containing the same RM system are no longer protected against infection. Therefore, many RM systems with different recognition sequences evolved through the billion years of bacteria/bacteriophage co-evolution. So far, over 2000 different RM systems are known and 700 different DNA MTases, which recognize and methylate almost 300 different DNA sequences [see http://www.neb.com/rebase (Roberts and Macelis, 2001)] have been sequenced. The diversification of RM systems conceptually can be compared with the diversification of MHC molecules in the evolution of the human immune system.

The role of DNA methylation in DNA repair and control of DNA replication is best understood in the *E. coli* Dam system (DNA adenine methylation), where DNA is modified at adenine residues in GATC sequences in both DNA strands by the Dam MTase (for a review, see Lobner-Olesen *et al.*, 2005). Similar systems are also present in other γ-proteobacteria. In these bacteria, the methylation status of GATC sites oscillates between the hemimethylation (GATC/G^{m}ATC) immediately after DNA replication and full methylation (G^{m}ATC/G^{m}ATC) after methylation by the Dam MTase, which usually happens 2–4 s after replication (Stancheva *et al.*, 1999). Within this short time span between DNA replication and Dam methylation, a directed repair of base mismatches caused by mis-incorporation of a base by DNA polymerase during replication is possible, because the methylation mark allows the distinction between the unmethylated daughter strand, which must be repaired, and the methylated original template strand, whose nucleotide sequence is correct. The methylation information is read by the MutH enzyme, which is a component of the MutHLS DNA repair system. In this system, MutS detects a mismatch after DNA synthesis. Then, in the presence of MutL, MutH scans for the nearest GATC site and specifically introduces a single strand cut in the unmethylated strand of hemimethylated GATC sites. This initiates the digestion of the daughter strand between the damaged sequence and the GATC site. Re-synthesis of the DNA uses the parental strand as template and, thereby, repairs the damage (for a review, see Li, 2008). In summary, in mismatch repair methylation of Dam sites encodes the information that the unmethylated strand is the newly synthesized daughter strand that needs to be repaired and the methylated strand is the parental strand that carries the correct sequence information.

As a second function, Dam methylation is used to couple the bacterial cell cycle to DNA replication. This is achieved because the re-methylation of the origin of replication is delayed for $\sim$20 min due to the SeqA protein binding to it and, thereby, preventing access of the Dam enzyme to this region of DNA. Since the hemimethylated origins of replication are not active, the delay in re-methylation serves to prevent a premature re-initiation of DNA replication (for a review, see Nielsen and Lobner-Olesen, 2008). The methylation of the origin of replication also plays a role in the segregation of the *E. coli* chromosome later in the cell cycle (Bach *et al.*, 2003; Bach and Skarstad, 2004).

Examples of gene regulation by the Dam MTase in *E. coli* were discovered several years ago (Low and Casadesus, 2008), when it was shown that the Pap operon in uropathogenic *E. coli* strains is regulated by differential DNA methylation. Analyses of the *E. coli* genome showed that there are stably unmethylated sets of GATC sites, whose position can vary with environmental conditions. Most likely, these sites are blocked by other proteins tightly bound to the DNA, such that the Dam enzyme can not gain access to these target sites. Recent whole genome expression profiles have revealed a larger number of genes whose expression is influenced by the Dam enzyme, by unknown mechanisms.

The unmethylated GATC sites that escape the general methylation are used to set up systems in which different genes are alternatively expressed, as exemplified in the Pap pili genes of uropathogenic *E. coli* which can be reversibly switched on or off (for reviews, see Hernday *et al.*, 2002, 2004). The Pap pili facilitate the colonization of the kidneys and represent an important pathogenicity factor of *E. coli*. The Pap promoter contains two GATC sites (a proximal and a distal one). Methylation of these sites interferes with the binding of the global regulator Lrp to these regions. Similarly, binding of Lrp to any of the sites prevents their methylation by Dam, generating a bistable system (i.e. a system having two stable states). In the OFF state, the proximal GATC site is bound by Lrp (and is therefore not available for methylation), which leads to the repression of pilin transcription. Conversely, methylation of the proximal GATC site by Dam prevents binding of Lrp to this site, leading to the activation of Pap transcription (ON state). The expression state is inherited through the DNA methylation and protein binding (hence DNA methylation serves as epigenetic mark), but it can change after DNA replication, leading to a switch in the expression pattern of daughter cells after cell division. This causes phase variations in the gene expression of bacteria that play important roles in infectivity.

In addition, some *E. coli* gene promoters such as the promoter of the transposase gene of the IS10 insertion element are induced in the hemimethylated state, which restricts their activity to the S-phase of *E. coli* (Roberts *et al.*, 1985). This effect could limit the effects of potential lethal integration events.

Adenine-N6 methylation has been shown to be involved in the pathogenicity of different bacteria (for reviews, see Low *et al.*, 2001; Heusipp *et al.*, 2007), including for example *Bordetella pertussis*, *E. coli*, *Salmonella thyphimurium* and *Neisseria meningitidis*. The role of Dam methylation in phase variation is the most likely reason why Dam negative strains of *S. thyphimurium* are not pathogenic and can serve as live vaccines.

A similar system of DNA methylation is present in *Caulobacter crescentus* and other α-proteobacteria (Collier *et al.*, 2007). Here, the CcrM MTase modifies adenine residues within GANTC sequences, where N is any nucleotide. Regulation of the cell cycle of *C. crescentus* is achieved through tight control of the transcription of essential transcriptional regulators including DnaA and CtrA and the CcrM MTase, expression of which is regulated by CcrM methylation. In summary, DNA replication is initiated by DnaA, which is expressed from methylated promoter. After replication, the promoter is converted into the hemimethylated state and inactivated. This happens early in the replication cycle, because the DnaA promoter is located close to the origin of replication. Remethylation does not occur at this stage, because the CcrM MTase is not expressed. For expression, the CcrM gene needs to be in the hemimethylated state (which happens after passage of the replication fork) and CtrA must be present. The CtrA protein is in an antagonistic relation with DnaA. It is induced by conversion of its promoter into the hemimethylated state, but this occurs late in DNA replication, because it is closer to the terminator. Thereby, the sequential changes in the methylation state of chromosomal DNA serve to couple the progression of DNA replication to the cell cycle and prevent the re-initiation events before the successful completion of the first replication round. In contrast to the Dam MTase, it has been shown that the CcrM enzyme is essential in α-proteobacteria (Stephens *et al.*, 1996). Since the group of α-proteobacteria contain important human pathogens and adenine-MTases appear not to exist in higher eukaryotes, the CcrM MTases could be attractive drug targets.

14.3 Structure and mechanism of prokaryotic DNA MTases

In this section, we briefly describe some features of DNA MTases, which explain how these fascinating enzymes are able to perform their complicated function. All DNA MTases use AdoMet as a methyl group donor, which has a high group transfer potential for methyl groups, because its activated methylsulfonium group can be attacked easily by a nucleophile which leads to the transfer of the methyl group (Figure 14.3). However, methylation of either the exocyclic amino groups of adenine and cytosine or the C5-position of cytosine is a difficult chemical task, because these positions are not intrinsically nucleophilic. To overcome this lack of reactivity, the two families of DNA MTases, N-MTases and C-MTases, have evolved highly sophisticated reaction mechanisms. Both mechanisms have in common the necessity for close contact between the enzyme active site and the target base. However, such proximity is not possible while the base is located within the DNA double helix. In a seminal structural study, Cheng, Roberts and co-workers demonstrated that the enzymes solve this contradiction by flipping the target base out of the DNA double helix during catalysis – an unprecedented, dramatic conformational change in the DNA (Klimasauskas *et al.*, 1994) (Figure 14.4). Since then, structural and biochemical studies have shown that base flipping is at least a two-step process. The base is first extruded from the DNA and then tightly contacted by the enzyme in a binding pocket of the catalytic domain, allowing base recognition and catalysis to occur.

The methylation of the C5-position of cytosine follows a two-step reaction mechanism that resembles the Michael addition (for reviews, see Cheng, 1995; Jeltsch, 2002). In the first step, a cysteine residue in the active site of the DNA MTase performs a nucleophilic attack on the C6-position of the target cytosine that is flipped out of the DNA and bound in a hydrophobic pocket of the enzyme. Thereby, a transient carbanion is generated, which is stabilized by a conserved glutamic acid residue that forms hydrogen bonds to the flipped base and protonates the N3 atom of the base. Since the carbanion has a high negative charge density at position 5, the activated base can now attack the methyl group of the cofactor, yielding a stable intermediate comprising the covalently linked complex of the methylated DNA and the MTase. The catalytic cysteine and glutamic acid residues are located in two highly conserved amino acid motifs found in all cytosine-C5

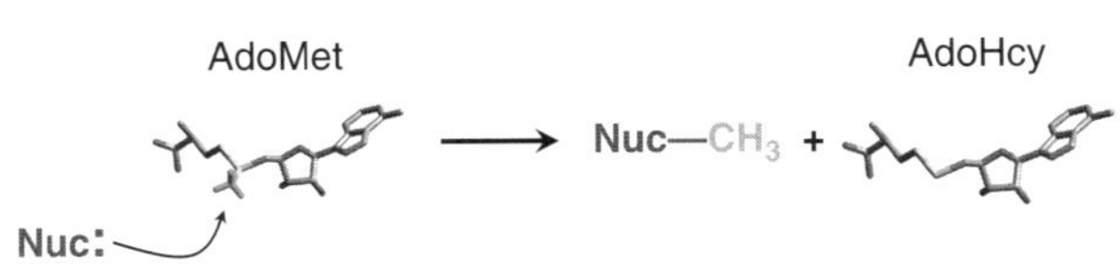

Figure 14.3 *General reaction catalyzed by methyltransferases. AdoMet and AdoHcy are shown in stick models colored by atom type, except that the activated methyl group is colored green. The reaction proceeds by an attack of a nucleophile on the methyl group, which leads to its transfer and the release of the cofactor product AdoHcy. See color plate section*

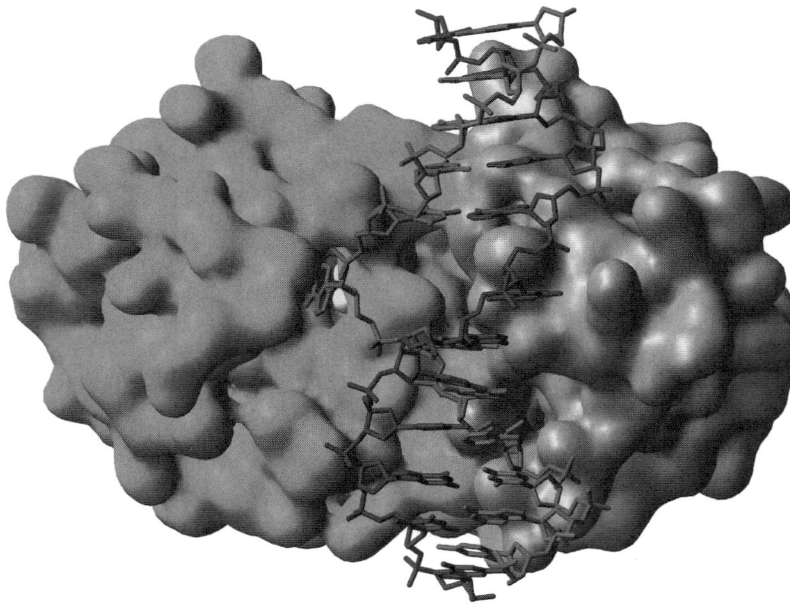

Figure 14.4 *Structure of the T4Dam DNA-(adenine-N6)-methyltransferase bound to a specific DNA molecule. The enzyme is shown in surface representation colored green and orange for its large and small domains. The two strands of the DNA are colored red and blue and shown as a stick model. AdoMet is rendered in space fill representation and colored yellow. Note the flipping of the central target base out of the DNA helix into a binding pocket of the enzyme, which harbors the catalytic center. See color plate section*

MTases, the PCQ motif and the ENV motif, also called motif IV and motif VI. The second step of the reaction resolves the covalent intermediate. It is initiated by deprotonation of C5 catalyzed by an unknown proton acceptor; deprotonation leads to the elimination of the cysteine residues and the re-establishment of aromaticity.

Methylation of either adenine or cytosine residues at their exocyclic amino group is chemically related, because in each case the substrate amino group is part of an electron-poor, heterocyclic, aromatic system. The free electron pairs of the amino groups are delocalized into the aromatic system and not available for nucleophilic attack on the AdoMet. The catalytic center of the MTases that act on exocyclic amino groups is formed by a (D/N/S)PP(Y/F/W) motif. The D/N/S side-chain oxygen and the backbone carbonyl group of the second proline residue of the active site tetrapeptide form hydrogen bonds to the amino group of the flipped adenine or cytosine target. Since the hydrogen acceptor groups are presented in a tetrahedral geometry, formation of the hydrogen bonds induces a change in hybridization of the amino group from sp^2 to sp^3. This localizes the free electron pair and makes it available for the nucleophilic attack on the methyl group of the AdoMet. It is interesting that HemK enzymes, which are conserved from bacteria to humans and strongly resemble adenine-N6 MTases in all their characteristic primary sequence motifs, are in fact glutamine-N5 methyltransferases (Heurgue-Hamard *et al.*, 2002; Nakahigashi *et al.*, 2002). This unexpected finding demonstrates that the (D/N/S)PP(Y/F/W) tetrapeptide is a versatile active site motif that supports methylation of amino groups that are conjugated to an electron-poor π-conjugation system, like the purine ring in adenine-N6 methylation, the pyrimidine ring in cytosine-N4 methylation or the carbonyl group in glutamine-N5 methylation.

All DNA MTases analyzed so far comprise two domains, one large and one small (for a review, see Cheng, 1995) (Figure 14.4). The large domain is evolutionarily conserved and contains the active site and the cofactor binding pocket. C-MTases and N-MTases show a striking similarity in the structures of their catalytic domains, with important catalytic residues [motif IV: PCQ in the case of cytosine-C5 MTases and (D/N/S)PP(Y/F/W) in the case of N-MTases] occupying equivalent positions in the 3D structures of both MTase families. This finding suggests an evolutionary relationship between the two classes of enzymes. The smaller domain of DNA MTases is more heterogeneous in size and structure. It contains most of the residues that mediate sequence-specific recognition of the DNA. Therefore, the variability of the target sequence is reflected by the low of amino acid sequence similarity of the small DNA MTase domains. It forms a battery of sequence-specific contacts to the bases of the target sequence, which mediate sequence-specific DNA interaction. DNA recognition is also supported by the detection of sequence-specific conformational preferences of the DNA in a process called indirect readout.

The localization of the specific target sites in the background of non-specific sites on the DNA is a challenging process. Like most other proteins or enzymes that interact with DNA in a sequence-specific manner, DNA MTases make use of facilitated diffusion to accelerate the search for their target sites. In this mechanism, the enzyme first binds to the DNA at a non-specific site in a very fast reaction that often is close to the theoretical limit for a bimolecular association process. Then, the enzyme follows the DNA in a one-dimensional random movement and searches for its target site.

Some of the MTases are able to introduce several methyl groups into one DNA molecule with several target sites in a processive reaction, whereas others (mostly the MTases that are parts of RM systems) need to dissociate from the DNA before a second turnover. Most likely, processive and distributive mechanisms are correlated with the biological roles played by the enzymes. To fulfill its function in an RM system, the MTase must not rapidly modify the incoming DNA at all targets sites, otherwise the phage DNA would be protected. The distributive reaction mechanism of the RM MTases efficiently prevents such complete methylation and ensures that any incoming phage DNA molecule remains vulnerable to the restriction enzyme for a sufficient period of time, because cleavage of one or a few sites is sufficient for inactivation of the phage. For the solitary MTases (not coupled with a restriction endonuclease), for example EcoDam, no such evolutionary pressure against processive methylation exists, because rapid methylation of the chromosomal DNA after replication is desirable.

14.4 The role of DNA methylation in higher eukaryotes

In higher eukaryotes, only cytosine-C5 methylation can be found in DNA and it mainly occurs at CG dinucleotide sequences. However, only certain CG sites are methylated, resulting in the establishment of a tissue- and cell type-specific pattern of methylation consisting of modified and unmodified CG sites. It has been estimated that approximately 60–90% of all CGs in the human genome are modified (3–8% of all cytosines) (for reviews, see Hermann *et al.*, 2004; Goll and Bestor, 2005; Klose and Bird, 2006). In plants, DNA methylation is additionally found at CNG sites and cytosines located in an asymmetric environment (Chan *et al.*, 2005). Thus, unlike in prokaryotes, where the DNA sequence alone determines the site of methylation, in eukaryotes a pattern of modified and non-modified recognition sites exists. If methylation takes place in both DNA strands at palindromic CG and CNG sites, DNA replication transforms the pattern of unmodified and fully methylated sites to a pattern comprising unmodified and hemimethylated CG sites. Therefore, after DNA replication, the information encoded in the pattern of DNA methylation is still available and the initial pattern can be re-established by a maintenance DNA MTase that specifically modifies hemimethylated target sites but does not react with unmethylated sites.

For CG methylation, this enzyme has been identified as Dnmt1 (for reviews, see Goll and Bestor, 2005; Jeltsch, 2006).

Together with covalent modifications of the histone proteins, DNA methylation constitutes so-called epigenetic information, that is, information which is heritable over cell divisions (occasionally even through the germ line) but is not encoded in the DNA sequence. Given these properties, epigenetic information is ideally suited to control processes such as cellular differentiation or development, where the developmental potential of the cell is stably confined (for a review, see Reik, 2007). Molecular epigenetics is a rapidly developing field with a great impact on gene regulation, development and important therapeutic perspectives (Allis *et al.*, 2007) [see also *Cell*, 2007, **128** (4), *Nature*, 2007, **447** (7143) and *Science*, 2001, **293** (5532), which are journal issues completely devoted to epigenetics or have epigenetics special sections]. Other epigenetic mechanisms include post-translational covalent modifications of the histone proteins (such as acetylation, methylation and phosphorylation), which influence the accessibility of the chromatin and modulate gene expression (Bannister and Kouzarides, 2005; Bernstein *et al.*, 2007; Kouzarides, 2007; Turner, 2007) and the expression of non-coding RNA molecules that lead to specific gene silencing (Bernstein and Allis, 2005; Grewal and Elgin, 2007; Henderson and Jacobsen, 2007). Although a detailed relation between DNA methylation, silencing by non-coding RNAs and histone modification is not entirely clear, evidence is accumulating that these processes are coupled and act in a synergistic fashion.

The presence of 5-methylcytosine in DNA has an important disadvantage, as it is potentially mutagenic, because it promotes deamination of the cytosine (Pfeifer *et al.*, 2000). Hydrolytic deamination of cytosine residues in double-stranded DNA occurs at a relatively slow rate with a half-life of about 30 000 years. It is initiated by a hydroxyl ion attack on the C4-position of cytosine bases that are protonated at the N3-position. However, 5-methylcytosines are deaminated 2–4 times more rapidly than cytosines (Lindahl, 1993; Shen *et al.*, 1994). The rate of deamination of cytosine and methylcytosine residues can be also accelerated by DNA MTases, in particular under conditions of low AdoMet (Shen *et al.*, 1992; Wyszynski *et al.*, 1994; Yebra and Bhagwat, 1995; Zingg *et al.*, 1996; Sharath *et al.*, 2000; Metivier *et al.*, 2008).This is conceivable, given that DNA MTases rotate the target base out of the DNA helix, such that it becomes more accessible and that protonation at N3 is part of the natural catalytic mechanism of cytosine-C5 methylation. The mutational damage caused by DNA methylation is augmented by the lower repair efficiency of T/G mismatches (arising from deamination of 5-methylcytosine), as compared with U/G mismatches (arising from deamination of unmethylated cytosine). This difference is due to the fact that uracil is a non-natural base in DNA and is efficiently recognized and excised by the repair enzyme uracil–DNA glycosylase (for a review, see Fromme *et al.*, 2004). In contrast, thymidine is a natural component of DNA that cannot be repaired by a global mechanism. However, at least two specialized repair enzymes exist for T/G mismatches in a CG context (for a review, see Walsh and Xu, 2006): one mismatch-specific thymine DNA glycosylase (TDG) and the methylcytosine binding protein MBD4. Despite the existence of these repair systems, methylation-mediated mutagenesis events apparently had a strong influence on the genome of vertebrates, because most CG sequences have been removed from the genome during evolution. In mammals, CG dinucleotides are 5–10-fold under-represented with respect to their expected normal frequency. No CG depletion is observed in CG islands that overlap with the annotated transcriptional start sites (TSS) of about 70% of all human genes (Saxonov *et al.*, 2006), including most housekeeping genes and tissue-specific genes. This can be explained, because the CG sites in CG islands are usually not methylated in the germline (Cross *et al.*, 1994), although examples of tissue-specific methylation of CG islands in developing cells accumulate. Deamination of 5-methylcytosine is also a predominant cause of mutations observed in somatic tissues. For example, methylated CG dinucleotides are the single most important mutational target in the p53 tumor suppressor gene (Rideout *et al.*, 1990), which is affected in approximately 50% of all human cancers. Given this mutational burden, the fact that DNA methylation is still observed in most animals, plants and fungi indicates that it must play important biological roles.

14.5 The role of DNA methylation in mammals

Methylation of CG sites in mammals is involved in gene regulation and methylation of CG sites in the promoter regions of genes leads to a reduction in gene expression. This regulatory function contributes to four major biological processes (Figure 14.5) (for reviews, see Hermann *et al.*, 2004; Goll and Bestor, 2005; Klose and Bird, 2006):

- epigenetic regulation of gene expression
- genomic imprinting
- X-chromosome inactivation
- protection against selfish genomic elements to preserve genomic stability.

Repression of gene expression by DNA methylation occurs at three levels of control: (1) several transcription factors, such as AP-2, c-Myc/Myn, E2F and NFκB, are not able to bind to methylated target

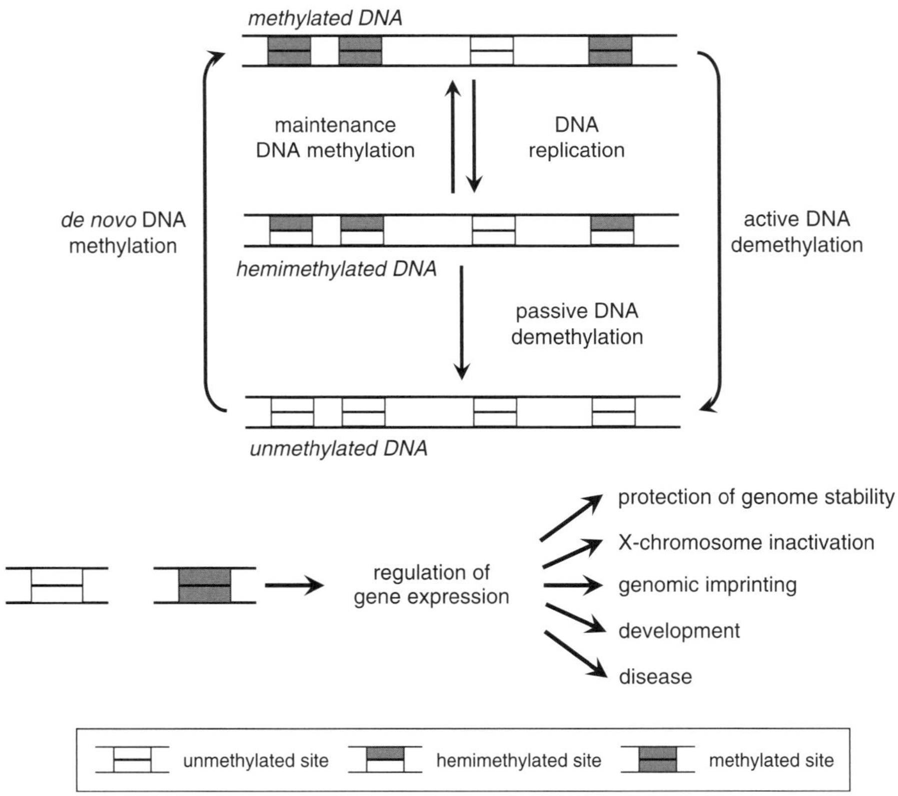

Figure 14.5 *Dynamics and roles of DNA methylation in higher eukaryotes*

sites; (2) DNA methylation recruits methylcytosine-binding proteins that act as repressors of gene transcription; and (3) DNA methylation triggers other chromatin changes such as histone deacetylation and thereby induces chromatin condensation, which leads to a strong and stable repression of gene expression.

Epigenetic gene regulation is a general tool that can be considered as an evolutionary device involved in many different biological functions. Most importantly, epigenetics controls the changes in gene expression during development. Since active transcription leads to the deposition of activating chromatin marks, whereas repression leads to chromatin condensation, the epigenetic marks serve as a cellular memory keeping track of transcriptional states. These changes are very important for the adaptation of cells to signals such as hormones, nutritional state and environment; they are also central to the processes occurring in the brain. In general, the epigenome functions at the interface of genome and environment, thereby controlling the specific response of the organism to its environment (for a review, see Jaenisch and Bird, 2003).

In mammals, genomic imprinting and X-chromosome inactivation are also mediated by epigenetic signals including DNA methylation. Imprinting refers to a small number of genes ($\sim$80) in mammals that carry a differential methylation mark that allows the paternal and maternal copies of these genes to be distinguished. Imprinted genes are only expressed from one chromosome; for example, only the maternal copy of the H19 and only the paternal copy of Igf2 are expressed (for reviews, see Reik *et al.*, 2001; Li, 2002; Delaval and Feil, 2004). The mechanism of imprinting of the H19/Igf2 pair has been understood in molecular terms: it involves an enhancer element, whose effect is directed to the nearby H19 promoter in the maternal genome. Its influence on the Igf2 gene is prevented by a chromatin boundary element. In the paternal genome, the H19 promoter sequence and the boundary element are methylated and thereby inactivated. Under these conditions, the enhancer acts on the promoter of Ifg2 gene, activating its expression. The imprint is transmitted as a certain pattern of methylation of imprinted genes, which is set in the gonads during spermatogenesis and oogenesis. After fertilization, the imprint persists in the somatic cells for the whole life of the individual. In contrast, in the germ cells it must be erased and reset, because the new imprint depends on the gender of the individual. Therefore, the paternal and maternal copies of imprinted genes will both obtain a female imprint in oocytes and a male imprint in sperms.

In mammals, females carry two X chromosomes whereas males have only one. Therefore, dose compensation is required for the genes encoded on the X chromosome to equalize their expression between the genders. To this end, in females, one X chromosome is inactivated in a process that involves specific expression of the Xist RNA from the inactivated chromosome, and also dense methylation and histone deacetylation of the inactivated chromosome (for reviews, see Heard and Disteche, 2006). The initial choice of which X chromosome will be inactivated is made in early embryogenesis. In embryonic tissues, one chromosome is selected for inactivation in a random fashion, whereas in extra embryonic tissues always the paternal chromosome is chosen for inactivation.

In many higher and lower eukaryotes, including mammals and plants, DNA methylation contributes to protection of the genome against selfish genetic elements. Eukaryotic genomes are challenged by different types of selfish genetic elements, such as transposons, retrotransposons and viruses, which constitute a large fraction of the human genome (for a review, see Jurka *et al.*, 2007). Since mobilization of transposons leads to genetic instability, stable silencing of these elements is crucial for life. Transposons and other repetitive DNA sequences usually are relatively rich in CG sequences and heavily methylated. Loss of DNA methylation at these sequences leads to an increase in transcription of transposons and recombination events, causing genomic instability. In this function, DNA methylation acts in concert with other mechanisms that also serve to protect the genome against parasitic DNA, like RNA interference (for reviews, see Matzke and Birchler, 2005; Golden *et al.*, 2008;) and pi-RNA (for a review, see Aravin *et al.*, 2007). At least in plants, RNA interference is also involved in targeting of DNA methylation (for a review, see Mathieu and Bender, 2004).

In mammals, the DNA methylation pattern in the organism is set during early embryogenesis (for a review, see Reik *et al.*, 2001). After fertilization, the methylation information of the paternal and maternal genome (which represents the functional state of oocyte and sperm, which are both highly specialized cell types) is almost completely erased to achieve totipotency. Before implantation, however, the methylation pattern is reset by the action of the *de novo* DNA MTases Dnmt3a and Dnmt3b. Later, the methylation pattern is propagated and maintained by the activity of the Dnmt1 enzyme, which shows a strong preference for methylation of hemimethylated CG sites. This maintenance step also depends on the targeting of Dnmt1 to the replication fork by interaction with PCNA and the targeting to hemimethylated DNA by UHRF1 (Jeltsch, 2008). The *de novo* enzymes also seem to play a role in the global maintenance of DNA methylation (Liang & Jones, 2009). In the primordial germ cells, another wave of demethylation and re-methylation occurs that leads to the gender-specific establishment of the methylation patterns of imprinted genes. This process depends on the *de novo* enzyme Dnmt3a and its regulator Dnmt3L.

Currently, it is unknown how the pattern of DNA methylation is generated and edited by a combination of specific *de novo* methylation and demethylation. Since DNA demethylation is much less well understood than DNA methylation, we focus here on the processes guiding the establishment of DNA methylation. It should be kept in mind, however, that the final DNA methylation pattern is the outcome of both of these counteracting processes. In principle, there are several alternative but compatible mechanisms that could guide DNA MTases to their target regions:

1. The MTases could have intrinsic specificity for certain target regions, as is the rule for prokaryotic DNA MTases. Examples of this mechanism include the preference of the *de novo* Dnmt3a DNA MTase for methylation of CG sites embedded in certain flanking sequences and for methylation of CG sites periodically placed with distances of 8–10 bp; both were shown to leave an imprint on genomic methylation patterns (Handa and Jeltsch, 2005; Jia *et al.*, 2007). However, sequence specificity comparable to prokaryotic enzymes has not been observed with any of the eukaryotic enzymes.
2. The MTases could be directed to the sites of methylation by other DNA binding proteins. Delivery of DNA MTases to target genes by an interaction with the sequence-specific transcription factors and DNA or chromatin-interacting proteins has already been demonstrated in several examples. Dnmt3a was shown to interact with the transcription factors Myc, PU.1, RP58 and p53, the retinoic acid receptor, Karopsi's sarcoma-associated herpes virus protein LANA and human papilloma virus protein E7.
3. Binding of other proteins could protect regions of the DNA from *de novo* methylation. This hypothesis could explain, for example, how CG islands escape methylation during embryonic development. Some experimental evidence supporting this scenario has been provided (Han *et al.*, 2001; Lin and Hsieh, 2001).
4. The accessibility of the chromatin is very important to control gene expression (Henikoff, 2008). Hence it could determine which regions of the DNA are subject to methylation or demethylation. However, this model cannot explain how the heterochromatic DNA, which is the most condensed, becomes highly methylated.
5. Specific histone modifications might attract DNA methylation. This model has been supported recently by the finding that Dnmt3L, a regulator of Dnmt3a, directly binds with its N-terminal domain to histone H3 tails that are unmethylated at K4 (Ooi *et al.*, 2007).
6. ncRNA could target DNA methylation (for a review, see Mathieu and Bender, 2004).

In agreement with the essential roles of epigenetic signaling in mammals, aberrant signaling has a profound impact on health and disease (for reviews, see Feinberg and Tycko, 2004; Robertson, 2005; Esteller, 2007; Feinberg, 2007; Jones and Baylin, 2007). Alterations in the DNA methylation pattern are frequently observed in most human cancers. Generally, a global hypomethylation of DNA in cancer cells is accompanied by a hypermethylation at specific loci. Both of these processes can have cancer-promoting effects: hypomethylation leads to genomic instability often observed in cancer cells and may lead to an activation of retrotransposons. In addition, the expression of oncogenes may become stimulated. Hypermethylation is often found in the promoter regions of tumor suppressor genes and there it has the same effect as a mutation in the gene itself. As the tumor suppressor gene products are often involved in cell-cycle regulation, cell signaling, apoptosis, cell adhesion, chromatin remodeling and DNA repair, the hypermethylation of the promoter regions of these key genes, leading to their inappropriate silencing, can provide a selective growth advantage to the tumor cells. Depending on the tumor type, epigenetic inactivation of tumor suppressor genes can be the predominant way of functional gene loss in tumor cells. It should be noted, however, that methylation defects in cancer cells, at least in principle, are reversible, which makes DNA methylation a promising target for a new generation of anti-cancer drugs. A genome-wide demethylation is also observed during aging and may contribute to the loss of gene regulation in aging cells.

Acknowledgments

Work in the authors' laboratory has been supported by the Deutsche Forschungsgemeinschaft and the BMBF.

References

1. Allis C.D., Jenuwein T., Reinberg D. (2007) *Epigenetics*, Cold Spring Harbor Laboratory Press, Cold Spring Harbor, NY.
2. Aravin A.A., Hannon G.J., Brennecke J. (2007) The Piwi–piRNA pathway provides an adaptive defense in the transposon arms race, *Science*, **318**, 761–764.
3. Bach T., Skarstad K. (2004) Re-replication from non-sequesterable origins generates three-nucleoid cells which divide asymmetrically, *Mol Microbiol*, **51**, 1589–1600.
4. Bach T., Krekling M.A., Skarstad K. (2003) Excess SeqA prolongs sequestration of oriC and delays nucleoid segregation and cell division, *EMBO J*, **22**, 315–323.
5. Bannister A.J., Kouzarides T. (2005) Reversing histone methylation, *Nature*, **436**, 1103–1106.
6. Bernstein B.E., Meissner A., Lander E.S. (2007) The mammalian epigenome, *Cell*, **128**, 669–681.
7. Bernstein E., Allis C.D. (2005) RNA meets chromatin, *Genes Dev*, **19**, 1635–1655.
8. Bestor T.H. (1988) Cloning of a mammalian DNA methyltransferase, *Gene*, **74**, 9–12.
9. Chan S.W., Henderson I.R., Jacobsen S.E. (2005) Gardening the genome: DNA methylation in Arabidopsis thaliana, Nat Rev Genet, **6**, 351–360.
10. Cheng X. (1995) Structure and function of DNA methyltransferases, *Annu Rev Biophys Biomol Struct*, **24**, 293–318.
11. Collier J., McAdams H.H., Shapiro L. (2007) A DNA methylation ratchet governs progression through a bacterial cell cycle, *Proc Natl Acad Sci USA*, **104**, 17111–17116.
12. Cross S.H., Charlton J.A., Nan X., Bird A.P. (1994) Purification of CpG islands using a methylated DNA binding column, *Nat Genet*, **6**, 236–244.
13. Delaval K., Feil R. (2004) Epigenetic regulation of mammalian genomic imprinting, *Curr Opin Genet Dev*, **14**, 188–195.
14. Esteller M. (2007) Cancer epigenomics: DNA methylomes and histone-modification maps, *Nat Rev Genet*, **8**, 286–298.
15. Feinberg A.P. (2007) Phenotypic plasticity and the epigenetics of human disease, *Nature*, **447**, 433–440.
16. Feinberg A.P., Tycko B. (2004) The history of cancer epigenetics, *Nat Rev Cancer*, **4**, 143–153.
17. Fromme J.C., Banerjee A., Verdine G.L. (2004) DNA glycosylase recognition and catalysis, *Curr Opin Struct Biol*, **14**, 43–49.
18. Gold M., Hurwitz J. (1964) The enzymatic methylation of ribonucleic acid and deoxyribonucleic acid. V. Purification and properties of the deoxyribonucleic acid-methylating activity of *Escherichia coli*, *J Biol Chem*, **239**, 3858–3865.

19. Golden D.E., Gerbasi V.R., Sontheimer E.J. (2008) An inside job for siRNAs, *Mol Cell*, **31**, 309–312.
20. Goll M.G., Bestor T.H. (2005) Eukaryotic cytosine methyltransferases, *Annu Rev Biochem*, **74**, 481–514.
21. Grewal S.I., Elgin S.C. (2007) Transcription and RNA interference in the formation of heterochromatin, *Nature*, **447**, 399–406.
22. Gruenbaum Y., Cedar H., Razin A. (1982) Substrate and sequence specificity of a eukaryotic DNA methylase, *Nature*, **295**, 620–622.
23. Han L., Lin I.G., Hsieh C.L. (2001) Protein binding protects sites on stable episomes and in the chromosome from *de novo* methylation, *Mol Cell Biol*, **21**, 3416–3424.
24. Handa V., Jeltsch A. (2005) Profound flanking sequence preference of Dnmt3a and Dnmt3b mammalian DNA methyltransferases shape the human epigenome, *J Mol Biol*, **348**, 1103–1112.
25. Heard E., Disteche C.M. (2006) Dosage compensation in mammals: fine-tuning the expression of the X chromosome, *Genes Dev*, **20**, 1848–1867.
26. Henderson I.R., Jacobsen S.E. (2007) Epigenetic inheritance in plants, *Nature*, **447**, 418–424.
27. Henikoff S. (2008) Nucleosome destabilization in the epigenetic regulation of gene expression, *Nat Rev Genet*, **9**, 15–26.
28. Hermann A., Gowher H., Jeltsch A. (2004) Biochemistry and biology of mammalian DNA methyltransferases, *Cell Mol Life Sci*, **61**, 2571–2587.
29. Hernday A., Krabbe M., Braaten B., Low D. (2002) Self-perpetuating epigenetic pili switches in bacteria, *Proc Natl Acad Sci USA*, **99** Suppl 4, 16470–16476.
30. Hernday A., Braaten B., Low D. (2004) The intricate workings of a bacterial epigenetic switch, *Adv Exp Med Biol*, **547**, 83–89.
31. Heurgue-Hamard V., Champ S., Engstrom A., Ehrenberg M., Buckingham R.H. (2002) The hemK gene in *Escherichia coli* encodes the N(5)-glutamine methyltransferase that modifies peptide release factors, *EMBO J*, **21**, 769–778.
32. Heusipp G., Falker S., Schmidt M.A. (2007) DNA adenine methylation and bacterial pathogenesis, *Int J Med Microbiol*, **297**, 1–7.
33. Hotchkiss R.D. (1948) The mode of action of chemotherapeutic agents, *Annu Rev Microbiol*, **2**, 183–214.
34. Jaenisch R., Bird A. (2003) Epigenetic regulation of gene expression: how the genome integrates intrinsic and environmental signals, *Nat Genet*, **33** Suppl, 245–254.
35. Jeltsch A. (2002) Beyond Watson and Crick: DNA methylation and molecular enzymology of DNA methyltransferases, *ChemBioChem*, **3**, 274–293.
36. Jeltsch A. (2006) On the enzymatic properties of Dnmt1: specificity, processivity, mechanism of linear diffusion and allosteric regulation of the enzyme, *Epigenetics*, **1**, 63–66.
37. Jeltsch A. (2008) Reading and writing DNA methylation, *Nat Struct Mol Biol*, **15**, 1003–1004.
38. Jia D., Jurkowska R.Z., Zhang X., Jeltsch A., Cheng X. (2007) Structure of Dnmt3a bound to Dnmt3L suggests a model for *de novo* DNA methylation, *Nature*, **449**, 248–251.
39. Jones P.A., Baylin S.B. (2007) The epigenomics of cancer, *Cell*, **128**, 683–692.
40. Jones P.A., Liang G. (2009) Rethinking how DNA methylation patterns are maintained, *Nat Rev Genet*, **10**, 805–811.
41. Jurka J., Kapitonov V.V., Kohany O., Jurka M.V. (2007) Repetitive sequences in complex genomes: structure and evolution, *Annu Rev Genomics Hum Genet*, **8**, 241–259.
42. Klimasauskas S., Kumar S., Roberts R.J., Cheng X. (1994) HhaI methyltransferase flips its target base out of the DNA helix, *Cell*, **76**, 357–369.
43. Klose R.J., Bird A.P. (2006) Genomic DNA methylation: the mark and its mediators, *Trends Biochem Sci*, **31**, 89–97.
44. Kouzarides T. (2007) Chromatin modifications and their function, *Cell*, **128**, 693–705.
45. Kuramochi-Miyagawa S., Watanabe T., Gotoh K., Totoki Y., Toyoda A., Ikawa M., Asada N., Kojima K., Yamaguchi Y., Ijiri T.W., Hata K., Li E., Matsuda Y., Kimura T., Okabe M., Sakaki Y., Sasaki H., Nakano T. (2008) DNA methylation of retrotransposon genes is regulated by Piwi family members MILI and MIWI2 in murine fetal testes, *Genes Dev*, **22**, 908–917.
46. Li E. (2002) Chromatin modification and epigenetic reprogramming in mammalian development, *Nat Rev Genet*, **3**, 662–673.
47. Li G.M. (2008) Mechanisms and functions of DNA mismatch repair, *Cell Res*, **18**, 85–98.

48. Lin I.G., Hsieh C.L. (2001) Chromosomal DNA demethylation specified by protein binding, *EMBO Rep*, **2**, 108–112.
49. Lindahl T. (1993) Instability and decay of the primary structure of DNA, *Nature*, **362**, 709–715.
50. Lobner-Olesen A., Skovgaard O., Marinus M.G. (2005) Dam methylation: coordinating cellular processes, *Curr Opin Microbiol*, **8**, 154–160.
51. Low D.A., Casadesus J. (2008) Clocks and switches: bacterial gene regulation by DNA adenine methylation, *Curr Opin Microbiol*, **11**, 106–112.
52. Low D.A., Weyand N.J., Mahan M.J. (2001) Roles of DNA adenine methylation in regulating bacterial gene expression and virulence, *Infect Immun*, **69**, 7197–7204.
53. Mathieu O., Bender J. (2004) RNA-directed DNA methylation, *J Cell Sci*, **117**, 4881–4888.
54. Matzke M.A., Birchler J.A. (2005) RNAi-mediated pathways in the nucleus, *Nat Rev Genet*, **6**, 24–35.
55. Metivier R., Gallais R., Tiffoche C., Le Peron C., R Jurkowska. Z., Carmouche R.P., Ibberson D., Barath P., Demay F., Reid G., Benes V., Jeltsch A., Gannon F., Salbert G. (2008) Cyclical DNA methylation of a transcriptionally active promoter, *Nature*, **452**, 45–50.
56. Nakahigashi K., Kubo N., Narita S., Shimaoka T., Goto S., Oshima T., Mori H., Maeda M., Wada C., Inokuchi H. (2002) HemK, a class of protein methyl transferase with similarity to DNA methyl transferases, methylates polypeptide chain release factors and hemK knockout induces defects in translational termination, *Proc Natl Acad Sci USA*, **99**, 1473–1478.
57. Nielsen O., Lobner-Olesen A. (2008) Once in a lifetime: strategies for preventing re-replication in prokaryotic and eukaryotic cells, *EMBO Rep*, **9**, 151–156.
58. Okano M., Xie S., Li E. (1998) Cloning and characterization of a family of novel mammalian DNA (cytosine-5) methyltransferases, *Nat Genet*, **19**, 219–220.
59. Ooi S.K., Qiu C., Bernstein E., Li K., Jia D., Yang Z., Erdjument-Bromage H., Tempst P., Lin S.P., Allis C.D., Cheng X., Bestor T.H. (2007) DNMT3L connects unmethylated lysine 4 of histone H3 to *de novo* methylation of DNA, *Nature*, **448**, 714–717.
60. Pfeifer G.P., Tang M., Denissenko M.F. (2000) Mutation hotspots and DNA methylation, *Curr Top Microbiol Immunol*, **249**, 1–19.
61. Pingoud A., Jeltsch A. (2001) Structure and function of type II restriction endonucleases, *Nucleic Acids Res*, **29**, 3705–3727.
62. Pingoud A., Fuxreiter M., Pingoud V., Wende W. (2005) Type II restriction endonucleases: structure and mechanism, *Cell Mol Life Sci*, **62**, 685–707.
63. Pradhan S., Talbot D., Sha M., Benner J., Hornstra L., Li E., Jaenisch R., Roberts R.J. (1997) Baculovirus-mediated expression and characterization of the full-length murine DNA methyltransferase, *Nucleic Acids Res*, **25**, 4666–4673.
64. Reik W. (2007) Stability and flexibility of epigenetic gene regulation in mammalian development, *Nature*, **447**, 425–432.
65. Reik W., Dean W., Walter J. (2001) Epigenetic reprogramming in mammalian development, *Science*, **293**, 1089–1093.
66. Rideout W.M. III, Coetzee G.A., Olumi A.F., Jones P.A. (1990) 5-Methylcytosine as an endogenous mutagen in the human LDL receptor and p53 genes, *Science*, **249**, 1288–1290.
67. Roberts D., Hoopes B.C., McClure W.R., Kleckner N. (1985) IS10 transposition is regulated by DNA adenine methylation, *Cell*, **43**, 117–130.
68. Roberts R.J., Macelis D. (2001) REBASE – restriction enzymes and methylases, *Nucleic Acids Res*, **29**, 268–269.
69. Robertson K.D. (2005) DNA methylation and human disease, *Nat Rev Genet*, **6**, 597–610.
70. Saxonov S., Berg P., Brutlag D.L. (2006) A genome-wide analysis of CpG dinucleotides in the human genome distinguishes two distinct classes of promoters, *Proc Natl Acad Sci USA*, **103**, 1412–1417.
71. Sharath A.N., Weinhold E., Bhagwat A.S. (2000) Reviving a dead enzyme: cytosine deaminations promoted by an inactive DNA methyltransferase and an *S*-adenosylmethionine analogue, *Biochemistry*, **39**, 14611–14616.
72. Shen J.C., Rideout W.M. III, Jones P.A. (1992) High frequency mutagenesis by a DNA methyltransferase, *Cell*, **71**, 1073–1080.
73. Shen J.C., Rideout W.M. III, Jones P.A. (1994) The rate of hydrolytic deamination of 5-methylcytosine in double-stranded DNA, *Nucleic Acids Res*, **22**, 972–976.

74. Stancheva I., Koller T., Sogo J.M. (1999) Asymmetry of Dam remethylation on the leading and lagging arms of plasmid replicative intermediates, *EMBO J*, **18**, 6542–6551.
75. Stephens C., Reisenauer A., Wright R., Shapiro L. (1996) A cell cycle-regulated bacterial DNA methyltransferase is essential for viability, *Proc Natl Acad Sci USA*, **93**, 1210–1214.
76. Turner B.M. (2007) Defining an epigenetic code, *Nat Cell Biol*, **9**, 2–6.
77. Walsh C.P., Xu G.L. (2006) Cytosine methylation and DNA repair, *Curr Top Microbiol Immunol*, **301**, 283–315.
78. Wion D., Casadesus J. (2006) N6-methyl-adenine: an epigenetic signal for DNA–protein interactions, *Nat Rev Microbiol*, **4**, 183–192.
79. Wyszynski M., Gabbara S., Bhagwat A.S. (1994) Cytosine deaminations catalyzed by DNA cytosine methyltransferases are unlikely to be the major cause of mutational hot spots at sites of cytosine methylation in Escherichia coli, Proc Natl Acad Sci USA, **91**, 1574–1578.
80. Yebra M.J., Bhagwat A.S. (1995) A cytosine methyltransferase converts 5-methylcytosine in DNA to thymine, *Biochemistry*, **34**, 14752–14757.
81. Zingg J.M., Shen J.C., Yang A.S., Rapoport H., Jones P.A. (1996) Methylation inhibitors can increase the rate of cytosine deamination by (cytosine-5)-DNA methyltransferase, *Nucleic Acids Res*, **24**, 3267–3275.

15

Frameworks for Programming RNA Devices

Maung Nyan Win, Joe C. Liang and Christina D. Smolke

15.1 Introduction

Engineered biological systems that process information, materials and energy hold great promise in providing solutions to many global challenges, including sustainability, renewable energy production, material and chemical synthesis and medical advancement. Our ability to understand and routinely engineer biological systems is limited by the tools available broadly to access, transmit and control molecular information encoded in the properties of various biological molecules, including small molecules, proteins and RNA. Synthetic biology is a rapidly growing field that involves the application of engineering principles to the design and construction of synthetic biological systems. Recent advances in synthetic biology have resulted in the engineering of complex biological systems that act as living chemical factories and exhibit programmed pattern formation (Basu *et al.*, 2005; Ro *et al.*, 2006). Synthetic biologists are also developing the engineering frameworks that support the reliable programming of biological function, including abstraction, standardization and modularity (Endy, 2005). These latter research activities are directed towards the goal of developing an engineering infrastructure and knowledge base that permit the rational design and rapid construction of increasingly complex, reliably operating biological systems.

An active area of research within synthetic biology has been the engineering of molecular components to exhibit new or enhanced activities. Although much of the traditional research effort has been directed towards the molecular design of protein-based components (Kortemme and Baker, 2004; Das and Baker, 2008), advances in RNA biology and engineering are fueling complementary efforts in functional RNA programming. Research in RNA synthetic biology is providing the foundational tools and knowledge base that will support broader efforts in encoding cellular information processing and control operations in synthetic RNA molecules and integrating these engineered components with biological networks to program higher level biological function.

The Chemical Biology of Nucleic Acids Edited by Günter Mayer

15.2 RNA provides a functionally diverse and programmable molecular substrate

Biological functions are encoded within the properties of the diverse molecules present in living organisms, including proteins, small molecules and nucleic acids. The biological functions of RNA can be grouped into protein-coding and nonprotein-coding, or noncoding, functions. Noncoding RNA exhibits diverse functional properties, such as gene-regulatory, ligand-binding and enzymatic properties. RNA molecules can also exhibit structural flexibility, which enables them to adopt dynamically different secondary and tertiary conformations and thereby exhibit allosteric properties. Due to their central location in the gene expression pathway, RNA molecules can regulate expression through diverse mechanisms, including transcription (Winkler *et al.*, 2003), translation (Winkler *et al.*, 2002), splicing (Cheah *et al.*, 2007) and decay (Winkler *et al.*, 2004). RNA molecules that implement sophisticated cellular information processing and control operations have also been characterized (Mandal *et al.*, 2004; Sudarsan *et al.*, 2006).

RNA is a powerful design substrate in synthetic biology due to its unique folding properties. RNA is composed of four nucleotide residues that interact through well-defined hydrogen-bond, base-stacking and electrostatic interactions. RNA secondary structure is dictated by these hydrogen-bond interactions, allowing the development of folding programs that predict the secondary structures and associated free energies of RNA molecules (Mathews, 2006). The energies involved in the folding of an RNA molecule are contributed to a lesser extent by its tertiary structure, such that the relationship between RNA sequence, structure and function is relatively accessible and predictable. These properties have allowed molecular engineers to design and construct a variety of functional RNAs that have been integrated into larger devices and systems to program cellular function (Isaacs *et al.*, 2006; Suess and Weigand, 2008).

15.3 RNA parts exhibit basic functions

RNA parts are genetic components composed of RNA molecules that are capable of performing basic biological functions such as gene regulation, ligand binding and directed conformation change. RNA parts can be generally grouped into three categories based on function – sensors, actuators and transmitters – where the function exhibited by an RNA part is determined by the intent of application specified by the user.

RNA detects signals through various binding events, including hybridization and tertiary interactions. RNA sensor parts can be utilized to detect diverse signals, including temperature, and a variety of molecular ligands such as small molecules, proteins and other nucleic acids. RNA senses temperature through the dynamic and temperature-dependent nature of nucleic acid hybridization interactions (Lai, 2003) and generally exhibits less structured conformations as a function of increasing temperature (Narberhaus *et al.*, 2006). RNA temperature sensor parts have been generated through selection/screening and rational design approaches (Yoshimatsu and Nagawa, 1989; Neupert *et al.*, 2008; Waldminghaus *et al.*, 2008). RNA senses molecular signals through direct binding interactions with the molecular target. RNA aptamers are the most common class of engineered RNA sensor parts capable of binding ligands with high affinities and specificities. RNA aptamers are typically generated *de novo* using an iterative *in vitro* selection strategy or SELEX (Systematic Evolution of Ligands by EXponential enrichment) (Ellington and Szostak, 1990; Tuerk and Gold, 1990). The flexibility of RNA as a sensor is highlighted by the diverse ligands against which synthetic RNA aptamers have been generated, including carbohydrates, proteins, peptides and small molecules (Hermann and Patel, 2000). In addition, RNA can bind nucleic acid ligands through Watson–Crick base-pairing, permitting the engineering of RNA sensors to nucleic acid ligands based on hybridization interactions.

RNA actuators control the activity of other biological molecules, thereby affecting responses in living systems. Advances in RNA biology and engineering have highlighted the variety of actuation functions that RNA can exhibit such as gene-expression regulation, post-translational regulation, directed localization and delivery to specific cell types. RNA parts that regulate gene expression are the most predominant class of RNA actuators. There are various types of RNA gene-expression actuators, where actuation can occur through different mechanisms, including transcription, translation, splicing and stability. These actuators can exhibit their regulatory effects either *in cis*, when the part is embedded within the target transcript, or *in trans*, when the part acts on a separate target RNA through binding interactions between the two molecules. The breadth of mechanisms through which RNA gene-regulatory parts can act has been covered in a recent review (Win *et al.*, 2009). In addition to more standard gene-regulatory elements, RNA aptamers have been shown to act as gene-regulatory parts when embedded within the 5′ untranslated region (UTR) of eukaryotic transcripts by adopting a more structured conformation upon ligand binding that inhibits ribosomal scanning and thus gene expression (Werstuck and Green, 1998; Harvey *et al.*, 2002; Suess *et al.*, 2003). Such RNA aptamer-based translational actuators encode two functions, sensing and actuation, and are therefore an example of a double-coding part.

A second class of RNA actuators function to regulate the activity of biological molecules. RNA aptamers can function as *trans*-acting RNA parts that actuate the activity of molecules post-translationally through binding interactions. A synthetic RNA aptamer can interact with chemical moieties important to the functional activity of the ligand, resulting in inhibition (Famulok *et al.*, 2001) or enhancement of activity (Babendure *et al.*, 2003). Intramers are a class of RNA aptamer-based post-translational activity actuators that are expressed intracellularly to inhibit the activity of various molecular ligands (Famulok *et al.*, 2001). For example, intramers against various viral proteins (Yamamoto *et al.*, 2000; Chaloin *et al.*, 2002; Nishikawa *et al.*, 2003; Kim and Jeong, 2004) have been shown to inhibit the activities of the target viral proteins in human cells, thereby reducing or inhibiting viral infectivity.

A third class of RNA actuators function to direct localization of biological molecules to different cellular compartments or regions, delivery to specific cells and colocalization of different molecules. Localization of transcripts is used to achieve higher local protein concentrations (Hazelrigg, 1998), whereas localization of gene-regulatory RNAs and protein-coding transcripts is used to achieve conditional or enhanced control of gene-regulatory functions (Kawamoto *et al.*, 2005). Synthetic RNA localization parts have either been adapted from naturally occurring RNA localization elements or generated as RNA aptamers to specific molecular targets. For example, heterologous transcripts harboring RNA zip-code parts within their 3′ UTRs have been shown to be localized to targeted compartments within the cytoplasm (Kislauskis *et al.*, 1993). In addition, such RNA zip-code parts have been integrated with RNA gene-regulatory parts to enhance gene-regulatory activities through colocalization (Lee *et al.*, 1999). Synthetic RNA scaffolds have also been built linking multiple RNA aptamers to localize DNA-binding and transcriptional activation proteins to regulate transcription initiation in yeast (Cassiday and Maher, 2001).

RNA transmitter parts are RNA sequences that translate an informational event (such as signal detection by a sensor) from one RNA part to another through conformational changes. One class of RNA transmitter parts, referred to as communication module parts, are based on a single sequence that can bind to a partially complementary target strand in multiple ways, where the energetics of strand binding allow the part to adopt different bound states through a helix slipping event (Soukup and Breaker, 1999). A second class of RNA transmitter parts, referred to as strand displacement parts, are based on two sequences that compete for binding to a target strand, where the energetics of each binding event allow the part to adopt different bound states through a competitive hybridization event (Mandal and Breaker, 2004).

15.4 RNA devices are composed of distinct parts and encode more complex functions

A device is a combination of refined parts that can perform a human-defined function (Endy, 2005). RNA devices are composed of multiple functionally distinct RNA parts and are capable of performing more complex biological functions as a result of the integrated component functions. Riboswitches are naturally occurring counterparts of RNA devices that link sensing and gene-regulatory functions (Mandal and Breaker, 2004). Riboswitches control gene expression in response to specific inputs and have been shown to act through diverse gene expression mechanisms, including transcription termination (Winkler *et al.*, 2003), translation initiation (Winkler *et al.*, 2002), splicing (Cheah *et al.*, 2007) and mRNA decay (Winkler *et al.*, 2004). Riboswitches that incorporate sensors that respond to temperature have also been characterized (Morita *et al.*, 1999; Nocker *et al.*, 2001), although the majority of characterized riboswitches respond to cellular metabolites, such as amino acids and cofactors. Riboswitches that incorporate multiple sensor and actuator domains through more complex architectures to perform higher order regulatory functions, such as multi-input signal integration and cooperativity, have also been characterized (Mandal *et al.*, 2004; Sudarsan *et al.*, 2006).

Building on these natural examples, engineers have built a variety of RNA devices that process and transmit molecular inputs to targeted gene expression outputs in a wide variety of cellular systems (Isaacs *et al.*, 2006; Suess and Weigand, 2008). The physical implementation of an RNA device as a component within a genetic construct is generally guided by its actuator part. Engineers have utilized various design strategies to build RNA devices from combinations of different parts and have employed both screening/selection and rational design approaches to generate functional devices. The different device designs will afford different advantages when linking to genetic systems and will exhibit different quantitative response properties. Therefore, the RNA device design strategy is an important consideration in ultimately linking to downstream applications.

15.5 RNA device design strategies based on the direct coupling of sensor and actuator parts

A direct coupling RNA device design strategy is one that links the sensor and actuator parts directly without the inclusion of a distinct transmitter part (Figure 15.1a). In this strategy, the binding information is typically transmitted between the sensor and actuator parts through conformational changes that are specific to each aptamer–ligand pair. As a result, the coupling of parts into a functional RNA device through this strategy can depend on mechanistic properties specific to the particular sensor and actuator.

The direct coupling strategy has been implemented based on the replacement of a portion of the actuator sequence with the sensor sequence. For example, a ribozyme-based RNA device was constructed by replacing a stem in the group I self-splicing intron from the bacteriophage T4 thymidylate synthase gene with the theophylline aptamer (Thompson *et al.*, 2002). The device was integrated into its native gene and demonstrated theophylline-dependent splicing and regulation of cell growth in an *Escherichia coli* thymidine auxotroph strain. In another example, an input-dependent transcriptional activator activity was built (Buskirk *et al.*, 2004) by replacing a stem in an RNA gene-regulatory part that functions as a transcriptional activator in yeast (Buskirk *et al.*, 2003) with the tetramethylrosamine (TMR) aptamer (Grate and Wilson, 1999). A functional TMR-responsive device was generated by randomizing several nucleotides in the region connecting the two parts and selecting for TMR-responsive sequences, which allowed the conformational change associated with TMR binding to activate the transcriptional activator and resulted in a 10-fold increase in expression in response to the input molecule.

Other examples of the direct coupling design strategy are based on the direct coupling of the sensor and actuator sequences. In one example, the theophylline aptamer was directly coupled to a small hairpin RNA (shRNA) actuator (An *et al.*, 2006) (Figure 15.1a). Processing of the shRNA sequence by Dicer was demonstrated to be inhibited by theophylline in a dose-dependent fashion in mammalian cells, resulting in the small-molecule regulation of RNA interference (RNAi)-based gene silencing. However, the functional activity of the device was shown to be very sensitive to the distance between the Dicer cleavage site and ligand binding site, as changes of even one base pair abolished function. This RNAi-based device was recently applied to the theophylline-dependent regulation of albumin

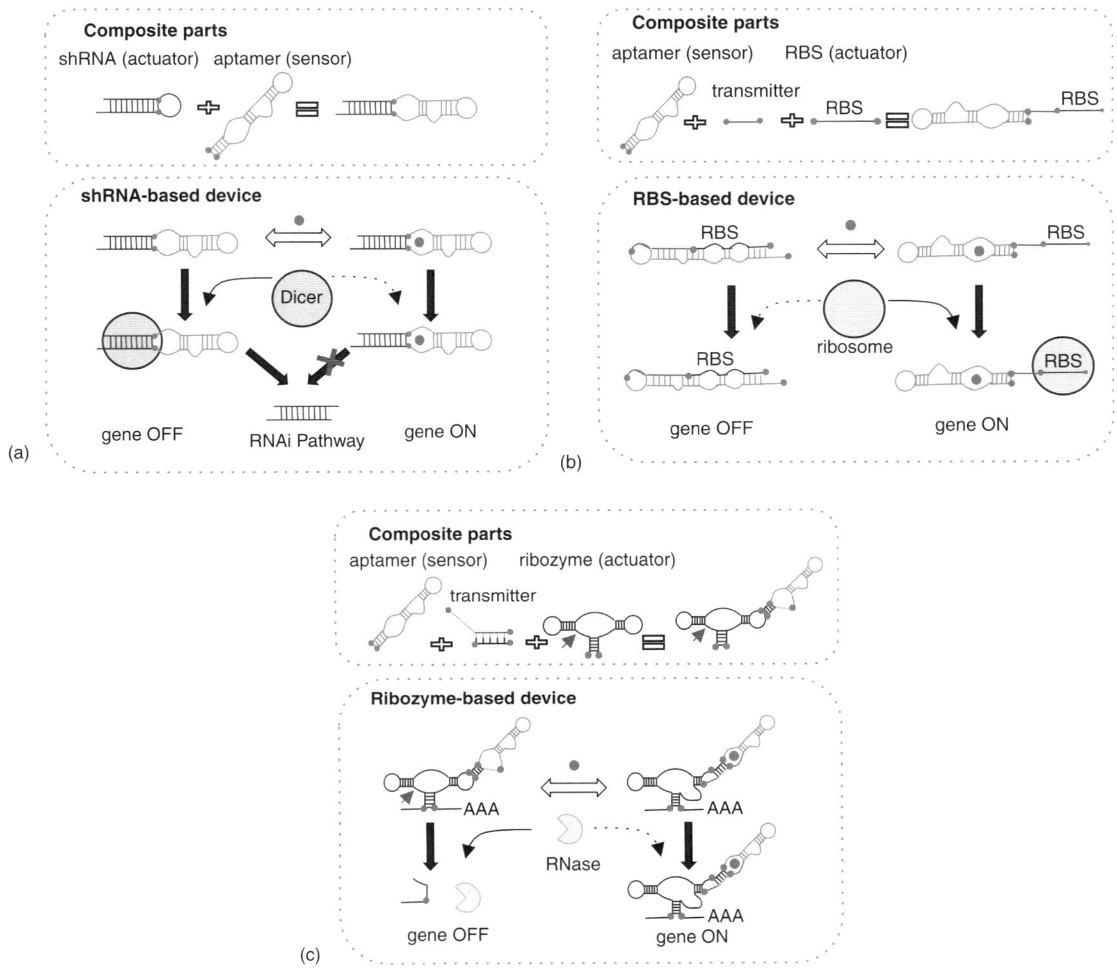

Figure 15.1 RNA device design strategies. (a) A device design strategy based on the direct coupling of sensor and actuator parts. In the example shown, an shRNA actuator part is linked directly to an RNA sensor part. Binding of the molecular input to the RNA sensor modulates cleavage by Dicer and subsequent RNAi-induced gene silencing. (b) A device design strategy based on the integration of an information transmission part between the sensor and actuator parts. In the example shown, an RBS actuator part is linked to an RNA sensor part through a linker sequence that acts as a transmitter part. Binding of input to the RNA device stabilizes the input-bound conformation in which the RBS is not occluded in a secondary structure, thereby increasing ribosome access. (c) A device design strategy based on the integration of an information transmission part that permits the modular assembly of sensor, actuator and information transmitter parts. In the example shown, a hammerhead ribozyme actuator part is linked to an RNA sensor part through a transmitter sequence that provides modular coupling between the two parts. Binding of the input to the RNA device stabilizes the input-bound conformation in which the ribozyme is inactive. All of the illustrated device designs result in gene expression increasing as a function of increasing input concentrations, thereby encoding gene expression ON switches. Reproduced from Win, M.N., Liang, J.C., and Smolke, C.D. (2009). Frameworks for programming biological function through RNA parts and devices. Chem Biol **16**, 298–310. Reprinted with permission from Elsevier © 2009. See color plate section

levels in liver cells (Tuleuova *et al.*, 2008). In another example, an RNA device was shown to regulate effectively splicing of a pre-mRNA in an input-dependent manner when the 5′ splice site was integrated within the stem of the tetracycline aptamer (Weigand and Suess, 2007).

15.6 RNA device design strategies based on the integration of a distinct information transmission function

A second type of device design strategy is one that links the sensor and actuator parts through a separate transmitter part (Figure 15.1b). The transmitter part translates information between the sensor and actuator through a change in conformation. Often this conformational change occurs at the level of secondary structure, such as helix slipping or strand displacement. As such, information is transmitted through a mechanism that is itself independent of input binding to the sensor, such that the RNA molecule adopts at least two different conformations associated with an active or inactive sensor. Input binding to the conformation harboring an active sensor shifts the distribution between these conformations to favor the input-bound state. By introducing an information transmission function that is distinct from the input binding event, devices built on this design strategy can achieve greater flexibility in the coupling of actuator and sensor parts.

Several examples of RNA devices that link aptamers to gene-regulatory parts that control the initiation of translation through linker sequences that perform information transmission functions have been described. These devices regulate the activity of the actuator by modulating its accessibility to translational machinery through hybridization interactions. For example, ribosome binding site (RBS)-based RNA devices were generated by coupling the theophylline aptamer to a RBS through a short linker region (Figure 15.1b). Randomized linker regions of various lengths were screened for those sequences that resulted in functional theophylline-responsive devices through colorimetric, flow cytometric and cell motility assays in *Escherichia coli* by altering the gene-regulatory output (Desai and Gallivan, 2004; Lynch *et al.*, 2007; Topp and Gallivan, 2008a,b; Lynch and Gallivan, 2009). The devices were proposed to function through modulating ribosome accessibility to the RBS via a strand displacement mechanism, where the RBS was partially base paired with the aptamer sequence in a gene expression OFF state or released from base pairing in the gene expression ON state in the input-bound device conformation. However, the aptamer sequence in the resulting devices encodes for both ligand binding and antisense functions such that the sequences of the sensor and RBS actuator are not independent of one another and cannot be independently modified while maintaining device function. A similar strategy was applied to generate two-input RBS-based RNA devices that exhibit logic operations (AND, NAND) by coupling the theophylline aptamer to a natural riboswitch element through a randomized linker region and selecting for desired function (Sharma *et al.*, 2008).

In another example (Suess *et al.*, 2004), a similar RBS-based device was constructed through the coupling of the theophylline aptamer to an RBS through a helix slipping-based transmitter part or communication module (Soukup and Breaker, 1999). These devices were proposed to function through small nucleotide shifts within the communication module that led to changes in the accessibility of the RBS and active state of the aptamer, resulting in theophylline-dependent regulation of gene expression. While the introduction of a communication module may impart some degree of sequence independence between the sensor and actuator, the function of such helix slipping-based transmitter parts was also shown to be dependent on the sensor sequence in a separate study (Win and Smolke, 2007).

Other examples of RNA devices built with separate transmitter parts couple multiple actuators into a single device. For example, two recent devices were designed to regulate ribosome access to the RBS through ribozyme cleavage (Ogawa and Maeda, 2008; Wieland and Hartig, 2008). The RBS and hammerhead ribozyme (hhRz) actuators were coupled in such a manner that the RBS was sequestered within the ribozyme stem structure and cleavage of the ribozyme resulted in separation of the two parts, thereby increasing ribosome access and thus expression levels. Ribozyme cleavage was in turn modulated through binding of theophylline to its aptamer, which was linked through a communication module to the ribozyme. In another example, a device was designed to regulate ribosome access to the RBS through *cis*- and *trans*-acting antisense parts (Isaacs *et al.*, 2004). The *cis*-acting antisense part was implemented to inhibit ribosome access to an RBS through direct hybridization interactions. The *trans*-acting antisense part was employed to hybridize with the *cis*-acting antisense part, thereby alleviating sequestration of the RBS and thus activating expression. Although this device design offers a greater level of flexibility in tailoring parts due to the isolation of binding events to hybridization interactions, the *cis*-acting antisense functions as both a transmitter and a sensor to the *trans*-acting antisense part. As a result, the sequence of the *cis*-acting antisense part is dependent on both the RBS and the *trans*-acting antisense part, thereby rendering interdependence among the parts.

15.7 Functional composition frameworks support device design strategies based on modular assembly of RNA parts

One of the goals of synthetic biology is to develop foundational technologies that make the engineering of biology easier and more reliable (Endy, 2005). Towards this goal, research efforts are being directed to the development of functional composition frameworks that support reliable device engineering. Functional composition frameworks are a type of device design strategy that supports the construction of devices through the modular assembly of distinct parts. As such, functional composition frameworks are characterized by functional modularity. Such frameworks support the efficient and reliable engineering of diverse device functions from a smaller number of refined parts through a plug-and-play type of strategy without having to redesign the device for every new instance or implementation of a device.

Functional composition frameworks were recently proposed for single-input, single-output RNA devices. In the proposed design strategies, functional modularity is achieved through the separation of functions (sensing, actuation and information transmission) into distinct and independent parts (Figures 15.1c and 15.2a). A common approach is the encoding of the information transmission function in a distinct transmitter part that employs a strand displacement event solely with the sequences of that transmitter part. In this design approach, the sequences of the sensor and actuator do not depend on one another and can therefore be changed independently. Therefore, although the functions of sensing and actuation frequently rely on more complex tertiary interactions, which are not accounted for in these first-generation frameworks, the integration of these functions is simplified via a distinct transmitter part that insulates part functions and controls the interactions between parts through predictive hybridization interactions.

In one example, a functional composition framework for RNA device design was proposed based on the assembly of three functional parts (Win and Smolke, 2007): a sensor, made of an RNA aptamer; an actuator, made of a self-cleaving hhRz; and a strand displacement-based transmitter part, made of a sequence that couples the sensor and actuator parts (Figure 15.1c). Competitive hybridization events localized within the transmitter part direct the distribution between two primary conformations. Several design choices were made in the implementation of the ribozyme-based device platform to support engineering properties such as portability, utility and composability. For example, the RNA device was integrated into the 3′ UTR of the target gene, where the integration of RNA devices is expected to have minimal nonspecific effects on gene expression and be extensible to multiple independent elements. In addition, ribozyme cleavage in the 3′ UTR inactivates a transcript and lowers gene expression independent of cell-specific machinery, thus allowing portability across different organisms. As another example, flanking sequences were included to insulate the device from surrounding gene-specific sequences that may disrupt the structure of the RNA device and therefore its activity to support the reliable coupling to other parts and devices.

Ribozyme-based devices were built in which the input-bound conformation was associated with the disruption or restoration of the catalytic core of the ribozyme, thus converting a molecular input to increased (ON switch) or decreased (OFF switch) expression, respectively, of a reporter gene in *Saccharomyces cerevisiae* (Figure 15.2b). The device response properties were shown to be tunable by altering nucleotide sequences within the transmitter part, which is proposed to alter the energetic and kinetic properties of the

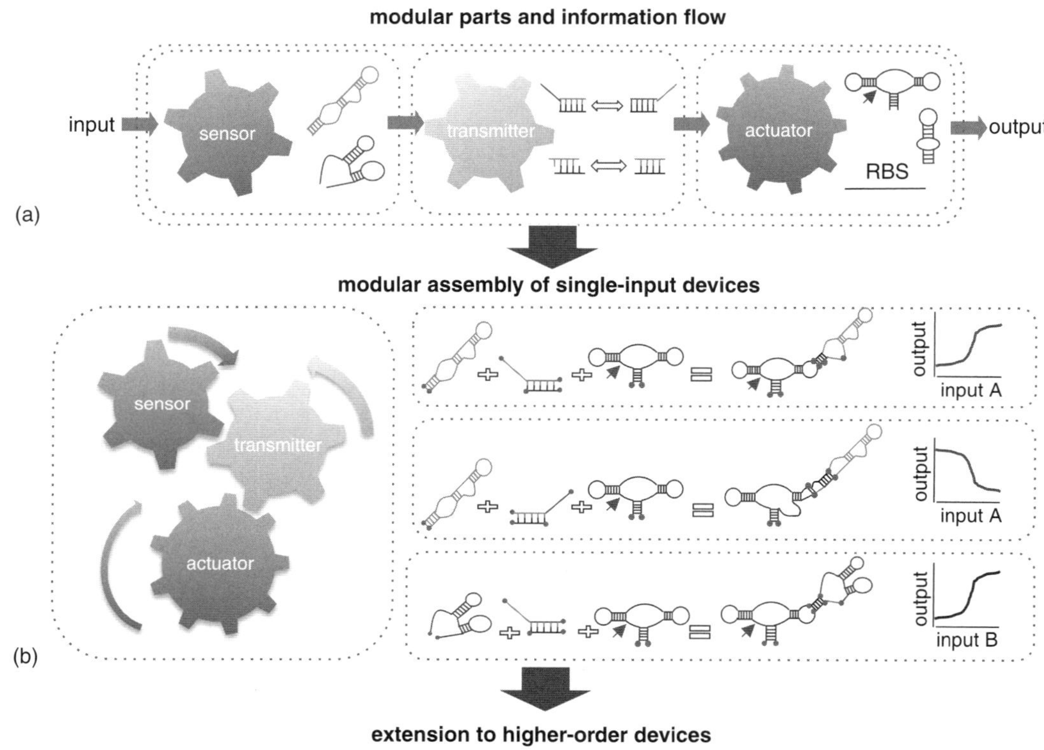

Figure 15.2 *Functional composition frameworks support the modular assembly of single-input, single-output RNA devices and extension to higher order devices. (a) A functional composition framework for assembling RNA devices from modular components. Information in the form of a molecular input is received by the sensor and transmitted by the transmitter to a regulated activity of the actuator, which in turn controls a gene expression event output. (b) Modular composition frameworks allow the modular assembly of single-input, single-output RNA devices from sensor, transmitter and actuator parts. In this example, the function of the RNA device is encoded in the transmitter part, such that alteration of the transmitter sequence results in devices that perform different computations. In addition, the input responsiveness of the device, or the particular input molecule to which the device responds, is encoded in the sensor part, such that direct replacement of the sensor part results in devices that process different inputs. (c) This single-input, single-output device framework can be directly extended to more complex devices encoding higher order functions. The computational operation encoded in the device is determined by the particular combination of parts (sensor, transmitter) and the device architecture. Reproduced from Win, M.N., Liang, J.C., and Smolke, C.D. (2009). Frameworks for programming biological function through RNA parts and devices.* Chem Biol ***16****, 298–310. Reprinted with permission from Elsevier © 2009. See color plate section*

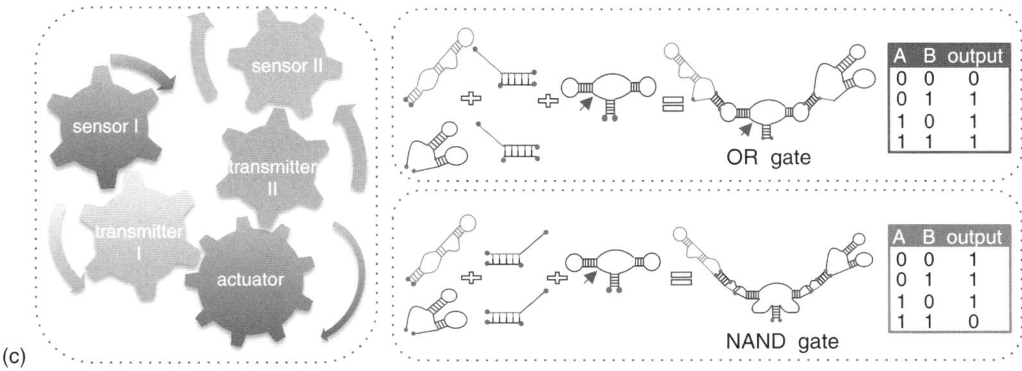

Figure 15.2 *(Continued)*

strand displacement event and therefore the ability of the device to distribute between and access different conformational states. The functional modularity of the proposed framework was demonstrated to support direct sensor replacement strategies, such that aptamer sequences to different molecular inputs (theophylline and tetracycline) were swapped into the framework and the function of the device was maintained while being made responsive to new small molecule inputs.

A similar design framework was used to build RNA devices that actuate through the RNAi pathway by coupling an aptamer to a shRNA through a distinct strand displacement-based transmitter part (Beisel *et al.*, 2008). shRNA-based devices were built in which the input-bound conformation was associated with the disruption of the shRNA stem, thereby inhibiting Dicer processing and subsequent gene silencing, converting a molecular input to increased (ON switch) expression of reporter and endogenous gene targets in mammalian cells. The functional modularity of the framework was further highlighted in this example, by demonstrating that direct replacement of aptamer sequences to different molecular inputs (theophylline, xanthine) in the platform maintained device function. In addition, this example demonstrated that the general framework was applicable to different actuation mechanisms, thereby supporting functional modularity of both the sensor and actuator parts. Other examples of RNA devices that follow this general framework have been implemented in yeast with *trans*-acting antisense actuators (Bayer and Smolke, 2005), further supporting the generality of this device design strategy.

15.8 Composition frameworks support extensions to higher order device functions

One of the important properties of a functional composition framework is its extensibility to the assembly of more complex devices that permit sophisticated information processing and control functions from basic sensing, actuation and transmitter parts. Such frameworks will provide general approaches for the forward engineering of multi-input devices that perform higher order functions and support the combinatorial assembly of many information processing, transduction and control devices from a smaller number of refined parts.

Recent work has described the extension of the functional composition framework described above for single-input, single-output ribozyme-based devices to the modular assembly of devices that perform higher order functions from a small set of refined sensor, hhRz actuator and transmitter parts (Win and Smolke, 2008) (Figure 15.2c). Three signal integration schemes based on single-layer device architectures were developed that involve the assembly of multiple single-input, single-output ribozyme-based devices and the coupling of multiple sensor–transmitter components to single ribozyme actuators. RNA devices that function as signal and bandpass filters, two-input logic gates (AND, NOR, NAND, OR gates) and cooperative control devices were built on the extended device framework. The variety of information processing operations demonstrated from a small number of parts through this extended framework highlights the utility and power of such modular assembly strategies. In addition, the described framework may be further extended to more complex device designs by combining multiple signal integration schemes within a single layer and implementing layered architectures (Rinaudo *et al.*, 2007).

15.9 Enabling technologies that support the engineering of RNA devices

The broad engineering of RNA devices tailored to different biological systems will require the development of several enabling technologies. These technologies are focused on the generation of well-characterized, refined libraries of RNA parts, generalizable approaches for integrating these parts into functioning devices and predictive tools for rationally programming or tuning the quantitative response properties of the engineered devices.

Strategies that support high-throughput and reliable RNA part generation, refinement and characterization are critical to RNA device engineering and therefore represent a basic enabling technology. RNA gene-regulatory actuator parts are commonly adapted from Nature with little modification or are designed from relatively simple hybridization-based design rules, although several efforts have focused on the *in vitro* evolution of RNA actuators exhibiting enhanced function (Vaish *et al.*, 1997; Conaty *et al.*, 1999; Persson *et al.*, 2002). RNA sensor parts based on aptamer sequences have been adapted from naturally occurring riboswitch elements (Nomura and Yokobayashi, 2007), but are more often generated *de novo* through *in vitro* selection strategies (Ellington and Szostak, 1990; Tuerk and Gold, 1990). The generation of RNA aptamers as sensor parts that exhibit desired *in vivo* affinities and specificities is currently a limiting step in RNA device design. *In vitro* selection strategies for DNA aptamers to proteins have been adapted to higher throughput formats based on automation (Cox *et al.*, 2002) and higher efficiency partitioning schemes such as capillary electrophoresis (Berezovski *et al.*, 2005; Drabovich *et al.*, 2005). However, similar improved aptamer generation schemes have not yet been demonstrated for RNA aptamers and other groups of target molecules such as small molecules.

Functional composition frameworks that support the rapid assembly of RNA devices from parts exhibiting basic functions represent another enabling technology supporting more efficient and reliable device design. Although first-generation frameworks have recently been proposed that highlight the design advantages afforded through such engineering design principles (Win and Smolke, 2007, 2008; Beisel *et al.*, 2008), much work remains in optimizing and refining such frameworks for RNA device design. As one example, effective composition frameworks require the careful refinement of parts, such that the parts can be more reliably integrated into the proposed framework. Therefore, large libraries of well-characterized, refined RNA parts that are compatible with device frameworks are needed to take full advantage of such composition frameworks. In addition, improved frameworks that address challenges in enhancing functional device performance (Win and Smolke, 2007; Beisel *et al.*, 2008), such as dynamic range and the insulation of part functions, are needed. The development of such next-generation frameworks will likely require the integration of recent technologies that allow the measurement of the kinetic folding and dynamic switching rates of RNA molecules (Greenleaf *et al.*, 2008), thereby advancing our understanding of RNA structure–function relationships. Finally, extended architectures (Deans *et al.*, 2007; Rinaudo *et al.*, 2007; Win and Smolke, 2008), including single- and multi-layered systems composed of heterogeneous components, are needed to support the engineering of more complex device functions, including signal restoration and amplification.

A third enabling technology is represented by strategies that support the engineering of parts within an RNA device platform. Such technologies can increase the efficiency of new device construction by removing the additional step of part integration and functional optimization within the context of the device. As one example, transmitter parts have been generated by applying *in vitro* (Soukup and Breaker, 1999) and *in vivo* screening strategies to small libraries within functional device platforms (Lynch *et al.*, 2007; Wieland and Hartig, 2008; Win and Smolke, 2007). The latter strategy provides the additional benefit of selecting for function within the desired cellular context, as parts that are generated *in vitro* do not necessarily translate to functional elements within the cellular environment (Link *et al.*, 2007; Win and Smolke, 2007). As such, *in vitro*-generated parts generally require secondary screening steps for *in vivo* function. Recently, cell-based screening strategies for functional transmitter parts have been adapted to higher throughput methods based on fluorescence-activated cell sorting (FACS) (Fowler *et al.*, 2008; Lynch and Gallivan, 2009). Researchers have also demonstrated the application of such cell-based screening methods to RNA gene-actuation parts within the device platform by extending the randomized sequence region into the actuator part (Lynch and Gallivan, 2009). A critical area of future research will be to apply these higher throughput, cell-based screening strategies to the generation of new sensor parts within device platforms.

Finally, computational tools that support the *in silico* design and programming of device function will provide a critical enabling technology to match quantitative device-response properties to the application-specific performance requirements of the genetic system of interest. Recent modeling tools have been described that provide early sequence-to-function computational frameworks for guiding the design and optimization of RNA devices (Beisel *et al.*, 2008). In addition, computational models have also been described recently that allow researchers to examine the effects of various parameters associated with RNA function on the resulting quantitative regulatory response properties (Beisel and Smolke, 2009). Such models provide important insights and guiding tools for advancing RNA device design, by elucidating the impact of different parameters, including RNA folding rates, gene-regulatory rates and ligand binding rates, on device performance. However, these studies have also highlighted the need for improved RNA folding algorithms that capture the kinetic properties and nature of *in vivo* RNA folding. The future refinement of *in silico* device design tools will require further insight into RNA structure–function relationships (Martick and Scott, 2006), kinetic and thermodynamic properties of RNA folding (Greenleaf *et al.*, 2008) and improved predictions of RNA secondary and tertiary structures (Parisien and Major, 2008).

15.10 Engineered systems composed of RNA parts and devices

RNA parts and devices are integrated into genetic circuits composed of heterogeneous parts, including proteins and small molecules, which comprise devices and systems exhibiting complex functions of interest to downstream applications. Numerous examples demonstrating the integration of RNA parts into heterogeneous genetic circuits have been described, while more recent examples have focused on the integration of RNA devices into circuits encoding functions relevant for biotechnological or medical applications. The functional diversity and programmability associated with RNA parts and devices make them powerful information processing and control molecules that can be implemented in broad application areas, including metabolic engineering, cellular biosensing, phenotype generation and therapeutics.

In the growing field of metabolic engineering, RNA parts have been implemented as molecular tools to tune enzyme levels, regulate pathway flux and optimize product yields. In one example, RNase activity actuators were integrated in intergenic regions of a multi-gene transcript in *E. coli* to direct the segmental processing and stability of the transcript, resulting in differential expression of the genes encoded in the construct (Smolke *et al.*, 2000). This regulatory strategy was applied to control the flux through a synthetic carotenoid pathway, where the production of β-carotene was modulated up to 300-fold relative to levels of lycopene, an intermediate metabolite in the pathway (Smolke *et al.*, 2001). Accumulation levels of intermediate pathway metabolites were varied through combinatorial implementation of RNase activity actuators exhibiting different regulatory strengths, thereby demonstrating the ability of RNA parts to modulate pathway enzyme levels, flux and metabolite levels. A modified strategy was developed based on the generation of libraries of combinations of RNase activity actuators, that were screened through two-color FACS for combinations that resulted in desired gene expression ratios (Pfleger *et al.*, 2006). This strategy was subsequently applied to a synthetic mevalonate pathway using a biosensor screening assay for mevalonate production, where selected intergenic regions optimized the relative expression of three pathway enzymes, resulting in a sevenfold increase in mevalonate levels.

RNA parts have also been implemented as regulatory components in circuits that have potential application in gene therapy. In a recent example, RNAi-based actuators were integrated as modular components into a genetic circuit that was employed as a tight and tunable transgenic regulatory system in mammalian cells (Deans *et al.*, 2007). The gene circuit coupled heterogeneous gene-regulatory actuator parts, repressor proteins and shRNAs to reduce synergistically the basal expression of the regulated gene from that obtained through any single actuator. In the OFF state of the circuit, the repressor protein LacI and the shRNA inhibited target gene expression through simultaneously targeting transcription and RNA stability, resulting in undetectable target protein levels. In the ON state of the circuit, shRNA expression was inhibited by the TetR repressor protein and a small-molecule effector, IPTG, was exogenously added to the system to disable the inhibitory function of LacI, thereby allowing target gene expression. The response of the genetic circuit was shown to be tunable with small-molecule concentrations and the utility of the circuit was demonstrated through the small molecule-responsive regulation of various genes affecting cell growth in different mammalian cells.

Recent work has also demonstrated the implementation of RNA devices in genetic circuits that are relevant to downstream biotechnological applications. For example, the application of RNA devices to the non-invasive detection of intracellular metabolite levels was recently demonstrated (Win and Smolke, 2007). Ribozyme-based RNA devices composed of a sensor responsive to a purine alkaloid xanthine were implemented as *in vivo* biosensors for the real-time detection of xanthine accumulation in *S. cerevisiae* by transmitting the metabolite binding event to a change in fluorescent protein levels. Genetic circuits that integrate RNA devices towards the regulation of cellular phenotypes and traits have also been demonstrated. In one example, an RBS-based RNA device was integrated into a genetic circuit in *E. coli* that enabled cells to detect, follow and locate precisely a molecular signal (Topp and Gallivan, 2007). The genetic circuit applied a theophylline-responsive RBS-based device to the regulation of a gene encoding a chemotaxis protein responsible for cell motility. Although this genetic circuit results in a slower response to artificial stimuli than to natural stimuli, which are recognized through binding to receptor proteins, the RNA device permits the reprogramming of cells to recognize new stimuli through genetic regulation of a single protein, providing an alternative to engineering new ligand specificity into the chemoreceptor proteins.

15.11 Conclusion

The thoughtful combination of foundational scientific discovery and engineering theory has resulted in significant advances in the design of functional RNA molecules. The field of RNA engineering has been fueled, influenced and inspired by discoveries and advances in the areas of RNA biology and *in vitro* nucleic acid circuits. The recent application of synthetic biology principles has resulted in more effective design strategies supporting the programming of RNA devices that perform complex information processing, transduction, communication and control functions in living systems. The integration of future scientific and technological advances will lead to enabling technologies supporting the reliable and efficient programming of RNA parts and devices and their integration as functional components into heterogeneous biological networks and systems. The resulting advances in our ability to transmit information to and from living systems, and implement control within cells themselves, will transform how we interact with and program biology.

References

An, C.I., Trinh, V.B. and Yokobayashi, Y. (2006). Artificial control of gene expression in mammalian cells by modulating RNA interference through aptamer-small molecule interaction. *RNA* **12**, 710–716.

Babendure, J.R., Adams, S.R. and Tsien, R.Y. (2003). Aptamers switch on fluorescence of triphenylmethane dyes. *J Am Chem Soc* **125**, 14716–14717.

Basu, S., Gerchman, Y., Collins, C.H., Arnold, F.H. and Weiss, R. (2005). A synthetic multicellular system for programmed pattern formation. *Nature* **434**, 1130–1134.

Bayer, T.S. and Smolke, C.D. (2005). Programmable ligand-controlled riboregulators of eukaryotic gene expression. *Nat Biotechnol* **23**, 337–343.

Beisel, C.L. and Smolke, C.D. (2009). Design principles for riboswitch function. *PLoS Comput Biol* **5**, e1000363.

Beisel, C.L., Bayer, T.S., Hoff, K.G. and Smolke, C.D. (2008). Model-guided design of ligand-regulated RNAi for programmable control of gene expression. *Mol Syst Biol* **4**, 224.

Berezovski, M., Drabovich, A., Krylova, S.M., Musheev, M., Okhonin, V., Petrov, A. and Krylov, S.N. (2005). Nonequilibrium capillary electrophoresis of equilibrium mixtures: a universal tool for development of aptamers. *J Am Chem Soc* **127**, 3165–3171.

Buskirk, A.R., Kehayova, P.D., Landrigan, A. and Liu, D.R. (2003). *In vivo* evolution of an RNA-based transcriptional activator. *Chem Biol* **10**, 533–540.

Buskirk, A.R., Landrigan, A. and Liu, D.R. (2004). Engineering a ligand-dependent RNA transcriptional activator. *Chem Biol* **11**, 1157–1163.

Cassiday, L.A. and Maher, L.J. III (2001). *In vivo* recognition of an RNA aptamer by its transcription factor target. *Biochemistry* **40**, 2433–2438.

Chaloin, L., Lehmann, M.J., Sczakiel, G. and Restle, T. (2002). Endogenous expression of a high-affinity pseudoknot RNA aptamer suppresses replication of HIV-1. *Nucleic Acids Res* **30**, 4001–4008.

Cheah, M.T., Wachter, A., Sudarsan, N. and Breaker, R.R. (2007). Control of alternative RNA splicing and gene expression by eukaryotic riboswitches. *Nature* **447**, 497–500.

Conaty, J., Hendry, P. and Lockett, T. (1999). Selected classes of minimised hammerhead ribozyme have very high cleavage rates at low Mg^{2+} concentration. *Nucleic Acids Res* **27**, 2400–2407.

Cox, J.C., Rajendran, M., Riedel, T., Davidson, E.A., Sooter, L.J., Bayer, T.S., Schmitz-Brown, M. and Ellington, A.D. (2002). Automated acquisition of aptamer sequences. *Comb Chem High Throughput Screen* **5**, 289–299.

Das, R. and Baker, D. (2008). Macromolecular modeling with rosetta. *Annu Rev Biochem* **77**, 363–382.

Deans, T.L., Cantor, C.R. and Collins, J.J. (2007). A tunable genetic switch based on RNAi and repressor proteins for regulating gene expression in mammalian cells. *Cell* **130**, 363–372.

Desai, S.K. and Gallivan, J.P. (2004). Genetic screens and selections for small molecules based on a synthetic riboswitch that activates protein translation. *J Am Chem Soc* **126**, 13247–13254.

Drabovich, A., Berezovski, M. and Krylov, S.N. (2005). Selection of smart aptamers by equilibrium capillary electrophoresis of equilibrium mixtures (ECEEM). *J Am Chem Soc* **127**, 11224–11225.

Ellington, A.D. and Szostak, J.W. (1990). *In vitro* selection of RNA molecules that bind specific ligands. *Nature* **346**, 818–822.

Endy, D. (2005). Foundations for engineering biology. *Nature* **438**, 449–453.

Famulok, M., Blind, M. and Mayer, G. (2001). Intramers as promising new tools in functional proteomics. *Chem Biol* **8**, 931–939.

Fowler, C.C., Brown, E.D. and Li, Y. (2008). A FACS-based approach to engineering artificial riboswitches. *ChemBioChem* **9**, 1906–1911.

Grate, D. and Wilson, C. (1999). Laser-mediated, site-specific inactivation of RNA transcripts. *Proc Natl Acad Sci USA* **96**, 6131–6136.

Greenleaf, W.J., Frieda, K.L., Foster, D.A., Woodside, M.T. and Block, S.M. (2008). Direct observation of hierarchical folding in single riboswitch aptamers. *Science* **319**, 630–633.

Harvey, I., Garneau, P. and Pelletier, J. (2002). Inhibition of translation by RNA–small molecule interactions. *RNA* **8**, 452–463.

Hazelrigg, T. (1998). The destinies and destinations of RNAs. *Cell* **95**, 451–460.

Hermann, T. and Patel, D.J. (2000). Adaptive recognition by nucleic acid aptamers. *Science* **287**, 820–825.

Isaacs, F.J., Dwyer, D.J., Ding, C., Pervouchine, D.D., Cantor, C.R. and Collins, J.J. (2004). Engineered riboregulators enable post-transcriptional control of gene expression. *Nat Biotechnol* **22**, 841–847.

Isaacs, F.J., Dwyer, D.J. and Collins, J.J. (2006). RNA synthetic biology. *Nat Biotechnol* **24**, 545–554.

Kawamoto, H., Morita, T., Shimizu, A., Inada, T. and Aiba, H. (2005). Implication of membrane localization of target mRNA in the action of a small RNA: mechanism of post-transcriptional regulation of glucose transporter in *Escherichia coli*. *Genes Dev* **19**, 328–338.

Kim, M.Y. and Jeong, S. (2004). Inhibition of the functions of the nucleocapsid protein of human immunodeficiency virus-1 by an RNA aptamer. *Biochem Biophys Res Commun* **320**, 1181–1186.

Kislauskis, E.H., Li, Z., Singer, R.H. and Taneja, K.L. (1993). Isoform-specific 3′-untranslated sequences sort alpha-cardiac and beta-cytoplasmic actin messenger RNAs to different cytoplasmic compartments. *J Cell Biol* **123**, 165–172.

Kortemme, T. and Baker, D. (2004). Computational design of protein–protein interactions. *Curr Opin Chem Biol* **8**, 91–97.

Lai, E.C. (2003). RNA sensors and riboswitches: self-regulating messages. *Curr Biol* **13**, R285–R291.

Lee, N.S., Bertrand, E. and Rossi, J. (1999). mRNA localization signals can enhance the intracellular effectiveness of hammerhead ribozymes. *RNA* **5**, 1200–1209.

Link, K.H., Guo, L., Ames, T.D., Yen, L., Mulligan, R.C. and Breaker, R.R. (2007). Engineering high-speed allosteric hammerhead ribozymes. *Biol Chem* **388**, 779–786.

Lynch, S.A. and Gallivan, J.P. (2009). A flow cytometry-based screen for synthetic riboswitches. *Nucleic Acids Res* **37**, 184–192.

Lynch, S.A., Desai, S.K., Sajja, H.K. and Gallivan, J.P. (2007). A high-throughput screen for synthetic riboswitches reveals mechanistic insights into their function. *Chem Biol* **14**, 173–184.

Mandal, M. and Breaker, R.R. (2004). Gene regulation by riboswitches. *Nat Rev Mol Cell Biol* **5**, 451–463.

Mandal, M., Lee, M., Barrick, J.E., Weinberg, Z., Emilsson, G.M., Ruzzo, W.L. and Breaker, R.R. (2004). A glycine-dependent riboswitch that uses cooperative binding to control gene expression. *Science* **306**, 275–279.

Martick, M. and Scott, W.G. (2006). Tertiary contacts distant from the active site prime a ribozyme for catalysis. *Cell* **126**, 309–320.

Mathews, D.H. (2006). Revolutions in RNA secondary structure prediction. *J Mol Biol* **359**, 526–532.

Morita, M.T., Tanaka, Y., Kodama, T.S., Kyogoku, Y., Yanagi, H. and Yura, T. (1999). Translational induction of heat shock transcription factor sigma32: evidence for a built-in RNA thermosensor. *Genes Dev* **13**, 655–665.

Narberhaus, F., Waldminghaus, T. and Chowdhury, S. (2006). RNA thermometers. *FEMS Microbiol Rev* **30**, 3–16.

Neupert, J., Karcher, D. and Bock, R. (2008). Design of simple synthetic RNA thermometers for temperature-controlled gene expression in *Escherichia coli*. *Nucleic Acids Res* **36**, e124.

Nishikawa, F., Kakiuchi, N., Funaji, K., Fukuda, K., Sekiya, S. and Nishikawa, S. (2003). Inhibition of HCV NS3 protease by RNA aptamers in cells. *Nucleic Acids Res* **31**, 1935–1943.

Nocker, A., Hausherr, T., Balsiger, S., Krstulovic, N.P., Hennecke, H. and Narberhaus, F. (2001). A mRNA-based thermosensor controls expression of rhizobial heat shock genes. *Nucleic Acids Res* **29**, 4800–4807.

Nomura, Y. and Yokobayashi, Y. (2007). Reengineering a natural riboswitch by dual genetic selection. *J Am Chem Soc* **129**, 13814–13815.

Ogawa, A. and Maeda, M. (2008). An artificial aptazyme-based riboswitch and its cascading system in *E. coli*. *ChemBioChem* **9**, 206–209.

Parisien, M. and Major, F. (2008). The MC-Fold and MC-Sym pipeline infers RNA structure from sequence data. *Nature* **452**, 51–55.

Persson, T., Hartmann, R.K. and Eckstein, F. (2002). Selection of hammerhead ribozyme variants with low Mg^{2+} requirement: importance of stem-loop II. *ChemBioChem* **3**, 1066–1071.

Pfleger, B.F., Pitera, D.J., Smolke, C.D. and Keasling, J.D. (2006). Combinatorial engineering of intergenic regions in operons tunes expression of multiple genes. *Nat Biotechnol* **24**, 1027–1032.

Rinaudo, K., Bleris, L., Maddamsetti, R., Subramanian, S., Weiss, R. and Benenson, Y. (2007). A universal RNAi-based logic evaluator that operates in mammalian cells. *Nat Biotechnol* **25**, 795–801.

Ro, D.K., Paradise, E.M., Ouellet, M., Fisher, K.J., Newman, K.L., Ndungu, J.M., Ho, K.A., Eachus, R.A., Ham, T.S., Kirby, J., *et al.* (2006). Production of the antimalarial drug precursor artemisinic acid in engineered yeast. *Nature* **440**, 940–943.

Sharma, V., Nomura, Y. and Yokobayashi, Y. (2008). Engineering complex riboswitch regulation by dual genetic selection. *J Am Chem Soc* **130**, 16310–16315

Smolke, C.D., Carrier, T.A. and Keasling, J.D. (2000). Coordinated, differential expression of two genes through directed mRNA cleavage and stabilization by secondary structures. *Appl Environ Microbiol* **66**, 5399–5405.

Smolke, C.D., Martin, V.J. and Keasling, J.D. (2001). Controlling the metabolic flux through the carotenoid pathway using directed mRNA processing and stabilization. *Metab Eng* **3**, 313–321.

Soukup, G.A. and Breaker, R.R. (1999). Engineering precision RNA molecular switches. *Proc Natl Acad Sci USA* **96**, 3584–3589.

Sudarsan, N., Hammond, M.C., Block, K.F., Welz, R., Barrick, J.E., Roth, A. and Breaker, R.R. (2006). Tandem riboswitch architectures exhibit complex gene control functions. *Science* **314**, 300–304.

Suess, B. and Weigand, J.E. (2008). Engineered riboswitches – Overview, problems and trends. *RNA Biol* **5**, 24–29.

Suess, B., Hanson, S., Berens, C., Fink, B., Schroeder, R. and Hillen, W. (2003). Conditional gene expression by controlling translation with tetracycline-binding aptamers. *Nucleic Acids Res* **31**, 1853–1858.

Suess, B., Fink, B., Berens, C., Stentz, R. and Hillen, W. (2004). A theophylline responsive riboswitch based on helix slipping controls gene expression *in vivo*. *Nucleic Acids Res* **32**, 1610–1614.

Thompson, K.M., Syrett, H.A., Knudsen, S.M. and Ellington, A.D. (2002). Group I aptazymes as genetic regulatory switches. *BMC Biotechnol* **2**, 21.

Topp, S. and Gallivan, J.P. (2007). Guiding bacteria with small molecules and RNA. *J Am Chem Soc* **129**, 6807–6811.

Topp, S. and Gallivan, J.P. (2008a). Random walks to synthetic riboswitches – a high-throughput selection based on cell motility. *ChemBioChem* **9**, 210–213.

Topp, S. and Gallivan, J.P. (2008b). Riboswitches in unexpected places – a synthetic riboswitch in a protein coding region. *RNA* **14**, 2498–2503.

Tuerk, C. and Gold, L. (1990). Systematic evolution of ligands by exponential enrichment: RNA ligands to bacteriophage T4 DNA polymerase. *Science* **249**, 505–510.

Tuleuova, N., An, C.I., Ramanculov, E., Revzin, A. and Yokobayashi, Y. (2008). Modulating endogenous gene expression of mammalian cells via RNA-small molecule interaction. *Biochem Biophys Res Commun* **376**, 169–173.

Vaish, N.K., Heaton, P.A. and Eckstein, F. (1997). Isolation of hammerhead ribozymes with altered core sequences by *in vitro* selection. *Biochemistry* **36**, 6495–6501.

Waldminghaus, T., Kortmann, J., Gesing, S. and Narberhaus, F. (2008). Generation of synthetic RNA-based thermosensors. *Biol Chem* **389**, 1319–1326.

Weigand, J.E. and Suess, B. (2007). Tetracycline aptamer-controlled regulation of pre-mRNA splicing in yeast. *Nucleic Acids Res* **35**, 4179–4185.

Werstuck, G. and Green, M.R. (1998). Controlling gene expression in living cells through small molecule–RNA interactions. *Science* **282**, 296–298.

Wieland, M. and Hartig, J.S. (2008). Improved aptazyme design and *in vivo* screening enable riboswitching in bacteria. *Angew Chem Int Ed* **47**, 2604–2607.

Win, M.N. and Smolke, C.D. (2007). From the cover: a modular and extensible RNA-based gene-regulatory platform for engineering cellular function. *Proc Natl Acad Sci USA* **104**, 14283–14288.

Win, M.N. and Smolke, C.D. (2008). Higher-order cellular information processing with synthetic RNA devices. *Science* **322**, 456–460.

Win, M.N., Liang, J.C. and Smolke, C.D. (2009). Frameworks for programming biological function through RNA parts and devices. *Chem Biol* **16**, 298–310.

Winkler, W., Nahvi, A. and Breaker, R.R. (2002). Thiamine derivatives bind messenger RNAs directly to regulate bacterial gene expression. *Nature* **419**, 952–956.

Winkler, W.C., Nahvi, A., Sudarsan, N., Barrick, J.E. and Breaker, R.R. (2003). An mRNA structure that controls gene expression by binding *S*-adenosylmethionine. *Nat Struct Biol* **10**, 701–707.

Winkler, W.C., Nahvi, A., Roth, A., Collins, J.A. and Breaker, R.R. (2004). Control of gene expression by a natural metabolite-responsive ribozyme. *Nature* **428**, 281–286.

Yamamoto, R., Katahira, M., Nishikawa, S., Baba, T., Taira, K. and Kumar, P.K. (2000). A novel RNA motif that binds efficiently and specifically to the Ttat protein of HIV and inhibits the *trans*-activation by Tat of transcription *in vitro* and *in vivo*. *Genes Cells* **5**, 371–388.

Yoshimatsu, T. and Nagawa, F. (1989). Control of gene expression by artificial introns in *Saccharomyces cerevisiae*. *Science* **244**, 1346–1348.

16

RNA as a Catalyst: The Diels–Alderase Ribozyme

Andres Jäschke

16.1 Introduction

One of the most fascinating questions at the interface of chemistry and biology is how enzymes achieve the catalysis of chemical reactions. These biocatalysts bring about tremendous rate accelerations under mild conditions with often extreme chemo-, regio- and stereoselectivities. For decades, generations of biochemists, enzymologists and structural biologists have been attracted to this phenomenon and, consequently, a fairly thorough understanding has been achieved regarding the catalytic mechanisms of protein enzymes.

Since the early 1980s, it has become clear that the exclusive assignment of enzymatic activity to the chemical substance class of proteins in incorrect, as ribonucleic acids were discovered that possess such activities [1,2]. Most of these naturally occurring 'ribozymes' catalyze (at least according to our present-day knowledge) the hydrolysis and formation of phosphodiester bonds and, for some of these systems, high-resolution crystal structures and mechanistic investigations provide a basis for an understanding of how these ribozymes work [3]. Combinatorial chemistry, on the other hand, has allowed the isolation (or generation) of RNA and also DNA catalysts with activities not known to exist in biology. It turns out that – although equipped with a much less varied arsenal of chemically different building blocks – nucleic acids can form tertiary structures with binding pockets and catalytic centers very much like proteins. RNA molecules containing the first artificial binding sites against human-chosen targets were presented in 1990, when, starting from synthetic combinatorial libraries, RNA molecules, so-called aptamers, were selected that featured binding properties similar to those of antibodies [4,5]. These were obtained by a new combinatorial technique termed SELEX (Systematic Evolution of Ligands by EXponential enrichment), and this technique eventually even produced RNAs with catalytic activities, neither of which were in any way related to the hitherto presumed biological functions of nucleic acids. Such SELEX experiments led to nucleic acids that

The Chemical Biology of Nucleic Acids Edited by Günter Mayer

catalyze a broad range of chemical transformations, ranging from cleavage of amide [6] and carboxylic ester bonds [7] and amide [8], C–C and C–S bond-forming reactions [9–11] to the catalysis of redox reactions [12,13] (for a recent review, see [14]).

The history of ribozymes, in the true sense of the word, is connected with the so-called RNA world [15,16], a hypothetic prebiotic stage in the evolution of life, where the majority of catalytic functions were effected by RNA and proteins did not yet exist. The idea of an RNA world alone implies that RNA should be practically omnipotent in the diversity of its catalytic potential, despite having been outperformed by proteins later.

The formation of C–C bonds is not only of great importance in the context of the RNA world hypothesis, but is also a quintessential task in biochemistry and organic synthesis. Consequently, several laboratories have attempted the development and characterization of nucleic acid catalysts for this purpose. The first successful example was a modified RNA (substituted with pyridine 'side-chains') that accelerated a Diels–Alder cycloaddition between tethered aliphatic dienes and maleimide dienophiles by about 700-fold [11], although not with multiple turnovers. Soon afterwards, the isolation of unmodified ribozymes for Diels–Alder cycloadditions of aromatic dienes was achieved by our group [9]. Other examples of RNA-catalyzed C–C bond formation include an aldol reaction between a tethered levulinic acid and a biotinylated benzaldehyde derivative [17], a Claisen condensation [18] and another Diels–Alder reaction catalyzed by native tRNA under high pressure [19], although these latter systems still await a detailed analytical characterization. Recently, even DNA was found to catalyze C–C bond formation by Diels–Alder reaction [20].

The Diels–Alder reaction is one of the most important C–C bond-forming processes in preparative organic chemistry. This transformation belongs to the class of pericyclic reactions and is a $[4\pi + 2\pi]$ cycloaddition, usually between an electron-rich 1,3-diene and an electron-deficient dienophile. The reaction creates two C–C bonds and up to four new stereocenters. In the course of the reaction, a six-membered carbocycle is formed. Because six π-electrons are involved, it has been viewed as proceeding through an aromatic transition state, which is energetically more favored than a nonconcerted reaction path.

For one of the Diels–Alderase ribozymes selected by Seelig and Jäschke [9], the three-dimensional structure could be determined by X-ray crystallography, both in the unbound state and in the complex with the reaction product of the Diels–Alder cycloaddition [21]. Thus, this ribozyme is the only RNA catalyst for small-molecule chemistry with a known spatial structure and, from extensive studies, it is arguably the best-characterized artificial ribozyme known to date. These data provide for the first time an insight into how a small RNA can accelerate reactions different from phosphodiester chemistry and what structural prerequisites are necessary.

This chapter discusses the current knowledge on these Diels–Alderase ribozymes, with special emphasis on crystallographic and complementary mechanistic investigations. Rather than following a chorological order, the chapter – after a brief discussion of the discovery and overall catalytic properties of the ribozymes – integrates data from over 10 different studies and systematically analyzes the structure and function of these ribozymes.

16.2 *In vitro* selection of Diels–Alderase ribozymes

The Diels–Alderase ribozymes were selected from synthetic combinatorial RNA libraries (pools), which typically contain $10^{14} - 10^{15}$ different sequences. These enormous complexities can only be handled since nucleic acid molecules are genetically encoded, that is, they carry the information for their own replication. This property allows for an iterative deconvolution of the libraries over several rounds. While *in vitro* selection of aptamers that bind immobilized Diels–Alder transition state analogs did not produce Diels–Alderase ribozymes [22], active catalysts were obtained by direct selection [9,11]. In both cases, tethered

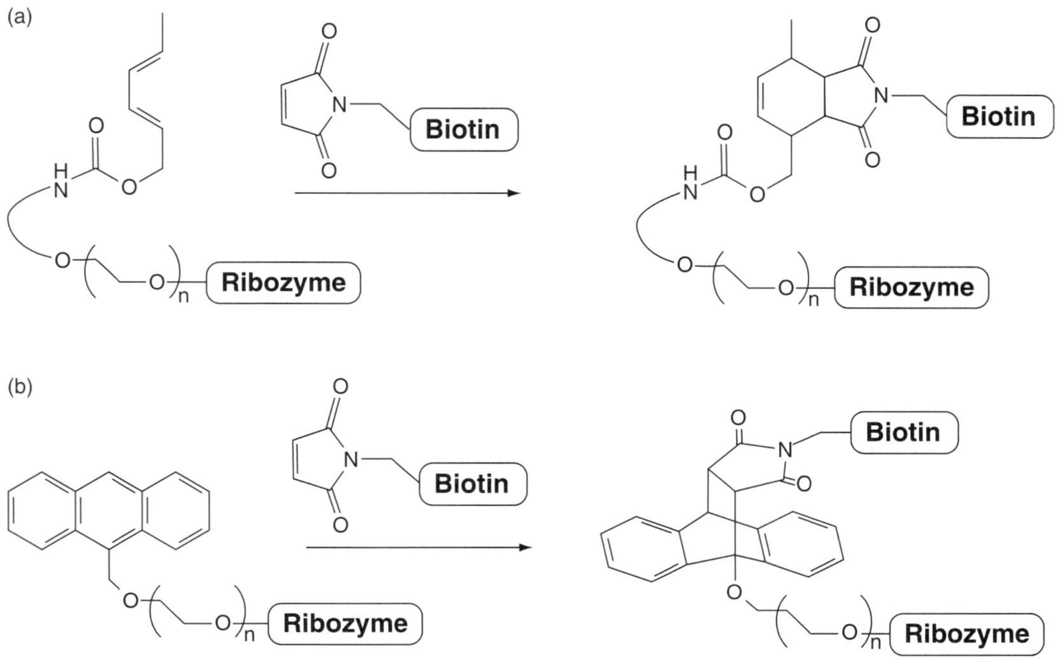

Figure 16.1 *RNA-catalyzed Diels–Alder reactions utilizing (a) tethered aliphatic [11] or (b) aromatic dienes [9]*

reactants were used (Figure 16.1). The diene (either an aliphatic hexadiene or an aromatic anthracene derivative) was covalently tethered [via inert poly(ethylene glycol) chains] to the members of the RNA library by chemo-enzymatic methods and the resulting library of RNA–tether–diene conjugates was then allowed to react with a biotinylated dienophile. RNA molecules that accelerated the reaction of the tethered diene thereby became tagged with the biotinyl residue and could be easily isolated by affinity chromatography and selectively amplified by reverse transcription and polymerase chain reaction (PCR). The resulting enriched DNA library was used as input for the next round of transcription, conjugation, reaction and selection and the cycle was repeated until active species dominated the library.

In the Seelig–Jäschke selection, 10 rounds of selection and amplification were carried out, starting with an RNA library of ~2×10^{14} sequences of 160 nucleotide length [9]. After tethering to anthracene, this library was initially allowed to react for 1 h with a 50-fold excess of biotinylated maleimide in buffered solution containing various metal ions and potential cofactors. Biotinylated RNA was recovered using immobilized streptavidin and the incorporation of ^{32}P-labeled nucleotides during transcription allowed for easy quantification. From the starting library and the enriched libraries from the first five iterations, about 0.1% of all conjugates were found to react with biotin maleimide (Figure 16.2), which is what could be expected from the uncatalyzed Diels–Alder reaction between biotin maleimide and various anthracene derivatives. In round 6, a significant increase in the fraction of biotinylated molecules occurred, indicating successful enrichment of catalytic species. To favor survival of only the most active catalysts, the selection pressure was gradually increased by reducing the reaction time from 1 h to 1 min and the maleimide concentration by a factor of 10. Furthermore, the diversity of the enriched pools was increased by random incorporation of point mutations by error-prone PCR. After 10 iterations, the enriched library showed a ~6500-fold rate acceleration, compared with the starting library (Figure 16.2).

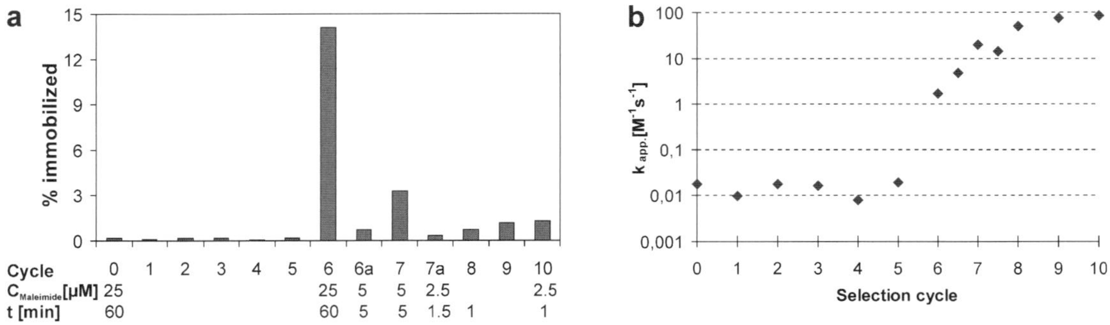

Figure 16.2 *Selection of Diels–Alderase ribozymes. (a) Enrichment of biotinylated RNA at each round of selection. Concentration of maleimide and reaction times are given below the abscissa. (b) Activity of the RNA library versus round of the selection [9]*

In order to isolate and characterize the individual members of this enriched pool, after reverse transcription and PCR amplification it was cloned into a plasmid (vector) that furthermore carried a gene for antibiotic resistance. *Escherichia coli* cells were then transformed with this vector and allowed to grow on agar plates in the presence of the antibiotic. Only plasmid-carrying bacteria could survive and formed colonies that were then individually picked, sequenced and the inserts analyzed for catalytic activity [9].

16.3 Sequence analysis – identification of a minimal motif

Fifty-five clones were submitted to sequencing, yielding 42 different sequences, 35 of which gave a significant rate acceleration in the same assay format that was used in the selection (anthracene covalently tethered, maleimide biotinylated). Sequence analysis by clustering algorithms revealed that these could be assigned to 16 independent sequence families. Although the sequences differ completely between the families, the members of each family differ from each other only by point mutations, which are likely due to PCR errors. This suggests that our starting library contained at least 16 different sequences (or ~1 out of 10^{13} randomly chosen sequences) capable of accelerating the Diels–Alder reaction. Apparently, catalysis of this reaction is only slightly more difficult for RNA to achieve than binding small molecule targets (such as ATP), for which a frequency of ~1 in 10^{10} – 10^{11} was reported [23].

Pairwise comparison of the consensus sequences of the individual families identified a common motif shared by 13 of the 16 families (Figure 16.3): the central element is an asymmetric internal bubble comprised of the two consensus sequences UGCCA and AAUACU, framed by two helices (helix II and III, light and dark gray, respectively). From helix II, another helix (helix I) continues in which one strand is formed by nucleotides from the conserved 5′-primer binding site. In all sequences containing the motif, the 5′-terminal GGAG is left (formally) unpaired. The first G is the site where the tethered anthracene is attached, suggesting an important structural role of this tetranucleotide. The respective ends of helix I and III are connected by loops L1 and L2 of varying size and sequence.

A small 49-mer RNA containing this motif only (with the ends of helix I and III closed by stabilizing tetranucleotide loops) accelerated the reaction between the tethered anthracene and biotin maleimide about 18 500-fold, confirming the findings of sequence analysis [9].

It should be noted that three of the selected catalytically active sequences (the ‘orphans’) did not contain the motif shown in Figure 16.3, and even re-analysis in hindsight, allowing for multiple mutations and forcing certain folding patterns, could not fit them to this motif. These orphans also bear no homology with each

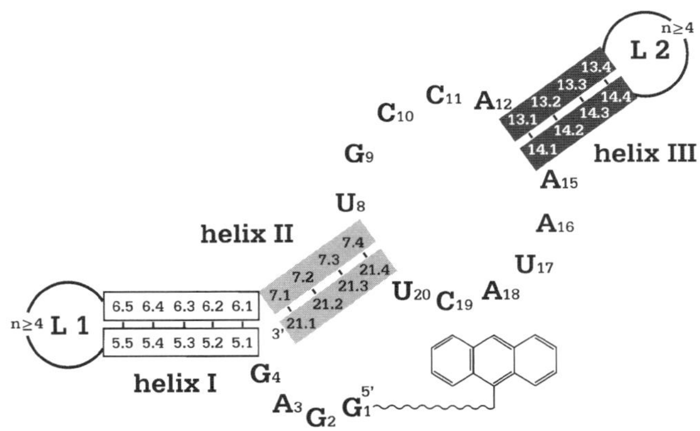

Figure 16.3 Secondary structure motif responsible for catalysis and numbering scheme [9]

other, indicating that there are at least four structurally different solutions for the catalysis of this Diels–Alder reaction. All characterization studies that follow have been performed on the minimal 49-mer derived from the conserved motif or on variants of this minimal ribozyme.

16.4 Overall catalytic properties

The ribozymes were selected to catalyze the Diels–Alder reaction between a covalently tethered anthracene and a biotinylated maleimide. Thus, the RNA–tether–anthracene conjugate is acting as enzyme and substrate at the same time (so-called *in cis* or self-modifying reaction), and this format does not allow for multiple turnovers. The minimal motif could be rationally converted into bi- and tripartite, highly active variants by formal strand breakage in one or both of the two loop regions and re-assembly of ribozymes from the two or three corresponding RNA fragments [9,24]. This format turned out to be useful for mutation studies. Finally, the selected Diels–Alderase ribozymes and the minimal 49-mer were shown to accelerate C–C bond formation in a true bimolecular fashion (*in trans* or true catalytic reaction, Figure 16.4) [25]. Substrate molecules as small as 9-hydroxymethylanthracene and *N*-ethylmaleimide are specifically recognized by the ribozymes, followed by conversion to the respective Diels–Alder products and product dissociation from the catalyst. The 49-mer ribozyme performs the reaction with fast multiple turnovers and a k_{cat} of 21 min^{-1} was measured. Saturation-type kinetics with respect to both reactants ($K_{M,Diene} = 370\ \mu M$, $K_{M,Dienophile} = 8$ mM) and also product inhibition were observed. The 'effective molarity' as a measure of the entropic gain of the ribozyme-catalyzed reaction was determined from the ratio k_{cat}/k_{uncat} to be 6.6 M, giving a lower estimate for the rate acceleration of 1100-fold [25]. These parameters compare favorably with those of catalytic Diels–Alderase antibodies, cyclodextrins and synthetic capsules [26]. From the measured parameters, these authors calculated an extra transition-state stabilization (relative to the ground state) of –1.1 kcal mol^{-1}, which is an order of magnitude lower than typical values for natural (protein) enzymes.

Another characteristic feature of enzymatic catalysis was demonstrated for the Diels–Alderase ribozymes, namely enantioselective bond formation. Whereas the uncatalyzed reaction produces racemic product mixtures, the ribozyme-catalyzed conversion shows an enantioselectivity of over 95% *ee* (enantiomeric excess). The enantioselectivity was shown to be primarily dependent on the size of the substituent at the anthracene ring system and the enantiomer of this ribozyme (chemically synthesized from the unnatural L-ribonucleotides) showed the opposite stereoselectivity (Figure 16.4) [25].

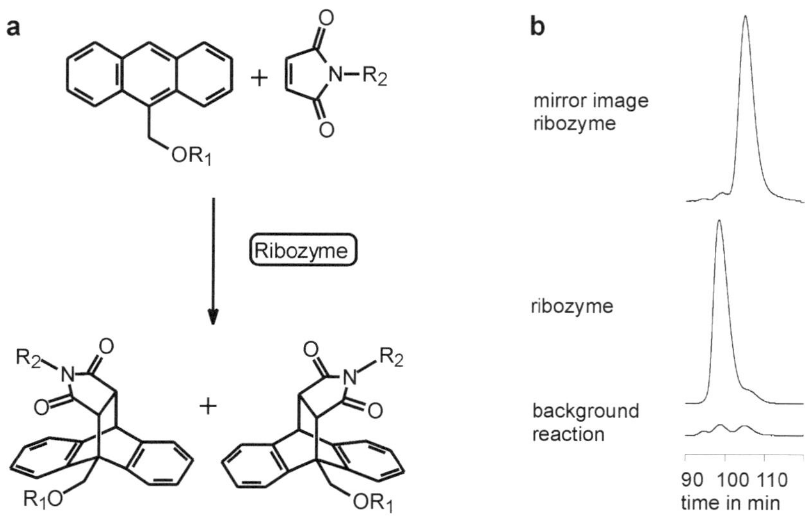

Figure 16.4 *Ribozyme catalysis* in trans *of a Diels–Alder reaction. (a) Chemistry of the reaction. (b) Chiral HPLC analysis of products obtained in reactions either without catalyst (background reaction), with the catalytic 49-mer minimum ribozyme or its synthetic mirror-image* L*-RNA version [$R_1 = (C_2H_4O)_6H$; $R_2 = (CH_2)_5COOCH_3$] [25]*

16.5 Overall structure of the ribozyme

After screening of several dozen constructs under hundreds of different conditions, the three-dimensional structure of the Diels–Alderase ribozyme could be solved both in the form of the apo-enzyme and in a complex with the Diels–Alder cycloaddition product [21]. The crystallization efforts succeeded with a bipartite Diels–Alderase ribozyme consisting of 38-mer and 11-mer RNA strands, with the shorter strand either unmodified or containing hexaethylene glycol (HEG)-linked cycloaddition product. The structures obtained provided details of the catalytic pocket, including bound product, surrounding RNA and coordinated cations.

The molecule adopts an overall topology that resembles the Greek letter lambda (λ) (Figure 16.5), featuring a special kind of three-way junction. Stems II and III are co-linearly stacked [best seen in panel (b)], bridged by a zipped-up asymmetric bubble, whereas stem I is extended by two base pairs and branches off at the bubble at a ~60° angle. In the center of the molecule there is extra electron density that is not related to nucleotides and can be traced back to the *S,S* stereoisomer of the Diels–Alder product, tightly bound into a pocket as a single, well-defined conformer (Figure 16.5c).

The 5′-GGAG tetranucleotide plays a critical role in shaping both the RNA scaffold and the catalytic pocket. G_1 and G_2 form Watson–Crick pairs with C_{11} and C_{10}, whereas A_3 and G_4 pair with U_{20} and C_{19}, thereby generating the nested pseudoknot topology within the RNA scaffold (dotted lines in Figure 16.5a and b). ‘Pseudoknot’ denotes a secondary structure resulting from nucleotides in the loop of an RNA stem–loop structure with nucleotides outside of the stem–loop. The 5′-terminal GGAG serves to clamp together the opposite sites of the asymmetric internal bubble and the direct connection of four helices without interjecting spacers can impose severe strain on the system. Such an architecture appears to be an efficient way to achieve a fairly dense and stable packing of helical elements. This pseudoknot formation was in agreement with earlier findings based on mutational and chemical probing data [24]. The asymmetric internal bubble is

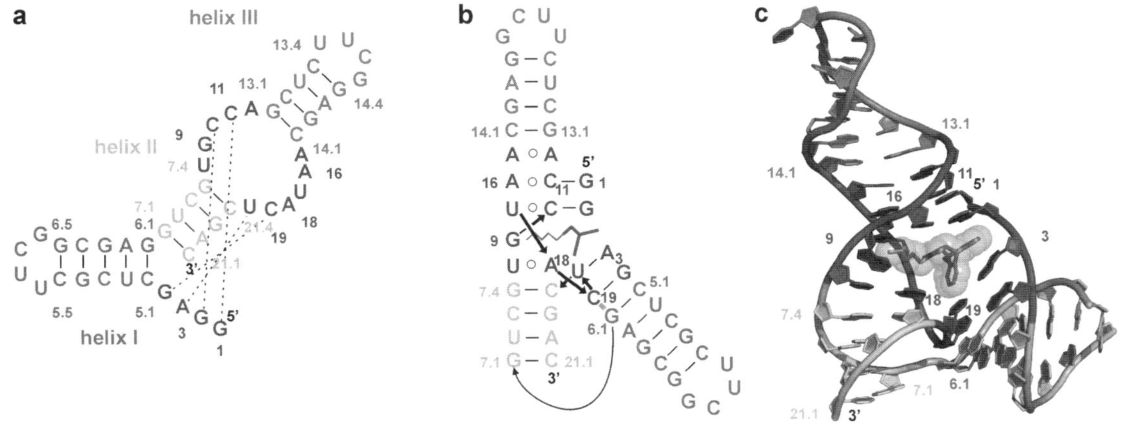

Figure 16.5 Crystal structure of the Diels–Alderase ribozyme. (a) Ribozyme secondary structure with helices I, II and III, the asymmetric bubble and the conserved 5′ end. Dotted lines represent pseudoknot Watson–Crick-type interactions. (b) Tertiary fold and (c) three-dimensional topology in the crystal structure of the ribozyme–product complex. One enantiomer of the Diels–Alder product is bound into the catalytic pocket of the ribozyme [21]. See color plate section

zipped up through various non-standard interactions between residues across its opposing strands. These involve a Watson–Crick-like $G_{12}{\cdot}A_{15}$ pair and a reversed Hoogsteen $U_8{\cdot}A_{18}$ pair, thereby extending both stems III and II respectively, by one step. Both pseudoknot minihelices are involved in higher order interactions: The $G_1G_2{:}C_{11}C_{10}$ minihelix attaches to the lower part of the bubble by formation of a $U_{17}{\cdot}(G_2{\cdot}C_{10})$ base triple and an $A_{16}{\cdot\cdot}(G_1{\cdot}C_{11})$ three-base platform (in a triple, the partners are close enough for H-bonding interaction, whereas in a platform they are not), whereas the $A_3G_4{:}U_{20}C_{19}$ minihelix stacks on top of helix I, with the two pyrimidine nucleotides being involved in an unusual sharp turn that spans the $A_{18}C_{19}U_{20}C_{21.4}$ segment. The G_9 residue is the only nucleotide in the center not formally involved in pairing, triple or platform interactions and apparently forms three hydrogen bonds along its Watson–Crick edge with the phosphate oxygens of A_{18} and the carbonyl group of the maleimide ring of the bound product, while its 2′-OH is hydrogen bonded with U_{17} [21].

16.6 Architecture of the catalytic pocket

The catalytic pocket is wedge-shaped and bracketed by the Watson–Crick $A_3{\cdot}U_{20}$ pair (bottom), the $U_{17}{\cdot}(G_2{\cdot}C_{10})$ triple (upper right) and the $U_8{\cdot}A_{18}$ reverse Hoogsteen pair (Figure 16.6, left) and is lined by an intricate network of hydrogen bonds (dotted lines). The HEG-linked bridgehead position of the bound Diels–Alder product is directed inwards into the RNA scaffold. Base G_9 is positioned opposite the opened edge of the pocket and shares the space with the side-chain projecting from the five-membered maleimide ring. The maleimide ring is stacked over C_{10} and apparently oriented through a pair of hydrogen bonds involving one of its two carbonyl oxygens, while its side-chain runs inside a hydrophobic canyon created by bases C_{10} (top), A_{18} (bottom) and G_9 (back). One of the six-membered aromatic rings of the product is sandwiched between purines G_2 and A_3, whereas the other is wedged between the base and sugar components of A_{18} and U_{20} (Figure 16.5). The catalytic pocket, which is primarily formed by base edges with a minimal

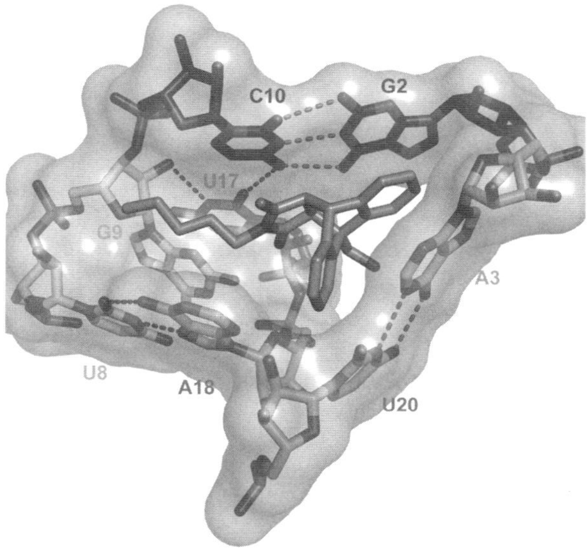

Figure 16.6 Surface representation of the catalytic pocket with stick representations of the nucleotides. See color plate section

contribution from the sugar–phosphate backbone, is accessible to the product from the front while a narrow orifice is visible in the back. It should be noted that the catalytic pocket contains not only a cavity to accommodate the cycloaddition product but also a direction-specific surface channel to accommodate the maleimide side-chain, which provides a convenient explanation for the observed stereoselectivity. There are no Mg^{2+} cations positioned within the immediate vicinity of the pocket that could participate in the catalytic process.

The structure of the free Diels–Alderase ribozyme was found to be virtually identical with that of the product complex, providing strong support for the concept of a preformed catalytic pocket, established by chemical probing [24]. The anthracene–RNA conjugate could also be crystallized; however, the anthracene module and its linker were found to be disordered in the crystal. The RNA mapped well with the other crystal structures.

The relevance of the crystal structure to catalysis in solution could be supported by photoaffinity crosslinking [27]. A Diels–Alder cycloaddition product was prepared from photoreactive *p*-azidobenzylmaleimide and was allowed to bind into the catalytic pocket. After irradiation, specific crosslinks were formed with residues in the direct vicinity. We identified two specific nucleotides, C10 and U17, that are crosslinked by the product analog and, according to the crystal structure, are positioned next to the maleimide side-chain. Which of these two nucleotides is predominantly crosslinked depends on the Mg^{2+} ion concentration, thereby revealing conformation-dependent interactions between ribozyme and probe.

16.7 Interactions between the ribozyme and its substrates and products

The crystallographic information about the architecture of the catalytic pocket could be compared with an earlier study that elucidated the interactions of the ribozyme with its substrates and products by chemical substitution analysis using 44 different, systematically varied analogs [28]. In that study, RNA–diene

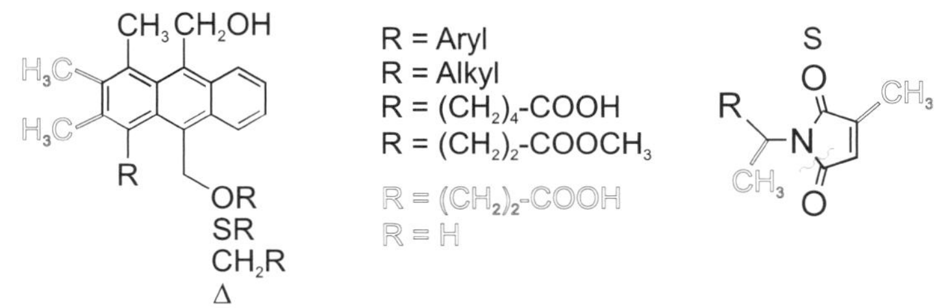

Figure 16.7 *Summarized structural requirements of dienes and dienophiles for acceptance by the Diels–Alderase ribozyme. Substitutions shown in black are tolerated, whereas those in empty letters are deleterious; Δ removal of whole substituent. For positions 1 and 8 in anthracene, large substituents [e.g. $CH_2(OCH_2CH_2)_6OH$] are tolerated [28].*

interaction was found to be governed by stacking interactions, whereas hydrogen bonding and metal ion coordination appeared to be less important. The diene has to be an anthracene derivative and substituents at defined positions are permitted, thereby shedding light on the geometry of the binding site (Figure 16.7). The crystal structure shows that there in no space to accommodate substituents at positions 2, 3, 6 and 7 (shown in red), as these would clash into the walls of the catalytic pocket. Some space for small substituents is found at positions 4, 5 and 10, pointing towards the back of the pocket. Large substituents are tolerated at positions 1, 8 and 9, pointing out of the pocket. Interestingly, the poly(ethylene glycol) tether used in the selection does not make any contribution to binding and can be removed or shifted to another position (e.g. position 1) without penalty.

The dienophile must be a five-membered maleimidyl ring with an unsubstituted reactive double bond. Substituents would again collide with the walls of the catalytic pocket. A hydrophobic side-chain (alkyl or aryl) makes a major contribution to RNA binding. Branching is not allowed at the alpha position (next to the maleimide ring), due to steric clash with the walls. The ribozyme distinguishes between different enantiomers of chiral substrates and accelerates cycloadditions with both enantio- and diastereoselectivity. The stereochemistry of the reaction is controlled by RNA–diene interactions. The RNA interacts strongly and stereoselectively with the cycloaddition products, requiring several structural features to be present. Strong and stereoselective product inhibition is observed [28].

The crystal structure shows three H-bonds between the reaction product and RNA. One that involves an ether oxygen of the poly(ethylene glycol) tether is irrelevant for catalysis as the tether can be removed entirely without effect on catalysis. The other two involve the maleimide's carbonyl oxygen and could be mechanistically meaningful. The importance of these bonds is, however, not yet completely established and atomic mutagenesis is currently used to probe these interactions.

16.8 Molecular determinants of stereoselectivity

As stated above, the size of the diene's substituent was identified as the major determinant of stereo-selectivity [28] and the diene was thought to enter the catalytic pocket with the sterically less demanding edge first, then reacting with the maleimide bound in one fixed orientation, while the opposite orientation of the anthracene was found to be disfavored. Rather than a typical enzyme pocket, the X-ray crystal structure of the ribozyme–Diels–Alder product complex surprisingly featured a catalytic center with access

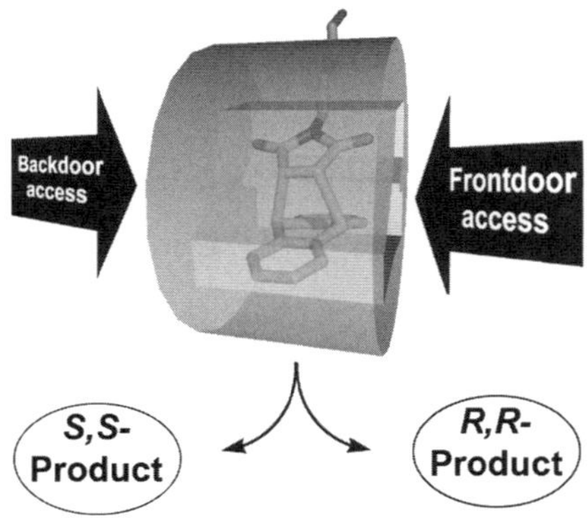

Figure 16.8 The active site of a Diels–Alderase ribozyme features two different 'doors' through which the diene may enter. Depending on which door is used, either of the two Diels–Alder product enantiomers is synthesized selectively. Enantioselectivity can be modulated by controlling access to the ribozyme's active site [29]

from both the front and back sides, with two openings of different size. In fact, the reaction product was bound inside the pocket with the sterically more demanding side first, contrary to what had been expected (Figure 16.6).[21] The only difference between these and the previous experiments was that the substrate specificity was investigated using free substrates (true catalysis, Figure 16.4a), whereas for co-crystallization, the Diels–Alder product was covalently linked to the ribozyme via an 18-atom flexible tether (tethered format, Figure 16.1b) which was attached to the RNA close to the back door. This raised the question of whether restriction of the substrate's translational and rotational mobility by tethering could force it to enter the catalytic pocket through the narrower, disfavored back door and thus to influence the stereochemistry of the reaction. The systematic investigation of this phenomenon provided direct chemical evidence that in the true catalytic reaction and the tethered reaction, the substrates use different approaches to the catalytic pocket, bind in different orientations and are converted to yield the opposite product enantiomers (Figure 16.8) [29].

16.9 Sequence conservation and tolerance for mutations

The intricate three-dimensional architecture suggests that certain nucleotides will be absolutely essential, while others can be easily replaced, which is in agreement with mutational data [24]. All proposed helices could be unambiguously supported by compensatory double mutations. The lower half of the asymmetric internal loop, the hexanucleotide $A_{15}A_{16}U_{17}A_{18}C_{19}U_{20}$, shows a significantly higher level of sequence variability than the upper pentanucleotide $U_8G_9C_{10}C_{11}A_{12}$ (Figure 16.9) and is more susceptible to both enzymatic and lead-induced hydrolysis. Three nucleotides of the internal loops were found to be absolutely invariant (at least at the single point mutation level), namely C_{10}, C_{11} (in the pentanucleotide) and U_{20} (in the hexanucleotide). The four 5′-terminal nucleotides $G_1G_2A_3G_4$ were also found to be highly conserved and the probing experiments indicated protection from modification under native conditions. Compensatory double mutations demonstrated that A_3G_4 pair with $U_{20}C_{19}$, whereas the probing data were suggestive of an interaction between G_1G_2 and $C_{11}C_{10}$. The probing data indicated a preformed tertiary structure (including a preformed catalytic pocket) which shows no major changes on

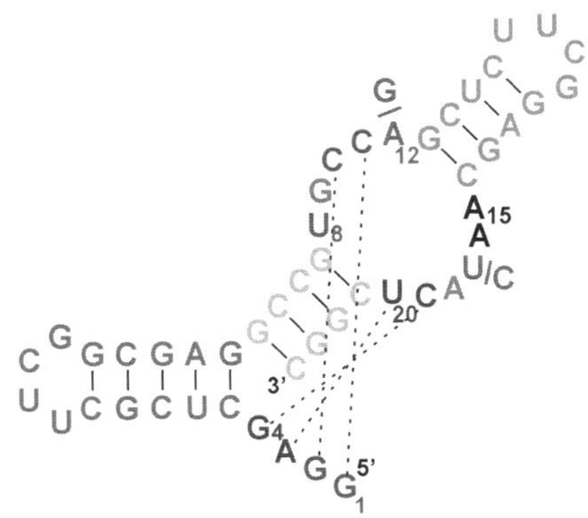

Figure 16.9 Mutation analysis of the Diels–Alderase ribozyme [24]. Green and blue stem–loop structures are completely variable in size and sequence, whereas the yellow helix is variable in sequence, but conserved in size. Nucleotides in red are invariant. Orange: highly conserved (over 90% activity reduction on mutation). Purple: pairwise complementary substitutions allowed. Gray: strong preference for two nucleotides. Black: variable (less than 50% activity reduction on mutation). Black dotted lines: Watson–Crick pseudoknot base pairs. See color plate section

substrate or product binding. Unlike many previously described aptamers, the Diels–Alderase ribozyme forms a stable, prestructured catalytic pocket which can accommodate the substrates and accelerate the reaction.

16.10 Conformational dynamics and the roles of metal ions

The investigation of this ribozyme's kinetics, reaction mechanism and folding by conventional chemical and biochemical methods revealed early complex relationships and suggested the existence of various conformational forms of this RNA molecule. In particular, it became apparent that the complex, mutual influences of substrate binding, Mg^{2+} ion binding, folding and ultimately catalysis cannot be easily deconvoluted. This situation prompted us to initiate detailed studies at the fundamental level of folding to establish a foundation that would allow extension to the catalytic mechanism in later stages. Single-molecule fluorescence resonance energy transfer (FRET) experiments were carried out using dye-labeled ribozyme molecules [30]. The results show that the Diels–Alderase ribozyme possesses three structurally different states that are distinguishable by their FRET efficiency distributions, the unfolded (U), intermediate (I) and folded (F) states. Mg^{2+} titration allowed the measurement of changes in their equilibrium populations and observation of the collapse of the intermediate state. Both effects were described with a thermodynamic model, revealing how the F state is stabilized by successive binding of Mg^{2+} ions. Furthermore, continuous fluctuations were observed between the I and F states on the 100 ms time-scale. The observation of conformational heterogeneity and continuous fluctuations between different states provides insights into the intriguing catalytic function of this ribozyme.

In the crystal structure of this ribozyme, eight Mg^{2+} ions were observed, of which two are suggested to be due to crystal packing. Interestingly, no Mg^{2+} ions were found in the immediate vicinity of the catalytic pocket, indicating that metal ions do not directly participate in catalysis. The interactions of the Diels–Alderase ribozyme with divalent metal ions in solution were furthermore elucidated by electron paramagnetic resonance

(EPR) spectroscopy using paramagnetic manganese instead of magnesium ions [31]. Manganese ion titrations revealed five high-affinity Mn^{2+} binding sites with an upper K_d of 0.6 μM. In order to characterize each binding site individually, EPR-silent Cd^{2+} ions were used to saturate the other binding sites. This cadmium-induced EPR silencing showed that the Mn^{2+} binding sites possess different affinities. In addition, these binding sites could be assigned to three different types, including inner-sphere, outer-sphere and an Mn^{2+} dimer. Based on simulations, the $Mn^{2+} - Mn^{2+}$ distance within the dimer was found to be ~6 Å, which is in good agreement with crystallographic data. The EPR spectroscopic characterization revealed no structural changes upon addition of a Diels–Alder product, supporting the concept of a preorganized catalytic pocket in the Diels–Alder ribozyme and the structural role of these ions.

16.11 Mechanistic considerations

The X-ray crystal structure of the ribozyme–product complex [21], coupled with extensive complementary chemical [28] and biochemical [24] experiments, suggests that the ribozyme should bind the diene and dienophile in a precisely defined steric orientation within a wedge-shaped catalytic pocket, thereby facilitating the reaction by reducing translational and rotational degrees of freedom. The anthracene substrate could be bound by stacking interactions between G2 and the A3–U20 pair. The reaction requires that the maleimide substrate be stacked on top of the anthracene and parallel to it, at a distance of about 3.5 Å. In this position, maleimide can form hydrogen bonds between its carbonyl oxygen and the exocyclic amino group of G9 and the 2′-OH of U17. The *N*-alkyl side-chain can be placed within the hydrophobic canyon (Figure 16.10), which would allow the approach of the maleimide to the bound anthracene only from one direction, thereby providing the stereoselectivity of the reaction. In the transition state of the

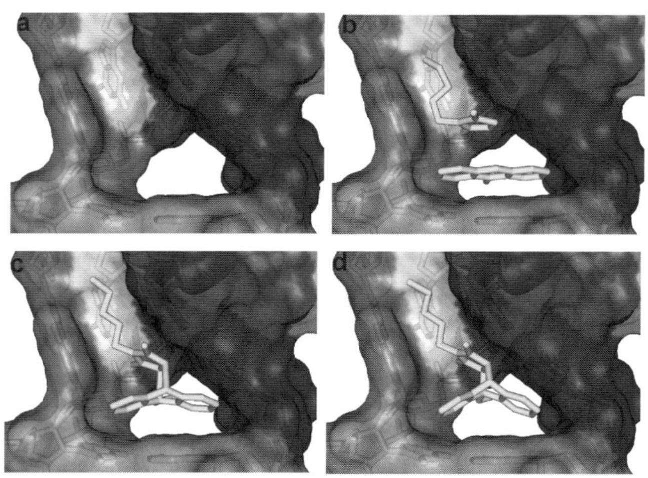

Figure 16.10 *Proposed model for the catalytic mechanism of the Diels–Alderase ribozyme. (a) Empty catalytic pocket. (b) Michaelis complex with both substrates bound. (c) Transition state inside the catalytic pocket. (d) Ribozyme-bound product. Panels (a) and (d) are derived directly from the crystallographic data; panels (b) and (c) were obtained by manually docking the two substrates and the transition state, respectively, into the pocket [21]. See color plate section*

reaction, two new single bonds are partly formed (bond lengths ~2.2 Å) and the anthracene ring system is bent out of planarity to yield an angle of about 154° (Figure 16.10), therefore further reducing the minimal unoccupied space underneath the bridgehead carbons seen in the structure of the ribozyme–product complex. Remarkably, the transition state has near-perfect shape complementarity with the catalytic pocket, a feature of prime importance in antibody catalysis of the Diels–Alder reactions [32,33]. In addition to steric factors, both stacking and hydrogen bonding may contribute to the energetics of the reaction. Stacking of the anthracene with nucleotides A3 and U20 could accelerate the reaction by increasing the diene's electron density, while hydrogen bonding of the maleimide would make it more electron deficient, thereby increasing catalytic reactivity. Hence it appears that ribozyme-based catalysis of Diels–Alder reactions reflects a combination of proximity, shape complementarity and energetic contributions to the catalytic process.

Computational analysis of the ribozyme-catalyzed Diels–Alder reaction confirmed the concerted nature of the cycloaddition and found that the proficiency of the RNA-catalyzed reaction originates from the active site holding the two reactants in reactive conformations, in which the reacting atoms are brought together at van der Waals distances and reactants approach each other at an appropriate angle [34].

16.12 Comparison with protein-catalyzed Diels–Alder reactions

Although there is still no unequivocal proof for the existence and relevance of Diels–Alderase enzymes in Nature, several catalytic antibodies have been generated that accelerate Diels–Alder and retro-Diels–Alder reactions. Several of these antibodies have been crystallized in complexes with either products or transition-state analogs [35–37]. Therefore, an opportunity exists to compare and contrast the structural and mechanistic principles associated with Diels–Alder reactions catalyzed by protein enzymes with their ribozyme counterparts.

Common features observed in these structures are the formation of hydrophobic cores lined predominantly by the side-chains of Tyr, Phe, Val and Trp residues and the stacking of one reactant over an aromatic residue, preferably tryptophan. Remarkably, the retro-Diels–Alderase catalytic antibody 10F11, the only antibody utilizing anthracene as the diene [36], forms a hydrophobic catalytic pocket with an overall shape similar to the Diels–Alder ribozyme pocket (Figure 16.11). The anthracene ring is stacked over a Trp residue, similar to the position of A3 in the ribozyme, with a few direct and water-mediated hydrogen bonds observed between heteroatoms on the bicyclic adduct and the amide backbone.

Macrophomate synthetase, a natural enzyme originally thought to be a Diels–Alderase, also possesses a mainly hydrophobic catalytic pocket and interacts with the substrates by hydrogen bonding. Major differences, however, lie in the utilization of a bound Mg^{2+} cation, which directly coordinates to the carbonyl oxygens of the dienophile and participates in the catalytic mechanism and in the generally lower importance of stacking in enzyme–substrate interactions [38,39]. Recent biochemical evidence, however, questions the use of a concerted Diels–Alder reaction pathway and suggests a two-step Michael–aldol mechanism instead [40].

Another Diels–Alder ribozyme requires both a covalent modification of the RNA and the presence of Cu^{2+} ions for catalytic activity, and apparently uses a different catalytic strategy [11]. The structural prerequisites, however, are as yet unknown.

The obvious similarities between antibody and RNA catalytic systems lead us to conclude that similar structural principles and catalytic mechanisms underlie macromolecule-mediated catalysis of Diels–Alder reactions. It is remarkable that RNA, although equipped with a much less varied arsenal of functional groups than proteins, has independently evolved matching criteria for generating catalytic pockets

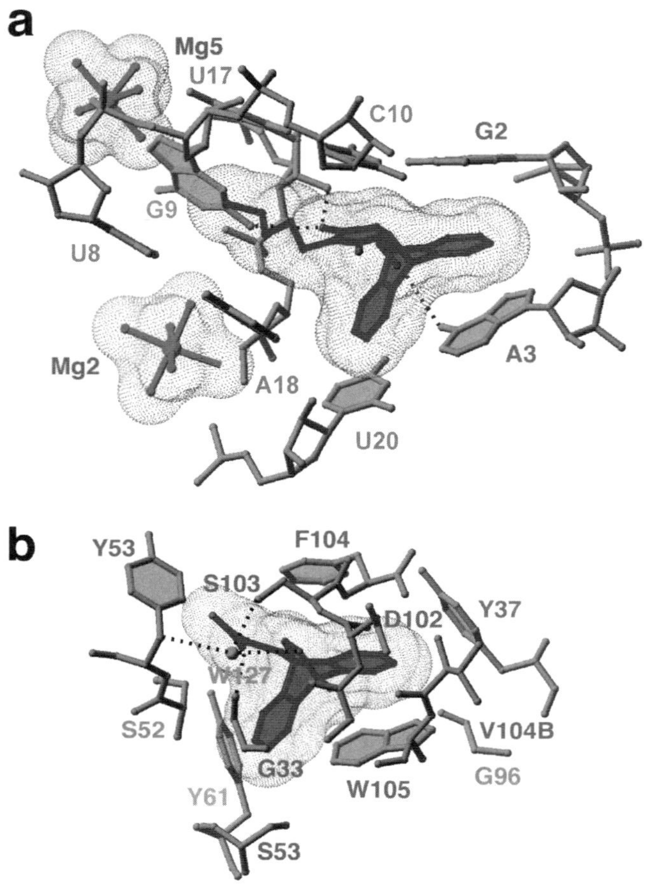

Figure 16.11 Comparison of RNA and protein catalytic sites for the Diels–Alder reaction. (a) Diels–Alderase ribozyme and (b) retro-Diels–Alderase antibody 10F11 [21]. See color plate section

capable of facilitating carbon–carbon bond-forming reactions, with comparable catalytic efficiency and enantioselectivity.

16.13 Miscellaneous

In contrast to protein enzymes, ribozymes can be easily converted into allosterically controlled catalysts by either rational design or combinatorial selection approaches [41]. The Diels–Alderase ribozyme could be brought under allosteric control of three different effectors: the purine alkaloid theophylline, the aminoglycoside tobramycin and a tumor-relevant mRNA sequence [42–44]. Attachment of the effector recognition domains (aptamers or complementary RNA sequences) could be achieved at different positions of the ribozymes. Allosteric activation factors between 20 and 2000 were achieved in all cases. The fluorescence properties of the anthracene substrates utilized by the Diels–Alderase allow for an easy optical readout without the need for labeling.

16.14 Conclusion

Despite their limited set of functional groups, ribozymes can accelerate complex organic transformations such as Diels–Alder reactions between small molecules in a way similar to protein enzymes or traditional chemical catalysts featuring multiple turnover, substrate specificity and stereoselectivity. The three-dimensional structure shows striking similarities with proteins evolved for similar reactions and the catalytic strategies used also appear to be similar.

References

1. C. Guerrier-Takada, K. Gardiner, T. Marsh, N. Pace and S. Altman, The RNA moiety of ribonuclease P is the catalytic subunit of the enzyme, *Cell*, **35**, 849–857 (1983).
2. K. Kruger, P.J. Grabowski, A.J. Zaug, J. Sands, D.E. Gottschling and T.R. Cech, Self-splicing RNA: autoexcision and autocyclization of the ribosomal RNA intervening sequence of *Tetrahymena*, *Cell*, **31**, 147–157 (1982).
3. J.C. Cochrane and S.A. Strobel, Catalytic strategies of self-cleaving ribozymes, *Acc. Chem. Res.*, **41**, 1027–1035 (2008).
4. A.D. Ellington and J.W. Szostak, *In vitro* selection of RNA molecules that bind specific ligands, *Nature*, **346**, 818–822 (1990).
5. C. Tuerk and L. Gold, Systematic evolution of ligands by exponential enrichment: RNA ligands to bacteriophage T4 DNA polymerase, *Science*, **249**, 505–510 (1990).
6. X. Dai, A. De Mesmaeker and G.F. Joyce, Cleavage of an amide bond by a ribozyme, *Science*, **267**, 237–240 (1995).
7. J.A. Piccirilli, T.S. McConnell, A.J. Zaug, H.F. Noller and T.R. Cech, Aminoacyl esterase activity of the *Tetrahymena* ribozyme, *Science*, **256**, 1420–1424 (1992).
8. T.W. Wiegand, R.C. Janssen and B.E. Eaton, Selection of RNA amide synthases, *Chem. Biol.*, **4**, 675–683 (1997).
9. B. Seelig and A. Jäschke, A small catalytic RNA motif with Diels–Alderase activity, *Chem. Biol.*, **6**, 167–176 (1999).
10. G. Sengle, A. Eisenfuhr, P.S. Arora, J.S. Nowick and M. Famulok, Novel RNA catalysts for the Michael reaction, *Chem. Biol.*, **8**, 459–473 (2001).
11. T.M. Tarasow, S.L. Tarasow and B.E. Eaton, RNA-catalysed carbon–carbon bond formation, *Nature*, **389**, 54–57 (1997).
12. S. Tsukiji, S.B. Pattnaik and H. Suga, An alcohol dehydrogenase ribozyme, *Nat. Struct. Biol.*, **10**, 713–717 (2003).
13. S. Tsukiji, S.B. Pattnaik and H. Suga, Reduction of an aldehyde by a NADH/Zn^{2+}-dependent redox active ribozyme, *J. Am. Chem. Soc.*, **126**, 5044–5045 (2004).
14. X. Chen, N. Li and A.D. Ellington, Ribozyme catalysis of metabolism in the RNA world, *Chem. Biodivers.*, **4**, 633–655 (2007).
15. W. Gilbert, The RNA world, *Nature*, **319**, 618 (1986).
16. C.R. Woese, *The Genetic Code: the Molecular Basis for Genetic Expression*, Harper and Row, New York, 1967.
17. S. Fusz, A. Eisenfuhr, S.G. Srivatsan, A. Heckel and M. Famulok, A ribozyme for the aldol reaction, *Chem. Biol.*, **12**, 941–950 (2005).
18. Y. Ryu, K.J. Kim, C.A. Roessner and A.I. Scott, Decarboxylative Claisen condensation catalyzed by *in vitro* selected ribozymes, *Chem. Commun.*, 1439–1441 (2006).
19. M. Mielcarek, M.Z. Barciszewska, P. Salanski, M. Stobiecki, J. Jurczak and J. Barciszewski, Native transfer RNA catalyzes Diels–Alder reaction, *Biochem. Biophys. Res. Commun.*, **294**, 145–148 (2002).
20. M. Chandra and S.K. Silverman, DNA and RNA can be equally efficient catalysts for carbon–carbon bond formation, *J. Am. Chem. Soc.*, **130**, 2936–2937 (2008).
21. A. Serganov, S. Keiper, L. Malinina, V. Tereshko, E. Skripkin, C. Hobartner, A. Polonskaia, A.T. Phan, R. Wombacher, R. Micura, Z. Dauter, A. Jäschke and D.J. Patel, Structural basis for Diels–Alder ribozyme-catalyzed carbon–carbon bond formation, *Nat. Struct. Mol. Biol.*, **12**, 218–224 (2005).
22. K.N. Morris, T.M. Tarasow, C.M. Julin, S.L. Simons, D. Hilvert and L. Gold, Enrichment for RNA molecules that bind a Diels–Alder transition state analog, *Proc. Natl. Acad. Sci. USA*, **91**, 13028–13032 (1994).

23. D.S. Wilson and J.W. Szostak, *In vitro* selection of functional nucleic acids, *Annu. Rev. Biochem.*, **68**, 611–647 (1999).
24. S. Keiper, D. Bebenroth, B. Seelig, E. Westhof and A. Jäschke, Architecture of a Diels–Alderase ribozyme with a preformed catalytic pocket, *Chem. Biol.*, **11**, 1217–1227 (2004).
25. B. Seelig, S. Keiper, F. Stuhlmann and A. Jäschke, Enantioselective ribozyme catalysis of a bimolecular cycloaddition reaction *Angew. Chem. Int. Ed.*, **39**, 4576–4579 (2000).
26. S.P. Kim, A.G. Leach and K.N. Houk, The origins of noncovalent catalysis of intermolecular Diels–Alder reactions by cyclodextrins, self-assembling capsules, antibodies and RNAses, *J. Org. Chem.*, **67**, 4250–4260 (2002).
27. R. Wombacher and A. Jäschke, Probing the active site of a Diels–Alderase ribozyme by photoaffinity cross-linking, *J. Am. Chem. Soc.*, **130**, 8594–8595 (2008).
28. F. Stuhlmann and A. Jäschke, Characterization of an RNA active site: interactions between a Diels–Alderase ribozyme and its substrates and products, *J. Am. Chem. Soc.*, **124**, 3238–3244 (2002).
29. R. Wombacher, S. Keiper, S. Suhm, A. Serganov, D.J. Patel and A. Jäschke, Control of stereoselectivity in an enzymatic reaction by backdoor access, *Angew. Chem. Int. Ed.*, **45**, 2469–2472 (2006).
30. A.Y. Kobitski, A. Nierth, M. Helm, A. Jäschke and G.U. Nienhaus, Mg^{2+}-dependent folding of a Diels–Alderase ribozyme probed by single-molecule FRET analysis, *Nucleic Acids Res.*, **35**, 2047–2059 (2007).
31. N. Kisseleva, S. Kraut, A. Jäschke and O. Schiemann, Characterizing multiple metal ion binding sites within a ribozyme by cadmium-induced EPR silencing, *HFSP J.*, **1**, 127–136 (2007).
32. J. Chen, Q. Deng, R. Wang, K. Houk and D. Hilvert, Shape complementarity, binding-site dynamics and transition state stabilization: a theoretical study of Diels–Alder catalysis by antibody 1E9, *ChemBioChem*, **1**, 255–261 (2000).
33. J. Xu, Q. Deng, J. Chen, K.N. Houk, J. Bartek, D. Hilvert and I.A. Wilson, Evolution of shape complementarity and catalytic efficiency from a primordial antibody template, *Science*, **286**, 2345–2348 (1999).
34. X. Zhang and T.C. Bruice, Diels–Alder ribozyme catalysis: a computational approach, *J. Am. Chem. Soc.*, **129**, 1001–1007 (2007).
35. A. Heine, E.A. Stura, J.T. Yli-Kauhaluoma, C. Gao, Q. Deng, B.R. Beno, K.N. Houk, K.D. Janda and I.A. Wilson, An antibody exo Diels–Alderase inhibitor complex at 1.95 angstrom resolution, *Science*, **279**, 1934–1940 (1998).
36. M. Hugot, N. Bensel, M. Vogel, M.T. Reymond, B. Stadler, J.L. Reymond and U. Baumann, A structural basis for the activity of retro-Diels–Alder catalytic antibodies: evidence for a catalytic aromatic residue, *Proc. Natl. Acad. Sci. USA*, **99**, 9674–9678 (2002).
37. F.E. Romesberg, B. Spiller, P.G. Schultz and R.C. Stevens, Immunological origins of binding and catalysis in a Diels–Alderase antibody, *Science*, **279**, 1929–1933 (1998).
38. T. Ose, K. Watanabe, T. Mie, M. Honma, H. Watanabe, M. Yao, H. Oikawa and I. Tanaka, Insight into a natural Diels–Alder reaction from the structure of macrophomate synthase, *Nature*, **422**, 185–189 (2003).
39. T. Ose, K. Watanabe, M. Yao, M. Honma, H. Oikawa and I. Tanaka, Structure of macrophomate synthase, *Acta. Crystallogr. D Biol. Crystallogr.*, **60**, 1187–1197 (2004).
40. J.M. Serafimov, D. Gillingham, S. Kuster and D. Hilvert, The putative Diels–Alderase macrophomate synthase is an efficient aldolase, *J. Am. Chem. Soc.*, **130**, 7798–7799 (2008).
41. S.M. Knudsen and A.D. Ellington. Aptazymes: allosteric ribozymes and deoxyribozymes as biosensors, in *The Aptamer Handbook*, ed. S. Klussmann, Wiley-VCH Verlag GmbH, Weinheim, pp. 290–310 (2006).
42. S. Amontov and A. Jäschke, Controlling the rate of organic reactions: rational design of allosteric Diels–Alderase ribozymes, *Nucleic Acids Res.*, **34**, 5032–5038 (2006).
43. M. Helm, M. Petermeier, B. Ge, R. Fiammengo and A. Jäschke, Allosterically activated Diels–Alder catalysis by a ribozyme, *J. Am. Chem. Soc.*, **127**, 10492–10493 (2005).
44. M. Petermeier and A. Jäschke, New theophylline-activated Diels–Alderase ribozymes by molecular engineering, *Org. Biomol. Chem.*, **7**, 288–292 (2009).

17

Evolving an Understanding of RNA Function by *In Vitro* Approaches

Qing Wang and Peter J. Unrau

17.1 Introduction

One of the central problems of our age is to explain how chemical and physical laws, combined with natural selection, have produced the biological complexity that surrounds us. RNA has played a fundamental role in the emergence and maintenance of this complexity, as evidenced by numerous naturally selected RNAs that perform a variety of biological tasks ranging from information transfer to catalysis. The discovery of self-splicing introns by Cech (1986) and the catalytic activity of RNase P by Altman (Guerrier-Takada *et al.*, 1983) in the 1980s demonstrated that RNAs could mediate chemical reactions just as protein enzymes do. Additional small self-cleaving *ribozymes*, such as the hammerhead ribozyme (Forster and Symons, 1987), the hairpin ribozyme (Feldstein *et al.*, 1989), the *Neurospora* VS ribozyme (Saville and Collins, 1990) and the HDV ribozyme (Wu *et al.*, 1989), which all play important roles in the maturation of viral genomes, added to this picture. In 2000, the crystal structure of the ribosome (Cech, 2000; Nissen *et al.*, 2000) demonstrated that the most central chemical reaction in current biology – peptide bond formation – is in fact RNA catalyzed. This discovery provided solid evidence for a bold prediction that RNA preceded the evolution of protein in what has come to be called the *'RNA world' hypothesis* (Orgel, 1968, 1986; Gilbert, 1986; Joyce, 2002).

The turn of the century also marked the discovery of *riboswitches* (Nahvi *et al.*, 2002) and RNA interference (Fire *et al.*, 1998), which together have dramatically altered our understanding of gene regulation. Riboswitches regulate enzymatic pathways by specifically recognizing small cellular metabolites (Mandal *et al.*, 2003); they modulate transcription and translation and are found in all three domains of life (Kubodera *et al.*, 2003; Sudarsan *et al.*, 2003; Tucker and Breaker, 2005; Winkler and Breaker, 2005). The structural and mechanistic flexibility exhibited by these allosteric molecules further demonstrate the evolutionary significance of RNA (Schwalbe *et al.*, 2007; Fernandez-Luna and Miranda-Rios, 2008).

The Chemical Biology of Nucleic Acids Edited by Günter Mayer

17.2 The interactions between an RNA system and its environment

Formally, the evolution of a functional RNA can be viewed as a *system* that interacts with its *selective environment* over some period of evolutionary time. From this perspective, the environment acts on *sequence space* to produce sequence variants that are selected based on their ability (fitness) to perform tasks that are in turn defined by interactions between the RNA system and the environment (Figure 17.1).

By defining the interactions possible between the environment and an RNA system, it is in principle possible to understand how optimization of RNA functionalities result in a particular level of fitness. For complex environments, such as those found in biology, this is challenging as fitness may be achieved by more than one functional strategy. A primary example of this is the selection of 'parasitic' sequences, such as plant viroids, that replicate using cellular RNA polymerases responsible for transcribing endogenous RNA populations. The existence of this parasitic strategy places selective pressure on the cellular environment and consequently prevents a straightforward analysis of host RNA function. Commonly, then RNA systems selected from complex environments possess a combination of functionalities that are required to satisfy multiple interactions with the environment (Figure 17.1, dashed lines). These independent selective constraints significantly complicate the analysis of biological RNA systems.

Since the 1990s, extensive information regarding RNA functionality has been gleaned from *in vitro selection* studies originally pioneered by the laboratories of Szostak, Gold and Joyce (Ellington and Szostak, 1990; Tuerk and Gold, 1990; Beaudry and Joyce, 1992). The artificial selection and design of ribozymes, *aptamers* and allosteric RNAs has provided much information about the relationship between

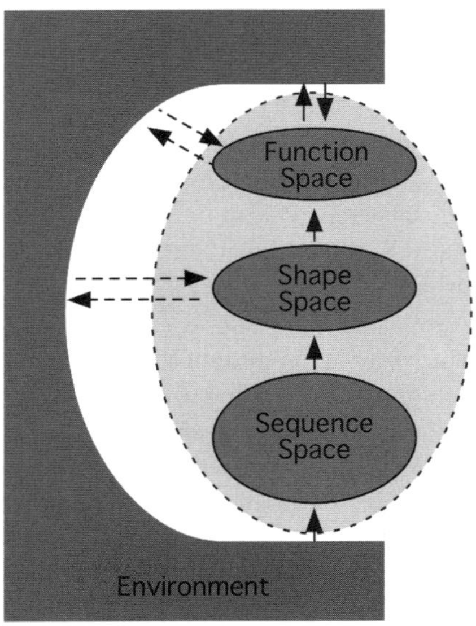

Figure 17.1 *A system-based approach for the analysis of functional RNA evolution. An RNA system (dotted oval) contains RNA sequences (as points in sequence space) that have definable shapes and functions (phenotypes) once the RNA system is placed into a specific environment (solid horseshoe). Interactions with this environment (paired arrows) define the selective pressure that drives the evolution of the RNA system. In this example, the environment is responsible for RNA replication (by, for example, a protein polymerase) and thus controls the sequence diversity available to be sampled (bottom vertical arrow). A complex environment can be difficult to characterize and may generate a multi-component selective pressure (dotted paired arrows). In contrast,* in vitro *selection can often be used to apply a particular pressure (solid paired arrows) making the relationship between selection pressure and RNA functionality simpler to understand*

RNA sequence and function. The great advantage of *in vitro* selection lies not only in its ability to produce completely new functional sequences, but also in the fact that the interactions between the RNA system and the environment can be specified and controlled (solid arrows in Figure 17.1). This makes it possible to relate the function of artificially selected RNAs more easily to their fitness.

17.3 Sequence and shape space: the arena for evolution of functional RNAs

Sequence space is an infinite-dimensional space, whose points uniquely specify the composition of every linear sequence within it (Schuster *et al.*, 1994) (Figure 17.2a). Generally, as the length of a polymer increases, the number of nearest neighbors increases in proportion to polymer length, while the number of possible sequences grows exponentially. For RNA, which in its unmodified form is a polymer comprised of four nucleotides, there are 4^n distinct sequences that can be formed at a given chain length n residues long. Each RNA sequence of this length has $8n + 4$ nearest neighbors that are defined by all possible one-base changes [mutation, $3n$; deletion, n; and insertion, $4(n + 1)$] possible relative to the sequence of interest. This concept of nearest neighbors allows a minimal integer distance to be defined between any two points in sequence space called the minimum *edit distance* [illustrated as the solid lines between two sequences $S_{(1)}$ and $S_{(2)}$ in Figure 17.2a].

Each point in sequence space can be mapped to a *shape space*. This space is defined only when environmental conditions are specified since RNA structure is sensitive to multitude of factors including small molecule interactions, metal ion concentrations, pH, and temperature (Figure 17.1). In addition, chaperone-type complexes in the environment can potentially influence RNA folding. For standard physiological conditions, the well-known canonical rules of base pairing hold and have led to thermodynamically based algorithms for the predication of RNA secondary structure (Mathews and Turner, 2006). These basic rules, when extended with more complex nucleotide interactions, have led to increased success in the prediction of not only secondary structure but also tertiary folds (St-Onge *et al.*, 2007; Parisien and Major, 2008).

Despite this progress, predicting RNA tertiary structure for anything but the simplest RNA folds is still very challenging. Significantly, many functional RNAs sample more than one configuration to achieve their activity [see Figure 17.2b, type (4)–(6)]. Recent advances in single-molecule techniques using fluorescence resonance energy transfer assay (FRET) and other advanced biophysical tools (Al-Hashimi and Walter, 2008) have provided insight into these dynamics. Allosteric RNAs, such as riboswitches, typically change conformation, in response to increasing/decreasing analyte concentrations [Figure 17.2b, type (4)] (Nahvi *et al.*, 2002; Mandal and Breaker, 2004b; Nudler, 2006). FRET has been used to resolve the multistep pathway of the hairpin ribozyme (Rueda *et al.*, 2004; Liu *et al.*, 2007) conformational dynamics in the HDV ribozyme core (Harris *et al.*, 2002) and recently the VS ribozyme (Pereira *et al.*, 2008), supporting a cyclical reaction trajectory during catalysis (Walter *et al.*, 2001). This mapping can also be extended to multi-component RNA systems, such as the ribosome, where the large and small subunits must act in concert to complete a cycle of peptide bond formation (Noller and Baucom, 2002; Rodnina *et al.*, 2002; Kiparisov *et al.*, 2006; Munro *et al.*, 2008). This thermodynamic understanding of RNA structural conformation during a functional work cycle provides a fruitful venue for the study of RNA structure and function (McDowell *et al.*, 2007; Al-Hashimi and Walter, 2008).

17.4 Sequence diversity and the modularity of functional RNA systems

The ability to find the fittest solution for a particular environmental condition is of central importance to evolution. The efficiency of this process is controlled by the distribution and structure of functional sequence in sequence space.

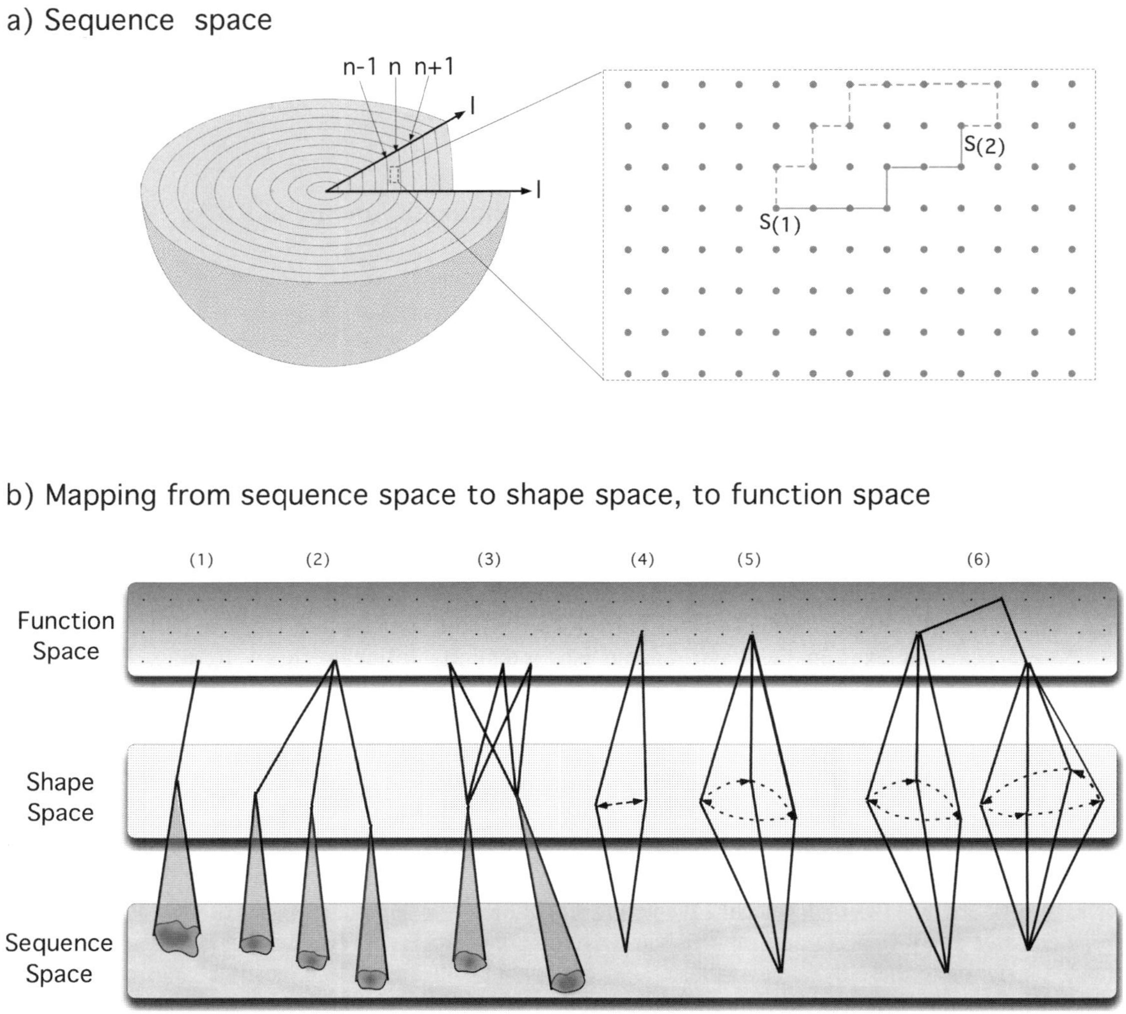

Figure 17.2 *Mapping sequence space to shape space and function space. (a) Sequence space. Sequence space can be visualized as an onion. Each layer of the onion contains* 4^n *distinct sequences and corresponds to sequences of length n. The right panel 'zooms in' to view a small region of sequence space. There are 3*n *nearest neighbors of sequence* S_1 *that have the same length as this sequence (are in the* n *shell of the onion) and that correspond to point mutations. There are a further 4*n + *4 nearest neighbors in the* n + *1 shell that correspond to all possible point insertions and* n *neighbors in the* n – *1 shell that correspond to the possible point deletions. As sequence space is infinite, there is an arbitrarily large set of paths that link any two sequence (*S_1*,* S_2*) in this space. Each path consists of a connected set of nearest neighbors as represented by the solid and dashed lines in the panel. The minimal edit distance is defined by the set of paths that have the smallest number of connections (here the solid trajectory). (b) Mapping between sequence, shape and function space. Shape and function space are defined when the environment is specified. Some of the common mapping relationships from sequence to shape to function space are shown: (1) one to one to one relationship: a set of sequences can fold into a common structure (one point in shape space) that performs a unique function (one point in function space). (2) Many to many to one relationship: many different sequence motifs (not directly connected in sequence space) can fold into distinct structures that all perform the same function. (3) Many to many to many relationship: a group of*

17.4.1 Sampling of sequence space has been low over evolutionary time

Over the entire history of this planet, evolution can only have sampled a tiny fraction of sequence space. The entire biomass of the planet likely reflects the existence of ~10^{31} microbial organisms (Whitman *et al.*, 1998), with in total perhaps 10^{38} independently evolving base pair (bp) of genomic sequence. The functional equivalence of most of this sequence must be vast, given the conservation of core metabolism; however, as an extreme position, if the entire planetary bacterial population replicated every hour and did so by generating a new and unique genome *each* time, the total sampled sequence diversity over the history of this planet (~4×10^{13} h) would reach an upper bound of no more than ~10^{51} bp. If transcribed from all registers into RNA strands having distinct lengths, this random string of genetic information would be sufficient to sample completely all RNA strings up to about 90 nucleotides (nt) long, but no more (this can be seen by the fact that the total number of sequences that are $\leq n$ residues long is given by $4^{n+1}/3$). We conclude from this calculation that it is entirely improbable that an exhaustive search of *RNA motifs* longer than 90 nt could have been performed by random processes on this planet. And yet large highly functional RNAs do exist biologically. The RNase P consensus sequence, which consists of more than 200 nt (Torres-Larios *et al.*, 2006), would be expected to have occurred spontaneously with a probability of ~10^{-120} and the probability of sampling the ribosomal RNAs by random sampling is entirely consistent with zero (Figure 17.3a). Clearly, then, evolutionary searches of sequence space must be organized by a set of principles that restrict and direct the search for function to a much more manageable dimension that permits sampling of large RNA motifs.

17.4.2 Random sampling

There are a variety of prebiotic (Orgel, 2004), biological and *in vitro* processes that can generate random sequence populations of nucleic acids. In appropriate conditions, a number of protein polymerases have been shown to be able to generate pseudo-random sequences (Biebricher and Orgel, 1973; Wettich and Biebricher, 2001). Artificially synthesized random sequence libraries from 20 to 276 nt long and containing up to ~10^{16} distinct random sequence variants have been used with great success to isolate functional RNAs (Bartel and Unrau, 1999; Fedor and Williamson, 2005; Chen *et al.*, 2007; Joyce, 2007; Stoltenburg *et al.*, 2007). These RNAs span a broad range of catalytic functions that include RNA cleavage and ligation, the synthesis and utilization of nucleotide-based cofactors and acyl-transfer reactions leading to peptide and protein synthesis (Bartel and Unrau, 1999; Fedor and Williamson, 2005; Chen *et al.*, 2007). The success of *in vitro* selections operationally demonstrates that the density of functional RNAs sequences commonly exceeds 1 in 10^{15} sequences as pools of artificial RNA molecules often contain numerous functional sequences (Bartel and Unrau, 1999; Chen *et al.*, 2007).

__Figure 17.2__ (Continued) sequences can fold into structures that perform different functions that are in common with a distinct second fold. For example, two capping ribozymes have distinct primary and secondary structures and yet perform the same set of chemical reactions. When the thermodynamic distribution of shapes is also considered as in (4), (5) and (6), a point in sequence space often samples more than one point in shape space, each representing one of the substates of the RNA system. The equilibrium between those states depends on environmental conditions. (4) Two states representing a conformational change before and after a binding event, such as occurs commonly with aptamers and some allosteric ribozymes. (5) Three structures representing three states of a ribozyme, before, during and after the transition state for a particular reaction. (6) Multi-component RNA systems, such as the ribosome, have subunits that each move through a set of states in order to complete a catalytic cycle. Here the large and small subunits of the ribosome provide discrete contributions to the overall function of peptide bond formation

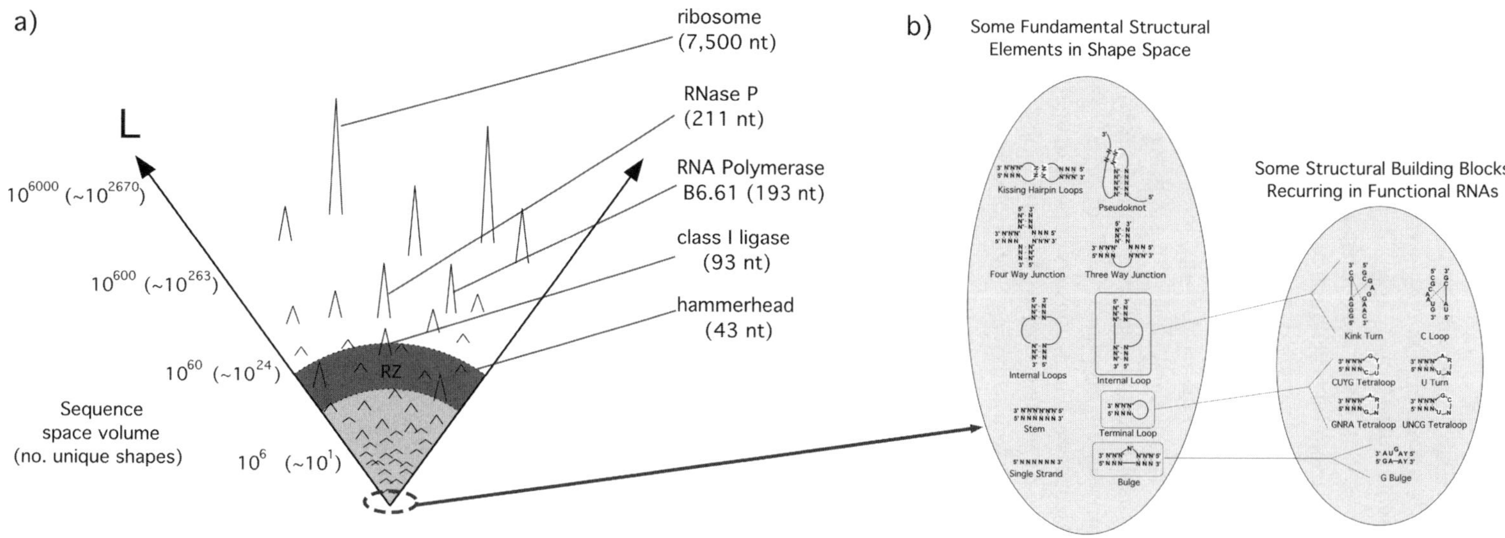

Figure 17.3 *The distribution and modularity of functional RNAs. (a) Functional RNA size and distribution in sequence space. Reproduced from Lehman, N. and Unrau, P.J. Recombination during* in vitro *evolution.* Journal of Molecular Evolution *2005,* ***61****, 245–252, with kind permission from Springer Science + Business Media ©2005. With increasing length, the number of possible sequences increases exponentially, but the density of distinct shapes, defined as the number of unique shapes (in parentheses) divided by the sequence space volume, decreases exponentially with increasing RNA length (Fontana, 2002). The height of each functional RNA peak indicates functional complexity, whereas the distance from the origin measures RNA length on a logarithmic scale. Small functional RNAs, such as hammerhead ribozyme core motif, are located in a region of sequence space where both the shape density in the sequence space is high and sampling can be exhaustive both naturally and* in vitro *(light gray shading). Large complex ribozymes such as class I ligase map to a region of sequence space with lower sequence density, that nevertheless in aggregate contains many more complex folds and functional sequence. Random sampling in this region of sequence space can never be exhaustive. Recombination of functional sequences in the dark gray area of this region (labeled RZ) appears likely to have played a very important role in the evolution of large and highly functional sequences, such as RNase P and the ribosome. The evolution of the artificial RNA polymerase ribozyme B6.61 has required recombination of the class I ligase core with a pool of random sequence elements in this size range. (b) RNA structures are built from simple and fundamental structural elements. The emergence of functional sequences is dramatically facilitated by the ease with which each fundamental structural element can be constructed from small easily sampled sequence elements.* *These foundational elements contribute to the basic overall modularity of RNA structure and function. Some common motifs (left oval) together with recurring motifs found in functional RNAs (right oval) are shown (Moore, 1999; Correll* et al.*, 2003; Hendrix* et al.*, 2005; Lescoute* et al.*, 2005)*

17.4.3 RNA modularity is a fundamental evolutionary principle

Natural and artificially selected RNA systems use modularity to achieve functionality. At the highest level, these modules provide subfunctions of the overall RNA system. The ribosome exhibits such large-scale modularity via the association of the large and small subunits with each subunit providing a discrete function, namely peptide bond formation (large subunit) and decoding (small subunit). In turn, the functions of the small and large subunits are provided by specific substructures found within each subunit (Allard *et al.*, 2000; Ban *et al.*, 2000; Schluenzen *et al.*, 2000). Similarly, the single-stranded RNase P RNA can be dissected into two major submotifs, one responsible for substrate recognition and the other catalysis. Satisfyingly two structurally distinct variants of the substrate recognition motif have been crystallized and both share nearly identical substrate interaction surfaces (Krasilnikov *et al.*, 2003, 2004). This demonstrates the functional and evolutionary independence of this motif from the enzymes catalytic domain (Torres-Larios *et al.*, 2006; Walker and Engelke, 2006). Even fairly small RNAs exhibit functional modularity. The *minimal motif* of self-cleaving hammerhead ribozyme (Wedekind and McKay, 1998) (Figure 17.4a) consists of two separate domains: a 'U turn' domain (C_3–A_6) and a domain (U_7–A_9 and G_{12}–A_{14}) that connects stems II and III by forming a continuous stack between these stems. Recent

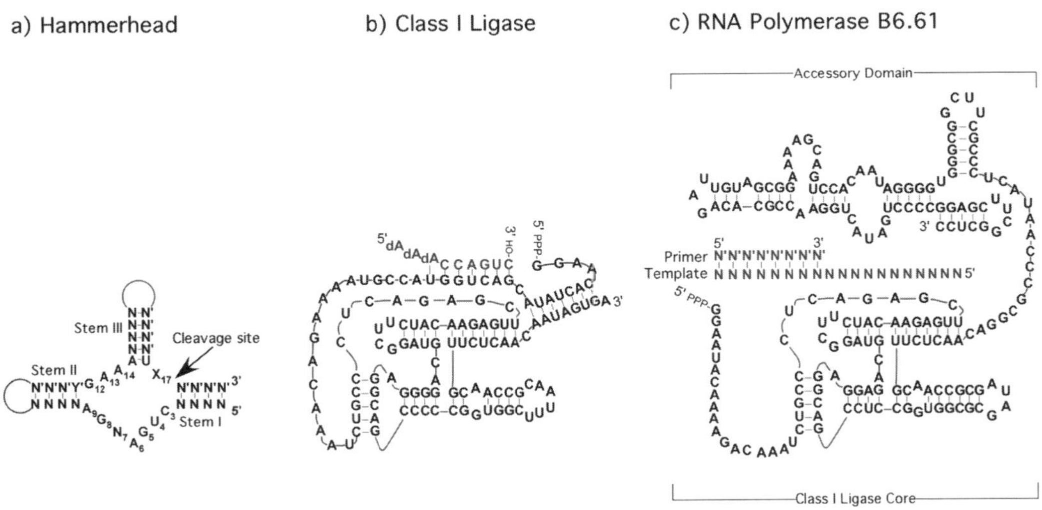

Figure 17.4 *Functional complexity requires increasingly complex RNA motifs. The secondary structures of three ribozymes, each having a unique evolution history, are shown. (a) The hammerhead ribozyme minimal motif (Wedekind and McKay, 1998). It consists of a catalytic core of 11 conserved nucleotides flanked by three helices. The hammerhead ribozyme motif has been easily discovered multiple times from natural and artificial selections for a self-cleaving RNA motif. An arrow indicates the cleavage site. (b) The structure of 119 nt class I ligase ribozyme (Bergman* et al.*, 2004). This ribozyme was originally isolated from an artificial random pool with the diversity of 10^{15}, catalyzing the formation of a 3′,5′-phosphodiester bond by the attack of a 3′-hydroxyl found on a well-positioned RNA strand (in gray) on an adjacent 5′-triphosphate found at the ribozyme terminus. (c) The structure of 193 nt B6.61 RNA polymerase ribozyme Q.S. Wang, L. Cheng and P.J. Unrau, unpublished data; Zaher and Unrau, 2007). This RNA polymerase ribozyme promotes template-dependent polymerization of a primer–template complex (shown in gray). Functionally it consists of two distinct structural domains, the class I ligase core and an accessory domain required for efficient NTP polymerization. With the increasing functional complexity represented in (a) to (c), the information needed to specify RNA function has grown exponentially. Nevertheless, the evolution of the complex RNA polymerase ribozyme is made possible by recombination of simpler motifs that are easily sampled from random sequence*

crystallographic data for a native hammerhead ribozyme displayed the importance of another domain formed by the peripheral interactions between stems I and stem II. This additional interaction dramatically lowers the magnesium requirements for cleavage (Martick and Scott, 2006; Nelson and Uhlenbeck, 2006). In addition, many allosteric ribozyme systems, which have an analyte detector motif coupled to a catalytic domain, have been engineered *in vitro* and have been found to exist naturally (Breaker, 2004). This breakdown of function into modular domains is consistent with a local optimization of structure to meet functional needs and suggests that Nature can efficiently construct complex macromolecular machines from much simpler evolutionary elements.

Both large and small modular RNA structures appear, based on a growing number of NMR and X-ray structures, to be themselves derived from a common and fundamental set of secondary and tertiary 'structural elements', such as helical stems, hairpins, internal loops, pseudoknots and kissing loops (Gan *et al.*, 2003; Hendrix *et al.*, 2005; Carothers *et al.*, 2006a; Leontis *et al.*, 2006) (Figure 17.3b, left oval), that together define many aspects of RNA three-dimensional structure. For example, 16S and 23S rRNAs are comprised of at least 210 such structural elements based on a quantification of modular sub-topologies (Pasquali *et al.*, 2005). Interestingly, structural elements with specific sequence constraints and shapes often recur in many natural and artificial functional RNAs and these elements dramatically simplify our understanding of basic RNA structure (Leontis and Westhof, 2003; Leontis *et al.*, 2006) (Figure 17.3b, right oval). These *elemental structural elements* are all exceedingly easy to find within sequence space by random sampling and are therefore likely to be commonly used in a broad range of evolutionary contexts. The fact that complex functional RNA systems are built up from functional subdomains that are in turn constructed from simple elemental structural elements in a hierarchy of increasing complexity has a number of profound evolutionary implications for the efficient sampling of sequence space and is the key to understanding the evolution of complex RNA systems.

17.4.4 The existence of common and rare motifs by random sampling

The knowledge of the modularity and redundancy of a motif, together with its minimal size, allow an estimation of its density in sequence space (Sabeti *et al.*, 1997). Interestingly, detailed analysis of *in vitro* selected functional RNAs has revealed motifs that are either fairly common (densities of 1 in 10^{12} or higher) or are exceedingly rare (densities less than 1 in 10^{20}). As these common and rare motifs have distinct evolutionary implications, we discuss one example of each type in further detail.

17.4.4.1 A common minimal cleavage motif: the hammerhead ribozyme

The self-cleaving hammerhead ribozyme is widely utilized by a variety of biological systems, including viroids and satellite RNA viruses of plants and animals, to mature RNA transcripts. Scrutinization of GenBank and EMBL databases has revealed several hundred hammerhead or hammerhead-like motifs widely spread in all life domains, including several families of repetitive sequence, in the satellite RNA, plant and animal viruses and in mRNAs (Ferbeyre *et al.*, 2000; Graf *et al.*, 2005). Most recently, a discontinuous form of this motif has been found in the 3′ UTRs of some rodent genes (Martick *et al.*, 2008).

In vitro isolation from an synthetic RNA pool containing 128 random positions demonstrated that, under near-physiological conditions, the 43 nt hammerhead ribozyme core motif (Figure 17.4a) is the most common RNA structure capable of self-cleavage at rates between 0.1 and 1.0 min^{-1} (Salehi-Ashtiani and Szostak, 2001). A comparable *in vitro* selection (Tang and Breaker, 2000) from a pool of 10^{14} RNAs with 40 nt random positions found one class (out of 12) of self-cleaving ribozyme corresponded to the hammerhead ribozyme motif. The hammerhead motif has been estimated to occur with a density in sequence space of about 1 in 10^{10} for random sequences ~100 nt long by a range of methodologies (Sabeti *et al.*, 1997;

Knight *et al.*, 2005). While this density is a function of random sequence length (Sabeti *et al.*, 1997), it suggests that RNAs occurring in this density range can be readily sampled by natural biological processes as evidenced by the biological popularity of this cleavage motif. Notably, a number of common riboswitches, such as the purine riboswitch (Serganov *et al.*, 2004), appear to have complexities that may only slightly exceed that of the hammerhead ribozyme. Satisfyingly, the fact that simple functional RNAs can be spontaneously selected when needed implies that RNAs of similar complexity could have existed very early in evolution.

17.4.4.2 Statistically rare motifs from random sequence: the class I ligase and nucleotide synthase ribozymes

At the other end of the spectrum, exceedingly improbable motifs have been isolated from random sequence. The class I ligase is the best characterized. This ribozyme has a complex and well characterized secondary and tertiary structure (Bergman *et al.*, 2004) (Figure 17.4b). The 93 nt catalytic motif has an exceedingly fast ligation activity that is comparable to highly evolved protein enzyme activities (Bartel and Szostak, 1993; Ekland *et al.*, 1995). The probability that this particular motif was present in Bartel and Szostak's original random sequence pool of ~10^{15} has been calculated to be approximately ~5×10^{-4} based on an analysis of the ribozymes motif structure (Ekland *et al.*, 1995) and, in contrast to motifs such as the hammerhead, appears to be a very lucky find. Since the discovery of the class I ligase motif, other ribozyme selections have also found uncommon ribozyme motifs. A pyrimidine nucleotide synthase, selected also from a very high diversity pool of random sequences, has a minimal motif of 124 nt that forms a complex pseudoknotted five helix structure, that also would not have been expected to occur with high probability (Chapple *et al.*, 2003). The inescapable conclusion, as noted by Bartel and Szostak after discovery of the class I ligase motif, is not that experimenters have been lucky, but rather that the density of complex ribozyme motifs must be relatively high in sequence space. Hence it might be reasonable to suppose that there are several thousand distinct ligase motifs having the complexity of the class I ligase that could have been potentially sampled instead of the class I ligase. This high density of complex functional RNA in sequence space suggests that the core functionalities for naturally occurring but complex motifs, such as RNase P and the hepatitis delta virus (HDV) self-cleaving ribozyme, may have emerged only once during natural selection, or that other processes, such as recombination, may have shaped their evolution (Figure 17.3a). Consistent with this idea, the HDV motif has been found to exist in the genome of primates, where it performs an unknown but presumably regulatory function. This motif has subsequently been exploited by a human-specific virus as a secondary evolutionary event (Salehi-Ashtiani *et al.*, 2006). Here, and in contrast with smaller motifs, phylogeny will continue to play a useful role in understanding the motif's evolution.

The notion that high densities of complex and distinct functional motifs exist in sequence space is entirely consistent with Schuster and Fontana's theoretical work demonstrating that a high density of RNA secondary structure folds exist for RNAs of intermediate length (Schuster *et al.*, 1994; Fontana and Schuster, 1998b; Fontana, 2002). Experimental explorations of sequence space have extended this seminal work to RNA function and indicate that a significant fraction of these folds have potentially interesting functional properties.

17.5 Fundamental mechanisms for sampling sequence space

There are two distinct mechanisms by which sequence space can be searched from a pre-existing RNA population and each plays a pivotal role in the evolution of functional RNA systems. The first is based on the generation of point mutants derived from a pre-existing RNA population. The second involves the recombination of preexisting RNA sequences to give rise to new recombinant molecules.

17.5.1 Local searches in sequence space: point mutation

The ability to move through sequence space from a point of defined function to one with increased or distinctly new function is of great practical and theoretical evolutionary interest. The experimentally determined values of functional density just discussed provide important clues about the structure of functional sequence space. Assuming a randomly selected starting position in sequence space that is 100 nt long, a sphere drawn about this point that has a radius of just seven point mutations/indels, contains in excess of 10^{15} sequences and should therefore reasonably be expected to contain perhaps several thousand of functional RNA sequences. This calculation implies that even though functional densities can appear fairly low, distinct functional sequences are not necessarily spaced far apart from one another in sequence space and might be reached by limited searches through sequence space.

Mutagenic polymerase chain reaction (PCR) and synthetic mutagenesis are the two main *in vitro* methods available to generate populations of mutant molecules clustered around a point in sequence space. Mutagenic PCR deliberately increases the error rate of naturally occurring DNA polymerases by adding metal ions such as manganese and altering dNTP concentrations (Kore *et al.*, 2000). Synthetically, mutations are introduced by coupling mixes of phosphoramidites at each nucleotide position. In this method, the mutation position and mutation rate can be completely controlled (Zaher and Unrau, 2005). Point deletions and insertions can also be introduced systematically using a partial reblocking/deblocking strategy on a DNA synthesizer (Chapple *et al.*, 2003). These methodologies result in sequence libraries, henceforth called mutation pools, that heavily sample sequence space in the close vicinity of the progenitor sequence.

The coverage of the mutation pool in sequence space (Figure 17.5a) depends on the mutation technique, the length of the progenitor sequence, and the total population size sampled. Take as a common example a library containing 10^{15} sequences, generated from a 100 nt RNA sequence having a 5% point mutation rate at all 100 positions. The probability of sampling all possible seven-point mutants in this library is nearly certain (as there are ~3.5×10^{13} such sequences). However, only about 82% of all sequences containing eight point mutations will be sampled and less 3% of the possible nine point mutants will be present in the pool. Therefore, this pool will sample exhaustively a disc on the $n = 100$ shell of sequence space. If point deletion and insertion were also included, this disc would extend to a true high-dimensional 'sphere' centered on the progenitor sequence; however, such libraries are technically much more challenging to create in an unbiased manner (Chapple *et al.*, 2003).

17.5.1.1 Enhancing an existing function

Typically, once a functional RNA is isolated from a random pool, a mutation pool covering the local sequence area around is generated to search for a locally optimal RNA solution [see $S_{(1*)}$ in Figure 17.5a]. For example, a pool based on a nucleotide synthase ribozyme called a15 was synthesized using a 20% mutation rate. Six cycles of re-selection resulted in isolate a.6.10 that was ~25 times more efficient. A second mutation pool with a diversity of ~5×10^{13} was then generated, which introduced systematic mutation and deletion in regions of sequence that lacked defined secondary structure conservation. The most efficient isolate was ~35-fold faster than a15 and resulted in the extraction of a 124 nt RNA motif (Chapple *et al.*, 2003).

Mutation pools have also been used to extend the function of RNA. For example, a pool of 10^{13} RNA sequences having a 5% mutation rate at 140 positions of the *Tetrahymena* group I ribozyme led to the

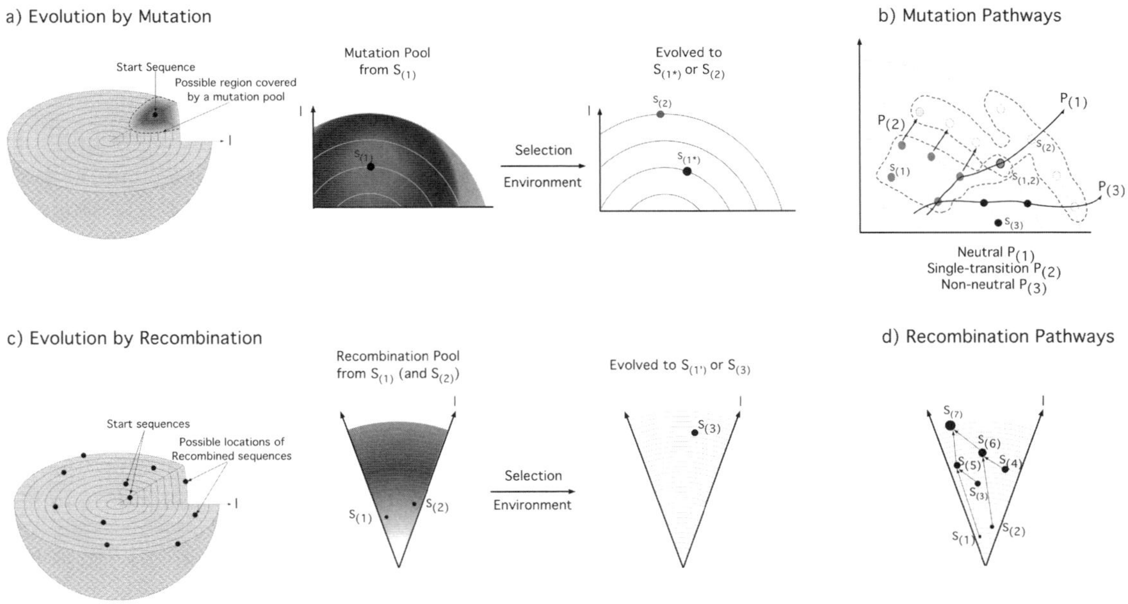

Figure 17.5 *Illustration of mutation and recombination based evolutionary mechanisms. The initial and final evolved sequences are illustrated as dots in sequence space, which is viewed either as half of an onion or as a horizontal slice. Distinct functions are indicated by numbers in parentheses following S and also by shading. The size of a point in sequence space represents increased functionality. (a) Local sampling by mutation. From a starting sequence $S_{(1)}$, point mutation densely samples the region of the sequence space proximal to it (darker area). Selection pressure generated by interaction(s) with the environment can lead to the isolation of either a more optimal sequence [$S_{(1*)}$] or a sequence with a new and presumably more optimal function [$S_{(2)}$]. (b) Evolutional pathways resulting from a series of point mutation events. Three different types of pathways linking three sequences of distinct function $S_{(1)}$ to $S_{(2)}$ and $S_{(3)}$ are shown: The neutral path (P_1) passes through a series of points that have function 1 before transitioning to function 2. Sometimes, a single point mutation can convert the function of a pre-existing RNA as indicated by the single-transition path (P_2). The common non-neutral pathway (P_3) has some points [such as $S_{(3)}$] that are inactive for both functions 1 and 2. (c) Global sampling by recombination. Recombinants generated from just two starting sequences [$S_{(1)}$ and $S_{(2)}$] can sample sequence space in regions impossible by point mutation. Large modular RNA $S_{(3)}$, created by recombination, can have the combined functionalities of $S_{(1)}$ and $S_{(2)}$ or may manifest a new functionality resulting from their combination. (d) Increasing functional complexity along a recombination pathway. The structurally and functionally complex but highly modular $S_{(7)}$ is too rare to be sampled by conventional random or mutation sampling. Its modularity was essential for its evolution and involved recombining the smaller $S_{(5)}$ and $S_{(6)}$ sequences. $S_{(5)}$ and $S_{(6)}$, in turn, have evolved by recombining from the simpler $S_{(1)}$, $S_{(2)}$, $S_{(3)}$ and $S_{(4)}$ sequences, which are easily sampled either at random or by point mutation*

isolation of variants able to utilize Ca^{2+} (Lehman and Joyce, 1993). In the case of RNA ligase ribozyme evolution, Joyce's laboratory designed an elegant *in vitro* continuous evolution system by coupling the ligation function with the RNA amplification. In this system, a serial dilution transfers performed either manually or by machine allows evolution in a continuous manner (Wright and Joyce, 1997; Johns and Joyce, 2005; Paegel and Joyce, 2008). With mutations generated both from an initial mutagenized pool and from the process of the continuous evolution, the class I ligase ribozyme has been evolved to operate at different ion or pH conditions and even undergo three successive nucleotidyl addition reactions (McGinness *et al.*, 2002; Kuhne and Joyce, 2003).

17.5.1.2 New functions from old: changing functionalities by short moves in sequence space

The mapping of RNA secondary structure as a function in sequence space had been studied theoretically based on the principle of RNA minimal folding energy. These studies indicate that there exist sequences with nearly all common structures in the close vicinity of any sequence (Schuster *et al.*, 1994; Fontana and Schuster, 1998a,b; Bornberg-Bauer and Chan, 1999). Experimental validation for this idea has been found by studies explicitly studying the folding of random sequence RNA (Schultes *et al.*, 2005). Such studies, together with simple estimates of functional sequence density discussed previously, suggest that functional RNAs, which comprise a subset of the folded RNAs, should have functionally active neighbors in their close proximity.

GTP *aptamers* have been isolated from a mutation pool derived from the sequence of a 40 nt ATP aptamer. A pool having a diversity of 10^{14} with a 15% mutation rate per position contained three distinct classes of GTP aptamers having on average a 24% difference from the original ATP aptamer in their primary sequence (Huang and Szostak, 2003). Most of these differences appeared to play roles either in disrupting the original secondary structure or in forming a new secondary structure. Similar findings have been made for more complex ribozyme functionalities. A library of sequences based on the Family A pyrimidine nucleotide synthase secondary structure yielded upon reselection many families of purine nucleotide synthase ribozymes. None of these ribozymes shared the global secondary structure of the original pyrimidine nucleotide synthase and yet all had a primary sequence that was similar to the progenitor sequence (Lau *et al.*, 2004). In another study, Curtis and co-workers successfully isolated more that 20 distinct kinase ribozymes from a mutation pool of an aminoacylase ribozyme, in which 65 of 90 positions were partially mutated at an average rate of ~11% per position. On average, only 14 mutational changes were needed to convert a 90 nt aminoacylase ribozyme into a kinase ribozyme (Curtis and Bartel, 2005). These findings conclusively demonstrate the close proximity of unrelated RNA functionalities in sequence space.

17.5.1.3 Neutral networks of functional sequence

The rules of secondary structure formation make most minimal motifs highly *degenerate* as the redundancy of the motif can be very high. This degeneracy defines a manifold or pseudo-continuous subregion of sequence space that has come to be called a 'neutral network'. Such networks are highly filamentous and allow considerable movement through sequence space by neutral mutations that sample the motif's degeneracy.

The close proximity between two distinct functional networks was elegantly demonstrated by Schultes and Bartel, who created a path of nearly neutral point mutations that walked both the hepatitis delta virus (HDV) self-cleaving ribozyme and the class III self-ligating ribozyme to a critical point that exhibited both cleavage and ligation activities albeit at a much lowered rate (Schultes and Bartel, 2000). This intersection sequence is ~40 mutational steps away (about half the length of the two ribozymes) from each of the initial ribozyme sequences [illustrated as $P_{(1)}$ in Figure 17.5b]. This intersection point is not expected to be particularly rare. In fact, it appears reasonable to assume that an extended interface between distinct neutral networks may be relatively common for moderately sized RNAs based on general properties of RNA motif degeneracy. A particularly good example of such an extended interface is provided by the guanine and adenine riboswitches in *Bacillus subtilis*. They share a high degree of homology in both their primary sequence and secondary structure. Breaker's laboratory has found that a single cytosine in the motif core defines the guanine-specific riboswitch. This residue when replaced with uridine completely changes specificity and creates an adenine-specific riboswitch (Mandal and Breaker, 2004a). The Watson–Crick pair formed between this specifying residue in the motif and the analyte makes possible accurate discrimination between guanine and adenine (Serganov *et al.*, 2004; Noeske *et al.*, 2005) and creates two neutral networks that are always precisely one mutation away from each other [illustrated as $P_{(2)}$ in Figure 17.5b].

The ability of two functional neutral networks to share an interface in sequence space raises the possibility that new functional RNA can arise from pre-existing folds without the need to carry an inactive intermediate sequence and by as little as a single point mutation. More specifically, the size of the interface between two functionalities relative to the size of each neutral network defines the spontaneous transition probability between the two functionalities at points in evolutionary time. Here in particular the rate of interconversion between the adenine and guanine riboswitches must be nearly equal over fairly short periods of evolutionary time. Consequently, any deviations from this equilibrium must reflect the relative selective fitness of each riboswitch as defined by the cellular environment and has important biological ramifications.

A combination of neutral drift through sequence space combined with selective pressure from the environment appears likely to have allowed the sampling of RNA functionalities of moderate size and with single domains of functionality (Lehman and Unrau, 2005). The evolution of RNA systems that have multiple subfunctions, however, is more complex and requires the sampling of larger RNA sequences. As the degeneracy of a particular functional RNA motif increases exponentially with RNA length, the amount of sampling required to find a new RNA subfunction can quickly become prohibitive, making other forms of sequence space sampling increasing competitive with point mutation (Figures 17.3a and 17.5).

17.5.2 Recombination

The second fundamental mechanism for the generation of new sequence diversity involves recombination (Figure 17.5c,d). While homologous recombination is biologically important, nonhomologous recombination appears likely to have been central for the discovery of large modular RNAs and can result from the reassortment of pre-existing sequences by either template switching during replication or strand cleavage and religation. From the evolutionary point of view, recombination is capable of moving large distances through sequence space by producing new sequences from a pre-existing population. This process allows the discovery of global optima that might be impossible or difficult to reach using only point mutation, and in particular appears to have played an important role in the discovery of large modular motifs past the recombination zone (Lehman and Unrau, 2005). Although mathematical descriptions of nonhomologous random recombination (NRR) during evolution are still in their infancy, several successful experimental investigations have provided insight into the importance of this form of recombination to evolutionary processes.

17.5.2.1 Efficiently finding a ribozyme minimal motif by NRR

Nonhomologous recombination has dramatically streamlined the search for ribozyme minimal core motifs. A Family B pyrimidine nucleotide synthase ribozyme originally 271 nt long was digested and religated together so as to construct a pool containing $<10^8$ recombinant molecules that included deletions, inversions and translocations (Wang and Unrau, 2005). During selection, a diverse set of variants having a similar catalytic rate but distinct lengths were found. The shortest isolate that was selected was 81 nt in length. The minimal motif, which folded into a secondary structure having three helix loops, consisted of four short sequence islands that were conserved within all of the selected isolates. This result clearly demonstrates how random nonhomologous recombination can sample sequence space efficiently so as to remove nonessential sequences without human intervention.

17.5.2.2 Joining two modules together by design

Small ribozymes have been engineered into riboswitches simply by fusing them to an aptamer motif. A good example is the artificial ATP-regulated hammerhead ribozyme (Tang and Breaker, 1997). This allosteric system exhibits a 180-fold inhibition in cleavage activity when ATP is present. Mutations in the aptamer domain that are expected to eliminate ATP binding or that increase the distance between aptamer and ribozyme domains result in a loss of ATP-specific allosteric control. To date, a versatile set of riboswitches have been discovered by a combination of rational design, *in vitro* selection and *in vivo* screening and are available to be used as novel molecular biology tools (Bauer and Suess, 2006).

17.5.2.3 Complex functions obtained by selection of new domains

Although point mutant libraries have been used with great success to select for optimized RNA functionalities, the sheer depth of sequence space makes it unlikely that such libraries contain RNAs that possesses distinct and correlated subfunctions without an increase in overall RNA length. With this idea in mind, appending random sequence modules on to an RNA motif having previously defined function allows the selection of new functionality (Johnston *et al.*, 2001; Voytek and Joyce, 2007).

The greatest triumph in this regard has been to convert a ligase ribozyme (Bartel and Szostak, 1993) into a polymerase (Johnston *et al.*, 2001) that in its present form can extend a primer–template by at least 20 nt (Zaher and Unrau, 2007) (Figure 17.4c). Template-based RNA polymerization requires an enzyme that can sequentially incorporate NTPs on to an elongating primer–template complex. This process has two fundamental steps: the appropriate recognition of the primer–template complex and phosphodiester bond formation between the primer and the correct NTP encoded by the template during each extension. Whereas the class I ligase performs very efficient phosphodiester bond formation, it exhibits only a limited ability to elongate a primer–template correctly (Ekland *et al.*, 1995; Johnston *et al.*, 2001). Selection for polymerase activity after appending a random sequence library to the 3′-terminus to the class I ligase produced a ribozyme possessing two domains that have recently been shown to be structurally distinct (Q.S. Wang, L. Cheng and P.J. Unrau, unpublished data) (Figure 17.4). The new 'accessory' domain enhances the functionality of the class I ligase core by interacting in a complex way with the primer--template complex so as to present it appropriately to the catalytic center of the ligase.

Recently, recombination has been successfully used to select a new ligase ribozyme with the necessary catalytic speed to compete with the class I ligase (Voytek and Joyce, 2007). A 35 nt accessory domain was added to a structural element of the DSL ligase ribozyme. After 21 rounds of stepwise *in vitro* selection, the resulting pool was capable of carrying out continuous evolution for 80 successive transfers and led to a ~10^5-fold improvement in catalytic efficiency. This new ligase motif will allow continuous evolution experiments to be performed for the first time that involve competition between two distinct ribozyme motifs.

The ability of *in vitro* selection to extend the function of a ribozyme in this modular fashion demonstrates experimentally the evolutionary potential of recombination to generate biologically relevant RNAs by similar sequential extensions (Figure 17.5d). In particular, complex naturally occurring ribozymes such as RNase P and the ribosome may have evolved from much simpler systems by such a mechanism (Lehman and Unrau, 2005).

17.6 Selection pressure, fitness and function

A range of successful *in vitro* selections have made it possible to understand how populations of random sequences respond to defined selective pressure. We consider selections for aptamer and ribozyme where fitness can be simply related to selected RNA functionality. When aptamer and ribozyme functionalities have been characterized, they exhibit interesting trends that are strongly related to the chemistry of the tasks they were required to perform so as to survive the selection process.

17.6.1 Aptamer affinity and specificity

The high affinity exhibited by aptamers towards their ligands can often approach that of antibodies, with dissociation constants (K_d) in the nanomolar to micromolar range (Stoltenburg *et al.*, 2007). However, depending on the physical and chemical properties of the ligand, aptamers can manifest a spectrum of specificities. At one extreme, aptamers that recognize a particular molecular submoiety have been found which indiscriminately recognize a range of related substrates. NTP aptamers that recognize the sugar–phosphate backbone region of the nucleotides and therefore bind to any of the four nucleotides with similar efficiency are a good example (Weill *et al.*, 2004). At the other end of the spectrum, aptamers that distinguish between closely related target pairs, such as theophylline and caffeine and D- and L-peptide (Jenison *et al.*, 1994; Michaud *et al.*, 2003), are known. In these studies, there is little evidence to suggest that higher specificity is well correlated with higher affinity. Carothers and co-workers examined the binding specificities of eleven structurally distinct GTP aptamers, which had binding affinities ranging from 8 μM to 9 nM (Carothers *et al.*, 2006b). Competitive binding studies with 16 different chemical analogs of GTP showed distinct patterns of specificity for each aptamer, but no general correlation between higher-affinity binding and higher-specificity was found.

17.6.2 Selective pressure and ribozyme functionality

Ribozyme fitness is also selected in response to an applied selective pressure. Generally, for a ribozyme reaction involving a free substrate this pressure is defined by two degrees of freedom: incubation time and substrate concentration. This selective pressure in turn results in the isolation of ribozymes whose *fitness* can be understood in terms of a particular catalytic (functional) strategy. If selection is performed with an excess of substrate or equivalently where binding affinity is guaranteed by base pairing to the substrate, then selective pressure is only strongly related to incubation time and selection favors ribozymes whose fitness is strongly related to k_{cat}. This can be seen as the fraction surviving becomes proportional to (tk_{cat}) under these conditions. This situation is commonly found during the selection of RNA ligase and cleavage ribozymes, which typically have very good substrate binding affinities as a consequence of the 'built-in' Watson–Crick recognition intrinsic to RNA systems (Bergman *et al.*, 2004; Przybilski and Hammann, 2006; Cochrane and Strobel, 2008).

In contrast, ribozymes selected where substrate concentrations and incubation times are both limiting can optimize k_{cat} and K_m independently so as to improve catalytic efficiency. For short enough times and low enough substrate concentrations, only the product of incubation time and substrate concentration can influence selective pressure. For ribozymes that follow Michaelis–Menten kinetics, survival is proportional to $[S]tk_{cat}/K_m$ and so the ratio between k_{cat} and K_m or the *efficiency* can be viewed as a direct measure of ribozyme fitness. Under these selection conditions, ribozymes with similar efficiency may manifest completely different functional strategies. Ribozymes that perform small-molecule chemistry therefore provide an ideal opportunity to study the optimization of fitness as maximizing binding and chemistry simultaneously may be difficult to perform.

17.6.2.1 A tradeoff between k_{cat} and K_m

The selection and characterization of purine and pyrimidine nucleotide synthase ribozymes (Unrau and Bartel, 1998; Lau *et al.*, 2004) yielded functional RNAs whose mechanisms spanned a continuum. Kinetic characterization of each of these ribozyme populations revealed that the aggregate pool efficiencies were fairly similar to the efficiencies measured by individual isolates from each selection (with purine synthase being 50–100 times more efficient). Further characterization of dominant ribozymes from both selections found that all the ribozymes exhibited good *substrate discrimination* (Unrau and Bartel, 1998; Lau *et al.*, 2004). However, whereas some ribozymes had measurable K_m values (Family A pyrimidine nucleotide synthase and RA purine nucleotide synthase), other ribozymes (Family B and C pyrimidine nucleotide synthase and MA purine nucleotide synthase) displayed unmeasurably high K_m values, suggesting very fast chemical rates. This range of catalytic strategy indicates that a compromise between the functionalities of substrate affinity and chemistry may exist in order to produce ribozymes of equivalent efficiencies. As small-molecule reactions require both functionalities to exist within a small region of space, the reason for this compromise may result from the inherent difficulty of placing the functional groups required for both activities within a small region of space.

17.6.2.2 Common nucleotide substrate handling by two dissimilar ribozymes

That small-molecule RNA functionality might be intrinsically limited by geometry was further reinforced by the characterization of two capping ribozymes, called Iso6 and 6.17. These ribozymes were isolated independently using distinct selection strategies (Huang and Yarus, 1997; Zaher *et al.*, 2006). Despite having distinct global secondary structures and metal ion requirements, both ribozymes catalyze a curiously indiscriminant capping reaction involving the attack of a nucleotide (any NMP, NDP, NTP, dNTP or ddNTP are taken as substrates) on to the α-phosphate of the triphosphate found at their 5′ termini. Although the ribozymes have very different k_{cat} values, each ribozyme has a chemical rate that is nearly completely independent of substrate type. More remarkably, the two ribozymes have virtually identical K_m values for nucleoside diphosphate and triphosphate substrates and both exhibit much lower binding affinities for nucleoside monophosphates (Zaher *et al.*, 2006) (Figure 17.6). This curious finding implies that the requirements of capping chemistry impose a common reaction mechanism on two otherwise independent points in sequence space [Figure 17.2b, type (3)]. Indeed characterizing the stereochemistry of the 6.17 ribozyme revealed a distinctive retaining mechanism that implicates a two-step reaction in the capping reaction. This mechanism is entirely consistent with the unusual reaction kinetics shared by both capping ribozymes (Zaher and Unrau, 2006). This mechanism requires two very symmetrical transition states that sandwich a transient ribozyme-substrate covalent intermediate. In all likelihood, the catalytic difficulty in achieving this mechanism has forced both ribozymes to focus all catalytic attention on only the phosphate moiety of their nucleotide substrates, giving rise to their curiously convergent catalytic properties.

17.7 Conclusion

We have presented evidence suggesting that the evolution of complex functional RNAs can most easily occur by taking advantage of the intrinsic modularity of RNA. A variety of evolutionary processes, in particular recombination, appear to be ideal for the generation of increasingly complex ribozymes with progressively expanding functionality. As such, it might be reasonable to expect that small-molecule ribozymes will be particularly sensitive from an evolutionary perspective to changes in their motif function and consequently their overall complexity.

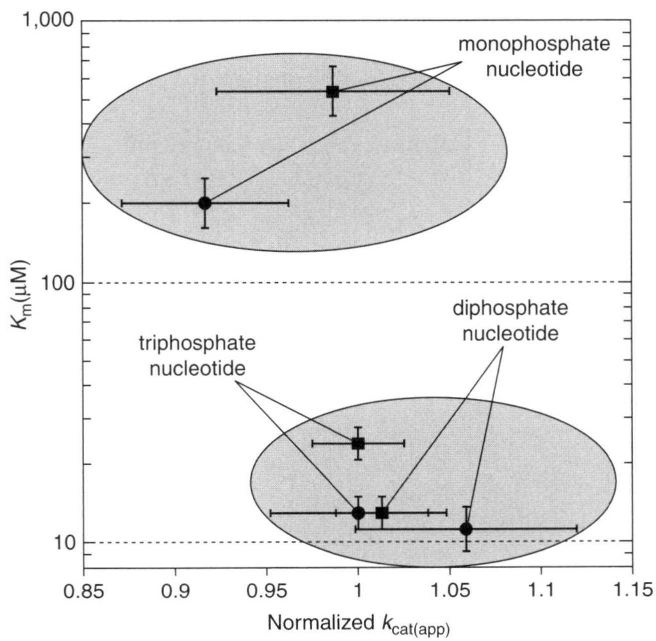

Figure 17.6 *Convergent functional outcomes for two RNA capping ribozymes. Isolated from two independent* in vitro *selections, the iso6 (circles) and c6.17 (squares) ribozymes have distinct metal ion requirements, secondary structures and* k_{cat} *values that differ by nearly two orders of magnitude. Nevertheless, nucleotide substrates are utilized in very similar patterns. Both ribozyme appear to share a complex two-step reaction mechanism that is independent of nucleotide base and sugar composition. Decreasing the phosphate chain length of the nucleotide substrate affects the substrate binding affinity* k_m *but leaves the reaction rate* $k_{cat(app)}$ *invariant for both ribozymes. Both ribozymes have similar* K_m *and normalized* $k_{cat(app)}$ *values for nucleotide triphosphate and diphosphates (bottom oval). The nucleotide monophosphates are much worse substrates for both ribozymes (top oval)*

If as Szostak and co-workers have hypothesized there is a general relationship between functional complexity and informational complexity (the amount of information required to specify an RNA structure) (Szostak, 2003; Carothers *et al.*, 2004; Hazen *et al.*, 2007), it will be very interesting to explore how evolutionary approaches are applied to the study of small-molecule RNA chemistry. This type of chemistry has particular relevance to theories of early evolution, which assume that RNA can perform such reactions with ease. A particularly important reaction in this respect is the chemistry of RNA replication. The evolution of a general RNA replicase ribozyme in the laboratory may possibly be within experimental reach given recent progress in the evolution of an RNA polymerase ribozyme. Such a replicase would be the ultimate tool for exploring RNA functionality in the future and a fundamental element for reproducing a hypothetical 'RNA world' in the laboratory.

17.8 Terminology

- The *'RNA world' hypothesis* claims that a major period of evolution dominated by RNA occurred prior to the emergence of protein catalysis. The primary evidence for this claim is that protein is synthesized by large and complex ribosomal machinery.

- *Riboswitches* are allosteric RNA molecules typically found in the 5′-UTR of a transcript that regulate either transcription or translation of the ORF.
- *Aptamers* are single-strand RNA or DNA sequences that can bind to their targets (ligands) with a high affinity. Many artificial aptamers have been isolated from random or mutation pool of nucleic acid using the method of 'systematic evolution of ligands by exponential enrichment' (SELEX) since first described in 1990 (Ellington and Szostak, 1990). Their ligands include proteins, peptides, complex molecules, drugs, small molecules and even metal ions.
- A *ribozyme* or catalytic RNA performs a chemical reaction. RNA cleavage and ligation are common activities performed by ribozymes, although many complex reactions have been isolated by *in vitro* selection.
- *In vitro selection* is an artificial procedure that allows the enrichment and amplification of functional RNAs.
- The *minimal motif* is defined by the smallest string of nucleotides required to generate the functionality of the motif and can generally be determined by understanding its domain structure and modularity. Normally this procedure is greatly facilitated by having a phylogenetic understanding of the RNA system together with an experimental confirmation of functionality for the minimal construct. A phylogeny also provides information that is required to estimate the *redundancy* of the motif.
- The organization of the minimal motif typically has a *modularity* that is defined by natural breakpoints in the linear organization where an arbitrary sequence can be inserted or removed without appreciably affecting the motif's functionality. Theoretically, high modularity can increase by several orders of magnitude the probability of finding a motif in random sequence and appears to be an important mechanism for sampling complex motifs from random sequence (Sabeti *et al.*, 1997; Knight and Yarus, 2003).
- *Redundancy (degeneracy).* Two RNA sequences are *functionally equivalent* if they are *indistinguishable* by the *environment*. This notion is analogous to the concept of equivalent microscopic states in thermodynamics. The total number of equivalent sequences that share a common *minimal motif* is defined to be the redundancy for that motif.
- *Phylogenetically*, a set of sequences is said to have a common *minimal motif* if *statistically* they share a pattern of primary sequence conservation that is unlikely to be due to chance. Such a pattern is commonly uncovered using a sequence alignment.
- *Fitness* defined for simple situations. Artificial selection for functional RNA typically uses *time* and *substrate concentration* to generate selection pressure: For well-behaved single-site aptamers, fitness is simply related to $1/K_D$ as the fraction surviving a selective step (assuming limiting amounts of substrate [S]) is proportional to $[S]/K_D$. For ribozyme systems that satisfy Michaelis–Menten kinetics, the fraction surviving an ideal selective step, when short times (t) and limiting amounts of substrate [S] are used, is equal to $[S]tk_{cat}/K_m$. In this situation, fitness is related to the system's *efficiency* or k_{cat}/K_m.

References

Al-Hashimi HM, Walter NG (2008) RNA dynamics: it is about time. *Curr Opin Struct Biol* **18**, 321–329.

Allard P, Rak AV, Wimberly BT, Clemons WM Jr, Kalinin A, Helgstrand M, Garber MB, Ramakrishnan V, Hard T (2000) Another piece of the ribosome: solution structure of S16 and its location in the 30S subunit. *Structure* **8**, 875–882.

Ban N, Nissen P, Hansen J, Moore PB, Steitz TA (2000) The complete atomic structure of the large ribosomal subunit at 2.4 Å resolution. *Science* **289**, 905–920.

Bartel DP, Szostak JW (1993) Isolation of new ribozymes from a large pool of random sequences. *Science* **261**, 1411–1418.

Bartel DP, Unrau PJ (1999) Constructing an RNA world. *Trends Cell Biol* **9**, M9–M13.

Bauer G, Suess B (2006) Engineered riboswitches as novel tools in molecular biology. *J Biotechnol* **124**, 4–11.

Beaudry AA, Joyce GF (1992) Directed evolution of an RNA enzyme. *Science* **257**, 635–641.

Bergman NH, Lau NC, Lehnert V, Westhof E, Bartel DP (2004) The three-dimensional architecture of the class I ligase ribozyme. *RNA* **10**, 176–184.

Biebricher CK, Orgel LE (1973) An RNA that multiplies indefinitely with DNA-dependent RNA polymerase: selection from a random copolymer. *Proc Natl Acad Sci USA* **70**, 934–938.

Bornberg-Bauer E, Chan HS (1999) Modeling evolutionary landscapes: mutational stability, topology and superfunnels in sequence space. *Proc Natl Acad Sci USA* **96**, 10689–10694.

Breaker RR (2004) Natural and engineered nucleic acids as tools to explore biology. *Nature* **432**, 838–845.

Carothers JM, Oestreich SC, Davis JH, Szostak JW (2004) Informational complexity and functional activity of RNA structures. *J Am Chem Soc* **126**, 5130–5137.

Carothers JM, Davis JH, Chou JJ, Szostak JW. (2006a). Solution structure of an informationally complex high-affinity RNA aptamer to GTP. *RNA* **12**, 567–579.

Carothers JM, Oestreich SC, Szostak JW. (2006b). Aptamers selected for higher-affinity binding are not more specific for the target ligand. *J Am Chem Soc* **128**, 7929–7937.

Cech TR (1986) Biologic catalysis by RNA. *Harvey Lect* **82**, 123–144.

Cech TR (2000) Structural biology. The ribosome is a ribozyme. *Science* **289**, 878–879.

Chapple KE, Bartel DP, Unrau PJ (2003) Combinatorial minimization and secondary structure determination of a nucleotide synthase ribozyme. *RNA* **9**, 1208–1220.

Chen X, Li N, Ellington AD (2007) Ribozyme catalysis of metabolism in the RNA world. *Chem Biodivers* **4**, 633–655.

Cochrane JC, Strobel SA (2008) Catalytic strategies of self-cleaving ribozymes. *Acc Chem Res* **41**, 1027–1035.

Correll CC, Beneken J, Plantinga MJ, Lubbers M, Chan YL (2003) The common and the distinctive features of the bulged-G motif based on a 1.04 Å resolution RNA structure. *Nucleic Acids Res* **31**, 6806–6818.

Curtis EA, Bartel DP (2005) New catalytic structures from an existing ribozyme. *Nat Struct Mol Biol* **12**, 994–1000.

Ekland EH, Szostak JW, Bartel DP (1995) Structurally complex and highly active RNA ligases derived from random RNA sequences. *Science* **269**, 364–370.

Ellington AD, Szostak JW (1990) *In vitro* selection of RNA molecules that bind specific ligands. *Nature* **346**, 818–822.

Fedor MJ, Williamson JR (2005) The catalytic diversity of RNAs. *Nat Rev* **6**, 399–412.

Feldstein PA, Buzayan JM, Bruening G (1989) Two sequences participating in the autolytic processing of satellite tobacco ringspot virus complementary RNA. *Gene* **82**, 53–61.

Ferbeyre G, Bourdeau V, Pageau M, Miramontes P, Cedergren R (2000) Distribution of hammerhead and hammerhead-like RNA motifs through the GenBank. *Genome Res* **10**, 1011–1019.

Fernandez-Luna MT, Miranda-Rios J (2008) Riboswitch folding: one at a time and step by step. **RNA Biol 5** (1), 1–4.

Fire A, Xu S, Montgomery MK, Kostas SA, Driver SE, Mello CC (1998) Potent and specific genetic interference by double-stranded RNA in *Caenorhabditis elegans*. *Nature* **391**, 806–811.

Fontana W (2002) Modelling 'evo-devo' with RNA. *Bioessays* **24**, 1164–1177.

Fontana W, Schuster P. (1998a). Continuity in evolution: on the nature of transitions. *Science* **280**, 1451–1455.

Fontana W, Schuster P. (1998b). Shaping space: the possible and the attainable in RNA genotype–phenotype mapping. *J Theor Biol* **194**, 491–515.

Forster AC, Symons RH (1987) Self-cleavage of virusoid RNA is performed by the proposed 55-nucleotide active site. *Cell* **50**, 9–16.

Gan HH, Pasquali S, Schlick T (2003) Exploring the repertoire of RNA secondary motifs using graph theory; implications for RNA design. *Nucleic Acids Res* **31**, 2926–2943.

Gilbert SD (1986) The RNA world. *Nature* **319**, 618.

Graf S, Przybilski R, Steger G, Hammann C (2005) A database search for hammerhead ribozyme motifs. *Biochem Soc Trans* **33**, 477–478.

Guerrier-Takada C, Gardiner K, Marsh T, Pace N, Altman S (1983) The RNA moiety of ribonuclease P is the catalytic subunit of the enzyme. *Cell* **35**, 849–857.

Harris DA, Rueda D, Walter NG (2002) Local conformational changes in the catalytic core of the *trans*-acting hepatitis delta virus ribozyme accompany catalysis. *Biochemistry* **41**, 12051–12061.

Hazen RM, Griffin PL, Carothers JM, Szostak JW (2007) Functional information and the emergence of biocomplexity. *Proc Natl Acad Sci USA* **104** Suppl 1, 8574–8581.

Hendrix DK, Brenner SE, Holbrook SR (2005) RNA structural motifs: building blocks of a modular biomolecule. *Q Rev Biophys* **38**, 221–243.

Huang F, Yarus M (1997) 5′-RNA self-capping from guanosine diphosphate. *Biochemistry* **36**, 6557–6563.

Huang Z, Szostak JW (2003) Evolution of aptamers with a new specificity and new secondary structures from an ATP aptamer. *RNA* **9**, 1456–1463.

Jenison RD, Gill SC, Pardi A, Polisky B (1994) High-resolution molecular discrimination by RNA. *Science* **263**, 1425–1429.

Johns GC, Joyce GF (2005) The promise and peril of continuous *in vitro* evolution. *J Mol Evol* **61**, 253–263.

Johnston WK, Unrau PJ, Lawrence MS, Glasner ME, Bartel DP (2001) RNA-catalyzed RNA polymerization: accurate and general RNA-templated primer extension. *Science* **292**, 1319–1325.

Joyce GF (2002) The antiquity of RNA-based evolution. *Nature* **418**, 214–221.

Joyce GF (2007) Forty years of *in vitro* evolution. *Angew Chem Int Ed* **46**, 6420–6436.

Kiparisov SV, Sergiev PV, Bogdanov AA, Dontsova OA (2006) [The structural changes in the ribosome during the elongation cycle]. *Mol Biol (Mosk)* **40**, 755–768.

Knight R, Yarus M (2003) Finding specific RNA motifs: function in a zeptomole world? *RNA* **9**, 218–230.

Knight R, De Sterck H, Markel R, Smit S, Oshmyansky A, Yarus M (2005) Abundance of correctly folded RNA motifs in sequence space, calculated on computational grids. *Nucleic Acids Res* **33**, 5924–5935.

Kore AR, Vaish NK, Morris JA, Eckstein F (2000) *In vitro* evolution of the hammerhead ribozyme to a purine-specific ribozyme using mutagenic PCR with two nucleotide analogues. *J Mol Biol* **301**, 1113–1121.

Krasilnikov AS, Yang X, Pan T, Mondragon A (2003) Crystal structure of the specificity domain of ribonuclease P. *Nature* **421**, 760–764.

Krasilnikov AS, Xiao Y, Pan T, Mondragon A (2004) Basis for structural diversity in homologous RNAs. *Science* **306**, 104–107.

Kubodera T, Watanabe M, Yoshiuchi K, Yamashita N, Nishimura A, Nakai S, Gomi K, Hanamoto H (2003) Thiamine-regulated gene expression of *Aspergillus oryzae* thiA requires splicing of the intron containing a riboswitch-like domain in the 5′-UTR. *FEBS Lett* **555**, 516–520.

Kuhne H, Joyce GF (2003) Continuous *in vitro* evolution of ribozymes that operate under conditions of extreme pH. *J Mol Evol* **57**, 292–298.

Lau MW, Cadieux KE, Unrau PJ (2004) Isolation of fast purine nucleotide synthase ribozymes. *J Am Chem Soc* **126**, 15686–15693.

Lehman N, Joyce GF (1993) Evolution *in vitro* of an RNA enzyme with altered metal dependence. *Nature* **361**, 182–185.

Lehman N, Unrau PJ (2005) Recombination during *in vitro* evolution. *J Mol Evol* **61**, 245–252.

Leontis NB, Westhof E (2003) Analysis of RNA motifs. *Curr Opin Struct Biol* **13**, 300–308.

Leontis NB, Lescoute A, Westhof E (2006) The building blocks and motifs of RNA architecture. *Curr Opin Struct Biol* **16**, 279–287.

Lescoute A, Leontis NB, Massire C, Westhof E (2005) Recurrent structural RNA motifs, isostericity matrices and sequence alignments. *Nucleic Acids Res* **33**, 2395–2409.

Liu S, Bokinsky G, Walter NG, Zhuang X (2007) Dissecting the multistep reaction pathway of an RNA enzyme by single-molecule kinetic 'fingerprinting'. *Proc Natl Acad Sci USA* **104**, 12634–12639.

Mandal M, Breaker RR. (2004a). Adenine riboswitches and gene activation by disruption of a transcription terminator. *Nat Struct Mol Biol* **11**, 29–35.

Mandal M, Breaker RR. (2004b). Gene regulation by riboswitches. *Nat Rev* **5**, 451–463.

Mandal M, Boese B, Barrick JE, Winkler WC, Breaker RR (2003) Riboswitches control fundamental biochemical pathways in *Bacillus subtilis* and other bacteria. *Cell* **113**, 577–586.

Martick M, Scott WG (2006) Tertiary contacts distant from the active site prime a ribozyme for catalysis. *Cell* **126**, 309–320.

Martick M, Horan LH, Noller HF, Scott WG (2008) A discontinuous hammerhead ribozyme embedded in a mammalian messenger RNA. *Nature* **454**, 899–902.

Mathews DH, Turner DH (2006) Prediction of RNA secondary structure by free energy minimization. *Curr Opin Struct Biol* **16**, 270–278.

McDowell SE, Spackova N, Sponer J, Walter NG (2007) Molecular dynamics simulations of RNA: an *in silico* single molecule approach. *Biopolymers* **85**, 169–184.

McGinness KE, Wright MC, Joyce GF (2002) Continuous *in vitro* evolution of a ribozyme that catalyzes three successive nucleotidyl addition reactions. *Chem Biol* **9**, 585–596.

Michaud M, Jourdan E, Villet A, Ravel A, Grosset C, Peyrin E (2003) A DNA aptamer as a new target-specific chiral selector for HPLC. *J Am Chem Soc* **125**, 8672–8679.

Moore PB (1999) Structural motifs in RNA. *Annu Rev Biochem* **68**, 287–300.

Munro JB, Vaiana A, Sanbonmatsu KY, Blanchard SC (2008) A new view of protein synthesis: mapping the free energy landscape of the ribosome using single-molecule FRET. *Biopolymers* **89**, 565–577.

Nahvi A, Sudarsan N, Ebert MS, Zou X, Brown KL, Breaker RR (2002) Genetic control by a metabolite binding mRNA. *Chem Biol* **9**, 1043.

Nelson JA, Uhlenbeck OC (2006) When to believe what you see. *Mol Cell* **23**, 447–450.

Nissen P, Hansen J, Ban N, Moore PB, Steitz TA (2000) The structural basis of ribosome activity in peptide bond synthesis. *Science* **289**, 920–930.

Noeske J, Richter C, Grundl MA, Nasiri HR, Schwalbe H, Wohnert J (2005) An intermolecular base triple as the basis of ligand specificity and affinity in the guanine- and adenine-sensing riboswitch RNAs. *Proc Natl Acad Sci USA* **102**, 1372–1377.

Noller HF, Baucom A (2002) Structure of the 70 S ribosome: implications for movement. *Biochem Soc Trans* **30**, 1159–1161.

Nudler E (2006) Flipping riboswitches. *Cell* **126**, 19–22.

Orgel LE (1968) Evolution of the genetic apparatus. *J Mol Biol* **38**, 381–393.

Orgel LE (1986) RNA catalysis and the origins of life. *J Theor Biol* **123**, 127–149.

Orgel LE (2004) Prebiotic chemistry and the origin of the RNA world. *Crit Rev Biochem Mol Biol* **39**, 99–123.

Paegel BM, Joyce GF (2008) Darwinian evolution on a chip. *PLoS Biol* **6**, e85.

Parisien M, Major F (2008) The MC-Fold and MC-Sym pipeline infers RNA structure from sequence data. *Nature* **452**, 51–55.

Pasquali S, Gan HH, Schlick T (2005) Modular RNA architecture revealed by computational analysis of existing pseudoknots and ribosomal RNAs. *Nucleic Acids Res* **33**, 1384–1398.

Pereira MJ, Nikolova EN, Hiley SL, Jaikaran D, Collins RA, Walter NG (2008) Single VS ribozyme molecules reveal dynamic and hierarchical folding toward catalysis. *J Mol Biol* **382**, 496–509.

Przybilski R, Hammann C (2006) The hammerhead ribozyme structure brought in line. *ChemBioChem* **7**, 1641–1644.

Rodnina MV, Daviter T, Gromadski K, Wintermeyer W (2002) Structural dynamics of ribosomal RNA during decoding on the ribosome. *Biochimie* **84**, 745–754.

Rueda D, Bokinsky G, Rhodes MM, Rust MJ, Zhuang X, Walter NG (2004) Single-molecule enzymology of RNA: essential functional groups impact catalysis from a distance. *Proc Natl Acad Sci USA* **101**, 10066–10071.

Sabeti PC, Unrau PJ, Bartel DP (1997) Accessing rare activities from random RNA sequences: the importance of the length of molecules in the starting pool. *Chem Biol* **4**, 767–774.

Salehi-Ashtiani K, Szostak JW (2001) *In vitro* evolution suggests multiple origins for the hammerhead ribozyme. *Nature* **414**, 82–84.

Salehi-Ashtiani K, Luptak A, Litovchick A, Szostak JW (2006) A genomewide search for ribozymes reveals an HDV-like sequence in the human CPEB3 gene. *Science* **313**, 1788–1792.

Saville BJ, Collins RA (1990) A site-specific self-cleavage reaction performed by a novel RNA in *Neurospora mitochondria. Cell* **61**, 685–696.

Schluenzen F, Tocilj A, Zarivach R, Harms J, Gluehmann M, Janell D, Bashan A, Bartels H, Agmon I, Franceschi F, Yonath A (2000) Structure of functionally activated small ribosomal subunit at 3.3 angstroms resolution. *Cell* **102**, 615–623.

Schultes EA, Bartel DP (2000) One sequence, two ribozymes: implications for the emergence of new ribozyme folds. *Science* **289**, 448–452.

Schultes EA, Spasic A, Mohanty U, Bartel DP (2005) Compact and ordered collapse of randomly generated RNA sequences. *Nat Struct Mol Biol* **12**, 1130–1136.

Schuster P, Fontana W, Stadler PF, Hofacker IL (1994) From sequences to shapes and back: a case study in RNA secondary structures. *Proc R Soc Lond B Biol Sci* **255**, 279–284.

Schwalbe H, Buck J, Furtig B, Noeske J, Wohnert J (2007) Structures of RNA switches: insight into molecular recognition and tertiary structure. *Angew Chem Int Ed* **46**, 1212–1219.

Serganov A, Yuan YR, Pikovskaya O, Polonskaia A, Malinina L, Phan AT, Hobartner C, Micura R, Breaker RR, Patel DJ (2004) Structural basis for discriminative regulation of gene expression by adenine- and guanine-sensing mRNAs. *Chem Biol* **11**, 1729–1741.

St-Onge K, Thibault P, Hamel S, Major F (2007) Modeling RNA tertiary structure motifs by graph-grammars. *Nucleic Acids Res* **35**, 1726–1736.

Stoltenburg R, Reinemann C, Strehlitz B (2007) SELEX – a (r)evolutionary method to generate high-affinity nucleic acid ligands. *Biomol Eng* **24**, 381–403.

Sudarsan N, Barrick JE, Breaker RR (2003) Metabolite-binding RNA domains are present in the genes of eukaryotes. *RNA* **9**, 644–647.

Szostak JW (2003) Functional information: molecular messages. *Nature* **423**, 689.

Tang J, Breaker RR (1997) Rational design of allosteric ribozymes. *Chem Biol* **4**, 453–459.

Tang J, Breaker RR (2000) Structural diversity of self-cleaving ribozymes. *Proc Natl Acad Sci USA* **97**, 5784–5789.

Torres-Larios A, Swinger KK, Pan T, Mondragon A (2006) Structure of ribonuclease P – a universal ribozyme. *Curr Opin Struct Biol* **16**, 327–335.

Tucker BJ, Breaker RR (2005) Riboswitches as versatile gene control elements. *Curr Opin Struct Biol* **15**, 342–348.

Tuerk C, Gold L (1990) Systematic evolution of ligands by exponential enrichment: RNA ligands to bacteriophage T4 DNA polymerase. *Science* **249**, 505–510.

Unrau PJ, Bartel DP (1998) RNA-catalysed nucleotide synthesis. *Nature* **395**, 260–263.

Voytek SB, Joyce GF (2007) Emergence of a fast-reacting ribozyme that is capable of undergoing continuous evolution. *Proc Natl Acad Sci USA* **104**, 15288–15293.

Walker SC, Engelke DR (2006) Ribonuclease P: the evolution of an ancient RNA enzyme. *Critl Rev Biochem Mol Biol* **41**, 77–102.

Walter NG, Harris DA, Pereira MJ, Rueda D (2001) In the fluorescent spotlight: global and local conformational changes of small catalytic RNAs. *Biopolymers* **61**, 224–242.

Wang QS, Unrau PJ (2005) Ribozyme motif structure mapped using random recombination and selection. *RNA* **11**, 404–411.

Wedekind JE, McKay DB (1998) Crystallographic structures of the hammerhead ribozyme: relationship to ribozyme folding and catalysis. *Annu Rev Biophys Biomol Struct* **27**, 475–502.

Weill L, Louis D, Sargueil B (2004) Selection and evolution of NTP-specific aptamers. *Nucleic Acids Res* **32**, 5045–5058.

Wettich A, Biebricher CK (2001) RNA species that replicate with DNA-dependent RNA polymerase from *Escherichia coli*. *Biochemistry* **40**, 3308–3315.

Whitman WB, Coleman DC, Wiebe WJ (1998) Prokaryotes: the unseen majority. *Proc Natl Acad Sci USA* **95**, 6578–6583.

Winkler WC, Breaker RR (2005) Regulation of bacterial gene expression by riboswitches. *Annu Rev Microbiol* **59**, 487–517.

Wright MC, Joyce GF (1997) Continuous *in vitro* evolution of catalytic function. *Science* **276**, 614–617.

Wu HN, Lin YJ, Lin FP, Makino S, Chang MF, Lai MM (1989) Human hepatitis delta virus RNA subfragments contain an autocleavage activity. *Proc Natl Acad Sci USA* **86**, 1831–1835.

Zaher HS, Unrau PJ (2005) *Nucleic Acid Library Construction Using Synthetic DNA Constructs*, Humana Press, Totowa, NJ.

Zaher HS, Unrau PJ (2006) A general RNA-capping ribozyme retains stereochemistry during cap exchange. *J Am Chem Soc* **128**, 13894–13900.

Zaher HS, Unrau PJ (2007) Selection of an improved RNA polymerase ribozyme with superior extension and fidelity. *RNA* **13**, 1017–1026.

Zaher HS, Watkins RA, Unrau PJ (2006) Two independently selected capping ribozymes share similar substrate requirements. *RNA* **12**, 1949–1958.

18

The Chemical Biology of Aptamers: Synthesis and Applications[1]

Günter Mayer and Bernhard Wulffen

18.1 Introduction

Aptamers are small, single-stranded nucleic acids that fold into a well-defined three-dimensional structure. They show a high affinity and specificity for their target molecules and inhibit their biological functions. Aptamers belong to the nucleic acids family and can be synthesized by chemical or enzymatic procedures or a combination of the two. They can, therefore, be considered as both chemical and biological substances, but certainly with chemical origin.

In 1990, three research groups independently reported the isolation of small nucleic acids with predefined functions given by their defined experimental set-ups: Ellington and Szostak reported on RNA molecules that bind to a small organic dye, which they denoted aptamers – a word chimera built up from the Latin expression *aptus* (to fit) and the Greek word *meros* (part) [1]. A second study by Tuerk and Gold described the selection of RNA molecules that bind to the bacteriophage T4 DNA polymerase and they termed this process SELEX [2]. Robertson and Joyce described the application of *in vitro* selection for the adaptation of the group I ribozyme so that it cleaved DNA rather than single-stranded RNA [3]. These ground-breaking studies already highlighted the power of *in vitro* selection for obtaining nucleic acid molecules with complex functions. Thereafter, various studies showed clearly that the aptamer technology fulfilled these expectations and proved that nucleic acid molecules can serve as sophisticated functional moieties, in addition to acting as blueprints of the genetic code or as rigid materials for nanoarchitectures [4–10]. Nucleic acid sequences have a defined function, as a result of their distinct 3D shape, and also carry the blueprint for their own synthesis in

[1] This chapter has been adapted from a recently published review written by G. Mayer essentially on the same topic (*Angew. Chem. Int. Ed.*, 2009, 48, 2672–2689), with permission from Wiley VCH Verlag GmbH & Co. KGaA.

The Chemical Biology of Nucleic Acids Edited by Günter Mayer

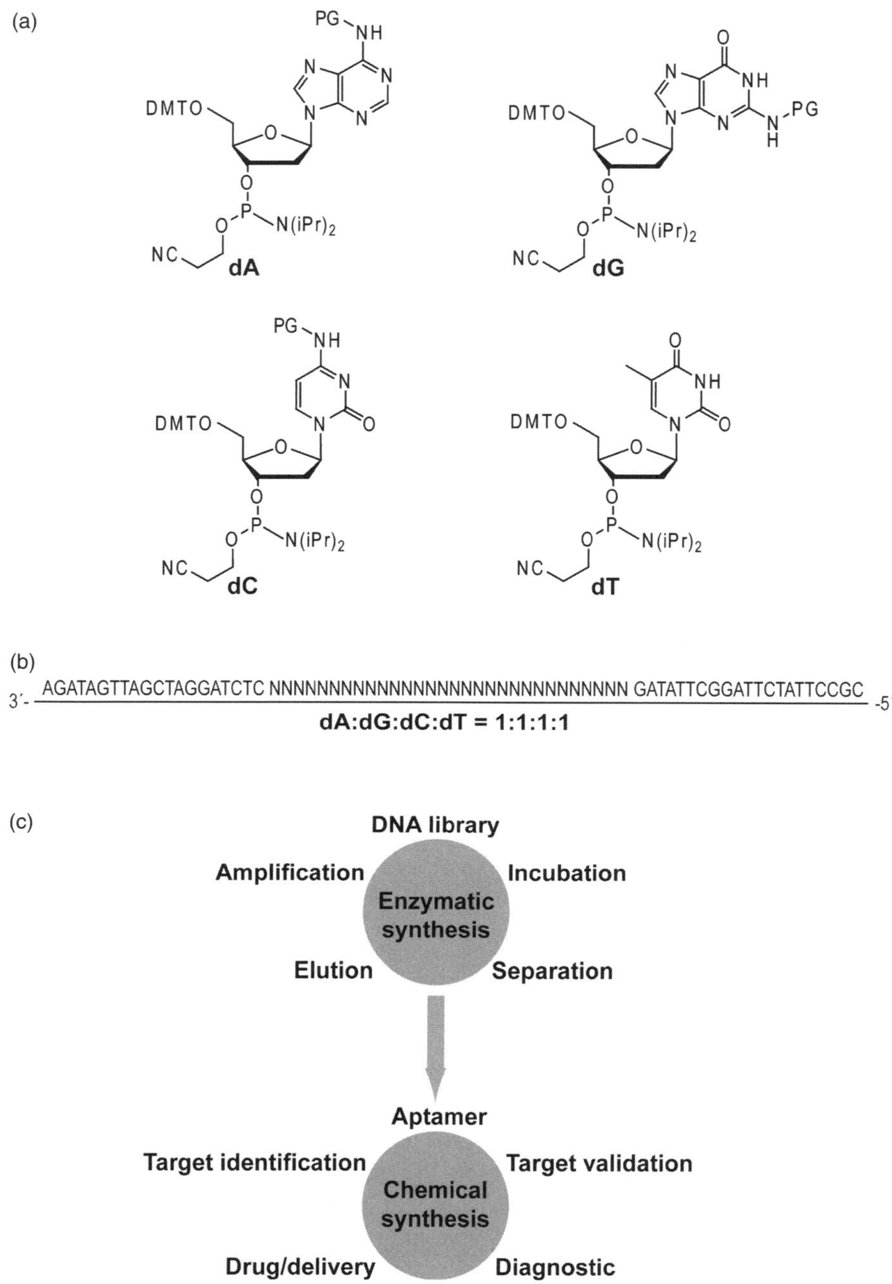

Figure 18.1 *(a) Dimethoxytrityl (DMT)-protected deoxynucleoside phosphoramidites for the solid-phase synthesis of DNA (PG = protecting group). (b) Nucleic acid libraries can be synthesized with defined primer binding sites flanking a random region (N). (c) Enzymatic methods are mandatory for the* in vitro *selection process (top circle), which consists of four key steps: incubation of the nucleic acid library (DNA library) with a target molecule of choice, separation of bound from non-bound nucleic acid species, elution of bound nucleic acid species and amplification of eluted nucleic acid species. Once an aptamer has been identified, it can be tailored for distinct applications (bottom circle), such as target validation, diagnostics, drug and delivery purposes and target identification*

their primary sequence. Consequently, chemical synthesis can be applied to equip an aptamer with a variety of additional functionalities, thereby tailoring it for diverse applications. Combinatorial solid-phase synthesis first produces a diverse library of nucleic acids, an *in vitro* selection process based on enzymatic methods then follows and the resulting aptamers can in turn be manipulated by chemical synthesis (Figure 18.1).

The early selection protocols targeting proteins were cumbersome and it frequently took several months to identify aptamers that specifically targeted the desired protein. Today, the selection process has been reduced to several days by elaborate handling protocols including methods based on single beads [11], capillary electrophoresis [12–16], surface plasmon resonance [17,18], high-performance liquid chromatography (HPLC) [19] and automated processes [20–22]. The development of nucleic acid libraries with superior replication properties, designed to avoid the accumulation of artifacts, such as molecular parasites, led to sophisticated selection procedures. The protocols that make use of robotic workstations are optimized to do without any purification steps of the nucleic acid intermediates through immobilization of a target molecule to magnetic beads [21,23]. By this means, the simultaneous selection against up to eight target molecules is possible. In addition to linear nucleic acid libraries (Figure 18.1b), structurally constrained libraries with predefined secondary structure elements have also been successfully applied for *in vitro* selection purposes [23,24]. These libraries have the advantage that truncation and structural characterization of a representative aptamer are straightforward [25]. In some cases, these libraries have been proven to require fewer selection cycles. Bugaut *et al.* demonstrated that chemical synthesis can be efficiently combined with SELEX procedures [26]. They synthesized a library containing 2′-amino-2′-deoxyuridine with 14 randomized positions. Prior to incubation with the target molecule, this library was post-transcriptionally modified with a set of three aldehydes. After incubation and elution of the bound RNA species, the aldehydes were removed to allow enzymatic replication and amplification of the isolated RNA molecules. In this chapter, the most recent developments in the aptamer field are summarized with emphasis on the modification of aptamers by enzymatic and chemical synthesis, with the ultimate goal of tailoring aptamers for specific applications. It should become clear how synthetic aptamers can be used for target identification and as validation tools and also for diagnostic and drug-discovery processes.

18.2 Aptamers – 'biological' with chemical origin

Aptamers are short, single-stranded nucleic acids with a defined three-dimensional shape that allows them to interact with high affinity with a target molecule. DNA and RNA polymerases are indispensable for the *in vitro* selection process since they provide a satisfactory means for the proper replication of selected sequences and for the mutual introduction of point mutations to obtain covariations of selected nucleic acids. Various modifications of nucleotides have been described which are compatible with the enzymatic steps of the *in vitro* selection procedure and thus enhance the chemical diversity and the biological properties of nucleic acid libraries (Figure 18.2) [27]. These modifications were introduced either at the phosphate/ribose backbone or at the nucleobases [28]. The replacement of the DNA phosphate backbone by a phosphorothioate has enhanced the stability against nucleases and the cellular availability of such molecules. In addition to antisense molecules, such as vitravene, aptamers with a phosphorothioate backbone have also been described [29,30]. However, the most prominent modification of aptamers is the derivatization of the 2′-ribose. This position contributes significantly to the stability of RNA aptamers and 2′-fluoro- and 2′-amino-2′-deoxypyrimidine triphosphates have frequently been used for the direct selection of nuclease-stabilized RNA aptamers. Chelliserrykattil and Ellington developed a variant of the T7 RNA polymerase that can use 2′-methoxypyrimidine nucleotide triphosphates and 2′-methoxyadenine nucleotide triphosphate as substrates for *in vitro* transcription [31,32]. Wengel and co-workers recently described the suitability of locked nucleic acid (LNA) triphosphates for polymerase chain reaction (PCR) and *in vitro* transcription [33,34]. LNAs bear a methylene ether bridge between the 2′-oxygen atom and the 4′-carbon atom. In this way, the 3′-carbon atom is 'locked' in the LNA

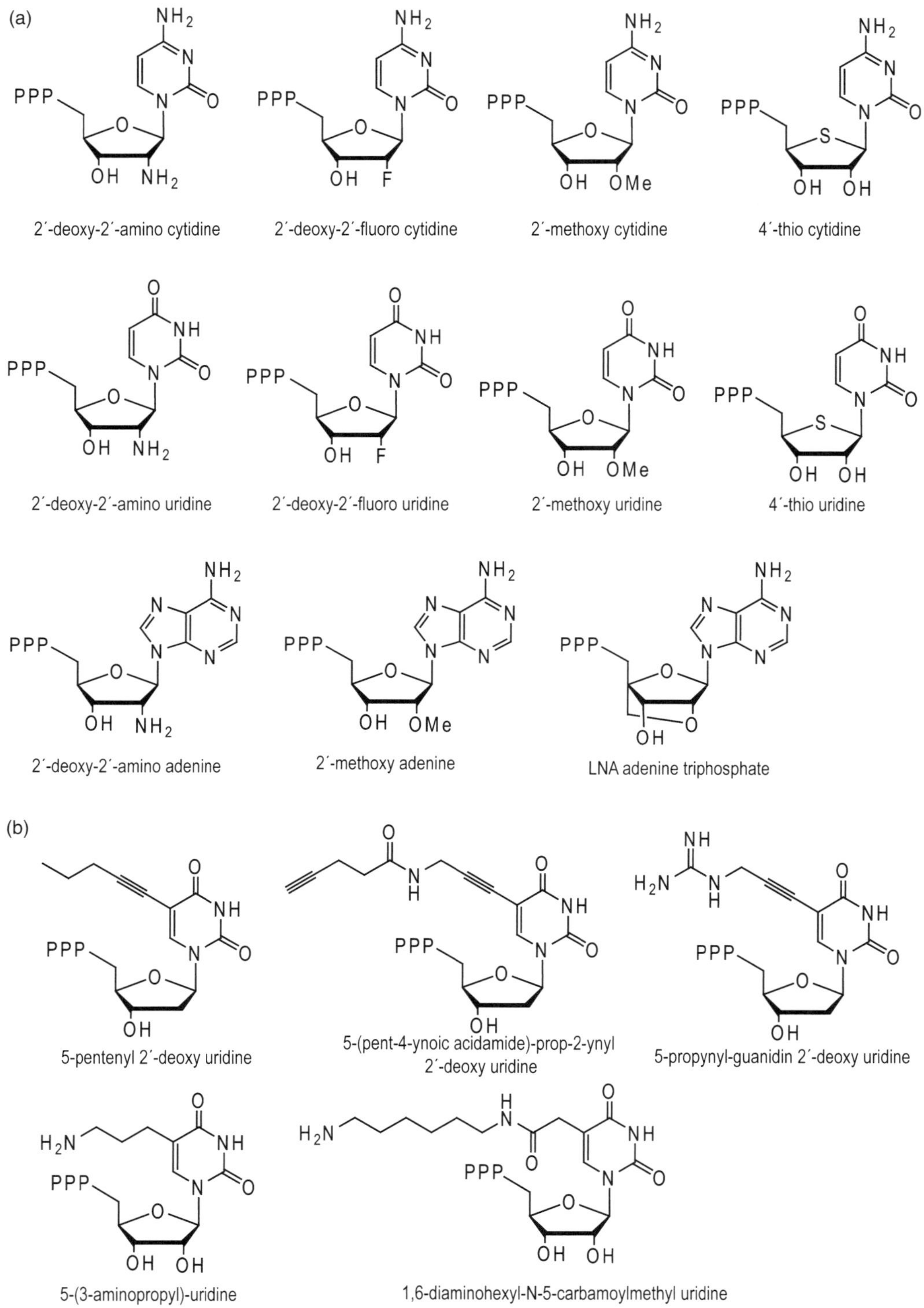

Figure 18.2 *Modifications of nucleotides at either the 2′-position (a) or the C5 position of uridine nucleotides (b) that are compatible with the enzymatic steps of the SELEX procedure (PPP = triphosphate)*

ribose in an *endo* conformation. The adaptation of LNAs for SELEX processes might pave the way to construct nucleic acid libraries with LNA building blocks and the *de novo* selection of LNA aptamers.

Modifications of the nucleobases that are compatible with enzymatic replication have also been reported, with modifications at the C5 position of uridine and deoxyuridine being the most prevalent [35–37]. In this way, distinct chemical moieties have been introduced in nucleic acid libraries and were applied for the *in vitro* selection of aptamers that require modification for target binding. Moreover, the introduction of alkyne groups opens up a route for the modification of aptamers at the completion of the selection, for example, through reaction with azide-derivatized compounds by copper(I)-catalyzed 1,3-dipolar Huisgen cycloadditions [38–41]. This allows an aptamer to be equipped with various functionalities and further adapted to individual applications. Our group has developed aptamers whose activity can be regulated by light (see also Chapter 13). This can be accomplished by the post-selective and site-specific incorporation of photolabile protecting groups, such as *o*-nitrophenylpropyl (NPP) and *o*-nitrophenylethyl (NPE), by solid-phase synthesis by employing modified nucleosides and phosphoramidite approaches [42]. Aptamers with photolabile protecting groups offer the possibility for the spatiotemporal regulation of aptamer activity and thus control over the activity of the target molecule. Aptamers, like nucleic acids, have the advantage that they can be designed to be either activated or deactivated by light [43–45]. This allows access to numerous light-regulated aptamer-based protein inhibitors.

18.2.1 Aptamers and their modifications for diagnostic purposes

Forcing aptamers to fluoresce under certain conditions permits their use as fluorescent probes. Several approaches have been described and excellent reviews dealing with this topic have been published [46–48]. Aptamers can be modified at their 5′-position by chemical synthesis, enzymatic synthesis or by a combination of the two [49,50]. DNA aptamers can be modified by the use of appropriate fluorescently labeled primer molecules and PCR amplification followed by single-strand displacement or directly by chemical synthesis. In the latter case, the introduction of a 5′-amino group represents a versatile anchor that can be further modified by using *N*-hydroxysuccinimide (NHS) or ethylenediaminecarbodiimide (EDC). In this way, the desired fluorescent molecule or any other functional moiety can be introduced (Figure 18.3) [51,52]. Fluorescing RNA aptamers can be synthesized directly by solid-phase synthesis. However, synthesis is limited since RNA molecules longer than 80 nucleotides are difficult to obtain. Alternatively, modifications of RNA molecules at the 5′ ends can be made enzymatically by *in vitro* transcription with T7 RNA polymerases in the presence of so-called initiator nucleotides, such as guanosine monophosphothioate (GMPS) [49,53,54]. The incorporation of GMPS at the 5′ end of RNA molecules permits the modification of RNA molecules with fluorescent moieties or other reporter molecules by iodoacetamido-based approaches (Figure 18.4) [49]. Recently, Jäschke and co-workers developed a novel initiator nucleotide that resulted in an aldehyde modification at the 5′ end of an enzymatically synthesized RNA, which can then be modified with amino- or hydrazine-functionalized groups (Figure 18.5) [55]. This approach allows the synthesis of a broader range of RNA molecules through a combination of chemical and enzymatic approaches and might be very useful for the introduction of diverse functional groups.

In addition to the modification of the 5′ end of RNA molecules, the 3′ ends can also be modified efficiently. One route employs redox reactions and derivatization with hydrazide or semithiocarbazide groups to introduce functional moieties at the 3′ end of RNA molecules [56]. Unfortunately, modification after selection can have a negative effect on the aptamer activities, thereby limiting their utility for diagnostic purposes. Approaches that circumvent this problem make use of aptamers that bind to small molecules, such as Malachite Green, whose fluorescent properties change in the aptamer-bound state [57–62]. The combination of such reporter aptamers with other functional nucleic acid subunits might pave the way to highly developed diagnostic tools. Excellent reports discussing the use of aptamer-based electronic sensors and

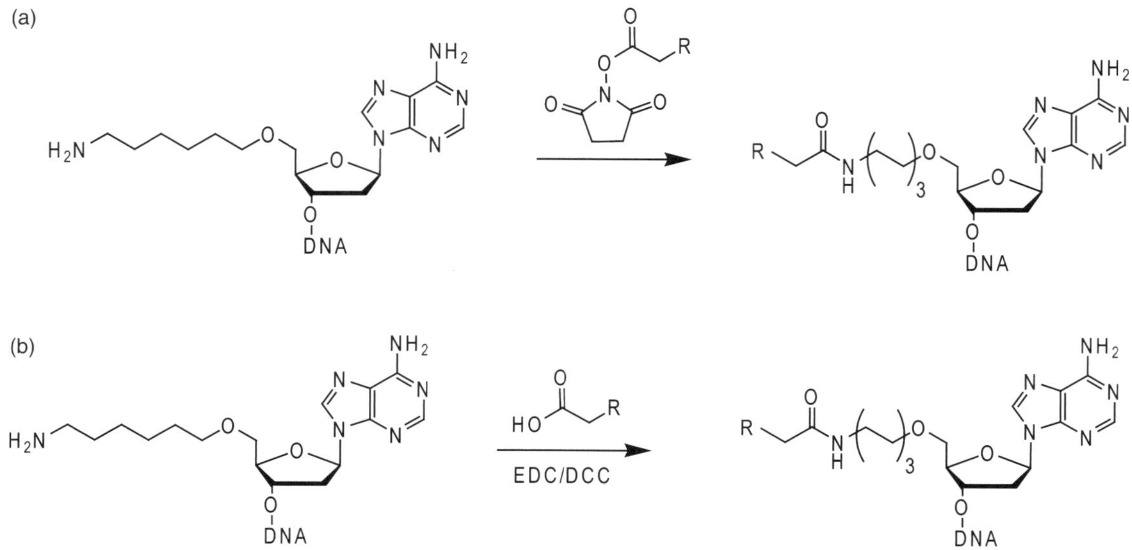

Figure 18.3 Derivatization of 5′-amino-modified DNA aptamers using NHS (a) or EDC/DCC chemistry (b)

microarrays have been published [63–72]. Gold and co-workers introduced the so-called photo-SELEX approach, which permits the photochemical cross-linking of an aptamer to a target protein through incorporated 5-bromodeoxyuridine nucleotides. This allows a significant enhancement of the signal-to-noise ratio in microarrays employing photosensitive aptamers as recognition elements [73]. An elegant approach in which structure switching aptamers are applied has been described by Li and co-workers [74,75]. These aptamers are designed to induce a change in the fluorescence intensity when complexed to the target molecule, thereby exploiting the adaptive binding nature of aptamers to receptor molecules [76,77]. Similarly to antibodies, modified aptamers have been used in diverse assays such as ELISA-like formats [78,79], Western blot analysis [80], capillary electrophoresis [14,81], flow cytometry [82], *in vivo* imaging [83], HPLC [19,84] and microarrays[73,85] that allow the sensitive detection of biomolecules.

An ingenious development in the field of aptamer diagnostics is the use of aptamers as both the recognition element and the template for amplification reactions by either rolling circle amplification [86,87] or aptamer-affinity PCR [88]. Landegren and co-workers adapted aptamers for assay formats based on proximity ligation, thus allowing the very sensitive detection of analytes in complex mixtures [89,90]. Proximity ligation relies on the ligation of DNA molecules located near each other because of receptor–ligand interaction, thus resulting in a new template which is sensitive for detection by PCR. However, no easy-to-use, commercially available clinical assay yet exists; this might be due to the different recognition properties of aptamers compared with the antibodies that are used daily in clinical practice: Aptamers require an intact and folded target structure to maintain high affinity and specific binding, whereas antibodies merely recognize a linear epitope, also in a slightly and totally denatured state of the target protein. Hence there are different prerequisites in target handling prior to the use of aptamers as detection reagents and the habits adopted in the use of highly established methods might be difficult to overcome. The detection of small molecules is a field in which aptamers will probably start to make a great impact, since in this case the handling of the target is not a major issue.

The combination of aptamers with quantum dot technology has provided novel advanced diagnostic assay formats [91–97]. For example, Levy *et al.* immobilized a thrombin-recognizing aptamer on quantum dots by

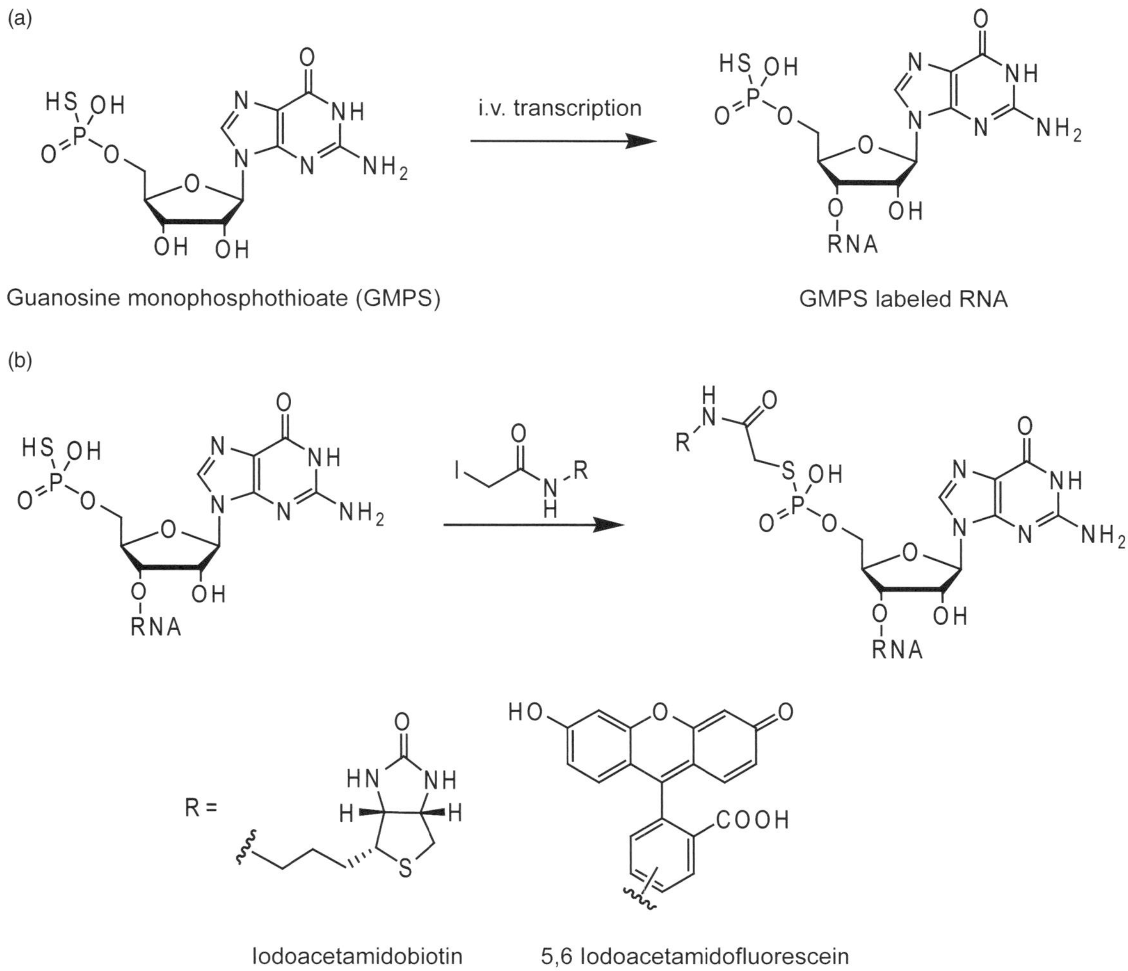

Figure 18.4 *(a) Introduction of guanosine monophosphothioate (GMPS) moieties at the 5′-position of RNA molecules by* in vitro *transcription. (b) GMPS-containing RNA can be further derivatized by iodoacetamido chemistry. Thereby, either biotin or fluorescent groups can be introduced*

employing a streptavidin–biotin approach [93]. An annealed complementary strand of the aptamer equipped with an eclipse quencher was efficiently displaced by the addition of thrombin, thus resulting in enhanced fluorescence of the quantum dots (Figure 18.6a) [98]. Liu and Lu described a similar approach: they derivatized quantum dots with two different short adapter oligodeoxynucleotides. The addition of an aptamer, which partially acts as a splint, causes the quantum dots to aggregate, thereby resulting in a quenched fluorescence signal. The addition of the aptamer ligand results in removal of the splint and thus disassociation of the quantum dots. This was quantified by a colorimetric assay (Figure 18.6b) [94,99,100].

In addition to the streptavidin–biotin approach, Lu and co-workers used thiol-modified aptamers to immobilize the aptamers efficiently on the Au surfaces of the quantum dots. They even went a step further and developed a simple dip-stick assay format based on quantum dot–aptamer conjugates for the rapid and easy detection of cocaine [100,101]. Stojanovic and Landry also used the same anti-cocaine aptamer and

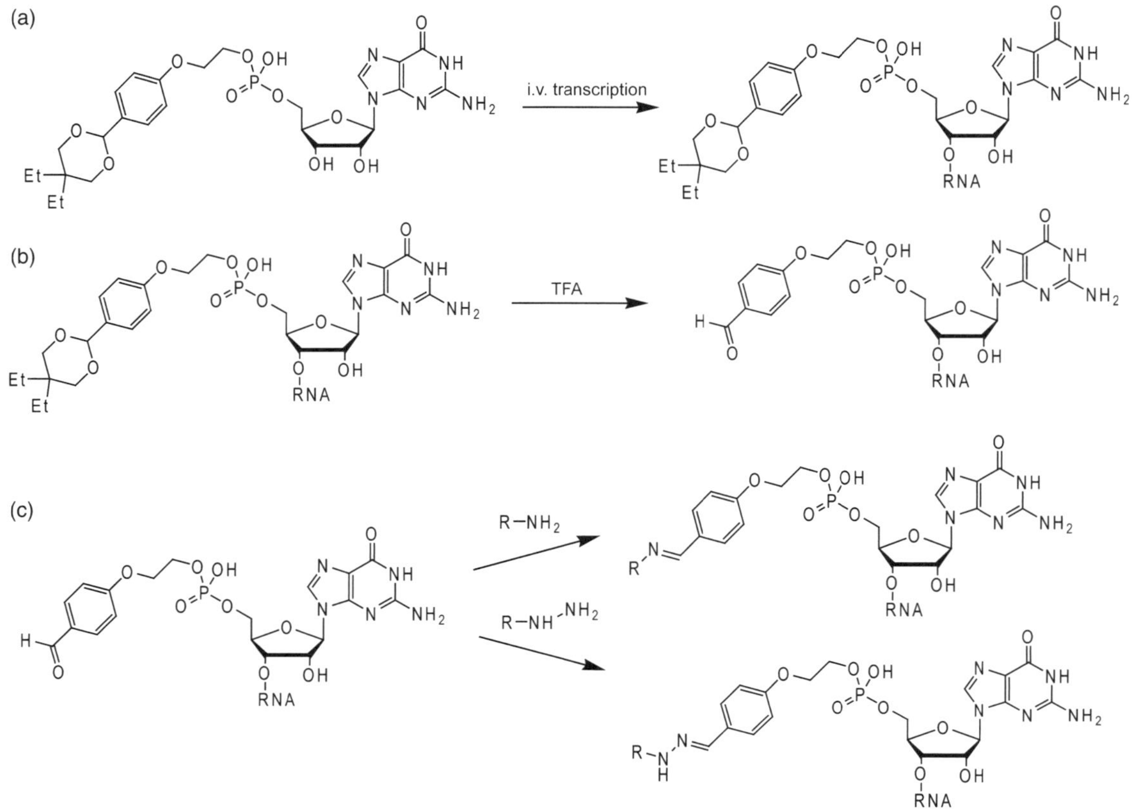

Figure 18.5 (a) The acetal-protected aldehyde–guanosine phosphate can be introduced at the 5′-end of RNA molecules during in vitro transcription. (b) Deprotection of the acetal by trifluoroacetic acid (TFA) releases the corresponding aldehyde, which can be subsequently modified by amino- or hydrazide-derivatized functional moieties (c)

developed a colorimetric detection assay. This assay is, in principle, based on a competition between cocaine and a cyanine dye prebound to the aptamer [95]. In an elegant approach, the antiprostate-specific membrane antigen (anti-PSMA) aptamer A10 has been coupled to quantum dots, whose fluorescence intensity was quenched by a doxorubicin (DOX) molecule that was also noncovalently associated with the aptamer (Figure 18.6c). This multifunctional complex has been shown to be useful for both the specific targeting of tumor cells expressing PSMA and the detection of these cells following DOX release [91]. Simultaneously, the cytostatic activity of the anthracycline drug DOX inhibits cell growth of the targeted cells [102]. Chen *et al.* coupled the 5′-amino-modified tenascin-C binding aptamer GBI-10 to quantum dots by using an EDC/sulfo-NHS approach to detect selectively and visualize glioma cells [103].

18.2.2 Modified aptamers as delivery vehicles

Aptamers with well-defined structure and function can be assembled into multifunctional molecules containing various functional groups. For example, aptamers targeting cells provide the means to develop drug-delivery vehicles that specifically address a certain malignant cell subtype. The synthesis of aptamers allows the manufacture of highly stable oligonucleotides with dramatically enhanced stability against degradation

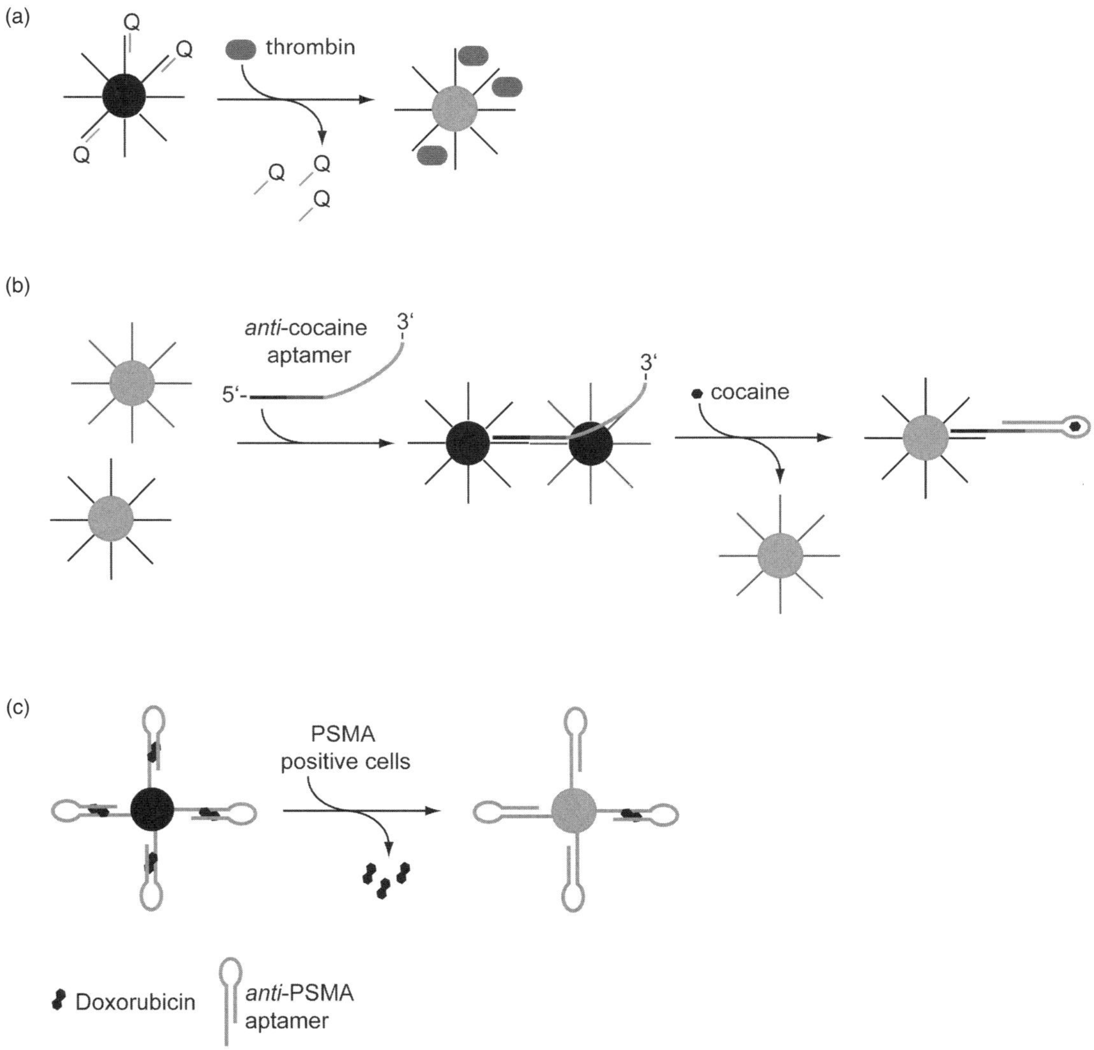

Figure 18.6 *Combination of the quantum dot and aptamer technology. (a) Thrombin-recognizing aptamers have been immobilized to quantum dots (black balls). The fluorescence of the quantum dot is quenched by a complementary oligodeoxynucleotide equipped with a quencher molecule (Q), which hybridizes to the aptamer sequence. Addition of thrombin induces release of the quencher and thus the fluorescence of the quantum dots (gray balls) is detectable. (b) Two different oligodeoxynucleotide sequences were immobilized of quantum dots (gray balls) and the anti-cocaine aptamer anneals to both sequences. Thereby the quantum dots are assembled in close proximity and quenching of fluorescence is induced (black balls). Addition of cocaine leads to the release of the aptamer and thus the disassembly of the quantum dots, resulting in enhanced fluorescence signaling (gray balls). (c) The anti-PSMA aptamer, in complex with doxorubicin (DOX), was immobilized on quantum dots (black ball), whereas DOX quenches the fluorescence of the quantum dots. The addition of PSMA-positive cells leads to a release of DOX and thus to enhanced fluorescence of the quantum dot (gray balls)*

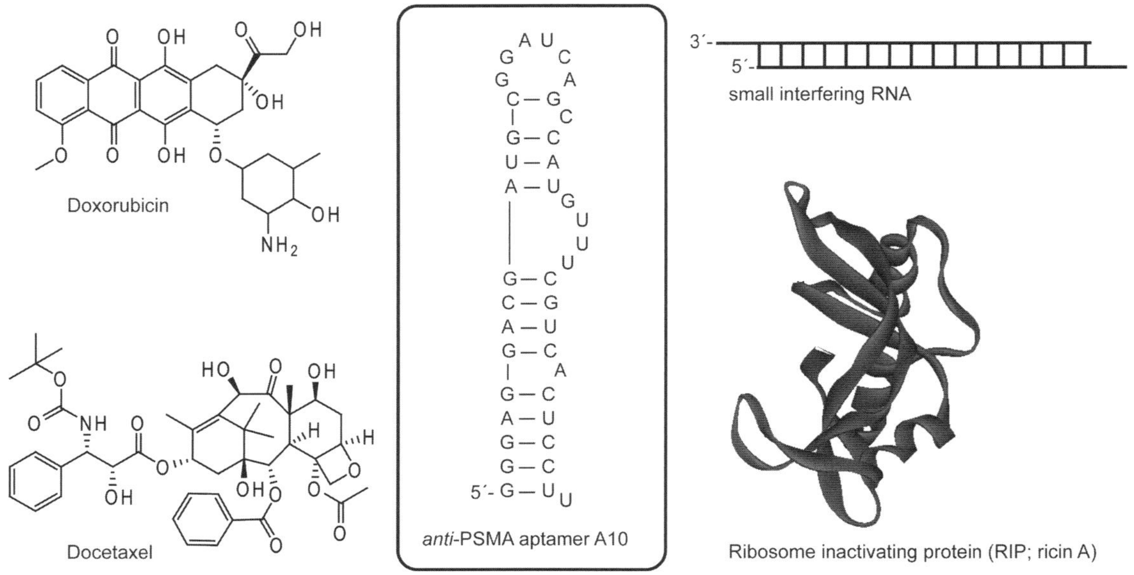

Figure 18.7 *The anti-PSMA aptamer A10 (boxed) has been conjugated to various agents that exert anti-tumor activity. All pyrimidine nucleotides are 2′-fluoro-2′-deoxynucleotides. In addition to coupling to the small molecules, doxorubicin and docetaxel, short interfering RNA (siRNA) molecules and the ribosome-inactivating protein gelonin were also attached. The 3D depicted structure represents ricin A, a protein which is highly homologous with gelonin and whose crystal structure was employed for structure superimposition of gelonin*

by nucleases. The bottom-up construction of functional molecular components with at least one aptamer building block allows access to multifunctional molecules. In this regard, aptamers have been applied as guiding modules in multifunctional complexes to target distinct cell subtypes and tissues. The most established and best characterized aptamer in this regard is the nucleic acid molecule A10, which binds to the prostate-specific membrane antigen (PSMA), a cell-surface molecule associated with the onset and progression of cancer [104]. Several studies have been concerned with the modification of this aptamer with additional molecules, thereby allowing their selective aptamer-triggered delivery to cancer cells (Figure 18.7). For example, polymer-coated nanoparticles with encapsulated chemotherapeutics (such as docetaxel) have been attached to the 5′-amino end of the aptamer by an NHS/EDC approach [105–107]. The systemic application of these complexes allows the efficient targeting of solid tumors. In a breakthrough study Langer, Farokzhad and co-workers showed the efficacy of such a chimeric aptamer cargo molecule in an *in vivo* xenograft rat tumor model system. The aptameric subunit A10 specifically localized the complex in a solid tumor where the adjacent chemotherapeutic induced a reduction in the size of the tumor up to total remission [106]. It is anticipated that such aptamer-based tumor-targeting systems will result in the unwanted side-effects of therapeutics being significantly reduced. In a second example, the same group directly complexed aptamer A10 with DOX, thereby obtaining so-called physical conjugates (noncovalent complexes between aptamers and DOX). Again, treatment of prostate cancer cells resulted in a significant reduction of tumor-cell proliferation [102].

In addition to chemotherapeutics, siRNA molecules can also be used to assemble chimeric siRNA–aptamer conjugates, thereby facilitating the cell-specific delivery of siRNA molecules. In this regard, siRNA molecules have been coupled to aptamer A10 either directly by nucleotidic extensions[108] or indirectly through the assembly of tetrameric streptavidin–biotin complexes consisting of two biotinylated aptamers and two biotinylated siRNA molecules per streptavidin moiety [109]. Both approaches were successful in the cell-specific siRNA-mediated reduction of the corresponding mRNA and protein levels. Aptamer–toxin conjugates were also generated and applied to give specificity for the toxin gelonin, a ribosome-inactivating protein [110]. Ferreira *et al.* reported the application of so-called 'phototoxic aptamers' for the specific entering and targeting of tumor cells [111]. They identified aptamers that bind to the *O*-glycan peptides present on the surface of cancer cells. Modification of the aptamers with the photosensitizer chlorine e6 yielded a bifunctional molecule that allows the targeting and destruction (upon light activation) of the targeted cell.

Recently, Rossi and co-workers added a second series of aptamers to the list of aptamer-based siRNA-delivering agents. They fused an anti-gp120-binding aptamer with siRNAs that are specific for the HIV transcripts tat and rev through a nucleotide linker [112,113]. Treatment of HIV-1-infected cells with these chimeras resulted in the specific down-regulation of the appropriate mRNA transcript. Several other aptamers have been described that specifically recognize cell-surface molecules. Of these, aptamers that target the avb3-integrin [114], the glutamate receptor channel GluR2 [115], VCAM-1 [116], epidermal growth factor receptor-3 [117,118] or the DC-SIGN[119] protein might be also suitable to generate sophisticated and cell-specific multifunctional tools for the delivery of therapeutically active agents.

18.3 Aptamers selected against unknown targets for the identification of biomolecules

An emerging direction in the aptamer field is the systematic identification of target molecules that are associated with distinct cellular states, either pathogenic or not. The identification of new therapeutically and diagnostically relevant biomolecules can be facilitated with the ultimate goal of developing an individualized medical approach. In this regard, aptamers that target a specific cell type or subpopulation of malignant cells (for example, tumor cells) are selected and characterized. After generation of the monoclonal aptamers, the cognate target molecules on the cell surfaces can be subsequently identified by employing aptamer-based pull-down protocols followed by sodium dodecyl sulfate polyacrylamide gel electrophoresis (SDS-PAGE), protease digestion and liquid chromatography–mass spectrometry (LC–MS) analysis [120,121]. The first example of such an approach was reported by Blank *et al.* [51]. In their study, aptamers with specificity for rat endothelial glioblastoma cells (YPEN) were identified. One aptamer has been shown, after modification with a fluorescent label at its 5′ end, to be efficient for staining tumor cells. LC–MS analysis identified the protein pigpen as the putative target of the aptamer. Pigpen is an endothelial proliferation marker protein that is expressed in the proliferating endothelium but its expression is down-regulated until confluency (full coverage of sample vial) is reached [122]. This finding indicates that pigpen is indeed a marker of the angiogenic state of endothelial cells, as already suggested by the aptamer-based approach introduced by Blank *et al.* [123]. A subsequent study by Gold's group further emphasizes the feasibility of the cell–SELEX approach for target identification. In this similar study, the authors identified a tumor cell-specific tenascin-C protein by an aptamer that had previously been selected to target specifically a different glioblastoma cell line [124].

SELEX procedures that target various eukaryotic cells, bacteria, parasites and viruses have since been described [125–128]. Anthrax spores and trypanosomes have been subjected to *in vitro* selection protocols aimed at the discovery of pathogen-neutralizing aptamers useful for therapeutic intervention. The aptamers were successfully applied to inhibit the virulence of the targeted microorganisms, although the cognate targets have not yet been identified. This is clearly an advantage of the aptamer technology, where cell-surface molecules and transmembrane receptors can be targeted within their native environment [129]. Cumbersome purification procedures are not required, which in turn lead to suboptimal activities and stabilities of the biomolecules. Taken together, the targeting of cells with SELEX allows a forward-genetic approach to be established based on a phenotype-correlated selection of aptamers (for example, a tumor cell).

Thereafter, the identification of the associated targets and antigens is possible. Recently, Tan and co-workers applied *in vitro* selection to identify single-stranded DNA (ssDNA) aptamers that target several immortalized tumor cell lines [130–133]. They employed a discriminating counter-SELEX protocol and enriched aptamers that specifically interact with a distinct tumor cell line and most remarkably also with primary cells from patients [134]. Aptamer-based pull-down analysis identified the membrane-bound heavy m chain of immunoglobulin as the putative target molecule presented on the Burkitt's lymphoma cell line (Ramos B-cell) used for the *in vitro* selection process; RTK7 was deconvulated as the target molecule of a second aptamer, selected to interact specifically with T-cell lymphomas [135,136]. Our group has successfully implemented the fluorescent-activated cell-sorting (FACS) technology in the *in vitro* selection process, thus allowing the selection of aptamers that not only possess high affinity, but also target a defined cellular phenotype in composite cell mixtures [137]. This approach might permit the direct targeting of subpopulations found in primary blood tumor cells and thus offers a route for the direct selection of patient-specific aptamers, thereby paving the way for individual diagnostics and therapeutics. Recently, Krylov and co-workers applied a similar whole-cell SELEX approach to mature and immature dendritic cells (DCs). Instead of using monoclonal aptamers, they employed the entire diverse enriched library as an affinity matrix for the simultaneous identification of target molecules by LC–MS analysis [138]. These target molecules might represent novel biomarkers associated with a distinct differentiation level of mature and immature cells.

The limited number of proteins which have been successfully identified by aptamer-based pull-down analysis demonstrates that aptamer selection involving living cells is a difficult task and is still in its infancy. Fitter and James showed that a subset of proteins, which are found in whole blood, for example, can also be employed directly as the target mixture for *in vitro* selection [139]. A similar selection protocol was used by Layzer and Sullenger, who selected aptamers against the γ-carboxyglutamic acid domain containing protein proteome from human plasma [140]. In both approaches, it was possible to enrich aptamers simultaneously against several targets. The limited number of possible targets means that the identification of the appropriate aptamer–target pairs might be simplified in these approaches compared with whole-cell–SELEX experiments. Wendel and co-workers identified aptamers that target stem cells and applied them for the isolation of the stem cells from composite mixtures, with the ultimate goal of developing functionalized materials with a preference for aggregating a certain cell subtype [141–145]. This approach could lead to the development of smart materials based on aptamers.

18.4 Aptamers for target validation

Aptamers are potent inhibitors of their target molecules and therefore they are particularly important for target validation and functional characterization. In the last decade, several studies have been described in which aptamers were employed to validate the function of a biomolecule in cell culture experiments and *in vivo* [146–148]. In the latter case, the stability and pharmacokinetic parameters of aptamers can be manipulated by site-specific chemical modifications, such as with 2′-fluoro, 2′-amino and 2′-methoxy nucleotides and the addition of CAP structures, such as 3′–3′-dT and poly(ethylene glycol) (PEG) [149,150]. This enables the aptamers to be subjected to versatile experimental set-ups and ultimately the use of aptamers as drugs. Aptamers offer a wide range of applications for target validation and the direct targeting of endogenous proteins. The synthesis of RNA aptamers by the cell's own transcription machinery is a great advantage and offers an alternative to transfection [147,151–154]. Recently, selection approaches based on the yeast three hybrid system have been described to assess the *in vivo* properties of *in vitro* selected RNA aptamers and to improve the activities of aptamers *in vivo* [155–159]. Lipid-based transfection experiments, which are used for the transmembrane location of plasmids and siRNA molecules, can also be used to deliver aptamers into the cytoplasm [160]. Other methodologies that might be useful for delivering aptamers inside cells include microinjection, receptor-mediated internalization and the construction of transgenic animals expressing the aptamer under the control of conditional promoters, as shown in a landmark study by Lis and co-workers [161]. The validation of intracellular target molecules with aptamers is a difficult task since this not only depends on the kinetic and thermodynamic properties of the aptamer, but also challenges the limitations of delivering aptamers into the cytoplasm. Each method has its own inherent limitations and makes optimization necessary with respect to the transfection efficiency (which also depends on the cell type used) and read-out format. Primary cells, tissues and organs in living animals are not particularly applicable to state-of-the-art lipid-mediated transfection protocols. However, the application of engineered recombinant viruses might be useful in these cases, but the labor-intensive procedures mean that such examples based on these approaches are rare [146,147,162]. Future examples are required to prove the general utility of viral delivery methods in this regard. An alternative approach might be an adaptation of the siRNA-transfection method. siRNA molecules have been successfully distributed into diverse tissues by equipping them with suitable delivery molecules such as peptides, lipoids, cholesterol and even cell-specific aptamers [108,163–167]. However, these approaches have not yet been applied for the delivery of aptamers. It would be of great interest to explore whether lipid- or peptide-mediated delivery is also feasible for the cellular delivery of aptamers, even in living organisms.

Similarly to off-target effects associated with siRNA approaches [168,169], the secondary endogenous targets of aptamers need to be addressed extensively in future experiments to allow the precise interpretation of aptamer-induced phenotypes. The comprehensive and reliable comparison of the aptamer technology with other techniques for validation of the target molecule, at least at a cellular level, makes the compilation of these data indispensable [170]. Data on the function of the target molecule at the protein level through the application of aptamers can be regarded as complementary to data derived from siRNA and genetic knockout studies. The combination of both approaches might give access to a precise analysis of the function of a biomolecule. In a recent study, Clary, Kontos and co-workers demonstrated that the synergistic effects of an aptamer targeting NF-κB can be exploited to suppress drug-resistant phenotypes of cancer cells that are no longer susceptible to DOX treatment [156,171,172]. DOX resistance has been shown to be caused by the activation of NF-κB, hence the anti-NF-κB aptamer had a beneficial effect. In a different study, Chan *et al.* established that the expression of the anti-NF-κB aptamer in combination with anti-NF-κB siRNA molecules leads to a quantitative inhibition of NF-κB function in mammalian (HeLa) cells. In contrast, the expression of only the aptamer or siRNA molecule leads to nonquantitative inhibition of the target [173].

The aptamer AGRO100, also known as AS1411, targets nucleolin and belongs to a family of G-rich oligodeoxynucleotides that are known to fold into a stable G-quartet structure and inhibit the proliferation of cancer cells [174,175]. The aptamer is undergoing clinical trials as an anticancer agent. A detailed analysis of its mode of action revealed that the aptamer inhibits cancer cell proliferation by forming a competition between the association of nucleolin with distinct co-factors, such as protein arginine methyltransferase 5 and NF-κB essential modulator (NEMO), thereby indicating that nucleolin is necessary for NF-κB regulation [176,177]. Furthermore, the treatment of breast cancer cells with AGRO100 resulted in an increased instability of the BCL-2 mRNA transcript [178]. The groups of Sullenger and Kontos at Duke University developed aptamers that either bind angiopoietin-1 but not angiopoietin-2 and *vice versa.* [179,180] These aptamers are useful for investigating the different functions of both angiopoietin molecules in cell culture and *in vivo*. These data demonstrate that a detailed analysis of the mode of action of an aptamer might uncover new insights into the molecule biology of the target molecule and pave the way for potential new targets for the development of novel therapeutics.

18.5 Drug discovery with aptamers

The modification of aptamers with fluorescing reporter molecules allows the establishment of high-throughput screening assays. In these formats, the aptamer is applied as a competitive reporter probe and small-molecule libraries are searched for compounds that are able to bind to the protein target and to compete with the aptamer for binding. The first report on using aptamers as competitors in screening assays was published in 2001. Green and co-workers employed an anti-PDGF aptamer and screened a small library of compounds which were already known to bind to and inhibit PDGF function, analogous to the parent aptamer [181,182]. Whereas that study used radioactively labeled aptamers, Famulok and co-workers employed so-called aptazymes for high-throughput screening [183,184]. Aptazymes are RNA molecules that contain two domains: one ribozyme domain which can be designed to undergo *cis* or *trans* cleavage and a second aptamer domain that binds to an effector molecule and thereby induces conformational changes within the ribozyme domain [185,186]. This set-up can be employed for the exogenous control of ribozyme activity and was used to screen a library of diverse antibiotics that compete with the anti-HIV-1 rev aptamer for binding to the rev protein. Remarkably, the compound identified, coumermycin A1, was shown to contain aptamer-inherited properties and was able to inhibit virus replication in cell cultures (Figure 18.8). HIV-1 reverse transcriptase inhibitors were identified by a very similar approach, but using a commercially available library of diverse small molecules (Figure 18.8) [187].

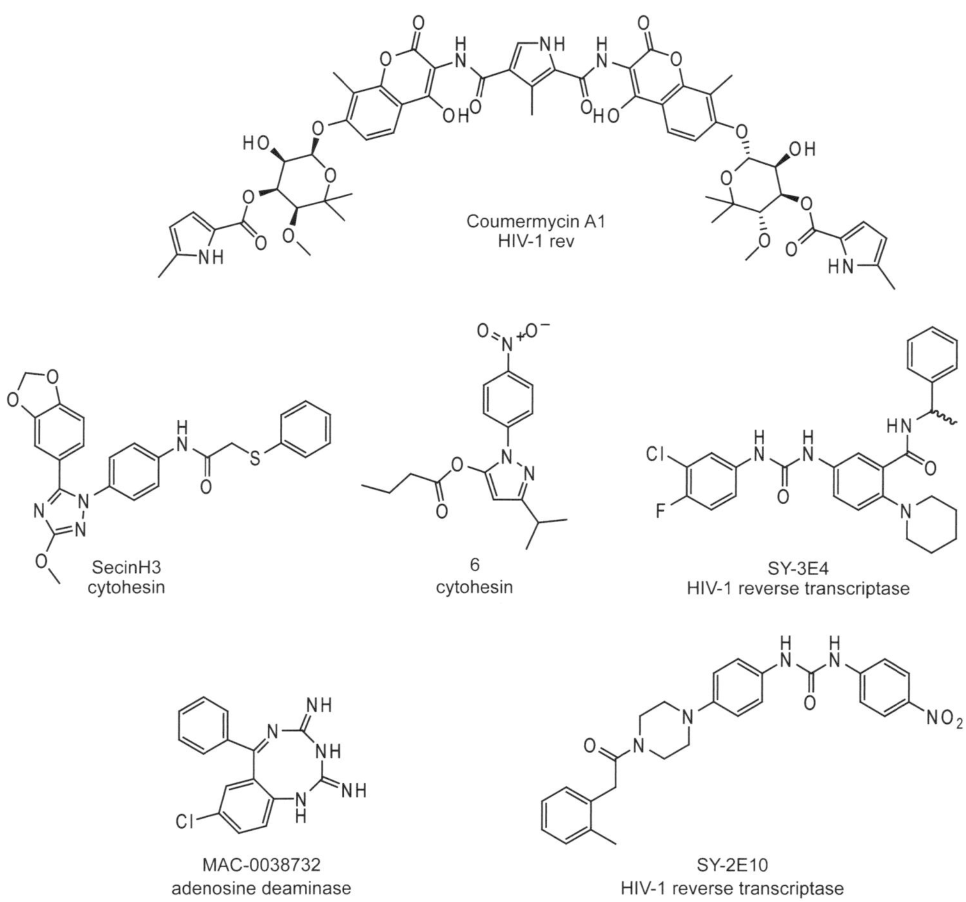

Figure 18.8 *Small organic molecules identified by aptamer-based high-throughput-like screening experiments. The name of the compound is given and below its primary biological target*

The compound SY-3E4 has been shown to inhibit virus replication in HIV-1 strains that are resistant to known reverse transcriptase inhibitors. Since the design of the aptazyme is rather complex and the future applications of aptazymes might focus on the regulation of gene expression [188–191], straightforward protocols are required that allow an aptamer to be used for general screening purposes. One such approach is based on fluorescence polarization and utilizes fluorescently labeled aptamers. In this method, the read-out signal is directly derived from the aptamer and strictly correlates with it being bound to the target molecule or not [192,193]. The application of such an approach led to the identification of the small molecule SecinH3, which binds to cytohesin and competes with the anti-cytohesin aptamer M69 (Figure 18.8) [146,194]. Moreover, the compound could be used to investigate further the function of cytohesins in animals [194,195]. We employed a fluorescence intensity-based approach, based on the detection of target-bound RNA molecules by ribogreen. This assay led to the identification of compound 6 (Figure 18.8) that also targets cytohesins [196].

In a different approach, Li, Brown and co-workers identified a new adenosine deaminase inhibitor [197]. Their screening assay was based on an aptamer that interacts with adenine, whereas the deaminated adenine (the product of deaminase enzymatic activity) was not recognized. The aptamer was thus used to detect enzymatic activity by differentially sensing the product and the reactant. A similar approach was applied by Srinivasan *et al.* [198]. They used an aptamer that recognizes ADP with a 330-fold greater affinity than adenosine triphosphate. This aptamer is a versatile reagent for detecting adenosine diphosphate, which is formed from adenosine triphosphate by kinase activity.

These examples illustrate that aptamers can be employed for the identification of small organic molecules that have protein-inhibiting properties. The most exciting approach is represented by the direct conversion of the chemical information stored in the structure of an aptamer into a small molecule by direct competition assays and fluorescent detection [192,199]. Compounds identified by these approaches displayed aptamer-similar activities and represent a direct link from macromolecular to small-molecule chemistry. An overview of novel protein inhibitors identified by aptamer-based approaches is given in Figure 18.8.

18.6 Aptamers as drugs

In addition to the application of aptamers for the discovery of small-molecule drug candidates, they can also be applied as drugs themselves [200]. Actually, this is one of the first areas that is envisioned to be ideally suited for aptamer applications. Since aptamers have restricted membrane-penetration capabilities, most aptamers for therapeutic purposes were selected to target extracellular proteins. As described in Section 18.2, aptamers with enhanced stability against nucleases can be selected directly by using nucleotides with modifications that are compatible with the enzymatic steps of the *in vitro* selection process. Alternatively, aptamers can be modified after the selection. In this way, they can be equipped with distinct chemical moieties to enhance their stability and improve their pharmacokinetics [149,150]. Healy *et al.* compared the pharmacokinetic and biodistribution of an anti-TGFb-2 aptamer equipped with different modifications, such as PEG20, PEG40, cholesterol or cell-permeating peptides (tat/antennapedia; Figure 18.9) [150]. The aptamer itself was modified with 2′-fluoro and 2′-methoxy groups, since in its unmodified version it was completely eliminated from the bloodstream 48 h following injection. Consistent with previous studies, conjugation with PEG led to an enhanced half-life. The cholesterol-modified aptamer was cleared from plasma, while the addition of cell-permeating peptides did not promote clearance. Conjugation with the antennapedia peptide resulted in a pronounced accumulation of the aptamer in the kidneys. All modifications led to a significant enrichment of the aptamer in the kidney, liver, spleen, heart and mediastinal lymph nodes. This example illustrates that aptamers can be rapidly modified with diverse functional groups, thereby modulating their pharmacokinetic properties. However, each aptamer–drug candidate has to be investigated

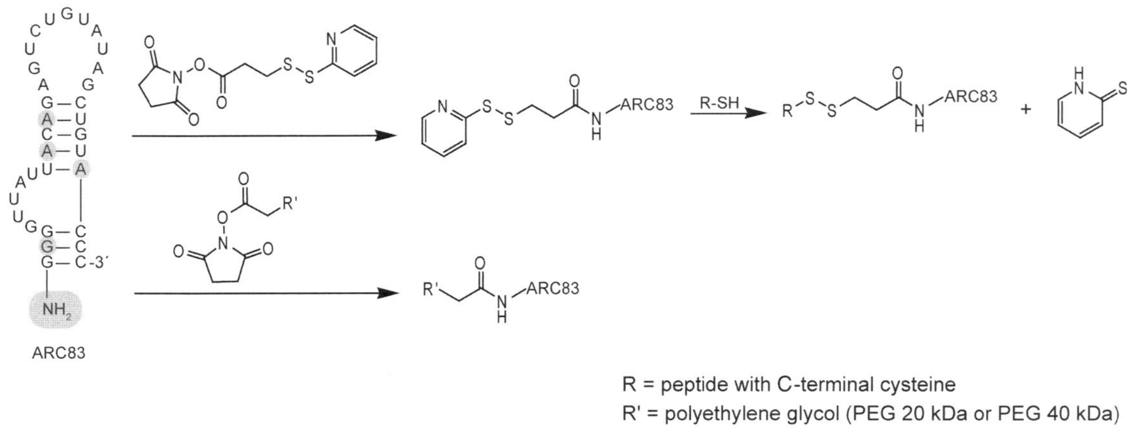

Figure 18.9 *The anti-TGFβ-2 aptamer ARC83 consists of 2′-fluoropyrimidine nucleotides and all purines, in addition to the ones highlighted with gray balls, are modified with 2′-OMe groups. The 5′-position bears an amino group allowing the introduction of further modifications. Therefore, either the commercially available N-succinismidyl 3-(2-pyridyldithio)propionate reagent (SPDP, Pierce) was used to couple peptides via C-terminal cysteine residues or NHS chemistry was applied to couple poly(ethylene glycol) moieties*

individually, since sequence-dependent variations of the biodistribution and pharmacokinetic are possible. One aptamer (Macugen) has been approved to date by the US Food and Drug Administration (FDA) and it is important for the field that others follow. Promisingly, a few aptamers and spiegelmers (RNA aptamers consisting of L-enantiomer building blocks) [201,202] are currently in clinical trials for different indications and disease patterns.

Apart from the treatment of cancer and cardiovascular diseases, the most prominent application of aptamers is as anticoagulants [193]. The first aptamer applied for this purpose is an anti-thrombin G-quadruplex-forming 15-mer oligodeoxynucleotide [203–205]. Clinical trials were initiated for the use of the aptamer, also termed ARC183 or HD1, during coronary artery bypass graft surgery (CABG). However, development was stopped after completion of the Phase I clinical trials because of a low dosage profile [206].

Improved anti-thrombin aptamers have been introduced recently by employing the concept of multivalency. In this regard, exosite I and exosite II binding aptamers have been interconnected by linker moieties, thus generating a bivalent aptamer with superior activities (Figure 18.10) [207–209]. We demonstrated that the bivalent anti-thrombin aptamer has outstanding anticoagulant activities and, advantageously, its activity can be inactivated efficiently by designed antisense molecules [210].

The concept of using complementary oligonucleotides as antidotes for aptamers was initially introduced by Sullenger and co-workers in 2002 [211–213]. They developed an aptamer that targets the blood coagulation factor IXa. A 5′-cholesterol and 3′-deoxythimidine CAP-modified variant of the anti-FIXa aptamer was later shown *in vivo* to be effective as an anticoagulant. Moreover, a 2′-methoxy-derivatized antisense strand, complementary to 17 nucleotides of the aptamer, was able to interfere with the aptamer activity, as shown in a murine model and also in humans [214–216]. Another aptamer that has been developed and already in clinical trials targets the von Willebrand factor (vWF). This aptamer functions as an antithrombotic reagent and is being assessed in the treatment of patients with acute coronary syndrome (ACS). Preliminary trials in humans fulfilled the expectations and showed inhibition of vWF-mediated platelet aggregation *in vivo* [217]. Furthermore, Oney *et al.* showed that a designed antisense molecule reverses the inhibitory activities of the anti-vWF aptamer [218]. Thus, the safety profile of aptamer-based drugs can be enhanced and adverse side-effects minimized. As stated earlier, macugen (or pegaptanib

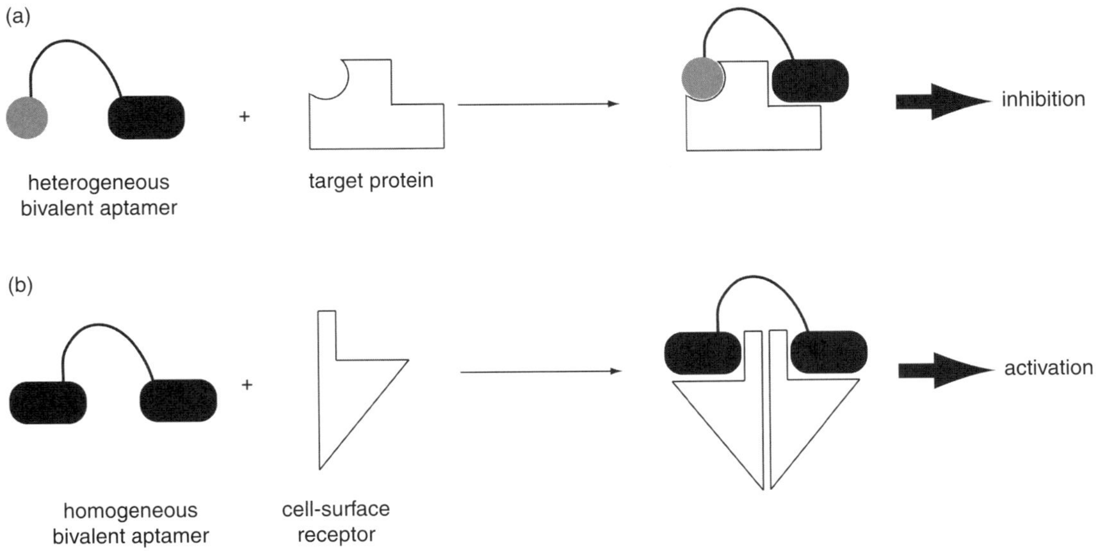

Figure 18.10 *Bivalent aptamers. (a) Heterobivalent aptamers (gray ball and black rectangle) that target a protein via two different domains can have improved inhibitory properties. (b) Homobivalent aptamers (black rectangle) that target cell-surface receptors can induce receptor dimerization and thus the activation of cell signaling*

sodium injection) is the sole aptamer which is approved by the FDA for the treatment of the wet form of age-related macula degeneration (AMD) [219]. Macugen targets the 165 amino acid splice isoform of the vascular-endothelial growth factor (VEGF) and its administration requires intravitreal injection of 6 nmol once every 6 weeks. The efficacy of macugen and its high specificity (also in the context of VEGF isoforms) give it a good safety profile. This is a factor of eminent importance since VEGF plays major roles in a wide range of physiological events.

Sullenger and co-workers introduced agonistic aptamers that are able to stimulate T cells, thus resulting in an efficient inhibition of tumor growth in mice. They identified an aptamer that targets 4-1BB, a co-stimulatory receptor which is up-regulated on activated T cells [220]. Agonistic antibodies targeting 4-1BB have proven successful in enhancing tumor immunity and tumor rejection in mice [221]. Similarly, a multivalent aptamer construct consisting of two identical aptamer domains assembled through a central stem structure was also shown to be effective in the stimulation of T cells and in inducing tumor rejection in mice (Figure 18.10). Recently, the same group expanded the bivalent agonistic aptamer approach to aptamers that target OX40, a member of the tumor necrosis factor receptor family of proteins [222]. These are interesting examples of how an aptamer that binds to cell-surface receptors can be transformed into an agonistic variant, based on a bivalent aptamer and induce clustering of the receptors, similarly to antibodies. Similarly, a study by Wang *et al.* described the identification of an aptamer targeting midkine, a heparin-binding growth factor. Application of the aptamer resulted in an expansion of regulatory T cells, thus inducing reduction of symptoms associated with experimental autoimmune encephalomyelitis (EAE), a model for the severe disease multiple sclerosis [223]. These examples demonstrate that aptamers and engineered variants thereof possess superior activities and might give access to sophisticated therapeutics. The low immunogenicity and cellular-targeting properties of aptamers leads to the hypothesis that they might represent suitable drugs with low side-effects.

18.7 Conclusion and outlook

Since the first description of aptamers in 1990, the field has developed into a mature technology, providing sophisticated tools for biological and medical sciences. The synthetic and enzymatic access to aptamers permits their site-specific and strategic modification and thus their adaptation to various applications such as target validation, as diagnostics and as drugs. Aptamers have the advantage that they do not induce immunological responses, at least at the level of antibody production. Moreover, aptamers can be used for the identification of cell-surface molecules associated with a pathogenic state of a cell; future studies will prove whether the aptamer technology can make significant contributions in this research field.

Aptamer technology thus represents an integrated technology platform. In contrast to other technologies based on nucleic acids, such as siRNA and antisense technologies, the identification of an aptamer is a labor-intensive procedure, consequently leading to the frequent use of these other technologies for target validation. A major challenge that needs to be overcome is the development of easy-to-use SELEX procedures, for example, the provision of ready-to-use 'SELEX kits' that allow the application of *in vitro* selection by any skilled researcher or technician. The characterization of an identified target gene by siRNA molecules needs to be verified by the application of inhibitors acting at the protein level, otherwise the interpretation of phenotypes might be suboptimal, because of off-target effects [224–226]. In this regard, aptamers and siRNA molecules represent complementary approaches whose combination might be a powerful tool.

Aptamers are excellent tools for chemical biology and the precise application and control of their activities will allow the accurate targeting of biomolecules and the high-resolution analysis of biomolecular function. Recent studies have used aptamer technology for the discovery of small organic molecules that possess aptamer-inherited properties. This might pave the way for the development of novel inhibitors of protein functions. The next few years will show whether aptamers will succeed as a novel drug class and as a diagnostic tool for daily clinical practice.

References

1. A. D. Ellington, J. W. Szostak, *Nature* 1990, **346**, 818.
2. C. Tuerk, L. Gold, *Science* 1990, **249**, 505.
3. D. L. Robertson, G. F. Joyce, *Nature* 1990, **344**, 467.
4. A. Heckel, M. Famulok, *Biochimie* 2008, **90**, 1096.
5. G. Mayer, D. Ackermann, N. Kuhn, M. Famulok, *Angew Chem Int Ed* 2008, **47**, 971.
6. G. Rasched, D. Ackermann, T. L. Schmidt, P. Broekmann, A. Heckel, M. Famulok, *Angew Chem Int Ed* 2008, **47**, 967.
7. C. Lin, X. Wang, Y. Liu, N. C. Seeman, H. Yan, *J Am Chem Soc* 2007, **129**, 14475.
8. P. W. Rothemund, *Nature* 2006, **440**, 297.
9. M. Famulok, A. Huttenhofer, *Biochemistry* 1996, **35**, 4265.
10. H. Schurer, K. Stembera, D. Knoll, G. Mayer, M. Blind, H. H. Forster, M. Famulok, P. Welzel, U. Hahn, *Bioorg Med Chem* 2001, **9**, 2557.
11. J. B. Tok, N. O. Fischer, *Chem Commun* 2008, 1883.
12. S. D. Mendonsa, M. T. Bowser, *Anal Chem* 2004, **76**, 5387.
13. S. D. Mendonsa, M. T. Bowser, *J Am Chem Soc* 2005, **127**, 9382.
14. M. Berezovski, A. Drabovich, S. M. Krylova, M. Musheev, V. Okhonin, A. Petrov, S. N. Krylov, *J Am Chem Soc* 2005, **127**, 3165.
15. A. Drabovich, M. Berezovski, S. N. Krylov, *J Am Chem Soc* 2005, **127**, 11224.
16. D. D. Buchanan, E. E. Jameson, J. Perlette, A. Malik, R. T. Kennedy, *Electrophoresis* 2003, **24**, 1375.
17. E. Kraus, W. James, A. N. Barclay, *J Immunol* 1998, **160**, 5209.

18. T. S. Misono, P. K. Kumar, *Anal Biochem* 2005, **342**, 312.
19. J. Muller, O. El-Maarri, J. Oldenburg, B. Potzsch, G. Mayer, *Anal Bioanal Chem* 2008, **390**, 1033.
20. D. Eulberg, K. Buchner, C. Maasch, S. Klussmann, *Nucleic Acids Res* 2005, **33**, e45.
21. J. C. Cox, A. D. Ellington, *Bioorg Med Chem* 2001, **9**, 2525.
22. A. Wochner, B. Cech, M. Menger, V. A. Erdmann, J. Glokler, *Biotechniques* 2007, **43**, 344.
23. G. Mayer, B. Wulffen, C. Huber, J. Brockmann, B. Flicke, L. Neumann, D. Hafenbradl, B. M. Klebl, M. J. Lohse, C. Krasel, M. Blind, *RNA* 2008, **14**, 524.
24. J. H. Davis, J. W. Szostak, *Proc Natl Acad Sci USA* 2002, **99**, 11616.
25. J.Hamm, *Nucleic Acids Res* 1996, **24**, 2220.
26. A. Bugaut, J. J. Toulme, B. Rayner, *Org Biomol Chem* 2006, **4**, 4082.
27. A. D. Keefe, S. T. Cload, *Curr Opin Chem Biol* 2008, **12**, 448.
28. R. M. Orr, *Curr Opin Mol Ther* 2001, **3**, 288.
29. X. Yang, S. E. Bassett, X. Li, B. A. Luxon, N. K. Herzog, R. E. Shope, J. Aronson, T. W. Prow, J. F. Leary, R. Kirby, A. D. Ellington, D. G. Gorenstein, *Nucleic Acids Res* 2002, **30**, e132.
30. J. T. Holmlund, *Ann N Y Acad Sci* 2003, **1002**, 244.
31. J. Chelliserrykattil, A. D. Ellington, *Nat Biotechnol* 2004, **22**, 1155.
32. R. Padilla, R. Sousa, *Nucleic Acids Res* 2002, **30**, e138.
33. R. N. Veedu, B. Vester, J. Wengel, *ChemBioChem* 2007, **8**, 490.
34. R. N. Veedu, B. Vester, J. Wengel, *J Am Chem Soc* 2008, **130**, 8124.
35. J. A. Latham, R. Johnson, J. J. Toole, *Nucleic Acids Res* 1994, **22**, 2817.
36. S. Jager, M. Famulok, *Angew Chem Int Ed* 2004, **43**, 3337.
37. S. Jager, G. Rasched, H. Kornreich-Leshem, M. Engeser, O. Thum, M. Famulok, *J Am Chem Soc* 2005, **127**, 15071.
38. N. Lin, J. Yan, Z. Huang, C. Altier, M. Li, N. Carrasco, M. Suyemoto, L. Johnston, S. Wang, Q. Wang, H. Fang, J. Caton-Williams, B. Wang, *Nucleic Acids Res* 2007, **35**, 1222.
39. J. F. Lutz, Z. Zarafshani, *Adv Drug Deliv Rev* 2008, **60**, 958.
40. P. M. Gramlich, S. Warncke, J. Gierlich, T. Carell, *Angew Chem Int Ed* 2008, **47**, 3442.
41. J. Gierlich, G. A. Burley, P. M. Gramlich, D. M. Hammond, T. Carell, *Org Lett* 2006, **8**, 3639.
42. G. Mayer, A. Heckel, *Angew Chem Int Ed* 2006, **45**, 4900.
43. A. Heckel, M. C. Buff, M. S. Raddatz, J. Muller, B. Potzsch, G. Mayer, *Angew Chem Int Ed* 2006, **45**, 6748.
44. A. Heckel, G. Mayer, *J. Am. Chem. Soc.* 2005, **127**, 822.
45. G. Mayer, L. Krock, V. Mikat, M. Engeser, A. Heckel, *ChemBioChem* 2005, **6**, 1966.
46. J. Hesselberth, M. P. Robertson, S. Jhaveri, A. D. Ellington, *J. Biotechnol.* 2000, **74**, 15.
47. J. R. Hesselberth, M. P. Robertson, S. M. Knudsen, A. D. Ellington, *Anal Biochem* 2003, **312**, 106.
48. M. Famulok, G. Mayer, *Curr. Top. Microbiol. Immunol.* 1999, **243**, 123.
49. G. Sengle, A. Jenne, P. S. Arora, B. Seelig, J. S. Nowick, A. Jaschke, M. Famulok, *Bioorg Med Chem* 2000, **8**, 1317.
50. B. Seelig, A. Jäschke, *Chem. Biol.* 1999, **6**, 167.
51. M. Blank, T. Weinschenk, M. Priemer, H. Schluesener, *J Biol Chem* 2001, **276**, 16464.
52. H. Ulrich, A. H. Martins, J. B. Pesquero, *Cytometry* 2004, **59A**, 220.
53. A. Eisenfuhr, P. S. Arora, G. Sengle, L. R. Takaoka, J. S. Nowick, M. Famulok, *Bioorg Med Chem* 2003, **11**, 235.
54. B. Seelig, A. Jaschke, *Bioconjug Chem* 1999, **10**, 371.
55. S. Pfander, R. Fiammengo, S. I. Kirin, N. Metzler-Nolte, A. Jaschke, *Nucleic Acids Res* 2007, **35**, e25.
56. F. Hansske, F. Cramer, *Methods Enzymol* 1979, **59**, 172.
57. C. Baugh, D. Grate, C. Wilson, *J. Mol. Biol.* 2000, **301**, 117.
58. D. Grate, C. Wilson, *Proc. Natl. Acad. Sci. USA* 1999, **96**, 6131.
59. M. Muller, J. E. Weigand, O. Weichenrieder, B. Suess, *Nucleic Acids Res* 2006, **34**, 2607.
60. S. Hanson, G. Bauer, B. Fink, B. Suess, *RNA* 2005, **11**, 503.
61. M. N. Stojanovic, P. de Prada, D. W. Landry, *J Am Chem Soc* 2001, **123**, 4928.
62. S. Sando, A. Narita, Y. Aoyama, *ChemBioChem* 2007, **8**, 1795.
63. C. A. Savran, S. M. Knudsen, A. D. Ellington, S. R. Manalis, *Anal Chem* 2004, **76**, 3194.
64. M. D. Schlensog, T. M. A. Gronewold, M. Tewes, M. Famulok, E. Quandt, *Sens Actuators B* 2004, **101**, 308.

65. H. Petach, R. Ostroff, C. Greef, G. M. Husar, *Methods Mol Biol* 2004, **264**, 101.
66. H. Petach, L. Gold, *Curr Opin Biotechnol* 2002, **13**, 309.
67. B. R. Baker, R. Y. Lai, M. S. Wood, E. H. Doctor, A. J. Heeger, K. W. Plaxco, *J Am Chem Soc* 2006, **128**, 3138.
68. M. Eisenstein, *Nat Methods* 2006, **3**, 244.
69. R. Kirby, E. J. Cho, B. Gehrke, T. Bayer, Y. S. Park, D. P. Neikirk, J. T. McDevitt, A. D. Ellington, *Anal Chem* 2004, **76**, 4066.
70. R. Y. Lai, K. W. Plaxco, A. J. Heeger, *Anal Chem* 2007, **79**, 229.
71. Y. Xiao, A. A. Lubin, A. J. Heeger, K. W. Plaxco, *Angew Chem Int Ed* 2005, **44**, 5456.
72. I. Willner, M. Zayats, *Angew Chem Int Ed* 2007, **46**, 6408.
73. C. Bock, M. Coleman, B. Collins, J. Davis, G. Foulds, L. Gold, C. Greef, J. Heil, J. S. Heilig, B. Hicke, M. N. Hurst, G. M. Husar, D. Miller, R. Ostroff, H. Petach, D. Schneider, B. Vant-Hull, S. Waugh, A. Weiss, S. K. Wilcox, D. Zichi, *Proteomics* 2004, **4**, 609.
74. R. Nutiu, J. M. Yu, Y. Li, *ChemBioChem* 2004, **5**, 1139.
75. R. Nutiu, Y. Li, *Chemistry* 2004, **10**, 1868.
76. T. Hermann, D. J. Patel, *Science* 2000, **287**, 820.
77. Y. Yang, M. Kochoyan, P. Burgstaller, E. Westhof, M. Famulok, *Science* 1996, **272**, 1343.
78. D. W. Drolet, L. Moon-McDermott, T. S. Romig, *Nat Biotechnol* 1996, **14**, 1021.
79. C. S. Ferreira, K. Papamichael, G. Guilbault, T. Schwarzacher, J. Gariepy, S. Missailidis, *Anal Bioanal Chem* 2008, **390**, 1039.
80. M. B. Murphy, S. T. Fuller, P. M. Richardson, S. A. Doyle, *Nucleic Acids Res* 2003, **31**, e110.
81. H. Zhang, X. F. Li, X. C. Le, *J Am Chem Soc* 2008, **130**, 34.
82. Y. F. Huang, H. T. Chang, W. Tan, *Anal Chem* 2008, **80**, 567.
83. J. Charlton, J. Sennello, D. Smith, *Chem Biol* 1997, **4**, 809.
84. M. Michaud, E. Jourdan, A. Villet, A. Ravel, C. Grosset, E. Peyrin, *J Am Chem Soc* 2003, **125**, 8672.
85. E. N. Brody, M. C. Willis, J. D. Smith, S. Jayasena, D. Zichi, L. Gold, *Mol Diagn* 1999, **4**, 381.
86. B. Schweitzer, S. Wiltshire, J. Lambert, S. O'Malley, K. Kukanskis, Z. Zhu, S. F. Kingsmore, P. M. Lizardi, D. C. Ward, *Proc Natl Acad Sci USA* 2000, **97**, 10113.
87. D. A. Di Giusto, W. A. Wlassoff, J. J. Gooding, B. A. Messerle, G. C. King, *Nucleic Acids Res* 2005, **33**, e64.
88. H. Zhang, Z. Wang, X. F. Li, X. C. Le, *Angew Chem Int Ed* 2006, **45**, 1576.
89. S. Fredriksson, M. Gullberg, J. Jarvius, C. Olsson, K. Pietras, S. M. Gustafsdottir, A. Ostman, U. Landegren, *Nat Biotechnol* 2002, **20**, 473.
90. S. M. Gustafsdottir, S. Wennstrom, S. Fredriksson, E. Schallmeiner, A. D. Hamilton, S. M. Sebti, U. Landegren, *Clin Chem* 2008, **54**, 1218.
91. V. Bagalkot, L. Zhang, E. Levy-Nissenbaum, S. Jon, P. W. Kantoff, R. Langer, O. C. Farokhzad, *Nano Lett* 2007, **7**, 3065.
92. J. A. Hansen, J. Wang, A. N. Kawde, Y. Xiang, K. V. Gothelf, G. Collins, *J Am Chem Soc* 2006, **128**, 2228.
93. M. Levy, S. F. Cater, A. D. Ellington, *ChemBioChem* 2005, **6**, 2163.
94. J. Liu, J. H. Lee, Y. Lu, *Anal Chem* 2007, **79**, 4120.
95. M. N. Stojanovic, D. W. Landry, *J Am Chem Soc* 2002, **124**, 9678.
96. A. P. Alivisatos, W. Gu, C. Larabell, *Annu Rev Biomed Eng* 2005, **7**, 55.
97. S. Dwarakanath, J. G. Bruno, A. Shastry, T. Phillips, A. A. John, A. Kumar, L. D. Stephenson, *Biochem Biophys Res Commun* 2004, **325**, 739.
98. L. C. Bock, L. C. Griffin, J. A. Latham, E. H. Vermaas, J. J. Toole, *Nature* 1992, **355**, 564.
99. J. Liu, Y. Lu, *Nat Protoc* 2006, **1**, 246.
100. J. Liu, D. Mazumdar, Y. Lu, *Angew Chem Int Ed* 2006, **45**, 7955.
101. M. Famulok, G. Mayer, *Nature* 2006, **439**, 666.
102. V. Bagalkot, O. C. Farokhzad, R. Langer, S. Jon, *Angew Chem Int Ed* 2006, **45**, 8149.
103. X. Chen, Y. Deng, Y. Lin, D. Pang, H. Qing, F. Qu, H. Xie, *Nanotechnology* 2008, **19**.
104. S. E. Lupold, B. J. Hicke, Y. Lin, D. S. Coffey, *Cancer Res* 2002, **62**, 4029.
105. O. C. Farokhzad, S. Jon, A. Khademhosseini, T. N. Tran, D. A. Lavan, R. Langer, *Cancer Res* 2004, **64**, 7668.
106. O. C. Farokhzad, J. Cheng, B. A. Teply, I. Sherifi, S. Jon, P. W. Kantoff, J. P. Richie, R. Langer, *Proc Natl Acad Sci USA* 2006, **103**, 6315.

107. J. Cheng, B. A. Teply, I. Sherifi, J. Sung, G. Luther, F. X. Gu, E. Levy-Nissenbaum, A. F. Radovic-Moreno, R. Langer, O. C. Farokhzad, *Biomaterials* 2007, **28**, 869.
108. J. O. McNamara, 2nd, E. R. Andrechek, Y. Wang, K. D. Viles, R. E. Rempel, E. Gilboa, B. A. Sullenger, P. H. Giangrande, *Nat Biotechnol* 2006, **24**, 1005.
109. T. C. Chu, K. Y. Twu, A. D. Ellington, M. Levy, *Nucleic Acids Res* 2006, **34**, e73.
110. T. C. Chu, J. W. Marks, 3rd, L. A. Lavery, S. Faulkner, M. G. Rosenblum, A. D. Ellington, M. Levy, *Cancer Res* 2006, **66**, 5989.
111. C. S. Ferreira, M. C. Cheung, S. Missailidis, S. Bisland, J. Gariepy, *Nucleic Acids Res* 2009, **37**, 866.
112. J. Zhou, P. Swiderski, H. Li, J. Zhang, C. P. Neff, R. Akkina, J. J. Rossi, *Nucleic Acids Res* 2009, **37**, 3094.
113. J. Zhou, H. Li, S. Li, J. Zaia, J. J. Rossi, *Mol Ther* 2008, **16**, 1481.
114. J. Mi, X. Zhang, P. H. Giangrande, J. O. McNamara II, S. M. Nimjee, S. Sarraf-Yazdi, B. A. Sullenger, B. M. Clary, *Biochem Biophys Res Commun* 2005, **338**, 956.
115. Z. Huang, W. Pei, S. Jayaseelan, H. Shi, L. Niu, *Biochemistry* 2007, **46**, 12648.
116. F. Chauveau, Y. Aissouni, J. Hamm, H. Boutin, D. Libri, F. Duconge, B. Tavitian, *Bioorg Med Chem Lett* 2007, **17**, 6119.
117. Y. Liu, C. T. Kuan, J. Mi, X. Zhang, B. M. Clary, D. D. Bigner, B. A. Sullenger, *Biol Chem* 2009, **390**, 137.
118. C. H. Chen, G. A. Chernis, V. Q. Hoang, R. Landgraf, *Proc Natl Acad Sci USA* 2003, **100**, 9226.
119. Y. Hui, L. Shan, Z. Lin-Fu, Z. Jian-Hua, *Mol Cell Biochem* 2007, **306**, 71.
120. J. A. Phillips, D. Lopez-Colon, Z. Zhu, Y. Xu, W. Tan, *Anal Chim Acta* 2008, **621**, 101.
121. S. M. Shamah, J. M. Healy, S. T. Cload, *Acc Chem Res* 2008, **41**, 130.
122. M. C. Alliegro, M. A. Alliegro, *J Biol Chem* 2002, **277**, 19037.
123. M. C. Alliegro, *Cell Biol Int* 2001, **25**, 577.
124. D. A. Daniels, H. Chen, B. J. Hicke, K. M. Swiderek, L. Gold, *Proc Natl Acad Sci USA* 2003, **100**, 15416.
125. J. G. Bruno, J. L. Kiel, *Biosens Bioelectron* 1999, **14**, 457.
126. F. Chen, J. Zhou, F. Luo, A. B. Mohammed, X. L. Zhang, *Biochem Biophys Res Commun* 2007, **357**, 743.
127. M. Homann, M. Lorger, M. Engstler, M. Zacharias, H. U. Goringer, *Comb Chem High Throughput Screen* 2006, **9**, 491.
128. M. Homann, H. U. Goringer, *Bioorg Med Chem* 2001, **9**, 2571.
129. L. Cerchia, F. Duconge, C. Pestourie, J. Boulay, Y. Aissouni, K. Gombert, B. Tavitian, V. de Franciscis, D. Libri, *PLoS Biol* 2005, **3**, e123.
130. H. W. Chen, C. D. Medley, K. Sefah, D. Shangguan, Z. Tang, L. Meng, J. E. Smith, W. Tan, *ChemMedChem* 2008, **3**, 991.
131. D. Shangguan, Y. Li, Z. Tang, Z. C. Cao, H. W. Chen, P. Mallikaratchy, K. Sefah, C. J. Yang, W. Tan, *Proc Natl Acad Sci USA* 2006, **103**, 11838.
132. D. Shangguan, L. Meng, Z. C. Cao, Z. Xiao, X. Fang, Y. Li, D. Cardona, R. P. Witek, C. Liu, W. Tan, *Anal Chem* 2008, **80**, 721.
133. Z. Tang, D. Shangguan, K. Wang, H. Shi, K. Sefah, P. Mallikratchy, H. W. Chen, Y. Li, W. Tan, *Anal Chem* 2007, **79**, 4900.
134. D. Shangguan, Z. C. Cao, Y. Li, W. Tan, *Clin Chem* 2007, **53**, 1153.
135. P. Mallikaratchy, Z. Tang, S. Kwame, L. Meng, D. Shangguan, W. Tan, *Mol Cell Proteomics* 2007, **6**, 2230.
136. D. Shangguan, Z. Cao, L. Meng, P. Mallikaratchy, K. Sefah, H. Wang, Y. Li, W. Tan, *J Proteome Res* 2008, **7**, 2133.
137. M. S. Raddatz, A. Dolf, E. Endl, P. Knolle, M. Famulok, G. Mayer, *Angew Chem Int Ed* 2008, **47**, 5190.
138. M. V. Berezovski, M. Lechmann, M. U. Musheev, T. W. Mak, S. N. Krylov, *J Am Chem Soc* 2008, **130**, 9137.
139. S. Fitter, R. James, *J Biol Chem* 2005, **280**, 34193.
140. J. M. Layzer, B. A. Sullenger, *Oligonucleotides* 2007, **17**, 1.
141. K. Guo, H. P. Wendel, L. Scheideler, G. Ziemer, A. M. Scheule, *J Cell Mol Med* 2005, **9**, 731.
142. K. T. Guo, R. SchAfer, A. Paul, A. Gerber, G. Ziemer, H. P. Wendel, *Stem Cells* 2006, **24**, 2220.
143. K. T. Guo, D. Scharnweber, B. Schwenzer, G. Ziemer, H. P. Wendel, *Biomaterials* 2007, **28**, 468.
144. J. Hoffmann, A. Paul, M. Harwardt, J. Groll, T. Reeswinkel, D. Klee, M. Moeller, H. Fischer, T. Walker, T. Greiner, G. Ziemer, H. P. Wendel, *J Biomed Mater Res A* 2008, **84**, 614.
145. R. Schafer, J. Wiskirchen, K. Guo, B. Neumann, R. Kehlbach, J. Pintaske, V. Voth, T. Walker, A. M. Scheule, T. O. Greiner, U. Hermanutz-Klein, C. D. Claussen, H. Northoff, G. Ziemer, H. P. Wendel, *Rofo* 2007, **179**, 1009.

146. G. Mayer, M. Blind, W. Nagel, T. Bohm, T. Knorr, C. L. Jackson, W. Kolanus, M. Famulok, *Proc Natl Acad Sci USA* 2001, **98**, 4961.
147. J. Mi, X. Zhang, Y. Liu, S. K. Reddy, Z. N. Rabbani, B. A. Sullenger, B. M. Clary, *Biochem Biophys Res Commun* 2007, **359**, 475.
148. M. Famulok, J. S. Hartig, G. Mayer, *Chem Rev* 2007, **107**, 3715.
149. T. G. McCauley, J. C. Kurz, P. G. Merlino, S. D. Lewis, M. Gilbert, D. M. Epstein, H. N. Marsh, *Pharm Res* 2006, **23**, 303.
150. J. M. Healy, S. D. Lewis, M. Kurz, R. M. Boomer, K. M. Thompson, C. Wilson, T. G. McCauley, *Pharm Res* 2004, **21**, 2234.
151. R. E. Martell, J. R. Nevins, B. A. Sullenger, *Mol Ther* 2002, **6**, 30.
152. H. K. Lee, H. Y. Kwak, J. Hur, I. A. Kim, J. S. Yang, M. W. Park, J. Yu, S. Jeong, *Cancer Res* 2007, **67**, 9315.
153. K. H. Choi, M. W. Park, S. Y. Lee, M. Y. Jeon, M. Y. Kim, H. K. Lee, J. Yu, H. J. Kim, K. Han, H. Lee, K. Park, W. J. Park, S. Jeong, *Mol Cancer Ther* 2006, **5**, 2428.
154. H. K. Lee, Y. S. Choi, Y. A. Park, S. Jeong, *Cancer Res* 2006, **66**, 10560.
155. J. Konig, C. Julius, S. Baumann, M. Homann, H. U. Goringer, M. Feldbrugge, *RNA* 2007, **13**, 614.
156. L. A. Cassiday, L. J. Maher III, *Proc Natl Acad Sci USA* 2003, **100**, 3930.
157. C. Lorenz, N. Piganeau, R. Schroeder, *Nucleic Acids Res* 2006, **34**, 334.
158. N. Piganeau, U. E. Schauer, R. Schroeder, *RNA* 2006, **12**, 177.
159. N. Piganeau, R. Schroeder, *Nat Protoc* 2006, **1**, 689.
160. M. G. Theis, A. Knorre, B. Kellersch, J. Moelleken, F. Wieland, W. Kolanus, M. Famulok, *Proc Natl Acad Sci USA* 2004, **101**, 11221.
161. H. Shi, B. E. Hoffman, J. T. Lis, *Proc Natl Acad Sci USA* 1999, **96**, 10033.
162. M. Blind, W. Kolanus, M. Famulok, *Proc Natl Acad Sci USA* 1999, **96**, 3606.
163. C. Wolfrum, S. Shi, K. N. Jayaprakash, M. Jayaraman, G. Wang, R. K. Pandey, K. G. Rajeev, T. Nakayama, K. Charrise, E. M. Ndungo, T. Zimmermann, V. Koteliansky, M. Manoharan, M. Stoffel, *Nat Biotechnol* 2007, **25**, 1149.
164. A. Akinc, A. Zumbuehl, M. Goldberg, E. S. Leshchiner, V. Busini, N. Hossain, S. A. Bacallado, D. N. Nguyen, J. Fuller, R. Alvarez, A. Borodovsky, T. Borland, R. Constien, A. de Fougerolles, J. R. Dorkin, K. Narayanannair Jayaprakash, M. Jayaraman, M. John, V. Koteliansky, M. Manoharan, L. Nechev, J. Qin, T. Racie, D. Raitcheva, K. G. Rajeev, D. W. Sah, J. Soutschek, I. Toudjarska, H. P. Vornlocher, T. S. Zimmermann, R. Langer, D. G. Anderson, *Nat Biotechnol* 2008, **26**, 561.
165. R. Rennert, I. Neundorf, A. G. Beck-Sickinger, *Adv Drug Deliv Rev* 2008, **60**, 485.
166. J. Soutschek, A. Akinc, B. Bramlage, K. Charisse, R. Constien, M. Donoghue, S. Elbashir, A. Geick, P. Hadwiger, J. Harborth, M. John, V. Kesavan, G. Lavine, R. K. Pandey, T. Racie, K. G. Rajeev, I. Rohl, I. Toudjarska, G. Wang, S. Wuschko, D. Bumcrot, V. Koteliansky, S. Limmer, M. Manoharan, H. P. Vornlocher, *Nature* 2004, **432**, 173.
167. P. Kumar, H. Wu, J. L. McBride, K. E. Jung, M. H. Kim, B. L. Davidson, S. K. Lee, P. Shankar, N. Manjunath, *Nature* 2007, **448**, 39.
168. E. Anderson, Q. Boese, A. Khvorova, J. Karpilow, *Methods Mol Biol* 2008, **442**, 45.
169. Y. Fedorov, E. M. Anderson, A. Birmingham, A. Reynolds, J. Karpilow, K. Robinson, D. Leake, W. S. Marshall, A. Khvorova, *RNA* 2006, **12**, 1188.
170. T. Mori, A. Oguro, T. Ohtsu, Y. Nakamura, *Nucleic Acids Res* 2004, **32**, 6120.
171. J. Mi, X. Zhang, Z. N. Rabbani, Y. Liu, S. K. Reddy, Z. Su, F. K. Salahuddin, K. Viles, P. H. Giangrande, M. W. Dewhirst, B. A. Sullenger, C. D. Kontos, B. M. Clary, *Mol Ther* 2008, **16**, 66.
172. J. Mi, X. Zhang, Z. N. Rabbani, Y. Liu, Z. Su, Z. Vujaskovic, C. D. Kontos, B. A. Sullenger, B. M. Clary, *Nucleic Acids Res* 2006, **34**, 3577.
173. R. Chan, M. Gilbert, K. M. Thompson, H. N. Marsh, D. M. Epstein, P. S. Pendergrast, *Nucleic Acids Res* 2006, **34**, e36.
174. V. Dapic, V. Abdomerovic, R. Marrington, J. Peberdy, A. Rodger, J. O. Trent, P. J. Bates, *Nucleic Acids Res* 2003, **31**, 2097.
175. C. R. Ireson, L. R. Kelland, *Mol Cancer Ther* 2006, **5**, 2957.
176. Y. Teng, A. C. Girvan, L. K. Casson, W. M. Pierce Jr, M. Qian, S. D. Thomas, P. J. Bates, *Cancer Res* 2007, **67**, 10491.

177. A. C. Girvan, Y. Teng, L. K. Casson, S. D. Thomas, S. Juliger, M. W. Ball, J. B. Klein, W. M. Pierce, Jr., S. S. Barve, P. J. Bates, *Mol Cancer Ther* 2006, **5**, 1790.
178. S. Soundararajan, W. Chen, E. K. Spicer, N. Courtenay-Luck, D. J. Fernandes, *Cancer Res* 2008, **68**, 2358.
179. R. R. White, J. A. Roy, K. D. Viles, B. A. Sullenger, C. D. Kontos, *Angiogenesis* 2008, **11**, 395.
180. R. R. White, S. Shan, C. P. Rusconi, G. Shetty, M. W. Dewhirst, C. D. Kontos, B. A. Sullenger, *Proc Natl Acad Sci USA* 2003, **100**, 5028.
181. L. S. Green, C. Bell, N. Janjic, *Biotechniques* 2001, **30**, 1094.
182. J. Floege, T. Ostendorf, U. Janssen, M. Burg, H. H. Radeke, C. Vargeese, S. C. Gill, L. S. Green, N. Janjic, *Am J Pathol* 1999, **154**, 169.
183. J. S. Hartig, S. H. Najafi-Shoushtari, I. Grune, A. Yan, A. D. Ellington, M. Famulok, *Nat Biotechnol* 2002, **20**, 717.
184. J. S. Hartig, M. Famulok, *Angew Chem Int Ed* 2002, **41**, 4263.
185. G. A. Soukup, R. R. Breaker, *Proc Natl Acad Sci USA* 1999, **96**, 3584.
186. A. Jenne, J. S. Hartig, N. Piganeau, A. Tauer, D. A. Samarsky, M. R. Green, J. Davies, M. Famulok, *Nat Biotechnol* 2001, **19**, 56.
187. S. Yamazaki, L. Tan, G. Mayer, J. S. Hartig, J. N. Song, S. Reuter, T. Restle, S. D. Laufer, D. Grohmann, H. G. Krausslich, J. Bajorath, M. Famulok, *Chem Biol* 2007, **14**, 804.
188. M. Wieland, J. S. Hartig, *Angew Chem Int Ed* 2008, **47**, 2604.
189. M. Wieland, J. S. Hartig, *ChemBioChem* 2008, **9**, 1873.
190. B. Suess, J. E. Weigand, *RNA Biol* 2008, **5**, 24.
191. T. S. Bayer, C. D. Smolke, *Nat Biotechnol* 2005, **23**, 337.
192. M. Hafner, E. Vianini, B. Albertoni, L. Marchetti, I. Grune, C. Gloeckner, M. Famulok, *Nat Protoc* 2008, **3**, 579.
193. G. Mayer, M. Famulok, *ChemBioChem* 2006, **7**, 602.
194. M. Hafner, A. Schmitz, I. Grune, S. G. Srivatsan, B. Paul, W. Kolanus, T. Quast, E. Kremmer, I. Bauer, M. Famulok, *Nature* 2006, **444**, 941.
195. B. Fuss, T. Becker, I. Zinke, M. Hoch, *Nature* 2006, **444**, 945.
196. Mayer G, Faulhammer D, Grättinger M, Fessele S and Blind M., *ChemBioChem* 2009, **10**, 1993.
197. N. H. Elowe, R. Nutiu, A. Allali-Hassani, J. D. Cechetto, D. W. Hughes, Y. Li, E. D. Brown, *Angew Chem Int Ed* 2006, **45**, 5648.
198. J. Srinivasan, S. T. Cload, N. Hamaguchi, J. Kurz, S. Keene, M. Kurz, R. M. Boomer, J. Blanchard, D. Epstein, C. Wilson, J. L. Diener, *Chem Biol* 2004, **11**, 499.
199. P. Porschewski, M. A. Grattinger, K. Klenzke, A. Erpenbach, M. R. Blind, F. Schafer, *J Biomol Screen* 2006, **11**, 773.
200. G. Mayer, A. Jenne, *BioDrugs* 2004, **18**, 351.
201. W. G. Purschke, D. Eulberg, K. Buchner, S. Vonhoff, S. Klussmann, *Proc Natl Acad Sci USA* 2006, **103**, 5173.
202. A. Vater, F. Jarosch, K. Buchner, S. Klussmann, *Nucleic Acids Res* 2003, **31**, e130.
203. S. M. Nimjee, C. P. Rusconi, R. A. Harrington, B. A. Sullenger, *Trends Cardiovasc Med* 2005, **15**, 41.
204. A. DeAnda, Jr., S. E. Coutre, M. R. Moon, C. M. Vial, L. C. Griffin, V. S. Law, M. Komeda, L. L. Leung, D. C. Miller, *Ann Thorac Surg* 1994, **58**, 344.
205. A. Joachimi, G. Mayer, J. S. Hartig, *J Am Chem Soc* 2007, **129**, 3036.
206. Archemix, press release, http://www.archemix.com/website/_popup_press_release.php?release=17, 2005.
207. J. Muller, B. Wulffen, B. Potzsch, G. Mayer, *ChemBioChem* 2007, **8**, 2223.
208. Y. Kim, Z. Cao, W. Tan, *Proc Natl Acad Sci USA* 2008, **105**, 5664.
209. L. Tian, T. Heyduk, *Biochemistry* 2009, **48**, 264.
210. J. Müller, D. Freitag, B. Pötzsch, G. Mayer, *J Thromb Haemost* 2008, **6**, 2105.
211. C. P. Rusconi, E. Scardino, J. Layzer, G. A. Pitoc, T. L. Ortel, D. Monroe, B. A. Sullenger, *Nature* 2002, **419**, 90.
212. C. P. Rusconi, J. D. Roberts, G. A. Pitoc, S. M. Nimjee, R. R. White, G. Quick Jr, E. Scardino, W. P. Fay, B. A. Sullenger, *Nat Biotechnol* 2004, **22**, 1423.
213. S. M. Nimjee, J. R. Keys, G. A. Pitoc, G. Quick, C. P. Rusconi, B. A. Sullenger, *Mol Ther* 2006, **14**, 408.
214. C. K. Dyke, S. R. Steinhubl, N. S. Kleiman, R. O. Cannon, L. G. Aberle, M. Lin, S. K. Myles, C. Melloni, R. A. Harrington, J. H. Alexander, R. C. Becker, C. P. Rusconi, *Circulation* 2006, **114**, 2490.

215. M. Y. Chan, M. G. Cohen, C. K. Dyke, S. K. Myles, L. G. Aberle, M. Lin, J. Walder, S. R. Steinhubl, I. C. Gilchrist, N. S. Kleiman, D. A. Vorchheimer, N. Chronos, C. Melloni, J. H. Alexander, R. A. Harrington, R. M. Tonkens, R. C. Becker, C. P. Rusconi, *Circulation* 2008, **117**, 2865.
216. M. Y. Chan, C. P. Rusconi, J. H. Alexander, R. M. Tonkens, R. A. Harrington, R. C. Becker, *J Thromb Haemost* 2008, **6**, 789.
217. J. C. Gilbert, T. DeFeo-Fraulini, R. M. Hutabarat, C. J. Horvath, P. G. Merlino, H. N. Marsh, J. M. Healy, S. Boufakhreddine, T. V. Holohan, R. G. Schaub, *Circulation* 2007, **116**, 2678.
218. S. Oney, S. M. Nimjee, J. Layzer, N. Que-Gewirth, D. Ginsburg, R. C. Becker, G. Arepally, B. A. Sullenger, *Oligonucleotides* 2007, **17**, 265.
219. R. S. Apte, *Expert Opin Pharmacother* 2008, **9**, 499.
220. J. O. McNamara, D. Kolonias, F. Pastor, R. S. Mittler, L. Chen, P. H. Giangrande, B. Sullenger, E. Gilboa, *J Clin Invest* 2008, **118**, 376.
221. I. Melero, W. W. Shuford, S. A. Newby, A. Aruffo, J. A. Ledbetter, K. E. Hellstrom, R. S. Mittler, L. Chen, *Nat Med* 1997, **3**, 682.
222. C. M. Dollins, S. Nair, D. Boczkowski, J. Lee, J. M. Layzer, E. Gilboa, B. A. Sullenger, *Chem Biol* 2008, **15**, 675.
223. J. Wang, H. Takeuchi, Y. Sonobe, S. Jin, T. Mizuno, S. Miyakawa, M. Fujiwara, Y. Nakamura, T. Kato, H. Muramatsu, T. Muramatsu, A. Suzumura, *Proc Natl Acad Sci USA* 2008, **105**, 3915.
224. A. C. Gavin, M. Bosche, R. Krause, P. Grandi, M. Marzioch, A. Bauer, J. Schultz, J. M. Rick, A. M. Michon, C. M. Cruciat, M. Remor, C. Hofert, M. Schelder, M. Brajenovic, H. Ruffner, A. Merino, K. Klein, M. Hudak, D. Dickson, T. Rudi, V. Gnau, A. Bauch, S. Bastuck, B. Huhse, C. Leutwein, M. A. Heurtier, R. R. Copley, A. Edelmann, E. Querfurth, V. Rybin, G. Drewes, M. Raida, T. Bouwmeester, P. Bork, B. Seraphin, B. Kuster, G. Neubauer, G. Superti-Furga, *Nature* 2002, **415**, 141.
225. A. C. Gavin, G. Superti-Furga, *Curr Opin Chem Biol* 2003, **7**, 21.
226. W. A. Weiss, S. S. Taylor, K. M. Shokat, *Nat Chem Biol* 2007, **3**, 739.

19

Nucleic Acids as Detection Tools

Jeffrey C.F. Lam, Sergio Aguirre and Yingfu Li

19.1 Introduction

The progress of biotechnology, especially towards nanobiotechnology, has been inspired from biological processes that have evolved over billions of years. Nucleic acids, long revered for their genetic coding properties, have now emerged as important materials for molecular diagnostic technologies. At the core of nucleic acid-based diagnostics is the fidelity of hybridization, a reversible thermodynamic process in which a nucleic acid molecule binds another nucleic acid molecule with the complementary sequence. This characteristic has been extensively explored in many areas of detection-oriented applications, to distinguish pathogenic species, to detect disease-related gene markers, to quantify the level of gene expression in cells and more.

In addition to the ability to form helical structures, nucleic acids are also known to have the ability to perform more complex tasks such as enzymatic catalysis and molecular recognition (ligand binding). The first discovery in this arena was made in the early 1980s when some natural RNA molecules, now known as ribozymes, were found to exhibit catalytic properties reminiscent of protein enzymes [1,2]. Years later, an elegant technique known as '*in vitro* selection' was invented and has since become a relatively routine process for engineering artificial DNA, RNA or modified nucleic acid molecules that can function as enzymes and receptors [3–6]. This has led to a colossal collection of functional nucleic acid sequences, many of which have been explored for bioanalytical applications.

Our pursuit towards a better quality of life has generated an increasing demand for new tools and techniques to diagnose disease accurately or to examine environmental conditions. The ideal detection device and assay must offer great selectivity, excellent sensitivity, ease of use and low cost of production. Nucleic acids can satisfy these requirements due to the following assets: (1) specific Watson–Crick base pairing under a wide range of conditions, (2) high stability and long shelf-life, (3) low cost of synthesis and (4) excellent adaptability to external modifications such as radiolabels and fluorescent and colorimetric dyes.

The Chemical Biology of Nucleic Acids Edited by Günter Mayer

In this chapter, we review some of the popular detection methods that utilize nucleic acids. First, we discuss assorted techniques in which nucleic acid probes are used to detect other nucleic acid sequences, followed by a discussion of functional nucleic acids and their applications as biosensors in conjunction with fluorescent, colorimetric, electrochemical and piezoelectric detection platforms.

19.2 Detection of nucleic acid targets by nucleic acid probes

Since many human diseases are linked to genetic perturbations, identification of specific mutations in patients has become a highly important practice in modern medicine. A great deal of DNA-based diagnostics pertain to identifying subtle differences in sequence content and their associated phenotypes. These differences are often the result of single nucleotide polymorphisms (SNPs) or other sequence alterations (such as point mutations, insertions and deletions). Since SNP is the most frequent form of genetic variation among individuals (roughly one for every 1000 nucleotides), many nucleic acid-based detection technologies are directed at identifying single-nucleotide differences [7,8].

Apart from local genetic mutations, structural and distant site changes of DNA sequence are also observed. These large genomic alterations include insertion–deletions, gene copy number variants (CNVs), inversions and translocations, which have been linked to a variety of genetic disorders [9]. For instance, CNVs are characterized by atypical gene content and/or gene deregulation associated with CHARGE syndrome and Parkinson's and Alzheimer's disease [8,10,11]. Currently, there are a plethora of methods for detecting local and structural genetic changes, but here we will only discuss a few major techniques.

19.2.1 *In situ* hybridization

In situ hybridization (ISH) is a method to visualize specific nucleic acid sequences using a reporter sequence [12,13]. It has been used for gene mapping, detection of chromosomal aberrations and structural analyses of cells [14,15]. Early assays employed radiolabeled probes, but fluorescence-based probes have become more popular as they are safer to use [16–18]. When fluorescent probes are used, the technique is often referred to as fluorescence *in situ* hybridization (FISH). Both ISH and FISH encompass three major steps: (1) sample preparation, (2) hybridization and (3) visualization (Figure 19.1).

Probes can be prepared by chemical synthesis or generated from host organisms (such as bacteria or yeast) by cloning. Extracted DNA can then be directly labeled with a fluorescent moiety or indirectly labeled with

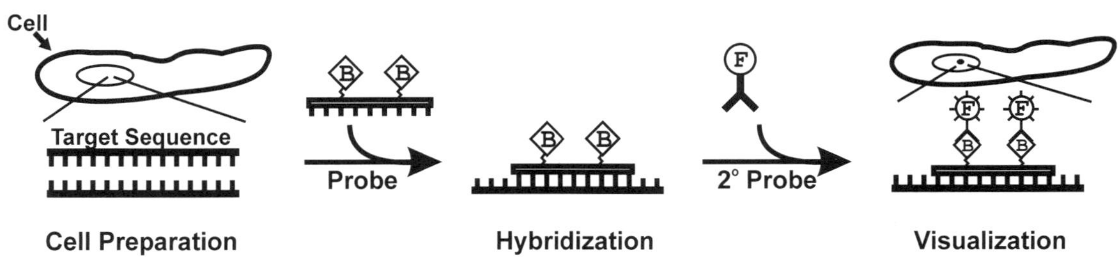

Figure 19.1 *Fluorescence* in situ *hybridization. Sample cells containing the target sequence are first fixed and permeabilized. Biotinylated probes (B) are subsequently added to the cells and the sample is heated and cooled to allow hybridization between the probe and its targeted sequence. Next, a secondary probe (2° probe) containing fluorescently modified streptavidin (F) is added, which binds biotinylated probes in the cells. The presence of target sequences can be detected by fluorescence microscopy*

biotin or digoxigenin for binding to secondary fluorescent reporters (Figure 19.1) [19]. Although FISH with directly labeled probes requires less experimental handling, the use of indirectly labeled probes allows for signal amplification where several layers of fluorescent antibodies are used concurrently to increase fluorescence intensity.

In the late 1980s, Raap and co-workers made an important advance where differentially labeled probes were used for target analyses by combinatorial labeling [20] or ratio labeling [21]. Combinatorial labeling uses a combination of fluorophores to identify multiple targets. For example, a probe labeled with FITC will fluoresce green and one labeled with AMCA will fluoresce blue, but a probe with both will fluoresce cyan. The combination of red, blue and green dyes has also provided yellow, magenta and white fluorescence. In ratio labeling, different probes are labeled with a combination of fluorophores, but the amount of the target is differentiated by the amount of fluorophore present in each probe. The use of multi-colored probes has given rise to several techniques, such as multiplex-FISH (M-FISH), spectral karyotyping (SKY) and combined binary ratio labeling (COBRA), which have been used to probe specific chromosomal regions [22–24] and study chromosomal characteristics [25–34].

19.2.2 Padlock probes and rolling circle amplification

Genotyping techniques such as ISH and FISH are widely used to determine physical locations of DNA and RNA in fixed cells. Their detection mechanisms are based on Watson–Crick base pairings where sensitivity and specificity are contingent on the strength of probe hybridization. As a result, FISH and ISH may not unambiguously resolve common genetic variations such as single-nucleotide polymorphisms. To overcome this problem, a class of oligonucleotide probes, called C-probes or padlock probes, were developed.

A padlock probe is a single-stranded oligonucleotide with two target-binding sequences at the 5′ and 3′ termini separated by a linker region (Figure 19.2) [35,36]. The two sequence elements bind the target sequence and position themselves for ligation by DNA ligase to create a circular DNA molecule. Since DNA ligase cannot perform the ligation reaction over a mismatched nucleotide, probe circularization will occur only if both probe sequences hybridize perfectly to its target. The resulting circular DNA molecule is topologically locked on to the target sequence, much like a padlock linked to a chain (Figure 19.2). Any non-specific (or non-circularized) probes can be thoroughly removed by stringent washing.

To detect bound circular probes, a number of methods have been reported. Initial efforts by Landegren and co-workers involved separation of circularized probes from unbound linear probes via denaturing polyacrylamide gels [35]. For *in situ* assays, fluorescently labeled antibodies can be used to recognize padlock probes tagged with specific antigens. Although strong fluorescence signals can be observed, it is difficult to achieve single-molecule sensitivity. To improve the detection limit, amplification strategies have also been implemented. The most popular format exploits bacteriophage DNA polymerases (e.g. ϕ29 DNA polymerase) that

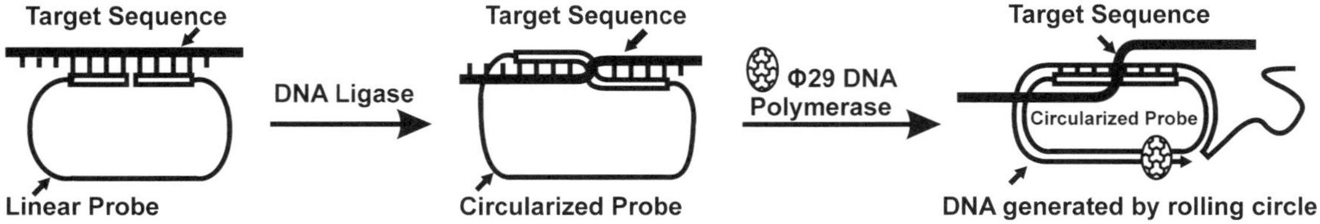

Figure 19.2 *Detection of target sequence via padlock probe and rolling circle amplification (RCA). The target sequence is hybridized with a linear probe, which is subsequently ligated with DNA ligase to create a circular probe. The locked circular probe is used as a template for RCA where the addition of ϕ29 DNA polymerase generates a long DNA strand with repeating sequence units*

uses circular DNA templates for amplification in a process known as rolling circle amplification (RCA) [37,38]. A bound primer complementary to the circular probe is extended by the polymerase over many revolutions to produce a long single-stranded DNA product thousands of times longer than the circular template (Figure 19.2). However, locked circular probes topologically prevent efficient amplification by polymerases. By creating a free 3′ end on the target strand using restriction enzymes that make a cut adjacent to the probe hybridization site, topological constraints can be relieved [39,40]. The resulting long DNA strands then can be easily detected by short complementary DNA probes labeled with colorimetric or fluorescent moieties [41].

To date, padlock probes have been used for numerous purposes, from SNP detection [35,41–43] to satellite DNA characterization [43] and the detection of pathogenic bacteria and viruses [44,45].

19.2.3 Dynamic allele-specific hybridization (DASH)

DASH is an SNP detection platform created in response to the establishment of several SNP databases in the 1990s [46]. Since then, it has evolved into a high-throughput fluorescence-based chip application [47]. The fundamental principle of DASH is real-time monitoring of allele-specific sequence differences using DNA thermal denaturation: First, a target DNA sequence is amplified by polymerase chain reaction (PCR), immobilized on a solid surface and denatured into single-stranded DNA (Figure 19.3). A complementary oligonucleotide probe is then added to hybridize with the target sequence at low temperature. A DNA intercalating dye (e.g. SYBR Green) which emits fluorescence proportional to the amount of duplex DNA is used to monitor the hybridization progress. The sample is steadily heated to induce denaturation while fluorescence is continuously measured. The rise in temperature induces a decrease in fluorescence as a result of probe–target dissociation. Any single-nucleotide mismatches can be characterized by significantly lower melting temperatures.

Generally, DASH permits allele identification with at least 99.9% accuracy [46]. Probe–target duplex melting temperatures can differ by 4 to 15 °C for matched and mismatched probes in homozygous and heterozygous samples. However, one of the major drawbacks of DASH is its low signal-to-noise ratio (SNR) produced by background fluorescence. In some cases, the target DNA may exhibit significant secondary structures that can interact with the dye to produce fluorescence in the absence of probes.

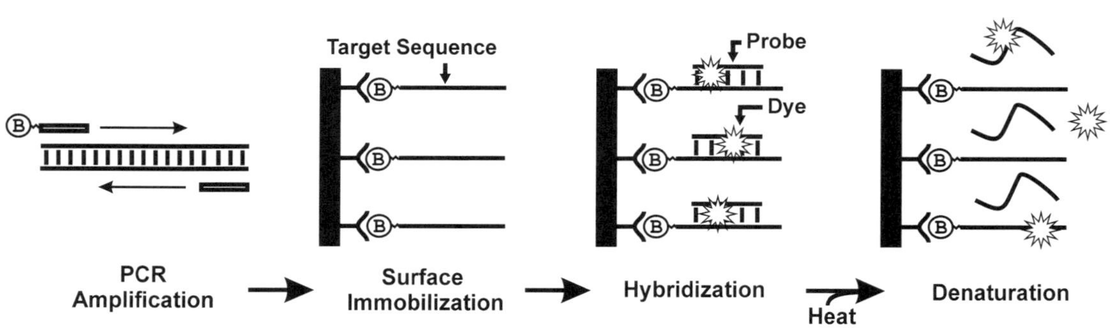

Figure 19.3 Dynamic allele-specific hybridization. Target sequences are amplified using polymerase chain reaction (PCR) with a biotinylated primer (B). The amplified sequences are isolated and immobilized on a streptavidin-coated surface and subsequently allowed to hybridize with complementary probes. A fluorescent DNA intercalating dye is then added. Depending on duplex stability, the release of dyes during thermal denaturation will provide information about sequence content (e.g. mismatches) of the sample (through monitoring changes in fluorescence intensity)

The use of induced fluorescence resonance energy transfer can enhance SNR and reduce background signals [48,49]. In this approach, the intercalating dye is used as the donor fluorophore whereas the oligonucleotide probe is chemically synthesized with an acceptor fluorophore. Hybridization of the probe with its target DNA will allow any intercalating dye to come into close proximity to the acceptor fluorophore-labeled probe, thus allowing efficient energy transfer between fluorophores while minimizing background interference.

19.2.4 Molecular beacons

The molecular beacon (MB) technology was devised in 1996 by Tyagi and Kramer, where dual-labeled fluorescent probes were used to report the presence of complementary nucleic acid sequences in solution [50]. An MB is a hairpin-shaped oligonucleotide containing a target-specific sequence flanked by two short complementary sequences that can hybridize to form a stem (Figure 19.4). Most probes are between 25 and 40 nucleotides long; the loop usually contains 15–30 nucleotides and is complementary to the target sequence. The stem of the hairpin may contain 5–10 base pairs. A fluorophore and a matching quencher are placed at the 5′ and 3′ termini of the stem and fluorescence signaling is mediated by fluorescence resonance energy transfer (FRET) (its mechanism will be discussed in Section 19.4). In the absence of a complementary target sequence, the fluorophore on the MB stem is quenched by the adjacent quencher. The binding of an MB to its target causes the stem to dissociate, thus increasing the distance between the fluorophore and quencher, leading to an increase in fluorescence intensity.

An MB can reversibly bind to and dissociate from its target, and this reversibility can significantly affect detection sensitivity and selectivity. The performance of an MB is dependent on several factors, such as temperature, pH and sequence content of the stem and loop [50,51]. Temperature can have very significant impact on MB performance since MB probes may possess different structural states at different temperatures [51]. At low temperature, the MB will exhibit strong intra-duplex stability even in the presence of target (Figure 19.4, left). Therefore, fluorescence enhancement is low due to intramolecular quenching by MBs that have not hybridized with target sequences. At optimal temperature, the MB will exhibit thermal instability with its

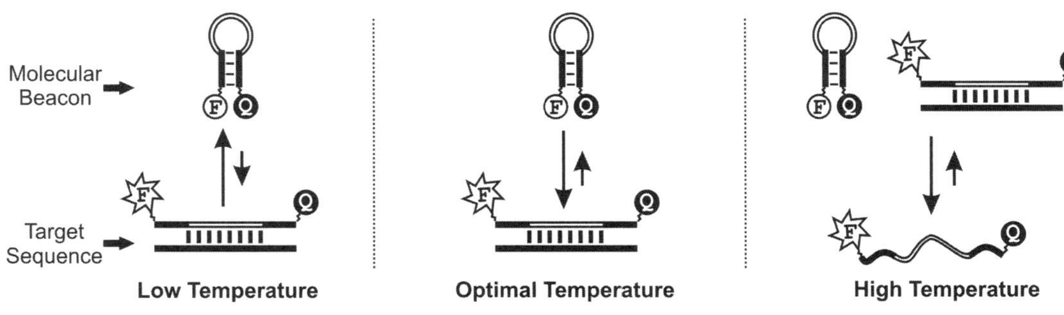

Figure 19.4 Temperature response of molecular beacons. The hairpin-shaped molecular beacon (MB) is modified with a fluorophore dye (F) and a quencher (Q) at its termini. The loop of the MB is used to bind a specific DNA sequence. Temperature can significantly influence the fluorescence intensity of an MB: at low temperatures, a predominant portion of MB remains in its native hairpin structure, preventing hybridization with the target sequence and lower fluorescence is observed. At optimal temperatures, the target sequence hybridizes to an MB where fluorophore–quencher pairs are separated, resulting in a high level of fluorescence. At high temperatures, the MB cannot maintain its native hairpin structure and a high level of fluorescence is observed even in the absence of target

intramolecular duplex and hybridize to the target sequence (Figure 19.4, middle), resulting in strong fluorescence signal. Finally, at higher temperature, internal stem of the MB and target–MB duplex are unstable, hence MBs remain in a linear conformation (Figure 19.4, right). In this case, fluorescence is observed as a result of forced thermal separation between dyes and not from target binding, which may lead to false-positive results.

Given their adaptability and simplicity, MBs have been used for many applications such as pathogen detection [52,53] and allele genotyping [54].

19.2.5 Array-based DNA technologies

DNA microarrays are a collection of sequence-specific nucleic acid probes strategically immobilized on a solid support to detect complementary DNA via hybridization. The first DNA arrays were known as dot blot arrays where radiolabeled DNA probes were used to quantify target DNAs on a nitrocellulose support [55,56]. Since then, several terminologies, including 'DNA arrays', 'gene chips' and 'biochips', have been used to describe this evolving niche of biotechnology [57].

Array-based hybridization chips can be fabricated using glass, silicon or plastic surfaces which can contain hundreds to thousands of micron-sized reactive zones. Early DNA arrays immobilized large amounts of PCR-amplified probes from plasmids, bacterial artificial chromosomes (BACs) or cDNA libraries for detection [58,59]. These large probes, often over kilobases in length, provide robust sensitivity and are particularly useful in genome-wide studies [60,61]. In later designs, oligonucleotide arrays were created using short synthetic DNA probes, which are more suitable for gene probing studies.

Today, most DNA microarrays contain nucleic acid probes that can hybridize with nucleic acid targets extracted and amplified from a cellular sample and modified with fluorescent groups (Figure 19.5). Hybridization between probe and target sequences with different levels of complementarity will generate different levels of fluorescence. Data acquisition and signal detection are attained by high-resolution laser scanning of the array and analyzed using integrated computer software. Probe–target duplexes with high complementarity will exhibit greater fluorescence signals compared with complexes with lesser complementarity. Tangible and reliable quantitative information from array-chip devices is contingent on the design and fabrication of probe surfaces, hybridization efficiency of the target and effectiveness of the signal transducer (i.e. fluorophore).

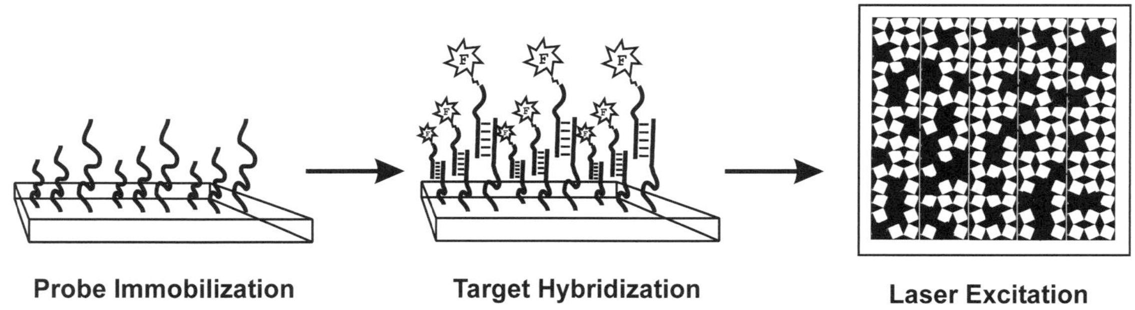

Figure 19.5 DNA microarrays. Different complementary probes are first spotted on a solid support in array format. Fluorophore-modified target sequences are added to allow hybridization with probes at specific array locations. Visualization of the array is achieved through laser excitation. The amount of a particular target sequence can be estimated because the intensity of fluorescence is correlated with the amount of target hybridized to surface probes at specific locations

For probe attachment on array surfaces, traditional immobilization methods have evolved to high-throughput techniques such as robotic spotting [62], inkjet printing [63] and light-directed deposition [64] with synthetic oligonucleotide probes or larger probes generated from BACs or cDNAs libraries [61,65]. Such advances have increased the research capacity to conduct high-throughput genotyping, sequencing or copy number analysis [66,67]. As a result, DNA microarray technology has become a popular method for global and large-scale genomic analyses.

19.3 Functional nucleic acids

As discussed earlier, nucleic acids can also perform catalysis and molecular recognition in addition to their roles in storage and transmission of genetic information in the cell. For instance, some naturally occurring RNA molecules, known as *ribozymes*, can catalyze RNA cleavage or peptidyl transfer reactions without proteins [68]. Similarly, synthetic catalytic DNA molecules, known as *deoxyribozymes* or *DNAzymes*, are able to catalyze reactions such as DNA or RNA cleavage [69,70], DNA phosphorylation [71–74], DNA ligation [75–81] and porphyrin metalation [82,83]. Aside from catalysis, nucleic acids may also act as molecular receptors (known as *aptamers*) and bind small molecules or whole cells with affinity similar antibodies.

Although some ribozymes are found in Nature, most ribozymes, deoxyribozymes and aptamers [collectively referred to as functional nucleic acids (FNAs)] are evolved in laboratories via a combinatorial technique known as *in vitro* selection or SELEX (Systematic Evolution of Ligands by EXponential enrichment), during which a pool of $10^{15} - 10^{18}$ random sequences are enriched through cycles of separation and amplification. Over the years, this technique has led to the creation of countless FNAs, many of which have been used to design biosensing devices.

19.3.1 Nucleic acid enzymes (NAEs)

Ribozymes and deoxyribozymes are often referred to as nucleic acid enzymes (NAEs). Their catalytic function is dependent on their ability to fold into well-defined tertiary structures, usually via Watson–Crick base pairing, non-canonical base pairing, base stacking and van der Waals interactions. Cofactors such as metal ions can also play a significant role in structural formation and/or catalysis [84,85]. Many existing NAEs facilitate RNA cleavage reactions by hydrolyzing a phosphodiester bond in the presence of specific metal ions such as Pb^{2+} and Hg^{2+} (Figure 19.6a) [86,87]. This metal-dependent property has been exploited in the design of metal ion sensors [87,88], nanomachines [89,90], computational logic gates [91] and therapeutic control agents [92,93].

19.3.2 Aptamers

In vitro selection has been used to engineer nucleic acids with ligand binding properties. This class of molecules are termed 'aptamers' (from the Latin word *aptus*, meaning 'to fit', and the Greek word *meros*, meaning particle) [3,94]. They are often regarded as nucleic acid equivalents of antibodies (Figure 19.6b). Binding of the target is usually mediated by molecular interactions such as base stacking, electrostatic interactions, van der Waals interactions and hydrogen bonding [95,96]. The first aptamers were composed strictly of RNA [3,5] while DNA-based [97] aptamers and those with chemically modified nucleotides were eventually developed [98]. To date, aptamers are known to bind nucleotides, amino acids, carbohydrates, polypeptides and even whole cells.

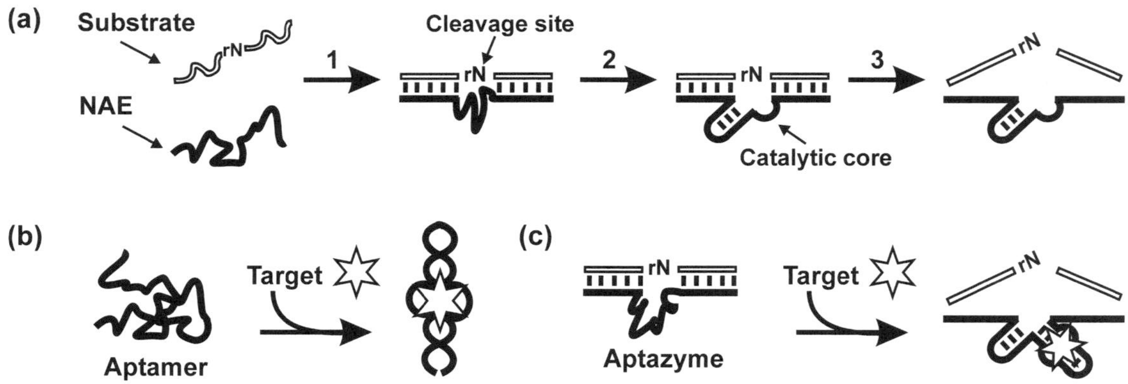

Figure 19.6 *Schematic of three classes of functional nucleic acids. (a) An RNA-cleaving nucleic acid enzyme (NAE) is hybridized to its substrate containing a single ribonucleotide (rN) where cleavage occurs upon folding of the NAE's catalytic core. (b) A target ligand (star) is recognized by its aptamer where the binding of the target triggers a conformational change or folding of the aptamer. (c) In the absence of the target, an RNA-cleaving aptazyme may be hybridized to its substrate but remains folded in a catalytically inactive conformation. The addition and subsequent binding of the target to the ligand binding site of the aptazyme induce folding of the aptazyme to a catalytically active conformation where RNA cleavage can occur*

19.3.3 Aptazymes

Aptazymes are allosteric (or effector-activated) nucleic acid enzymes. They can be engineered via semi-rational or rational designs. In the rational approach, an existing aptamer is linked to an NAE via a 'communication module'. In the absence of a target, the aptazyme holds the NAE domain in an inactive conformation via sequence-specific sequestration (Figure 19.6c). Addition of target causes a structural rearrangement and restores catalytic activity. Although numerous aptazymes have been created via rational design [99–101], the main limitation is that both NAE and aptamer structural foreknowledge are required to engineer a viable sensor. In the semi-rational approach, a random-sequence domain is placed alongside an aptamer and only molecules with catalytic activity are selected by SELEX. Alternatively, known NAEs can be linked with a random-sequence domain and molecules with target-dependent activity are identified. By applying semi-rational or rational designs, DNA and RNA aptazymes for targets such as nucleotides [100–102], nucleic acids [103–106], secondary messengers [107,108], cofactors [102,109,110] and proteins [111] have been created.

19.4 Fluorescent nucleic acid sensors

The use of fluorescence is widely popular in nucleic acid sensors for several reasons: (1) a large selection of fluorophores for nucleic acid conjugation, (2) minimal health risks in handling fluorophores, (3) instruments capable of detecting fluorescence at ultralow concentrations, (4) portability of instruments for on-site detection and (5) relatively long shelf-life of fluorophores.

In most fluorescent nucleic acid sensors, signaling depends on changes in fluorescence intensity regulated by FRET. When a donor fluorophore is excited (by light), the absorbed energy can be partially transferred to an acceptor fluorophore or quencher via non-radiative dipole–dipole oscillations. The energy received by the acceptor is then released via its characteristic emission fluorescence or it can be entirely quenched. The amount of energy transfer between donor to acceptor, also known as FRET efficiency (E), is inversely proportional to the distance between the dyes defined by the equation $E = 1/[1 + (r/R_0)^6]$, where R_0 is the Förster distance and r is the physical distance between the donor and acceptor. Since E is dependent on the sixth power of interfluorophore distance, a small increase in r (typically between 10 and 100 nm) will yield a significant change in fluorescence. This is exploited in nucleic acid sensors by carefully placing dyes such that changes in nucleic acid structure will result in a change in fluorescence [112]. The following subsections discuss how various fluorescence signaling strategies are employed in these sensors.

19.4.1 Detection using fluorescent nucleic acid enzymes

Since the majority of fluorescent NAE sensors encompass NAEs with RNA-cleaving activity, this section will discuss the design, modification and characterization of RNA-cleaving NAEs for fluorescent sensing. Conceptually, NAE cleavage activity is coupled to a change in fluorescence by attaching a fluorescence signaling module containing a donor fluorophore (F) and a quencher (Q). When the substrate is cleaved by NAE, the signaling components are separated, thus causing an increase in fluorescence. However, since NAE catalysis is dependent on its correct folding, modifying an NAE with a fluorophore and quencher may hinder its function. Consequently, it is often necessary to assess suitable fluorophore positions when constructing fluorescent NAE sensors.

Unlike molecular beacons, where fluorophores are fixed at terminal ends of an oligonucleotide, different dye arrangements can be used in fluorescence signaling NAEs (Figure 19.7). For instance, the F and Q can be placed in the following manners: (1) F at one end of the substrate and Q at the end of the NAE [113–116], (2) F and Q at both ends of the substrate [117–121], (3) F and Q at opposing ends of the substrate along with a second Q at the end of the NAE [122,123], (4) F and Q juxtaposed to the cleavage site [124–129] or (5) F at the end of the substrate along with NAEs chemically conjugated to a solid support with quenching ability (e.g. gold surface) [130,131].

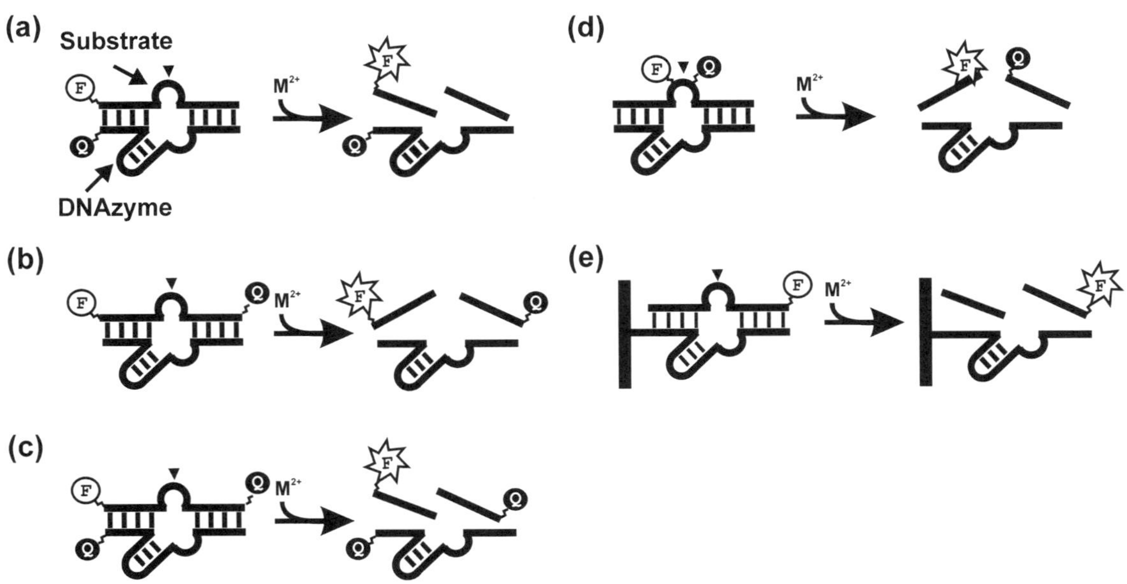

Figure 19.7 Various dye arrangements of fluorescence signaling NAEs. (a)–(e) In each panel, the reaction scheme of NAE-mediated substrate cleavage is represented. The location of fluorophores (with low fluorescence) and quenchers are illustrated by F and Q circles, respectively. Fluorophore with high emission of fluorescence (e.g. due to a large inter-fluorophore–quencher distance) are represented by F-labeled stars. Each cleavage site is marked by the filled triangle

In addition to the aforementioned strategies, fluorescent NAE sensors can be derived by *in vitro* selection using substrates pre-attached with F and Q adjacent to the cleavage site (Figure 19.7d). Such NAEs are able to accommodate bulky moieties while performing RNA cleavage efficiently. In a study by Li's group, a fluorescence-signaling NAE named DET22-18 with impressive catalytic efficiency and fluorescence signal enhancement was generated [125]. Using the same approach but under different pH conditions, additional nucleic acid sensors were derived [124]. These fluorescence signaling NAEs all required F and Q for substrate cleavage, suggesting that catalysis was partially mediated by the signaling module [132].

To date, the use of fluorescence signaling NAEs have been primarily focused on, but not limited to, the detection of metal ions. For example, Lu and co-workers developed several highly sensitive fluorogenic assays for metal ion detection. They engineered a fluorescent Pb^{2+} sensor using the 8–17 RNA-cleaving NAE which exhibited a 4-fold signal enhancement and a detection limit of 10 nM. This was lower than the US Environmemntal Protection Agency (EPA)-defined toxicological limit [133]. Recently, they also reported a fluorescent NAE sensor for uranyl ion (UO_2^{2+}) with ~15-fold fluorescence enhancement and a detection limit of 45 pM (which is ~10 times more sensitive than mass spectrometry) [122]. Apart from lead and uranium, similar fluorescent NAE sensors have also been created for Hg^{2+}, Cu^{2+}, Mn^{2+}, Ni^{2+}, Co^{2+} and Cd^{2+} [87,127,134].

19.4.2 Detection using aptamers

Aptamers have been selected for a plethora of targets ranging from metal ions to whole cells. Some aptamers exhibit an affinity for their target 10 000 times greater than its structural analog [135,136]. To date, numerous aptamers have been employed for fluorescence-based biosensing applications. Some strategies are similar in principle to those used in fluorescence signaling NAEs, whereas others were specifically developed for aptamers.

19.4.2.1 Fluorescently labeled aptamers

Since aptamers undergo conformational change in the presence of a target and many fluorophores are sensitive to environmental changes, labeling an aptamer with a fluorophore can offer a convenient way to visualize target binding. Early designs relied on covalent attachment of a fluorophore to the aptamer [137–140]. However, fluorescence enhancement was highly dependent on the type of fluorophore used and its location on the aptamer, hence only a modest signal enhancement could be observed (up to 4-fold). Woodbury and co-workers improved on this method by incorporating a fluorogenic nucleotide analog to the aptamer sequence [141]. Since the quantum yield of the analog is dependent on nucleobase-stacking interactions, a 30-fold signal enhancement could be achieved with this strategy [142,143].

Fluorescence signaling aptamers can also be constructed by employing fluorophores and quenchers. For example, a cocaine aptamer was separated and labeled with an F and Q moiety (Figure 19.8a) [144]. In the absence of cocaine, the separated strands remain distant and only high fluorescence is observed. In the presence of a target, the two fragments reform the native aptamer, bringing the fluorophore and quencher in close proximity, thus resulting in fluorescence reduction. This design was able to detect 1 μM cocaine. The same method was demonstrated using PDGF, thrombin and L-arginamide aptamers [145–151].

Fluorescent aptamer sensors can also be designed based on the ability that an aptamer can adopt two different conformations: a duplex structure with a complementary oligonucleotide in the absence of target or a tertiary structure when bound to its target. This approach is termed 'structure-switching aptamers' [152]. The original design uses a 5′-extended ATP aptamer that can hybridize to both a fluorophore-labeled complementary strand (F-DNA) and a Q-labeled complementary oligonucleotide (Q-DNA) (Figure 19.8b). In the absence of ATP, the three strands come together to form a duplex. Since F and Q are positioned next to

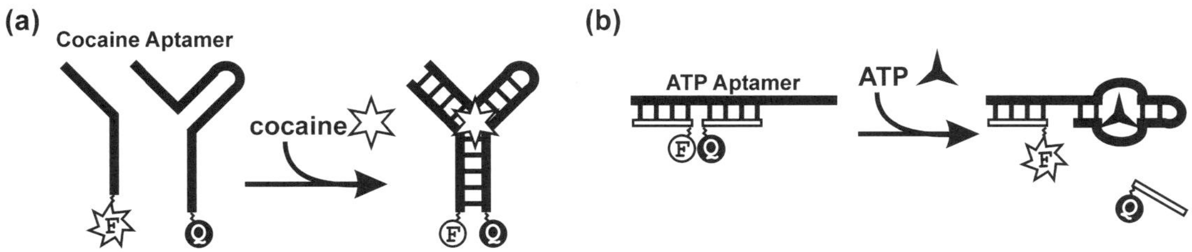

Figure 19.8 Fluorescence signaling aptamers. (a) A cocaine aptamer is divided and labeled with a fluorophore and quencher (F and Q, respectively). In the absence of cocaine, the two aptamer portions remain separated where high fluorescence intensity is detected. Upon the addition of cocaine, the two labeled strands come together, resulting in a decrease in fluorescence. (b) An ATP aptamer is hybridized to two short F- and Q-labeled complementary sequences (open bars). In the absence of ATP, the sequences form a stable duplex with the aptamer where F and Q are in close proximity and fluorescence is quenched. The interaction between ATP and the aptamer causes a conformational change resulting in the release of Q-labeled sequence and an increase in fluorescence intensity

each other, fluorophore emission is quenched. The addition of ATP promotes the switch from duplex structure to a target–aptamer complex structure, accompanied by the departure of the Q-DNA and an increase in fluorescence (Figure 19.8b). Further characterization experiments revealed that the aptamers' ability to 'structure switch' was greatly dependent on temperature; only at optimal temperature was there an evident signal in the presence of ATP. This is due to the greater stability of DNA duplex at lower temperatures where aptamer structure switching becomes increasingly difficult. To overcome temperature-dependent effects, the length of Q-DNA was tuned to allow structure switching in real time at a given temperature.

19.4.2.2 Aptamers with intercalating fluorescent dyes

Aptamers contain duplexes that are formed or dissociated in the presence of target. It is known that many fluorescent dyes can intercalate duplex structures to produce an increase in fluorescence. When these dyes are used with aptamers, binding of a target can induce a structural change in the aptamer, thus affecting the dye–aptamer interaction and fluorescence intensity.

Given that the largest change in fluorescence coincides with the greatest displacement of dyes from the aptamer, a short oligonucleotide complementary to the aptamer sequence can be used to provide additional dye binding sites. The presence of target will interact with the aptamer and release the short oligonucleotide and intercalated dye, resulting in a decrease in fluorescence. Dong and co-workers employed this strategy using the thrombin aptamer and ethidium bromide dyes to construct a fluorescent thrombin sensor [153].

To create a sensor that exhibits an *increase* in fluorescence upon target binding, strong fluorescence quenchers such as gold nanoparticles (AuNPs) have been employed to modulate intercalating dye fluorescence. For instance, AuNPs functionalized with the PDGF aptamer produced minimal fluorescence in the absence of a target due to quenching effects between the dye, DMDAP (*N*,*N*-dimethyl-2,7-diazapyrenium), and the nanoparticle surface. The addition of PDGF displaced the dye from AuNPs, producing a 40-fold increase in fluorescence [154].

19.4.3 Detection using aptazymes

Signaling principles in fluorescent aptazymes are analogous to fluorescent NAE sensors. There are several approaches to synergize fluorescence signaling with target binding and catalysis; the most common is by

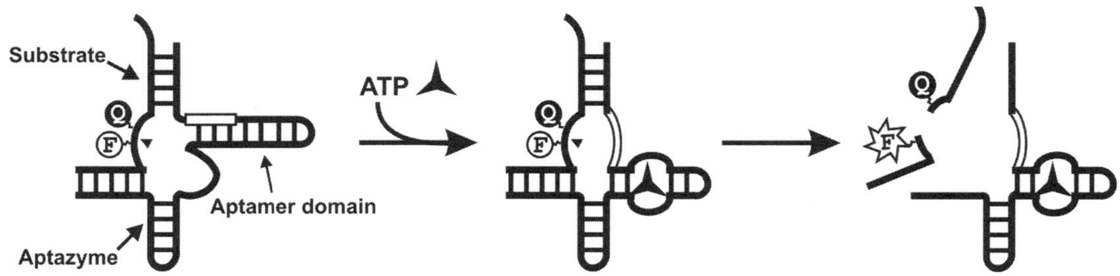

Figure 19.9 Fluorescence signaling aptazyme. In the absence of the target (ATP) the catalytic core of the aptazyme (represented by open bars) is restricted from accessing its substrate by hybridizing to complementary nucleotides in the aptamer domain. However, the presence of ATP induces folding of the aptamer to release the catalytic core for substrate cleavage, resulting in fluorescence enhancement

rational design, where an F and a Q moiety are placed at compatible locations on the aptazyme. The presence of target switches the NAE from an inactive to an active conformation, allowing catalysis to occur. This is detected when the dyes are separated by aptazyme cleavage and an increase in fluorescence is observed. Using this strategy, aptazymes for proteins and other small molecules such as ADP [155], caffeine and aspartame [156] have been converted to fluorescence signaling reporters.

Li and co-workers created fluorescence signaling aptazymes by converting existing fluorescent NAEs (see Section 19.4.1) [157] into aptazymes. Since the signaling module which often interferes with NAE activity is already embedded in the reporter, allosteric binding is simply integrated by appending a desirable aptamer. In this study, an ATP aptamer was fused to a fluorescent RNA-cleaving NAE via rational design (Figure 19.9). The aptamer domain sequestered the catalytic core of the NAE in the absence of ATP and the addition of target led to structural refolding of the aptamer, thus 'releasing' the catalytic core and restoring function to the NAE domain.

Apart from RNA cleavage, a fluorescent aptazyme with ligase activity has also been reported [158]: an ATP-dependent NAE was immobilized on a silica surface. The presence of ATP activated the aptazyme to self-ligate with a substrate to produce a circular DNA probe. ϕ29 DNA polymerase was then added to utilize the circular probe as a template to generate a long, single-stranded DNA chain with repeating sequences. The addition of fluorophore-labeled oligonucleotides complementary to the DNA chain allowed a large fluorescence signal to be observed.

Aside from detection, fluorescent aptazymes can be used for inhibitor screening and protein–protein interaction studies. For example, Famulok and co-workers converted an HIV1-Rev protein-responsive aptazyme to a fluorescence signaling aptazyme to screen for other interacting protein partners [159]. Small molecules that interact with Rev proteins can also be screened by observing molecules that disrupted protein binding to the aptamer domain and modulated Rev-responsive aptazyme activity [160]. These examples suggest the possibility of using fluorescence signaling aptazymes for high-throughput screening in drug discovery or to reveal unknown protein binders.

19.5 Colorimetric nucleic acid sensors

Colorimetric sensors are commonly employed since color changes can be readily detected by the naked eye. This is especially convenient for qualitative analysis in rural areas where access to instrumentation is limited. On the other hand, if quantitative measurements are required, a simple spectrophotometer can be used. Since nucleic acids do not absorb visible light, color-reporting methods (described below) must be implemented.

19.5.1 Common color-reporting methods in nucleic acid sensors

19.5.1.1 Colorimetric dyes

In colorimetric nucleic acid sensors, color-producing molecules such as natural or synthetic dyes found in food and fabric coloring are utilized. In most systems, the chosen dye interacts with nucleic acids to transduce a binding event into an observable color change. This is similar to the mechanism employed by nucleic acid stains that intercalate between stacked bases to produce an increase in fluorescence [161].

19.5.1.2 Gold nanoparticles (AuNPs)

Gold nanoparticles are frequently used in colorimetric sensors because they exhibit different colors in aggregated or dispersed states [162,163]. This distance-dependent phenomenon, known as surface plasmon resonance (SPR), is due to coherent oscillation of nanoparticle surface electrons induced by an incident electromagnetic field [164]. In the case of gold nanoparticles (AuNPs), when visible light is shone, a specific resonant wavelength is absorbed by AuNPs to induce surface electron oscillations. The resultant oscillation generates electromagnetic radiation (light) at the same frequency as the oscillating electrons. This effect is highly dependent on the size, shape and distance between nanoparticles. For example, when AuNPs are far apart (large inter-particle distance), an absorption in the green spectrum is present and AuNPs appear red. Conversely, when AuNPs are close together due to nanoparticle aggregation (small inter-particle distance), the absorption shifts toward a longer wavelength and the solution appears purple (Figure 19.10a). This distance-dependent property is widely exploited in colorimetric nucleic acid sensors.

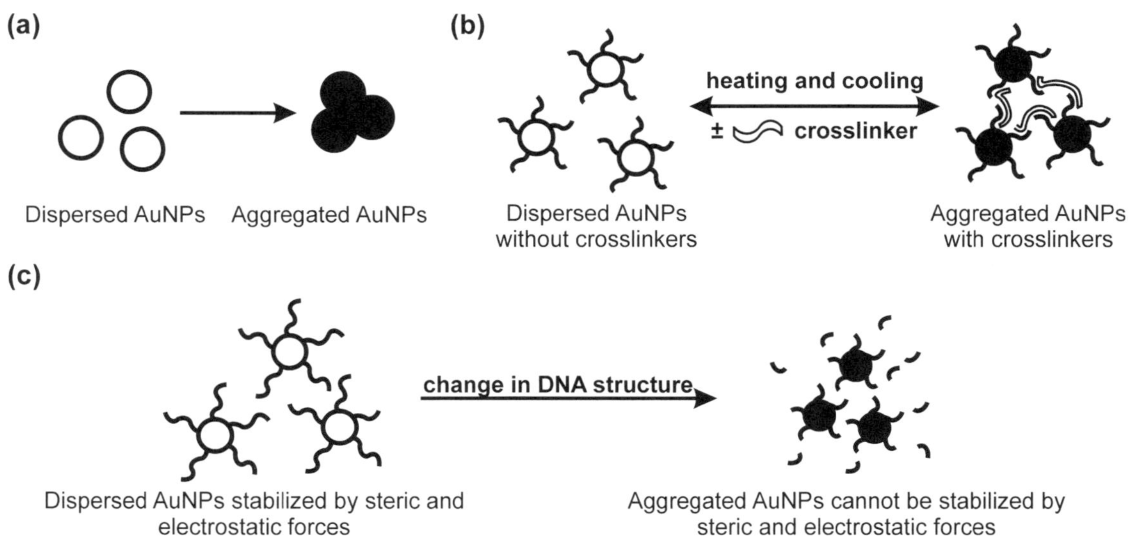

Figure 19.10 *Three modes of gold nanoparticle (AuNP) aggregation. (a) Bare (unmodified) gold nanoparticles exhibit a red color when well dispersed in solution. Conversely, the aggregation of gold nanoparticles by decreasing the interparticle distance can be characterized by a purple color. (b) The aggregation of nanoparticles is reversibly modulated by heating and cooling duplexes between DNA-functionalized nanoparticles and its complementary crosslinkers. (c) Structural changes of DNA on the surface of AuNP affect surface steric and electrosteric forces which can induce nanoparticle aggregation. In this panel, DNA on AuNPs is digested by nuclease and nanoparticles are destabilized by the change in surface DNA structure*

19.5.1.3 Organic polymers

The use of organic polymers as a color-reporting group has also been reported [165,166]. For instance, water-soluble cationic polymers (e.g. polythiophene) can act as a stain to transduce optical signals based on nucleic acid structures and electrostatic interactions. The polymer forms a complex with anionic single-stranded DNA. Any nucleic acid structural change induces a conformational change in the polymer, which in turn causes a color change.

Another polymer-based approach uses artificial polypeptides as a color reporting group [167,168]. Peptides with modified groups (e.g. *n*-hexyl-L-glutamate and ϵ-benzyloxycarbonyl-L-lysine) are graft-polymerized on a silicon surface to produce multilayers of polypeptides. Depending on their orientation, the thickness of the film will differ and the reflection of incident light on the film will be altered, thus producing a color change. Since polypeptide films are cationic, nucleic acids can be immobilized on their surface. Any change in structure will also disturb the molecular orientation and thickness of the peptide layer, translating to a color change.

19.5.2 Colorimetric assays involving nucleic acid enzymes

19.5.2.1 Assays by crosslinking aggregation of AuNPs

Gold nanoparticles exhibit distance-dependent colorimetric properties which can be modulated by nucleic acids grafted on an AuNP surface and DNA crosslinkers in solution (Figure 19.10b): Dispersed nanoparticles can be aggregated via hybridization between surface DNA and DNA crosslinkers, resulting in a red-to-purple color change. Conversely, when crosslinking duplexes are broken up and nanoparticles are redispersed, a purple-to-red color change is observed.

In the first NAE-based AuNP colorimetric sensor, nanoparticles are assembled with linker DNAs containing terminal sequences complementary to surface-bound oligonucleotides. A single ribonucleotide is also placed in the middle of each DNA crosslinker for RNA cleavage by a Pb^{2+}-dependent NAE (Figure 19.11a) [169]. To detect the presence of Pb^{2+}, pre-assembled DNA-AuNPs containing RNA–substrate linkers and Pb^{2+}-specific NAEs are heated and cooled. If Pb^{2+} is present, substrate linkers are cleaved and the reassembly of AuNPs becomes impossible, hence nanoparticles remain red (Figure 19.11b). If Pb^{2+} is not present, substrate linkers can reassemble nanoparticles to produce a purple color. Using this method, a detection limit of ~100 nM was reported.

The main disadvantage of this system is the need to preassemble nanoparticles prior to each assay. Since only one identical DNA sequence was grafted on all AuNPs, hybridization between DNA-AuNPs and substrate linkers will only assemble in a head-to-tail manner, which suffers from steric effects between nanoparticles (Figure 19.11a) [170]. By grafting two different DNA sequences on AuNPs, nanostructures can be aligned in a tail-to-tail arrangement, which can be dispersed at constant temperature without additional heating and cooling procedures.

19.5.2.2 Assays by non-crosslinking aggregation of AuNPs

In a second type of AuNP-based sensor, nanoparticle stability is modulated by changing electrolyte concentration [5]. When non-functionalized (or bare) AuNPs are exposed to high salt concentrations, nanoparticles are unstable and quickly aggregate to induce a color change. On the other hand, DNA in DNA-functionalized AuNPs can stabilize nanoparticles via steric and electrosteric forces, thus tolerating higher salt concentrations than non-functionalized nanoparticles. However, physical or chemical changes on surface DNA can affect AuNP stability, causing particle aggregation or dispersion (Figure 19.10c). Since this mechanism does not rely on DNA linkers, it is often referred to as a *non-crosslinking* mechanism.

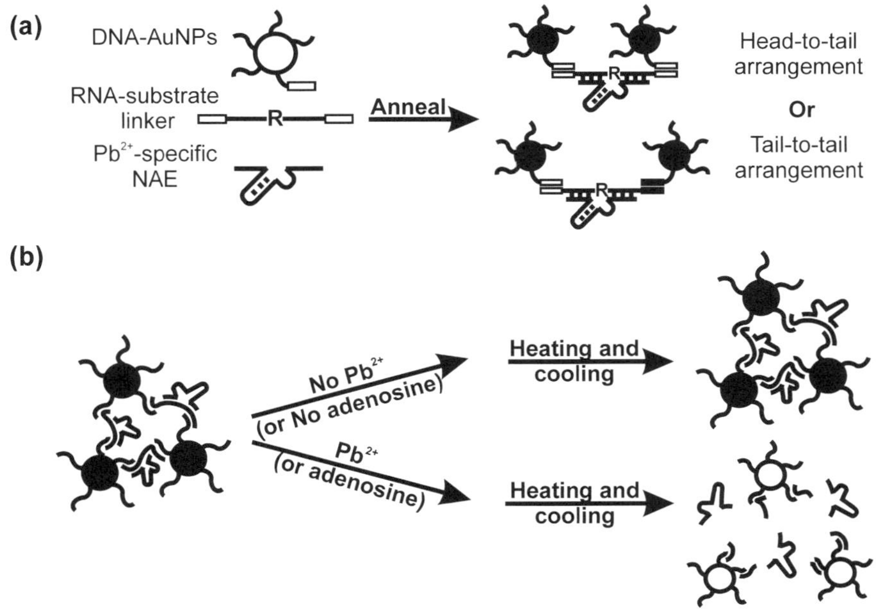

Figure 19.11 Various nanoparticle preassembly arrangements and detection mechanism for AuNP-based colorimetric lead sensor. (a) Nanoparticles in the lead sensor can be preassembled by annealing DNA-AuNPs with RNA–substrate linkers and Pb^{2+}-specific NAEs in a head-to-tail arrangement where only one set of linker sequence is used (opened bars). Alternatively, a tail-to-tail arrangement can be achieved by using two sets of crosslinking sequences (open and filled bars). (b) In this NAE-based AuNP lead sensor, the presence of lead induces Pb^{2+}-specific cleavage of crosslinkers by NAEs. When DNA-AuNPs are heated and cooled, cleaved crosslinkers can no longer reassemble nanoparticles again, hence a purple-to-red color change is detected. In the absence of lead, crosslinkers remain intact and reassembly of nanoparticles is possible and no color change is observed

Recently, non-crosslinking-based AuNP aggregation was exploited in an NAE-based colorimetric sensor [171]. DNA oligonucleotides containing an RNA linkage are grafted on an AuNP surface to function as substrates for Pb^{2+}-dependent NAEs. Prior to each assay, salt concentrations are adjusted so that AuNPs remain well dispersed. In the presence of Pb^{2+}, substrate cleavage results in a significant loss of negative charge from the nanoparticle surface, causing interparticle electrostatic and steric repulsion forces to decrease. AuNPs are no longer stable at the provided salt concentration, which results in colloid aggregation. Compared with the previous system (Section 19.5.2.1), non-crosslinking-based assays begin with well-dispersed nanoparticles which allow better substrate accessibility by NAEs to produce faster color changes than crosslinking-based assays.

19.5.3 Colorimetric assays involving aptamers

19.5.3.1 Assays by colorimetric dye

In a study by Stojanovic and Landry, a cocaine sensor was constructed using DNA aptamers and colorimetric dyes [172]. The aptamer was previously known to form a three-way junction motif (three Watson–Crick stems connected by a common branch point) and the analyte is thought to bind to a hydrophobic pocket in the aptamer. It was speculated that certain dyes may also form a complex with the aptamer; therefore, 35 dyes were screened for binding in the absence of cocaine. A cyanine dye (diethylthiotricarbocyanine iodide) was found to form a monomeric complex with the aptamer but produced a large spectral shift in the presence of cocaine (Figure 19.12a). This assay was able to detect cocaine concentrations from 2 to 600 μM.

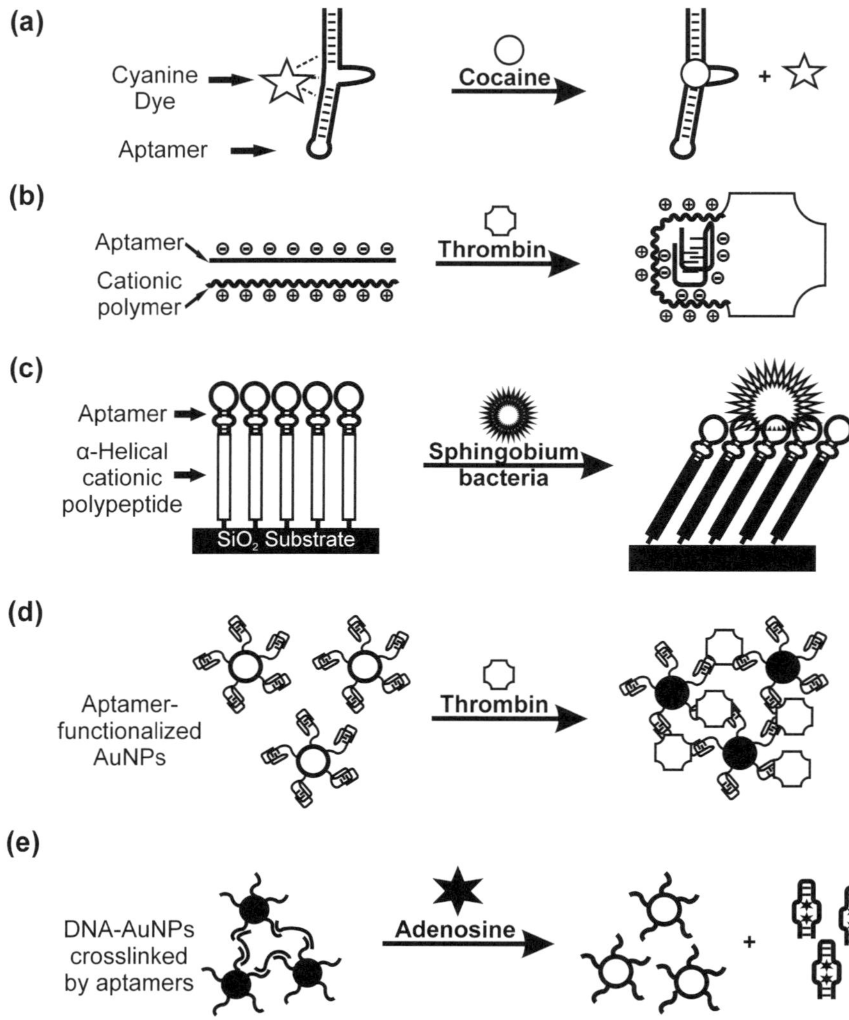

Figure 19.12 Schematics of various aptamer-based colorimetric sensors. (a) Detection of cocaine via dissociation of bound dyes from aptamer. (b) Detection of thrombin via structural change of cationic polymers. (c) Detection of bacteria based on conformational changes of cationic polypeptides. (d) Detection of thrombin via gold nanoparticles crosslinked by target molecule and its aptamer. (e) Detection of adenosine via dehybridization and folding of anti-adenosine aptamer crosslinked between nanoparticles. The details of each sensor are described in the main text

19.5.3.2 Assays by conjugated cationic polymers

In a study by Ho and co-workers, cationic polythiophene was used as a 'stain' to bind the thrombin aptamer via electrostatic interaction (Figure 19.12b). The flexible backbone of the polymer was highly conjugated and an optical change occurred when a different structure was formed. In the absence of thrombin, the DNA aptamer was unfolded and the polymer remained highly conjugated in a planar structure to exhibit a red–violet color (Figure 19.12b). In the presence of thrombin, the aptamer adopted a compact quadruplex structure and the cationic polymer wrapped around the folded aptamer and disrupted its backbone configuration [173,174]. This results in an orange color which can be monitored using a spectrophotometer.

19.5.3.3 Assays by artificial polypeptides

In artificial polypeptide-based assays, polypeptides with modified amino acids are densely grafted on a solid surface, forming many rod-like α-helices (Figure 19.12c). Depending on the thickness and number of layers, incident light will be interfered with, diffracted or scattered and the reflected light will have different colors (Figure 19.12c). This is known as *structural color* because light is reflected based on nanostructures and does not require color pigments. In a sensor reported by Yokogawa and co-workers, lysine-based polymers made from ϵ-benzyloxycarbonyl-L-lysine are immobilized on a silica substrate [167,168]. To modulate the polymer structure, an RNA aptamer for *Sphingobium* bacteria is absorbed on the cationic polypeptide surface. The binding of target to RNA aptamer structurally reorganizes surface polypeptides, resulting in a color change which can be detected by visible-reflective analysis using a spectrophotometer.

19.5.3.4 Assays by crosslinking aggregation of AuNPs

Some analytes may have distinct binding sites for two different aptamers which can be used as linkers for AuNP aggregation [174–177]. In this design, AuNPs are functionalized with two different aptamer sequences and can be crosslinked in the presence of a target (Figure 19.12d) [178]. However, it was found that low amounts of analyte could not effectively assemble nanoparticles to produce a color change. Instead, some aggregates were used as *seeds* to promote additional assembly. This is done by separating aggregated from non-aggregated nanoparticles, then placing aggregates in a new solution of dispersed AuNPs. Since this only works for an analyte with two (or more) distinct aptamer binding sites, only colorimetric sensors for thrombin and PDGF were constructed (Figure 19.12d).

Aptamers can also be used as crosslinkers to modulate nanoparticle assembly in a different way. A linker containing an aptamer sequence and AuNPs functionalized with DNA complementary to the linker's 5′ and 3′ ends are preassembled by heating and cooling (Figure 19.12e). Addition of target causes the aptamer to dissociate from nanoparticle assembly to form an aptamer–target complex. AuNP aggregates are redispersed, causing a purple-to-red color change. Using this approach, sensors for adenosine and cocaine have been reported [179].

19.5.3.5 Assays by non-crosslinking aggregation of AuNPs

Aptamers can also modulate nanoparticle stability via non-crosslinking aggregation mechanisms. This was demonstrated by Li and co-workers using adenosine aptamers [180]. Short complementary DNAs grafted on an AuNP surface were hybridized with adenosine aptamers (Figure 19.13a). The anionic DNA duplexes stabilize AuNPs under high salt concentrations. When adenosine is added, the aptamer strands dissociate from AuNPs and bind to their target (Figure 19.13a). The surface charge of AuNPs is reduced and colloid stability can no longer be maintained under the same ionic environment, hence AuNPs aggregate and a red-to-purple color change is detected.

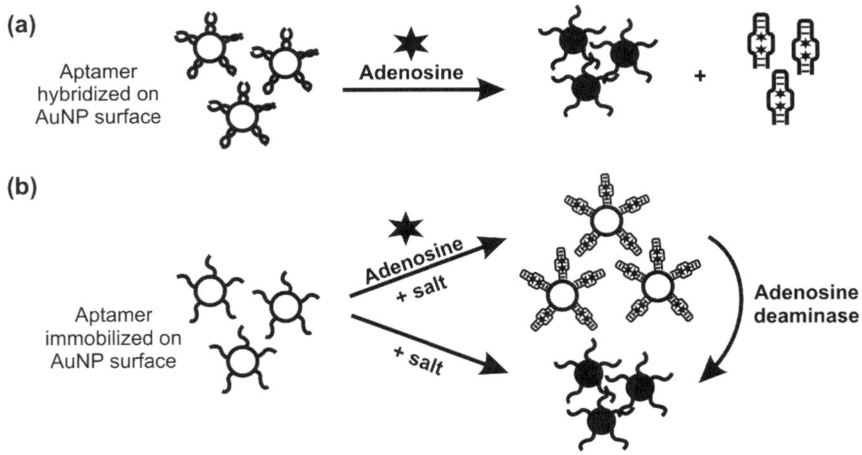

Figure 19.13 Colorimetric detection of adenosine using AuNPs with non-crosslinking aggregation. (a) Adenosine-binding aptamers initially hybridized with DNA-functionalized AuNPs are released in the presence of adenosine, causing destabilization and subsequent aggregation of nanoparticles, which can be visualized by a red-to-purple color change. (b) Under high salt conditions, the aptamers immobilized on the AuNP surface are well dispersed in the presence of target due to steric and electrostatic stabilization by folded aptamers. In the absence of target, AuNPs cannot be stabilized by unfolded aptamers under high salt conditions and a color change is detected. Similarly, the conversion of aptamer-bound adenosine into inosine by adenosine deaminase causes the unfolding of aptamers on the AuNP surface and subsequent aggregation of nanoparticles under the same salt condition

In a study by Maeda and co-workers, it was reported that the molecular conformation on a nanoparticle surface can modulate colloidal stability [181]. For instance, single-stranded DNA on AuNP is more resistant towards salt-induced aggregation than double-stranded DNA, even when both are identical in length. These observations suggest that DNA aptamers on a nanoparticle surface can be used to control colloid stability.

Li and co-workers studied the influence of aptamer folding on nanoparticle stability [182]. Adenosine aptamers are immobilized on an AuNP surface and salt is added to the solution. In the absence of a target, the high salt concentration overcomes DNA's repulsive force and nanoparticles aggregate (Figure 19.13b). However, in the presence of adenosine, folded aptamers on the colloid surface stabilize nanoparticles against aggregation and no color change is observed.

A unique characteristic of this platform is that the assay is reversible since aptamers can be folded and denatured repeatedly. In the reverse process, adenosine deaminase, an enzyme which converts adenosine to inosine, is added to well-dispersed nanoparticles which are stabilized by tethered aptamer–adenosine complexes (Figure 19.13b). Since inosine does not bind to the aptamer, the conversion of a target to a non-target molecule causes the aptamer to unfold on the gold surface. This change in DNA structure results in nanoparticle destabilization and the opposite red-to-purple color change is observed.

19.5.4 Colorimetric assays involving aptazymes

In an aptazyme-based colorimetric sensor, adenosine-dependent RNA-cleaving aptazymes were preassembled with AuNPs and DNA crosslinkers containing a ribonucleotide similar to the NAE-based assay described previously (See section 19.5.2.1) [183,184]. In the presence of adenosine, the aptamer domain binds to its target, allowing NAE to cleave the ribonucleotide linkage embedded within each crosslinker (Figure 19.11b). Consequently, a purple-to-red color change is observed since nanoparticles are no longer held together by crosslinkers. In the absence of adenosine, the aptamer domain remains unfolded, preventing NAE in the aptazyme from cleaving crosslinkers. As a result, the nanoparticle assembly remains aggregated after heating and cooling and no color change is detected.

19.6 Electrochemical nucleic acid sensors

19.6.1 General considerations

In electrochemical nucleic acid sensors, signal transduction is mediated by nucleic acid-functionalized electrodes. The interaction between an aptamer and a target is measured by an electric signal proportional to the amount of target present. This signal can be transduced either via labeled or label-free methods. In the latter, strong redox molecules are bound intrinsically or tethered to aptamers through attachment groups. The redox molecules report on the conformational state of the aptamer on an electrode surface. In the label-free approach, the sensor relies strictly on intrinsic electrical changes from aptamer–target interactions between DNA and the electrode.

To immobilize nucleic acids on electrodes, the simplest way is to attach probes directly to surface materials through electrostatic adsorption. Although uncomplicated, this method results in non-specific interactions which may affect signal reliability. Therefore, the most frequent form of immobilization is by attaching thiolated nucleic acids on a gold surface [185]. Any non-specific interactions between thiolated probes and the gold surface are removed by mercaptohexanol treatment, by which non-specifically bound DNAs are displaced [186,187].

Since aptamers require space for proper folding, the surface density and probe-to-electrode distance may have profound effects on detection capacity [188,189]. Probes typically range between 15 and 100 nucleotides in length and have 3–6 carbon linkers to avoid surface interference. Reports characterizing probe density effects revealed that a surface coverage of 5×10^{12} molecules cm^{-2} is sufficient to avoid target repulsion [190].

19.6.2 Electrochemical sensors using label-free nucleic acids

In label-free nucleic acid electrochemical sensors, the signal is dependent on the structural rigidity of the target, the probe–target complex or oxidation of nucleotide bases in the probe [191,192]. The sensor is constructed using electrochemical impedance spectroscopy (EIS), where nucleic acid probe structures are monitored. This is based on dielectric changes on the electrode produced by probe–target interactions [193]. A high-frequency alternating current is used to discriminate between native probes or probe–target complexes. In the presence of a target, a phase lag is induced by poor migration of redox molecules to the electrode surface and an increase in resistance is detected. Using this design, O'Sullivan and co-workers achieved low detection limits of thrombin using ferricyanide in a reusable sensor (Figure 19.14a) [194]. The aptamer–thrombin complex retarded electron transfer to the gold surface and an increase in transfer resistance was recorded. To regenerate the sensor, high salt concentrations (e.g. 2 M NaCl) were introduced to dissociate the complex and resistance measurements returned to basal levels. Using the same design, electrochemical sensors for PDGF [195], AMP [196] and lysozyme [197] with detection limits from low nanomolar to micromolar concentrations have been reported.

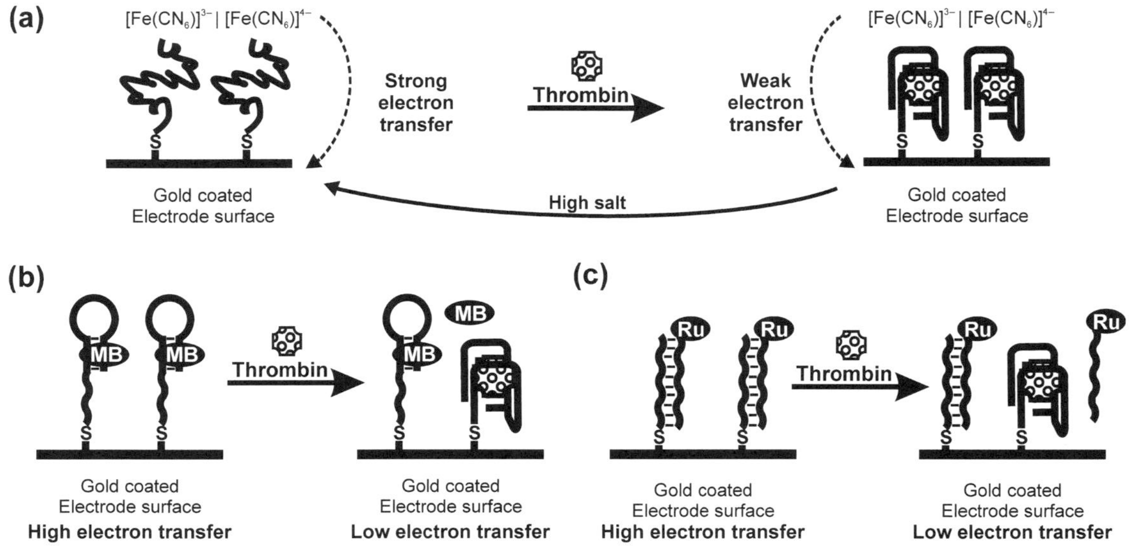

Figure 19.14 Various electrochemical detection methods for thrombin using aptamers. In the panels, thrombin-binding aptamers are immobilized on a gold-coated electrode and the binding between thrombin and the aptamer can be detected by different mechanisms. (a) The binding is detected by a reduction in electron transfer between ferricyanide and the gold-coated electrode. (b) The binding displaces a duplex binding dye [e.g. methylene blue (MB)] from the aptamer, consequently decreasing electron transfer to the electrode surface. (c) $Ru(bpy)_3^{2+}$-conjugated complementary strands are initially hybridized with the aptamer in the absence of thrombin. The folding of the aptamer results in the displacement of $Ru(bpy)_3^{2+}$-labeled strands, accompanied by a reduction in electron transfer to the gold electrode

19.6.3 Electrochemical sensors using labeled nucleic acids

Electrochemical sensors with labeled moieties have also been exploited as an effective biosensing mechanism. In this method, the signal is dependent on the redox group. A strong redox molecule can be either tethered to the DNA probe, or placed in solution if it can preferentially bind a specific DNA conformation that can be altered by ligand binding. An early electrochemical nucleic acid sensor employed a minor groove binding label, $Co(Phen)_3^{3+}$, to distinguish between single- and double-stranded DNA [198,199]. Another method used methylene blue for the detection of viral DNA from hepatitis B. In both cases, DNA duplex promoted the association of reducing agents, thus increasing its proximity to the electrode surface to generate a signal. Such electrochemical sensors have also been developed to detect gene mutations related to cancer [200,201].

Kim and co-workers developed a comparable method for thrombin detection using its DNA aptamer. The aptamer was designed to have a hairpin structure and the presence of thrombin disrupted the hairpin to release previously bound intercalating dyes (Figure 19.14b) [202]. A similar strategy was used in a cocaine sensor where the presence of target caused a methylene blue moiety to move closer to the electrode surface for enhanced electron transfer [203,204].

Other studies used modified nucleotides as redox groups. For example, Wang and co-workers utilized DNA probes that were synthesized with inosine bases instead of guanine bases to detect hybridization of complementary sequences [205,206]. In this system, inosine forms hydrogen bonds with cytosine and displays oxidation peaks that are away from those of guanine. The guanine oxidation peaks appear only upon annealing to target sequences, and are measured using chronopotentiometry. Sequences that share less complementarity with the probe display diminished oxidation peaks. Although high background signals were initially observed due to non-specific interaction between guanine-containing target sequences and the graphite electrode, magnetic beads were subsequently applied to remove non-specifically bound sequences and a detection limit in the picomolar range was achieved [207].

Researchers have also engineered electrochemical sandwich assays to generate extremely sensitive devices. Fang and co-workers utilized an aptamer structure-switching approach for the detection of thrombin [208]. They immobilized the thrombin aptamer on gold electrodes with a short complementary oligonucleotide sequence that was tethered to silica particles packed with a strong reducing agent, $Ru(bpy)_3^{2+}$ (Figure 19.14c). Upon target binding, the aptamer switches from its duplex conformation to a complex structure with thrombin, which causes the release of oligonucleotides from the electrode surface and generates a decrease in electrochemiluminescent signal. Such electrochemical sandwich assays have been reported to achieve detection limits in the femto- to nanomolar range [208–211].

19.7 Piezoelectric nucleic acid sensors

Mechanical stress applied to the surface of certain crystals can generate an electric potential proportional to the mechanical stress (known as the *piezoelectric effect*). Materials that exhibit piezoelectric properties include lithium niobate, aluminum nitride, zinc oxide, tourmaline and lead zirconate titanate. Two popular forms of piezoelectric biosensors for the detection of DNA and other non-nucleic acid targets are the quartz crystal microbalance (QCM) and the cantilever; both are operated by applying an alternating electric field to a crystal or ceramic to produce a defined oscillation pattern. The sensor resonates when the excitation frequency is equivalent to the mechanical frequency of the material and will change when probes interact with the target [212].

The ability to immobilize nucleic acid probes while maintaining their binding properties is critical to the performance of QCMs. An optimal immobilization method is dependent on probe characteristics and also the transducing material. To date, covalent attachment on pre-functionalized surfaces, affinity immobilization and self-assembly have been shown to be successful [213]. Most piezoelectric crystals are first functionalized with gold and the resulting gold-coated piezoelectric surface is further modified by direct chemisorption of thiol-modified probes. Suitable amounts of thiolated nucleic acid probes can be attached to the gold surface for optimal hybridization. Secondary thiols such as mercaptohexanol can be used to reduce non-specific binding and adjust surface probe densities [214]. Alternatively, the surface can be modified with streptavidin so that biotinylated probes can be used.

19.7.1 Quartz crystal-based piezoelectric sensors

Piezoelectric quartz crystals have been used in microbalances and microviscometers for thin-film deposition control, etching studies, aerosol mass measurements and system contamination studies. The first piezoelectric nucleic acid sensor was based on hybridization of oligonucleotides to their complementary sequence on a QCM [215]. It was able to differentiate point mutations in DNA where mismatched duplexes generated lower oscillating frequency shifts than perfectly complementary sequences. Prohaska and co-workers extended this technique for non-nucleic acid detection by designing a sensor using an aptamer that binds immunoglobulin E (IgE) (Figure 19.15a) [216]. They used a gold-coated QCM functionalized with the IgE aptamer or anti-IgE antibody in a flow-through injection system [217]. Both the aptamer and antibody can exhibit a detection limit of $\sim$0.5 nM. However, the QCM using IgE aptamer could achieve a wider detection range than the antibody device because nucleic acids can be easily immobilized to higher surface densities. The performance of the aptamer sensor can be further improved by optimizing the experimental conditions (such as ionic strength and pH of the buffer) and the optimized sensor is compatible with heterogeneous samples such as plasma and meat extracts [218–220].

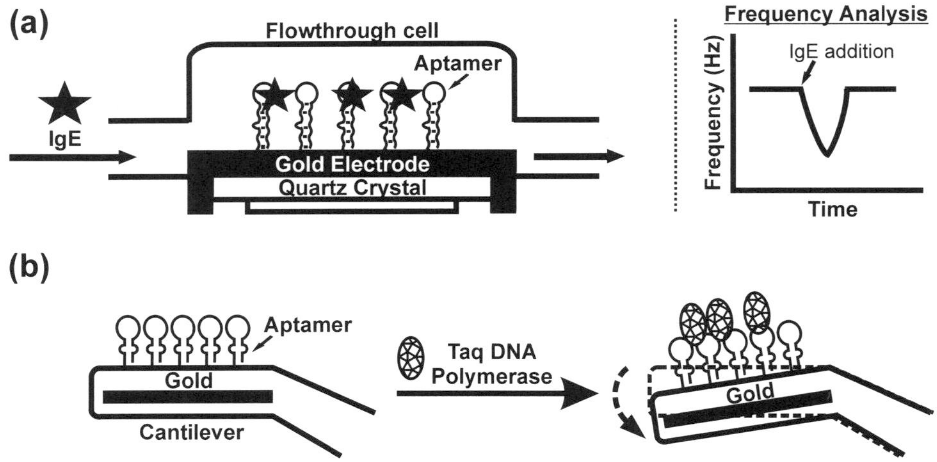

Figure 19.15 Piezoelectric nucleic acid sensors. (a) Schematic of a flow-through cell for IgE detection. The binding of IgE to the aptamer grafted on a gold electrode can be detected by changes in the oscillation frequency of a quartz crystal over time. (b) Aptamer molecules for Taq DNA polymerase are immobilized on a gold-coated piezoelectric cantilever. The binding of the polymerase to the aptamer causes a mass change and subsequent bending of the cantilever, which is measured using a laser beam

19.7.2 Piezoelectric cantilevers

Piezoelectric cantilever technology has been successfully applied to detect IgE, HIV Tat proteins [219], thrombin [218], and DNA [221,222]. In these sensors, the interaction between immobilized nucleic acid probes and their targets alters the surface stress of the cantilever. The bending of cantilever is monitored with a laser beam. For instance, Manalis and co-workers immobilized the aptamer that binds Taq DNA polymerase via standard gold–thiol chemistry to create a cantilever sensor (Figure 19.15b) [223]. The polymerase induced bending of cantilever by 3–32 nm depending on the density of aptamers on gold surface. Since target binding on a cantilever was comparable to aptamer performance in solution, this suggests that immobilization of aptamers on cantilevers has minimal interference on detection.

19.8 Conclusion and perspectives

We have discussed many examples in which nucleic acids are utilized as functional probes for biomolecular detection. The straightforward approach takes advantage of predictable Watson–Crick interactions where nucleic acid targets are detected by DNA probes in a sequence-specific manner. By combining this property with existing platforms such as fluorescence signaling and solid-phase immobilization, it has allowed researchers to detect individual nucleic acid sequence *in situ* or to examine multiple sequence copies in a high-throughput fashion. In both cases, high detection sensitivity (e.g. detection of a single molecule) and high selectivity (e.g. discrimination of single nucleotide mismatch) can be achieved.

Apart from detecting nucleic acids through Watson–Crick hybridization, the discovery of FNA has broadened the potential of nucleic acids as a diagnostic tool. For instance, many *in vitro* selection experiments have generated RNA and DNA species which are capable of metal-specific catalysis and such molecules have been demonstrated to detect various metal pollutants in the environment. Similarly,

SELEX experiments have produced a large array of aptamers capable of recognizing a wide range of targets from small molecules to whole cells. These molecular receptors have become increasingly popular in assay and sensor development. Finally, by linking aptamers to nucleic acid enzymes, numerous aptazymes have been created. These molecules have also provided a useful, ligand-responsive and catalytic platform for detection based applications.

The progress of FNA-based detection technologies has been phenomenal to date: FNAs have been widely coupled to a plethora of detection strategies to generate fluorescent, colorimetric, electrochemical and piezoelectric sensing devices. However, the current bottleneck in broadening the utility of these sensors is the availability of novel FNA motifs, which is often limited by labor-intensive and time-consuming SELEX procedures. Therefore, it is important to focus future efforts on crafting facile or high-throughput techniques to expand the current repertoire of FNAs.

References

1. K. Kruger, P.J. Grabowski *et al.*, Self-splicing RNA: autoexcision and autocyclization of the ribosomal RNA intervening sequence of *Tetrahymena*, *Cell*, **31**, 147–157 (1982).
2. C. Guerrier-Takada, K. Gardiner *et al.*, The RNA moiety of ribonuclease P is the catalytic subunit of the enzyme, *Cell*, **35**, 849–857 (1983).
3. A.D. Ellington and J.W. Szostak, *In vitro* selection of RNA molecules that bind specific ligands, *Nature*, **346**, 818–822 (1990).
4. D.L. Robertson and G.F. Joyce, Selection *in vitro* of an RNA enzyme that specifically cleaves single-stranded DNA, *Nature*, **344**, 467–468 (1990).
5. C. Tuerk and L. Gold, Systematic evolution of ligands by exponential enrichment: RNA ligands to bacteriophage T4 DNA polymerase, *Science*, **249**, 505–510 (1990).
6. S.W. Santoro and G.F. Joyce, A general purpose RNA-cleaving DNA enzyme, *Proc Natl Acad Sci USA*, **94**, 4262–4266 (1997).
7. N. Risch and K. Merikangas, The future of genetic studies of complex human diseases, *Science*, **273**, 1516–1517 (1996).
8. J.N. Hirschhorn and M.J. Daly, Genome-wide association studies for common diseases and complex traits, *Nat Rev Genet*, **6**, 95–108 (2005).
9. L. Feuk, C.R. Marshall *et al.*, Structural variants: changing the landscape of chromosomes and design of disease studies, *Hum Mol Genet*, **15** Spec. No. 1, R57–R66 (2006).
10. J.R. Lupski, R.M. de Oca-Luna *et al.*, DNA duplication associated with Charcot-Marie-Tooth disease type 1a, *Cell*, **66**, 219–232 (1991).
11. R. Redon, S. Ishikawa *et al.*, Global variation in copy number in the human genome, *Nature*, **444**, 444–454 (2006).
12. J.G. Gall and M.L. Pardue, Formation and detection of RNA–DNA hybrid molecules in cytological preparations, *Proc Natl Acad Sci USA*, **63**, 378–383 (1969).
13. M.L. Pardue and J.G. Gall, Molecular hybridization of radioactive DNA to the DNA of cytological preparations, *Proc Natl Acad Sci USA*, **64**, 600–604 (1969).
14. L.A. Brown and D. Huntsman, Fluorescent *in situ* hybridization on tissue microarrays: challenges and solutions, *J Mol Histol*, **38**, 151–157 (2007).
15. J.A. Fletcher, DNA *in situ* hybridization as an adjunct in tumor diagnosis, *Am J Clin Pathol*, **112**, S11–18 (1999).
16. D. Pinkel, T. Straume *et al.*, Cytogenetic analysis using quantitative, high-sensitivity, fluorescence hybridization, *Proc Natl Acad Sci USA*, **83**, 2934–2938 (1986).
17. T. Cremer, J. Landegent *et al.*, Detection of chromosome aberrations in the human interphase nucleus by visualization of specific target dnas with radioactive and non-radioactive *in situ* hybridization techniques: diagnosis of trisomy 18 with probe L1.84, *Hum Genet*, **74**, 346–352 (1986).
18. J. Nath and K.L. Johnson, A review of fluorescence *in situ* hybridization (FISH): current status and future prospects, *Biotechnol Histochem*, **75**, 54–78 (2000).

19. J.H. Knoll, P. Lichter *et al.*, *In situ* hybridization and detection using nonisotopic probes, *Curr Protoc Mol Biol*, Ch. 14, Unit 14 17 (2007).
20. P.M. Nederlof, D. Robinson *et al.*, Three-color fluorescence *in situ* hybridization for the simultaneous detection of multiple nucleic acid sequences, *Cytometry*, **10**, 20–27 (1989).
21. P.M. Nederlof, S. van der Flier *et al.*, Fluorescence ratio measurements of double-labeled probes for multiple *in situ* hybridization by digital imaging microscopy, *Cytometry*, **13**, 839–845 (1992).
22. C. Fauth, H. Zhang *et al.*, A new strategy for the detection of subtelomeric rearrangements, *Hum Genet*, **109**, 576–583 (2001).
23. J. Brown, K. Saracoglu, S. Uhrig, M.R. Speicher, R. Eils and L. Kearney, Subtelomeric chromosome rearrangements are detected using an innovative 12-color FISH assay (M-TEL), *Nat Med*, **7**, 497–501 (2001).
24. S.J. Knight, C.M. Lese *et al.*, An optimized set of human telomere clones for studying telomere integrity and architecture, *Am J Hum Genet*, **67**, 320–332 (2000).
25. J. Bayani and J. Squire, Multi-color FISH techniques, *Curr Protoc Cell Biol*, Ch. 22, Unit 22 **25** (2004).
26. K. Junker, T. Fritsch *et al.*, Multicolor fluorescence *in situ* hybridization (M-FISH) on cells from urine for the detection of bladder cancer, *Cytogenet Genome Res*, **114**, 279–283 (2006).
27. M.I. Stamouli, A.D. Panani *et al.*, Detection of genetic alterations in primary bladder carcinoma with dual-color and multiplex fluorescence *in situ* hybridization, *Cancer Genet Cytogenet*, **149**, 107–113 (2004).
28. K. Mrozek, M. Iliszko *et al.*, Spectral karyotyping reveals 17;22 fusions in a cytogenetically atypical dermatofibrosarcoma protuberans with a large marker chromosome as a sole abnormality, *Genes Chromosomes Cancer*, **31**, 182–186 (2001).
29. F.F. Zhang, J.L. Murata-Collins *et al.*, Twenty-four-color spectral karyotyping reveals chromosome aberrations in cytogenetically normal acute myeloid leukemia, *Genes Chromosomes Cancer*, **28**, 318–328 (2000).
30. A. Ohsaka, K. Otsubo *et al.*, Spectral karyotyping and fluorescence *in situ* hybridization analyses identified a novel three-way translocation involving inversion 16 in therapy-related acute myeloid leukemia M4Eo, *Cancer Genet Cytogenet*, **184**, 113–118 (2008).
31. S. Mergenthaler-Gatfield, W. Holzgreve *et al.*, Spectral karyotyping (SKY): applications in prenatal diagnostics, *Methods Mol Biol*, **444**, 3–26 (2008).
32. H. Engels, A. Ehrbrecht *et al.*, Comprehensive analysis of human subtelomeres with combined binary ratio labelling fluorescence *in situ* hybridisation, *Eur J Hum Genet*, **11**, 643–651 (2003).
33. H.J. Tanke, J. Wiegant *et al.*, New strategy for multi-colour fluorescence *in situ* hybridisation: COBRA: combined binary ratio labelling, *Eur J Hum Genet*, **7**, 2–11 (1999).
34. F. Darroudi, V. Bezrookove *et al.*, Insights into the sites of x ray and neutron induced chromosomal aberrations in human lymphocytes using COBRA–MFISH, *Radiat Prot Dosimetry*, **99**, 189–192 (2002).
35. M. Nilsson, H. Malmgren *et al.*, Padlock probes: circularizing oligonucleotides for localized DNA detection, *Science*, **265**, 2085–2088 (1994).
36. D.Y. Zhang, M. Brandwein *et al.*, Amplification of target-specific, ligation-dependent circular probe, *Gene*, **211**, 277–285 (1998).
37. A. Fire and S.Q. Xu, Rolling replication of short DNA circles, *Proc Natl Acad Sci USA*, **92**, 4641–4645 (1995).
38. D.Y. Liu, S.L. Daubendiek *et al.*, Rolling circle DNA synthesis: small circular oligonucleotides as efficient templates for DNA polymerases, *J Am Chem Soc*, **118**, 1587–1594 (1996).
39. A.T. Christian, M.S. Pattee *et al.*, Detection of DNA point mutations and mRNA expression levels by rolling circle amplification in individual cells, *Proc Natl Acad Sci USA*, **98**, 14238–14243 (2001).
40. C. Larsson, J. Koch *et al.*, *In situ* genotyping individual DNA molecules by target-primed rolling-circle amplification of padlock probes, *Nat Methods*, **1**, 227–232 (2004).
41. P.M. Lizardi, X. Huang *et al.*, Mutation detection and single-molecule counting using isothermal rolling-circle amplification, *Nat Genet*, **19**, 225–232 (1998).
42. A.F. Faruqi, S. Hosono *et al.*, High-throughput genotyping of single nucleotide polymorphisms with rolling circle amplification, *BMC Genomics*, **2**, 4 (2001).
43. M. Nilsson, K. Krejci *et al.*, Padlock probes reveal single-nucleotide differences, parent of origin and *in situ* distribution of centromeric sequences in human chromosomes 13 and 21, *Nat Genet*, **16**, 252–255 (1997).

44. W. Zhang, M. Cohenford *et al.*, Detection of *Chlamydia trachomatis* by isothermal ramification amplification method: a feasibility study, *J Clin Microbiol*, **40**, 128–132 (2002).
45. A. Rector, G.D. Bossart *et al.*, Characterization of a novel close-to-root papillomavirus from a Florida manatee by using multiply primed rolling-circle amplification: *Trichechus manatus latirostris* papillomavirus type 1, *J Virol*, **78**, 12698–12702 (2004).
46. J.A. Prince, L. Feuk *et al.*, Robust and accurate single nucleotide polymorphism genotyping by dynamic allele-specific hybridization (DASH): design criteria and assay validation, *Genome Res*, **11**, 152–162 (2001).
47. A. Russom, S. Haasl *et al.*, Genotyping by dynamic heating of monolayered beads on a microheated surface, *Electrophoresis*, **25**, 3712–3719 (2004).
48. W.M. Howell, M. Jobs *et al.*, IFRET: An improved fluorescence system for DNA-melting analysis, *Genome Res*, **12**, 1401–1407 (2002).
49. M. Jobs, W.M. Howell *et al.*, DASH-2: flexible, low-cost and high-throughput SNP genotyping by dynamic allele-specific hybridization on membrane arrays, *Genome Res*, **13**, 916–924 (2003).
50. S. Tyagi and F.R. Kramer, Molecular beacons: probes that fluoresce upon hybridization, *Nat Biotechnol*, **14**, 303–308 (1996).
51. G. Bonnet, S. Tyagi *et al.*, Thermodynamic basis of the enhanced specificity of structured DNA probes, *Proc Natl Acad Sci USA*, **96**, 6171–6176 (1999).
52. J.A. Vet, A.R. Majithia *et al.*, Multiplex detection of four pathogenic retroviruses using molecular beacons, *Proc Natl Acad Sci USA*, **96**, 6394–6399 (1999).
53. S. Sandhya, W. Chen *et al.*, Molecular beacons: a real-time polymerase chain reaction assay for detecting *Escherichia coli* from fresh produce and water, *Anal Chim Acta*, **614**, 208–212 (2008).
54. L.G. Kostrikis, S. Tyagi *et al.*, Spectral genotyping of human alleles, *Science*, **279**, 1228–1229 (1998).
55. E.M. Southern, Detection of specific sequences among DNA fragments separated by gel electrophoresis, *J Mol Biol*, **98**, 503–517 (1975).
56. F.C. Kafatos, C.W. Jones *et al.*, Determination of nucleic acid sequence homologies and relative concentrations by a dot hybridization procedure, *Nucleic Acids Res*, **7**, 1541–1552 (1979).
57. J. Wang, From DNA biosensors to gene chips, *Nucleic Acids Res*, **28**, 3011–3016 (2000).
58. D. Pinkel, R. Segraves *et al.*, High resolution analysis of DNA copy number variation using comparative genomic hybridization to microarrays, *Nat Genet*, **20**, 207–211 (1998).
59. J.R. Pollack, T. Sorlie *et al.*, Microarray analysis reveals a major direct role of DNA copy number alteration in the transcriptional program of human breast tumors, *Proc Natl Acad Sci USA*, **99**, 12963–12968 (2002).
60. A.M. Snijders, N. Nowak *et al.*, Assembly of microarrays for genome-wide measurement of DNA copy number, *Nat Genet*, **29**, 263–264 (2001).
61. A.S. Ishkanian, C.A. Malloff *et al.*, A tiling resolution DNA microarray with complete coverage of the human genome, *Nat Genet*, **36**, 299–303 (2004).
62. D. Shalon, S.J. Smith *et al.*, A DNA microarray system for analyzing complex DNA samples using two-color fluorescent probe hybridization, *Genome Res*, **6**, 639–645 (1996).
63. T.R. Hughes, M. Mao *et al.*, Expression profiling using microarrays fabricated by an ink-jet oligonucleotide synthesizer, *Nat Biotechnol*, **19**, 342–347 (2001).
64. A.C. Pease, D. Solas *et al.*, Light-generated oligonucleotide arrays for rapid DNA sequence analysis, *Proc Natl Acad Sci USA*, **91**, 5022–5026 (1994).
65. H. Fiegler, R. Redon *et al.*, Accurate and reliable high-throughput detection of copy number variation in the human genome, *Genome Res*, **16**, 1566–1574 (2006).
66. A.L. Beaudet and J.W. Belmont, Array-based DNA diagnostics: let the revolution begin, *Annu Rev Med*, **59**, 113–129 (2008).
67. N. Waddell, Microarray-based DNA profiling to study genomic aberrations, *IUBMB Life*, **60**, 437–440 (2008).
68. J.A. Doudna and T.R. Cech, The chemical repertoire of natural ribozymes, *Nature*, **418**, 222–228 (2002).
69. N. Carmi, S.R. Balkhi *et al.*, Cleaving DNA with DNA, *Proc Natl Acad Sci USA*, **95**, 2233–2237 (1998).
70. S.K. Silverman, *In vitro* selection, characterization and application of deoxyribozymes that cleave RNA, *Nucleic Acids Res*, **33**, 6151–6163 (2005).
71. Y. Li and R.R. Breaker, Phosphorylating DNA with DNA, *Proc Natl Acad Sci USA*, **96**, 2746–2751 (1999).

72. W. Wang, L.P. Billen *et al.*, Sequence diversity, metal specificity and catalytic proficiency of metal-dependent phosphorylating DNA enzymes, *Chem Biol*, **9**, 507–517 (2002).
73. J.C. Achenbach, G.A. Jeffries *et al.*, Secondary-structure characterization of two proficient kinase deoxyribozymes, *Biochemistry*, **44**, 3765–3774 (2005).
74. S.A. McManus and Y. Li, A deoxyribozyme with a novel guanine quartet-helix pseudoknot structure, *J Mol Biol*, **375**, 960–968 (2008).
75. B. Cuenoud and J.W. Szostak, A DNA metalloenzyme with DNA ligase activity, *Nature*, **375**, 611–614 (1995).
76. M. Levy and A.D. Ellington, Selection of deoxyribozyme ligases that catalyze the formation of an unnatural internucleotide linkage, *Bioorg Med Chem*, **9**, 2581–2587 (2001).
77. M. Levy and A.D. Ellington, *In vitro* selection of a deoxyribozyme that can utilize multiple substrates, *J Mol Evol*, **54**, 180–190 (2002).
78. A. Sreedhara, Y. Li *et al.*, Ligating DNA with DNA, *J Am Chem Soc*, **126**, 3454–3460 (2004).
79. A. Flynn-Charlebois, Y. Wang *et al.*, Deoxyribozymes with 2′–5′ RNA ligase activity, *J Am Chem Soc*, **125**, 2444–2454 (2003).
80. A. Flynn-Charlebois, T.K. Prior *et al.*, *In vitro* evolution of an RNA-cleaving DNA enzyme into an RNA ligase switches the selectivity from 3′–5′ to 2′–5′, *J Am Chem Soc*, **125**, 5346–5350 (2003).
81. W.E. Purtha, R.L. Coppins *et al.*, General deoxyribozyme-catalyzed synthesis of native 3′–5′ RNA linkages, *J Am Chem Soc*, **127**, 13124–13125 (2005).
82. Y. Li and D. Sen, Toward an efficient dnazyme, *Biochemistry*, **36**, 5589–5599 (1997).
83. Y. Li and D. Sen, A catalytic DNA for porphyrin metalation, *Nat Struct Biol*, **3**, 743–747 (1996).
84. D.A. Baum and S.K. Silverman, Deoxyribozymes: useful DNA catalysts *in vitro* and *in vivo*, *Cell Mol Life Sci*, (2008).
85. J.C. Achenbach, W. Chiuman *et al.*, Dnazymes: from creation *in vitro* to application *in vivo*, *Curr Pharm Biotechnol*, **5**, 321–336 (2004).
86. R.R. Breaker and G.F. Joyce, A DNA enzyme that cleaves RNA, *Chem Biol*, **1**, 223–229 (1994).
87. J. Liu and Y. Lu, Rational design of 'turn-on' allosteric dnazyme catalytic beacons for aqueous mercury ions with ultrahigh sensitivity and selectivity, *Angew Chem Int Ed*, **46**, 7587–7590 (2007).
88. J. Liu and Y. Lu, Colorimetric biosensors based on DNAzyme-assembled gold nanoparticles, *J Fluoresc*, **14**, 343–354 (2004).
89. Y. Chen, M. Wang *et al.*, An autonomous DNA nanomotor powered by a DNA enzyme, *Angew Chem Int Ed*, **43**, 3554–3557 (2004).
90. Y. Tian, Y. He *et al.*, A DNAzyme that walks processively and autonomously along a one-dimensional track, *Angew Chem Int Ed*, **44**, 4355–4358 (2005).
91. M.N. Stojanovic, S. Semova *et al.*, Deoxyribozyme-based ligase logic gates and their initial circuits, *J Am Chem Soc*, **127**, 6914–6915 (2005).
92. S. Chakraborti and A.C. Banerjea, Inhibition of HIV-1 gene expression by novel DNA enzymes targeted to cleave HIV-1 TAR RNA: potential effectiveness against all HIV-1 isolates, *Mol Ther*, **7**, 817–826 (2003).
93. A.V. Vlassov, H. Ilves *et al.*, Inhibition of hepatitis C IRES-mediated gene expression by 8–17 deoxyribozymes in human tissue culture cells, *Dokl Biochem Biophys*, **410**, 257–259 (2006).
94. R. Stoltenburg, C. Reinemann *et al.*, SELEX – a (r)evolutionary method to generate high-affinity nucleic acid ligands, *Biomol Eng*, **24**, 381–403 (2007).
95. S.D. Jayasena, Aptamers: an emerging class of molecules that rival antibodies in diagnostics, *Clin Chem*, **45**, 1628–1650 (1999).
96. T. Hermann and D.J. Patel, Adaptive recognition by nucleic acid aptamers, *Science*, **287**, 820–825 (2000).
97. A.D. Ellington and J.W. Szostak, Selection *in vitro* of single-stranded DNA molecules that fold into specific ligand-binding structures, *Nature*, **355**, 850–852 (1992).
98. L.S. Green, D. Jellinek *et al.*, Nuclease-resistant nucleic acid ligands to vascular permeability factor/vascular endothelial growth factor, *Chem Biol*, **2**, 683–695 (1995).
99. G.A. Soukup and R.R. Breaker, Design of allosteric hammerhead ribozymes activated by ligand-induced structure stabilization, *Structure*, **7**, 783–791 (1999).
100. J. Tang and R.R. Breaker, Rational design of allosteric ribozymes, *Chem Biol*, **4**, 453–459 (1997).

101. J. Tang and R.R. Breaker, Mechanism for allosteric inhibition of an ATP-sensitive ribozyme, *Nucleic Acids Res*, **26**, 4214–4221 (1998).
102. M.P. Robertson and A.D. Ellington, Design and optimization of effector-activated ribozyme ligases, *Nucleic Acids Res*, **28**, 1751–1759 (2000).
103. H. Porta and P.M. Lizardi, An allosteric hammerhead ribozyme, *Biotechnology (N Y)*, **13**, 161–164 (1995).
104. T. Kuwabara, M. Warashina *et al.*, A novel allosterically *trans*-activated ribozyme, the maxizyme, with exceptional specificity *in vitro* and *in vivo*, *Mol Cell*, **2**, 617–627 (1998).
105. M.P. Robertson and A.D. Ellington, *In vitro* selection of an allosteric ribozyme that transduces analytes to amplicons, *Nat Biotechnol*, **17**, 62–66 (1999).
106. Y. Komatsu, S. Yamashita *et al.*, Construction of new ribozymes requiring short regulator oligonucleotides as a cofactor, *J Mol Biol*, **299**, 1231–1243 (2000).
107. M. Koizumi, G.A. Soukup *et al.*, Allosteric selection of ribozymes that respond to the second messengers cGMP and cAMP, *Nat Struct Biol*, **6**, 1062–1071 (1999).
108. G.A. Soukup, E.C. DeRose *et al.*, Generating new ligand-binding RNAs by affinity maturation and disintegration of allosteric ribozymes, *RNA*, **7**, 524–536 (2001).
109. M. Araki, Y. Okuno *et al.*, Allosteric regulation of a ribozyme activity through ligand-induced conformational change, *Nucleic Acids Res*, **26**, 3379–3384 (1998).
110. G.A. Soukup and R.R. Breaker, Engineering precision RNA molecular switches, *Proc Natl Acad Sci USA*, **96**, 3584–3589 (1999).
111. D.Y. Wang and D. Sen, Rationally designed allosteric variants of hammerhead ribozymes responsive to the HIV-1 TAT protein, *Comb Chem High Throughput Screen*, **5**, 301–312 (2002).
112. V.V. Didenko, DNA probes using fluorescence resonance energy transfer (FRET): designs and applications, *Biotechniques*, **31**, 1106–1116, 1118, 1120–1121 (2001).
113. T.A. Perkins, D.E. Wolf *et al.*, Fluorescence resonance energy transfer analysis of ribozyme kinetics reveals the mode of action of a facilitator oligonucleotide, *Biochemistry*, **35**, 16370–16377 (1996).
114. N.G. Walter and J.M. Burke, Real-time monitoring of hairpin ribozyme kinetics through base-specific quenching of fluorescein-labeled substrates, *RNA*, **3**, 392–404 (1997).
115. Y. Lu, J. Liu *et al.*, New highly sensitive and selective catalytic DNA biosensors for metal ions, *Biosens Bioelectron*, **18**, 529–540 (2003).
116. T.A. Perkins and J. Goodchild, Using fluorescence resonance energy transfer to investigate hammerhead ribozyme kinetics, *Methods Mol Biol*, **74**, 241–251 (1997).
117. M.N. Stojanovic, P. de Prada *et al.*, Homogeneous assays based on deoxyribozyme catalysis, *Nucleic Acids Res*, **28**, 2915–2918 (2000).
118. D. Vitiello, D.B. Pecchia *et al.*, Intracellular ribozyme-catalyzed *trans*-cleavage of RNA monitored by fluorescence resonance energy transfer, *RNA*, **6**, 628–637 (2000).
119. A. Jenne, Gmelin, W., Raffler, N. and Famulok, M., Real-time characterization of ribozymes by fluorecence resonance energy transfer (FRET), *Angew. Chem. Int. Ed.*, **38**, 1300–1303 (1999).
120. K.K. Singh, R. Parwaresch *et al.*, Rapid kinetic characterization of hammerhead ribozymes by real-time monitoring of fluorescence resonance energy transfer (FRET), *RNA*, **5**, 1348–1356 (1999).
121. A. Jenne, J.S. Hartig *et al.*, Rapid identification and characterization of hammerhead-ribozyme inhibitors using fluorescence-based technology, *Nat Biotechnol*, **19**, 56–61 (2001).
122. J. Liu, A.K. Brown *et al.*, A catalytic beacon sensor for uranium with parts-per-trillion sensitivity and millionfold selectivity, *Proc Natl Acad Sci USA*, **104**, 2056–2061 (2007).
123. J. Liu and Y. Lu, Improving fluorescent DNAzyme biosensors by combining inter- and intramolecular quenchers, *Anal Chem*, **75**, 6666–6672 (2003).
124. Z. Liu, S.H. Mei *et al.*, Assemblage of signaling DNA enzymes with intriguing metal-ion specificities and pH dependences, *J Am Chem Soc*, **125**, 7539–7545 (2003).
125. S.H. Mei, Z. Liu *et al.*, An efficient RNA-cleaving DNA enzyme that synchronizes catalysis with fluorescence signaling, *J Am Chem Soc*, **125**, 412–420 (2003).
126. W. Chiuman and Y. Li, Evolution of high-branching deoxyribozymes from a catalytic DNA with a three-way junction, *Chem Biol*, **13**, 1061–1069 (2006).

127. Y. Shen, G. Mackey *et al.*, Entrapment of fluorescence signaling DNA enzymes in sol–gel-derived materials for metal ion sensing, *Anal Chem*, **79**, 3494–3503 (2007).
128. W. Chiuman and Y. Li, Revitalization of six abandoned catalytic DNA species reveals a common three-way junction framework and diverse catalytic cores, *J Mol Biol*, **357**, 748–754 (2006).
129. W. Chiuman and Y. Li, Efficient signaling platforms built from a small catalytic DNA and doubly labeled fluorogenic substrates, *Nucleic Acids Res*, **35**, 401–405 (2007).
130. C.B. Swearingen, D.P. Wernette *et al.*, Immobilization of a catalytic DNA molecular beacon on au for Pb(II) detection, *Anal Chem*, **77**, 442–448 (2005).
131. D.P. Wernette, C.B. Swearingen *et al.*, Incorporation of a DNAzyme into Au-coated nanocapillary array membranes with an internal standard for Pb(II) sensing, *Analyst*, **131**, 41–47 (2006).
132. S.A. Kandadai and Y. Li, Characterization of a catalytically efficient acidic RNA-cleaving deoxyribozyme, *Nucleic Acids Res*, **33**, 7164–7175 (2005).
133. J. Li and Y. Lu, A highly sensitive and selective catalytic DNA biosensor for lead ions, *J Am Chem Soc*, **122**, 10466–10467 (2000).
134. J. Liu and Y. Lu, A DNAzyme catalytic beacon sensor for paramagnetic Cu^{2+} ions in aqueous solution with high sensitivity and selectivity, *J Am Chem Soc*, **129**, 9838–9839 (2007).
135. R.D. Jenison, S.C. Gill *et al.*, High-resolution molecular discrimination by RNA, *Science*, **263**, 1425–1429 (1994).
136. A. Geiger, P. Burgstaller *et al.*, RNA aptamers that bind L-arginine with sub-micromolar dissociation constants and high enantioselectivity, *Nucleic Acids Res*, **24**, 1029–1036 (1996).
137. E.J. Merino and K.M. Weeks, Facile conversion of aptamers into sensors using a 2′-ribose-linked fluorophore, *J Am Chem Soc*, **127**, 12766–12767 (2005).
138. N. Kamekawa, Y. Shimomura *et al.*, Pyrene-modified DNA aptamer as a fluorescent biosensor with high affinity and specificity for ATP sensing, *Chem Let.*, **35**, 660 (2006).
139. S.D. Jhaveri, R. Kirby *et al.*, Designed signaling aptamers that transduce molecular recognition to changes in fluorescence intensity, *J Am Chem Soc*, **122**, 2469–2473 (2000).
140. K. Yamana, Y. Ohtani *et al.*, Bis-pyrene labeled DNA aptamer as an intelligent fluorescent biosensor, *Bioorg Med Chem Lett*, **13**, 3429–3431 (2003).
141. E. Katilius, Z. Katiliene and N.W. Woodbury, Signaling aptamers created using fluorescent nucleotide analogues, *Anal Chem*, **78**, 6484–6489 (2006).
142. M.E. Hawkins, Fluorescent pteridine nucleoside analogs: a window on DNA interactions, *Cell Biochem Biophys*, **34**, 257–281 (2001).
143. M.J. Rist and J.P. Marino, Fluorescent nucleotide base analogs as probes of nucleic acid structure, dynamics and interactions, *Curr Org Chem*, **6**, 775–793 (2002).
144. M.N. Stojanovic, P. de Prada *et al.*, Fluorescent sensors based on aptamer self-assembly, *J Am Chem Soc*, **122**, 11547–11548 (2000).
145. N. Hamaguchi, A. Ellington *et al.*, Aptamer beacons for the direct detection of proteins, *Anal Biochem*, **294**, 126–131 (2001).
146. D.P. Morse, Direct selection of RNA beacon aptamers, *Biochem Biophys Res Commun*, **359**, 94–101 (2007).
147. H. Ozaki, A. Nishihira *et al.*, Biomolecular sensor based on fluorescence-labeled aptamer, *Bioorg Med Chem Lett*, **16**, 4381–4384 (2006).
148. H. Urata, K. Nomura *et al.*, Fluorescent-labeled single-strand ATP aptamer DNA: chemo- and enantio-selectivity in sensing adenosine, *Biochem Biophys Res Commun*, **360**, 459–463 (2007).
149. C.J. Yang, S. Jockusch *et al.*, Light-switching excimer probes for rapid protein monitoring in complex biological fluids, *Proc Natl Acad Sci USA*, **102**, 17278–17283 (2005).
150. A. Ono and H. Togashi, Highly selective oligonucleotide-based sensor for mercury(II) in aqueous solutions, *Angew Chem Int Ed*, **33**, 4300–4302 (2004).
151. S. Nagatoishi, T. Nojima *et al.*, Fluorescence energy transfer probes based on the guanine quadruplex formation for the fluorometric detection of potassium ion, *Anal Chim Acta*, **581**, 125–131 (2007).
152. R. Nutiu and Li, Y., Structure-switching signaling aptamers, *J Am Chem Soc*, **125**, 4771–4778 (2003).
153. B. Li, H. Wei and S. Dong, Sensitive detection of protein by an aptamer-based label-free fluorescing molecular switch, *Chem Commun*, 73–75 (2007).

154. C. Huang, S.H. Chiu, Y.F. Huang and H.T. Chang, Aptamer-functionalized gold nanoparticles for turn-on light switch detection of platelet-derived growth factor, *Anal Chem*, **79**, 4798–4804 (2007).
155. J. Srinivasan, S.T. Cload *et al.*, ADP-specific sensors enable universal assay of protein kinase activity, *Chem Biol*, **11**, 499–508 (2004).
156. A. Ferguson, R.M. Boomer *et al.*, A novel strategy for selection of allosteric ribozymes yields riboreporter sensors for caffeine and aspartame, *Nucleic Acids Res*, **32**, 1756–1766 (2004).
157. Y. Shen, W. Chiuman *et al.*, Catalysis and rational engineering of *trans*-acting pH6DZ1, an RNA-cleaving and fluorescence-signaling deoxyribozyme with a four-way junction structure, *ChemBioChem*, **7**, 1343–1348 (2006).
158. E.J. Cho, L. Yang *et al.*, Using a deoxyribozyme ligase and rolling circle amplification to detect a non-nucleic acid analyte, ATP, *J Am Chem Soc*, **127**, 2022–2023 (2005).
159. J.S. Hartig, S.H. Najafi-Shoushtari *et al.*, Protein-dependent ribozymes report molecular interactions in real time, *Nat Biotechnol*, **20**, 717–722 (2002).
160. S. Yamazaki, L. Tan *et al.*, Aptamer displacement identifies alternative small-molecule target sites that escape viral resistance, *Chem Biol*, **14**, 804–812 (2007).
161. E. Tuite and J.M. Kelly, The interaction of methylene-blue, azure-B and thionine with DNA – formation of complexes with polynucleotides and mononucleotides as model systems, *Biopolymers*, **35**, 419–433 (1995).
162. S.K. Ghosh and T. Pal, Interparticle coupling effect on the surface plasmon resonance of gold nanoparticles: from theory to applications, *Chem Rev*, **107**, 4797–4862 (2007).
163. M.C. Daniel and D. Astruc, Gold nanoparticles: assembly, supramolecular chemistry, quantum-size-related properties and applications toward biology, catalysis and nanotechnology, *Chem Rev*, **104**, 293–346 (2004).
164. K. Aslan, J.R. Lakowicz *et al.*, Plasmon light scattering in biology and medicine: new sensing approaches, visions and perspectives, *Curr Opin Chem Biol*, **9**, 538–544 (2005).
165. H.A. Ho, M. Bera-Aberem *et al.*, Optical sensors based on hybrid DNA/conjugated polymer complexes, *Chemistry*, **11**, 1718–1724 (2005).
166. H.A. Ho and M. Leclerc, Optical sensors based on hybrid aptamer/conjugated polymer complexes, *J Am Chem Soc*, **126**, 1384–1387 (2004).
167. T.A. Kinoshita, S. Hayashi *et al.*, Preparation of a structural color forming system by polypeptide-based LB films, *J PhotochemPhotobiol A Chem*, **145**, 101–106 (2001).
168. M. Sivakumar, R. Tominaga *et al.*, Studies on visual sensor from self-assembled polypeptides, *Sci Technol Adv Mater*, **6**, 91–96 (2005).
169. J.W. Liu and Y. Lu, A colorimetric lead biosensor using DNAzyme-directed assembly of gold nanoparticles, *J Am Chem Soc*, **125**, 6642–6643 (2003).
170. J. Liu and Y. Lu, Accelerated color change of gold nanoparticles assembled by DNAzymes for simple and fast colorimetric Pb^{2+} detection, *J Am Chem Soc*, **126**, 12298–12305 (2004).
171. W. Zhao, J.C. Lam *et al.*, Enzymatic cleavage of nucleic acids on gold nanoparticles: a generic platform for facile colorimetric biosensors, *Small*, **4**, 810–816 (2008).
172. M.N. Stojanovic and D.W. Landry, Aptamer-based colorimetric probe for cocaine, *J Am Chem Soc*, **124**, 9678–9679 (2002).
173. K.Y. Wang, S. McCurdy *et al.*, A DNA aptamer which binds to and inhibits thrombin exhibits a new structural motif for DNA, *Biochemistry*, **32**, 1899–1904 (1993).
174. K. Padmanabhan, K.P. Padmanabhan *et al.*, The structure of alpha-thrombin inhibited by a 15-mer single-stranded DNA aptamer, *J Biol Chem*, **268**, 17651–17654 (1993).
175. L.C. Bock, L.C. Griffin *et al.*, Selection of single-stranded DNA molecules that bind and inhibit human thrombin, *Nature*, **355**, 564–566 (1992).
176. R.F. Macaya, J.A. Waldron *et al.*, Structural and functional characterization of potent antithrombotic oligonucleotides possessing both quadruplex and duplex motifs, *Biochemistry*, **34**, 4478–4492 (1995).
177. C.C. Huang, Y.F. Huang *et al.*, Aptamer-modified gold nanoparticles for colorimetric determination of platelet-derived growth factors and their receptors, *Anal Chem*, **77**, 5735–5741 (2005).
178. V. Pavlov, Y. Xiao *et al.*, Aptamer-functionalized au nanoparticles for the amplified optical detection of thrombin, *J Am Chem Soc*, **126**, 11768–11769 (2004).

179. J. Liu and Y. Lu, Fast colorimetric sensing of adenosine and cocaine based on a general sensor design involving aptamers and nanoparticles, *Angew Chem Int Ed*, **45**, 90–94 (2005).
180. W. Zhao, W. Chiuman *et al.*, Simple and rapid colorimetric biosensors based on DNA aptamer and noncrosslinking gold nanoparticle aggregation, *ChemBioChem*, **8**, 727–731 (2007).
181. K. Sato, K. Hosokawa *et al.*, Rapid aggregation of gold nanoparticles induced by non-cross-linking DNA hybridization, *J Am Chem Soc*, **125**, 8102–8103 (2003).
182. W. Zhao, W. Chiuman *et al.*, DNA aptamer folding on gold nanoparticles: from colloid chemistry to biosensors, *J Am Chem Soc*, **130**, 3610–3618 (2008).
183. D.Y. Wang, B.H.Y. Lai *et al.*, A general strategy for effector-mediated control of RNA-cleaving ribozymes and DNA enzymes, *J Mol Biol*, **318**, 33–43 (2002).
184. J.W. Liu and Y. Lu, Adenosine-dependent assembly of aptazyme-functionalized gold nanoparticles and its application as a colorimetric biosensor, *Anal Chem*, **76**, 1627–1632 (2004).
185. K.J. Odenthal and J.J. Gooding, An introduction to electrochemical DNA biosensors, *Analyst*, **132**, 603–610 (2007).
186. T.M. Herne, M.J. Tarlov, Characterization of DNA probes immobilized on gold surfaces, *J Am Chem Soc*, **119**, 8916–8920 (1997).
187. A.B. Steel, T.M. Herne *et al.*, Electrochemical quantitation of DNA immobilized on gold, *Anal Chem*, **70**, 4670–4677 (1998).
188. J. Watterson, P.A.E. Piunno and U.J. Krull, Practical physical aspects of interfacial nucleic acid oligomer hybridisation for biosensor design, *Anal Chim Acta*, **469**, 115–127 (2002).
189. E.L. Wong, E. Chow and J.J. Gooding, DNA recognition interfaces: the influence of interfacial design on the efficiency and kinetics of hybridization, *Langmuir*, **21**, 6957–6965 (2005).
190. K.A. Peterlinz and R.M. Georgiadis, Observation of hybridization and dehybridization of thiol-tethered DNA using two-color surface plasmon resonance spectroscopy, *J Am Chem Soc*, **119**, 3401–3402 (1997).
191. F.J. Mearns, E.L. Wong, K. Short, D.B. Hibbert and J.J. Gooding, DNA biosensor concepts based on a change in the DNA persistence length upon hybridization, *Electroanalysis*, **18**, 1971–1981 (2006).
192. J.J. Gooding, A. Chou *et al.*, The ion gating effect: using a change in flexibility to allow label free electrochemical detection of DNA hybridisation, *Chem Commun*, 1938–1939 (2003).
193. K. Eugenii and I. Willner, Probing biomolecular interactions at conductive and semiconductive surfaces by impedance spectroscopy: routes to impedimetric immunosensors, DNA-sensors and enzyme biosensors, *Electroanalysis*, **15**, 913–947 (2003).
194. A.E. Radi, J.L. Acero Sanchez *et al.*, Reusable impedimetric aptasensor, *Anal Chem*, **77**, 6320–6323 (2005).
195. W. Liao and X.T. Cui, Reagentless aptamer based impedance biosensor for monitoring a neuro-inflammatory cytokine PDGF, *Biosens Bioelectron*, **23**, 218–224 (2007).
196. M. Zayats, Y. Huang *et al.*, Label-free and reagentless aptamer-based sensors for small molecules, *J Am Chem Soc*, **128**, 13666–13667 (2006).
197. M.C. Rodriguez, A.N. Kawde *et al.*, Aptamer biosensor for label-free impedance spectroscopy detection of proteins based on recognition-induced switching of the surface charge, *Chem Commun*, 4267–4269 (2005).
198. K.M. Millan and S.R. Mikkelsen, Sequence-selective biosensor for DNA based on electroactive hybridization indicators, *Anal Chem*, **65**, 2317–2323 (1993).
199. K.M. Millan, A. Saraullo *et al.*, Voltammetric DNA biosensor for cystic fibrosis based on a modified carbon paste electrode, *Anal Chem*, **66**, 2943–2948 (1994).
200. V. Pavlov, I. Willner *et al.*, Amplified detection of telomerase activity using electrochemical and quartz crystal microbalance measurements, *Biosens Bioelectron*, **20**, 1011–1021 (2004).
201. P.M. Armistead and H.H. Thorp, Electrochemical detection of gene expression in tumor samples: overexpression of Rak nuclear tyrosine kinase, *Bioconjug Chem*, **13**, 172–176 (2002).
202. G.S. Bang, S. Cho *et al.*, A novel electrochemical detection method for aptamer biosensors, *Biosens Bioelectron*, **21**, 863–870 (2005).
203. M.N. Stojanovic, P. de Prada *et al.*, Aptamer-based folding fluorescent sensor for cocaine, *J Am Chem Soc*, **123**, 4928–4931 (2001).
204. B.R. Baker, R.Y. Lai *et al.*, An electronic, aptamer-based small-molecule sensor for the rapid, label-free detection of cocaine in adulterated samples and biological fluids, *J Am Chem Soc*, **128**, 3138–3139 (2006).

205. J. Wang, G. Rivas, J.R. Fernandes, J.L. Lopez, M. Jiang and R. Waymire, Indicator-free electrochemical DNA hybridization biosensor, *Anal Chim Acta*, **375**, 197–206 (1998).
206. J. Wang and A.N. Kawde, Pencil-based renewable biosensor for label-free electrochemical detection of DNA hybridization, *Anal Chim Acta*, **431**, 219–224 (2001).
207. J. Wang, A.N. Kawde *et al.*, Magnetic bead-based label-free electrochemical detection of DNA hybridization, *Analyst*, **126**, 2020–2024 (2001).
208. X. Wang, J. Zhou *et al.*, Detection of thrombin using electrogenerated chemiluminescence based on $Ru(bpy)_3^{2+}$-doped silica nanoparticle aptasensor via target protein-induced strand displacement, *Anal Chim Acta*, **598**, 242–248 (2007).
209. S. Centi, S. Tombelli *et al.*, Aptamer-based detection of plasma proteins by an electrochemical assay coupled to magnetic beads, *Anal Chem*, **79**, 1466–1473 (2007).
210. R. Polsky, R. Gill *et al.*, Nucleic acid-functionalized Pt nanoparticles: catalytic labels for the amplified electrochemical detection of biomolecules, *Anal Chem*, **78**, 2268–2271 (2006).
211. K. Ikebukuro, C. Kiyohara *et al.*, Novel electrochemical sensor system for protein using the aptamers in sandwich manner, *Biosens Bioelectron*, **20**, 2168–2172 (2005).
212. G.A. Campbell and R. Mutharasan, Monitoring of the self-assembled monolayer of 1-hexadecanethiol on a gold surface at nanomolar concentration using a piezo-excited millimeter-sized cantilever sensor, *Langmuir*, **21**, 11568–11573 (2005).
213. F. Lucarelli, S. Tombelli *et al.*, Electrochemical and piezoelectric DNA biosensors for hybridisation detection, *Anal Chim Acta*, **609**, 139–159 (2008).
214. T.M. Herne and M.J. Tarlov, Characterization of DNA probes immobilized on gold surfaces, *J Am Chem Soc*, **119**, 8916–8920 (1997).
215. S. Tombelli, M. Minunni *et al.*, Piezoelectric biosensors: strategies for coupling nucleic acids to piezoelectric devices, *Methods*, **37**, 48–56 (2005).
216. M. Liss, B. Petersen *et al.*, An aptamer-based quartz crystal protein biosensor, *Anal Chem*, **74**, 4488–4495 (2002).
217. T.W. Wiegand, P.B. Williams *et al.*, High-affinity oligonucleotide ligands to human IgE inhibit binding to Fc epsilon receptor I, *J Immunol*, **157**, 221–230 (1996).
218. T. Hianik, V. Ostatna *et al.*, Detection of aptamer–protein interactions using QCM and electrochemical indicator methods, *Bioorg Med Chem Lett*, **15**, 291–295 (2005).
219. M. Minunni, S. Tombelli *et al.*, Development of biosensors with aptamers as bio-recognition element: the case of HIV-1 TAT protein, *Biosens Bioelectron*, **20**, 1149–1156 (2004).
220. A. Bini, M. Minunni *et al.*, Analytical performances of aptamer-based sensing for thrombin detection, *Anal Chem*, **79**, 3016–3019 (2007).
221. P. Vattanaviboon, K. Sangseekhiow *et al.*, Detection and haplotype differentiation of Southeast Asian alpha-thalassemia using polymerase chain reaction and a piezoelectric biosensor immobilized with a single oligonucleotide probe, *Transl Res*, **151**, 246–254 (2008).
222. K. Rijal and R. Mutharasan, PEMC-based method of measuring DNA hybridization at femtomolar concentration directly in human serum and in the presence of copious noncomplementary strands, *Anal Chem*, **79**, 7392–7400 (2007).
223. C.A. Savran, S.M. Knudsen *et al.*, Micromechanical detection of proteins using aptamer-based receptor molecules, *Anal Chem*, **76**, 3194–3198 (2004).

20

Bacterial Riboswitch Discovery and Analysis

Tyler D. Ames and Ronald R. Breaker

20.1 RNA molecules as sensors and switches

Numerous recent studies have revealed that metabolite-sensing riboswitches are commonly used by bacteria to sense small-molecule ligands and control gene expression [1–8]. Interestingly, the notion that RNA molecules could function to control the expression of metabolic genes was proposed at the dawn of the gene regulation field by Jacob and Monod, who were speculating on the mechanism of *lac* operon regulation in *Escherichia coli* [9]. However, the discovery that the *lac* repressor was made of protein and not nucleic acid [10] suggested that other soon-to-be identified regulatory factors would also be made of protein.

As the decades passed, researchers accumulated additional examples of protein factors in bacteria that control gene expression in response to small-molecule binding. Although these findings further solidified the view that metabolite sensing and gene control were dominated by proteins, there were two developments in nucleic acids science that caused some researchers to reconsider RNA as a medium for metabolite sensing and regulation.

The first advance began in the early 1980s with the discovery of ribozymes [11–13]. Although nearly all known natural ribozymes catalyze phosphoester transfer or hydrolysis reactions, some of these accelerate reactions with rate enhancements that approach those of protein enzymes. The second advance began in the early 1990s with the use of directed evolution strategies [14–16] to create aptamers that bound to small organic compounds.

Together, these findings strongly supported speculation that RNA molecules were functioning both as an information storage system and as the biochemically active components of 'RNA World' organisms [17–19], long before the emergence of proteins. If RNA World organisms of ever-increasing complexity needed to sense and respond to chemical cues, then it seems reasonable to speculate that natural RNA aptamers would have been employed billions of years ago to function as sub-structures of RNA switches.

The hypothesis that RNA can serve as a robust medium for molecular switch construction was further confirmed by using modular rational design and directed evolution methods to create numerous allosteric

The Chemical Biology of Nucleic Acids Edited by Günter Mayer

ribozymes that function as ligand-responsive RNA switches [20–22]. This technology has now advanced sufficiently such that researchers are able to produce examples of designer aptamer and allosteric ribozyme constructs that generate robust changes in gene expression levels in response to small-molecule binding [23–27]. The increasing diversity of RNA switches being engineered in the laboratory would have provided convincing evidence of their usefulness in regulating gene expression even if the existence of natural riboswitches had yet to be established.

20.2 Solving gene control mysteries

Our path to riboswitch discovery and analysis passed through the areas of ribozyme biochemistry and directed evolution of nucleic acids noted above. It had previously been established that guanosine or many of its 5′-derivatized analogs are selectively recognized by group I ribozymes without the need for protein factors [11,28]. However, unlike most known riboswitches, the guanosine ligand serves as a substrate for the self-splicing ribozyme and becomes attached to the excised intron.

In some instances, however, group I ribozymes may indeed be riboswitches because self-splicing induced by metabolite binding results in the formation of a contiguous open reading frame that permits protein expression. It is also apparent how the metabolite substrate for group I ribozymes could be altered [29] or how the ribozyme itself can be mutated to accept altered substrates [28,30]. This latter mechanism could have naturally been used to diversify hypothetical riboswitches based on self-splicing to yield gene control elements that are responsive to a greater array of compounds.

We first explored the functional diversity of RNA switches by engineering allosteric ribozymes [20]. By the early 2000s, we and others had created a sizeable collection of allosteric ribozymes that selectively responded to targets ranging from metal ions and small molecules to nucleic acids and proteins [21,22]. Some of these engineered RNAs exhibit orders of magnitude discrimination against close ligand analogs and typically undergo alterations in ribozyme rate enhancements of between 100- and 10 000-fold. Furthermore, these engineered RNA switches could be used to report accurately the concentration of natural ligands in complex biological mixtures [31]. These findings strongly suggested that RNA World organisms would have carried RNA domains capable of serving as natural metabolite sensors and that it was likely that at least some modern organisms would still make use of such riboswitches to control gene expression.

We speculated that, if riboswitches were as common as we believed they should be, it was likely that researchers already had encountered examples in their gene control studies over the previous four decades. Armed with this hypothesis and our background in ribozyme engineering, we searched for publications that described gene regulation observations wherein a protein regulatory factor was notably absent. Two long-standing genetic mysteries involving vitamin B_{12} or cyanocobalamin [32] and riboflavin [33] were of particular interest. In both instances, there was evidence for ligand-mediated gene control, the existence of a conserved RNA sequence and structure, and a notable lack of a protein regulatory factor.

The data emerging from RNA engineering studies fostered speculation that RNA may play a more active role in metabolite sensing and gene control [34]. Furthermore, genetic studies such as those noted above [32,33] and a similar finding with thiamin-mediated regulation [35,36] caused researchers to suggest that these RNAs and others may directly bind small molecules. Despite the existence of engineered RNA switches and intriguing genetic data hinting at the existence of natural riboswitches, few would have accepted the riboswitch hypothesis, for two reasons. First, the field lacked an experimentally validated mechanism for how metabolite binding could modulate gene expression. Second, and perhaps most importantly, there was no definitive proof that the reported riboswitch candidates could even bind ligands in the absence of proteins. Demonstrating these capabilities would show that RNA is the dominant component of these regulatory systems that are capable of selectively recognizing small molecules and using these binding events to bring about the molecular changes necessary to modulate gene expression.

A historical parallel occurred when the first ribozymes were described. Researchers became convinced that RNA was able to catalyze chemical transformations once activity was measured in the complete absence of protein [11,12]. Therefore, our original efforts to establish the existence of riboswitches [37–39] included experiments designed to provide biochemical proof for metabolite binding in the complete absence of proteins and to provide both biochemical and genetic proof for gene control mechanisms. These [37–39] and related studies demonstrated that the genetic mysteries surrounding vitamin B_{12}, riboflavin and thiamin were due to widespread riboswitch classes that respond to adenosylcobalamin (AdoCbl), flavin mononucleotide (FMN) and thiamin pyrophosphate (TPP), respectively. With the similar validation of riboswitch classes for purines [40], *S*-adenosylmethionine [41–43] (SAM), lysine [44,45] and molybdenum cofactor [46] (Moco), to date a total of seven pre-existing genetic mysteries have been solved by riboswitch discoveries.

20.3 Common riboswitch architectures

In their simplest form, riboswitches carry two functional domains: an aptamer that selectively binds to a target metabolite and an expression platform that interprets the ligand binding status of the aptamer and communicates this status to other gene expression components of the cell. Most riboswitch expression platforms control transcription termination or translation initiation [47] (Figure 20.1). Simple Watson–Crick base-paired structures are usually involved in evaluating the occupancy status of the aptamer and likewise are used to control gene expression. As a result, the base-pairing interactions important for expression platform function frequently can be predicted using algorithms that evaluate the thermodynamic stability of RNA secondary structures.

Expression platforms that control transcription termination frequently exploit intrinsic transcription terminators, which comprise a strong base-paired stem followed by a run of approximately six uridine residues [48,49]. If this stem begins to form as RNA polymerase is just completing transcription of the U residues, transcription will terminate due to separation of the hybrid between the RNA transcript and the DNA template strand. The aptamer domain can control transcription termination by binding ligand and controlling the formation of the terminator stem to either repress or activate gene expression. Similarly, expression platforms that control translation initiation frequently make use of base-pairing interactions either to expose or to occlude access to the purine-rich ribosome binding site (RBS) [50].

Most riboswitches repress gene expression (genetic 'OFF' switches) when a ligand binds, although a few of examples of riboswitches that activate gene expression (genetic 'ON' switches) have been identified [51,52]. The distributional bias in favor of OFF switches is not because RNA is less capable of forming expression platforms that turn gene expression on when ligand binds, but because biological systems have a more frequent need to repress expression of metabolic genes when adequate concentrations of the corresponding metabolite have been reached. Repressing gene expression when a ligand binds can be achieved by sequestering an anti-terminator segment within the aptamer domain [39,53] or by using nucleotides of the RBS to form part of the aptamer domain [46]. In contrast, activation of gene expression can be achieved by sequestering within the aptamer domain a stretch of nucleotides otherwise used to form a terminator stem [54] or by controlling access to an anti-anti-RBS stem [55].

Since expression platforms can control gene expression via multiple mechanisms and use simple secondary structure alterations to commit to 'OFF' or 'ON' states, these domains can vary widely in sequence and structure between riboswitches. In contrast, riboswitches can be readily classified based on the striking sequence and structural conservation of their aptamer domains [47]. The extensive level of conservation amongst the aptamer domains of members of a particular metabolite-binding riboswitch class is easily rationalized by recognizing that

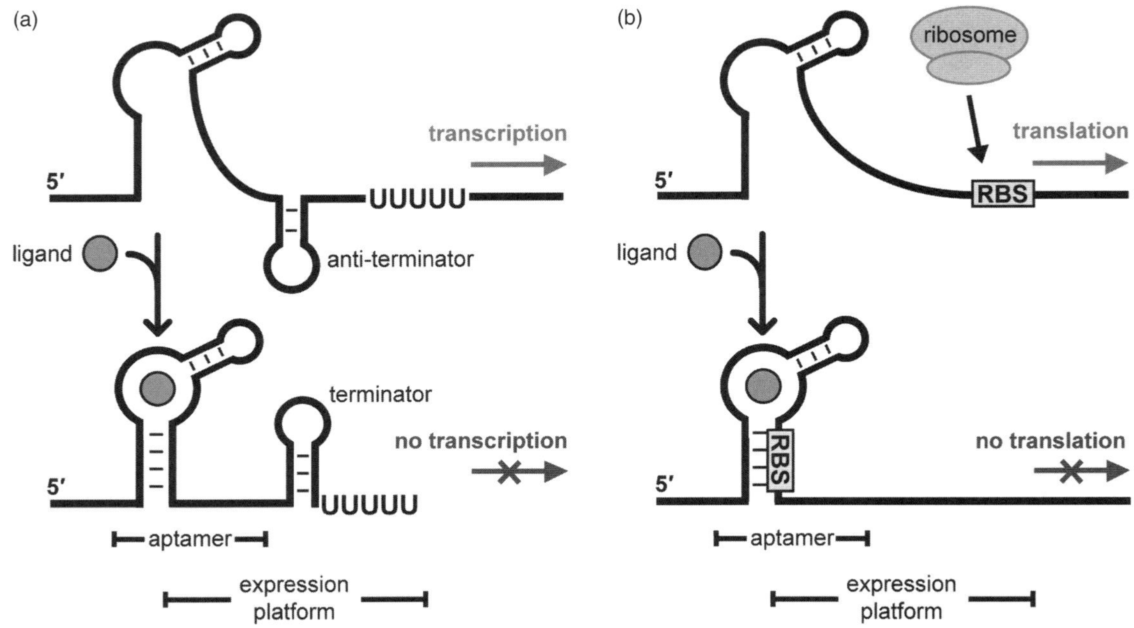

Figure 20.1 *Common riboswitch gene control mechanisms in bacteria. (a) Expression platforms that control transcription termination sometimes carry an intrinsic transcription terminator stem and can form a competing antiterminator stem. A genetic OFF switch is depicted wherein ligand binding favors formation of the terminator stem. (b) Expression platforms that control translation initiation commonly carry an anti-RBS stem and can form a competing anti-anti-RBS stem. A genetic OFF switch is depicted wherein ligand binding favors formation of the anti-RBS stem. Reproduced from J. N. Kim and R. R. Breaker, Purine sensing by riboswitches,* Biol. Cell*, 2008,* ***100****, 1–11, by permission of Portland Press*

RNA must use only four types of nucleotides to form a binding pocket for a compound that never changes through evolution. This places extraordinary evolutionary restrictions on variation of the aptamer domain for riboswitches compared with RNAs that are forming complexes with protein factors [56–59] or other RNAs [60–62], both which could mutate and thereby allow its partner RNA to mutate in response.

20.4 Biochemical and genetic validation of riboswitch candidates

Given the distinguishing functional and architectural features of metabolite-binding riboswitches, there are several experimental challenges that confront researchers who seek to validate the existence of novel riboswitch classes. Although the order of experimental analysis can vary, the following key aspects of a complete proof of riboswitch function should be sought.

Initial efforts to confirm riboswitch function involve the identification of a candidate RNA sequence that has characteristics expected for metabolite binding and gene control. It is becoming increasingly unlikely that new riboswitch candidates will be found by surveying decades-old literature. Rather, given the explosion of available genomic sequence data, new candidate riboswitch classes are more likely to be found by employing powerful bioinformatics tools to search for structured non-coding RNAs. Details regarding the strategies and tools used to identify novel riboswitch candidates will be discussed in greater detail below.

Once a riboswitch candidate has been identified, the genomic distribution of representatives of the candidate can be assessed to provide a rationale for gene control function. Riboswitches are almost always associated with a particular metabolic pathway in diverse organisms and are almost exclusively located in the 5′ UTRs of the genes they control. Such assessments also can provide clues to help solve which ligand triggers gene regulation. For example, if the candidate always appears in the 5′ UTR of a gene coding for the biosynthesis or transport of an essential metabolite, then the biosynthesis or transport product is most likely to be the riboswitch ligand.

Genomic distribution assessments can be very straightforward for most riboswitches that sense fundamental metabolites. In contrast, class I SAM-sensing riboswitches [41–43] control many genes whose protein products catalyze a wide array of reactions involving general sulfur metabolism, although the riboswitch frequently is also found associated with SAM synthase genes [63,64]. Similarly, riboswitches that sense cyclic diguanosine-5′-monophosphate (c-di-GMP) are associated with a great diversity of genes whose protein products can differ dramatically in function between even closely related organisms [52]. This diversity is an outcome of the role of c-di-GMP as a bacterial second messenger that controls a wide range of physiological responses [65,66].

An important aspect of any riboswitch proof is to assess whether the RNA alone can selectively bind a target ligand. This does not mean that the action of a metabolite-responsive riboswitch precludes the involvement of protein factors in all instances. Rather, demonstration of selective complex formation between the RNA and its ligand in the absence of protein factors conclusively rules out the possibility that a key aspect of the gene control process – ligand sensing – is carried out by an undiscovered protein factor. Perhaps there will be instances discovered wherein a riboswitch selectively binds its metabolite target only when bound by a protein factor that, for example, assists in RNA folding. However, demonstration of direct binding of ligand by RNA alone proves that there is no need to invoke the involvement of protein factors in the function of the aptamer.

Demonstrations of direct binding of ligand by riboswitch aptamers have been achieved by using a variety of techniques. We routinely employ a method called in-line probing [67–69], which takes advantage of the natural instability of RNA 5′-phosphoester linkages to evaluate binding affinity and RNA structural changes (Figure 20.2). Specifically, nucleophilic attack by the 2′-oxygen of a nucleotide on the neighboring phosphorus center causes scission of the phosphoester bond linking the phosphate to the nucleotide downstream. This reaction yields cleaved RNA fragments carrying either 2′,3′-cyclic phosphate or 5′-hydroxyl termini. Cleavage proceeds more quickly for RNA linkages that position the attacking oxygen nucleophile in close proximity to the phosphorus center electrophile. However, the reaction yields measurable amounts of cleaved products only if the 2′-oxygen nucleophile, the phosphorus center and the 5′-oxygen leaving group also form an in-line geometry [70,71]. To accelerate the speed of spontaneous cleavage, in-line probing assays are performed under slightly elevated pH conditions ($pH \approx 8.3$) to increase the frequency of 2′-oxyanion formation. This deprotonated form of the functional group is a much stronger nucleophile than the corresponding hydroxyl group.

In-line probing reactions are incubated under desired conditions for a time that yields an average of less than one cleavage event per RNA molecule. The use of 5′-^{32}P-labeled RNA permits the detection of only the 5′ cleavage products on separation by gel electrophoresis. RNA linkages that are joining nucleotides involved in forming typical secondary structures will rarely undergo cleavage because A-form RNA structures preclude the formation of the in-line orientation necessary for the reaction. Likewise, tertiary-structured regions of an RNA are also unlikely to adopt an in-line geometry. Therefore, radioactive product bands corresponding to cleavage at linkages in highly structured regions of an RNA will be greatly reduced in intensity compared with product bands from regions that are relatively unstructured and thus more likely to sample an in-line geometry.

Comparison of the banding patterns resulting from in-line probing assays conducted with or without the addition of a candidate ligand can reveal changes in the shape of the RNA that are induced by ligand binding. These shape changes cannot be so subtle that the speed at which an internucleotide linkage spontaneously cleaves is unchanged. However, ligand binding commonly affects the stability of nucleotides that the ligand

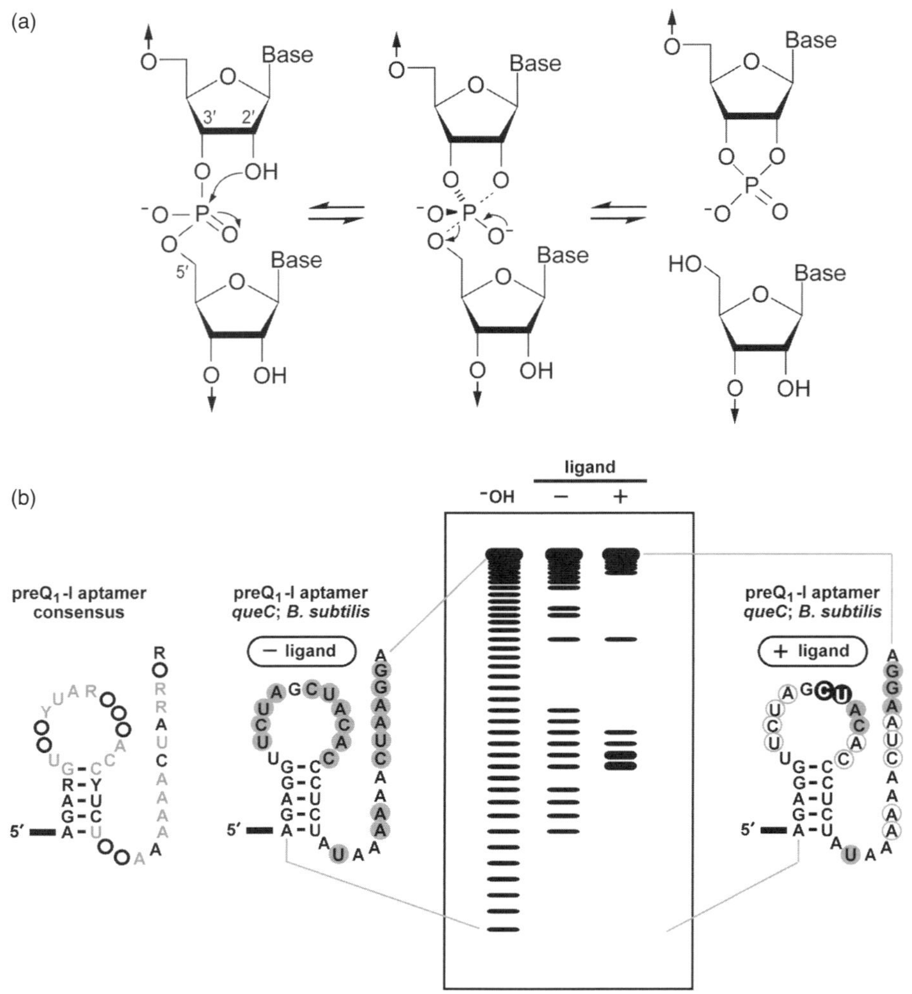

Figure 20.2 *Schematic representation of in-line probing assay results and interpretation. (a) Mechanism of spontaneous RNA cleavage. (b) Typical in-line probing results. Left: consensus sequence and structure of a type II preQ1 riboswitch aptamer. Gray nucleotides identify positions that are conserved in 95% or more of the known representatives. Circles represent any nucleotide; R and Y represent purine and pyrimidine nucleotides, respectively. Right: schematic depiction of the products of spontaneous RNA cleavage generated during in-line probing and separated by polyacrylamide gel electrophoresis. The top band in each lane is the uncleaved 5′-^{32}P-labeled precursor RNA. Partial digestion with alkali ($^{-}$OH) yields products representing cleavage at all possible linkages. The pattern of product bands generated during incubation in the absence of ligand reveal regions of the aptamer that are largely unstructured (filled gray circles). This pattern changes if the RNA is incubated with saturating amounts of the appropriate ligand, preQ$_1$. Specifically, open gray circles identify nucleotide linkages that cleave more slowly (become more structured), black circles identify nucleotide linkages that undergo more rapid cleavage (become more unstructured or are held in an in-line attack geometry) and filled gray circles represent measurable but unchanged extent of cleavage. Actual in-line probing data for this RNA [111] are consistent with the structural features of this aptamer class determined by X-ray and NMR methods [137–139]*

directly contacts, in addition to causing distal rearrangements in aptamer structure. These structural changes are easily observed by analyzing and quantitating cleavage band patterns. Indeed, the dissociation constant (K_D) values for aptamer–ligand complexes can be assessed simply by conducting several parallel in-line probing assays using various concentrations of ligand with an RNA concentration that is below the K_D value [68].

Numerous other approaches have been used to assess the binding characteristics of riboswitch aptamers. A structural probing assay called SHAPE [72] has been used on riboswitch aptamers with results [73,74] similar to those produced from in-line probing assays. SHAPE takes advantage of structure-dependent changes in the reactivity of 2′-hydroxyl groups to chemical agents, wherein modification is assessed by reverse transcription of chemically treated RNA templates. Changes in RNA structure caused by ligand binding will affect the extent of chemical alteration, which blocks reverse transcription at the site of modification. Chemical agents that have long been used to modify other functional groups on RNA in a structure-dependent fashion also have been used to assess ligand binding [75].

RNA structural changes can also be evaluated by using oligonucleotides and nucleases as probes. For example, binding of ligand to an FMN riboswitch has been established by using RNase H activity to assess whether a DNA oligonucleotide was able to base pair to the aptamer when incubated in the absence or presence of ligand [53]. RNase H will cleave the RNA strand of an RNA–DNA hybrid helix. Therefore, detection of cleaved RNA indicates that the RNA structure may have not bound a ligand because nucleotides that otherwise would be occupied in forming the bound conformation remain accessible to oligonucleotide binding. Researchers have also used the activity of RNases that do not rely on DNA probes, but whose activities are influenced by RNA structure, to reveal shape changes induced by ligand binding [76,77]. One concern about probing methods involving protein reagents is the remote possibility that proteins added intentionally or as contaminants may be responsible for any ligand-mediated effects observed. This concern can be eliminated by using protein-free assays that employ nucleic acids [75,78] or chemical reagents [72,75] alone as structural probes.

All of the methods discussed so far can yield both confirmation of ligand binding and estimations of binding affinity. These methods can also provide information on the locations of shape changes that are induced by ligand binding. Other methods have been used to validate binding without requiring measurable folding changes to occur in the RNA. The *glmS* riboswitch class senses glucosamine-6-phosphate (GlcN6P) and is distinct from other riboswitches because it operates as a cofactor-dependent self-cleaving ribozyme [79–81]. This 'ribozyme riboswitch' also appears to form nearly identical structures both in the presence and in the absence of ligand [79–82], which precludes the use of analytical methods that rely on shape change. However, simple ribozyme cleavage assays can be used to assess the K_D values for GlcN6P or of analogs that also can promote ribozyme action [79,83–86].

Methods also exist for assessing ligand binding that are applicable to any riboswitch aptamer, regardless of its shape change potential. If a radiolabeled ligand is available, then methods such as equilibrium dialysis [37,38] can be employed. In a two-chambered equilibrium dialysis system, aptamers that bind a ligand will shift the distribution of the radiolabeled ligand to favor the chamber containing RNA. However, if the chamber carries mutant RNAs that cannot bind ligand or if sufficient amounts of an unlabeled competitor ligand is present, then the radioactive ligand will distribute equally. Scatchard analyses can also be conducted using equilibrium dialysis data gathered with riboswitch aptamers [43] to yield information on affinity and stoichiometry of the RNA–ligand interaction. Similar information, along with detailed thermodynamic parameters, can be gained by employing isothermal titration calorimetry [87,88]. However, relatively large amounts of RNA are required for each analysis, which may be restrictive if K_D values are to be estimated for many analogs of the natural ligand.

Examination of the mechanisms by which ligand binding controls gene expression can involve both *in vitro* and *in vivo* assays. Many demonstrations of riboswitch-mediated transcription termination have been published, wherein the amount of full-length transcript is compared with the amount of product that terminates via the action of an intrinsic transcription terminator stem [39,41–45,52,53,89,90]. Similarly, ribosome binding assays using nitrocellulose filter binding [91] or using reverse transcriptase primer extension [32,92] techniques have been conducted to determine whether the presence of ligand mediates access to RBS elements proposed to be important for riboswitches that control translation initiation. Each of these assays can provide evidence indicating that a riboswitch operates by transcription termination or translation initiation control, but the extent of termination or ribosome binding modulation may not reflect the extent of gene control or the concentration of compound required to trigger the riboswitch in cells. The factors that confound these *in vitro* assays will be discussed in greater detail in a later section [88,89,93].

If few or no biochemical data are available on ligand binding and expression platform function for a candidate riboswitch, genetic experiments can help reveal riboswitch ligands and mechanisms. These analyses are greatly facilitated by the use of riboswitch–reporter mRNA fusions that reflect gene expression control of the natural riboswitch representative. Translational fusions are created by joining a DNA template for a 5′ UTR encompassing a candidate riboswitch and the first few codons of its associated open reading frame (ORF) to a DNA coding for a reporter gene. The fusion is made such that either transcription control or translation control mechanisms used by the expression platform will regulate reporter gene expression. In contrast, a transcriptional fusion is made from similar DNAs wherein the 5′ UTR fragment is trimmed to exclude its natural RBS and start codon and this fragment is joined to a reporter gene construct that retains its own RBS and start codon. This construct will only respond to ligand if the expression platform controls transcription termination or some other mechanism that affects mRNA stability.

Once a useful riboswitch–reporter fusion construct has been created, experiments can be conducted to search for a ligand, to examine gene control mechanisms and to gain a more accurate determination of the extent of gene regulation brought about by the riboswitch. Riboswitch-triggering ligands can be sought by conducting genetic screens for mutations in metabolic genes that disregulate reporter expression [94]. Likewise, known strains with metabolic pathway disruptions can be examined. Alternatively, candidate compounds could be supplemented to a minimal medium used to culture cells harboring the reporter construct.

The mechanism of gene control can be assessed in cells by site-directed mutagenesis of the aptamer and expression platform portions of the riboswitch–reporter fusion. These experiments must be planned carefully because aptamer and expression platform features commonly overlap. Mutations that eliminate base-pairing interactions in aptamer stems usually disrupt ligand binding. Subsequent introduction of compensatory mutations that restore pairing, but result in a different sequence within the paired structure, usually restore ligand binding activity and gene control [37,38,40,43,52,54,95]. However, if these nucleotides mutated in the aptamer are also involved in forming structures critical for expression platform function, then compensatory mutations may need to be made elsewhere in the expression platform to retain activity.

20.5 Bioinformatics approaches to riboswitch discovery

All of the currently known riboswitch classes exhibit a common set of characteristics that can be exploited to guide the search for new classes. In particular, extensive sequence conservation, conserved secondary and tertiary structure features and distinctive genomic distributions can be used to identify candidate riboswitches. Computer algorithms that search through the growing expanse of genomic sequence data are proving to be very effective tools for discovering new riboswitches and other structured RNA motifs.

The first bioinformatics algorithms dedicated to finding novel riboswitch classes centered on the identification of highly conserved stretches of nucleotide sequences in the non-coding or intergenic regions (IGRs) of bacterial genomes. This was carried out by using blastn [96] searches to compare the nucleotide sequence of each IGR of *Bacillus subtilis* with all other IGRs from this organism and 90 other fully sequenced bacterial species [97]. This produced a collection of candidate riboswitches that had at least one representative in *B. subtilis*. In later studies, the IGRs of all sequenced species were compared with each other [98,99]. Clusters of similar sequences that showed evidence of structural conservation were designated candidate motifs, while groups of sequences that were very small in length, known or hypothesized protein-binding elements or comprised of repetitive AT-rich sequences were rejected. Once a candidate riboswitch class was identified, the functions of the gene products coded immediately downstream of the candidate were assigned using the COG database [100]. Additional sequence conservation among members of candidate motifs were mapped using the FASTA software package [101], which serves as a more sensitive search algorithm but could only be applied in practice on data sets smaller than those used in genome-wide searches due to its relatively slow speed compared with blastn.

Consensus sequence and structure models were then constructed manually, refined and used as a guide to search for additional representatives of the candidate riboswitch class via algorithms such as sequencesniffer [97]. Candidate riboswitches to be biochemically or genetically tested were best chosen from the collection of motifs that are predicted to have complex-folded structures and are commonly located upstream of metabolic genes. Candidate ligands for testing were chosen on the basis of the functions of the protein products whose genes were associated with the RNA domains, as these functions provided clues regarding riboswitch ligand identities.

The methods described above were successfully employed to discover several riboswitch classes. However, the use of improved search algorithms that automatically build consensus sequence and structure models can significantly aid searching through larger datasets and reveal rarer candidate riboswitches. A computational pipeline was developed to identify more efficiently RNAs that shared regions of nucleotide sequence and that share secondary structure features [102]. The computational demands for this type of search are substantial and therefore limiting the size of the sequence database to be examined is desirable. In addition to limiting the search to non-coding portions of genomes, an additional simplification can be achieved by identifying homologous genes across available bacterial genomic sequences using the National Center for Biotechnology Information (NCBI) Conserved Domain Database (CDD) [103]. If a structured RNA functions as a riboswitch, then it is highly likely to control homologous genes in related bacterial species.

This computational pipeline was employed to search IGRs $5'$ of similar sets of genes for evidence of conserved structured RNAs using the program CMfinder [104]. To expand these motifs to include riboswitch examples that were not associated with the initial set of genes, the RAVENNA program [105] was used to scan full prokaryotic genomes for additional RNA sequences to add to the motif. The extra sequences were used to refine the motifs using CMfinder and these revised motifs were in turn used in the RAVENNA program to identify even more examples. This process was iteratively employed to search exhaustively for representatives of the candidate riboswitch class within any genomic DNA database.

The computational pipeline described above has uncovered numerous previously undiscovered RNA motifs [99], including riboswitches that sense c-di-GMP [52], *S*-adenosylhomocysteine [95] (SAH), Moco [46], a second prequeuosine 1 ($preQ_1$) riboswitch class [106] and a host of other conserved non-coding RNAs. Other algorithms have been developed that take advantage of the characteristics of individual riboswitch classes or of riboswitches as a group to search for additional representatives of known riboswitch classes or for novel candidates [107–109].

20.6 Many bacterial riboswitch classes remain to be discovered

One way to understand better the functional potential of structured RNAs is to recognize more fully the diversity of molecules such as ribozymes and riboswitches that exists in biological systems. Most riboswitch classes being identified yield knowledge of new-found aptamer structures and expand the number of ligands known to be bound by RNA. Also, some ligands such as SAM [110] and $preQ_1$ [106,111] are sensed by multiple riboswitch classes, suggesting that there may also be multiple riboswitch architectures for other metabolites. Comparisons of the diversity and distributions of the known riboswitch classes already can be made to help estimate how many riboswitch classes remain to be discovered. However, before this can be done, we need to point out some challenges in creating a clear definition for what constitutes a distinct riboswitch class.

The problem of riboswitch classification could be approached from two perspectives. Since the aptamer domain is the most conserved portion of a riboswitch, sequence and secondary-structure features could be used to define each class. This criterion can be applied to SAM-I [41–43], SAM-II [98] and SAM-III [91,112] riboswitches to yield three classes that indeed use very different three-dimensional architectures to bind the same coenzyme [113–115]. However, bioinformatics searches have revealed the existence of RNA motifs called SAM-IV [116] and SAM-V [117] that were segregated differently from the known riboswitch classes. Although these two SAM-responsive riboswitch 'classes' do have distinctive sequence and structural features, on closer inspection, SAM-IV carries the same ligand-binding core as SAM-I, whereas SAM-V carries the same core as SAM-II. Although bioinformatics search algorithms used to discover many of these RNAs yielded five distinct groupings, the representatives could be clustered into three classes based on the sequences and structures of the ligand binding pockets alone.

Riboswitch classification could also be guided by the ligands they sense. The simplest approach could be the assignment of 'class' status for each collection of RNAs that bind to a different ligand. If this criterion were applied to riboswitches that selectively sense and respond to guanine [40], adenine [51] and 2′-deoxyguanosine [118], then three distinct classes would be defined. However, changing only one C-to-U mutation in a guanine riboswitch aptamer changes the ligand-binding preference to adenine [51], which is permitted because this nucleotide forms a Watson–Crick base pair with the ligand. As expected, the global tertiary structures of guanine- and adenine-sensing aptamers are near identical, which would otherwise be indicative of a single aptamer class. Although 11 mutations to otherwise well-conserved guanine aptamer nucleotides are observed in natural 2′-deoxyguanosine riboswitch aptamers [118], only a few changes to the core are needed to create an RNA that strongly discriminates against guanine and favors 2′-deoxyguanosine [73]. Again, the global tertiary structures are similar, which would support grouping all these aptamers into a single structural class [119].

If we apply the most conservative class assignment criterion where RNA groups are combined based on structural similarity regardless of ligand specificity, then there have been 15 confirmed classes of riboswitches reported. Another five RNA classes have been identified with structures and genomic distributions which suggest that they are strong candidates for novel riboswitches. Other weaker candidates also exist, but are not included in the analyses described below. The relative abundance of each riboswitch class was estimated by determining the number of representatives for each class present in bacteria whose genomes have been sequenced (Figure 20.3). As reported previously [47], there are only a few riboswitch classes known that are common to many bacteria. Interestingly, three riboswitch candidates (*yybP*, *ydaO* and *ykkC*) [97], each with hundreds of representatives identified, remain unvalidated.

Separating the data on riboswitch copy number into the years in which each riboswitch class was reported reveals that the riboswitch classes discovered first are between 10- and 100-fold more common than riboswitch classes reported more recently Figure 20.4). Nearly all recently discovered riboswitch classes were identified using bioinformatics search strategies. Since these algorithms employ comparative sequence analysis, riboswitches with more representatives in the genomic databases are more likely to be found. Exceedingly rare riboswitch classes will therefore be discovered only when they are present at a certain threshold number that is reached only when the total size of the genomic DNA databases increases sufficiently.

Some common riboswitch classes may not have been detected if they are very small or have very few strictly conserved features. Also, if the genomic sequence data currently available are not representative of

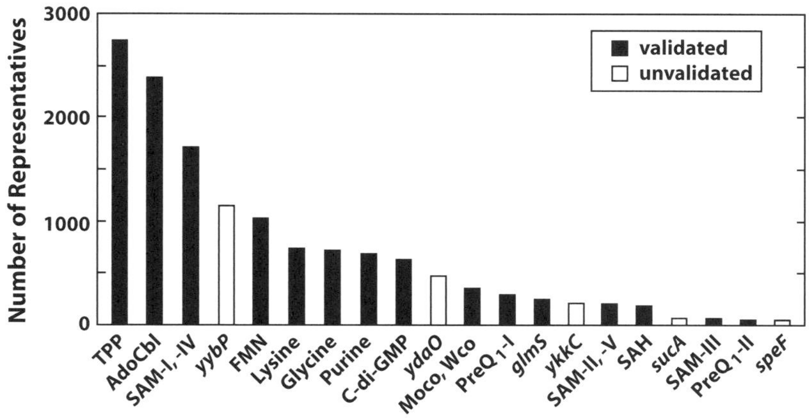

Figure 20.3 *Plot of the number of representatives for each riboswitch class identified in the collection of bacteria whose genomes have been fully sequenced*

the genomic DNA for the vast number of bacteria awaiting sequence analysis, then additional common riboswitch classes may still be found. However, the current trend (Figure 20.4) strongly suggests that newly found riboswitch classes will be progressively rarer. Given the difficulty in finding uncommon riboswitch classes, the current collection is likely to under-represent these rare classes. If true, then the data correspond well to a power law distribution (Figure 20.5) where there is a log–linear relationship between the number of riboswitches for a class and its rank order in abundance. Data following a power law relationship should be proportional regardless of the scale examined. Although the right-most data points do not maintain the same slope as other points, this is likely due to the fact that the collection of riboswitches is under-represented in rare classes. If riboswitches are distributed according to a power law relationship, then hundreds of riboswitch classes (albeit progressively rarer) may be discovered as bacterial sequence data expand.

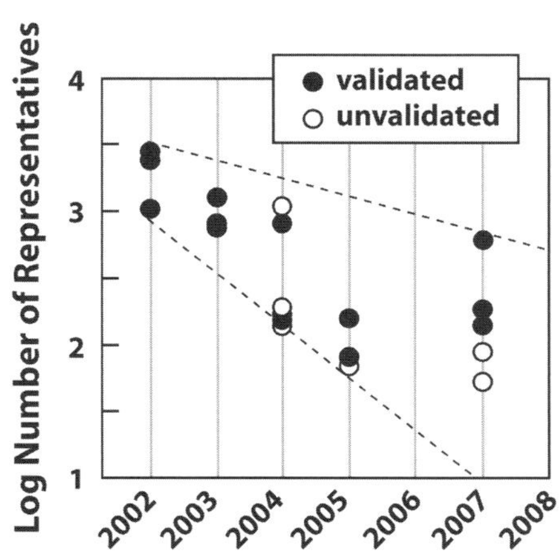

Figure 20.4 *Plot of the logarithm of the number of representatives for each riboswitch class versus the year they were reported to function as riboswitches*

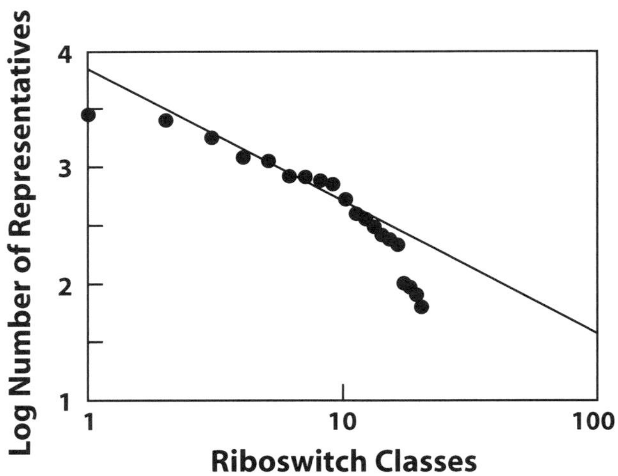

Figure 20.5 *Plot of the logarithm of the number of representatives for each riboswitch class versus the rank order in abundance for each riboswitch class (1 is the most common riboswitch class, TPP). The data resemble a power law relationship:* $\log y = k \log x + \log a$*, where* y *is the number of representatives,* x *is the rank order of the riboswitch class,* k *is the slope of the resulting line and* a *is the scaling factor. The right-most portion of the plot is the poorest fit for a power law relationship, which corresponds to the dataset that is most likely to be incomplete due to the difficulty in discovering the rarest riboswitch classes*

20.7 Structural and mechanistic complexity of riboswitches

Most riboswitches are each composed of a single aptamer that controls the function of a single expression platform. This simple arrangement of functional domains appears to be sufficient for many gene control challenges faced by bacteria. However, single aptamers can yield only one type of dose–response curve, which is limiting (Figure 20.6). The gene control output must perfectly reflect the one-to-one binding interaction between the aptamer and its corresponding ligand. Thus, the dynamic range of the ligand concentration (Figure 20.6a) that causes the majority of simple riboswitch-mediated gene control inherently will cover approximately two orders of magnitude [54,120,121] (Figure 20.6b).

In contrast, the extent of gene expression control will be dictated in part by the extent to which each aptamer transcribed actually folds and functions in a timely manner. Also, gene expression dynamic range will be affected by the function of the expression platform, which for example could terminate transcription only occasionally when a ligand binds or it could imperfectly control translation initiation. *In vitro* transcription termination assays with riboswitches that carry an intrinsic terminator stem as part of their expression platform are usually observed to control transcription imperfectly. However, conclusions from these assays should be drawn cautiously as variables such as changes in nucleoside-5′-triphosphate (NTP) concentrations [89,120], the use of surrogate RNA polymerases from other bacterial species or viruses [39,89,120] or the addition of other transcription factors [89] can have significant effects on the extent of transcription termination observed.

In some instances, imperfect termination even when ligand is at saturating concentrations is likely due to the fact that the assay conditions do not perfectly reflect the normal cellular environment. This may be the case for TPP riboswitches from *Bacillus anthracis* that exhibit incomplete transcription termination in test-tube assays, but exhibit essentially complete repression of gene expression in cells [120]. However, a transcription-controlling FMN riboswitch in *B. subtilis* was recently observed to repress incompletely

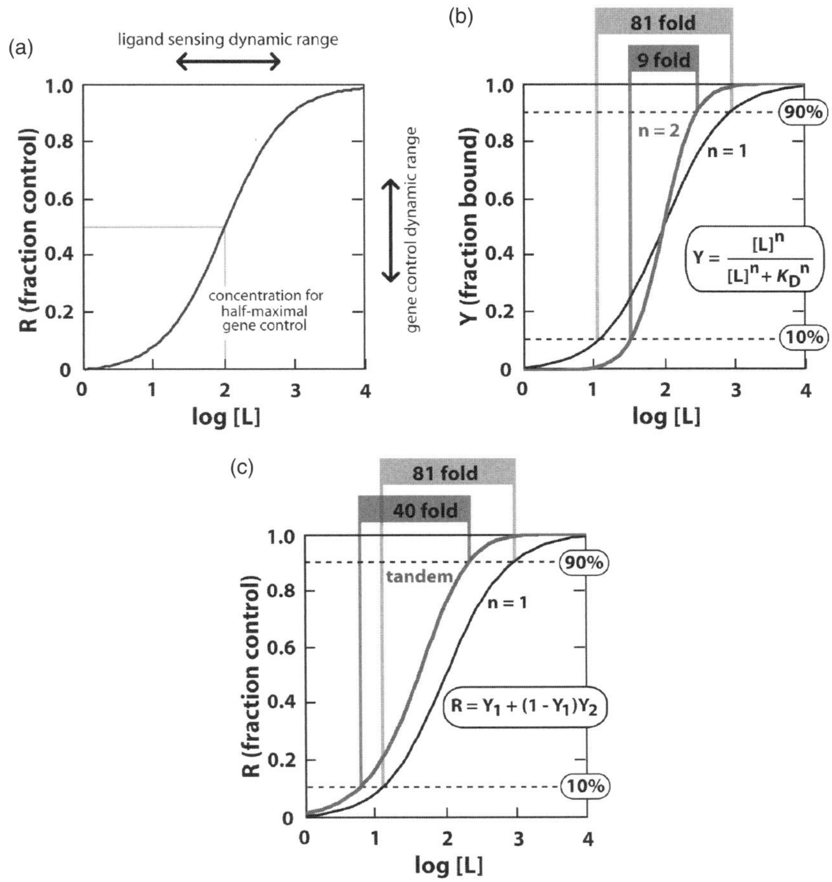

Figure 20.6 *Dynamic ranges for simple, cooperative and tandem riboswitches. (a) Plot of the fraction of riboswitches that are regulated (R) for a genetic ON switch versus the logarithm of the ligand concentration (L) in arbitrary units. This graphic represents a perfect riboswitch that fully regulates gene expression (from zero to maximum levels) as the aptamer becomes saturated. (b) Plot of the fraction of aptamers (Y) bound to ligand for a simple riboswitch (n = 1) and a riboswitch with two perfectly cooperative aptamers (n = 2) versus ligand concentration. Curves will follow the Hill equation where* n *is the Hill coefficient and* K_D *is the apparent dissociation constant. The changes in ligand concentration required to progress from 10% to 90% of the aptamers bound to ligand are noted. (c) Plot of the fraction of riboswitches controlled by ligand binding versus the logarithm of the ligand concentration for a simple switch (n = 1) and for a tandem arrangement involving two independently functioning riboswitches that respond to the same ligand with similar* K_D *values. Function of the tandem riboswitch arrangement follows the equation depicted. Adapted from R. Welz and R. R. Breaker, Ligand binding and gene control characteristics of tandem riboswitches in* Bacillus anthracis, *RNA, 2007,* ***13****, 573–582 and reproduced by permission of Cold Spring Harbor Laboratory Press*

reporter gene expression with either its natural ligand or an analog of FMN called roseoflavin [122]. This riboswitch represses only about 80% of its maximum gene expression, suggesting that approximately 20% of the time the riboswitch either fails to bind FMN or ligand binding fails to trigger expression platform function. Similarly, a TPP riboswitch that controls alternative splicing in plants [123–125] suppresses only about 80% of its maximum gene expression [124]. Perhaps these riboswitches have been tuned through evolution to function imperfectly, which permits the organisms to maintain a basal level of gene expression regardless of how much ligand is present.

It is tempting for many to assume that the binding affinity of a riboswitch aptamer correlates directly with the concentration of ligand needed to trigger riboswitch function in cells. If this were always true, then one could use any riboswitch as a probe to establish the *in vivo* concentration of specific metabolites. Unfortunately, the aptamers of most riboswitches that control transcription termination may not reach thermodynamic equilibrium with their ligands, but rather their function may be driven by the kinetics of other processes including RNA folding, ligand association and RNA transcription [88,89,93,126].

One disadvantage to organisms that carry kinetically driven riboswitches is that the riboswitch requires a ligand concentration that is higher than the K_D of the aptamer alone to half-maximally trigger gene regulation [89]. The riboswitch could adapt to sense lower concentrations of ligand by binding the ligand more quickly. Interestingly, the K_D values established for some riboswitch aptamers are in the mid-picomolar range [120,122,127]. Since typical bacterial cells such as *Escherichia coli* have a cell volume of approximately 1 fl, the concentration of a single free metabolite molecule in a cell of this size is equivalent to approximately 60 nM. Therefore, some riboswitch aptamers have K_D values that are nearly 1000-fold better than needed to respond to the presence of one free ligand molecule. Perhaps the riboswitch aptamer has evolved to associate rapidly with ligand, whereas the rate of dissociation may be fairly slow and biologically inconsequential.

In at least one way, kinetically driven riboswitches have a substantial advantage over thermodynamically driven riboswitches or protein genetic factors. The concentration of ligand needed to trigger simple gene control elements that are thermodynamically driven can only be changed by the cell through mutation of the ligand binding site, thereby changing the K_D of the receptor for its target metabolite. In contrast, simple kinetically driven riboswitches can undergo adjustment of the ligand concentration needed to control gene expression by accruing aptamer mutations that affect the rate of ligand association, by accruing mutations in the riboswitch RNA or in RNA polymerase that change the kinetics of transcription or by changes in other cellular conditions that alter transcription speed [89,93,126].

Despite the points discussed above, riboswitches with a single aptamer and a single expression platform are limited in function. However, combining the components of riboswitches or stacking riboswitches in tandem can yield higher ordered functions that are used by bacteria to take on more challenging gene control tasks [126–128]. Two aptamers that bind ligand cooperatively can be integrated with a single expression platform to create a riboswitch that can control gene expression in response to smaller changes in ligand concentration (increased 'digital' character) [54] (Figure 20.6b). Members of the glycine riboswitch class are the only representatives known that exploit a tandem aptamer arrangement and these generate Hill coefficients measured between 1.4 and 1.6.

A second type of more digital riboswitch has been observed with tandem TPP riboswitches from *B. anthracis* and also for several other classes of riboswitches [120,126]. In the tandem TPP riboswitch example examined in detail [120], two complete and independently functioning riboswitches that are tuned to respond to the same ligand concentration yield a dose–response curve that is steeper than a single riboswitch (Figure 20.6c). Although the reduction in the dynamic range for the ligand is much smaller compared with a digital switch based on cooperative aptamers, this arrangement does not require interplay between the two riboswitches and may be more likely to emerge during evolution.

Mixing representatives of more than one riboswitch class in the same mRNA adds another layer of sophistication to the function of simple riboswitches. One such mixed tandem riboswitch arrangement has been examined from the *metE* gene of *Bacillus clausii* [127]. This system carries a riboswitch for SAM followed by a riboswitch for AdoCbl, wherein both function independently to terminate transcription when their ligands bind. This arrangement represses gene expression when either SAM or AdoCbl are at high concentrations and thus functions like a two-input NOR Boolean logic gate [129]. In contrast to electronic logic gates, these simple riboswitches do not work with perfect digital control and so the corresponding genetic output will match the NOR truth table entries only at extremely low and extremely high ligand concentrations. Intermediate ligand concentrations residing within the dynamic ranges of the riboswitches will yield intermediate gene expression levels.

As noted above, most riboswitches whose mechanisms for gene control have been proven or predicted appear to regulate transcription termination or translation initiation. However, these two major mechanisms of gene control could be more similar than first presumed. Organisms that commonly employ riboswitches that control translation initiation also appear to be more dependent on the action of the transcription terminator protein Rho [47]. This protein is known to bind to nascent mRNAs that are not being actively translated and cause transcription termination. Perhaps some riboswitches that preclude rapid initiation of translation indirectly cause transcription termination by facilitating the action of Rho. If true, this would explain how some riboswitches that appear to control translation initiation of the first gene in a multigene operon could serve as efficient gene control elements and prevent wasteful transcription. Again, these riboswitches would be dependent on the kinetics of translation initiation and transcription factor binding. However, there is evidence that at least one riboswitch responsive to adenine that controls translation initiation can interact with ligands in a time frame to be under thermodynamic control [130].

It is very unlikely that all the mechanisms for riboswitch-mediated gene control have been recognized. The diversity of rare mechanisms possible is reflected in part by the discovery of the self-cleaving ribozyme–riboswitch that senses GlcN6P [79–81]. Initially, the existence of a riboswitch that uses self-cleavage was puzzling for two reasons. First, how does RNA cleavage upstream of the coding region control expression of the gene? Second, why would a riboswitch exploit an RNA-cleavage mechanism for gene control when there are other, perhaps simpler, mechanisms that could be exploited?

The first question was answered by experiments revealing the involvement of a protein-based RNase that selectively degrades mRNAs that have undergone GlcN6P-mediated ribozyme cleavage [131]. The second question has not been definitively addressed and so we are left only with speculation. It seems reasonable to expect that, in most instances, a simple aptamer that interfaces with a common expression platform would demand a smaller length of RNA with fewer conserved nucleotides compared with an aptamer that controls the function of a ribozyme. The diversity of sequences and structures that expression platforms can use to bring about transcription or translation control is great and therefore expression platforms can rapidly change in nucleotide sequence through evolution compared with more strictly conserved aptamer domains. In contrast, fusion of aptamer and ribozyme domains usually would require strictly conserved sequence and structural features to be present in both domains. Perhaps this ribozyme–riboswitch structure emerged because the aptamer and ribozyme domains are not separated and so this relatively small RNA that selectively binds ligand and cleaves RNA was comparable in size and information content to other more typical riboswitches.

Another possibility for the presence of a ribozyme–riboswitch arrangement is that cells need to more strongly discriminate against closely related analogs of GlcN6P than is possible by simple molecular recognition alone. In this regard, it is interesting that a natural analog of GlcN6P, glucose-6-phosphate (Glc6P) does not trigger self-cleavage activity of the ribozyme [79,83,84]. However, the compound is bound by the ribozyme [80,83] at concentrations only about 10-fold lower than that of GlcN6P [83]. Although the relative concentrations of these two compounds in bacterial cells is not known, it may be that the level of Glc6P is sufficiently high that it can occupy the binding site for GlcN6P, but does not trigger gene repression because it cannot activate the ribozyme. Since the amine group on GlcN6P is essential for ribozyme action [79,83,84] and appears to serve a direct role in the cleavage reaction [80,81], the ribozyme activity may be essential for ensuring that only the correct metabolite triggers gene control.

On rare occasions, bioinformatics analyses reveal the existence of riboswitch aptamers that are not located in the 5′ UTRs of mRNAs, but are located in positions that do not allow them to regulate expression of a covalently linked ORF. In some instances, the aptamers are predicted to be expressed independently, although their possible functions remain to be determined. One example of a SAM riboswitch located downstream of an operon coding for genes involved in sulfur metabolism was predicted [47,64] to function as an interfering transcriptional unit. This mechanism was validated experimentally [132] and provides the first known instance of a riboswitch that can control the expression of a gene to which it is not physically tethered. Most likely, many other examples of *in trans* riboswitch-mediated control of gene expression exist, but these could be far more difficult to predict than those that use *cis*-acting or antisense-based mechanisms.

20.8 Conclusion

Some widespread riboswitch classes, such as those that sense TPP, AdoCbl and FMN, could be modern relics that had their origin in organisms from the RNA World. Regardless, modern examples of riboswitches exhibit characteristics that are sufficient to compete successfully with protein factors to meet many challenging molecular sensing and gene control needs. The current collection of riboswitches likely represents only a small portion of the true diversity of metabolite-sensing RNAs in bacteria and we should be able to access many undiscovered bacterial riboswitches by using bioinformatics search strategies that are already being employed.

Given the diversity of riboswitches already known to be present in bacteria, it may seem surprising that so few riboswitches have been reported in species of archaea and eukaryota. Representatives of TPP riboswitches have been identified in some archaeal and eukaryotic species [123,133]. The TPP riboswitches in eukaryotes are present in introns, where they control alternative splicing in response to ligand binding [124,125,133,134].

It is intriguing to find that TPP riboswitches, the most common in bacteria, have representation in species from all three domains of life. These representatives could be direct descendants from TPP riboswitches present in the last common ancestor of modern cells. However, its prevalence in bacteria also means that this RNA had an increased probability of entering these other domains of life via horizontal gene transfer. TPP riboswitch aptamers are also relatively modest in size and smaller aptamers with few conserved sequences and structural features have a greater chance of emerging anew via evolution.

The existence of only one validated riboswitch class in eukaryotes does not mean that these organisms are entirely devoid of other examples. With additional evidence, a putative arginine-sensing riboswitch in fungi could become the second such RNA class to be validated [135]. The apparent absence of other examples of bacterial riboswitches in eukaryotes could be explained in several ways. For example, eukaryotes may have less of a need to sense and respond to the same metabolites that are so important for bacteria to monitor. Or perhaps protein factors have more fully replaced metabolite-sensing RNAs. Given the greater diversity of non-coding RNA expression in eukaryotes, each cell is sampling large amounts of RNA 'sequence space' and this could facilitate the emergence of entirely new classes of riboswitches that look nothing like their bacterial counterparts. Application of increasingly powerful bioinformatics and genetics search strategies should reveal any conserved RNA elements that could be the missing riboswitch candidates in eukaryotes.

References

1. W.C. Winkler and R.R. Breaker (2003) Genetic control by metabolite-binding riboswitches. *ChemBioChem* **4**, 1024–1032.
2. M. Mandal and R.R. Breaker (2004) Gene regulation by riboswitches. *Nat. Rev. Mol. Cell Biol.* **5**, 451–463.
3. E. Nudler and A.S. Mironov (2004) The riboswitch control of bacterial metabolism. *Trends Biochem. Sci.* **29**, 11–17.

4. J.K. Soukup and G.A. Soukup (2004) Riboswitches exert genetic control through metabolite-induced conformational change. *Curr. Opin. Struct. Biol.* **14**, 344–349.
5. B.J. Tucker and R.R. Breaker (2005) Riboswitches as versatile gene control elements. *Curr. Opin. Struct. Biol.* **15**, 342–348.
6. W.C. Winkler and R.R. Breaker (2005) Regulation of bacterial gene expression by riboswitches. *Annu. Rev. Microbiol.* **59**, 487–517.
7. R.L. Coppins, K.B. Hall and E.A. Groisman (2007) The intricate world of riboswitches. *Curr. Opin. Microbiol.* **10**, 176–181.
8. A. Roth and R.R. Breaker (2009) The structural and functional diversity of metabolite-binding riboswitches. *Annu. Rev. Biochem.* **78**, 305–334.
9. F. Jacob and J.Monod (1961) Genetic regulatory mechanisms in the synthesis of proteins. *J. Mol. Biol.* **3**, 318–356.
10. W. Gilbert and B. Müller-Hill (1966) Isolation of the LAC repressor. *Proc. Natl. Acad. Sci. USA* **56**, 1891–1898.
11. K. Kruger, P.J. Grabowski, A.J. Zaug, J. Sands, D.E. Gottschling and T.R. Cech (1982) Self-splicing RNA: autoexcision and autocyclization of the ribosomal RNA intervening sequence of *Tetrahymena*. *Cell* **31**, 147–157.
12. C. Guerrier-Takada, K. Gardiner, T. Marsh, N. Pace and S. Altman (1983) The RNA moiety of ribonuclease P is the catalytic subunit of the enzyme. *Cell* **35**, 849–857.
13. E.A. Doherty and J.A. Doudna (2000) Ribozyme structures and mechanisms. *Annu. Rev. Biochem.* **69**, 597–615.
14. G.F. Joyce (1989) Amplification, mutation and selection of catalytic RNA. *Gene* **82**, 83–87.
15. C. Tuerk and L. Gold (1990) Systematic evolution of ligands by exponential enrichment: RNA ligands to bacteriophage T4 DNA polymerase. *Science* **249**, 505–510.
16. A.D. Ellington and J.W. Szostak (1990) *In vitro* selection of RNA molecules that bind specific ligands. *Nature* **346**, 818–822.
17. G.F. Joyce (1991) The rise and fall of the RNA World. *New Biol.* **3**, 399–407.
18. X. Chen, N. Li and A.D. Ellington (2007) Ribozyme catalysis of metabolism in the RNA World. *Chem. Biodivers.* **4**, 633–655.
19. T.R. Cech (2009) Crawling out of the RNA World. *Cell* **136**, 599–602.
20. J. Tang and R.R. Breaker (1997) Rational design of allosteric ribozymes. *Chem. Biol.* **4**, 453–459.
21. R.R. Breaker (2002) Engineered allosteric ribozymes as biosensor components. *Curr. Opin. Biotechnol.* **13**, 31–39.
22. S.K. Silverman (2003) Rube Goldberg goes (ribo)nuclear? Molecular switches and sensors made from RNA. *RNA* **9**, 377–383.
23. G. Werstuck and M.R. Green (1998) Controlling gene expression in living cells through small molecule-RNA interactions. *Science* **282**, 296–298.
24. J.P. Gallivan (2007) Toward reprogramming bacteria with small molecules and RNA. *Curr. Opin. Chem. Biol.* **11**, 612–619.
25. S. Topp and J.P. Gallivan (2007) Guiding bacteria with small molecules and RNA. *J. Am. Chem. Soc.* **129**, 6807–6811.
26. J. E. Weigand and B. Suess (2007) Tetracycline aptamer-controlled regulation of pre-mRNA splicing in yeast. *Nucleic Acids Res.* **35**, 4179–4185.
27. B. Suess and J.E. Weigand (2008) Engineered riboswitches: overview, problems and trends. *RNA Biol.* **5**, 24–29.
28. F. Michel, M. Hanna, R. Green, D.P. Bartel and J.W. Szostak (1989) The guanosine binding site of the *Tetrahymena* ribozyme. *Nature* **342**, 391–395.
29. R.R. Breaker and G.F. Joyce (1995) Self-incorporation of coenzymes by ribozymes. *J. Mol. Evol.* **40**, 551–558.
30. M.D. Been and A.T. Perotta (1991) Group I intron self-splicing with adenosine: evidence for a single nucleoside-binding site. *Science* **252**, 434–437.
31. S. Seetharaman, M. Zivarts, N. Sudarsan and R.R. Breaker (2001) Immobilized RNA switches for the analysis of complex chemical and biological mixtures. *Nat. Biotechnol.* **19**, 336–341.
32. X. Nou and R.J. Kadner (2000) Adenosylcobalamin inhibits ribosome binding to *btuB* RNA. *Proc. Natl. Acad. Sci. USA* **97**, 7190–7195.
33. M.S. Gelfand, A.A. Mironov, J. Jomantas, Y.I. Kozlov and D.A. Perumov (1999) A conserved RNA structure element involved in the regulation of bacterial riboflavin synthesis genes. *Trends Genet.* **15**, 439–442.
34. L. Gold, D. Brown, Y.-Y. He, T. Shtatland, B.S. Singer and Y. Wu. (1997) From oligonucleotide shapes to genomic SELEX: novel biological regulatory loops. *Proc. Natl. Acad. USA* **94**, 59–64.

35. J. Miranda-Ríos, M.Navarro and M.Soberón (2001) A conserved RNA structure (thi box) is involved in regulation of thiamine biosynthetic gene expression in bacteria. *Proc. Natl. Acad. Sci. USA* **98**, 9736–9741.
36. G.D. Stormo and Y. Ji (2001) Do mRNAs act as direct sensors of small molecules to control their expression? *Proc. Natl. Acad. Sci. USA* **98**, 9465–9467.
37. A. Nahvi, N. Sudarsan, M.S. Ebert, X. Zou, K.L. Brown and R.R. Breaker (2002) Genetic control by a metabolite binding mRNA. *Chem. Biol.* **9**, 1043–1049.
38. W. Winkler, A. Nahvi and R.R. Breaker (2002) Thiamine derivatives bind messenger RNAs directly to regulate bacterial gene expression. *Nature* **419**, 952–956.
39. W.C. Winkler, S. Cohen-Chalamish and R.R. Breaker (2002) An mRNA structure controls gene expression by binding FMN. *Proc. Natl. Acad. Sci. USA* **99**, 15908–15913.
40. M. Mandal, B. Boese, J.E. Barrick, W. C. Winkler and R.R. Breaker (2003) Riboswitches control fundamental biochemical pathways in *Bacillus subtilis* and other bacteria. *Cell* **113**, 577–586.
41. B.A. McDaniel, F.J. Grundy, I. Artsimovitch, T.M. Henkin (2003) Transcription termination control of the S box system: direct measurement of *S*-adenosylmethionine by the leader RNA. *Proc. Natl. Acad. Sci. USA* **100**, 3083–3088.
42. V. Epshtein, A.S. Mironov and E. Nudler (2003) The riboswitch-mediated control of sulfur metabolism in bacteria. *Proc. Natl. Acad. Sci. USA* **100**, 5052–5026.
43. W.C. Winkler, A. Nahvi, N. Sudarsan, J.E. Barrick and R.R. Breaker (2003) An mRNA structure that controls gene expression by binding *S*-adenosylmethionine. *Nat. Struct. Biol.* **10**, 701–707.
44. F.J. Grundy, S.C. Lehman and T.M. Henkin (2003) The L box regulon: lysine sensing by leader RNAs of bacterial lysine biosynthesis genes. *Proc. Natl. Acad. Sci. USA* **100**, 12057–12062.
45. N. Sudarsan, J.K. Wickiser, S. Nakamura, M.S. Ebert and R.R. Breaker (2003) An mRNA structure in bacteria that controls gene expression by binding lysine. *Genes Dev.* **17**, 2688–2697.
46. E.E. Regulski, R.H. Moy, Z. Weinberg, J.E. Barrick, Z. Yao, W.L. Ruzzo and R.R. Breaker (2008) A widespread riboswitch candidate that controls bacterial genes involved in molybdenum cofactor and tungsten cofactor metabolism. *Mol. Microbiol.* **68**, 918–932.
47. J.E. Barrick and R.R. Breaker (2007) The distributions, mechanisms, and structures of metabolite-binding riboswitches. *Genome Biol.* **8**, R239.
48. W.S. Yarnell and J.W. Roberts (1999) Mechanism of intrinsic transcription termination and antitermination. *Science* **284**, 611–615.
49. I. Gusarov and E. Nudler (1999) The mechanism of intrinsic transcription termination. *Mol. Cell* **3**, 495–504.
50. J. Shine and L. Dalgarno (1975) Determinant of cistron specificity in bacterial ribosomes. *Nature* **254**, 34–38.
51. M. Mandal and R.R. Breaker (2004) Adenine riboswitches and gene activation by disruption of a transcription terminator. *Nat. Struct. Mol. Biol.* **11**, 29035.
52. N. Sudarsan, E.R. Lee, Z. Weinberg, R.H. Moy, J.N. Kim, K.H. Link, R.R. Breaker (2008) Riboswitches in eubacteria sense the second messenger cyclic di-GMP. *Science* **321**, 411–413.
53. A.S. Mironov, I. Gusarov, R. Rafikov, L.E. Lopez, K. Shatalin, R.A. Kreneva, D.A. Perumov and E. Nudler (2002) Sensing small molecules by nascent RNA: a mechanism to control transcription in bacteria. *Cell* **111**, 747–756.
54. M. Mandal, M. Lee, J.E. Barrick, Z. Weinberg, G.M. Emilsson, W.L. Ruzzo and R.R. Breaker (2004) A glycine-dependent riboswitch that uses cooperative binding to control gene expression. *Science* **306**, 275–279.
55. A.G. Vitreschak, D.A. Rodionov, A.A. Mironov and M. S. Gelfand (2003) Regulation of the vitamin B_{12} metabolism and transport in bacteria by a conserved RNA structural element. *RNA* **9**, 1084–1097.
56. O. Amster-Choder (2005) The *bgl* sensory system: a transmembrane signaling pathway controlling transcriptional antitermination. *Curr. Opin. Microbiol.* **8**, 127–134.
57. A. Gutiérrez, C. Yanofsky and E. Mereno (2007) Comparison of tryptophan biosynthetic operon regulation in different Gram-positive bacterial species. *Trends Genet.* **23**, 422–426.
58. T. Kumarevel (2007) Structural insights of HutP-mediated regulation of transcription of the hut operon in *Bacillus subtilis*. *Biophys. Chem.* **128**, 1–12.
59. K.M. Wassarman (2007) 6S RNA: a regulator of transcription. *Mol. Microbiol.* **65**, 1425–1431.
60. T.M. Henkin and F.J. Grundy (2006) Sensing metabolic signals with nascent RNA transcripts: the T box and S box riboswitches as paradigms. *Cold Spring Harbor Symp. Quant. Biol.* **71**, 231–237.

61. S. Gottesman (2005) Micros for microbes: non-coding regulatory RNAs in bacteria. *Trends Genet.* **21**, 399–404.
62. L.S. Waters and G. Storz (2009) Regulatory RNAs in bacteria. *Cell* **136**, 615–628.
63. F.J. Grundy and T.M. Henkin (1998) The S box regulon: a new global transcription termination system for methionine and cysteine biosynthesis genes in Gram-positive bacteria. *Mol. Microbiol.* **30**, 737–749.
64. D.A. Rodionov, A. G. Vitreschak, A. A. Mironov and M. S. Gelfand (2004) Comparative genomics of the methionine metabolism in Gram-positive bacteria: a variety of regulatory systems. *Nucleic Acids Res.* **32**, 3340–3353.
65. R. Tamayo, J.T. Pratt and A. Camilli (2007) Roles of cyclic diguanylate in the regulation of bacterial pathogenesis. *Annu. Rev. Microbiol.* **61**, 131–148.
66. R. Hengge (2009) Principles of c-di-GMP signaling in bacteria. *Nat. Rev. Microbiol.* **7**, 263–273.
67. G.A. Soukup and R.R. Breaker (1999) Relationship between internucleotide linkage geometry and the stability of RNA. *RNA* **5**, 1308–1325.
68. G.A. Soukup, E.C. DeRose, M. Koizumi and R.R. Breaker (2001) Generating new ligand-binding RNAs by affinity maturation and disintegration of allosteric ribozymes. *RNA* **7**, 524–536.
69. E.E. Regulski and R.R. Breaker (2008) In-line probing analysis of riboswitches. *Methods Mol. Biol.* **419**, 53–67.
70. F.H. Westheimer (1968) Pseudo-rotation in the hydrolysis of phosphate esters. *Acc. Chem. Res.* **1**, 70–78.
71. D.A. Usher (1969) On the mechanism of ribonuclease action. *Proc. Natl. Acad. Sci. USA* **62**, 661–667.
72. E.J. Merino, K.A. Wilkinson, J.L. Coughlan and K.M. Weeks (2005) RNA structure analysis at single nucleotide resolution by selective 2′-hydroxyl acylation and primer extension (SHAPE). *J. Am. Chem. Soc.* **127**, 4223–4231.
73. A.L. Edwards and R.T. Batey (2009) A structural basis for the recognition of 2′-deoxyguanosine by the purine riboswitch. *J. Mol. Biol.* **385**, 938–948.
74. C.D. Stoddard, S.D. Gilbert and R.T. Batey (2008) Ligand-dependent folding of the three-way junction in the purine riboswitch. *RNA* **14**, 675–684.
75. A. Rentmeister, G. Mayer, N. Kuhn and M. Famulok (2008) Secondary structures and functional requirements for thiM riboswitches from *Desulfovibrio vulgaris*, *Erwinia cartovora* and *Rhodobacter spheroides*. *Biol. Chem.* **389**, 127–134.
76. A. Serganov, A. Polonskaia, A.T. Phan, R.R. Breaker and D.J. Patel (2006) Structural basis for gene regulation by a thiamine pyrophosphate-sensing riboswitch. *Nature* **441**, 1167–1171.
77. A. Serganov, L. Huang and D.J. Patel (2008) Structural insights into amino acid binding and gene control by a lysine riboswitch. *Nature* **455**, 1263–1267.
78. G. Mayer, M.-S. L. Raddatz, J.D. Grunwald and M. Famulok (2007) RNA ligands that distinguish metabolite-induced conformations in the TPP riboswitch. *Angew. Chem. Int. Ed.* **46**, 557–560.
79. W.C. Winkler, A. Nahvi, A. Roth, J.A. Collins and R.R. Breaker (2004) Control of gene expression by a natural metabolite-responsive ribozyme. *Nature* **428**, 281–286.
80. D.J. Kline and A.R. Ferré-D'Amaré (2006) Structural basis of *glmS* ribozyme activation by glucosamine-6-phosphate. *Science* **313**, 1752–1756.
81. J.C. Cochrane, S.V. Lipchock and S.A. Strobel (2007) Structural investigation of the GlmS ribozyme bound to its catalytic cofactor. *Chem. Biol.* **14**, 97–105.
82. K.J. Hampel and M.M. Tinsley (2006) Evidence for preorganization of the *glmS* ribozyme ligand binding pocket. *Biochemistry* **45**, 7861–7871.
83. T.J. McCarthy, M.A. Plog, S.A. Floy, J.A. Jansen, J.K. Soukup and G.A. Soukup (2005) Ligand requirements for *glmS* ribozyme self-cleavage. *Chem. Biol.* **12**, 1221–1226.
84. J. Lim, B.C. Grove, A. Roth and R.R. Breaker (2006) Characteristics of ligand recognition by a *glmS* self-cleaving ribozyme. *Angew. Chem. Int. Ed.* **45**, 6689–6693.
85. G. Mayer and M. Famulok (2006) High-throughput-compatible assay for *glmS* riboswitch metabolite dependence. *ChemBioChem* **7**, 602–604.
86. K. Blount, I. Puskarz, R. Penchovsky and R. Breaker (2006) Development and application of a high-throughput assay for *glmS* ribozyme activators. *RNA Biol.* **3**, 77–81.
87. R.T. Batey, S.D. Gilbert and R.K. Montange (2004) Structure of a natural guanine-responsive riboswitch complexed with the metabolite hypoxanthine. *Nature* **432**, 411–415.
88. S.D. Gilbert, C.D. Stoddard, S.J. Wise and R.T. Batey (2006) Thermodynamic and kinetic characterization of ligand binding to the purine riboswitch aptamer domain. *J. Mol. Biol.* **359**, 754–768.

89. J.K. Wickiser, W.C. Winkler, R.R. Breaker and D.M. Crothers (2005) The speed of RNA transcription and metabolite binding kinetics operate an FMN riboswitch. *Mol. Cell* **18**, 49–60.
90. S. Blouin and D.A. Lafontaine (2007) A loop–loop interaction and a K-turn motif located in the lysine aptamer domain are important for the riboswitch gene regulation control. *RNA* **13**, 1256–1267.
91. R.T. Fuchs, F.J. Grundy and T.M. Henkin (2007) *S*-Adenosylmethionine directly inhibits binding of 30S ribosomal subunits to the S_{MK} box translational riboswitch RNA. *Proc. Natl. Acad. Sci. USA* **104**, 4876–4880.
92. N. Ontiveros-Palacios, A.M. Smith, F.J. Grundy, M. Soberon, T.M. Henkin and J. Miranda-Rios (2008) Molecular basis of gene regulation by the THI-box riboswitch. *Mol. Microbiol.* **76**, 793–803.
93. J.K. Wickiser, M.T. Cheah, R.R. Breaker and D.M. Crothers (2005) The kinetics of ligand binding by an adenine-sensing riboswitch. *Biochemistry* **44**, 13404–13414.
94. B.A. McDaniel, F.J. Grundy, V.P. Kurlekar, J. Tomsic and T.M. Henkin (2006) Identification of a mutant in the *Bacillus subtilis S*-adenosylmethionine synthetase gene that results in derepression of S-box gene expression. *J. Bacteriol.* **188**, 3674–3681.
95. J.X. Wang, E.R. Lee, D.R. Morales, J. Lim, R.R. Breaker (2008) Riboswitches that sense *S*-adenosylhomocysteine and activate genes involved in coenzyme recycling. *Mol. Cell* **29**, 691–702.
96. S.F. Altschul, W. Gish, W. Miller, E.W. Myers and D. J. Lipman (1990) Basic local alignment search tool. *J. Mol. Biol.* **215**, 403–410.
97. J.E. Barrick, K.A. Corbino, W.C. Winkler, A. Nahvi, M. Mandal, J. Collins, M. Lee, A. Roth, N. Sudarsan, I. Jona, J.K. Wickiser and R. R. Breaker (2004) New RNA motifs suggest an expanded scope for riboswitches in bacterial genetic control. *Proc. Natl. Acad. Sci. USA* **101**, 6421–6426.
98. K. Corbino, J.E. Barrick, J. Lim, R. Welz, B.J. Tucker, I. Puskarz, M. Mandal, N.D. Rudnick and R.R. Breaker (2005) Evidence for a second class of *S*-adenosylmethionine riboswitches and other regulatory RNA motifs in alpha-proteobacteria. *Genome Biol.* **6**, R70.
99. Z. Weinberg, J.E. Barrick, Z. Yao, A. Roth, J.N. Kim, J. Gore, J.X. Wang, E.R. Lee, K.F. Block, N. Sudarsan, S. Neph, M. Tompa, W.L. Ruzzo and R.R. Breaker (2007) Identification of 22 candidate structured RNAs in bacteria using the CMfinder comparative genomics pipeline. *Nucleic Acids Res.* **35**, 4809–4819.
100. R.L. Tatusov, D.A. Natale, I.V. Garkavtsev, T.A. Tatusova, U.T. Shankayaram, B.S. Rao, B. Kiryutin, M.Y. Galperin, N.D. Fedorova and E.V. Koonin (2001) The COG database: new developments in phylogenetic classification of proteins from complete genomes. *Nucleic Acids Res.* **29**, 22–28.
101. D.J. Lipman and W.R. Pearson (1985) Rapid and sensitive protein similarity searches. *Science* **227**, 1435–1441.
102. Z. Yao, J. Barrick, Z. Weinberg, S. Neph, R. Breaker, M. Tompa and W.L. Ruzzo (2007) A computational pipeline for high-throughput discovery of *cis*-regulatory noncoding RNA in prokaryotes. *PLoS Comput. Biol.* **3**, e126.
103. A. Marchler-Bauer, J.B. Anderson, P.F. Cherukuri, C. DeWeese-Scott, L.Y. Geer *et al.* (2005) CDD: a conserved domain database for protein classification. *Nucleic Acids Res.* **33**, D192–D196.
104. Z. Yao, Z. Weinberg and W.L. Ruzzo (2006) CMfinder – a covariance model based RNA motif finding algorithm. *Bioinformatics* **22**, 445–452.
105. Z. Weinberg and W.L. Ruzzo (2004) Exploiting conserved structure for faster annotation of non-coding RNAs without loss of accuracy. *Bioinformatics* **20**, i334–i341.
106. M.M. Meyer, A. Roth, S.M. Chervin, G.A. Garcia and R.R. Breaker (2008) Confirmation of a second natural $preQ_1$ aptamer class in *Streptococcaceae*. *RNA* **14**, 685–695.
107. P. Bengert and T. Dandekar (2004) Riboswitch finder – a tool for identification of riboswitch RNAs. *Nucleic Acids Res.* **32**, W154–159.
108. S. Zhang, I. Borovok, Y. Aharonowitz, R. Sharan and V. Bafna (2006) A sequence-based filtering method for ncRNA identification and its application to searching for riboswitch elements. *Bioinformatics* **22**, e557–565.
109. X Xu, Y. Ji and G.D. Stormo (2009) Discovering *cis*-regulatory RNAs in *Shewanella* genomes by support vector machines. *PLoS Comput. Biol.* **5**, e10000338.
110. J.X. Wang and R.R. Breaker (2008) Riboswitches that sense *S*-adenosylmethionine and *S*-adenosylhomocysteine. *Biochem. Cell Biol.* **86**, 157–168.
111. A. Roth, W.C. Winkler, E.E. Regulski, B.W.K. Lee, J. Lim, I. Jona, J.E. Barrick, A. Ritwik, J.N. Kim, R. Welz, D. Iwata-Reuyl and R.R. Breaker (2007) A riboswitch selective for the queuosine precursor $preQ_1$ contains as unusually small aptamer domain. *Nat. Struct. Mol. Biol.* **14**, 308–317.

112. R.T. Fuchs, F.J. Grundy and T.M. Henkin (2006) The S(MK) box is a new SAM-binding RNA for translational regulation of SAM synthase. *Nat. Struct. Mol. Biol.* **13**, 226–233.
113. R.K. Montange and R.T. Batey (2006) Structure of the *S*-adenosylmethionine riboswitch regulatory mRNA element. *Nature* **441**, 1172–1175.
114. S.D. Gilbert, R.P. Rambo, D. Van Tyne and R.T. Batey (2008) Structure of the SAM-II riboswitch bound to *S*-adenosylmethionine. *Nat. Struct. Mol. Biol.* **15**, 177–182.
115. C. Lu, A.M. Smith, R.T. Fuchs, F. Ding, K. Rajashankar, T.M. Henkin and A. Ke. (2008) Crystal structures of the SAMIII/S(MK) riboswitch reveal the SAM-dependent translation inhibition mechanism. *Nat. Struct. Mol. Biol.* **15**, 1076–1083.
116. Z. Weinberg, E.E. Regulski, M.C. Hammond, J.E. Barrick, Z. Yao, W.L. Ruzzo and R.R. Breaker (2008) The aptamer core of SAM-IV riboswitches mimics the ligand-binding site of SAM-I riboswitches. *RNA* **14**, 822–828.
117. M.M. Meyer, T.D. Ames, D.P. Smith, Z. Weinberg, M.S. Schwalbach, S.J. Giovannoni and R.R. Breaker (2009) Identification of candidate structured RNAs in the marine organism '*Candidatus* Pelagibacter ubique'. *BMC Genomics* **10**, 268.
118. J.N. Kim, A. Roth and R.R. Breaker (2007) Guanine riboswitch variants from *Mesoplasma florum* selectively recognize 2′-deoxyguanosine. *Proc. Natl. Acad. Sci. USA* **104**, 16092–16097.
119. J.N. Kim and R.R. Breaker (2008) Purine sensing by riboswitches. *Biol. Cell* **100**, 1–11.
120. R. Welz and R.R. Breaker (2007) Ligand binding and gene control characteristics of tandem riboswitches in *Bacillus anthracis*. *RNA* **13**, 573–582.
121. K.H. Link and R.R. Breaker (2009) Engineering ligand-responsive gene control elements: lessons learned from natural riboswitches. *Gene Ther.* **16**, 1189–1201.
122. E.R. Lee, K.F. Blount and R.R. Breaker (2009) Roseoflavin is a natural antibacterial compound that binds to FMN riboswitches and regulates gene expression. *RNA Biol.* **6**, 187–194.
123. N. Sudarsan, J.E. Barrick and R.R. Breaker (2003) Metabolite binding RNA domains are present in the genes of eukaryotes. *RNA* **9**, 644–647.
124. A. Wachter, M. Tunc-Ozdemir, B.C. Grove, P.J. Green, D.K. Shintani and R.R. Breaker (2007) Riboswitch control of gene expression in plants by splicing and alternative 3′ end processing of mRNAs. *Plant Cell* **19**, 3437–3450.
125. S. Bocobza, A. Adato, T. Mandal, M. Shapira, E. Nudler and A. Aharoni (2007) Riboswitch-dependent gene regulation and its evolution in the plant kingdom. *Genes Dev.* **21**, 2874–2879.
126. R.R. Breaker (2008) Complex riboswitches. *Science* **319**, 1795–1797.
127. N. Sudarsan, M.C. Hammond, K.F. Block, R. Welz, J.E. Barrick, A. Roth and R.R. Breaker (2006) Tandem riboswitch architectures exhibit complex gene control functions. *Science* **314**, 300–304.
128. C.D. Stoddard and R.T. Batey (2006) Mix-and-match riboswitches. *ACS Chem. Biol.* **15**, 751–754.
129. J.R. Gregg (1998) *Ones and Zeros: Understanding Boolean Algebra, Digital Circuits, and the Logic of Sets*. IEEE Press, New York, pp. 77–100.
130. R. Rieder, K. Lang, D. Graber and R. Micura (2007) Ligand-induced folding of the adenosine deaminase A-riboswitch and implications on riboswitch translation control. *ChemBioChem* **25**, 896–902.
131. J.A. Collins, I. Irnov, S. Baker and W.C. Winkler (2007) Mechanism of mRNA destabilization by the *glmS* ribozyme. *Genes Dev.* **21**, 3356–3368.
132. G. André, S. Even, H. Putzer, P. Burquiére, C. Croux, A. Danchin, I. Martin-Verstraete and O. Soutourina (2008) S-box and T-box riboswitches and antisense RNA control a sulfur metabolic operon of *Clostridium acetobutylicum*. *Nucleic Acids Res.* **36**, 5955–5969.
133. T. Kubodera, M. Watanabe, K. Yoshiuchi, N. Yamashita, A. Nishimura, S. Nakai, K. Gomi and H. Hanamoto (2003) Thiamine-regulated gene expression of *Aspergillus oryzae thiA* requires splicing of the intron containing a riboswitch-like domain in the 5′-UTR. *FEBS Lett.* **555**, 516–520.
134. M.T. Cheah, A. Wachter, N. Sudarsan and R.R. Breaker (2007) Control of alternative RNA splicing and gene expression by eukaryotic riboswitches. *Nature* **447**, 497–500.
135. P. Borsuk, A. Przykorska, K. Blachnio, M. Koper, J.M. Pawlowicz, M. Pekala and P. Weglenski (2007) L-Arginine influences the structure and function of arginase mRNA in *Aspergillus nidulans*. *Biol. Chem.* **388**, 135–144.

136. M.T. Croft, M. Moulin, M.E. Webb and A.G. Smith (2007) Thiamine biosynthesis in algae is regulated by riboswitches. *Proc. Natl. Acad. Sci. USA* **104**, 20770–20775.
137. D.J. Kline, T.E. Edwards and A.R. Ferré-D'Amare (2009) Cocrystal structue of a class I preQ_1 riboswitch reveals a pseudoknot recognizing an essential hypermodified nucleobase. *Nat. Struct. Mol. Biol.* **16**, 343–344.
138. R.C. Spitale, A.T. Torelli, J. Krucinska, V. Bandarian and J.E. Wedekind (2009) The structural basis for recognition of the preQ_0 metabolite by an unusually small riboswitch aptamer domain. *J. Biol. Chem.* **284**, 11012–11016.
139. M. Kang, R. Peterson and J. Feigon (2009) Structural insights into riboswitch control of the biosynthesis of queuosine, a modified nucleotide found in the anticodon of tRNA. *Mol. Cell* **33**, 784–790.

Index

Note: Page numbers in *italic* refer to figures, schemes or tables.

The Chemical Biology of Nucleic Acids Edited by Günter Mayer